Basic heat transfer relations

Fourier's law of heat conduction:

$$q_x = -kA\frac{\partial T}{\partial x}$$

Convection heat transfer from a surface:

$$q = hA(T_{\text{surface}} - T_{\text{free stream}}) \quad \text{for exterior flows}$$

$$q = hA(T_{\text{surface}} - T_{\text{fluid bulk}}) \quad \text{for flow in channels}$$

Forced convection: $\text{Nu} = f(\text{Re}, \text{Pr})$ (Chapters 5 and 6, Tables 5-2 and 6-8)

Free convection: $\text{Nu} = f(\text{Gr}, \text{Pr})$ (Chapter 7, Table 7-4)

$$\text{Re} = \frac{\rho u x}{\mu} \qquad \text{Gr} = \frac{\rho^2 g\beta\,\Delta T x^3}{\mu^2} \qquad \text{Pr} = \frac{c_p\mu}{k}$$

x = characteristic dimension

Radiation heat transfer (Chapter 8)

Black body emissive power, $\dfrac{\text{energy emitted by black body}}{\text{area}\cdot\text{time}} = \sigma T^4$

$$\text{Radiosity} = \frac{\text{energy leaving surface}}{\text{area}\cdot\text{time}}$$

$$\text{Irradiation} = \frac{\text{energy incident on surface}}{\text{area}\cdot\text{time}}$$

Radiation shape factor F_{mn} = fraction of energy leaving surface m and arriving at surface n

Reciprocity relation: $A_m F_{mn} = A_n F_{nm}$

Radiation heat transfer from surface with area A_1, emissivity ϵ_1, and temperature $T_1(K)$ to large enclosure at temperature T_2:

$$q = \sigma A_1 \epsilon_1 (T_1^4 - T_2^4)$$

LMTD method for heat exchangers (Section 10-5):

$$q = UAF\,\Delta T_m$$

where F = factor for specific heat exchanger; ΔT_m = LMTD for counterflow double-pipe heat exchanger with same inlet temperatures

Effectiveness-NTU method for heat exchangers (Section 10-6, Table 10-3):

$$\epsilon = \frac{\text{Temperature difference for fluid with minimum value of } \dot{m}c}{\text{Largest temperature difference in heat exchanger}}$$

$$\text{NTU} = \frac{UA}{C_{\min}} \qquad \epsilon = f(\text{NTU}, C_{\min}/C_{\max})$$

See List of Symbols on page xvii for definitions of terms.

HEAT TRANSFER

HEAT TRANSFER

Seventh Edition

J. P. HOLMAN
Professor of Mechanical Engineering
Southern Methodist University

McGraw-Hill Publishing Company
New York St. Louis San Francisco Auckland
Bogotá Caracas Hamburg Lisbon London
Madrid Mexico Milan Montreal New Delhi
Oklahoma City Paris San Juan São Paulo
Singapore Sydney Tokyo Toronto

THE COVER

Plate Heat Exchanger
Courtesy *Alfa-Laval Company*

Heat Transfer

2 3 4 5 6 7 8 9 0 DOCDOC 9 4 3 2 1 0

P/N 029638-3
PART OF
ISBN 0-07-909388-4

This book was set in Times Roman by General Graphic Services, Inc.
The editors were Lyn Beamesderfer and Scott Amerman;
the production supervisor was Leroy A. Young.
The cover was designed by Karen Quigley.
R. R. Donnelley & Sons Company was printer and binder.

Library of Congress Cataloging-in-Publication Data

Holman, J. P. (Jack Philip)
Heat Transfer / J. P. Holman.—7th ed.
p. cm.
Includes bibliographical references.
ISBN 0-07-909388-4
1. Heat—Transmission. 1. Title.
QC320.H64 1990
621.402′2—dc20 89-13168

ABOUT THE AUTHOR

JACK P. HOLMAN received his Ph.D. in mechanical engineering from Oklahoma State University in 1958. After two years active duty as a research scientist in the Air Force Aerospace Research Laboratory he joined the faculty of Southern Methodist University, where he is presently Brown Foundation Professor of Mechanical Engineering.

During his tenure at Southern Methodist University he has nine times been voted the Outstanding Engineering Faculty Member by the student body in a poll conducted annually. He has been active on many committees and has held administrative positions as Director of the Thermal and Fluid Sciences Center, Head of Civil and Mechanical Engineering Department, and Assistant Provost for Instructional Media.

As a principal investigator for research sponsored by the Atomic Energy Commission, National Science Foundation, NASA, and the Environmental Protection Agency, he has published extensively in such journals as *Industrial and Engineering Chemistry, International Journal of Heat and Mass Transfer, Journal of the Aerospace Sciences,* and others.

His three widely used textbooks, *Heat Transfer,* 1963 (7th ed. 1990), *Experimental Methods for Engineers,* 1966 (5th ed. 1989), and *Thermodynamics,* 1969 (4th ed. 1988), all published by the McGraw-Hill Publishing Company, have been translated into Spanish, Chinese, Japanese, Korean, and Portuguese and are distributed world wide. Dr. Holman is the consulting editor for the McGraw-Hill Series in Mechanical Engineering and also consults for industry in the fields of energy conservation and energy systems.

A member of the American Society of Engineering Education, he is past Chairman of the National Mechanical Engineering Division and past Chairman of the ASME Region X Mechanical Engineering Department Heads. Dr. Holman is a registered professional engineer in the state of Texas and received the *Mechanical Engineer of the Year* award by the North Texas Section of the American Society of Mechanical Engineers in 1971.

Dr. Holman is also the recipient of the *George Westinghouse Award* from the American Society of Engineering Education for distinguished contributions to Engineering Education (1972), the *James Harry Potter Gold Medal* for contributions to thermodynamics from ASME (1986), and the *Worcester Reed Warner Gold Medal* for outstanding contributions to the permanent literature of engineering from ASME (1987). He is a Fellow of ASME.

CONTENTS

PREFACE

This book presents an elementary treatment of the principles of heat transfer. As a text it contains sufficient material for a one-semester course which may be presented at the junior level, or higher, depending on individual course objectives. A background in ordinary differential equations is helpful for proper understanding of the material. Although some familiarity with fluid mechanics will aid in the convection discussions, it is not essential. The concepts of thermodynamic energy balances are also useful in the various analytical developments.

Presentation of the subject follows classical lines of separate discussions for conduction, convection, and radiation, although it is emphasized that the physical mechanism of convection heat transfer is one of conduction through the stationary fluid layer near the heat transfer surface. Throughout the book emphasis has been placed on physical understanding while, at the same time, relying on meaningful experimental data in those circumstances which do not permit a simple analytical solution.

Conduction is treated from both the analytical and the numerical viewpoint, so that the reader is afforded the insight which is gained from analytical solutions as well as the important tools of numerical analysis which must often be used in practice. A similar procedure is followed in the presentation of convection heat transfer. An integral analysis of both free- and forced-convection boundary layers is used to present a physical picture of the convection process. From this physical description inferences may be drawn which naturally lead to the presentation of empirical and practical relations for calculating convection heat-transfer coefficients. Because it provides an easier instruction vehicle than other methods, the radiation-network method is used extensively in the introduction of analysis of radiation systems, while a more generalized formulation is given later.

Systems of nonlinear equations requiring iterative solutions are also discussed in the conduction and radiation chapters.

The log-mean-temperature-difference and effectiveness approaches are presented in heat-exchanger analysis since both are in wide use and each offers its own advantages to the designer. A brief introduction to diffusion and mass transfer is presented in order to acquaint the reader with these processes and to establish more firmly the important analogies between heat, mass, and momentum transfer.

A number of special topics are discussed in Chapter 12 which give added flavor to the basic material of the preceding chapters.

Problems are included at the end of each chapter. Some of these problems are of a routine nature to familiarize the student with the numerical manipulations and orders of magnitude of various parameters which occur in the subject of heat transfer. Other problems extend the subject matter by requiring students to apply the basic principles to new situations and develop their own equations. Both types of problems are important.

The subject of heat transfer is not static. New developments occur quite regularly, and better analytical solutions and empirical data are continuously made available to the professional in the field. Because of the huge amount of information which is available in the research literature, the beginning student could easily be overwhelmed if too many of the nuances of the subject were displayed and expanded. The book is designed to serve as an elementary text, so the author has assumed a role of interpreter of the literature with those findings and equations being presented which can be of immediate utility to the reader. It is hoped that the student's attention is called to more extensive works in a sufficient number of instances to emphasize the greater depth which is available on most of the subjects of heat transfer. For the serious student, then, the end-of-chapter references offer an open door to the literature of heat transfer which can pyramid upon further investigation.

A textbook in its seventh edition obviously reflects many compromises and evolutionary processes over the years. This book is no exception. While the basic physical mechanisms of heat transfer have not changed, analytical techniques and experimental data are constantly being revised and improved. One objective of this new edition is to keep the exposition up to date with recent information while still retaining a simple approach which can be understood by the beginning student.

The computer is now the preferred vehicle for solution of many heat-transfer problems. Personal computers with either local software or communication links offer the engineer ample power for the solution of most problems. Despite the ready availability of this computing power I have resisted the temptation to include specific computer programs for two reasons: (1) each computer installation is somewhat different in its input-output capability and (2) a number of programs for microcomputers in a menu-driven format are already on the scene. The central issue here has been directed toward problem setup which can be adapted to any computational facility.

For those persons wishing to exploit the convenience of the microcomputer, a software package developed by Professor Allan D. Kraus, of the Naval Postgraduate School, has been included as Appendix D. A disk containing the programs will be found on the inside back cover. Appendix D contains the necessary documentation, examples, and problems for use of the programs. Some open-ended design problems are included to take advantage of the power of the computer. Note that the body of the text *does not require use of these computer programs*. On the other hand, intelligent use of the programs requires an understanding of the subject of heat transfer. References to appropriate sections of the text are therefore given in Appendix D.

The SI (metric) system of units is the primary one for the text. Because the Btu-ft-pound system is still in wide use, answers and intermediate steps to examples are occasionally stated in these units. A few examples and problems are completely in English units. Some figures have dual coordinates that show both systems of units. These displays will enable the student to develop a ''bilingual'' capability during the period before full metric conversion is achieved.

In this edition minor modifications and adjustments have been made along with the inclusion of the heat-transfer software package. Many new problems have been added so that the instructor and student may now choose from over 1000 problems of varying complexity. The open-ended design problems associated with the heat-transfer software are an important part of these additions.

It is not possible to cover all the topics in this book in either a quarter or semester term course, but it is hoped that the variety of topics and problems will provide the necessary flexibility for many applications.

McGraw-Hill and I would like to thank the following reviewers for their many helpful comments and suggestions: J. Benjamin Austin, Bucknell University; Roger Carlson, Auburn University; Young Cho, Drexel University; Ronald Mussulman, Cal Poly—San Luis Obispo; Douglas J. Nelson, Virginia Polytechnic Institute and State University; Eugene E. Niemi, Jr., University of Lowell; Brian Vick, Virginia Polytechnic Institute and State University; and Paul H. Zang, GMI Engineering and Management Institute.

With a book at this stage of revision the list of other people who have been generous with their comments and suggestions has grown very long indeed. Rather than risk omission of a single name, I hope that a grateful general acknowledgment will express my sincere gratitude for these persons' help and encouragement.

J. P. Holman

LIST OF SYMBOLS

a	Local velocity of sound
a	Attenuation coefficient (Chap. 8)
A	Area
A	Albedo (Chap. 8)
A_m	Fin profile area (Chap. 2)
B	Magnetic field strength
c	Specific heat, usually kJ/kg · °C
C	Concentration (Chap. 11)
C_D	Drag coefficient, defined by Eq. (6-13)
C_f	Friction coefficient, defined by Eq. (5-52)
c_p	Specific heat at constant pressure, usually kJ/kg · °C
c_v	Specific heat at constant volume, usually kJ/kg · °C
d	Diameter
D	Depth or diameter
D	Diffusion coefficient (Chap. 11)
D_H	Hydraulic diameter, defined by Eq. (6-14)
e	Internal energy per unit mass, usually kJ/kg
E	Internal energy, usually kJ
E	Emissive power, usually W/m^2 (Chap. 8)
E_{b0}	Solar constant (Chap. 8)
$E_{b\lambda}$	Blackbody emmissive power per unit wavelength, defined by Eq. (8-12)
E	Electric field vector
f	Friction factor, defined by Eq. (5-107) or Eq. (10-29)
F	Force, usually N
F_{m-n} or F_{mn}	Radiation shape factor for radiation from surface m to surface n
g	Acceleration of gravity
g_c	Conversion factor, defined by Eq. (1-14)
$G = \dfrac{\dot{m}}{A}$	Mass velocity
G	Irradiation (Chap. 8)
h	Heat-transfer coefficient, usually W/m^2 · °C
$\bar{h}$	Average heat-transfer coefficient
h_D	Mass-transfer coefficient, usually m/h
h_{fg}	Enthalpy of vaporization, kJ/kg
h_r	Radiation heat-transfer coefficient (Chap. 8)

H	Magnetic field intensity
i	Enthalpy, usually kJ/kg
I	Intensity of radiation
I	Solar insolation (Chap. 8)
I_0	Solar insolation at outer edge of atmosphere
J	Radiosity (Chap. 8)
J	Current density
k	Thermal conductivity, usually W/m · °C
k_e	Effective thermal conductivity of enclosed spaces (Chap. 7)
k_λ	Scattering coefficient (Chap. 8)
L	Length
L_c	Corrected fin length (Chap. 2)
m	Mass
$\dot{m}$	Mass rate of flow
M	Molecular weight (Chap. 11)
n	Molecular density
n	Turbidity factor, defined by Eq. (8-120)
N	Molal diffusion rate, moles per unit time (Chap. 11)
p	Pressure, usually N/m², Pa
P	Perimeter
q	Heat-transfer rate, kJ per unit time
q''	Heat flux, kJ per unit time per unit area
$\dot{q}$	Heat generated per unit volume
$\bar{q}_{m,n}$	Residual of a node, used in relaxation method (Chaps. 3,4)
Q	Heat, kJ
r	Radius or radial distance
r	Recovery factor, defined by Eq. (5-120)
R	Fixed radius
R	Gas constant
R_{th}	Thermal resistance, usually °C/W
s	A characteristic dimension (Chap. 4)
S	Molecular speed ratio (Chap. 12)
S	Conduction shape factor, usually m
t	Thickness, applied to fin problems (Chap. 2)
t, T	Temperature
u	Velocity
v	Velocity
v	Specific volume usually m³/kg
V	Velocity
V	Molecular volume (Chap. 11)
W	Weight, usually N
x, y, z	Space coordinates in cartesian system
$\alpha = \frac{k}{\rho c}$	Thermal diffusivity, usually m²/s
α	Absorptivity (Chap. 8)
α	Accommodation coefficient (Chap. 12)
α	Solar altitude angle, deg (Chap. 8)
β	Volume coefficient of expansion, 1/K
β	Temperature coefficient of thermal conductivity, 1/°C
$\gamma = \frac{c_p}{c_v}$	Isentropic exponent, dimensionless
Γ	Condensate mass flow per unit depth of plate (Chap. 9)
δ	Hydrodynamic-boundary-layer thickness
δ_t	Thermal-boundary-layer thickness
ϵ	Heat-exchanger effectiveness
ϵ	Emissivity
ϵ_H, ϵ_M	Eddy diffusivity of heat and momentum (Chap. 5)
$\zeta = \frac{\delta_t}{\delta}$	Ratio of thermal-boundary-layer thickness to hydrodynamic-boundary-layer thickness
η	Similarity variable, defined by Eq. (B-6)
η_f	Fin efficiency, dimensionless
θ	Angle in spherical or cylindrical coordinate system
θ	Temperature difference, $T - T_{\text{reference}}$

	The reference temperature is chosen differently for different systems (see Chaps. 2 to 4)
λ	Wavelength
λ	Mean free path (Chap. 12)
μ	Dynamic viscosity
ν	Kinematic viscosity
ν	Frequency of radiation (Chap. 8)
ρ	Density, usually kg/m^3
ρ	Reflectivity (Chap. 8)
ρ_e	Charge density
σ	Electrical conductivity
σ	Stefan-Boltzmann constant
σ	Surface tension of liquid-vapor interface (Chap. 9)
τ	Time
τ	Shear stress between fluid layers
τ	Transmissivity (Chap. 8)
ϕ	Angle in spherical or cylindrical coordinate system
ψ	Stream function

Dimensionless Groups

$\mathrm{Bi} = \dfrac{hs}{k}$	Biot number
$\mathrm{Ec} = \dfrac{u_\infty^2}{c_p(T_\infty - T_w)}$	Eckert number
$\mathrm{Fo} = \dfrac{\alpha\tau}{s^2}$	Fourier number
$\mathrm{Gr} = \dfrac{g\beta(T_w - T_\infty)x^3}{\nu^2}$	Grashof number
$\mathrm{Gr}^* = \mathrm{Gr\,Nu}$	Modified Grashof number for constant heat flux
$\mathrm{Gz} = \mathrm{Re\,Pr}\,\dfrac{d}{L}$	Graetz number
$\mathrm{Kn} = \dfrac{\lambda}{L}$	Knudsen number
$\mathrm{Le} = \dfrac{\alpha}{D}$	Lewis number (Chap. 11)
$\mathrm{M} = \dfrac{u}{a}$	Mach number
$\mathrm{N} = \dfrac{\sigma B_y^2 x}{\rho u_\infty}$	Magnetic-influence number
$\mathrm{Nu} = \dfrac{hx}{k}$	Nusselt number
$\overline{\mathrm{Nu}} = \dfrac{\bar{h}x}{k}$	Average Nusselt number
$\mathrm{Pe} = \mathrm{Re\,Pr}$	Peclet number
$\mathrm{Pr} = \dfrac{c_p\mu}{k}$	Prandtl number
$\mathrm{Ra} = \mathrm{Gr\,Pr}$	Rayleigh number
$\mathrm{Re} = \dfrac{\rho u x}{\mu}$	Reynolds number
$\mathrm{Sc} = \dfrac{\nu}{D}$	Schmidt number (Chap. 11)
$\mathrm{Sh} = \dfrac{h_D x}{D}$	Sherwood number (Chap. 11)
$\mathrm{St} = \dfrac{h}{\rho c_p u}$	Stanton number
$\overline{\mathrm{St}} = \dfrac{\bar{h}}{\rho c_p u}$	Average Stanton number

Subscripts

Subscript	Meaning
aw	Adiabatic wall conditions
b	Refers to blackbody conditions (Chap. 8)
b	Evaluated at bulk conditions
d	Based on diameter

f	Evaluated at film conditions
g	Saturated vapor conditions (Chap. 9)
i	Initial or inlet conditions
L	Based on length of plate
m	Mean flow conditions
m, n	Denotes nodal positions in numerical solution (see Chaps. 3, 4)
0	Denotes stagnation flow conditions (Chap. 5) or some initial condition at time zero
r	At specified radial position
s	Evaluated at condition of surroundings
x	Denotes some local position with respect to x coordinate
w	Evaluated at wall conditions
*	(Superscript) Properties evaluated at reference temperature, given by Eq. (5-124)
∞	Evaluation at free-stream conditions

HEAT TRANSFER

INTRODUCTION

Heat transfer is that science which seeks to predict the energy transfer which may take place between material bodies as a result of a temperature difference. Thermodynamics teaches that this energy transfer is defined as heat. The science of heat transfer seeks not merely to explain how heat energy may be transferred, but also to predict the rate at which the exchange will take place under certain specified conditions. The fact that a heat-transfer *rate* is the desired objective of an analysis points out the difference between heat transfer and thermodynamics. Thermodynamics deals with systems in equilibrium; it may be used to predict the amount of energy required to change a system from one equilibrium state to another; it may not be used to predict how fast a change will take place since the system is not in equilibrium during the process. Heat transfer supplements the first and second principles of thermodynamics by providing additional experimental rules which may be used to establish energy-transfer rates. As in the science of thermodynamics, the experimental rules used as a basis of the subject of heat transfer are rather simple and easily expanded to encompass a variety of practical situations.

As an example of the different kinds of problems which are treated by thermodynamics and heat transfer, consider the cooling of a hot steel bar which is placed in a pail of water. Thermodynamics may be used to predict the final equilibrium temperature of the steel bar–water combination. Thermodynamics will not tell us how long it takes to reach this equilibrium condition or what the temperature of the bar will be after a certain length of time before the equilibrium condition is attained. Heat transfer may be used to predict the temperature of both the bar and the water as a function of time.

Most readers will be familiar with the terms used to denote the three modes of heat transfer: conduction, convection, and radiation. In this chapter we seek to explain the mechanism of these modes qualitatively so that each may be considered in its proper perspective. Subsequent chapters treat the three types of heat transfer in detail.

1-1 CONDUCTION HEAT TRANSFER

When a temperature gradient exists in a body, experience has shown that there is an energy transfer from the high-temperature region to the low-temperature region. We say that the energy is transferred by conduction and that the heat-transfer rate per unit area is proportional to the normal temperature gradient:

$$\frac{q}{A} \sim \frac{\partial T}{\partial x}$$

When the proportionality constant is inserted,

$$q = -kA \frac{\partial T}{\partial x} \tag{1-1}$$

where q is the heat-transfer rate and $\partial T/\partial x$ is the temperature gradient in the direction of the heat flow. The positive constant k is called the *thermal conductivity* of the material, and the minus sign is inserted so that the second principle of thermodynamics will be satisfied; i.e., heat must flow downhill on the temperature scale, as indicated in the coordinate system of Fig. 1-1. Equation (1-1) is called Fourier's law of heat conduction after the French mathematical physicist Joseph Fourier, who made very significant contributions to the analytical treatment of conduction heat transfer. It is important to note that Eq. (1-1) is the defining equation for the thermal conductivity and that k has the units of watts per meter per Celsius degree in a typical system of units in which the heat flow is expressed in watts.

We now set ourselves the problem of determining the basic equation which governs the transfer of heat in a solid, using Eq. (1-1) as a starting point.

Consider the one-dimensional system shown in Fig. 1-2. If the system is in a steady state, i.e., if the temperature does not change with time, then the problem is a simple one, and we need only integrate Eq. (1-1) and substitute the appropriate values to solve for the desired quantity. However, if the temperature of the solid is changing with time, or if there are heat sources or sinks within the solid, the situation is more complex. We consider the general case where the temperature may be changing with time and heat sources may be

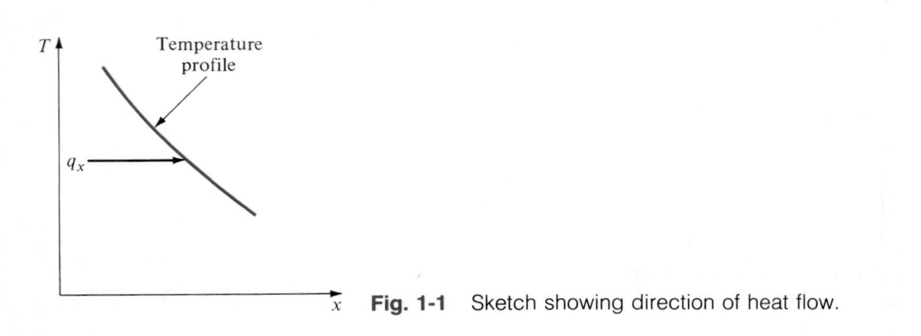

Fig. 1-1 Sketch showing direction of heat flow.

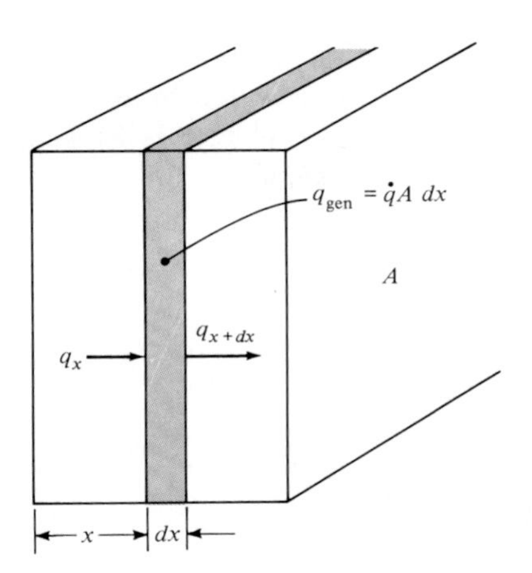

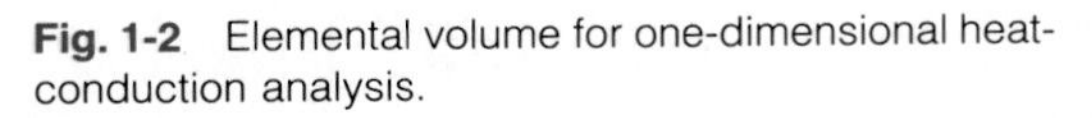

Fig. 1-2 Elemental volume for one-dimensional heat-conduction analysis.

present within the body. For the element of thickness dx the following energy balance may be made:

Energy conducted in left face + heat generated within element
= change in internal energy + energy conducted out right face

These energy quantities are given as follows:

$$\text{Energy in left face} = q_x = -kA\frac{\partial T}{\partial x}$$

$$\text{Energy generated within element} = \dot{q}A\ dx$$

$$\text{Change in internal energy} = \rho c A\frac{\partial T}{\partial \tau}\ dx$$

$$\text{Energy out right face} = q_{x+dx} = -kA\frac{\partial T}{\partial x}\bigg]_{x+dx}$$

$$= -A\left[k\frac{\partial T}{\partial x} + \frac{\partial}{\partial x}\left(k\frac{\partial T}{\partial x}\right)dx\right]$$

where $\dot{q}$ = energy generated per unit volume, W/m^3
c = specific heat of material, J/kg·°C
ρ = density, kg/m^3

Combining the relations above gives

$$-kA\frac{\partial T}{\partial x} + \dot{q}A\ dx = \rho c A\frac{\partial T}{\partial \tau}\ dx - A\left[k\frac{\partial T}{\partial x} + \frac{\partial}{\partial x}\left(k\frac{\partial T}{\partial x}\right)dx\right]$$

or

$$\frac{\partial}{\partial x}\left(k\frac{\partial T}{\partial x}\right) + \dot{q} = \rho c\frac{\partial T}{\partial \tau} \tag{1-2}$$

This is the one-dimensional heat-conduction equation. To treat more than one-dimensional heat flow, we need consider only the heat conducted in and out of a unit volume in all three coordinate directions, as shown in Fig. 1-3a. The energy balance yields

$$q_x + q_y + q_z + q_{\text{gen}} = q_{x+dx} + q_{y+dy} + q_{z+dz} + \frac{dE}{d\tau}$$

and the energy quantities are given by

$$q_x = -k\,dy\,dz\,\frac{\partial T}{\partial x}$$

$$q_{x+dx} = -\left[k\frac{\partial T}{\partial x} + \frac{\partial}{\partial x}\left(k\frac{\partial T}{\partial x}\right)dx\right]dy\,dz$$

$$q_y = -k\,dx\,dz\,\frac{\partial T}{\partial y}$$

$$q_{y+dy} = -\left[k\frac{\partial T}{\partial y} + \frac{\partial}{\partial y}\left(k\frac{\partial T}{\partial y}\right)dy\right]dx\,dz$$

$$q_z = -k\,dx\,dy\,\frac{\partial T}{\partial z}$$

$$q_{z+dz} = -\left[k\frac{\partial T}{\partial z} + \frac{\partial}{\partial z}\left(k\frac{\partial T}{\partial z}\right)dz\right]dx\,dy$$

$$q_{\text{gen}} = \dot{q}\,dx\,dy\,dz$$

$$\frac{dE}{d\tau} = \rho c\,dx\,dy\,dz\,\frac{\partial T}{\partial \tau}$$

so that the general three-dimensional heat-conduction equation is

$$\frac{\partial}{\partial x}\left(k\frac{\partial T}{\partial x}\right) + \frac{\partial}{\partial y}\left(k\frac{\partial T}{\partial y}\right) + \frac{\partial}{\partial z}\left(k\frac{\partial T}{\partial z}\right) + \dot{q} = \rho c\frac{\partial T}{\partial \tau} \tag{1-3}$$

For constant thermal conductivity Eq. (1-3) is written

$$\frac{\partial^2 T}{\partial x^2} + \frac{\partial^2 T}{\partial y^2} + \frac{\partial^2 T}{\partial z^2} + \frac{\dot{q}}{k} = \frac{1}{\alpha}\frac{\partial T}{\partial \tau} \tag{1-3a}$$

where the quantity $\alpha = k/\rho c$ is called the *thermal diffusivity* of the material. The larger the value of α, the faster heat will diffuse through the material. This may be seen by examining the quantities which make up α. A high value of α could result either from a high value of thermal conductivity, which would indicate a rapid energy-transfer rate, or from a low value of the thermal heat capacity ρc. A low value of the heat capacity would mean that less of the energy moving through the material would be absorbed and used to raise the temperature of the material; thus more energy would be available for further transfer. Thermal diffusivity α has units of square meters per second.

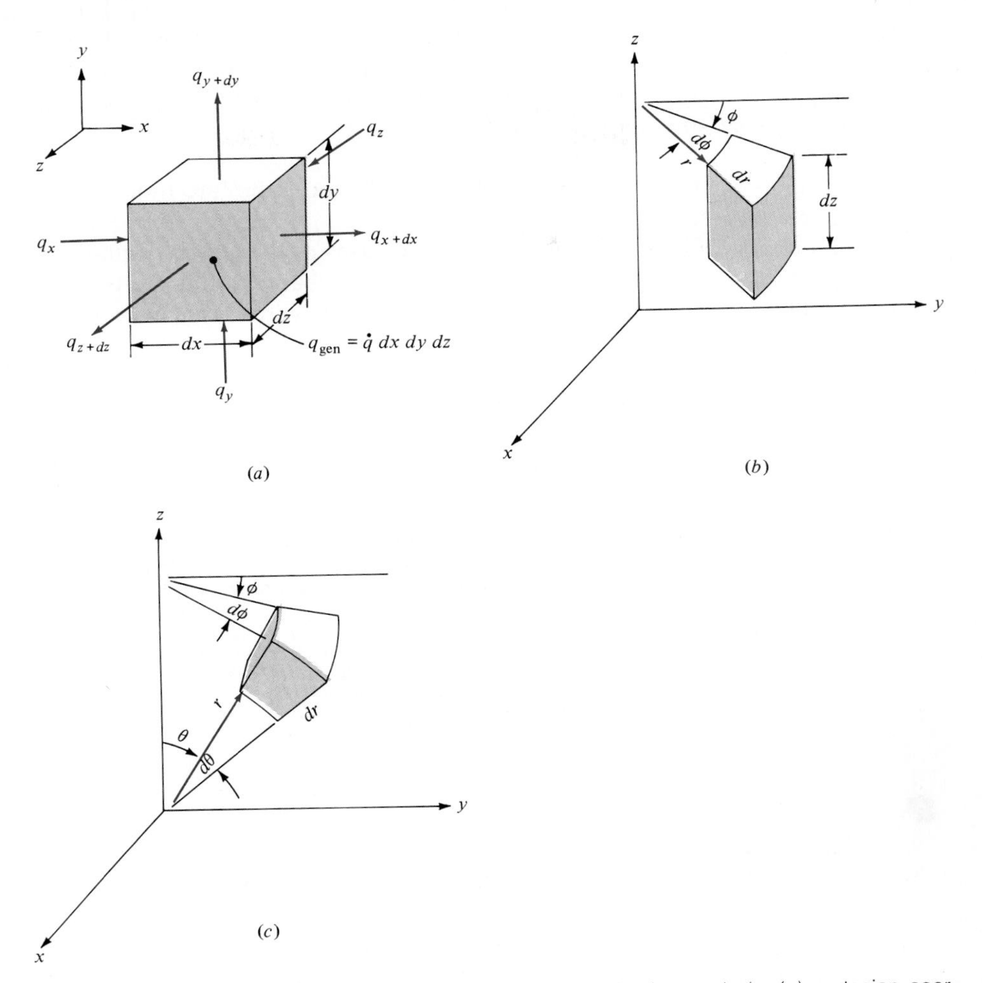

Fig. 1-3 Elemental volume for three-dimensional heat-conduction analysis: (*a*) cartesian coordinates; (*b*) cylindrical coordinates; (*c*) spherical coordinates.

In the derivations above, the expression for the derivative at $x + dx$ has been written in the form of a Taylor-series expansion with only the first two terms of the series employed for the development.

Equation (1-3*a*) may be transformed into either cylindrical or spherical coordinates by standard calculus techniques. The results are as follows:

Cylindrical Coordinates:

$$\frac{\partial^2 T}{\partial r^2} + \frac{1}{r}\frac{\partial T}{\partial r} + \frac{1}{r^2}\frac{\partial^2 T}{\partial \phi^2} + \frac{\partial^2 T}{\partial z^2} + \frac{\dot{q}}{k} = \frac{1}{\alpha}\frac{\partial T}{\partial \tau} \tag{1-3b}$$

Spherical Coordinates:

$$\frac{1}{r}\frac{\partial^2}{\partial r^2}(rT) + \frac{1}{r^2 \sin\theta}\frac{\partial}{\partial\theta}\left(\sin\theta\,\frac{\partial T}{\partial\theta}\right) + \frac{1}{r^2 \sin^2\theta}\frac{\partial^2 T}{\partial\phi^2} + \frac{\dot{q}}{k} = \frac{1}{\alpha}\frac{\partial T}{\partial\tau} \tag{1-3c}$$

The coordinate systems for use with Eqs. (1-3*b*) and (1-3*c*) are indicated in Fig. 1-3*b* and *c*, respectively.

Many practical problems involve only special cases of the general equations listed above. As a guide to the developments in future chapters, it is worthwhile to show the reduced form of the general equations for several cases of practical interest.

Steady-State One-Dimensional Heat Flow (No Heat Generation):

$$\frac{d^2T}{dx^2} = 0 \tag{1-4}$$

Note that this equation is the same as Eq. (1-1) when q = constant.

Steady-State One-Dimensional Heat Flow in Cylindrical Coordinates (No Heat Generation):

$$\frac{d^2T}{dr^2} + \frac{1}{r}\frac{dT}{dr} = 0 \tag{1-5}$$

Steady-State One-Dimensional Heat Flow with Heat Sources:

$$\frac{d^2T}{dx^2} + \frac{\dot{q}}{k} = 0 \tag{1-6}$$

Two-Dimensional Steady-State Conduction without Heat Sources:

$$\frac{\partial^2 T}{\partial x^2} + \frac{\partial^2 T}{\partial y^2} = 0 \tag{1-7}$$

1-2 THERMAL CONDUCTIVITY

Equation (1-1) is the defining equation for thermal conductivity. On the basis of this definition, experimental measurements may be made to determine the thermal conductivity of different materials. For gases at moderately low temperatures, analytical treatments in the kinetic theory of gases may be used to predict accurately the experimentally observed values. In some cases, theories are available for the prediction of thermal conductivities in liquids and solids, but in general, many open questions and concepts still need clarification where liquids and solids are concerned.

The mechanism of thermal conduction in a gas is a simple one. We identify the kinetic energy of a molecule with its temperature; thus, in a high-temperature region, the molecules have higher velocities than in some lower-temperature region. The molecules are in continuous random motion, colliding with one another and exchanging energy and momentum. The molecules have this random motion whether or not a temperature gradient exists in the gas. If a molecule moves from a high-temperature region to a region of lower temperature, it transports kinetic energy to the lower-temperature part of the system and gives up this energy through collisions with lower-energy molecules.

Table 1-1 lists typical values of the thermal conductivities for several materials to indicate the relative orders of magnitude to be expected in practice. More complete tabular information is given in Appendix A. In general, the thermal conductivity is strongly temperature-dependent.

We noted that thermal conductivity has the units of watts per meter per Celsius degree when the heat flow is expressed in watts. Note that a heat *rate* is involved, and the numerical value of the thermal conductivity indicates how fast heat will flow in a given material. How is the rate of energy transfer taken into account in the molecular model discussed above? Clearly, the faster the molecules move, the faster they will transport energy. Therefore the thermal conductivity of a gas should be dependent on temperature. A simplified analytical treatment shows the thermal conductivity of a gas to vary with the square root of the absolute temperature. (It may be recalled that the velocity of sound in a gas varies with the square root of the absolute temperature; this velocity is approximately the mean speed of the molecules.) Thermal conductivities of some typical gases are shown in Fig. 1-4. For most gases at moderate pressures the thermal conductivity is a function of temperature alone. This means that the gaseous data for 1 atmosphere (atm), as given in Appendix A, may be used for a rather wide range of pressures. When the pressure of the gas becomes of the order of its critical pressure or, more generally, when non-ideal-gas behavior is encountered, other sources must be consulted for thermal-conductivity data.

The physical mechanism of thermal-energy conduction in liquids is qualitatively the same as in gases; however, the situation is considerably more complex because the molecules are more closely spaced and molecular force fields exert a strong influence on the energy exchange in the collision process. Thermal conductivities of some typical liquids are shown in Fig. 1-5.

In the English system of units heat flow is expressed in British thermal units per hour (Btu/h), area in square feet, and temperature in degrees Fahrenheit. Thermal conductivity will then have units of Btu/h · ft · °F.

Thermal energy may be conducted in solids by two modes: lattice vibration and transport by free electrons. In good electrical conductors a rather large number of free electrons move about in the lattice structure of the material. Just as these electrons may transport electric charge, they may also carry thermal energy from a high-temperature region to a low-temperature region, as in the case of gases. In fact, these electrons are frequently referred to as the *electron gas*. Energy may also be transmitted as vibrational energy in the lattice structure of the material. In general, however, this latter mode of energy transfer is not as large as the electron transport, and for this reason good

Table 1-1 Thermal Conductivity of Various Materials at 0°C

Material	*Thermal conductivity k*	
	W/m · °C	Btu/h · ft · °F
Metals:		
Silver (pure)	410	237
Copper (pure)	385	223
Aluminum (pure)	202	117
Nickel (pure)	93	54
Iron (pure)	73	42
Carbon steel, 1% C	43	25
Lead (pure)	35	20.3
Chrome-nickel steel (18% Cr, 8% Ni)	16.3	9.4
Nonmetallic solids:		
Quartz, parallel to axis	41.6	24
Magnesite	4.15	2.4
Marble	2.08–2.94	1.2–1.7
Sandstone	1.83	1.06
Glass, window	0.78	0.45
Maple or oak	0.17	0.096
Sawdust	0.059	0.034
Glass wool	0.038	0.022
Liquids:		
Mercury	8.21	4.74
Water	0.556	0.327
Ammonia	0.540	0.312
Lubricating oil, SAE 50	0.147	0.085
Freon 12, CCl_2F_2	0.073	0.042
Gases:		
Hydrogen	0.175	0.101
Helium	0.141	0.081
Air	0.024	0.0139
Water vapor (saturated)	0.0206	0.0119
Carbon dioxide	0.0146	0.00844

electrical conductors are almost always good heat conductors, viz., copper, aluminum, and silver, and electrical insulators are usually good heat insulators. Thermal conductivities of some typical solids are shown in Fig. 1-6. Other data are given in Appendix A.

The thermal conductivities of various insulating materials are also given in Appendix A. Some typical values are 0.038 W/m · °C for glass wool and 0.78 W/m · °C for window glass. At high temperatures, the energy transfer through

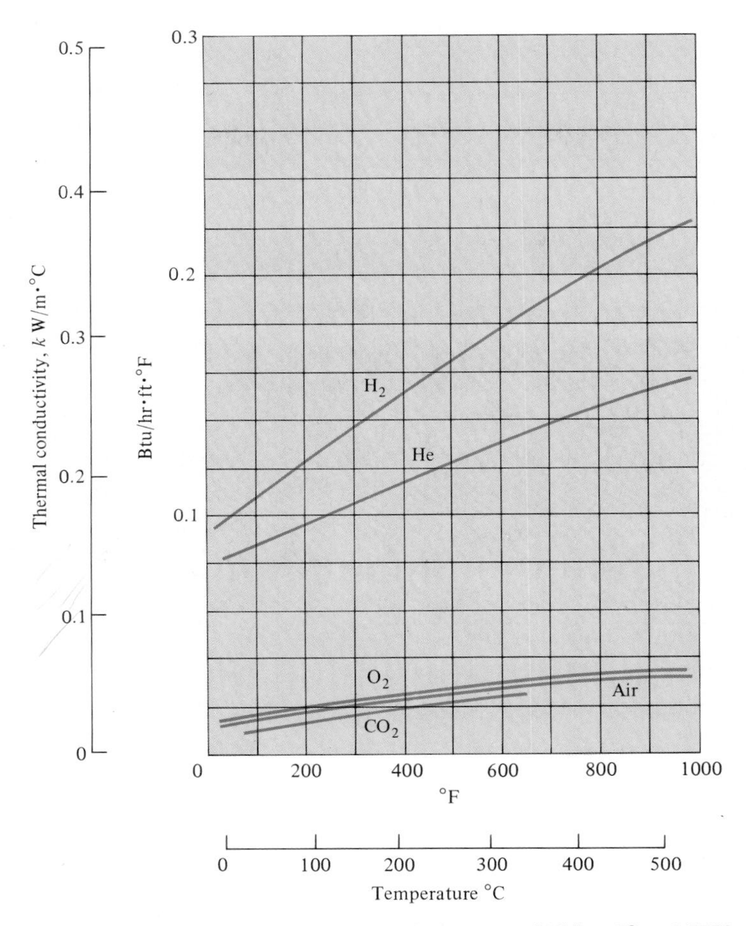

Fig. 1-4 Thermal conductivities of some typical gases [1 W/m · °C = 0.5779 Btu/h · ft · °F].

insulating materials may involve several modes: conduction through the fibrous or porous solid material; conduction through the air trapped in the void spaces; and, at sufficiently high temperatures, radiation.

An important technical problem is the storage and transport of cryogenic liquids like liquid hydrogen over extended periods of time. Such applications have led to the development of *superinsulations* for use at these very low temperatures (down to about −250°C). The most effective of these superinsulations consists of multiple layers of highly reflective materials separated by insulating spacers. The entire system is evacuated to minimize air conduction, and thermal conductivities as low as 0.3 mW/m · °C are possible. A convenient summary of the thermal conductivities of insulating materials at cryogenic

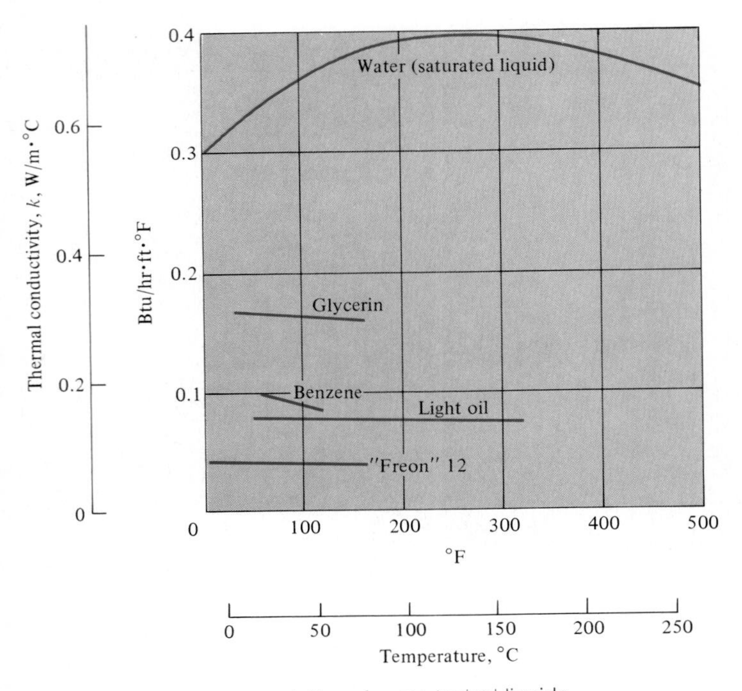

Fig. 1-5 Thermal conductivities of some typical liquids.

temperatures is given in Fig. 1-7. Further information on multilayer insulation is given in Refs. 3 and 2.

1-3 CONVECTION HEAT TRANSFER

It is well known that a hot plate of metal will cool faster when placed in front of a fan than when exposed to still air. We say that the heat is convected away, and we call the process *convection heat transfer*. The term *convection* provides the reader with an intuitive notion concerning the heat-transfer process; however, this intuitive notion must be expanded to enable one to arrive at anything like an adequate analytical treatment of the problem. For example, we know that the velocity at which the air blows over the hot plate obviously influences the heat-transfer rate. But does it influence the cooling in a linear way; i.e., if the velocity is doubled, will the heat-transfer rate double? We should suspect that the heat-transfer rate might be different if we cooled the plate with water instead of air, but, again, how much difference would there be? These questions may be answered with the aid of some rather basic analyses presented in later chapters. For now, we sketch the physical mechan-

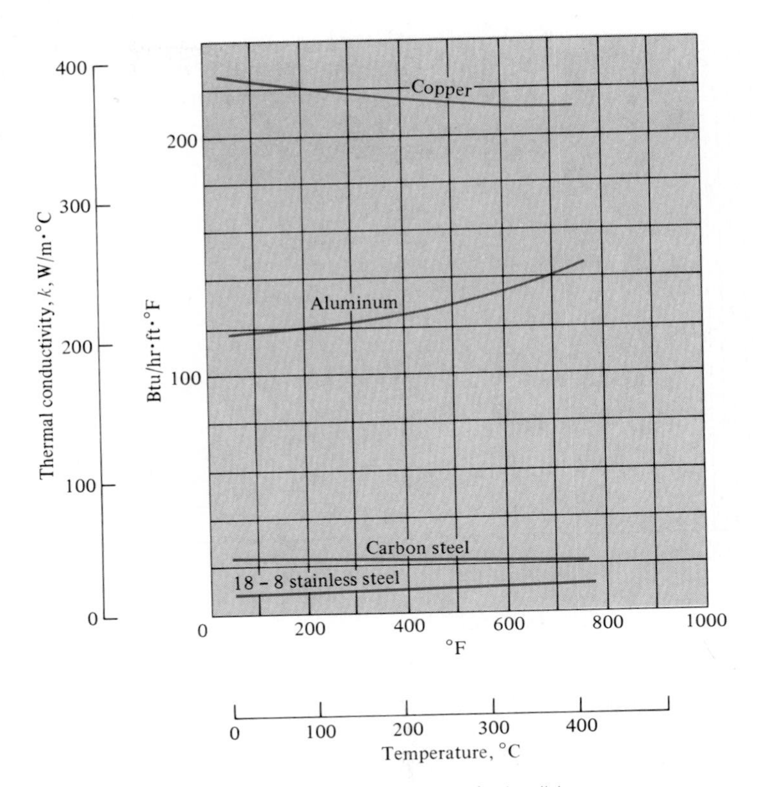

Fig. 1-6 Thermal conductivities of some typical solids.

ism of convection heat transfer and show its relation to the conduction process.

Consider the heated plate shown in Fig. 1-8. The temperature of the plate is T_w, and the temperature of the fluid is T_∞. The velocity of the flow will appear as shown, being reduced to zero at the plate as a result of viscous action. Since the velocity of the fluid layer at the wall will be zero, the heat must be transferred only by conduction at that point. Thus we might compute the heat transfer, using Eq. (1-1), with the thermal conductivity of the fluid and the fluid temperature gradient at the wall. Why, then, if the heat flows by conduction in this layer, do we speak of *convection* heat transfer and need to consider the velocity of the fluid? The answer is that the temperature gradient is dependent on the rate at which the fluid carries the heat away; a high velocity produces a large temperature gradient, and so on. Thus the temperature gradient at the wall depends on the flow field, and we must develop in our later analysis an expression relating the two quantities. Nevertheless, it must be remembered that the physical mechanism of heat transfer at the wall is a conduction process.

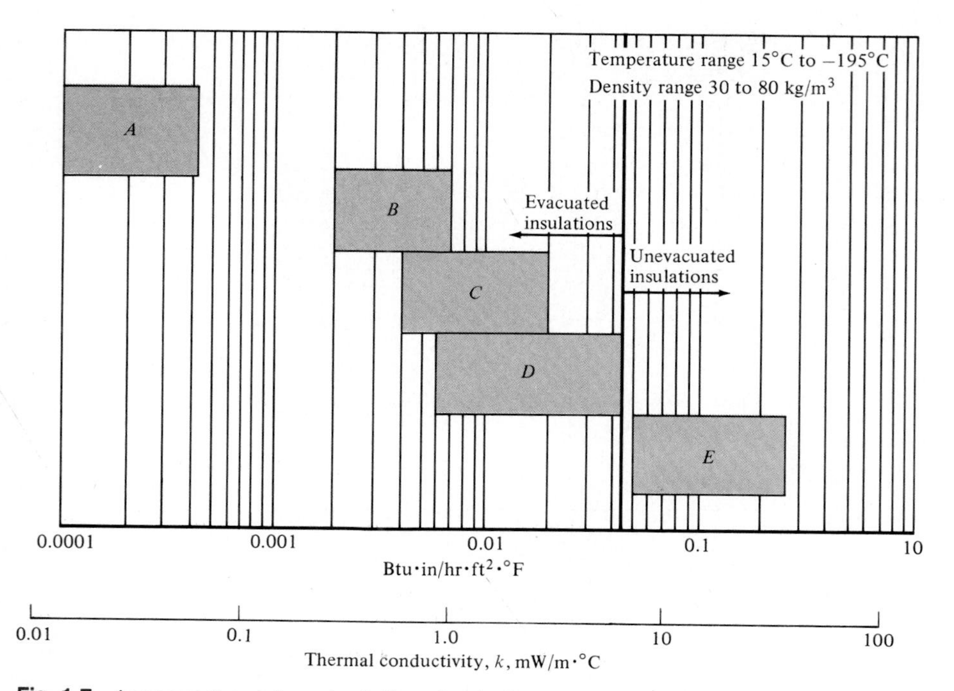

Fig. 1-7 Apparent thermal conductivities of typical cryogenic insulation material: (*a*) multilayer insulations; (*b*) opacified powders; (*c*) glass fibers; (*d*) powders; (*e*) foams, powders, and fibers, according to Ref. 1. [1 Btu in/h · ft² · °F = 144 mW/m · °C]

To express the overall effect of convection, we use Newton's law of cooling:

$$q = hA\,(T_w - T_\infty) \tag{1-8}$$

Here the heat-transfer rate is related to the overall temperature difference between the wall and fluid and the surface area A. The quantity h is called the *convection heat-transfer coefficient,* and Eq. (1-8) is the defining equation. An analytical calculation of h may be made for some systems. For complex situations it must be determined experimentally. The heat-transfer coefficient is

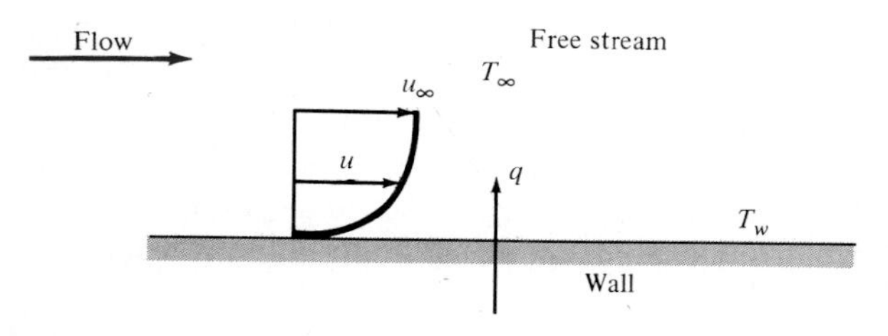

Fig. 1-8 Convection heat transfer from a plate.

sometimes called the *film conductance* because of its relation to the conduction process in the thin stationary layer of fluid at the wall surface. From Eq. (1-8) we note that the units of h are in watts per square meter per Celsius degree when the heat flow is in watts.

In view of the foregoing discussion, one may anticipate that convection heat transfer will have a dependence on the viscosity of the fluid in addition to its dependence on the thermal properties of the fluid (thermal conductivity, specific heat, density). This is expected because viscosity influences the velocity profile and, correspondingly, the energy-transfer rate in the region near the wall.

If a heated plate were exposed to ambient room air without an external

Table 1-2 Approximate Values of Convection Heat-Transfer Coefficients

	h	
Mode	$W/m^2 \cdot °C$	$Btu/h \cdot ft^2 \cdot °F$
Free convection, $\Delta T = 30°C$		
Vertical plate 0.3 m [1 ft] high in air	4.5	0.79
Horizontal cylinder, 5-cm diameter, in air	6.5	1.14
Horizontal cylinder, 2-cm diameter, in water	890	157
Forced convection		
Airflow at 2 m/s over 0.2-m square plate	12	2.1
Airflow at 35 m/s over 0.75-m square plate	75	13.2
Air at 2 atm flowing in 2.5-cm-diameter tube at 10 m/s	65	11.4
Water at 0.5 kg/s flowing in 2.5-cm-diameter tube	3500	616
Airflow *across* 5-cm-diameter cylinder with velocity of 50 m/s	180	32
Boiling water		
In a pool or container	2500–35,000	440–6200
Flowing in a tube	5000–100,000	880–17,600
Condensation of water vapor, 1 atm		
Vertical surfaces	4000–11,300	700–2000
Outside horizontal tubes	9500–25,000	1700–4400

source of motion, a movement of the air would be experienced as a result of the density gradients near the plate. We call this *natural,* or *free,* convection as opposed to *forced* convection, which is experienced in the case of the fan blowing air over a plate. Boiling and condensation phenomena are also grouped under the general subject of convection heat transfer. The approximate ranges of convection heat-transfer coefficients are indicated in Table 1-2.

1-4 RADIATION HEAT TRANSFER

In contrast to the mechanisms of conduction and convection, where energy transfer through a material medium is involved, heat may also be transferred through regions where a perfect vacuum exists. The mechanism in this case is electromagnetic radiation. We shall limit our discussion to electromagnetic radiation which is propagated as a result of a temperature difference; this is called *thermal radiation.*

Thermodynamic considerations show* that an ideal thermal radiator, or *blackbody,* will emit energy at a rate proportional to the fourth power of the absolute temperature of the body and directly proportional to its surface area. Thus

$$q_{\text{emitted}} = \sigma A T^4 \tag{1-9}$$

where σ is the proportionality constant and is called the Stefan-Boltzmann constant with the value of 5.669×10^{-8} W/m² · K⁴. Equation (1-9) is called the Stefan-Boltzmann law of thermal radiation, and it applies only to blackbodies. It is important to note that this equation is valid only for thermal radiation; other types of electromagnetic radiation may not be treated so simply.

Equation (1-9) governs only radiation *emitted* by a blackbody. The net radiant *exchange* between two surfaces will be proportional to the difference in absolute temperatures to the fourth power; i.e.,

$$\frac{q_{\text{net exchange}}}{A} \propto \sigma(T_1^4 - T_2^4) \tag{1-10}$$

We have mentioned that a blackbody is a body which radiates energy according to the T^4 law. We call such a body *black* because black surfaces, like as a piece of metal covered with carbon black, approximate this type of behavior. Other types of surfaces, like a glossy painted surface or a polished metal plate, do not radiate as much energy as the blackbody; however, the total radiation emitted by these bodies still generally follows the T_1^4 proportionality. To take account of the "gray" nature of such surfaces we introduce another factor into Eq. (1-9), called the emissivity ϵ, which relates the radiation of the "gray" surface to that of an ideal black surface. In addition, we must

*See, for example, J. P. Holman, "Thermodynamics," 4th ed., p. 705, McGraw-Hill Book Company, New York, 1988.

take into account the fact that not all the radiation leaving one surface will reach the other surface since electromagnetic radiation travels in straight lines and some will be lost to the surroundings. We therefore introduce two new factors in Eq. (1-9) to take into account both situations, so that

$$q = F_\epsilon F_G \sigma A \, (T_1^4 - T_2^4) \tag{1-11}$$

where F_ϵ is the emissivity function and F_G is the geometric "view factor" function. The determination of the form of these functions for specific configurations is the subject of a subsequent chapter. It is important to alert the reader at this time, however, to the fact that these functions usually are not independent of one another as indicated in Eq. (1-11).

Radiation in an Enclosure

A simple radiation problem is encountered when we have a heat transfer surface at temperature T_1 completely enclosed by a much larger surface maintained at T_2. We will show in Chap. 8 that the net radiant exchange in this case can be calculated with

$$q = \epsilon_1 \sigma A_1 \, (T_1^4 - T_2^4) \tag{1-12}$$

Values of ϵ are given in Appendix A.

Radiation heat-transfer phenomena can be exceedingly complex, and the calculations are seldom as simple as implied by Eq. (1-11). For now, we wish to emphasize the difference in physical mechanism between radiation heat-transfer and conduction-convection systems. In Chap. 8 we examine radiation in detail.

1-5 DIMENSIONS AND UNITS

In this section we outline the systems of units which are used throughout the book. One must be careful not to confuse the meaning of the terms *units* and *dimensions*. A dimension is a physical variable used to specify the behavior or nature of a particular system. For example, the length of a rod is a dimension of the rod. In like manner, the temperature of a gas may be considered one of the thermodynamic dimensions of the gas. When we say the rod is so many meters long, or the gas has a temperature of so many degrees Celsius, we have given the units with which we choose to measure the dimension. In our development of heat transfer we use the dimensions

L = length

M = mass

F = force

τ = time

T = temperature

All the physical quantities used in heat transfer may be expressed in terms of these fundamental dimensions. The units to be used for certain dimensions are selected by somewhat arbitrary definitions which usually relate to a physical phenomenon or law. For example, Newton's second law of motion may be written

$$\text{Force} \sim \text{time rate of change of momentum}$$

$$F = k\frac{d(mv)}{d\tau}$$

where k is the proportionality constant. If the mass is constant,

$$F = kma \tag{1-13}$$

where the acceleration is $a = dv/d\tau$. Equation (1-11) is usually written

$$F = \frac{1}{g_c}ma \tag{1-14}$$

with $1/g_c = k$. Equation (1-14) is used to define our systems of units for mass, force, length, and time. Some typical systems of units are

1. 1-pound force will accelerate a 1-lb mass 32.17 ft/s^2.
2. 1-pound force will accelerate a 1-slug mass 1 ft/s^2.
3. 1-dyne force will accelerate a 1-g mass 1 cm/s^2.
4. 1-newton force will accelerate a 1-kg mass 1 m/s^2.
5. 1-kilogram force will accelerate a 1-kg mass 9.806 m/s^2.

The 1-kg force is sometimes called a *kilopond* (kp).

Since Eq. (1-14) must be dimensionally homogeneous, we shall have a different value of the constant g_c for each of the unit systems in items 1 to 5 above. These values are

1. $g_c = 32.17\ \text{lb}_m \cdot \text{ft/lb}_f \cdot \text{s}^2$
2. $g_c = 1\ \text{slug} \cdot \text{ft/lb}_f \cdot \text{s}^2$
3. $g_c = 1\ \text{g} \cdot \text{cm/dyn} \cdot \text{s}^2$
4. $g_c = 1\ \text{kg} \cdot \text{m/N} \cdot \text{s}^2$
5. $g_c = 9.806\ \text{kg}_m \cdot \text{m/kg}_f \cdot \text{s}^2$

It matters not which system of units is used so long as it is consistent with the above definitions.

Work has the dimensions of a product of force times a distance. Energy has the same dimensions. The units for work and energy may be chosen from any of the systems used above, and would be

1. $lb_f \cdot ft$

2. $lb_f \cdot ft$

3. dyn · cm = 1 erg

4. N · m = 1 joule (J)

5. $kg_f \cdot m$ = 9.806 J

In addition, we may use the units of energy which are based on thermal phenomena:

1 Btu will raise 1 lb_m of water 1°F at 68°F.

1 cal will raise 1 g of water 1°C at 20°C.

1 kcal will raise 1 kg of water 1°C at 20°C.

Some conversion factors for the various units of work and energy are

$$1 \text{ Btu} = 778.16 \text{ lb}_f \cdot \text{ft}$$
$$1 \text{ Btu} = 1055 \text{ J}$$
$$1 \text{ kcal} = 4182 \text{ J}$$
$$1 \text{ lb}_f \cdot \text{ft} = 1.356 \text{ J}$$
$$1 \text{ Btu} = 252 \text{ cal}$$

Other conversion factors are given in Appendix A.

The weight of a body is defined as the force exerted on the body as a result of the acceleration of gravity. Thus

$$W = \frac{g}{g_c} m \tag{1-15}$$

where W is the weight and g is the acceleration of gravity. Note that the weight of a body has the dimensions of a force. We now see why systems 1 and 5 above were devised; 1 lb_m will weigh 1 lb_f at sea level, and 1 kg_m will weigh 1 kg_f.

Temperature conversions are performed with the familiar formulas

$$°F = \tfrac{9}{5}°C + 32$$
$$°R = °F + 459.69$$
$$K = °C + 273.16$$
$$°R = \tfrac{9}{5} K$$

Unfortunately, *all* the above unit systems are used in various places through-

Table 1-3 Multiplier Factors for SI Units

Multiplier	*Prefix*	*Abbreviation*
10^{12}	tera	T
10^{9}	giga	G
10^{6}	mega	M
10^{3}	kilo	k
10^{2}	hecto	h
10^{-2}	centi	c
10^{-3}	milli	m
10^{-6}	micro	μ
10^{-9}	nano	n
10^{-12}	pico	p
10^{-18}	atto	a

out the world. While the food-pound force, pound mass, second, degree Fahrenheit, Btu system is still widely used in the United States, there is increasing impetus to institute the SI (Système International d'Unités) units as a worldwide standard. In this system, the fundamental units are meter, newton, kilogram mass, second, and degrees Celsius; a "thermal" energy unit is not used; i.e., the joule (newton-meter) becomes the energy unit used throughout. The watt (joules per second) is the unit of power in this system. In the SI system, the standard units for thermal conductivity would become

$$k \text{ in W/m} \cdot {}^\circ\text{C}$$

and the convection heat-transfer coefficient would be expressed as

$$h \text{ in W/m}^2 \cdot {}^\circ\text{C}$$

Because SI units are so straightforward we shall use them as the standard in this text, with intermediate steps and answers in examples also given parenthetically in the Btu–pound mass system. A worker in heat transfer must obtain a feel for the order of magnitudes in both systems. In the SI system the concept of g_c is not normally used, and the newton is *defined* as

$$1 \text{ N} = 1 \text{ kg} \cdot \text{m/s}^2 \tag{1-16}$$

Even so, one should keep in mind the physical relation between force and mass as expressed by Newton's second law of motion.

The SI system also specifies standard multiples to be used to conserve space when numerical values are expressed. They are summarized in Table 1-3. Standard symbols for quantities normally encountered in heat transfer are summarized in Table 1-4. Conversion factors are given in Appendix A.

■ **EXAMPLE 1-1** Conduction through copper plate

One face of a copper plate 3 cm thick is maintained at 400°C, and the other face is maintained at 100°C. How much heat is transferred through the plate?

Table 1-4 SI Quantities Used in Heat Transfer

Quantity	*Unit abbreviation*
Force	N (newton)
Mass	kg (kilogram mass)
Time	s (second)
Length	m (meter)
Temperature	°C or K
Energy	J (joule)
Power	W (watt)
Thermal conductivity	W/m · °C
Heat-transfer coefficient	W/m^2 · °C
Specific heat	J/kg · °C
Heat flux	W/m^2

Solution

From Appendix A the thermal conductivity for copper is 370 W/m · °C at 250°C. From Fourier's law

$$\frac{q}{A} = -k\frac{dT}{dx}$$

Integrating gives

$$\frac{q}{A} = -k\frac{\Delta T}{\Delta x} = \frac{-(370)(100 - 400)}{3 \times 10^{-2}} = 3.7 \text{ MW/m}^2 \quad [1.172 \times 10^6 \text{ Btu/h} \cdot \text{ft}^2]$$

■ **EXAMPLE 1-2** Convection calculation

Air at 20°C blows over a hot plate 50 by 75 cm maintained at 250°C. The convection heat-transfer coefficient is 25 W/m^2 · °C. Calculate the heat transfer.

Solution

From Newton's law of cooling

$$\begin{aligned} q &= hA\,(T_w - T_\infty) \\ &= (25)(0.50)(0.75)(250 - 20) \\ &= 2.156 \text{ kW} \quad [7356 \text{ Btu/h}] \end{aligned}$$

■ **EXAMPLE 1-3** Multimode heat transfer

Assuming that the plate in Ex. 1-2 is made of carbon steel (1%) 2 cm thick and that 300 W is lost from the plate surface by radiation, calculate the inside plate temperature.

Solution

The heat conducted through the plate must be equal to the sum of convection and radiation heat losses:

$$q_{\text{cond}} = q_{\text{conv}} + q_{\text{rad}}$$

$$-kA\frac{\Delta T}{\Delta x} = 2.156 + 0.3 = 2.456 \text{ kW}$$

$$\Delta T = \frac{(-2456)(0.02)}{(0.5)(0.75)(43)} = -3.05°\text{C} \quad [-5.49°\text{C}]$$

where the value of k is taken from Table 1-1. The inside plate temperature is therefore

$$T_i = 250 + 3.05 = 253.05°\text{C}$$

■ **EXAMPLE 1-4** Heat source and convection

An electric current is passed through a wire 1 mm in diameter and 10 cm long. The wire is submerged in liquid water at atmospheric pressure, and the current is increased until the water boils. For this situation $h = 5000 \text{ W/m}^2 \cdot °\text{C}$, and the water temperature will be 100°C. How much electric power must be supplied to the wire to maintain the wire surface at 114°C?

Solution

The total convection loss is given by Eq. (1-8):

$$q = hA\,(T_w - T_\infty)$$

For this problem the surface area of the wire is

$$A = \pi dL = \pi(1 \times 10^{-3})(10 \times 10^{-2}) = 3.142 \times 10^{-4} \text{ m}^2$$

The heat transfer is therefore

$$q = (5000 \text{ W/m}^2 \cdot °\text{C})(3.142 \times 10^{-4} \text{ m}^2)(114 - 100) = 21.99 \text{ W} \quad [75.03 \text{ Btu/h}]$$

and this is equal to the electric power which must be applied.

■ **EXAMPLE 1-5** Radiation heat transfer

Two infinite black plates at 800 and 300°C exchange heat by radiation. Calculate the heat transfer per unit area.

Solution

Equation (1-10) may be employed for this problem, so we find immediately

$$\begin{aligned} q/A &= \sigma(T_1^4 - T_2^4) \\ &= (5.669 \times 10^{-8})(1073^4 - 573^4) \\ &= 69.03 \text{ kW/m}^3 \; [21{,}884 \text{ Btu/h} \cdot \text{ft}^2] \end{aligned}$$

■ **EXAMPLE 1-6** Total heat loss

A horizontal steel pipe having a diameter of 5 cm is maintained at a temperature of 50°C in a large room where the air and wall temperature are at 20°C. The surface emissivity of the steel may be taken as 0.8. Using the data of Table 1-2 calculate the total heat lost by the pipe per unit length.

Solution

The total heat loss is the sum of convection and radiation. From Table 1-2 we see that an estimate for the heat transfer coefficient for *free* convection with this geometry and air is $h = 6.5\ \text{W/m}^2 \cdot {}^\circ\text{C}$. The surface area is $\pi\, d\, L$, so the convection loss per unit length is

$$q/L]_{\text{conv}} = h(\pi d)\,(T_w - T_\infty)$$
$$= (6.5)(\pi)(0.05)(50 - 20) = 30.63\ \text{W/m}$$

The pipe is a body surrounded by a large enclosure so the radiation heat transfer can be calculated from Eq. (1-12). With $T_1 = 50^\circ\text{C} = 323^\circ\text{K}$ and $T_2 = 20^\circ\text{C} = 293^\circ\text{K}$, we have

$$q/L = \epsilon_1\,(\pi d_1)\sigma(T_1^4 - T_2^4)$$
$$= (0.8)(\pi)(0.05)(5.669 \times 10^{-8})(323^4 - 293^4)$$
$$= 25.04\ \text{W/m}$$

The total heat loss is therefore

$$q/L]_{\text{tot}} = q/L]_{\text{conv}} + q/L]_{\text{rad}}$$
$$= 30.63 + 25.04 = 55.67\ \text{W/m}$$

In this example we see that the convection and radiation are about the same. To neglect either would be a serious mistake.

1-6 COMPUTER SOLUTION OF HEAT-TRANSFER PROBLEMS

Many practical problems in heat transfer are solved with computer techniques. In the chapters which follow, many examples will be presented to illustrate setups for computer solutions, and many end-of-chapter problems will pose interesting questions to be answered by the reader. But this is not to play down analysis or appreciation of the fundamental principles of the subject. These principles are even more important in computer work because the engineer must be extra careful about input variables and have a notion of the correct magnitude of the output from the computer.

The reader may note the absence of specific computer programs throughout the text. There are many ways to approach a solution and the assumption is made that the reader is "computer literate" and can handle the solution once a basic algorithm is laid out. Thus, our general approach to computer-related problems is to:

1. Set up the problem with diagrams and equations.
2. Set the objectives of the analysis.
3. Lay out a computation algorithm, where appropriate.
4. Obtain a solution.
5. If necessary, consider and evaluate alternative solutions.

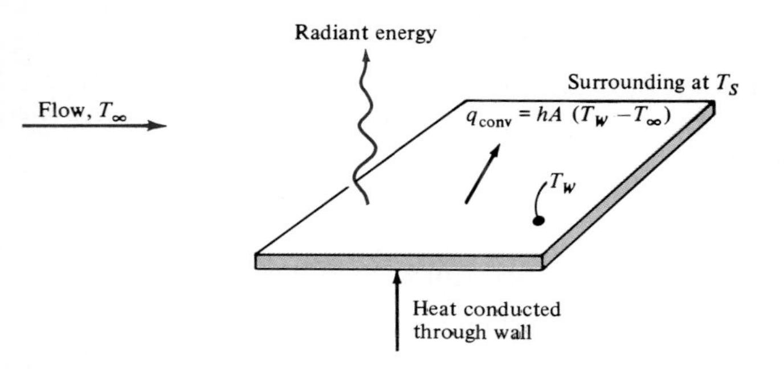

Fig. 1-9 Combination of conduction, convection, and radiation heat transfer.

1-7 SUMMARY

We may summarize our introductory remarks very simply. Heat transfer may take place by one or more of three modes: conduction, convection, and radiation. It has been noted that the physical mechanism of convection is related to the heat conduction through the thin layer of fluid adjacent to the heat-transfer surface. In both conduction and convection Fourier's law is applicable, although fluid mechanics must be brought into play in the convection problem in order to establish the temperature gradient.

Radiation heat transfer involves a different physical mechanism—that of propagation of electromagnetic energy. To study this type of energy transfer we introduce the concept of an ideal radiator, or blackbody, which radiates energy at a rate proportional to its absolute temperature to the fourth power.

It is easy to envision cases in which all three modes of heat transfer are present, as in Fig. 1-9. In this case the heat conducted through the plate is removed from the plate surface by a combination of convection and radiation. An energy balance would give

$$-kA\left.\frac{dT}{dy}\right]_{wall} = hA\ (T_w - T_\infty) + F_\epsilon F_G \sigma A\ (T_w^4 - T_s^4)$$

where T_s = temperature of surroundings
T_w = surface temperature
T_∞ = fluid temperature

To apply the science of heat transfer to practical situations, a thorough knowledge of all three modes of heat transfer must be obtained.

REVIEW QUESTIONS

1 Define thermal conductivity.

2 Define the convection heat-transfer coefficient.

3 Discuss the mechanism of thermal conduction in gases and solids.

4 Discuss the mechanism of heat convection.

5 What is the order of magnitude for the convection heat transfer coefficient in free convection? Forced convection? Boiling?

6 When may one expect radiation heat transfer to be important?

7 Name some good conductors of heat; some poor conductors.

8 What is the order of magnitude of thermal conductivity for (*a*) metals, (*b*) solid insulating materials, (*c*) liquids, (*d*) gases?

9 Suppose a person stated that heat can not be transfered in a vacuum. How do you respond?

10 Review any standard text on thermodynamics and define: (*a*) heat, (*b*) internal energy, (*c*) work, (*d*) enthalpy.

11 Define and discuss g_c.

PROBLEMS

1-1 If 3 kW is conducted through a section of insulating material 1.0 m^2 in cross section and 2.5 cm thick and the thermal conductivity may be taken as 0.2 W/m · °C, compute the temperature difference across the material.

1-2 A temperature difference of 85°C is impressed across a fiber-glass layer of 13 cm thickness. The thermal conductivity of the fiber glass is 0.035 W/m · °C. Compute the heat transferred through the material per hour per unit area.

1-3 A truncated cone 30 cm high is constructed of aluminum. The diameter at the top is 7.5 cm, and the diameter at the bottom is 12.5 cm. The lower surface is maintained at 93°C; the upper surface, at 540°C. The other surface is insulated. Assuming one-dimensional heat flow, what is the rate of heat transfer in watts?

1-4 The temperatures on the faces of a plane wall 15 cm thick are 370 and 93°C. The wall is constructed of a special glass with the following properties: $k = 0.78$ W/m · °C, $\rho = 2700$ kg/m³, $c_p = 0.84$ kJ/kg · °C. What is the heat flow through the wall at steady-state conditions?

1-5 A certain superinsulation material having a thermal conductivity of 2×10^{-4} W/m · °C is used to insulate a tank of liquid nitrogen that is maintained at −320°F; 85.8 Btu is required to vaporize each pound mass of nitrogen at this temperature. Assuming that the tank is a sphere having an inner diameter (ID) of 2 ft, estimate the amount of nitrogen vaporized per day for an insulation thickness of 1.0 in and an ambient temperature of 70°F. Assume that the outer temperature of the insulation is 70°F.

1-6 Rank the following materials in order of (*a*) transient response and (*b*) steady-state conduction. Taking the material with the highest rank, give the other materials as a percentage of the maximum: aluminum, copper, silver, iron, lead, chrome steel (18% Cr, 8% Ni), magnesium. What do you conclude from this ranking?

1-7 A 50-cm-diameter pipeline in the Arctic carries hot oil at 30°C and is exposed to a surrounding temperature of $-20°C$. A special powder insulation 5 cm thick surrounds the pipe and has a thermal conductivity of 7 mW/m · °C. The convection heat-transfer coefficient on the outside of the pipe is 12 W/m^2 · °C. Estimate the energy loss from the pipe per meter of length.

1-8 A 5-cm layer of loosely packed asbestos is placed between two plates at 100 and 200°C. Calculate the heat transfer across the layer.

1-9 A certain insulation has a thermal conductivity of 10 mW/m · °C. What thickness is necessary to effect a temperature drop of 500°C? What would be the heat flow under these conditions?

1-10 Assuming that the heat transfer to the sphere is Prob. 1-5 occurs by free convection with a heat-transfer coefficient of 2.7 W/m^2 · °C, calculate the temperature difference between the outer surface of the sphere and the environment.

1-11 Two perfectly black surfaces are constructed so that all the radiant energy leaving a surface at 800°C reaches the other surface. The temperature of the other surface is maintained at 250°C. Calculate the heat transfer between the surfaces per hour and per unit area of the surface maintained at 800°C.

1-12 Two very large parallel planes having surface conditions which very nearly approximate those of a blackbody are maintained at 1100 and 425°C, respectively. Calculate the heat transfer by radiation between the planes per unit time and per unit surface area.

1-13 Calculate the radiation heat exchange in 1 day between two black planes having the area of the surface of a 2-ft-diameter sphere when the planes are maintained at -320 and 70°F. What does this calculation indicate in regard to Prob. 1-5?

1-14 Two infinite black plates at 500 and 100°C exchange heat by radiation. Calculate the heat-transfer rate per unit area. If another perfectly black plate is placed between the 500 and 100°C plates, by how much is the heat transfer reduced? What is the temperature of the center plate?

1-15 Water flows at the rate of 0.5 kg/s in a 2.5-cm-diameter tube having a length of 3 m. A constant heat flux is imposed at the tube wall so that the tube wall temperature is 40°C higher than the water temperature. Calculate the heat transfer and estimate the temperature rise in the water. The water is pressurized so that boiling can not occur.

1-16 Steam at 1 atm pressure ($T_{sat} = 100°C$) is exposed to a 30-by-30-cm vertical square plate which is cooled such that 3.78 kg/h is condensed. Calculate the plate temperature. Consult steam tables for any necessary properties.

1-17 Boiling water at 1 atm may require a surface heat flux of 3×10^4 Btu/h · ft^2 for a surface temperature of 232°F. What is the value of the heat-transfer coefficient?

1-18 A small radiant heater has metal strips 6 mm wide with a total length of 3 m. The surface emissivity of the strips is 0.85. To what temperature must the strips be heated if they are to dissipate 1600 W of heat to a room at 25°C?

1-19 Calculate the energy emitted by a blackbody at 1000°C.

1-20 If the radiant flux from the sun is 1350 W/m^2, what would be its equivalent blackbody temperature?

1-21 A 4.0-cm-diameter sphere is heated to a temperature of 150°C and is enclosed in a large room at 20°C. Calculate the radiant heat loss if the surface emissivity is 0.65.

1-22 A flat wall is exposed to an environmental temperature of 38°C. The wall is covered with a layer of insulation 2.5 cm thick whose thermal conductivity is 1.4 W/m · °C, and the temperature of the wall on the inside of the insulation is 315°C. The wall loses heat to the environment by convection. Compute the value of the convection heat-transfer coefficient which must be maintained on the outer surface of the insulation to ensure that the outer-surface temperature does not exceed 41°C.

1-23 Consider a wall heated by convection on one side and cooled by convection on the other side. Show that the heat-transfer rate through the wall is

$$q = \frac{T_1 - T_2}{1/h_1A + \Delta x/kA + 1/h_2A}$$

where T_1 and T_2 are the fluid temperatures on each side of the wall and h_1 and h_2 are the corresponding heat-transfer coefficients.

1-24 One side of a plane wall is maintained at 100°C, while the other side is exposed to a convection environment having $T = 10$°C and $h = 10$ W/m² · °C. The wall has $k = 1.6$ W/m · °C and is 40 cm thick. Calculate the heat-transfer rate through the wall.

1-25 How does the free-convection heat transfer from a vertical plate compare with pure conduction through a vertical layer of air having a thickness of 2.5 cm and a temperature difference the same at $T_w - T_\infty$? Use information from Table 1-2.

1-26 A $\frac{1}{4}$-in steel plate having a thermal conductivity of 25 Btu/h · ft · °F is exposed to a radiant heat flux of 1500 Btu/h · ft² in a vacuum space where the convection heat transfer is negligible. Assuming that the surface temperature of the steel exposed to the radiant energy is maintained at 100°F, what will be the other surface temperature if all the radiant energy striking the plate is transferred through the plate by conduction?

1-27 A solar radiant heat flux of 700 W/m² is absorbed in a metal plate which is perfectly insulated on the back side. The convection heat-transfer coefficient on the plate is 11 W/m² · °C, and the ambient air temperature is 30°C. Calculate the temperature of the plate under equilibrium conditions.

1-28 A 5.0-cm-diameter cylinder is heated to a temperature of 200°C, and air at 30°C is forced across it at a velocity of 50 m/s. If the surface emissivity is 0.7, calculate the total heat loss per unit length if the walls of the enclosing room are at 10°C. Comment on this calculation.

1-29 A vertical square plate, 30 cm on a side, is maintained at 50°C and exposed to room air at 20°C. The surface emissivity is 0.8. Calculate the total heat lost by both sides of the plate.

1-30 A black 20-by-20-cm plate has air forced over it at a velocity of 2 m/s and a temperature of 0°C. The plate is placed in a large room whose walls are at 30°C. The back side of the plate is perfectly insulated. Calculate the temperature of the plate resulting from the convection-radiation balance. Use information from Table 1-2. Are you surprised at the result?

1-31 Two large black plates are separated by a vacuum. On the outside of one plate is a convection environment of $T = 80°C$ and $h = 100\ W/m^2 \cdot °C$, while the outside of the other plate is exposed to 20°C and $h = 15\ W/m^2 \cdot °C$. Make an energy balance on the system and determine the plate temperatures. For this problem $F_G = F_\epsilon = 1.0$.

1-32 Using the basic definitions of units and dimensions given in Sec. 1-5, arrive at expressions (*a*) to convert joules to British thermal units, (*b*) to convert dyne-centimeters to joules, (*c*) to convert British thermal units to calories.

1-33 Beginning with the three-dimensional heat-conduction equation in cartesian coordinates [Eq. (1-3*a*)], obtain the general heat-conduction equation in cylindrical coordinates [Eq. (1-3*b*)].

1-34 A woman informs an engineer that she frequently feels cooler in the summer when standing in front of an open refrigerator. The engineer tells her that she is only "imagining things" because there is no fan in the refrigerator to blow the cool air over her. A lively argument ensues. Whose side of the argument do you take? Why?

1-35 A woman informs her engineer husband that "hot water will freeze faster than cold water." He calls this statement nonsense. She answers by saying that she has actually timed the freezing process for ice trays in the home refrigerator and found that hot water does indeed freeze faster. As a friend, you are asked to settle the argument. Is there any logical explanation for the woman's observation?

1-36 An air-conditioned classroom in Texas is maintained at 72°F in the summer. The students attend classes in shorts, sandals, and skimpy shirts and are quite comfortable. In the same classroom during the winter, the same students wear wool slacks, long-sleeve shirts, and sweaters and are equally comfortable with the room temperature maintained at 75°F. Assuming that humidity is not a factor, explain this apparent anomaly in "temperature comfort."

1-37 Write the simplified heat-conduction equation for (*a*) steady one-dimensional heat flow in cylindrical coordinates in the *azimuth* (ϕ) direction and (*b*) steady one-dimensional heat flow in spherical coordinates in the azimuth (ϕ) direction.

1-38 A vertical cylinder 6 ft tall and 1 ft in diameter might be used to approximate a man for heat-transfer purposes. Suppose the surface temperature of the cylinder is 78°F, $h = 2\ Btu/h \cdot ft^2 \cdot °F$, the surface emissivity is 0.9, and the cylinder is placed in a large room where the air temperature is 68°F and the wall temperature is 45°F. Calculate the heat lost from the cylinder. Repeat for a wall temperature of 80°F. What do you conclude from these calculations?

■ REFERENCES

1. Glaser, P. E., I. A. Black, and P. Doherty: Multilayer Insulation, *Mech. Eng.*, August 1965, p. 23.
2. Barron, R.: "Cryogenic Systems," McGraw-Hill Book Company, New York, 1967.
3. Dewitt, W. D., N. C. Gibbon, and R. L. Reid,: Multifoil Type Thermal Insulation, *IEEE Trans. Aerosp. Electron. Syst.*, vol. 4, no. 5, suppl. pp. 263–271, 1968.

STEADY-STATE CONDUCTION—ONE DIMENSION

2-1 INTRODUCTION

We now wish to examine the applications of Fourier's law of heat conduction to calculation of heat flow in some simple one-dimensional systems. Several different physical shapes may fall in the category of one-dimensional systems: cylindrical and spherical systems are one-dimensional when the temperature in the body is a function only of radial distance and is independent of azimuth angle or axial distance. In some two-dimensional problems the effect of a second-space coordinate may be so small as to justify its neglect, and the multidimensional heat-flow problem may be approximated with a one-dimensional analysis. In these cases the differential equations are simplified, and we are led to a much easier solution as a result of this simplification.

2-2 THE PLANE WALL

First consider the plane wall where a direct application of Fourier's law [Eq. (1-1)] may be made. Integration yields

$$q = -\frac{kA}{\Delta x}(T_2 - T_1) \tag{2-1}$$

when the thermal conductivity is considered constant. The wall thickness is Δx, and T_1 and T_2 are the wall-face temperatures. If the thermal conductivity varies with temperature according to some linear relation $k = k_0(1 + \beta T)$, the resultant equation for the heat flow is

$$q = -\frac{k_0 A}{\Delta x}\left[(T_2 - T_1) + \frac{\beta}{2}(T_2^2 - T_1^2)\right] \tag{2-2}$$

If more than one material is present, as in the multilayer wall shown in Fig. 2-1, the analysis would proceed as follows: The temperature gradients in the three materials are shown, and the heat flow may be written

$$q = -k_A A \frac{T_2 - T_1}{\Delta x_A} = -k_B A \frac{T_3 - T_2}{\Delta x_B} = -k_C A \frac{T_4 - T_3}{\Delta x_C}$$

Note that the heat flow must be the same through all sections.

Solving these three equations simultaneously, the heat flow is written

$$q = \frac{T_1 - T_4}{\Delta x_A/k_A A + \Delta x_B/k_B A + \Delta x_C/k_C A} \tag{2-3}$$

At this point we retrace our development slightly to introduce a different conceptual viewpoint for Fourier's law. The heat-transfer rate may be considered as a flow, and the combination of thermal conductivity, thickness of material, and area as a resistance to this flow. The temperature is the potential, or driving, function for the heat flow, and the Fourier equation may be written

$$\text{Heat flow} = \frac{\text{thermal potential difference}}{\text{thermal resistance}} \tag{2-4}$$

a relation quite like Ohm's law in electric-circuit theory. In Eq. (2-1) the thermal resistance is $\Delta x/kA$, and in Eq. (2-3) it is the sum of the three terms in the denominator. We should expect this situation in Eq. (2-3) because the three walls side by side act as three thermal resistances in series. The equivalent electric circuit is shown in Fig. 2-1*b*.

The electrical analogy may be used to solve more complex problems involving both series and parallel thermal resistances. A typical problem and its analogous electric circuit are shown in Fig. 2-2. The one-dimensional heat-flow equation for this type of problem may be written

$$q = \frac{\Delta T_{\text{overall}}}{\Sigma R_{\text{th}}} \tag{2-5}$$

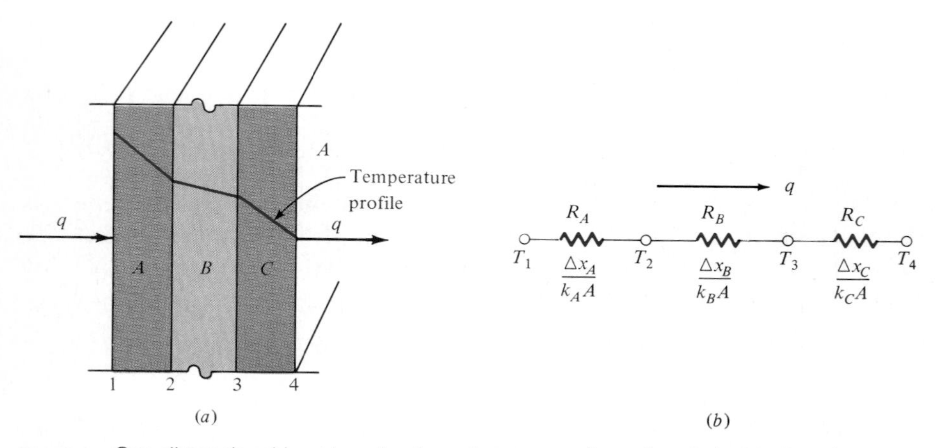

Fig. 2-1 One-dimensional heat transfer through a composite wall and electrical analog.

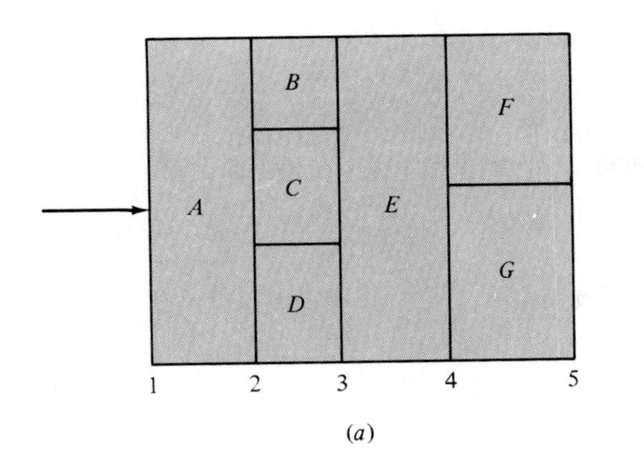

(a)

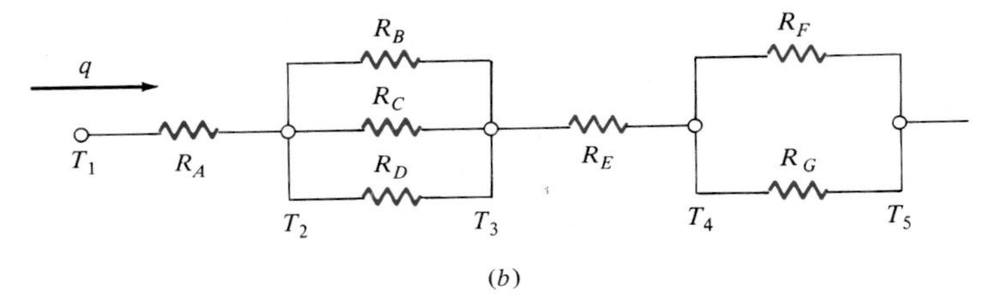

(b)

Fig. 2-2 Series and parallel one-dimensional heat transfer through a composite wall and electrical analog.

where the R_{th} are the thermal resistances of the various materials. The units for the thermal resistance are °C/W or °F · h/Btu.

It is well to mention that in some systems like that in Fig. 2-2 two-dimensional heat flow may result if the thermal conductivities of materials B, C, and D differ by an appreciable amount. In these cases other techniques must be employed to effect a solution.

2-3 INSULATION AND *R* VALUES

In Chap. 1 we noted that the thermal conductivities for a number of insulating materials are given in Appendix A. In classifying the performance of insulation, it is a common practice in the building industry to use a term called the *R value*, which is defined as

$$R = \frac{\Delta T}{q/A} \tag{2-6}$$

The units for R are °C · m²/W or °F · ft² · h/Btu. Note that this differs from the thermal-resistance concept discussed above in that a heat flow *per unit area* is used.

At this point it is worthwhile to classify insulation materials in terms of their application and allowable temperature ranges. Table 2-1 furnishes such information and may be used as a guide for the selection of insulating materials.

Table 2-1 Insulation Types and Applications

Type	Temperature range, °C	Thermal conductivity, mW/m · °C	Density, kg/m³	Application
1 Linde evacuated superinsulation	−240–1100	0.0015–0.72	Variable	Many
2 Urethane foam	−180–150	16–20	25–48	Hot and cold pipes
3 Urethane foam	−170–110	16–20	32	Tanks
4 Cellular glass blocks	−200–200	29–108	110–150	Tanks and pipes
5 Fiber-glass blanket for wrapping	−80–290	22–78	10–50	Pipe and pipe fittings
6 Fiber-glass blankets	−170–230	25–86	10–50	Tanks and equipment
7 Fiber-glass preformed shapes	−50–230	32–55	10–50	Piping
8 Elastomeric sheets	−40–100	36–39	70–100	Tanks
9 Fiber-glass mats	60–370	30–55	10–50	Pipe and pipe fittings
10 Elastomeric preformed shapes	−40–100	36–39	70–100	Pipe and fittings
11 Fiber glass with vapor barrier blanket	−5–70	29–45	10–32	Refrigeration lines
12 Fiber glass without vapor barrier jacket	to 250	29–45	24–48	Hot piping
13 Fiber-glass boards	20–450	33–52	25–100	Boilers, tanks, heat exchangers
14 Cellular glass blocks and boards	20–500	29–108	110–150	Hot piping
15 Urethane foam blocks and boards	100–150	16–20	24–65	Piping
16 Mineral fiber preformed shapes	to 650	35–91	125–160	Hot piping
17 Mineral fiber blankets	to 750	37–81	125	Hot piping
18 Mineral wool blocks	450–1000	52–130	175–290	Hot piping
19 Calcium silicate blocks, boards	230–1000	32–85	100–160	Hot piping, boilers, chimney linings
20 Mineral fiber blocks	to 1100	52–130	210	Boilers and tanks

2-4 RADIAL SYSTEMS—CYLINDERS

Consider a long cylinder of inside radius r_i, outside radius r_o, and length L, such as the one shown in Fig. 2-3. We expose this cylinder to a temperature differential $T_i - T_o$ and ask what the heat flow will be. For a cylinder with length very large compared to diameter, it may be assumed that the heat flows only in a radial direction, so that the only space coordinate needed to specify the system is r. Again, Fourier's law is used by inserting the proper area relation. The area for heat flow in the cylindrical system is

$$A_r = 2\pi rL$$

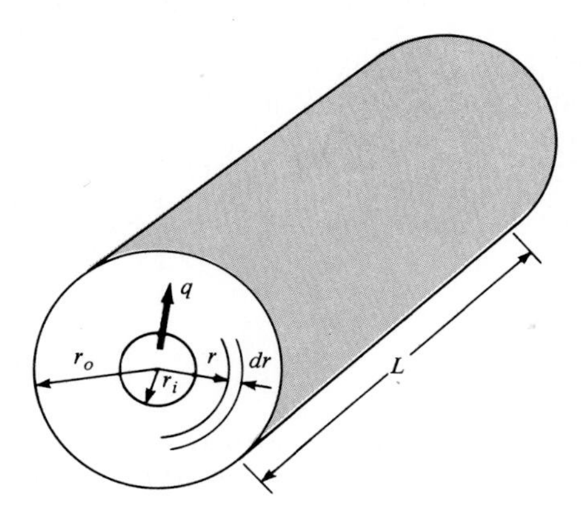

Fig. 2-3 One-dimensional heat flow through a hollow cylinder and electrical analog.

so that Fourier's law is written

$$q_r = -kA_r \frac{dT}{dr}$$

or (2-7)

$$q_r = -2\pi krL \frac{dT}{dr}$$

with the boundary conditions

$$T = T_i \qquad \text{at } r = r_i$$

$$T = T_o \qquad \text{at } r = r_o$$

The solution to Eq. (2-7) is

$$q = \frac{2\pi kL(T_i - T_o)}{\ln(r_o/r_i)} \tag{2-8}$$

and the thermal resistance in this case is

$$R_{th} = \frac{\ln(r_o/r_i)}{2\pi kL}$$

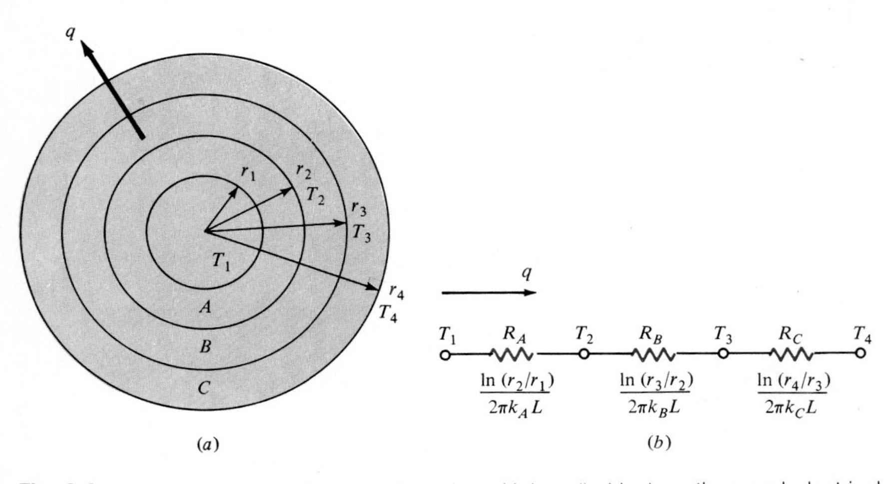

Fig. 2-4 One-dimensional heat flow through multiple cylindrical sections and electrical analog.

The thermal-resistance concept may be used for multiple-layer cylindrical walls just at it was used for plane walls. For the three-layer system shown in Fig. 2-4 the solution is

$$q = \frac{2\pi L(T_1 - T_4)}{\ln (r_2/r_1)/k_A + \ln (r_3/r_2)/k_B + \ln (r_4/r_3)/k_C} \tag{2-9}$$

The thermal circuit is shown in Fig. 2-4*b*.

Spherical systems may also be treated as one-dimensional when the temperature is a function of radius only. The heat flow is then

$$q = \frac{4\pi k(T_i - T_o)}{1/r_i - 1/r_o} \tag{2-10}$$

The derivation of Eq. (2-10) is left as an exercise.

■ **EXAMPLE 2-1** Multilayer conduction

An exterior wall of a house may be approximated by a 4-in layer of common brick [k = 0.7 W/m · °C] followed by a 1.5-in layer of gypsum plaster [k = 0.48 W/m · °C]. What thickness of loosely packed rock-wool insulation [k = 0.065 W/m · °C] should be added to reduce the heat loss (or gain) through the wall by 80 percent?

Solution

The overall heat loss will be given by

$$q = \frac{\Delta T}{\Sigma R_{\text{th}}}$$

Because the heat loss with the rock-wool insulation will be only 20 percent (80 percent reduction) of that before insulation

$$\frac{q \text{ with insulation}}{q \text{ without insulation}} = 0.2 = \frac{\Sigma R_{\text{th}} \text{ without insulation}}{\Sigma R_{\text{th}} \text{ with insulation}}$$

We have for the brick and plaster, for unit area,

$$R_b = \frac{\Delta x}{k} = \frac{(4)(0.0254)}{0.7} = 0.145 \text{ m}^2 \cdot {}^\circ\text{C/W}$$

$$R_p = \frac{\Delta x}{k} = \frac{(1.5)(0.0254)}{0.48} = 0.079 \text{ m}^2 \cdot {}^\circ\text{C/W}$$

so that the thermal resistance without insulation is

$$R = 0.145 + 0.079 = 0.224 \text{ m}^2 \cdot {}^\circ\text{C/W}$$

Then $$R \text{ with insulation} = \frac{0.224}{0.2} = 1.122 \text{ m}^2 \cdot {}^\circ\text{C/W}$$

and this represents the sum of our previous value and the resistance for the rock wool

$$1.122 = 0.224 + R_{rw}$$

$$R_{rw} = 0.898 = \frac{\Delta x}{k} = \frac{\Delta x}{0.065}$$

so that $$\Delta x_{rw} = 0.0584 \text{ m} = 2.30 \text{ in}$$

■ **EXAMPLE 2-2** Multilayer cylindrical system

A thick-walled tube of stainless steel [18% Cr, 8% Ni, $k = 19$ W/m · °C] with 2-cm inner diameter (ID) and 4-cm outer diameter (OD) is covered with a 3-cm layer of asbestos insulation [$k = 0.2$ W/m · °C]. If the inside wall temperature of the pipe is maintained at 600°C, calculate the heat loss per meter of length.

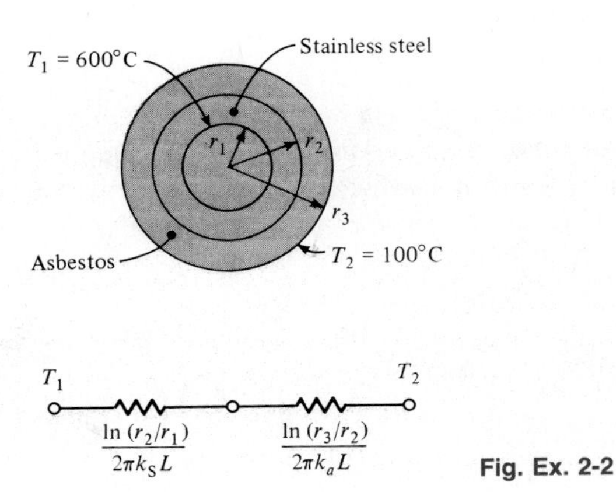

Fig. Ex. 2-2

Solution

The accompanying figure shows the thermal network for this problem. The heat flow is given by

$$\frac{q}{L} = \frac{2\pi(T_1 - T_2)}{\ln(r_2/r_1)/k_s + \ln(r_3/r_2)/k_a} = \frac{2\pi(600 - 100)}{(\ln 2)/19 + (\ln\frac{5}{2})/0.2} = 680 \text{ W/m}$$

Convection Boundary Conditions

We have already seen in Chap. 1 that convection heat transfer can be calculated from

$$q_{\text{conv}} = hA\,(T_w - T_\infty)$$

An electric-resistance analogy can also be drawn for the convection process by rewriting the equation as

$$q_{\text{conv}} = \frac{T_w - T_\infty}{1/hA} \tag{2-11}$$

where now the $1/hA$ term becomes the convection resistance.

2-5 THE OVERALL HEAT-TRANSFER COEFFICIENT

Consider the plane wall shown in Fig. 2-5 exposed to a hot fluid A on one side and a cooler fluid B on the other side. The heat transfer is expressed by

$$q = h_1A(T_A - T_1) = \frac{kA}{\Delta x}(T_1 - T_2) = h_2A\,(T_2 - T_B)$$

The heat-transfer process may be represented by the resistance network in Fig. 2-5b, and the overall heat transfer is calculated as the ratio of the overall temperature difference to the sum of the thermal resistances:

$$q = \frac{T_A - T_B}{1/h_1A + \Delta x/kA + 1/h_2A} \tag{2-12}$$

Observe that the value $1/hA$ is used to represent the convection resistance. The overall heat transfer by combined conduction and convection is frequently expressed in terms of an overall heat-transfer coefficient U, defined by the relation

$$q = UA\,\Delta T_{\text{overall}} \tag{2-13}$$

where A is some suitable area for the heat flow. In accordance with Eq. (2-12), the overall heat-transfer coefficient would be

$$U = \frac{1}{1/h_1 + \Delta x/k + 1/h_2}$$

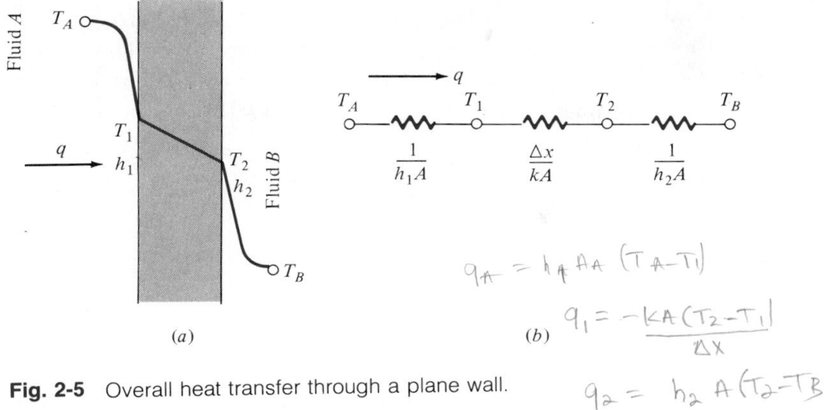

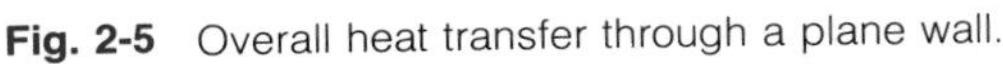

Fig. 2-5 Overall heat transfer through a plane wall.

For a hollow cylinder exposed to a convection environment on its inner and outer surfaces, the electric-resistance analogy would appear as in Fig. 2-6 where, again, T_A and T_B are the two fluid temperatures. Note that the area for convection is not the same for both fluids in this case, these areas depending on the inside tube diameter and wall thickness. In this case the overall heat transfer would be expressed by

$$q = \frac{T_A - T_B}{\dfrac{1}{h_i A_i} + \dfrac{\ln (r_o/r_i)}{2\pi k L} + \dfrac{1}{h_o A_o}} \tag{2-14}$$

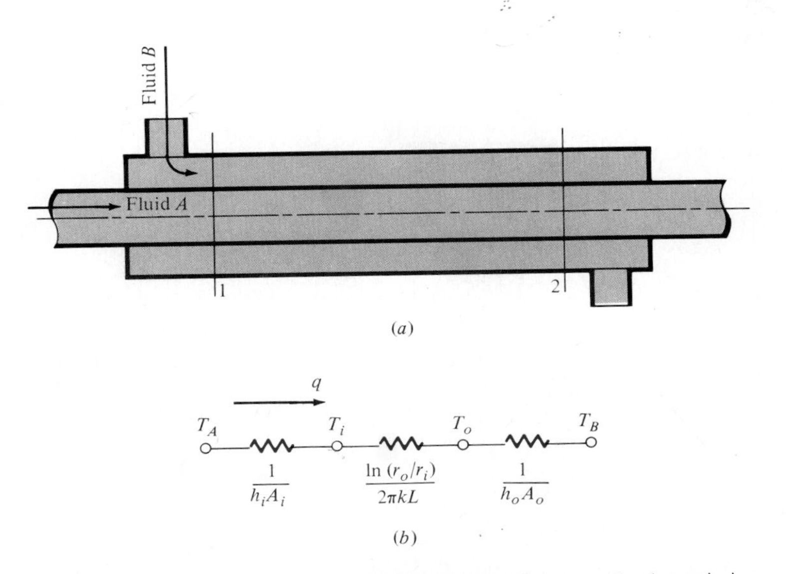

Fig. 2-6 Resistance analogy for hollow cylinder with convection boundaries.

in accordance with the thermal network shown in Fig. 2-6. The terms A_i and A_o represent the inside and outside surface areas of the inner tube. The overall heat-transfer coefficient may be based on either the inside or the outside area of the tube. Accordingly,

$$U_i = \frac{1}{\dfrac{1}{h_i} + \dfrac{A_i \ln (r_o/r_i)}{2\pi kL} + \dfrac{A_i}{A_o}\dfrac{1}{h_o}} \tag{2-15}$$

$$U_o = \frac{1}{\dfrac{A_o}{A_i}\dfrac{1}{h_i} + \dfrac{A_o \ln (r_o/r_i)}{2\pi kL} + \dfrac{1}{h_o}} \tag{2-16}$$

Calculations of the convection heat-transfer coefficients for use in the overall heat-transfer coefficient are made in accordance with the methods described in later chapters. Some typical values of the overall heat-transfer coefficient are given in Table 10-1.

2-6 CRITICAL THICKNESS OF INSULATION

Let us consider a layer of insulation which might be installed around a circular pipe, as shown in Fig. 2-7. The inner temperature of the insulation is fixed at T_i, and the outer surface is exposed to a convection environment at T_∞. From the thermal network the heat transfer is

$$q = \frac{2\pi L(T_i - T_\infty)}{\dfrac{\ln (r_o/r_i)}{k} + \dfrac{1}{r_o h}} \tag{2-17}$$

Now let us manipulate this expression to determine the outer radius of insulation r_o which will maximize the heat transfer. The maximization condition is

$$\frac{dq}{dr_o} = 0 = \frac{-2\pi L(T_i - T_\infty)\left(\dfrac{1}{kr_o} - \dfrac{1}{hr_o^2}\right)}{\left[\dfrac{\ln (r_o/r_i)}{k} + \dfrac{1}{r_o h}\right]^2}$$

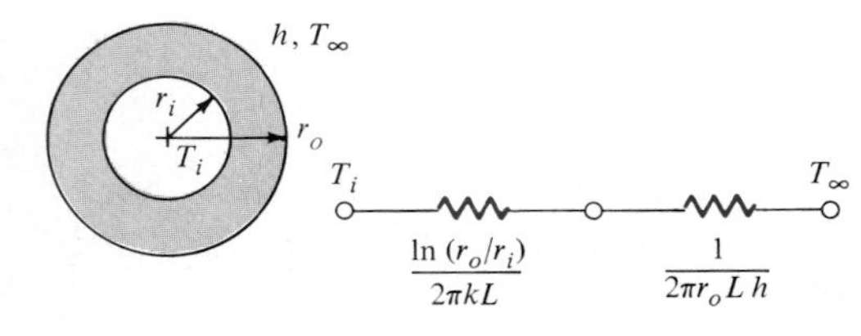

Fig. 2-7 Critical insulation thickness.

which gives the result

$$r_o = \frac{k}{h} \tag{2-18}$$

Equation (2-18) expresses the critical-radius-of-insulation concept. If the outer radius is less than the value given by this equation, then the heat transfer will be *increased* by adding more insulation. For outer radii greater than the critical value an increase in insulation thickness will cause a decrease in heat transfer. The central concept is that for sufficiently small values of h the convection heat loss may actually increase with the addition of insulation because of increased surface area.

■ **EXAMPLE 2-3** Critical insulation thickness

Calculate the critical radius of insulation for asbestos [$k = 0.17$ W/m · °C] surrounding a pipe and exposed to room air at 20°C with $h = 3.0$ W/m² · °C. Calculate the heat loss from a 200°C, 5.0-cm-diameter pipe when covered with the critical radius of insulation and without insulation.

Solution

From Eq. (2-18) we calculate r_o as

$$r_o = \frac{k}{h} = \frac{0.17}{3.0} = 0.0567 \text{ m} = 5.67 \text{ cm}$$

The inside radius of the insulation is 5.0/2 = 2.5 cm, so the heat transfer is calculated from Eq. (2-17) as

$$\frac{q}{L} = \frac{2\pi(200 - 20)}{\dfrac{\ln(5.67/2.5)}{0.17} + \dfrac{1}{(0.0567)(3.0)}} = 105.7 \text{ W/m}$$

Without insulation the convection from the outer surface of the pipe is

$$\frac{q}{L} = h(2\pi r)(T_i - T_o) = (3.0)(2\pi)(0.025)(200 - 20) = 84.8 \text{ W/m}$$

So, the addition of 3.17 cm (5.67 − 2.5) of insulation actually *increases* the heat transfer by 25 percent.

■ 2-7 HEAT-SOURCE SYSTEMS

A number of interesting applications of the principles of heat transfer are concerned with systems in which heat may be generated internally. Nuclear reactors are one example; electrical conductors and chemically reacting systems are others. At this point we shall confine our discussion to one-dimensional systems, or, more specifically, systems where the temperature is a function of only one space coordinate.

Plane Wall with Heat Sources

Consider the plane wall with uniformly distributed heat sources shown in Fig. 2-8. The thickness of the wall in the x direction is $2L$, and it is assumed that the dimensions in the other directions are sufficiently large that the heat flow may be considered as one-dimensional. The heat generated per unit volume is $\dot{q}$, and we assume that the thermal conductivity does not vary with temperature. This situation might be produced in a practical situation by passing a current through an electrically conducting material. From Chap. 1, the differential equation which governs the heat flow is

$$\frac{d^2T}{dx^2} + \frac{\dot{q}}{k} = 0 \qquad (2\text{-}19)$$

For the boundary conditions we specify the temperatures on either side of the wall, i.e.,

$$T = T_w \qquad \text{at } x = \pm L \qquad (2\text{-}20)$$

The general solution to Eq. (2-19) is

$$T = -\frac{\dot{q}}{2k}x^2 + C_1x + C_2 \qquad (2\text{-}21)$$

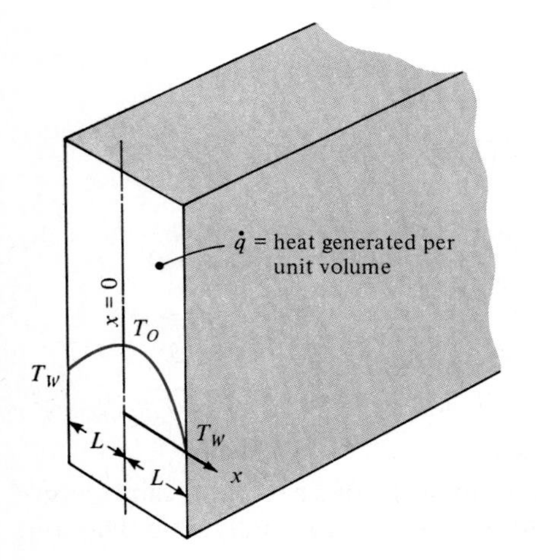

Fig. 2-8 Sketch illustrating one-dimensional conduction problem with heat generation.

Because the temperature must be the same on each side of the wall, C_1 must be zero. The temperature at the midplane is denoted by T_0 and from Eq. (2-21)

$$T_0 = C_2$$

The temperature distribution is therefore

$$T - T_0 = -\frac{\dot{q}}{2k} x^2 \tag{2-22a}$$

or

$$\frac{T - T_0}{T_w - T_0} = \left(\frac{x}{L}\right)^2 \tag{2-22b}$$

a parabolic distribution. An expression for the midplane temperature T_0 may be obtained through an energy balance. At steady-state conditions the total heat generated must equal the heat lost at the faces. Thus

$$2\left(-kA \left.\frac{dT}{dx}\right]_{x=L}\right) = \dot{q}A\,2L$$

where A is the cross-sectional area of the plate. The temperature gradient at the wall is obtained by differentiating Eq. (2-22b):

$$\left.\frac{dT}{dx}\right]_{x=L} = \left.(T_w - T_0)\left(\frac{2x}{L^2}\right)\right]_{x=L} = (T_w - T_0)\frac{2}{L}$$

Then

$$-k(T_w - T_0)\frac{2}{L} = \dot{q}L$$

and

$$T_0 = \frac{\dot{q}L^2}{2k} + T_w \tag{2-23}$$

This same result could be obtained by substituting $T = T_w$ at $x = L$ into Eq. (2-22a).

The equation for the temperature distribution could also be written in the alternative form

$$\frac{T - T_w}{T_0 - T_w} = 1 - \frac{x^2}{L^2} \tag{2-22c}$$

■ 2-8 CYLINDER WITH HEAT SOURCES

Consider a cylinder of radius R with uniformly distributed heat sources and constant thermal conductivity. If the cylinder is sufficiently long that the temperature may be considered a function of radius only, the appropriate differential equation may be obtained by neglecting the axial, azimuth, and time-dependent terms in Eq. (1-3*b*),

$$\frac{d^2T}{dr^2} + \frac{1}{r}\frac{dT}{dr} + \frac{\dot{q}}{k} = 0 \tag{2-24}$$

The boundary conditions are

$$T = T_w \qquad \text{at } r = R$$

and heat generated equals heat lost at the surface:

$$\dot{q}\pi R^2 L = -k2\pi RL \left.\frac{dT}{dr}\right]_{r=R}$$

Since the temperature function must be continuous at the center of the cylinder, we could specify that

$$\frac{dT}{dr} = 0 \qquad \text{at } r = 0$$

However, it will not be necessary to use this condition since it will be satisfied automatically when the two boundary conditions are satisfied.

We rewrite Eq. (2-24)

$$r\frac{d^2T}{dr^2} + \frac{dT}{dr} = \frac{-\dot{q}r}{k}$$

and note that

$$r\frac{d^2T}{dr^2} + \frac{dT}{dr} = \frac{d}{dr}\left(r\frac{dT}{dr}\right)$$

Then integration yields

$$r\frac{dT}{dr} = \frac{-\dot{q}r^2}{2k} + C_1$$

and

$$T = \frac{-\dot{q}r^2}{4k} + C_1 \ln r + C_2$$

From the second boundary condition above,

$$\left.\frac{dT}{dr}\right]_{r=R} = \frac{-\dot{q}R}{2k} = \frac{-\dot{q}R}{2k} + \frac{C_1}{R}$$

Thus

$$C_1 = 0$$

We could also note that C_1 must be zero because at $r = 0$ the logarithm function becomes infinite.

From the first boundary condition,

$$T = T_w = \frac{-\dot{q}R^2}{4k} + C_2 \qquad \text{at } r = R$$

so that

$$C_2 = T_w + \frac{\dot{q}R^2}{4k}$$

The final solution for the temperature distribution is then

$$T - T_w = \frac{\dot{q}}{4k}(R^2 - r^2) \tag{2-25a}$$

or, in dimensionless form,

$$\frac{T - T_w}{T_0 - T_w} = 1 - \left(\frac{r}{R}\right)^2 \tag{2-25b}$$

where T_0 is the temperature at $r = 0$ and is given by

$$T_0 = \frac{\dot{q}R^2}{4k} + T_w \tag{2-26}$$

It is left as an exercise to show that the temperature gradient at $r = 0$ is zero.

For a hollow cylinder with uniformly distributed heat sources the appropriate boundary conditions would be

$$T = T_i \qquad \text{at } r = r_i \text{ (inside surface)}$$

$$T = T_o \qquad \text{at } r = r_o \text{ (outside surface)}$$

The general solution is still

$$T = -\frac{\dot{q}r^2}{4k} + C_1 \ln r + C_2$$

Application of the new boundary conditions yields

$$T - T_o = \frac{\dot{q}}{4k}(r_o^2 - r^2) + C_1 \ln \frac{r}{r_o} \tag{2-27}$$

where the constant C_1 is given by

$$C_1 = \frac{T_i - T_o + \dot{q}(r_i^2 - r_o^2)/4k}{\ln (r_i/r_o)} \tag{2-28}$$

■ **EXAMPLE 2-4** Heat source with convection

A current of 200 A is passed through a stainless-steel wire [k = 19 W/m · °C] 3 mm in diameter. The resistivity of the steel may be taken as 70 μΩ · cm, and the length of the wire is 1 m. The wire is submerged in a liquid at 110°C and experiences a convection heat-transfer coefficient of 4 kW/m² · °C. Calculate the center temperature of the wire.

Solution

All the power generated in the wire must be dissipated by convection to the liquid:

$$P = I^2R = q = hA\,(T_w - T_\infty) \qquad (a)$$

The resistance of the wire is calculated from

$$R = \rho \frac{L}{A} = \frac{(70 \times 10^{-6})(100)}{\pi(0.15)^2} = 0.099\ \Omega$$

where ρ is the resistivity of the wire. The surface area of the wire is $\pi\, dL$, so from Eq. (a),

$$(200)^2(0.099) = 4000\pi(3 \times 10^{-3})(1)(T_w - 110) = 3960\ \text{W}$$

and

$$T_w = 215°\text{C} \quad [419°\text{F}]$$

The heat generated per unit volume $\dot{q}$ is calculated from

$$P = \dot{q}V = \dot{q}\pi r^2 L$$

so that

$$\dot{q} = \frac{3960}{\pi(1.5 \times 10^{-3})^2(1)} = 560.2\ \text{MW/m}^3 \quad [5.41 \times 10^7\ \text{Btu/h} \cdot \text{ft}^3]$$

Finally, the center temperature of the wire is calculated from Eq. (2-26):

$$T_0 = \frac{\dot{q}r_o^2}{4k} + T_w = \frac{(5.602 \times 10^8)(1.5 \times 10^{-3})^2}{(4)(19)} + 215 = 231.6°\text{C} \quad [449°\text{F}]$$

2-9 CONDUCTION-CONVECTION SYSTEMS

The heat which is conducted through a body must frequently be removed (or delivered) by some convection process. For example, the heat lost by conduction through a furnace wall must be dissipated to the surroundings through convection. In heat-exchanger applications a finned-tube arrangement might be used to remove heat from a hot liquid. The heat transfer from the liquid to the finned tube is by convection. The heat is conducted through the material and finally dissipated to the surroundings by convection. Obviously, an analysis of combined conduction-convection systems is very important from a practical standpoint.

We shall defer part of our analysis of conduction-convection systems to Chap. 10 on heat exchangers. For the present we wish to examine some simple extended-surface problems. Consider the one-dimensional fin exposed to a surrounding fluid at a temperature T_∞ as shown in Fig. 2-9. The temperature of the base of the fin is T_0. We approach the problem by making an energy balance on an element of the fin of thickness dx as shown in the figure. Thus

Energy in left face = energy out right face + energy lost by convection

The defining equation for the convection heat-transfer coefficient is recalled as

$$q = hA(T_w - T_\infty) \tag{2-29}$$

where the area in this equation is the surface area for convection. Let the cross-sectional area of the fin be A and the perimeter be P. Then the energy quantities are

$$\text{Energy in left face} = q_x = -kA\frac{dT}{dx}$$

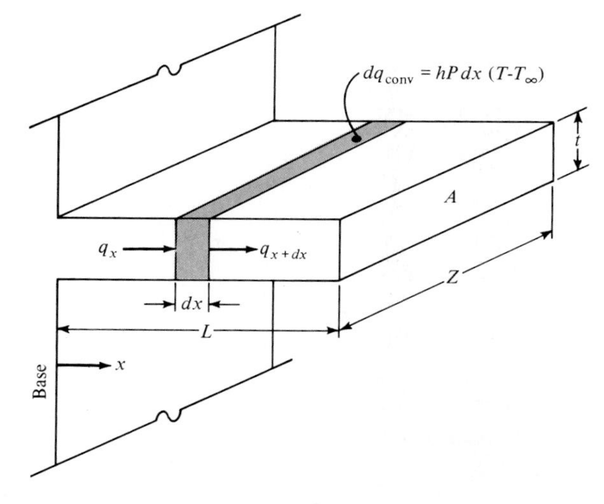

Fig. 2-9 Sketch illustrating one-dimensional conduction and convection through a rectangular fin.

$$\text{Energy out right face} = q_{x+dx} = -kA \left.\frac{dT}{dx}\right]_{x+dx}$$

$$= -kA\left(\frac{dT}{dx} + \frac{d^2T}{dx^2}\,dx\right)$$

$$\text{Energy lost by convection} = hP\,dx(T - T_\infty)$$

Here it is noted that the differential surface area for convection is the product of the perimeter of the fin and the differential length dx. When we combine the quantities, the energy balance yields

$$\frac{d^2T}{dx^2} - \frac{hP}{kA}(T - T_\infty) = 0 \tag{2-30a}$$

Let $\theta = T - T_\infty$. Then Eq. (2-30a) becomes

$$\frac{d^2\theta}{dx^2} - \frac{hP}{kA}\theta = 0 \tag{2-30b}$$

One boundary condition is

$$\theta = \theta_0 = T_0 - T_\infty \qquad \text{at } x = 0$$

The other boundary condition depends on the physical situation. Several cases may be considered:

CASE 1 The fin is very long, and the temperature at the end of the fin is essentially that of the surrounding fluid.

CASE 2 The fin is of finite length and loses heat by convection from its end.

CASE 3 The end of the fin is insulated so that $dT/dx = 0$ at $x = L$.

If we let $m^2 = hP/kA$, the general solution for Eq. (2-30b) may be written

$$\theta = C_1 e^{-mx} + C_2 e^{mx} \tag{2-31}$$

For case 1 the boundary conditions are

$$\theta = \theta_0 \qquad \text{at } x = 0$$

$$\theta = 0 \qquad \text{at } x = \infty$$

and the solution becomes

$$\frac{\theta}{\theta_0} = \frac{T - T_\infty}{T_0 - T_\infty} = e^{-mx} \tag{2-32}$$

For case 3 the boundary conditions are

$$\theta = \theta_0 \qquad \text{at } x = 0$$

$$\frac{d\theta}{dx} = 0 \qquad \text{at } x = L$$

Thus
$$\theta_0 = C_1 + C_2$$
$$0 = m(-C_1 e^{-mL} + C_2 e^{mL})$$

Solving for the constants C_1 and C_2, we obtain

$$\frac{\theta}{\theta_0} = \frac{e^{-mx}}{1 + e^{-2mL}} + \frac{e^{mx}}{1 + e^{2mL}} \tag{2-33a}$$

$$= \frac{\cosh [m(L - x)]}{\cosh mL} \tag{2-33b}$$

The hyperbolic functions are defined as

$$\sinh x = \frac{e^x - e^{-x}}{2} \qquad \cosh x = \frac{e^x + e^{-x}}{2}$$

$$\tanh x = \frac{\sinh x}{\cosh x} = \frac{e^x - e^{-x}}{e^x + e^{-x}}$$

The solution for case 2 is more involved algebraically, and the result is

$$\frac{T - T_\infty}{T_0 - T_\infty} = \frac{\cosh m(L - x) + (h/mk) \sinh m(L - x)}{\cosh mL + (h/mk) \sinh mL} \tag{2-34}$$

All of the heat lost by the fin must be conducted into the base at $x = 0$. Using the equations for the temperature distribution, we can compute the heat loss from

$$q = -kA \left. \frac{dT}{dx} \right]_{x=0}$$

An alternative method of integrating the convection heat loss could be used:

$$q = \int_0^L hP(T - T_\infty)\, dx = \int_0^L hP\theta\, dx$$

In most cases, however, the first equation is easier to apply. For case 1,

$$q = -kA(-m\theta_0 e^{-m(0)}) = \sqrt{hPkA}\, \theta_0 \tag{2-35}$$

For case 3,

$$q = -kA\theta_0 m \left(\frac{1}{1 + e^{-2mL}} - \frac{1}{1 + e^{+2mL}} \right)$$
$$= \sqrt{hPkA}\, \theta_0 \tanh mL \tag{2-36}$$

The heat flow for case 2 is

$$q = \sqrt{hPkA}\, (T_0 - T_\infty) \frac{\sinh mL + (h/mk) \cosh mL}{\cosh mL + (h/mk) \sinh mL} \tag{2-37}$$

In the above development it has been assumed that the substantial temperature gradients occur only in the x direction. This assumption will be satisfied if the fin is sufficiently thin. For most fins of practical interest the error introduced by this assumption is less than 1 percent. The overall accuracy of practical fin calculations will usually be limited by uncertainties in values of the convection coefficient h. It is worthwhile to note that the convection coefficient is seldom uniform over the entire surface, as has been assumed above. If severe non-uniform behavior is encountered, numerical finite-difference techniques must be employed to solve the problem. Such techniques are discussed in Chap. 3.

2-10 FINS

In the foregoing development we derived relations for the heat transfer from a rod or fin of uniform cross-sectional area protruding from a flat wall. In practical applications, fins may have varying cross-sectional areas and may be attached to circular surfaces. In either case the area must be considered as a variable in the derivation, and solution of the basic differential equation and the mathematical techniques become more tedious. We present only the results for these more complex situations. The reader is referred to Refs. 1 and 8 for details on the mathematical methods used to obtain the solutions.

To indicate the effectiveness of a fin in transferring a given quantity of heat, a new parameter called *fin efficiency* is defined by

$$\text{Fin efficiency} = \frac{\text{actual heat transferred}}{\begin{array}{c}\text{heat which would be transferred}\\ \text{if entire fin area were}\\ \text{at base temperature}\end{array}} = \eta_f$$

For case 3 above, the fin efficiency becomes

$$\eta_f = \frac{\sqrt{hPkA}\,\theta_0 \tanh mL}{hPL\theta_0} = \frac{\tanh mL}{mL} \tag{2-38}$$

The fins discussed above were assumed to be sufficiently deep that the heat flow could be considered one-dimensional. The expression for mL may be written

$$mL = \sqrt{\frac{hP}{kA}}L = \sqrt{\frac{h(2z + 2t)}{kzt}}L$$

where z is the depth of the fin and t is the thickness. Now, if the fin is sufficiently deep, the term $2z$ will be large compared with $2t$, and

$$mL = \sqrt{\frac{2hz}{ktz}}L = \sqrt{\frac{2h}{kt}}L$$

Multiplying numerator and denominator by $L^{1/2}$ gives

$$mL = \sqrt{\frac{2h}{kLt}}L^{3/2}$$

Lt is the profile area of the fin, which we define as

$$A_m = Lt$$

so that

$$mL = \sqrt{\frac{2h}{kA_m}}L^{3/2} \tag{2-39}$$

We may therefore use the expression in Eq. (2-39) to compute the efficiency of a fin with insulated tip as given by Eq. (2-38).

Harper and Brown [2] have shown that the solution in case 2 above may be expressed in the same form as Eq. (2-38) when the length of the fin is extended by one-half the thickness of the fin. A corrected length L_c is then used in all the equations which apply for the case of the fin with an insulated tip. Thus

$$L_c = L + \frac{t}{2} \tag{2-40}$$

The error which results from this approximation will be less than 8 percent when

$$\left(\frac{ht}{2k}\right)^{1/2} \leq \frac{1}{2} \tag{2-41}$$

If a cylindrical spine extends from a wall as shown in Fig. 2-10*g*, the corrected fin length is calculated from

$$L_c = L + \frac{\pi d^2/4}{\pi d} = L + d/4 \tag{2-42}$$

Examples of other types of fins are shown in Fig. 2-10 according to Ref. 8. Figure 2-11 presents a comparison of the efficiencies of a triangular fin and a

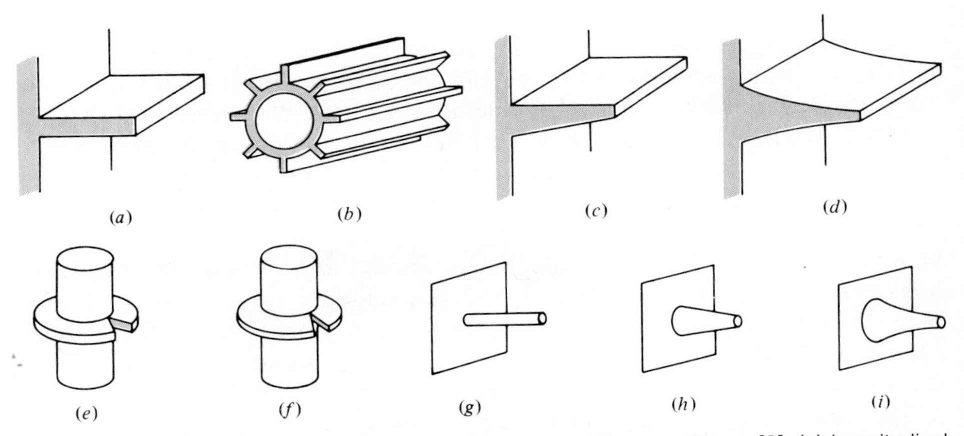

Fig. 2-10 Different types of finned surfaces, according to Kern and Kraus [8]: (*a*) Longitudinal fin of rectangular profile; (*b*) cylindrical tube equipped with fins of rectangular profile; (*c*) longitudinal fin of trapezoidal profile; (*d*) longitudinal fin of parabolic profile; (*e*) cylindrical tube equipped with radial fin of rectangular profile; (*f*) cylindrical tube equipped with radial fin of truncated conical profile; (*g*) cylindrical spine; (*h*) truncated conical spine; (*i*) parabolic spine.

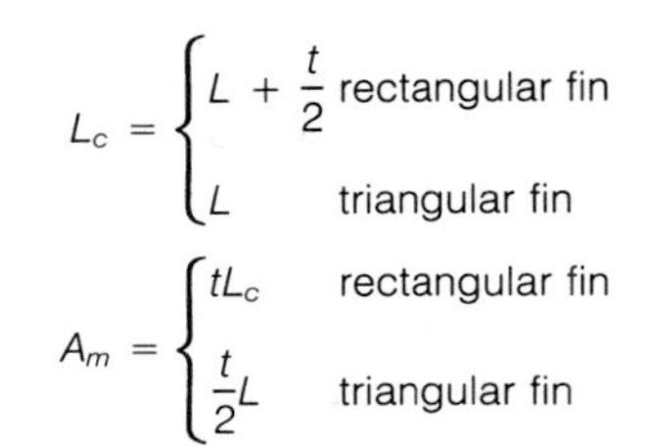

$$L_c = \begin{cases} L + \dfrac{t}{2} & \text{rectangular fin} \\ L & \text{triangular fin} \end{cases}$$

$$A_m = \begin{cases} tL_c & \text{rectangular fin} \\ \dfrac{t}{2}L & \text{triangular fin} \end{cases}$$

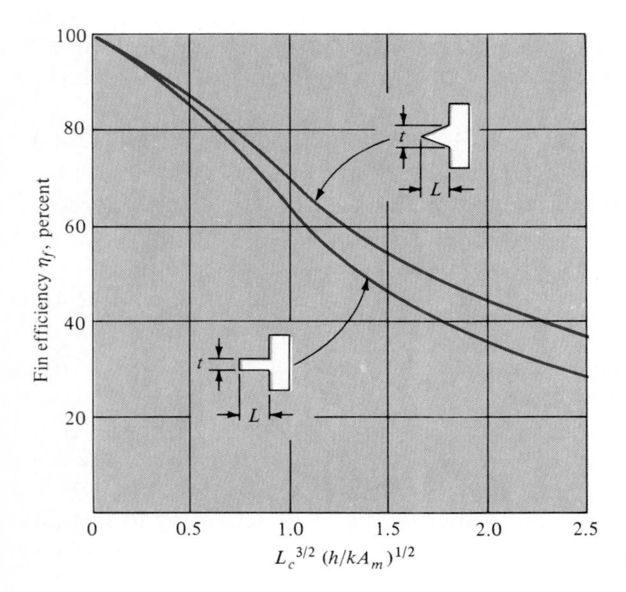

Fig. 2-11 Efficiencies of rectangular and triangular fins.

straight rectangular fin corresponding to case 2. Figure 2-12 shows the efficiencies of circumferential fins of rectangular cross-sectional area. Notice that the corrected fin lengths L_c and profile area A_m have been used in Figs. 2-11 and 2-12. We may note that as $r_{2c}/r_1 \rightarrow 1.0$, the efficiency of the circumferential fin becomes identical to that of the straight fin of rectangular profile.

It is interesting to note that the fin efficiency reaches its maximum value for the trivial case of $L = 0$, or no fin at all. Therefore, we should not expect to be able to maximize fin performance with respect to fin length. It is possible, however, to maximize the efficiency with respect to the quantity of fin material (mass, volume, or cost), and such a maximization process has rather obvious economic significance. We have not discussed the subject of radiation heat transfer from fins. The radiant transfer is an important consideration in a number of applications, and the interested reader should consult Siegel and Howell [9] for information on this subject.

In some cases a valid method of evaluating fin performance is to compare the heat transfer with the fin to that which would be obtained without the fin.

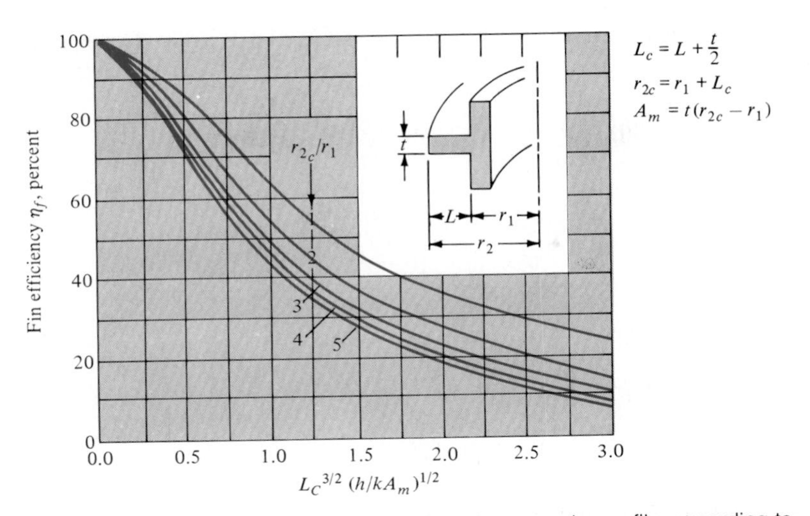

Fig. 2-12 Efficiencies of circumferential fins of rectangular profile, according to Ref. 3.

The ratio of these quantities is

$$\frac{q \text{ with fin}}{q \text{ without fin}} = \frac{\eta_f A_f h \theta_0}{h A_b \theta_0}$$

where A_f is the total surface area of the fin and A_b is the base area. For the insulated tip fin described by Eq. (2-36),

$$A_f = PL$$

$$A_b = A$$

and the heat ratio would become

$$\frac{q \text{ with fin}}{q \text{ without fin}} = \frac{\tanh mL}{\sqrt{hA/kP}}$$

This is sometimes called the *fin effectiveness*.

☐ Conditions When Fins Do Not Help

At this point we should remark that the installation of fins on a heat-transfer surface will not necessarily increase the heat-transfer rate. If the value of h, the convection coefficient, is large, as it is with high-velocity fluids or boiling liquids, the fin may produce a reduction in heat transfer because the conduction resistance then represents a larger impediment to the heat flow than the convection resistance. To illustrate the point, consider a stainless-steel pin fin which has $k = 16$ W/m · °C, $L = 10$ cm, $d = 1$ cm and which is exposed to a boiling-water convection situation with $h = 5000\ \text{W/m}^2 \cdot °\text{C}$. From Eq. (2-36) we can compute

$$\frac{q \text{ with fin}}{q \text{ without fin}} = \frac{\tanh mL}{\sqrt{hA/kP}}$$

$$= \frac{\tanh\left\{\left[\dfrac{5000\pi(1 \times 10^{-2})(4)}{16\pi(1 \times 10^{-2})^2}\right]^{1/2} (10 \times 10^{-2})\right\}}{\left[\dfrac{5000\pi(1 \times 10^{-2})^2}{(4)(16)\pi(1 \times 10^{-2})}\right]^{1/2}}$$

$$= 1.13$$

Thus, this rather large pin produces an increase of only 13 percent in the heat transfer.

Still another method of evaluating fin performance is discussed in Prob. 2-68. Kern and Kraus [8] give a very complete discussion of extended-surface heat transfer. Some photographs of different fin shapes used in electronic cooling applications are shown in Fig. 2-13.

■ EXAMPLE 2-5 Influence of thermal conductivity on fin temperature profiles

Compare the temperature distributions in a straight fin of rectangular profile having a thickness of 2 cm and a length of 10 cm and exposed to a convection environment with $h = 25\ \text{W/m}^2 \cdot °\text{C}$, for three fin materials: copper [$k = 385\ \text{W/m} \cdot °\text{C}$], stainless steel [$k = 17\ \text{W/m} \cdot °\text{C}$], and glass [$k = 0.8\ \text{W/m} \cdot °\text{C}$]. Also compare the relative heat flows and fin efficiencies.

Solution

We have

$$\frac{hP}{kA} = \frac{(25)\pi(0.02)}{k\pi(0.01)^2} = \frac{5000}{k}$$

The terms of interest are therefore

Material	$\frac{hP}{kA}$	m	mL
Copper	12.99	3.604	0.3604
Stainless steel	294.1	17.15	1.715
Glass	6250	79.06	7.906

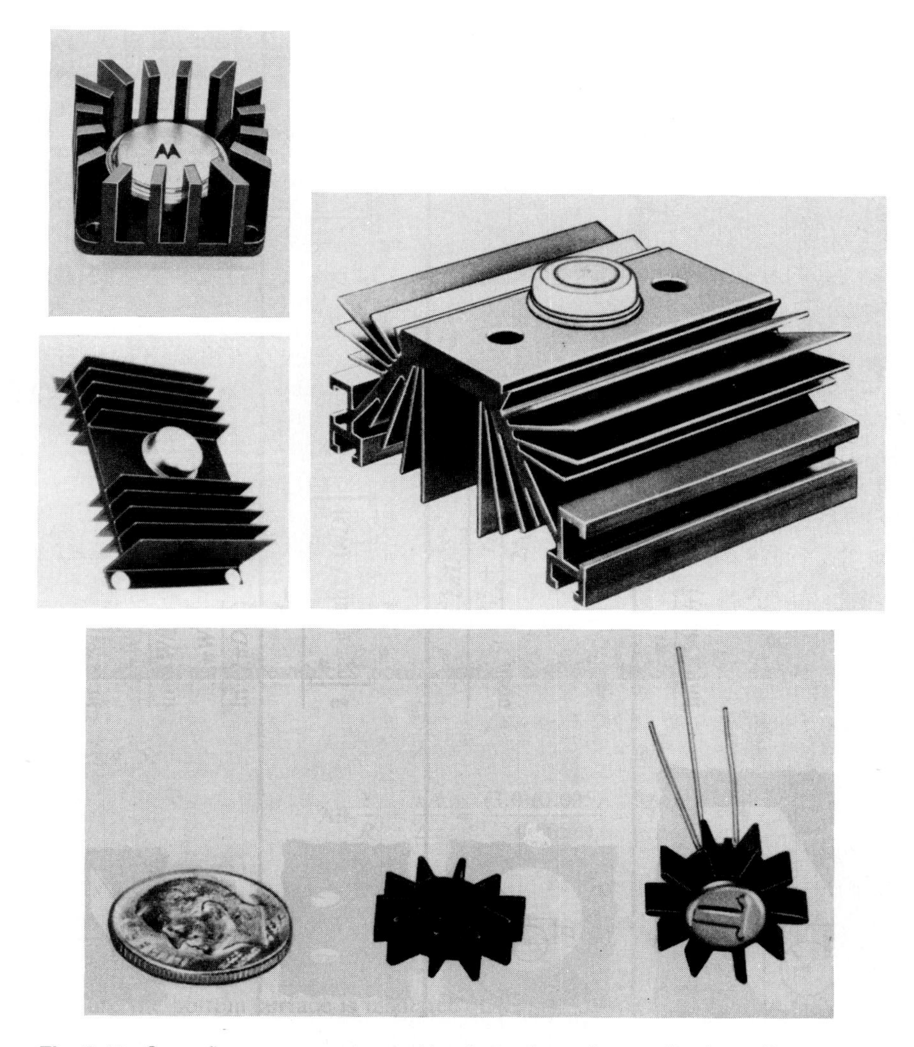

Fig. 2-13 Some fin arrangements used in electronic cooling applications. (Courtesy Wakefield Engineering Inc., Wakefield, Mass.)

These values may be inserted into Eq. (2-33*a*) to calculate the temperatures at different x locations along the rod, and the results are shown in the accompanying figure. We notice that the glass behaves as a "very long" fin, and its behavior could be calculated from Eq. (2-32). The fin efficiencies are calculated from Eq. (2-38) by using the corrected length approximation of Eq. (2-40). We have

$$L_c = L + \frac{t}{2} = 10 + \frac{2}{2} = 11 \text{ cm} \quad [4.33 \text{ in}]$$

The parameters of interest for the heat-flow and efficiency comparisons are now tabulated as

Material	*hPkA*	mL_c
Copper	0.190	0.3964
Stainless steel	0.0084	1.8865
Glass	3.9×10^{-4}	8.697

To compare the heat flows we could either calculate the values from Eq. (2-36) for a unit value of θ_0 or observe that the fin efficiency gives a relative heat-flow comparison because the maximum heat transfer is the same for all three cases; i.e., we are dealing with the same fin size, shape, and value of h. We thus calculate the values of η_f from Eq. (2-38) and the above values of mL_c.

Material	η_f	*q relative to copper, %*
Copper	0.951	100
Stainless steel	0.506	53.6
Glass	0.115	12.1

The temperature profiles in the accompanying figure can be somewhat misleading. The glass has the steepest temperature gradient at the base, but its much lower value of k produces a lower heat-transfer rate.

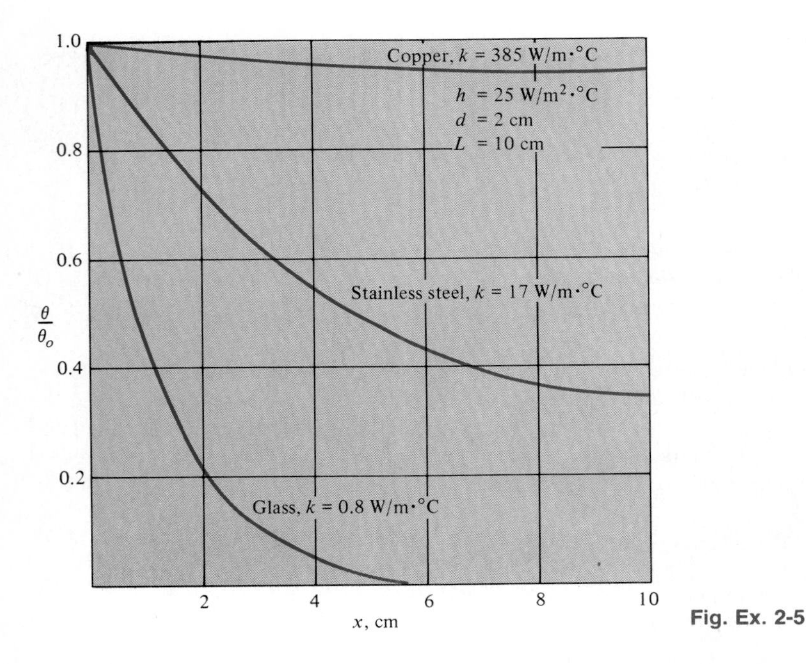

Fig. Ex. 2-5

■ **EXAMPLE 2-6**

An aluminum fin [k = 200 W/m · °C] 3.0 mm thick and 7.5 cm long protrudes from a wall, as in Fig. 2-9. The base is maintained at 300°C, and the ambient temperature is 50°C with h = 10 W/m² · °C. Calculate the heat loss from the fin per unit depth of material.

Solution

We may use the approximate method of solution by extending the fin a fictitious length $t/2$ and then computing the heat transfer from a fin with insulated tip as given by Eq. (2-36). We have

$$L_c = L + t/2 = 7.5 + 0.15 = 7.65 \text{ cm} \quad [3.01 \text{ in}]$$

$$m = \sqrt{\frac{hP}{kA}} = \left[\frac{h(2z + 2t)}{ktz}\right]^{1/2} \approx \sqrt{\frac{2h}{kt}}$$

when the fin depth $z \gg t$. So,

$$m = \left[\frac{(2)(10)}{(200)(3 \times 10^{-3})}\right]^{1/2} = 5.774$$

From Eq. (2-36), for an insulated-tip fin

$$q = (\tanh mL_c)\sqrt{hPkA}\,\theta_0$$

For a 1-m depth

$$A = (1)(3 \times 10^{-3}) = 3 \times 10^{-3} \text{ m}^2 \quad [4.65 \text{ in}^2]$$

and

$$q = (5.774)(200)(3 \times 10^{-3})(300 - 50) \tanh [(5.774)(0.0765)]$$
$$= 359 \text{ W/m} \quad [373.5 \text{ Btu/h} \cdot \text{ft}]$$

■ **EXAMPLE 2-7**

Aluminum fins 1.5 cm wide and 1.0 mm thick are placed on a 2.5-cm-diameter tube to dissipate the heat. The tube surface temperature is 170°C, and the ambient-fluid temperature is 25°C. Calculate the heat loss per fin for h = 130 W/m² · °C. Assume k = 200 W/m · °C for aluminum.

Solution

For this example we can compute the heat transfer by using the fin-efficiency curves in Fig. 2-12. The parameters needed are

$$L_c = L + t/2 = 1.5 + 0.05 = 1.55 \text{ cm}$$

$$r_1 = 2.5/2 = 1.25 \text{ cm}$$

$$r_{2c} = r_1 + L_c = 1.25 + 1.55 = 2.80 \text{ cm}$$

$$r_{2c}/r_1 = 2.80/1.25 = 2.24$$

$$A_m = t\,(r_{2c} - r_1) = (0.001)(2.8 - 1.25)(10^{-2}) = 1.55 \times 10^{-5} \text{ m}^2$$

$$L_c^{3/2}\left(\frac{h}{kA_m}\right)^{1/2} = (0.0155)^{3/2}\left[\frac{130}{(200)(1.55 \times 10^{-5})}\right]^{1/2} = 0.396$$

From Fig. 2-12 η_f = 82 percent. The heat which would be transferred if the entire fin were at the base temperature is (both sides of fin exchanging heat)

$$q_{max} = 2\pi(r_{2c}^2 - r_1^2)h(T_0 - T_\infty)$$
$$= 2\pi(2.8^2 - 1.25^2)(10^{-4})(130)(170 - 25)$$
$$= 74.35 \text{ W} \quad [253.7 \text{ Btu/h}]$$

The actual heat transfer is then the product of the heat flow and the fin efficiency:

$$q_{act} = (0.82)(74.35) = 60.97 \text{ W} \quad [208 \text{ Btu/h}]$$

■ **EXAMPLE 2-8** Rod with heat sources

A rod containing uniform heat sources per unit volume $\dot{q}$ is connected to two temperatures as shown in the accompanying figure. The rod is also exposed to an environment with convection coefficient h and temperature T_∞. Obtain an expression for the temperature distribution in the rod.

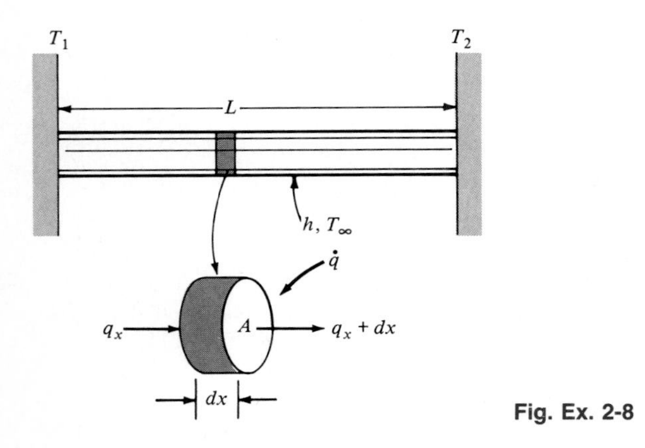

Fig. Ex. 2-8

Solution

We first must make an energy balance on the element of the rod shown, similar to that used to derive Eq. (2-30). We have

Energy in left face + heat generated in element
= energy out right face + energy lost by convection

or
$$-kA\frac{dT}{dx} + \dot{q}A\,dx = -kA\left(\frac{dT}{dx} + \frac{d^2T}{dx^2}dx\right) + hP\,dx\,(T - T_\infty)$$

Simplifying, we have

$$\frac{d^2T}{dx^2} - \frac{hP}{kA}(T - T_\infty) + \frac{\dot{q}}{k} = 0 \qquad (a)$$

or, with $\theta = T - T_\infty$ and $m^2 = hP/kA$

$$\frac{d^2\theta}{dx} + m^2\theta + \frac{\dot{q}}{k} = 0 \qquad (b)$$

We can make a further variable substitution as

$$\theta' = \theta - \dot{q}/km^2$$

so that our differential equation becomes

$$\frac{d^2\theta'}{dx^2} - m^2\theta' = 0 \qquad (c)$$

which has the general solution

$$\theta' = c_1 e^{-mx} + c_2 e^{mx} \qquad (d)$$

The two end temperatures are used to establish the boundary conditions:

$$\theta' = \theta_1' = T_1 - T_\infty - \dot{q}/km^2 = C_1 + C_2$$

$$\theta' = \theta_2' = T_2 - T_\infty - \dot{q}/km^2 = C_1 e^{-mL} + C_2 e^{mL}$$

Solving for the constants C_1 and C_2 gives

$$\theta' = \frac{(\theta_1' e^{2mL} - \theta_2' e^{mL})e^{-mx} + (\theta_2' e^{mL} - \theta_2')e^{mx}}{e^{2mL} - 1} \qquad (e)$$

2-11 THERMAL CONTACT RESISTANCE

Imagine two solid bars brought into contact as indicated in Fig. 2-14, with the sides of the bars insulated so that heat flows only in the axial direction. The materials may have different thermal conductivities, but if the sides are insulated, the heat flux must be the same through both materials under steady-state conditions. Experience shows that the actual temperature profile through the two materials varies approximately as shown in Fig. 2-14*b*. The temperature

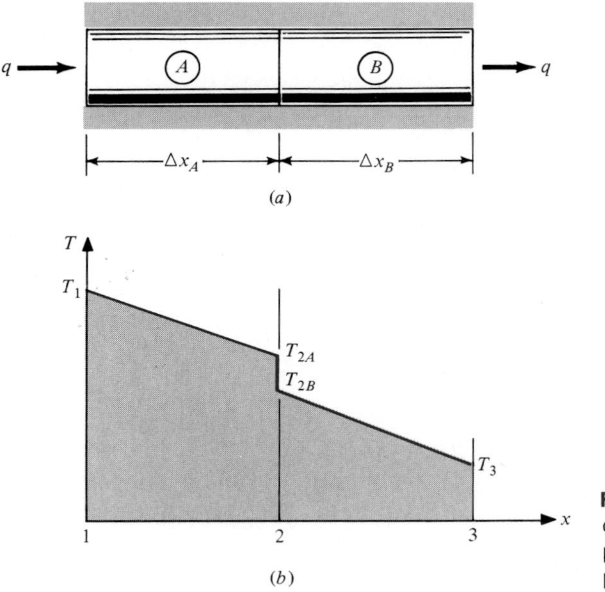

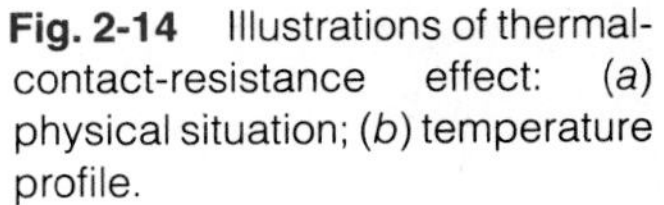
Fig. 2-14 Illustrations of thermal-contact-resistance effect: (*a*) physical situation; (*b*) temperature profile.

drop at plane 2, the contact plane between the two materials, is said to be the result of a *thermal contact resistance*. Performing an energy balance on the two materials, we obtain

$$q = k_A A \frac{T_1 - T_{2A}}{\Delta x_A} = \frac{T_{2A} - T_{2B}}{1/h_c A} = k_B A \frac{T_{2B} - T_3}{\Delta x_B}$$

or
$$q = \frac{T_1 - T_3}{\Delta x_A/k_A A + 1/h_c A + \Delta x_B/k_B A} \tag{2-43}$$

where the quantity $1/h_c A$ is called the thermal contact resistance and h_c is called the contact coefficient. This factor can be extremely important in a number of applications because of the many heat-transfer situations which involve mechanical joining of two materials.

The physical mechanism of contact resistance may be better understood by examining a joint in more detail, as shown in Fig. 2-15. The actual surface roughness is exaggerated to implement the discussion. No real surface is perfectly smooth, and the actual surface roughness is believed to play a central role in determining the contact resistance. There are two principal contributions to the heat transfer at the joint:

1. The solid-to-solid conduction at the spots of contact
2. The conduction through entrapped gases in the void spaces created by the contact

The second factor is believed to represent the major resistance to heat flow, because the thermal conductivity of the gas is quite small in comparison to that of the solids.

Designating the contact area by A_c and the void area by A_v, we may write for the heat flow across the joint

$$q = \frac{T_{2A} - T_{2B}}{L_g/2k_A A_c + L_g/2k_B A_c} + k_f A_v \frac{T_{2A} - T_{2B}}{L_g} = \frac{T_{2A} - T_{2B}}{1/h_c A}$$

where L_g is the thickness of the void space and k_f is the thermal conductivity of the fluid which fills the void space. The *total* cross-sectional area of the bars is A. Solving for h_c, the contact coefficient, we obtain

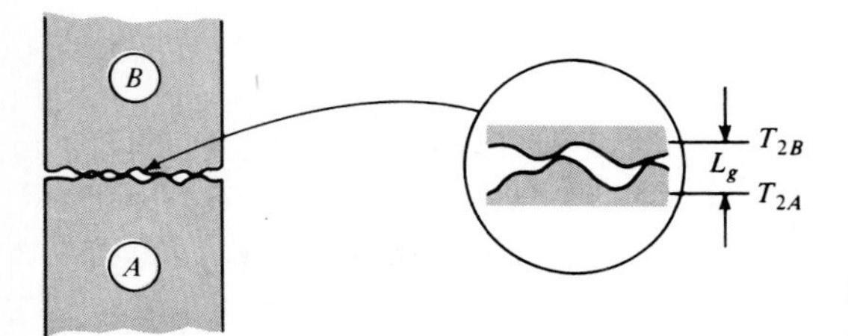

Fig. 2-15 Joint-roughness model for analysis of thermal contact resistance.

$$h_c = \frac{1}{L_g}\left(\frac{A_c}{A}\frac{2k_A k_B}{k_A + k_B} + \frac{A_v}{A}k_f\right) \qquad (2\text{-}44)$$

In most instances, air is the fluid filling the void space and k_f is small compared with k_A and k_B. If the contact area is small, the major thermal resistance results from the void space. The main problem with this simple theory is that it is extremely difficult to determine effective values of A_c, A_v, and L_g for surfaces in contact.

From the above physical model, we may tentatively conclude:

1. The contact resistance should increase with a decrease in the ambient gas pressure when the pressure is decreased below the value where the mean free path of the molecules is large compared with a characteristic dimension of the void space, since the effective thermal conductance of the entrapped gas will be decreased for this condition.
2. The contact resistance should be decreased for an increase in the joint pressure since this results in a deformation of the high spots of the contact surfaces, thereby creating a greater contact area between the solids.

A very complete survey of the contact-resistance problem is presented in Refs. 4, 6, 7, 10, 11. Unfortunately, there is no satisfactory theory which will predict thermal contact resistance for all types of engineering materials, nor have experimental studies yielded completely reliable empirical correlations.

Table 2-2 Contact Conductance of Typical Surfaces

	Roughness		*Temperature,*	*Pressure,*	$1/h_c$	
Surface type	μ*in*	μ*m*	*°C*	*atm*	**h · ft² · °F/Btu**	**m² · °C/W × 10⁴**
416 Stainless, ground, air	100	2.54	90–200	3–25	0.0015	2.64
304 Stainless, ground, air	45	1.14	20	40–70	0.003	5.28
416 Stainless, ground, with 0.001-in brass shim, air	100	2.54	30–200	7	0.002	3.52
Aluminum, ground, air	100	2.54	150	12–25	0.0005	0.88
	10	0.25	150	12–25	0.0001	0.18
Aluminum, ground, with 0.001-in brass shim, air	100	2.54	150	12–200	0.0007	1.23
Copper, ground, air	50	1.27	20	12–200	0.00004	0.07
Copper, milled-air	150	3.81	20	10–50	0.0001	0.18
Copper, milled, vacuum	10	0.25	30	7–70	0.0005	0.88

This is understandable because of the many complex surface conditions which may be encountered in practice.

Radiation heat transfer across the joint can also be important when high temperatures are encountered. This energy transfer may be calculated by the methods discussed in Chap. 8.

For design purposes the contact conductance values given in Table 2-2 may be used in the absence of more specific information. Thermal contact resistance can be reduced markedly, perhaps as much as 75 percent, by the use of a "thermal grease" like Dow 340.

EXAMPLE 2-9

Two 3.0-cm-diameter 304 stainless-steel bars, 10 cm long, have ground surfaces and are exposed to air with a surface roughness of about 1 μm. If the surfaces are pressed together with a pressure of 50 atm and the two-bar combination is exposed to an overall temperature difference of 100°C, calculate the axial heat flow and temperature drop across the contact surface.

Solution

The overall heat flow is subject to three thermal resistances, one conduction resistance for each bar, and the contact resistance. For the bars

$$R_{\text{th}} = \frac{\Delta x}{kA} = \frac{(0.1)(4)}{(16.3)\pi(3 \times 10^{-2})^2} = 8.679°\text{C/W}$$

From Table 2-2 the contact resistance is

$$R_c = \frac{1}{h_c A} = \frac{(5.28 \times 10^{-4})(4)}{\pi(3 \times 10^{-2})^2} = 0.747°\text{C/W}$$

The total thermal resistance is therefore

$$\Sigma R_{\text{th}} = (2)(8.679) + 0.747 = 18.105$$

and the overall heat flow is

$$q = \frac{\Delta T}{\Sigma R_{\text{th}}} = \frac{100}{18.105} = 5.52 \text{ W} \quad [18.83 \text{ Btu/h}]$$

The temperature drop across the contact is found by taking the ratio of the contact resistance to the total thermal resistance:

$$\Delta T_c = \frac{R_c}{\Sigma R_{\text{th}}} \Delta T = \frac{(0.747)(100)}{18.105} = 4.13°\text{C} \quad [39.43°\text{F}]$$

REVIEW QUESTIONS

1 What is meant by the term *one-dimensional* when applied to conduction problems?

2 What is meant by thermal resistance?

3 Why is the one-dimensional heat-flow assumption important in the analysis of fins?

4 Define fin efficiency.

5 Why is the insulated-tip solution important for the fin problems?

6 What is meant by thermal contact resistance? Upon what parameters does this resistance depend?

■ PROBLEMS

2-1 A wall 2 cm thick is to be constructed from material which has an average thermal conductivity of 1.3 W/m · °C. The wall is to be insulated with material having an average thermal conductivity of 0.35 W/m·°C, so that the heat loss per square meter will not exceed 1830 W. Assuming that the inner and outer surface temperatures of the insulated wall are 1300 and 30°C, calculate the thickness of insulation required.

2-2 A certain material 2.5 cm thick, with a cross-sectional area of 0.1 m^2, has one side maintained at 35°C and the other at 95°C. The temperature at the center plane of the material is 62°C, and the heat flow through the material is 1 kW. Obtain an expression for the thermal conductivity of the material as a function of temperature.

2-3 A composite wall is formed of a 2.5-cm copper plate, a 3.2-mm layer of asbestos, and a 5-cm layer of fiber glass. The wall is subjected to an overall temperature difference of 560°C. Calculate the heat flow per unit area through the composite structure.

2-4 Find the heat transfer per unit area through the composite wall sketched. Assume one-dimensional heat flow.

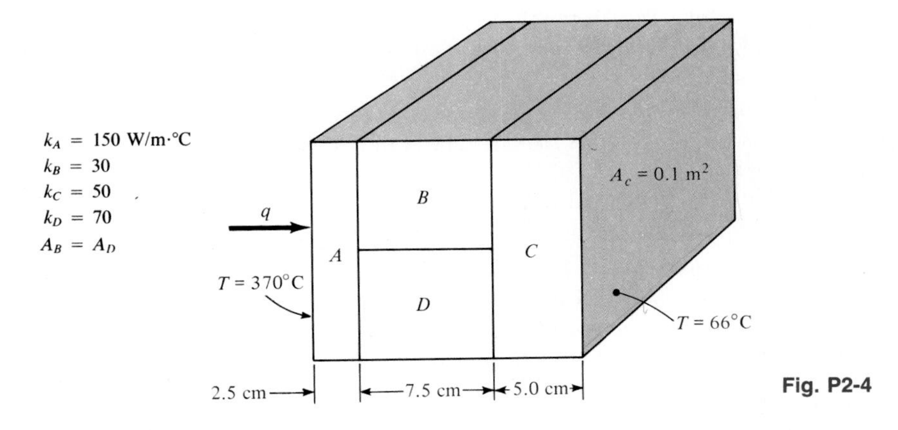

Fig. P2-4

2-5 One side of a copper block 5 cm thick is maintained at 260°C. The other side is covered with a layer of fiber glass 2.5 cm thick. The outside of the fiber glass is maintained at 38°C, and the total heat flow through the copper-fiber-glass combination is 44 kW. What is the area of the slab?

2-6 An outside wall for a building consists of a 10-cm layer of common brick and a 2.5-cm layer of fiber glass [k = 0.05 W/m · °C]. Calculate the heat flow through the wall for a 45°C temperature differential.

2-7 One side of a copper block 4 cm thick is maintained at 175°C. The other side is covered with a layer of fiber glass 1.5 cm thick. The outside of the fiber glass is maintained at 80°C, and the total heat flow through the composite slab is 300 W. What is the area of the slab?

2-8 A plane wall is constructed of a material having a thermal conductivity that varies as the square of temperature according to the relation $k = k_0(1 + \beta T^2)$. Derive an expression for the heat transfer in such a wall.

2-9 A certain material has a thickness of 30 cm and a thermal conductivity of 0.04 W/m · °C. At a particular instant in time the temperature distribution with x, the distance from the left face, is $T = 150x^2 - 30x$, where x is in meters. Calculate the heat flow rates at $x = 0$ and $x = 30$ cm. Is the solid heating up or cooling down?

2-10 A wall is constructed of 2.0 cm of copper, 3.0 mm of asbestos sheet [$k = 0.166$ W/m · °C], and 6.0 cm of fiber glass. Calculate the heat flow per unit area for an overall temperature difference of 500°C.

2-11 A certain building wall consists of 6.0 in of concrete [$k = 1.2$ W/m · °C], 2.0 in of fiber-glass insulation, and $\frac{3}{8}$ in of gypsum board [$k = 0.05$ W/m · °C]. The inside and outside convection coefficients are 2.0 and 7.0 Btu/h · ft² · °F, respectively. The outside air temperature is 20°F, and the inside temperature is 72°F. Calculate the overall heat-transfer coefficient for the wall, the R value, and the heat loss per unit area.

2-12 A wall is constructed of a section of stainless steel [$k = 16$ W/m · °C] 4.0 mm thick with identical layers of plastic on both sides of the steel. The overall heat-transfer coefficient, considering convection on both sides of the plastic, is 120 W/m² · °C. If the overall temperature difference across the arrangement is 60°C, calculate the temperature difference across the stainless steel.

2-13 An ice chest is constructed of styrofoam [$k = 0.033$ W/m · °C] with inside dimensions of 25 by 40 by 100 cm. The wall thickness is 5.0 cm. The outside of the chest is exposed to air at 25°C with $h = 10$ W/m² · °C. If the chest is completely filled with ice, calculate the time for the ice to completely melt. State your assumptions. The heat of fusion for water is 330 kJ/kg.

2-14 A spherical tank, 1 m in diameter, is maintained at a temperature of 120°C and exposed to a convection environment. With $h = 25$ W/m² · °C and $T_\infty = 15$°C, what thickness of urethane foam should be added to ensure that the outer temperature of the insulation does not exceed 40°C? What percentage reduction in heat loss results from installing this insulation?

2-15 A hollow sphere is constructed of aluminum with an inner diameter of 4 cm and an outer diameter of 8 cm. The inside temperature is 100°C and the outer temperature is 50°C. Calculate the heat transfer.

2-16 Suppose the sphere in problem 2-15 is covered with a 1-cm layer of an insulating material having $k = 50$ mW/m · °C and the outside of the insulation is exposed to an environment with $h = 20$ W/m² · °C and $T_\infty = 10$°C. The inside of the sphere remains at 100°C. Calculate the heat transfer under these conditions.

2-17 In Appendix A, dimensions of standard steel pipe are given. Suppose a 3-in schedule 80 pipe is covered with 1 in of an insulation having $k = 60$ mW/m · °C and the outside of the insulation is exposed to an environment having $h = 10$ W/m² · °C and $T_\infty = 20$°C. The temperature of the inside of the pipe is 250°C. For unit length of the pipe calculate (*a*) overall thermal resistance and (*b*) heat loss.

2-18 A steel pipe with a 5-cm OD is covered with a 6.4-mm asbestos insulation [$k = 0.096$ Btu/h · ft · °F] followed by a 2.5-cm layer of fiber-glass insulation [$k = 0.028$ Btu/h · ft · °F]. The pipe-wall temperature is 315°C, and the outside insulation temperature is 38°C. Calculate the interface temperature between the asbestos and fiber glass.

2-19 Derive an expression for the thermal resistance through a hollow spherical shell of inside radius r_i and outside radius r_o having a thermal conductivity k.

2-20 A 1.0-mm-diameter wire is maintained at a temperature of 400°C and exposed to a convection environment at 40°C with $h = 120$ W/m² · °C. Calculate the thermal conductivity which will just cause an insulation thickness of 0.2 mm to produce a "critical radius." How much of this insulation must be added to reduce the heat transfer by 75 percent from that which would be experienced by the bare wire?

2-21 A 2.0-in schedule 40 steel pipe (see Appendix A) has $k = 27$ Btu/h · ft · °F. The fluid inside the pipe has $h = 30$ Btu/h · ft² · °F, and the outer surface of the pipe is covered with 0.5-in fiber-glass insulation with $k = 0.023$ Btu/h · ft · °F. The convection coefficient on the outer insulation surface is 2.0 Btu/h · ft · °F. The inner fluid temperature is 320°F and the ambient temperature is 70°F. Calculate the heat loss per foot of length.

2-22 Derive a relation for the critical radius of insulation for a sphere.

2-23 A cylindrical tank 80 cm in diameter and 2.0 m high contains water at 80°C. The tank is 90 percent full, and insulation is to be added so that the water temperature will not drop more than 2°C per hour. Using the information given in this chapter, specify an insulating material and calculate the thickness required for the specified cooling rate.

2-24 A hot steam pipe having an inside surface temperature of 250°C has an inside diameter of 8 cm and a wall thickness of 5.5 mm. It is covered with a 9-cm layer of insulation having $k = 0.5$ W/m · °C, followed by a 4-cm layer of insulation having $k = 0.25$ W/m · °C. The outside temperature of the insulation is 20°C. Calculate the heat lost per meter of length. Assume $k = 47$ W/m · °C for the pipe.

2-25 A house wall may be approximated as two 1.2-cm layers of fiber insulating board, a 8.0-cm layer of loosely packed asbestos, and a 10-cm layer of common brick. Assuming convection heat-transfer coefficients of 15 W/m² · °C on both sides of the wall, calculate the overall heat-transfer coefficient for this arrangement.

2-26 Calculate the R value for the following insulations: (*a*) urethane foam, (*b*) fiber-glass mats, (*c*) mineral wool blocks, (*d*) calcium silicate blocks.

2-27 An insulation system is to be selected for a furnace wall at 1000°C using first a layer of mineral wool blocks followed by fiber-glass boards. The outside of the insulation is exposed to an environment with $h = 15$ W/m² · °C and $T_\infty = 40$°C. Using the data of Table 2-1 calculate the thickness of each insulating material such that the interface temperature is not greater than 400°C and the outside temperature is not greater than 55°C. Use mean values for the thermal conductivities. What is the heat loss in this wall in watts per square meter?

2-28 Derive an expression for the temperature distribution in a plane wall having uniformly distributed heat sources and one face maintained at a temperature T_1 while the other face is maintained at a temperature T_2. The thickness of the wall may be taken as $2L$.

2-29 Derive an expression for the temperature distribution in a plane wall in which distributed heat sources vary according to the linear relation

$$\dot{q} = \dot{q}_w[1 + \beta(T - T_w)]$$

where $\dot{q}_w$ is a constant and equal to the heat generated per unit volume at the wall temperature T_w. Both sides of the plate are maintained at T_w, and the plate thickness is $2L$.

2-30 A plane wall 6.0 cm thick generates heat internally at the rate of 0.3 MW/m³. One side of the wall is insulated, and the other side is exposed to an environment at 93°C. The convection heat-transfer coefficient between the wall and the environment is 570 W/m² · °C. The thermal conductivity of the wall is 21 W/m · °C. Calculate the maximum temperature in the wall.

2-31 Consider a shielding wall for a nuclear reactor. The wall receives a gamma-ray flux such that heat is generated within the wall according to the relation

$$\dot{q} = \dot{q}_0 e^{-ax}$$

where $\dot{q}_0$ is the heat generation at the inner face of the wall exposed to the gamma-ray flux and a is a constant. Using this relation for heat generation, derive an expression for the temperature distribution in a wall of thickness L, where the inside and outside temperatures are maintained at T_i and T_0, respectively. Also obtain an expression for the maximum temperature in the wall.

2-32 Repeat Prob. 2-31, assuming that the outer surface is adiabatic while the inner surface temperature is maintained at T_i.

2-33 Rework Prob. 2-29 assuming that the plate is subjected to a convection environment on both sides of temperature T_∞ with a heat-transfer coefficient h. T_w is now some reference temperature not necessarily the same as the surface temperature.

2-34 Heat is generated in a 2.5-cm-square copper rod at the rate of 35.3 MW/m³. The rod is exposed to a convection environment at 20°C, and the heat-transfer coefficient is 4000 W/m² · °C. Calculate the surface temperature of the rod.

2-35 A plane wall of thickness $2L$ has an internal heat generation which varies according to $\dot{q} = \dot{q}_0 \cos ax$, where $\dot{q}_0$ is the heat generated per unit volume at the center of the wall ($x = 0$) and a is a constant. If both sides of the wall are maintained at a constant temperature of T_w, derive an expression for the total heat loss from the wall per unit surface area.

2-36 A certain semiconductor material has a conductivity of 0.0124 W/cm · °C. A rectangular bar of the material has a cross-sectional area of 1 cm² and a length of 3 cm. One end is maintained at a 300°C and the other end at 100°C, and the bar carries a current of 50 A. Assuming the longitudinal surface is insulated, calculate the midpoint temperature in the bar. Take the resistivity as $1.5 \times 10^{-3} \Omega \cdot \text{cm}$.

2-37 The temperature distribution in a certain plane wall is

$$\frac{T - T_1}{T_2 - T_1} = C_1 + C_2x^2 + C_3x^3$$

where T_1 and T_2 are the temperatures on each side of the wall. If the thermal conductivity of the wall is constant and the wall thickness is L, derive an expression for the heat generation per unit volume as a function of x, the distance from the plane where $T = T_1$. Let the heat generation rate be $\dot{q}_0$ at $x = 0$.

2-38 Electric heater wires are installed in a solid wall having a thickness of 8 cm and $k = 2.5$ W/m · °C. The right face is exposed to an environment with $h = 50$ W/m² · °C and $T_\infty = 30$°C, while the left face is exposed to $h = 75$ W/m² · °C and $T_\infty = 50$°C. What is the maximum allowable heat generation rate such that the maximum temperature in the solid does not exceed 300°C?

2-39 A 3.0-cm-thick plate has heat generated uniformly at the rate of 5×10^5 W/m³. One side of the plate is maintained at 200°C and the other side at 50°C. Calculate the temperature at the center of the plate for $k = 20$ W/m · °C.

2-40 Heat is generated uniformly in a stainless steel plate having $k = 20$ W/m · °C. The thickness of the plate is 1.0 cm and the heat generation rate is 500 MW/m³. If the two sides of the plate are maintained at 100 and 200°C respectively, calculate the temperature at the center of the plate.

2-41 A plate having a thickness of 4.0 mm has an internal heat generation of 200 MW/m³ and a thermal conductivity of 25 W/m · °C. One side of the plate is insulated and the other side is maintained at 100°C. Calculate the maximum temperature in the plate.

2-42 A 3.2-mm-diameter stainless-steel wire 30 cm long has a voltage of 10 V impressed on it. The outer surface temperature of the wire is maintained at 93°C. Calculate the center temperature of the wire. Take the resistivity of the wire as 70 $\mu\Omega \cdot$ cm and the thermal conductivity as 22.5 W/m · °C.

2-43 The heater wire of Ex. 2-4 is submerged in a fluid maintained at 93°C. The convection heat-transfer coefficient is 5.7 kW/m² · °C. Calculate the center temperature of the wire.

2-44 An electric current is used to heat a tube through which a suitable cooling fluid flows. The outside of the tube is covered with insulation to minimize heat loss to the surroundings, and thermocouples are attached to the outer surface of the tube to measure the temperature. Assuming uniform heat generation in the tube, derive an expression for the convection heat-transfer coefficient on the inside of the tube in terms of the measured variables: voltage E, current I, outside tube wall temperature T_0, inside and outside radii r_i and r_o, tube length L, and fluid temperature T_f.

2-45 Derive an expression for the temperature distribution in a sphere of radius r with uniform heat generation $\dot{q}$ and constant surface temperature T_w.

2-46 A stainless-steel sphere [$k = 16$ W/m · °C] having a diameter of 4 cm is exposed to a convection environment at 20°C, $h = 15$ W/m² · °C. Heat is generated uniformly in the sphere at the rate of 1.0 MW/m³. Calculate the steady-state temperature for the center of the sphere.

2-47 An aluminum-alloy electrical cable has $k = 190$ W/m · °C, a diameter of 30 mm, and carries an electric current of 230 A. The resistivity of the cable is 2.9 $\mu\Omega$ · cm, and the outside surface temperature of the cable is 180°C. Calculate the maximum temperature in the cable if the surrounding air temperature is 15°C.

2-48 Derive an expression for the temperature distribution in a hollow cylinder with heat sources which vary according to the linear relation

$$\dot{q} = a + br$$

with $\dot{q}_i$ the generation rate per unit volume at $r = r_i$. The inside and outside temperatures are $T = T_i$ at $r = r_i$ and $T = T_o$ at $r = r_o$.

2-49 The outside of a copper wire having a diameter of 2 mm is exposed to a convection environment with $h = 5000$ W/m² · °C and $T_\infty = 100$°C. What current must be passed through the wire to produce a center temperature of 150°C? Repeat for an aluminum wire of the same diameter.

2-50 A hollow tube having an inside diameter of 2.5 cm and a wall thickness of 0.4 mm is exposed to an environment at $h = 100$ W/m² · °C and $T_\infty = 40$°C. What heat generation rate in the tube will produce a maximum tube temperature of 250°C for $k = 24$ W/m · °C?

2-51 Water flows on the inside of a steel pipe with an ID of 2.5 cm. The wall thickness is 2 mm, and the convection coefficient on the inside is 500 W/m² · °C. The convection coefficient on the outside is 12 W/m² · °C. Calculate the overall heat-transfer coefficient. What is the main determining factor for U?

2-52 The pipe in Prob. 2-51 is covered with a layer of asbestos [$k = 0.18$ W/m · °C] while still surrounded by a convection environment with $h = 12$ W/m² · °C. Calculate the critical insulation radius. Will the heat transfer be increased or decreased by adding an insulation thickness of (*a*) 0.5 mm, (*b*) 10 mm?

2-53 Calculate the overall heat-transfer coefficient for Prob. 2-4.

2-54 Calculate the overall heat-transfer coefficient for Prob. 2-5.

2-55 Air at 120°C in a thin-wall stainless-steel tube with $h = 65$ W/m² · °C. The inside diameter of the tube is 2.5 cm and the wall thickness is 0.4 mm. $k = 18$ W/m · °C for the steel. The tube is exposed to an environment with $h = 6.5$ W/m² · °C and $T_\infty = 15$°C. Calculate the overall heat transfer coefficient and the heat loss per meter of length. What thickness of an insulation having $k = 40$ mW/m · °C should be added to reduce the heat loss by 90 percent?

2-56 An insulating glass window is constructed of two 5-mm glass plates separated by an air layer having a thickness of 4 mm. The air layer may be considered stagnant so that pure conduction is involved. The convection coefficients for the inner and outer surfaces are 12 and 50 W/m² · °C respectively. Calculate the overall heat transfer coefficient for this arrangement, and the R value. Repeat the calculation for a single glass plate 5 mm thick.

2-57 A wall consists of a 1-mm layer of copper, a 4-mm layer of 1 percent carbon steel, a 1-cm layer of asbestos sheet, and 10 cm of fiber-glass blanket. Calculate the overall heat-transfer coefficient for this arrangement. If the two outside surfaces are at 10 and 150°C, calculate each of the interface temperatures.

2-58 A thin rod of length L has its two ends connected to two walls which are maintained at temperatures T_1 and T_2, respectively. The rod loses heat to the environment at T_∞ by convection. Derive an expression (*a*) for the temperature distribution in the rod and (*b*) for the total heat lost by the rod.

2-59 A rod of length L has one end maintained at temperature T_0 and is exposed to an environment of temperature T_∞. An electrical heating element is placed in the rod so that heat is generated uniformly along the length at a rate $\dot{q}$. Derive an expression (*a*) for the temperature distribution in the rod and (*b*) for the total heat transferred to the environment. Obtain an expression for the value of $\dot{q}$ which will make the heat transfer zero at the end which is maintained at T_0.

2-60 One end of a copper rod 30 cm long is firmly connected to a wall which is maintained at 200°C. The other end is firmly connected to a wall which is maintained at 93°C. Air is blown across the rod so that a heat-transfer coefficient of 17 W/m² · °C is maintained. The diameter of the rod is 12.5 mm. The temperature of the air is 38°C. What is the net heat lost to the air in watts?

2-61 Verify the temperature distribution for case 2 in Sec. 2-9, i.e., that

$$\frac{T - T_\infty}{T_0 - T_\infty} = \frac{\cosh m(L - x) + (h/mk) \sinh m(L - x)}{\cosh mL + (h/mk) \sinh mL}$$

Subsequently show that the heat transfer is

$$q = \sqrt{hPkA}\,(T_0 - T_\infty)\frac{\sinh mL + (h/mk) \cosh mL}{\cosh mL + (h/mk) \sinh mL}$$

2-62 An aluminum rod 2.5 cm in diameter and 15 cm long protrudes from a wall which is maintained at 260°C. The rod is exposed to an environment at 16°C. The convection heat-transfer coefficient is 15 W/m² · °C. Calculate the heat lost by the rod.

2-63 Derive Eq. (2-35) by integrating the convection heat loss from the rod of case 1 in Sec. 2-9.

2-64 Derive Eq. (2-36) by integrating the convection heat loss from the rod of case 3 in Sec. 2-9.

2-65 A long, thin copper rod 6.4 mm in diameter is exposed to an environment at 20°C. The base temperature of the rod is 150°C. The heat-transfer coefficient between the rod and the environment is 24 W/m² · °C. Calculate the heat given up by the rod.

2-66 A very long copper rod [k = 372 W/m · °C] 2.5 cm in diameter has one end maintained at 90°C. The rod is exposed to a fluid whose temperature is 40°C. The heat-transfer coefficient is 3.5 W/m² · °C. How much heat is lost by the rod?

2-67 An aluminum fin 1.6 mm thick is placed on a circular tube with 2.5-cm OD. The fin is 6.4 mm long. The tube wall is maintained at 150°C, the environment temperature is 15°C, and the convection heat-transfer coefficient is 23 W/m² · C. Calculate the heat lost by the fin.

2-68 The total efficiency for a finned surface may be defined as the ratio of the total heat transfer of the combined area of the surface and fins to the heat which would be transferred if this total area were maintained at the base temperature T_0. Show that this efficiency can be calculated from

$$\eta_t = 1 - \frac{A_f}{A}(1 - \eta_f)$$

where η_t = total efficiency
A_f = surface area of all fins
A = total heat-transfer area, including fins and exposed tube or other surface
η_f = fin efficiency

2-69 A triangular fin of stainless steel (18% Cr, 8% Ni) is attached to a plane wall maintained at 460°C. The fin thickness is 6.4 mm, and the length is 2.5 cm. The environment is at 93°C, and the convection heat-transfer coefficient is 28 W/m² · °C. Calculate the heat lost from the fin.

2-70 A 2.5-cm-diameter tube has circumferential fins of rectangular profile spaced at 9.5-mm increments along its length. The fins are constructed of aluminum and are 0.8 mm thick and 12.5 mm long. The tube wall temperature is maintained at 200°C, and the environment temperature is 93°C. The heat-transfer coefficient is 110 W/m² · °C. Calculate the heat loss from the tube per meter of length.

2-71 A circumferential fin of rectangular cross section surrounds a 2.5-cm-diameter tube. The length of the fin is 6.4 mm, and the thickness is 3.2 mm. The fin is constructed of mild steel. If air blows over the fin so that a heat-transfer coefficient of 28 W/m² · °C is experienced and the temperatures of the base and air are 260 and 93°C, respectively, calculate the heat transfer from the fin.

2-72 A straight rectangular fin 2.0 cm thick and 14 cm long is constructed of steel and placed on the outside of a wall maintained at 200°C. The environment temperature is 15°C, and the heat-transfer coefficient for convection is 20 W/m² · °C. Calculate the heat lost from the fin per unit depth.

2-73 An aluminum fin 1.6 mm thick surrounds a tube 2.5 cm in diameter. The length of the fin is 12.5 mm. The tube-wall temperature is 200°C, and the environment temperature is 20°C. The heat-transfer coefficient is 60 W/m² · °C. What is the heat lost by the fin?

2-74 Obtain an expression for the optimum thickness of a straight rectangular fin for a given profile area. Use the simplified insulated-tip solution.

2-75 Derive a differential equation (do not solve) for the temperature distribution in a straight triangular fin. For convenience take the coordinate axis as shown and assume one-dimensional heat flow.

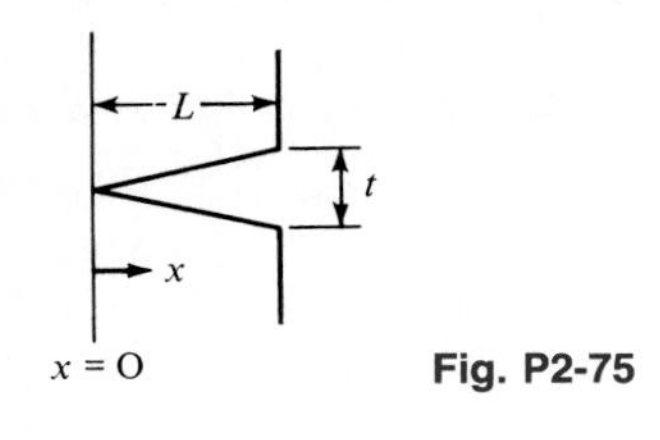

Fig. P2-75

2-76 A long stainless-steel rod [$k = 16$ W/m · °C] has a square cross section 12.5 by 12.5 mm and has one end maintained at 250°C. The heat-transfer coefficient is 40 W/m² · °C, and the environment temperature is 90°C. Calculate the heat lost by the rod.

2-77 A straight fin of rectangular profile is constructed of duralumin (94% Al, 3% Cu) with a thickness of 2.4 mm. The fin is 19 mm long, and it is subjected to a convection environment with $h = 85$ W/m² · °C. If the base temperature is 90°C and the environment is at 25°C, calculate the heat transfer per unit length of fin.

2-78 A certain internal-combustion engine is air-cooled and has a cylinder constructed of cast iron [$k = 35$ Btu/h · ft · °F]. The fins on the cylinder have a length of $\frac{5}{8}$ in and thickness of $\frac{1}{8}$ in. The convection coefficient is 12 Btu/h · ft² · °F. The cylinder diameter is 4 in. Calculate the heat loss per fin for a base temperature of 450°F and environment temperature of 100°F.

2-79 A 1.6-mm-diameter stainless-steel rod [$k = 22$ W/m · °C] protrudes from a wall maintained at 49°C. The rod is 12.5 mm long, and the convection coefficient is 570 W/m² · °C. The environment temperature is 25°C. Calculate the temperature of the tip of the rod. Repeat the calculation for $h = 200$ and 1200 W/m² · °C.

2-80 A 2-cm-diameter glass rod 6 cm long [$k = 0.8$ W/m · °C] has a base temperature of 100°C and is exposed to an air convection environment at 20°C. The temperature at the tip of the rod is measured as 35°C. What is the convection heat-transfer coefficient? How much heat is lost by the rod?

2-81 A straight rectangular fin has a length of 2.0 cm and a thickness of 1.5 mm. The thermal conductivity is 55 W/m · °C, and it is exposed to a convection environment at 20°C and $h = 500$ W/m² · °C. Calculate the maximum possible heat loss for a base temperature of 200°C. What is the actual heat loss?

2-82 A straight rectangular fin has a length of 3.5 cm and a thickness of 1.4 mm. The thermal conductivity is 55 W/m · °C. The fin is exposed to a convection environment at 20°C and $h = 500$ W/m² · °C. Calculate the maximum possible heat loss for a base temperature of 150°C. What is the actual heat loss for this base temperature?

2-83 A circumferential fin of rectangular profile is constructed of 1 percent carbon steel and attached to a circular tube maintained at 150°C. The diameter of the tube is 5 cm, and the length is also 5 cm with a thickness of 2 mm. The surrounding air is maintained at 20°C and the convection heat-transfer coefficient may be taken as 100 W/m² · °C. Calculate the heat lost from the fin.

2-84 A circumferential fin of rectangular profile is constructed of aluminum and surrounds a 3-cm-diameter tube. The fin is 2 cm long and 1 mm thick. The tube wall temperature is 200°C, and the fin is exposed to a fluid at 20°C with a convection heat-transfer coefficient of 80 W/m² · °C. Calculate the heat loss from the fin.

2-85 A 1.0-cm-diameter steel rod ($k = 20$ W/m² · °C) is 20 cm long. It has one end maintained at 50°C and the other at 100°C. It is exposed to a convection environment at 20°C with $h = 85$ W/m² · °C. Calculate the temperature at the center of the rod.

2-86 A straight rectangular fin of steel (1% C) is 2.6 cm thick and 17 cm long. It is placed on the outside of a wall which is maintained at 230°C. The surrounding air temperature is 25°C,and the convection heat-transfer coefficient is 23 W/m² · °C. Calculate the heat lost from the fin per unit depth and the fin efficiency.

2-87 A straight fin having a triangular profile has a length of 5 cm and a thickness of 4 mm and is constructed of a material having $k = 23$ W/m · °C. The fin is exposed to surroundings with a convection coefficient of 20 W/m² · °C and a temperature of 40°C. The base of the fin is maintained at 200°C. Calculate the heat lost per unit depth of fin.

2-88 A circumferential aluminum fin is installed on a 1-in-diameter tube. The length of the fin is 0.5 in and the thickness is 1.0 mm. It is exposed to a convection environment at 30°C with a convection coefficient of 56 W/m² · °C. The base temperature is 125°C. Calculate the heat lost by the fin.

2-89 A circumferential fin of rectangular profile is constructed of stainless steel (18% Cr, 8% Ni). The thickness of the fin is 2.0 mm, the inside radius is 2.0 cm, and the length is 8.0 cm. The base temperature is maintained at 135°C and the fin is exposed to a convection environment at 15°C with $h = 20$ W/m² · °C. Calculate the heat lost by the fin.

2-90 A rectangular fin has a length of 2.5 cm and thickness of 1.1 mm. The thermal conductivity is 55 W/m · °C. The fin is exposed to a convection environment at 20°C and $h = 500$ W/m² · °C. Calculate the heat loss for a base temperature of 125°C.

2-91 A 1.0-mm-thick aluminum fin surrounds a 2.5-cm-diameter tube. The length of the fin is 1.25 cm. The fin is exposed to a convection environment at 30°C with $h = 75$ W/m² · °C. The tube surface is maintained at 100°C. Calculate the heat lost by the fin.

2-92 A glass rod having a diameter of 1 cm and length of 5 cm is exposed to a convection environment at a temperature of 20°C. One end of the rod is maintained at a temperature of 180°C. Calculate the heat lost by the rod if the convection heat-transfer coefficient is 15 W/m² · °C.

2-93 A stainless steel rod has a square cross-section measuring 1 by 1 cm. The rod length is 8 cm, and $k = 18$ W/m² · °C. The base temperature of the rod is 300°C. The rod is exposed to a convection environment at 50°C with $h = 45$ W/m² · °C. Calculate the heat lost by the rod and the fin efficiency.

2-94 Copper fins with a thickness of 1.0 mm are installed on a 2.5-cm-diameter tube. The length of each fin is 12 mm. The tube temperature is 250°C and the fins are exposed to air at 30°C with a convection heat-transfer coefficient of 120 W/m² · °C. Calculate the heat lost by each fin.

2-95 A straight fin of rectangular profile is constructed of stainless steel (18% Cr, 8% Ni) and has a length of 5 cm and a thickness of 2.5 cm. The base temperature is maintained at 100°C and the fin is exposed to a convection environment at 20°C with $h = 47$ W/m² · °C. Calculate the heat lost by the fin per meter of depth, and the fin efficiency.

2-96 A circumferential fin of rectangular profile is constructed of Duralumin and surrounds a 3-cm-diameter tube. The fin is 3 cm long and 1 mm thick. The tube wall temperature is 200°C, and the fin is exposed to a fluid at 20°C with a convection heat-transfer coefficient of 80 W/m² · °C. Calculate the heat loss from the fin.

2-97 A circular fin of rectangular profile is attached to a 3.0-cm-diameter tube maintained at 100°C. The outside diameter of the fin is 9.0 cm and the fin thickness is 1.0 mm. The environment has a convection coefficient of 50 W/m² · C and a temperature of 30°C. Calculate the thermal conductivity of the material for a fin efficiency of 60 percent.

2-98 A circumferential fin of rectangular profile having a thickness of 1.0 mm and a length of 2.0 cm is placed on a 2.0-cm-diameter tube. The tube temperature is 150°C, the environment temperature is 20°C, and h = 200 W/m² · °C. The fin is aluminum. Calculate the heat lost by the fin.

2-99 Two 1-in-diameter bars of stainless steel [k = 17 W/m · °C] are brought into end-to-end contact so that only 0.1 percent of the cross-sectional area is in contact at the joint. The bars are 7.5 cm long and subjected to an axial temperature difference of 300°C. The roughness depth in each bar ($L_g/2$) is estimated to be 1.3 μm. The surrounding fluid is air, whose thermal conductivity may be taken as 0.035 W/m · °C for this problem. Estimate the value of the contact resistance and the axial heat flow. What would the heat flow be for a continuous 15-cm stainless-steel bar?

2-100 When the *joint pressure* for two surfaces in contact is increased, the high spots of the surfaces are deformed so that the contact area A_c is increased and the roughness depth L_g is decreased. Discuss this effect in the light of the presentation of Sec. 2-11. (Experimental work shows that joint conductance varies almost directly with pressure.)

2-101 Two aluminum plates 5 mm thick with a ground roughness of 100 μin are bolted together with a contact pressure of 20 atm. The overall temperature difference across the plates is 80°C. Calculate the temperature drop across the contact joint.

2-102 Fins are frequently installed on tubes by a press-fit process. Consider a circumferential aluminum fin having a thickness of 1.0 mm to be installed on a 2.5-cm-diameter aluminum tube. The fin length is 1.25 cm, and the contact conductance may be taken from Table 2-2 for a 100-μin ground surface. The convection environment is at 20°C, and h = 125 W/m² · °C. Calculate the heat transfer for each fin for a tube wall temperature of 200°C. What percentage reduction in heat transfer is caused by the contact conductance?

2-103 An aluminum fin is attached to a transistor which generates heat at the rate of 300 mW. The fin has a total surface area of 9.0 cm² and is exposed to surrounding air at 27°C. The contact conductance between transistor and fin is 0.9×10^{-4} m² · °C/W, and the contact area is 0.5 cm². Estimate the temperature of the transistor, assuming the fin is uniform in temperature.

2-104 Suppose you have a choice between a straight triangular or rectangular fin constructed of aluminum with a base thickness of 3.0 mm. The convection coefficient is 50 W/m² · °C. Select the fin with the least weight for a given heat flow.

2-105 Consider aluminum circumferential fins with r_1 = 1.0 cm, r_2 = 2.0 cm, and thicknesses of 1.0, 2.0, and 3.0 mm. The convection coefficient is 160 W/m² · °C. Compare the heat transfers for six 1.0-mm fins, three 2.0-mm fins, and two 3.0-mm fins. What do you conclude? Repeat for h = 320 W/m² · °C.

2-106 A plane wall 20 cm thick with uniform internal heat generation of 200 kW/m³ is exposed to a convection environment on both sides at 50°C with h = 400 W/m² · °C. Calculate the center temperature of the wall for k = 20 W/m · °C.

2-107 Suppose the wall of Prob. 2-106 is only 10 cm thick and has one face insulated. Calculate the maximum temperature in the wall assuming all the other conditions are the same. Comment on the results.

2-108 A straight aluminum fin of triangular profile has a base maintained at 200°C and is exposed to a convection environment at 25°C with h = 45 W/m² · °C. The fin has a length of 8 mm and a thickness of 2.0 mm. Calculate the heat lost per unit depth of fin.

2-109 One hundred circumferential aluminum fins of rectangular profile are mounted on a 1.0-m tube having a diameter of 2.5 cm. The fins are 1 cm long and 2.0 mm thick. The base temperature is 180°C, and the convection environment is at 20°C with h = 50 W/m² · °C. Calculate the total heat lost from the finned-tube arrangement over the 1.0-m length.

2-110 "Pin fins" of aluminum are to be compared in terms of their relative performance as a function of diameter. Three "pins" having diameters of 2, 5, and 10 mm with a length of 5 cm are exposed to a convection environment with T_∞ = 20°C, and h = 40 W/m² · °C. The base temperature is 200 °C. Calculate the heat transfer for each pin. How does it vary with pin diameter?

2-111 Calculate the heat transfer per unit mass for the pin fins in Prob. 2-110. How does it vary with diameter?

2-112 The cylindrical segment shown has a thermal conductivity of 100 W/m · °C. The inner and outer radii are 1.5 and 1.7 cm, respectively, and the surfaces are insulated. Calculate the circumferential heat transfer per unit depth for an imposed temperature difference of 50°C. What is the thermal resistance?

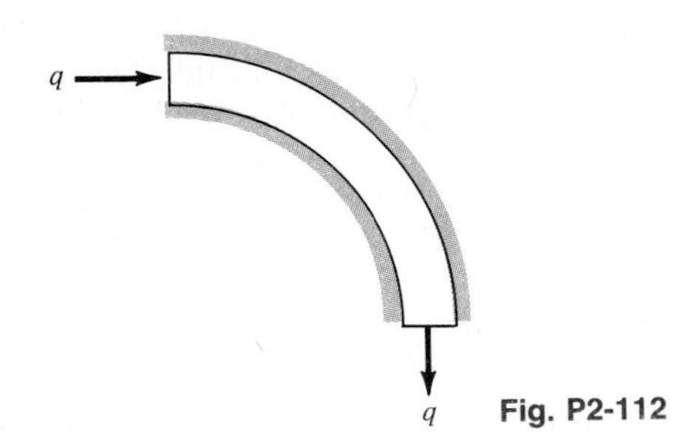

Fig. P2-112

2-113 A straight rectangular fin has a length of 1.5 cm and a thickness of 1.0 mm. The convection coefficient is 20 W/m² · °C. Compare the heat-transfer rates for aluminum and magnesium fins.

2-114 Suppose both fins in Prob. 2-113 are to dissipate the same heat. Which would be lower in weight? Assume that the thickness is the same for both fins but adjust the lengths until the heat transfers are equal.

2-115 The truncated hollow cone shown is used in laser-cooling applications and is constructed of copper with a thickness of 0.5 mm. Calculate the thermal resistance for one-dimensional heat flow. What would be the heat transfer for a temperature difference of 300°C?

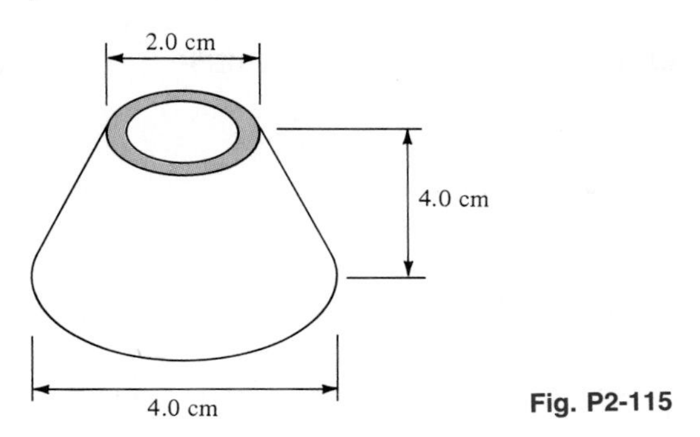

Fig. P2-115

■ REFERENCES

1 Schneider, P. J.: "Conduction Heat Transfer," Addison-Wesley Publishing Company, Inc., Reading, Mass., 1955.

2 Harper, W. B., and D. R. Brown: Mathematical Equations for Heat Conduction in the Fins of Air-cooled Engines, *NACA Rep.* 158, 1922.

3 Gardner, K. A.: Efficiency of Extended Surfaces, *Trans. ASME,* vol. 67, pp. 621–631, 1945.

4 Moore, C. J.: Heat Transfer across Surfaces in Contact: Studies of Transients in One-dimensional Composite Systems, *Southern Methodist Univ., Thermal/Fluid Sci. Ctr. Res. Rep.* 67-2, Dallas, Tex., March 1967.

5 Ybarrondo, L. J., and J. E. Sunderland: Heat Transfer from Extended Surfaces, *Bull. Mech. Eng. Educ.,* vol. 5, pp. 229–234, 1966.

6 Moore, C. J., Jr., H. A. Blum, and H. Atkins: Subject Classification Bibliography for Thermal Contact Resistance Studies, *ASME Pap.* 68-WA/HT-18, December 1968.

7 Clausing, A. M.: Transfer at the Interface of Dissimilar Metals: The Influence of Thermal Strain, *Int. J. Heat Mass Transfer,* vol. 9, p. 791, 1966.

8 Kern, D. Q., and A. D. Kraus: "Extended Surface Heat Transfer," McGraw-Hill Book Company, New York, 1972.

9 Siegel, R., and J. R. Howell: "Thermal Radiation Heat Transfer," 2d ed., McGraw-Hill Book Company, New York, 1980.

10 Fried E.: Thermal Conduction Contribution to Heat Transfer at Contacts, "Thermal Conductivity," (R. P. Tye, ed.) vol. 2, Academic Press, Inc., New York, 1969.

11 Fletcher, L. S.: Recent Developments in Contact Conductance Heat Transfer, *J. Heat Transfer*, vol. 110, no. 4(B), p. 1059, Nov. 1988.

STEADY-STATE CONDUCTION—MULTIPLE DIMENSIONS

3-1 INTRODUCTION

In Chap. 2 steady-state heat transfer was calculated in systems in which the temperature gradient and area could be expressed in terms of one space coordinate. We now wish to analyze the more general case of two-dimensional heat flow. For steady state with no heat generation, the Laplace equation applies.

$$\frac{\partial^2 T}{\partial x^2} + \frac{\partial^2 T}{\partial y^2} = 0 \tag{3-1}$$

assuming constant thermal conductivity. The solution to this equation may be obtained by analytical, numerical, or graphical techniques.

The objective of any heat-transfer analysis is usually to predict heat flow or the temperature which results from a certain heat flow. The solution to Eq. (3-1) will give the temperature in a two-dimensional body as a function of the two independent space coordinates x and y. Then the heat flow in the x and y directions may be calculated from the Fourier equations

$$q_x = -kA_x \frac{\partial T}{\partial x} \tag{3-2}$$

$$q_y = -kA_y \frac{\partial T}{\partial y} \tag{3-3}$$

These heat-flow quantities are directed either in the x direction or in the y direction. The total heat flow at any point in the material is the resultant of the q_x and q_y at that point. Thus the total heat-flow vector is directed so that it is perpendicular to the lines of constant temperature in the material, as shown in Fig. 3-1. So if the temperature distribution in the material is known, we may easily establish the heat flow.

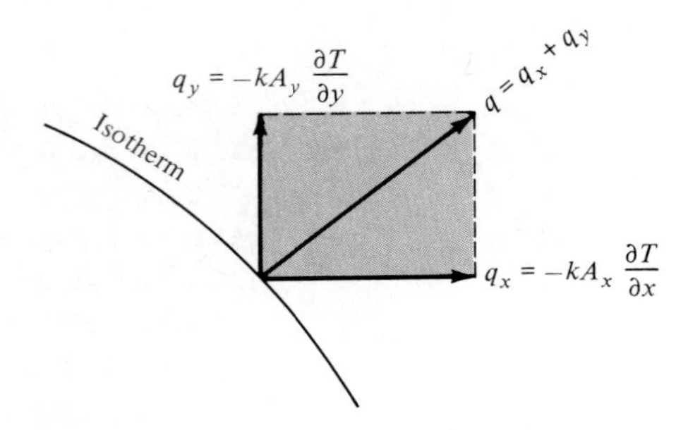

Fig. 3-1 Sketch showing the heat flow in two dimensions.

3-2 MATHEMATICAL ANALYSIS OF TWO-DIMENSIONAL HEAT CONDUCTION

We first consider an analytical approach to a two-dimensional problem and then indicate the numerical and graphical methods which may be used to advantage in many other problems. It is worthwhile to mention here that analytical solutions are not always possible to obtain; indeed, in many instances they are very cumbersome and difficult to use. In these cases numerical techniques are frequently used to advantage. For a more extensive treatment of the analytical methods used in conduction problems, the reader may consult Refs. 1, 2, 12, and 13.

Consider the rectangular plate shown in Fig. 3-2. Three sides of the plate are maintained at the constant temperature T_1, and the upper side has some temperature distribution impressed upon it. This distribution could be simply a constant temperature or something more complex, such as a sine-wave distribution. We shall consider both cases.

To solve Eq. (3-1), the separation-of-variables method is used. The essential point of this method is that the solution to the differential equation is assumed to take a product form

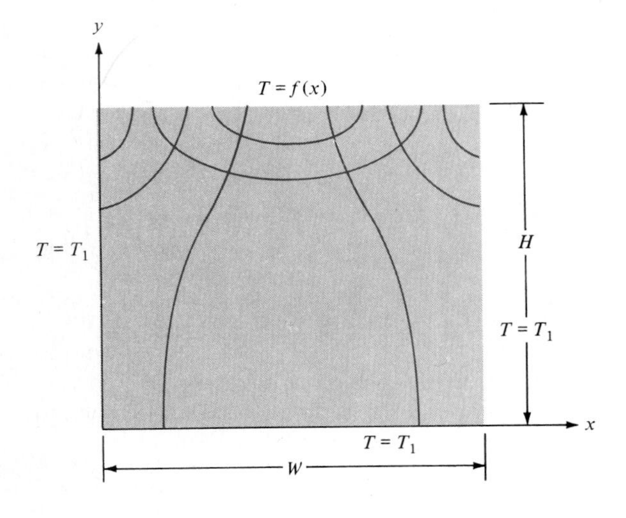

Fig. 3-2 Isotherms and heat flow lines in a rectangular plate.

$$T = XY \qquad \text{where} \qquad X = X(x) \qquad \text{(3-4)}$$
$$Y = Y(y)$$

The boundary conditions are then applied to determine the form of the functions X and Y. The basic assumption as given by Eq. (3-4) can be justified *only* if it is possible to find a solution of this form which satisfies the boundary conditions.

First consider the boundary conditions with a sine-wave temperature distribution impressed on the upper edge of the plate. Thus

$$T = T_1 \text{ at } y = 0$$
$$T = T_1 \text{ at } x = 0 \qquad \text{(3-5)}$$
$$T = T_1 \text{ at } x = W$$
$$T = T_m \sin\left(\frac{\pi x}{W}\right) + T_1 \qquad \text{at } y = H$$

where T_m is the amplitude of the sine function. Substituting Eq. (3-4) in (3-1) gives

$$-\frac{1}{X}\frac{d^2X}{dx^2} = \frac{1}{Y}\frac{d^2Y}{dy^2} \qquad \text{(3-6)}$$

Observe that each side of Eq. (3-6) is independent of the other because x and y are independent variables. This requires that each side be equal to some constant. We may thus obtain two ordinary differential equations in terms of this constant,

$$\frac{d^2X}{dx^2} + \lambda^2 X = 0 \qquad \text{(3-7)}$$

$$\frac{d^2Y}{dy^2} - \lambda^2 Y = 0 \qquad \text{(3-8)}$$

where λ^2 is called the *separation constant*. Its value must be determined from the boundary conditions. Note that the form of the solution to Eqs. (3-7) and (3-8) will depend on the sign of λ^2; a different form would also result if λ^2 were zero. The only way that the correct form can be determined is through an application of the boundary conditions of the problem. So we shall first write down all possible solutions and then see which one fits the problem under consideration.

For $\lambda^2 = 0$:

$$X = C_1 + C_2 x$$
$$Y = C_3 + C_4 y \qquad \text{(3-9)}$$
$$T = (C_1 + C_2 x)(C_3 + C_4 y)$$

This function cannot fit the sine-function boundary condition, so that the $\lambda^2 = 0$ solution may be excluded.

For $\lambda^2 < 0$:

$$X = C_5 e^{-\lambda x} + C_6 e^{\lambda x}$$
$$Y = C_7 \cos \lambda y + C_8 \sin \lambda y \tag{3-10}$$
$$T = (C_5 e^{-\lambda x} + C_6 e^{\lambda x})(C_7 \cos \lambda y + C_8 \sin \lambda y)$$

Again, the sine-function boundary condition cannot be satisfied, so this solution is excluded also.

For $\lambda^2 > 0$:

$$X = C_9 \cos \lambda x + C_{10} \sin \lambda x$$
$$Y = C_{11} e^{-\lambda y} + C_{12} e^{\lambda y} \tag{3-11}$$
$$T = (C_9 \cos \lambda x + C_{10} \sin \lambda x)(C_{11} e^{-\lambda y} + C_{12} e^{\lambda y})$$

Now, it is possible to satisfy the sine-function boundary condition; so we shall attempt to satisfy the other conditions. The algebra is somewhat easier to handle when the substitution

$$\theta = T - T_1$$

is made. The differential equation and the solution then retain the same form in the new variable θ, and we need only transform the boundary conditions. Thus

$$\begin{aligned} \theta &= 0 && \text{at } y = 0 \\ \theta &= 0 && \text{at } x = 0 \\ \theta &= 0 && \text{at } x = W \\ \theta &= T_m \sin \frac{\pi x}{W} && \text{at } y = H \end{aligned} \tag{3-12}$$

Applying these conditions, we have

$$0 = (C_9 \cos \lambda x + C_{10} \sin \lambda x)(C_{11} + C_{12}) \tag{a}$$
$$0 = C_9(C_{11} e^{-\lambda y} + C_{12} e^{\lambda y}) \tag{b}$$
$$0 = (C_9 \cos \lambda W + C_{10} \sin \lambda W)(C_{11} e^{-\lambda y} + C_{12} e^{\lambda y}) \tag{c}$$
$$T_m \sin \frac{\pi x}{W} = (C_9 \cos \lambda x + C_{10} \sin \lambda x)(C_{11} e^{-\lambda H} + C_{12} e^{\lambda H}) \tag{d}$$

Accordingly,

$$C_{11} = -C_{12}$$
$$C_9 = 0$$

and from (*c*),

$$0 = C_{10} C_{12} \sin \lambda W (e^{\lambda y} - e^{-\lambda y})$$

This requires that

$$\sin \lambda W = 0 \tag{3-13}$$

Recall that λ was an undetermined separation constant. Several values will satisfy Eq. (3-13), and these may be written

$$\lambda = \frac{n\pi}{W} \tag{3-14}$$

where n is an integer. The solution to the differential equation may thus be written as a sum of the solutions for each value of n. This is an infinite sum, so that the final solution is the infinite series

$$\theta = T - T_1 = \sum_{n=1}^{\infty} C_n \sin \frac{n\pi x}{W} \sinh \frac{n\pi y}{W} \tag{3-15}$$

where the constants have been combined and the exponential terms converted to the hyperbolic function. The final boundary condition may now be applied:

$$T_m \sin \frac{\pi x}{W} = \sum_{n=1}^{\infty} C_n \sin \frac{n\pi x}{W} \sinh \frac{n\pi H}{W}$$

which requires that $C_n = 0$ for $n > 1$. The final solution is therefore

$$T = T_m \frac{\sinh (\pi y/W)}{\sinh (\pi H/W)} \sin \left(\frac{\pi x}{W}\right) + T_1 \tag{3-16}$$

The temperature field for this problem is shown in Fig. 3-2. Note that the heat-flow lines are perpendicular to the isotherms.

We now consider the set of boundary conditions

$$\begin{aligned} T &= T_1 \qquad \text{at } y = 0 \\ T &= T_1 \qquad \text{at } x = 0 \\ T &= T_1 \qquad \text{at } x = W \\ T &= T_2 \qquad \text{at } y = H \end{aligned}$$

Using the first three boundary conditions, we obtain the solution in the form of Eq. (3-15):

$$T - T_1 = \sum_{n=1}^{\infty} C_n \sin \frac{n\pi x}{W} \sinh \frac{n\pi y}{W} \tag{3-17}$$

Applying the fourth boundary condition gives

$$T_2 - T_1 = \sum_{n=1}^{\infty} C_n \sin \frac{n\pi x}{W} \sinh \frac{n\pi H}{W} \tag{3-18}$$

This is a Fourier sine series, and the values of the C_n may be determined by expanding the constant temperature difference $T_2 - T_1$ in a Fourier series over the interval $0 < x < W$. This series is

$$T_2 - T_1 = (T_2 - T_1) \frac{2}{\pi} \sum_{n=1}^{\infty} \frac{(-1)^{n+1} + 1}{n} \sin \frac{n\pi x}{W} \tag{3-19}$$

Upon comparison of Eq. (3-18) with Eq. (3-19), we find that

$$C_n = \frac{2}{\pi}(T_2 - T_1)\frac{1}{\sinh(n\pi H/W)}\frac{(-1)^{n+1}+1}{n}$$

and the final solution is expressed as

$$\frac{T - T_1}{T_2 - T_1} = \frac{2}{\pi}\sum_{n=1}^{\infty}\frac{(-1)^{n+1}+1}{n}\sin\frac{n\pi x}{W}\frac{\sinh(n\pi y/W)}{\sinh(n\pi H/W)} \tag{3-20}$$

An extensive study of analytical techniques used in conduction heat transfer requires a background in the theory of orthogonal functions. Fourier series are one example of orthogonal functions, as are Bessel functions and other special functions applicable to different geometries and boundary conditions. The interested reader may consult one or more of the conduction heat-transfer texts listed in the references for further information on the subject.

3-3 GRAPHICAL ANALYSIS

Consider the two-dimensional system shown in Fig. 3-3. The inside surface is maintained at some temperature T_1, and the outer surface is maintained at T_2. We wish to calculate the heat transfer. Isotherms and heat-flow lines have been

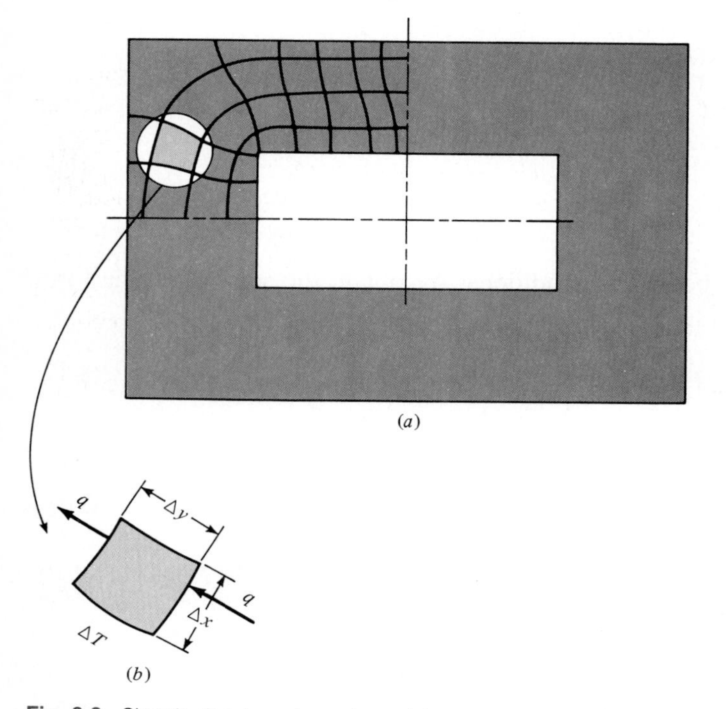

Fig. 3-3 Sketch showing element used for curvilinear-square analysis of two-dimensional heat flow.

sketched to aid in this calculation. The isotherms and heat-flow lines form groupings of curvilinear figures like that shown in Fig. 3-3*b*. The heat flow across this curvilinear section is given by Fourier's law, assuming unit depth of material:

$$q = -k\,\Delta x(1)\frac{\Delta T}{\Delta y} \tag{3-21}$$

This heat flow will be the same through each section within this heat-flow lane, and the total heat flow will be the sum of the heat flows through all the lanes. If the sketch is drawn so that $\Delta x \cong \Delta y$, the heat flow is proportional to the ΔT across the element and, since this heat flow is constant, the ΔT across each element must be the same within the same heat-flow lane. Thus the ΔT across an element is given by

$$\Delta T = \frac{\Delta T_{\text{overall}}}{N}$$

where N is the number of temperature increments between the inner and outer surfaces. Furthermore, the heat flow through each lane is the same since it is independent of the dimensions Δx and Δy when they are constructed equal. Thus we write for the total heat transfer

$$q = \frac{M}{N}k\,\Delta T_{\text{overall}} = \frac{M}{N}k(T_2 - T_1) \tag{3-22}$$

where M is the number of heat-flow lanes. So, to calculate the heat transfer, we need only construct these curvilinear-square plots and count the number of temperature increments and heat-flow lanes. Care must be taken to construct the plot so that $\Delta x \approx \Delta y$ and the lines are perpendicular.

The accuracy of this method is dependent entirely on the skill of the person sketching the curvilinear squares. Even a crude sketch, however, can frequently help to give fairly good estimates of the temperatures that will occur in a body; and these estimates may then be refined with numerical techniques discussed in Sec. 3-5. An electrical analogy may be employed to sketch the curvilinear squares, as discussed in Sec. 3-9.

The graphical method presented here is mainly of historical interest to show the relation of heat-flow lanes and isotherms. It may not be expected to be used for the solution of many practical problems.

3-4 THE CONDUCTION SHAPE FACTOR

In a two-dimensional system where only two temperature limits are involved, we may define a conduction shape factor S such that

$$q = kS\,\Delta T_{\text{overall}} \tag{3-23}$$

The values of S have been worked out for several geometries and are summarized in Table 3-1. A very comprehensive summary of shape factors for a

Table 3-1 Conduction Shape Factors, Summarized from Refs. 6 and 7

Physical system	*Schematic*	*Shape factor*	*Restrictions*
Isothermal cylinder of radius r buried in semi-infinite medium having isothermal surface	Isothermal; D; r; L	$\dfrac{2\pi L}{\cosh^{-1}(D/r)}$	$L \gg r$
		$\dfrac{2\pi L}{\ln(2D/r)}$	$L \gg r$ $D > 3r$
		$\dfrac{2\pi L}{\ln\dfrac{L}{r}\left[1 - \dfrac{\ln(L/2D)}{\ln(L/r)}\right]}$	$D \gg r$ $L \gg D$
Isothermal sphere of radius r buried in infinite medium	r	$4\pi r$	
Isothermal sphere of radius r buried in semi-infinite medium having isothermal surface	Isothermal; D; r	$\dfrac{4\pi r}{1 - r/2D}$	
Conduction between two isothermal cylinders buried in infinite medium	r_1; r_2; D	$\dfrac{2\pi L}{\cosh^{-1}\left(\dfrac{D^2 - r_1^2 - r_2^2}{2r_1r_2}\right)}$	$L \gg r$ $L \gg D$

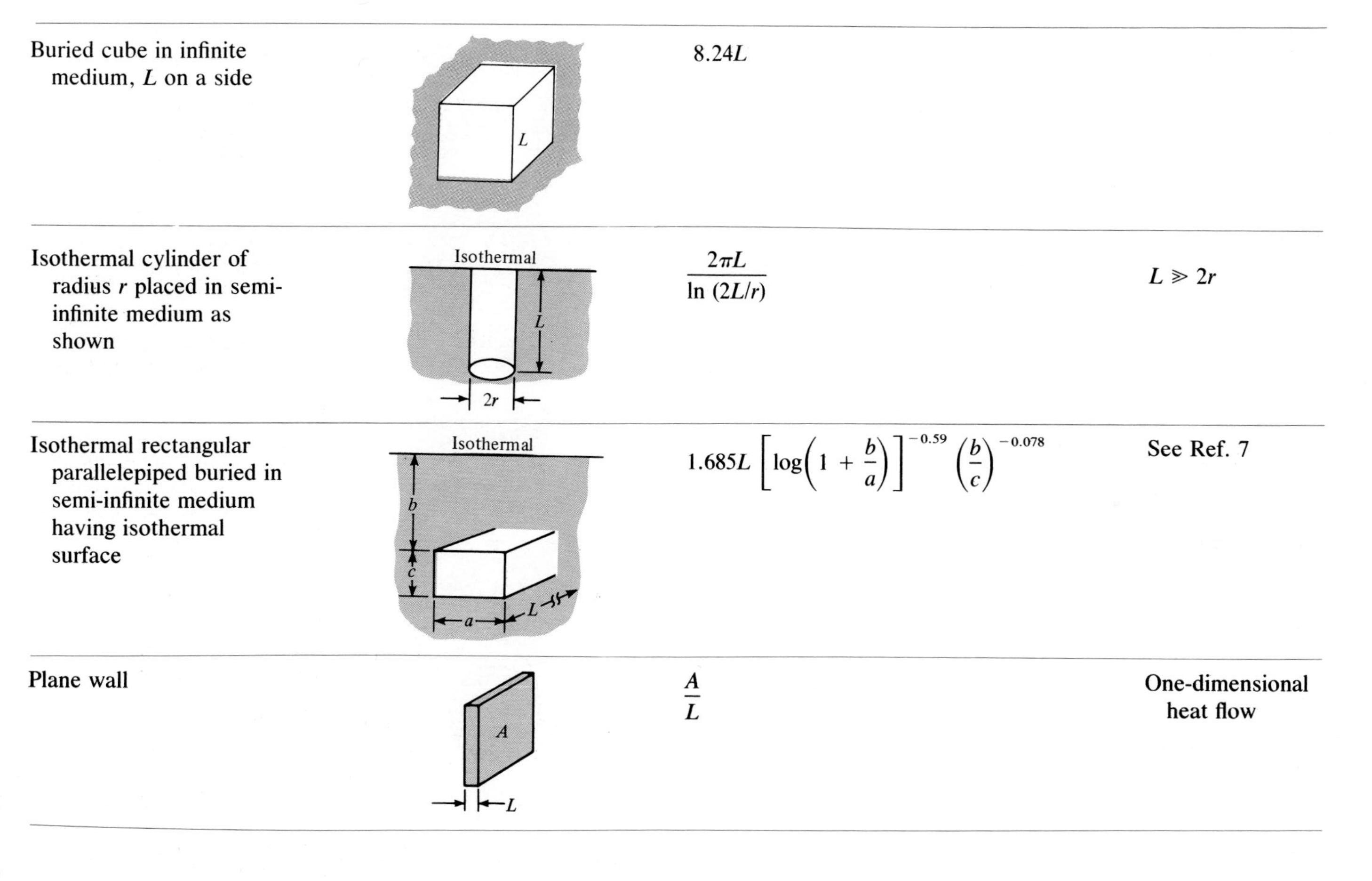

Buried cube in infinite medium, L on a side		$8.24L$	
Isothermal cylinder of radius r placed in semi-infinite medium as shown		$\dfrac{2\pi L}{\ln (2L/r)}$	$L \gg 2r$
Isothermal rectangular parallelepiped buried in semi-infinite medium having isothermal surface		$1.685L\left[\log\left(1+\dfrac{b}{a}\right)\right]^{-0.59}\left(\dfrac{b}{c}\right)^{-0.078}$	See Ref. 7
Plane wall		$\dfrac{A}{L}$	One-dimensional heat flow

Table 3-1 Conduction Shape Factors, Summarized from Refs. 6 and 7 (*Continued*)

Physical system	*Schematic*	*Shape factor*	*Restrictions*
Hollow cylinder, length L	r_i, r_o	$\frac{2\pi L}{\ln (r_o/r_i)}$	$L \gg r$
Hollow sphere	r_i, r_o	$\frac{4\pi r_o r_i}{r_o - r_i}$	
Thin horizontal disk buried in semi-infinite medium with isothermal surface	Isothermal; $2r$; D	$4r$ $8r$	$D = 0$ $D \gg 2r$
Hemisphere buried in semi-infinite medium	r	$2\pi r$	
Isothermal sphere buried in semi-infinite medium with insulated surface	Insulated; D; T_∞; r	$\frac{4\pi r}{1 + r/2D}$	
Two isothermal spheres buried in infinite medium	r_1; r_2; D	$\frac{4\pi}{\frac{r_2}{r_1}\left[1 - \frac{(r_1/D)^4}{1 - (r_2/D)^2}\right] - \frac{2r_2}{D}}$	$D > 5r_{max}$

Thin rectangular plate of length L, buried in semi-infinite medium having isothermal surface	Isothermal	$\dfrac{\pi W}{\ln(4W/L)}$ $\dfrac{2\pi W}{\ln(4W/L)}$ $\dfrac{2\pi W}{\ln(2\pi D/L)}$	$D = 0$ $W > L$ $D \gg W$ $W > L$ $W \gg L$ $D > 2W$
Parallel disks buried in infinite medium		$\dfrac{4\pi}{2\left[\dfrac{\pi}{2} - \tan^{-1}(r/D)\right]}$	$D > 5r$ r/D in radians
Eccentric cylinders of length L		$\dfrac{2\pi L}{\cosh^{-1}\left(\dfrac{r_1^2 + r_2^2 - D^2}{2r_1 r_2}\right)}$	$L \gg r_2$
Cylinder centered in a square of length L		$\dfrac{2\pi L}{\ln(0.54W/r)}$	$L \gg W$

Table 3-1 Conduction Shape Factors, Summarized from Refs. 6 and 7 (*Continued*)

Physical system	*Schematic*	*Shape factor*	*Restrictions*
Horizontal cylinder centered in infinite plate	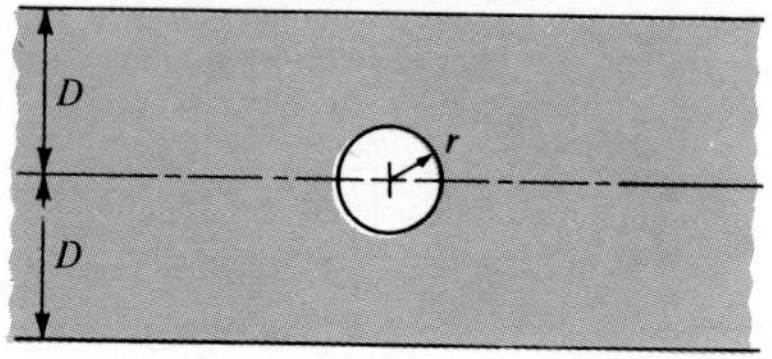 D, D, r	$\dfrac{2\pi L}{\ln (4D/r)}$	

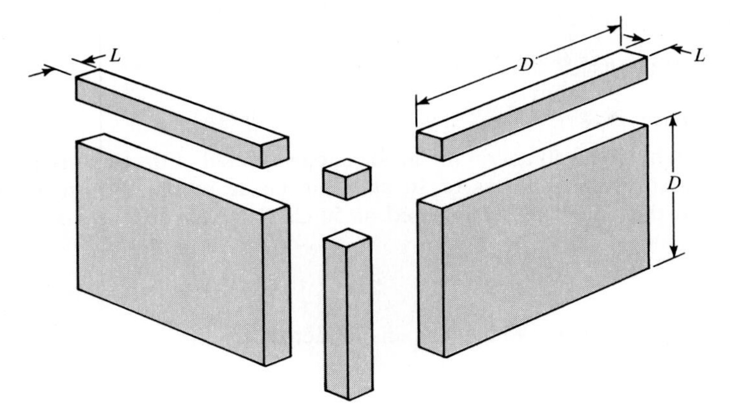

Fig. 3-4 Sketch illustrating dimensions for use in calculating three-dimensional shape factors.

large variety of geometries is given by Hahne and Grigull [18, 23]. Note that the inverse hyperbolic cosine can be calculated from

$$\cosh^{-1}x = \ln\,(x \pm \sqrt{x^2 - 1})$$

For a three-dimensional wall, as in a furnace, separate shape factors are used to calculate the heat flow through the edge and corner sections. When all the interior dimensions are greater than one-fifth of the wall thickness,

$$S_{\text{wall}} = \frac{A}{L} \qquad S_{\text{edge}} = 0.54D \qquad S_{\text{corner}} = 0.15L$$

where A = area of wall
L = wall thickness
D = length of edge

These dimensions are illustrated in Fig. 3-4. Note that the shape factor per unit depth is given by the ratio M/N when the curvilinear-squares method is used for calculations. The use of the shape factor for calculation purposes is illustrated in Examples 3-1 and 3-2.

■ **EXAMPLE 3-1** Buried pipe

A horizontal pipe 15 cm in diameter and 4 m long is buried in the earth at a depth of 20 cm. The pipe-wall temperature is 75°C, and the earth surface temperature is 5°C. Assuming that the thermal conductivity of the earth is 0.8 W/m · °C, calculate the heat lost by the pipe.

Solution

We may calculate the shape factor for this situation using the equation given in Table 3-1. Since $D < 3r$,

$$S = \frac{2\pi L}{\cosh^{-1}(D/r)} = \frac{2\pi(4)}{\cosh^{-1}(20/7.5)} = 15.35 \text{ m}$$

The heat flow is calculated from

$$q = kS\,\Delta T = (0.8)(15.35)(75 - 5) = 859.6\ \text{W} \quad [2933\ \text{Btu/h}]$$

EXAMPLE 3-2 Cubical furnace

A small cubical furnace 50 by 50 by 50 cm on the inside is constructed of fireclay brick [$k = 1.04$ W/m · °C] with a wall thickness of 10 cm. The inside of the furnace is maintained at 500°C, and the outside is maintained at 50°C. Calculate the heat lost through the walls.

Solution

We compute the total shape factor by adding the shape factors for the walls, edges, and corners:

Walls: $$S = \frac{A}{L} = \frac{(0.5)(0.5)}{0.1} = 2.5\ \text{m}$$

Edges: $$S = 0.54D = (0.54)(0.5) = 0.27\ \text{m}$$

Corners: $$S = 0.15L = (0.15)(0.1) = 0.015\ \text{m}$$

There are six wall sections, twelve edges, and eight corners, so that the total shape factor is

$$S = (6)(2.5) + (12)(0.27) + (8)(0.015) = 18.36\ \text{m}$$

and the heat flow is calculated as

$$q = kS\,\Delta T = (1.04)(18.36)(500 - 50) = 8.592\ \text{kW} \quad [29{,}320\ \text{Btu/h}]$$

3-5 NUMERICAL METHOD OF ANALYSIS

An immense number of analytical solutions for conduction heat-transfer problems have been accumulated in the literature over the past 100 years. Even so, in many practical situations the geometry or boundary conditions are such that an analytical solution has not been obtained at all, or if the solution has been developed, it involves such a complex series solution that numerical evaluation becomes exceedingly difficult. For such situations the most fruitful approach to the problem is one based on finite-difference techniques, the basic principles of which we shall outline in this section.

Consider a two-dimensional body which is to be divided into equal increments in both the x and y directions, as shown in Fig. 3-5. The nodal points are designated as shown, the m locations indicating the x increment and the n locations indicating the y increment. We wish to establish the temperatures at any of these nodal points within the body, using Eq. (3-1) as a governing condition. Finite differences are used to approximate differential increments in the temperature and space coordinates; and the smaller we choose these finite increments, the more closely the true temperature distribution will be approximated.

The temperature gradients may be written as follows:

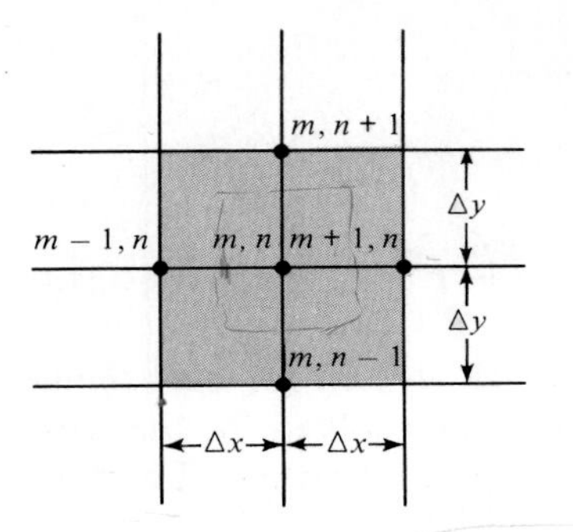

Fig. 3-5 Sketch illustrating nomenclature used in two-dimensional numerical analysis of heat conduction.

$$\left.\frac{\partial T}{\partial x}\right]_{m+1/2,n} \approx \frac{T_{m+1,n} - T_{m,n}}{\Delta x}$$

$$\left.\frac{\partial T}{\partial x}\right]_{m-1/2,n} \approx \frac{T_{m,n} - T_{m-1,n}}{\Delta x}$$

$$\left.\frac{\partial T}{\partial y}\right]_{m,n+1/2} \approx \frac{T_{m,n+1} - T_{m,n}}{\Delta y}$$

$$\left.\frac{\partial T}{\partial y}\right]_{m,n-1/2} \approx \frac{T_{m,n} - T_{m,n-1}}{\Delta y}$$

$$\left.\frac{\partial^2 T}{\partial x^2}\right]_{m,n} \approx \frac{\left.\dfrac{\partial T}{\partial x}\right]_{m+1/2,n} - \left.\dfrac{\partial T}{\partial x}\right]_{m-1/2,n}}{\Delta x} = \frac{T_{m+1,n} + T_{m-1,n} - 2T_{m,n}}{(\Delta x)^2}$$

$$\left.\frac{\partial^2 T}{\partial y^2}\right]_{m,n} \approx \frac{\left.\dfrac{\partial T}{\partial y}\right]_{m,n+1/2} - \left.\dfrac{\partial T}{\partial y}\right]_{m,n-1/2}}{\Delta y} = \frac{T_{m,n+1} + T_{m,n-1} - 2T_{m,n}}{(\Delta y)^2}$$

Thus the finite-difference approximation for Eq. (3-1) becomes

$$\frac{T_{m+1,n} + T_{m-1,n} - 2T_{m,n}}{(\Delta x)^2} + \frac{T_{m,n+1} + T_{m,n-1} - 2T_{m,n}}{(\Delta y)^2} = 0$$

If $\Delta x = \Delta y$, then

$$T_{m+1,n} + T_{m-1,n} + T_{m,n+1} + T_{m,n-1} - 4T_{m,n} = 0 \qquad (3\text{-}24)$$

Since we are considering the case of constant thermal conductivity, the heat flows may all be expressed in terms of temperature differentials. Equation (3-24) states very simply that the net heat flow into any node is zero at steady-state conditions. In effect, the numerical finite-difference approach replaces the continuous temperature distribution by fictitious heat-conducting rods connected between small nodal points which do not generate heat.

We can also devise a finite-difference scheme to take heat generation into account. We merely add the term $\dot{q}/k$ into the general equation and obtain

$$\frac{T_{m+1,n} + T_{m-1,n} - 2T_{m,n}}{(\Delta x)^2} + \frac{T_{m,n+1} + T_{m,n-1} - 2T_{m,n}}{(\Delta y)^2} + \frac{\dot{q}}{k} = 0$$

Then for a square grid in which $\Delta x = \Delta y$,

$$T_{m+1,n} + T_{m-1,n} + T_{m,n+1} + T_{m,n-1} + \frac{\dot{q}(\Delta x)^2}{k} - 4T_{m,n} = 0 \quad \text{(3-24a)}$$

To utilize the numerical method, Eq. (3-24) must be written for each node within the material and the resultant system of equations solved for the temperatures at the various nodes. A very simple example is shown in Fig. 3-6, and the four equations for nodes 1, 2, 3, and 4 would be

$$100 + 500 + T_2 + T_3 - 4T_1 = 0$$

$$T_1 + 500 + 100 + T_4 - 4T_2 = 0$$

$$100 + T_1 + T_4 + 100 - 4T_3 = 0$$

$$T_3 + T_2 + 100 + 100 - 4T_4 = 0$$

These equations have the solution

$$T_1 = T_2 = 250°\text{C} \qquad T_3 = T_4 = 150°\text{C}$$

Of course, we could recognize from symmetry that $T_1 = T_2$ and $T_3 = T_4$ and would then only need two nodal equations,

$$100 + 500 + T_3 - 3T_1 = 0$$

$$100 + T_1 + 100 - 3T_3 = 0$$

Once the temperatures are determined, the heat flow may be calculated from

$$q = \sum k \, \Delta x \frac{\Delta T}{\Delta y}$$

where the ΔT is taken at the boundaries. In the example the heat flow may be calculated at either the 500°C face or the three 100°C faces. If a sufficiently fine grid is used, the two values should be very nearly the same. As a matter

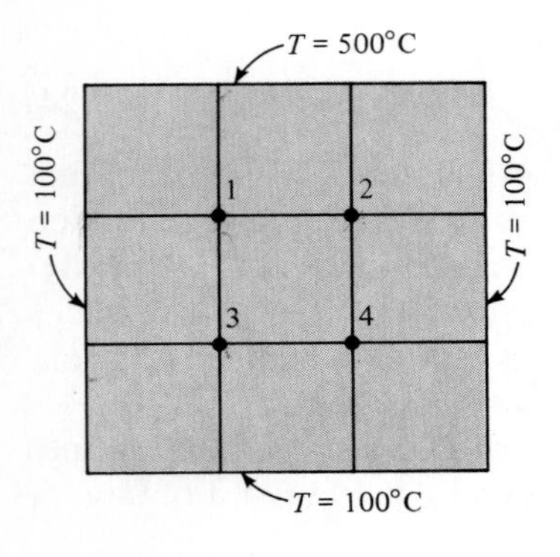

Fig. 3-6 Four-node problem.

of general practice, it is usually best to take the arithmetic average of the two values for use in the calculations. In the example the two calculations yield:

500°C face:

$$q = -k \frac{\Delta y}{\Delta x}[(250 - 500) + (250 - 500)] = 500k$$

100°C face:

$$q = -k \frac{\Delta y}{\Delta x}[(250 - 100) + (150 - 100) + (150 - 100) + (150 - 100) + (150 - 100) + (250 - 100)] = -500k$$

and the two values agree in this case. The calculation of the heat flow in cases in which curved boundaries or complicated shapes are involved is treated in Refs. 2, 3, and 15.

When the solid is exposed to some convection boundary condition, the temperatures at the surface must be computed differently from the method given above. Consider the boundary shown in Fig. 3-7. The energy balance on node (m, n) is

$$-k\,\Delta y \frac{T_{m,n} - T_{m-1,n}}{\Delta x} - k\frac{\Delta x}{2}\frac{T_{m,n} - T_{m,n+1}}{\Delta y} - k\frac{\Delta x}{2}\frac{T_{m,n} - T_{m,n+1}}{\Delta y} = h\Delta y(T_{m,n} - T_\infty)$$

If $\Delta x = \Delta y$, the boundary temperature is expressed in the equation

$$T_{m,n}\left(\frac{h\,\Delta x}{k} + 2\right) - \frac{h\,\Delta x}{k}T_\infty - \frac{1}{2}(2T_{m-1,n} + T_{m,n+1} + T_{m,n-1}) = 0 \qquad (3\text{-}25)$$

An equation of this type must be written for each node along the surface shown in Fig. 3-7. So when a convection boundary condition is present, an equation like (3-25) is used at the boundary and an equation like (3-24) is used for the interior points.

Equation (3-25) applies to a plane surface exposed to a convection boundary

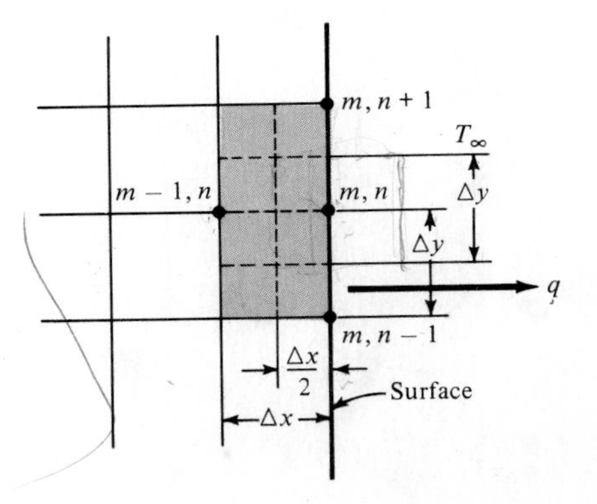

Fig. 3-7 Nomenclature for nodal equation with convective boundary condition.

condition. It will not apply for other situations, such as an insulated wall or a corner exposed to a convection boundary condition. Consider the corner section shown in Fig. 3-8. The energy balance for the corner section is

$$-k\frac{\Delta y}{2}\frac{T_{m,n}-T_{m-1,n}}{\Delta x}-k\frac{\Delta x}{2}\frac{T_{m,n}-T_{m,n-1}}{\Delta y} = h\frac{\Delta x}{2}(T_{m,n}-T_\infty)+h\frac{\Delta y}{2}(T_{m,n}-T_\infty)$$

If $\Delta x = \Delta y$,

$$2T_{m,n}\left(\frac{h\,\Delta x}{k}+1\right)-2\frac{h\,\Delta x}{k}T_\infty-(T_{m-1,n}+T_{m,n-1})=0 \qquad (3\text{-}26)$$

Other boundary conditions may be treated in a similar fashion, and a convenient summary of nodal equations is given in Table 3-2 for different geometrical and boundary situations. Situations *f* and *g* are of particular interest since they provide the calculation equations which may be employed with curved boundaries, while still using uniform increments in Δx and Δy.

EXAMPLE 3-3

Consider the square of Fig. 3-9. The left face is maintained at 100°C and the top face at 500°C, while the other two faces are exposed to an environment at 100°C:

$$h = 10\ \text{W/m}^2\cdot{}^\circ\text{C} \qquad \text{and} \qquad k = 10\ \text{W/m}\cdot{}^\circ\text{C}$$

The block is 1 m square. Compute the temperatures of the various nodes as indicated in Fig. 3-9 and the heat flows at the boundaries.

Solution

The nodal equation for nodes 1, 2, 4, and 5 is

$$T_{m+1,n}+T_{m-1,n}+T_{m,n+1}+T_{m,n-1}-4T_{m,n}=0$$

The equation for nodes 3, 6, 7, and 8 is given by Eq. (3-25), and the equation for 9 is given by Eq. (3-26):

$$\frac{h\,\Delta x}{k}=\frac{(10)(1)}{(3)(10)}=\frac{1}{3}$$

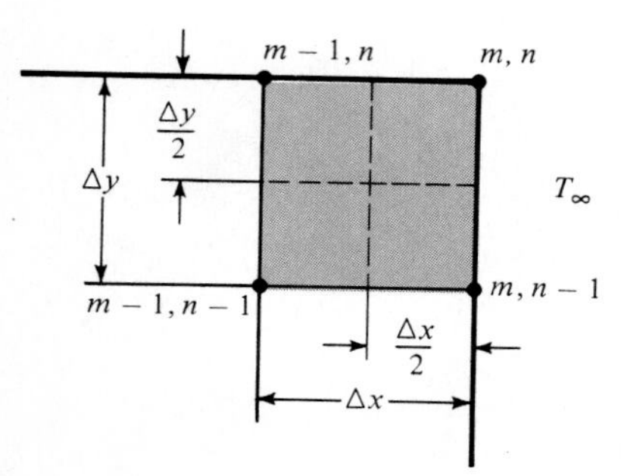

Fig. 3-8 Nomenclature for nodal equation with convection at a corner section.

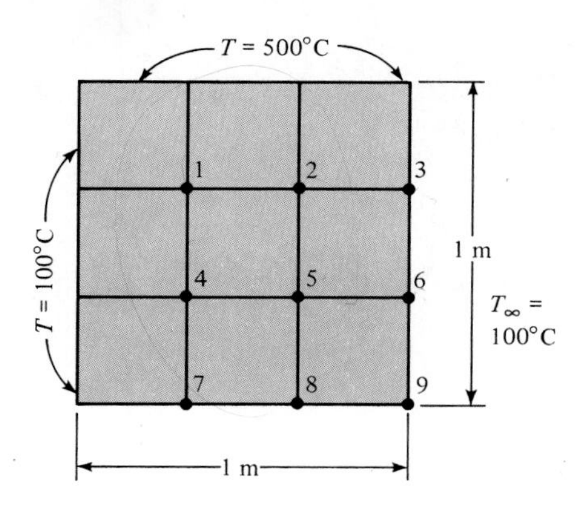

Fig. 3-9 Nomenclature for Example 3-3.

The equations for nodes 3 and 6 are thus written

$$2T_2 + T_6 + 567 - 4.67T_3 = 0$$

$$2T_5 + T_3 + T_9 + 67 - 4.67T_6 = 0$$

The equations for nodes 7 and 8 are given by

$$2T_4 + T_8 + 167 - 4.67T_7 = 0$$

$$2T_5 + T_7 + T_9 + 67 - 4.67T_8 = 0$$

and the equation for node 9 is

$$T_6 + T_8 + 67 - 2.67T_9 = 0$$

We thus have nine equations and nine unknown nodal temperatures. We shall discuss solution techniques shortly, but for now we just list the answers:

Node	*Temperature*, °C
1	280.67
2	330.30
3	309.38
4	192.38
5	231.15
6	217.19
7	157.70
8	184.71
9	175.62

The heat flows at the boundaries are computed in two ways: as conduction flows for the 100 and 500°C faces and as convection flows for the other two faces. For the 500°C face, the heat flow *into* the face is

$$q = \sum k\,\Delta x\,\frac{\Delta T}{\Delta y} = (10)[500 - 280.67 + 500 - 330.30 + (500 - 309.38)(\tfrac{1}{2})]$$

$$= 4843.4 \text{ W/m}$$

Table 3-2 Summary of Nodal Formulas for Finite-Difference Calculations (Dashed Lines Indicate Element Volume.)†

Physical situation	*Nodal equation for equal increments in x and y (second equation in situation is in form for Gauss-Seidel iteration)*
(*a*) Interior node	$0 = T_{m+1,n} + T_{m,n+1} + T_{m-1,n} + T_{m,n-1} - 4T_{m,n}$ $T_{m,n} = (T_{m+1,n} + T_{m,n+1} + T_{m-1,n} + T_{m,n-1})/4$
(*b*) Convection boundary node	$0 = \frac{h\,\Delta x}{k}T_\infty + \frac{1}{2}(2T_{m-1,n} + T_{m,n+1} + T_{m,n-1}) - \left(\frac{h\,\Delta x}{k} + 2\right)T_{m,n}$ $T_{m,n} = \frac{T_{m-1,n} + (T_{m,n+1} + T_{m,n-1})/2 + \mathrm{Bi}T_\infty}{2 + \mathrm{Bi}}$ $\mathrm{Bi} = \frac{h\,\Delta x}{k}$

(*c*) Exterior corner with convection boundary

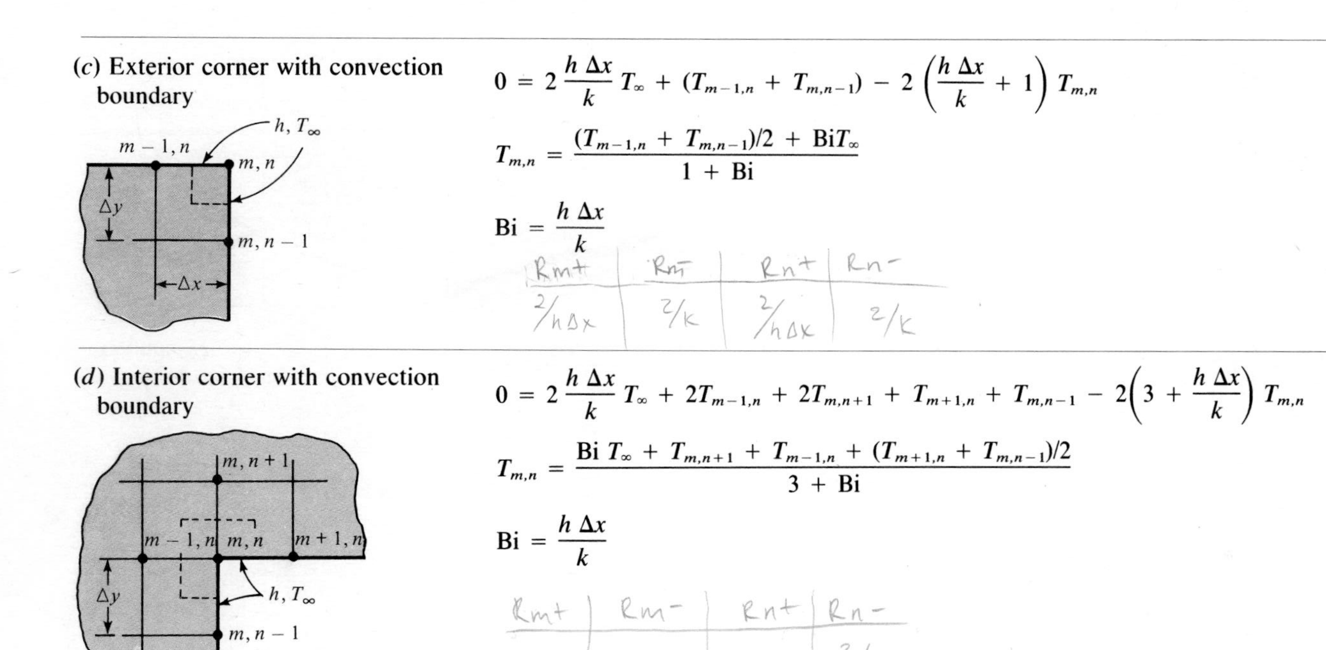

$$0 = 2\frac{h\,\Delta x}{k}T_\infty + (T_{m-1,n} + T_{m,n-1}) - 2\left(\frac{h\,\Delta x}{k} + 1\right)T_{m,n}$$

$$T_{m,n} = \frac{(T_{m-1,n} + T_{m,n-1})/2 + \mathrm{Bi}T_\infty}{1 + \mathrm{Bi}}$$

$$\mathrm{Bi} = \frac{h\,\Delta x}{k}$$

Rm+	Rm-	Rn+	Rn-
2/hΔx	2/k	2/hΔx	2/k

(*d*) Interior corner with convection boundary

$$0 = 2\frac{h\,\Delta x}{k}T_\infty + 2T_{m-1,n} + 2T_{m,n+1} + T_{m+1,n} + T_{m,n-1} - 2\left(3 + \frac{h\,\Delta x}{k}\right)T_{m,n}$$

$$T_{m,n} = \frac{\mathrm{Bi}\,T_\infty + T_{m,n+1} + T_{m-1,n} + (T_{m+1,n} + T_{m,n-1})/2}{3 + \mathrm{Bi}}$$

$$\mathrm{Bi} = \frac{h\,\Delta x}{k}$$

Rm+	Rm-	Rn+	Rn-
2/k	1/k	1/k	2/k

Table 3-2 Summary of Nodal Formulas for Finite-Difference Calculations (Dashed Lines Indicate Element Volume.) (*Continued*)

Physical situation	*Nodal equation for equal increments in x and y (second equation in situation is in form for Gauss-Seidel iteration)*
(*e*) Insulated boundary	$0 = T_{m,n+1} + T_{m,n-1} + 2T_{m-1,n} - 4T_{m,n}$ $T_{m,n} = (T_{m,n+1} + T_{m,n-1} + 2T_{m-1,n})/4$
(*f*) Interior node near curved boundary‡	$0 = \dfrac{2}{b(b+1)} T_2 + \dfrac{2}{a+1} T_{m+1,n} + \dfrac{2}{b+1} T_{m,n-1} + \dfrac{2}{a(a+1)} T_1 - 2\left(\dfrac{1}{a} + \dfrac{1}{b}\right) T_{m,n}$

Rm+	Rm−	Rn+	Rn−
∞	1/k	2/k	2/k

Rm+	Rm−	Rn+	Rn−

(g) Boundary node with convection along curved boundary—node 2 for (f) above§	$$0 = \frac{b}{\sqrt{a^2 + b^2}} T_1 + \frac{b}{\sqrt{c^2 + 1}} T_3 + \frac{a + 1}{b} T_{m,n} + \frac{h \, \Delta x}{k} (\sqrt{c^2 + 1} + \sqrt{a^2 + b^2}) T_\infty - \left[\frac{b}{\sqrt{a^2 + b^2}} + \frac{b}{\sqrt{c^2 + 1}} + \frac{a + 1}{b} + (\sqrt{c^2 + 1} + \sqrt{a^2 + b^2}) \frac{h \, \Delta x}{k} \right] T_2$$

†Convection boundary may be converted to insulated surface by setting $h = 0$ (Bi = 0).
‡This equation is obtained by multiplying the resistance formulation by $4/(a + 1)(b + 1)$.
§This relation is obtained by dividing the resistance formulation by 2.

The heat flow *out* of the 100°C face is

$$q = \sum k\,\Delta y \frac{\Delta T}{\Delta x} = (10)[280.67 - 100 + 192.38 - 100 + (157.70 - 100)(\tfrac{1}{2})]$$

$$= 3019 \text{ W/m}$$

The heat flow *out* the right face is given by the convection relation

$$q = \sum h\,\Delta y(T - T_\infty)$$

$$= (10)(\tfrac{1}{3})[309.38 - 100 + 217.19 - 100 + (175.62 - 100)(\tfrac{1}{2})]$$

$$= 1214.6 \text{ W/m}$$

Finally, the heat flow *out* the bottom face is

$$q = \sum h\,\Delta x(T - T_\infty)$$

$$= (10)(\tfrac{1}{3})[(100 - 100)(\tfrac{1}{2}) + 157.70 - 100 + 184.71 - 100 + (175.62 - 100)(\tfrac{1}{2})]$$

$$= 600.7 \text{ W/m}$$

The total heat flow out is

$$q_{\text{out}} = 3019 + 1214.6 + 600.7 = 4834.3 \text{ W/m}$$

This compares favorably with the 4843.4 W/m conducted into the top face.

Solution Techniques

From the foregoing discussion we have seen that the numerical method is simply a means of approximating a continuous temperature distribution with the finite nodal elements. The more nodes taken, the closer the approximation; but, of course, more equations mean more cumbersome solutions. Fortunately, computers and even programmable calculators have the capability to obtain these solutions very quickly.

In practical problems the selection of a large number of nodes may be unnecessary because of uncertainties in boundary conditions. For example, it is not uncommon to have uncertainties in h, the convection coefficient of ± 15 to 20 percent.

The nodal equations may be written as

$$\begin{aligned}
a_{11}T_1 + a_{12}T_2 + \cdots + a_{1n}T_n &= C_1 \\
a_{21}T_1 + a_{22}T_2 + \cdots \qquad\qquad &= C_2 \\
a_{31}T_1 + \cdots \qquad\qquad\qquad\quad &= C_3 \\
\cdots\cdots\cdots\cdots\cdots\cdots\cdots\cdots\cdots & \\
a_{n1}T_1 + a_{n2}T_2 + \cdots + a_{nn}T_n &= C_n
\end{aligned} \tag{3-27}$$

where $T_1, T_2, \ldots, T_n$ are the unknown nodal temperatures. By using the matrix notation

$$[A] = \begin{bmatrix} a_{11} & a_{12} & \cdots & a_{1n} \\ a_{21} & a_{22} & \cdots & \\ a_{31} & & \cdots & \\ \cdots & \cdots & \cdots & \cdots \\ a_{n1} & a_{n2} & \cdots & a_{nn} \end{bmatrix} \qquad [C] = \begin{bmatrix} C_1 \\ C_2 \\ \cdot \\ \cdot \\ \cdot \\ C_n \end{bmatrix} \qquad [T] = \begin{bmatrix} T_1 \\ T_2 \\ \cdot \\ \cdot \\ \cdot \\ T_n \end{bmatrix}$$

Eq. (3-27) can be expressed as

$$[A][T] = [C] \tag{3-28}$$

and the problem is to find the inverse of $[A]$ such that

$$[T] = [A]^{-1}[C] \tag{3-29}$$

Designating $[A]^{-1}$ by

$$[A]^{-1} = \begin{bmatrix} b_{11} & b_{12} & \cdots & b_{1n} \\ b_{21} & & \cdots & \\ \cdots & \cdots & \cdots & \cdots \\ b_{n1} & b_{n2} & \cdots & b_{nn} \end{bmatrix}$$

the final solutions for the unknown temperatures are written in expanded form as

$$\begin{aligned} T_1 &= b_{11}C_1 + b_{12}C_2 + \cdots + b_{1n}C_n \\ T_2 &= b_{21}C_1 + \cdots \\ &\cdots\cdots\cdots\cdots\cdots\cdots \\ T_n &= b_{n1}C_1 + b_{n2}C_2 + \cdots + b_{nn}C_n \end{aligned} \tag{3-30}$$

Clearly, the larger the number of nodes, the more complex and time-consuming the solution, even with a high-speed computer. For most conduction problems the matrix contains a large number of zero elements so that some simplification in the procedure is afforded. For example, the matrix notation for the system of Example 3-3 would be

$$\begin{bmatrix} -4 & 1 & 0 & 1 & 0 & 0 & 0 & 0 & 0 \\ 1 & -4 & 1 & 0 & 1 & 0 & 0 & 0 & 0 \\ 0 & 2 & -4.67 & 0 & 0 & 1 & 0 & 0 & 0 \\ 1 & 0 & 0 & -4 & 1 & 0 & 1 & 0 & 0 \\ 0 & 1 & 0 & 1 & -4 & 1 & 0 & 1 & 0 \\ 0 & 0 & 1 & 0 & 2 & -4.67 & 0 & 0 & 1 \\ 0 & 0 & 0 & 2 & 0 & 0 & -4.67 & 1 & 0 \\ 0 & 0 & 0 & 0 & 2 & 0 & 1 & -4.67 & 1 \\ 0 & 0 & 0 & 0 & 0 & 1 & 0 & 1 & -2.67 \end{bmatrix} \begin{bmatrix} T_1 \\ T_2 \\ T_3 \\ T_4 \\ T_5 \\ T_6 \\ T_7 \\ T_8 \\ T_9 \end{bmatrix} = \begin{bmatrix} -600 \\ -500 \\ -567 \\ -100 \\ 0 \\ -67 \\ -167 \\ -67 \\ -67 \end{bmatrix}$$

We see that because of the structure of the equations the coefficient matrix is very sparse. For this reason iterative methods of solution may be very efficient.

The Gauss-Seidel method is one which we shall discuss later. An old method suitable for hand calculations with a small number of nodes is called the *relaxation method*. In this technique the nodal equation is set equal to some residual $\bar{q}_{m,n}$ (instead of zero) and the following calculation procedure followed:

1. Values of the nodal temperatures are assumed.
2. The value of the residual for each node is calculated from the respective equation and the assumed temperatures.
3. The residuals are "relaxed" to zero by changing the assumptions of the nodal temperatures. The largest residuals are usually relaxed first.
4. As each nodal temperature is changed, a new residual must be calculated for connecting nodes.
5. The procedure is continued until the residuals are sufficiently close to zero.

In Table 3-3 a relaxation solution for the system of Fig. 3-6 is shown. For the most part, the relaxation method would be employed as an expedient vehicle only when a computer was not readily available.

Other methods of solution include a transient analysis carried through to steady state (see Chap. 4), direct elimination (Gauss elimination [9]), or more sophisticated iterative techniques [14]. A number of large computer programs are available for the solution of heat-transfer problems. Kern and Kraus [19] present both steady-state and transient programs which can handle up to 300 nodes. A general circuit-analysis progam applicable to heat-transfer problems is available in Ref. 17, and most computer centers have some kind of in-house program available for heat-transfer computations. Further information on numerical techniques is given in Refs. 11 to 19.

Table 3-3 Relaxation Table for System of Fig. 3-6

T_1	$\bar{q}_1$	T_2	$\bar{q}_2$	T_3	$\bar{q}_3$	T_4	$\bar{q}_4$
300	−100	300	−100	200	−100	200	−100
275	0		−125		−125		
	−30	270	−5				−130
			−45		−165	160	30
	−70			160	−5		−10
255	10		−65		−25		
	0	260	−25				−20
	−5			155	−5		−25
	−15	250	15				−35
			5		−15	150	5
	−20			150	5		0
250	0		0		0		

3-6 NUMERICAL FORMULATION IN TERMS OF RESISTANCE ELEMENTS

Up to this point we have shown how conduction problems can be solved by finite-difference approximations to the differential equations. An equation is formulated for each node and the set of equations solved for the temperatures throughout the body. In formulating the equations we could just as well have used a resistance concept for writing the heat transfer between nodes. Designating our node of interest with the subscript i and the adjoining nodes with subscript j, we have the general-conduction-node situation shown in Fig. 3-10. At steady state the net heat input to node i must be zero or

$$q_i + \sum_j \frac{T_j - T_i}{R_{ij}} = 0 \tag{3-31}$$

where q_i is the heat delivered to node i by heat generation, radiation, etc. The R_{ij} can take the form of convection boundaries, internal conduction, etc., and Eq. (3-31) can be set equal to some residual for a relaxation solution or to zero for treatment with matrix methods.

No new information is conveyed by using a resistance formulation, but some workers may find it convenient to think in these terms. When a numerical solution is to be performed which takes into account property variations, the resistance formulation is particularly useful.

For convenience of the reader Table 3-4 lists the resistance elements which correspond to the nodes in Table 3-2. Note that all resistance elements are for unit depth of material and $\Delta x = \Delta y$. The nomenclature for the table is that R_{m+} refers to the resistance on the positive x side of node (m, n), R_{n-} refers to the resistance on the negative y side of node (m, n), and so on.

The resistance formulation is also useful for numerical solution of complicated three-dimensional shapes. The volume elements for the three common coordinate systems are shown in Fig. 3-11, and internal nodal resistances for each system are given in Table 3-5. The nomenclature for the (m, n, k) subscripts is given at the top of the table, and the plus or minus sign on the resistance subscripts designates the resistance in a positive or negative direction from the central node (m, n, k). The elemental volume ΔV is also indicated for each

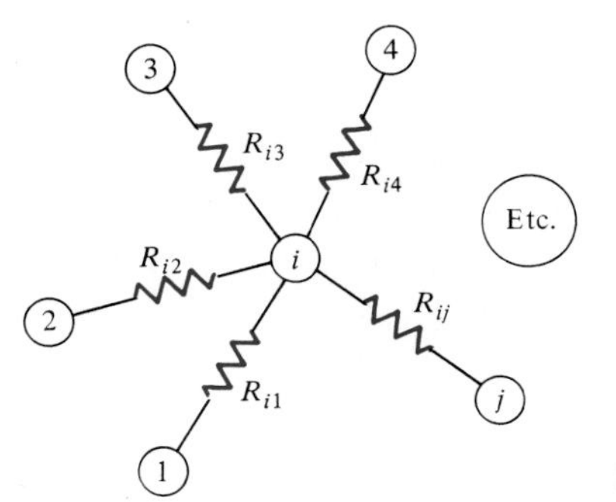

Fig. 3-10 General conduction node.

Table 3-4 Resistances for Nodes of Table 3-2 $\Delta x = \Delta y$, $\Delta z = 1$

Physical situation	R_{m+}	R_{m-}	R_{n+}	R_{n-}	ΔV
(*a*) Interior node	$\frac{1}{k}$	$\frac{1}{k}$	$\frac{1}{k}$	$\frac{1}{k}$	$(\Delta x)^2$
(*b*) Convection boundary	$\frac{1}{h\Delta x}$	$\frac{1}{k}$	$\frac{2}{k}$	$\frac{2}{k}$	$\frac{(\Delta x)^2}{2}$
(*c*) Exterior corner, convection	$\frac{2}{h\Delta x}$	$\frac{2}{k}$	$\frac{2}{h\Delta x}$	$\frac{2}{k}$	$\frac{(\Delta x)^2}{4}$
(*d*) Interior corner, convection†	$\frac{2}{k}$	$\frac{1}{k}$	$\frac{1}{k}$	$\frac{2}{k}$	$\frac{3(\Delta x)^2}{4}$
(*e*) Insulated boundary	∞	$\frac{1}{k}$	$\frac{2}{k}$	$\frac{2}{k}$	$\frac{(\Delta x)^2}{2}$
(*f*) Interior node near curved boundary	$\frac{2}{(b+1)k}$ to node $(m+1, n)$	$\frac{2a}{(b+1)k}$ to node 1	$\frac{2b}{(a+1)k}$ to node 2	$\frac{2}{(a+1)k}$ to node $(m, n-1)$	$0.25(1+a)(1+b)(\Delta x)^2$
(*g*) Boundary node with curved boundary node 2 for (*f*) above	$R_{23} = \frac{2\sqrt{c^2+1}}{bk}$ $R_{21} = \frac{2\sqrt{a^2+b^2}}{bk}$ $R_{2-\infty} = \frac{2}{h\Delta x(\sqrt{c^2+1} + \sqrt{a^2+b^2})}$ $R_{n-} = \frac{2b}{k(a+1)}$ to node (m, n)				$\Delta V = 0.125[(2+a)+c](\Delta x)^2$

† Also $R_{\infty} = 1/h\Delta x$ for convection to T_{∞}.

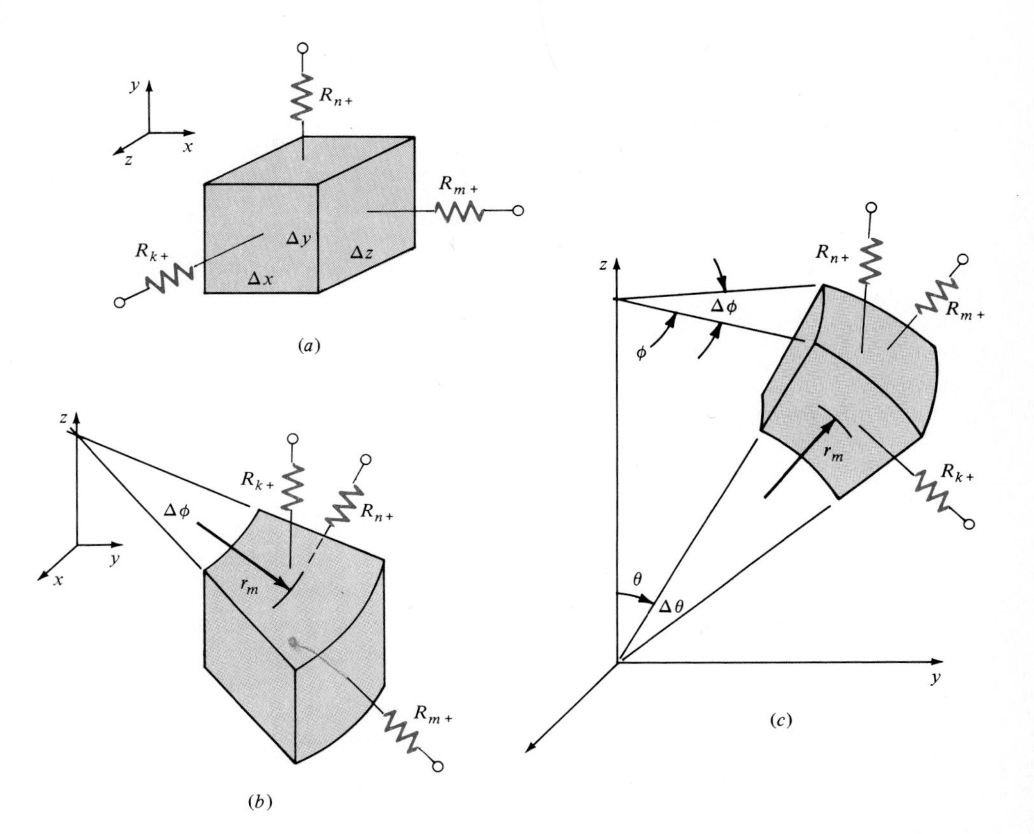

Fig. 3-11 Volume and resistance elements: (*a*) cartesian, (*b*) cylindrical, and (*c*) spherical coordinate systems.

coordinate system. We note, of course, that in a practical problem the coordinate increments are frequently chosen so that $\Delta x = \Delta y = \Delta z$, etc., and the resistances are simplified.

3-7 GAUSS-SEIDEL ITERATION

When the number of nodes is very large, an iterative technique may frequently yield a more efficient solution to the nodal equations than a direct matrix inversion. One such method is called the *Gauss-Seidel iteration* and is applied in the following way. From Eq. (3-31) we may solve for the temperature T_i in terms of the resistances and temperatures of the adjoining nodes T_j as

$$T_i = \frac{q_i + \sum_j (T_j/R_{ij})}{\sum_j (1/R_{ij})} \tag{3-32}$$

Table 3-5 Internal Nodal Resistances for Different Coordinate Systems

	Cartesian	*Cylindrical*	*Spherical*
Nomenclature for increments	x, m y, n z, k	r, m ϕ, n z, k	r, m ϕ, n θ, k
Volume element ΔV	$\Delta x\ \Delta y\ \Delta z$	$r_m\ \Delta r\ \Delta\phi\ \Delta z$	$r_m^2 \sin\theta\ \Delta r\ \Delta\phi\ \Delta\theta$
R_{m+}	$\dfrac{\Delta x}{\Delta y\ \Delta z\ k}$	$\dfrac{\Delta r}{(r_m + \Delta r/2)\ \Delta\phi\ \Delta z\ k}$	$\dfrac{\Delta r}{(r_m + \Delta r/2)^2 \sin\theta\ \Delta\phi\ \Delta\theta\ k}$
R_{m-}	$\dfrac{\Delta x}{\Delta y\ \Delta z\ k}$	$\dfrac{\Delta r}{(r_m - \Delta r/2)\ \Delta\phi\ \Delta z\ k}$	$\dfrac{\Delta r}{(r_m + \Delta r/2) \sin\theta\ \Delta\phi\ \Delta\theta\ k}$
R_{n+}	$\dfrac{\Delta y}{\Delta x\ \Delta z\ k}$	$\dfrac{r_m\ \Delta\phi}{\Delta r\ \Delta z\ k}$	$\dfrac{\Delta\phi \sin\theta}{\Delta r\ \Delta\theta\ k}$
R_{n-}	$\dfrac{\Delta y}{\Delta x\ \Delta z\ k}$	$\dfrac{r_m\ \Delta\phi}{\Delta r\ \Delta z\ k}$	$\dfrac{\Delta\phi \sin\theta}{\Delta r\ \Delta\theta\ k}$
R_{k+}	$\dfrac{\Delta z}{\Delta x\ \Delta y\ k}$	$\dfrac{\Delta z}{r_m\ \Delta\phi\ \Delta r\ k}$	$\dfrac{\Delta\theta}{\sin(\theta + \Delta\theta/2)\ \Delta r\ \Delta\phi\ k}$
R_{k-}	$\dfrac{\Delta z}{\Delta x\ \Delta y\ k}$	$\dfrac{\Delta z}{r_m\ \Delta\phi\ \Delta r\ k}$	$\dfrac{\Delta\theta}{\sin(\theta - \Delta\theta/2)\ \Delta r\ \Delta\phi\ k}$

The Gauss-Seidel iteration makes use of the difference equations expressed in the form of Eq. (3-32) through the following procedure.

1. An initial set of values for the T_i is assumed. This initial assumption can be obtained through any expedient method. For a large number of nodes to be solved on a computer the T_i's are frequently assigned a zero value to start the calculation.
2. Next, the new values of the nodal temperatures T_i are calculated according to Eq. (3-32), *always using the most recent values of the* T_j.
3. The process is repeated until successive calculations differ by a sufficiently small amount. In terms of a computer program, this means that a test will be inserted to stop the calculations when

$$|T_{i_{n+1}} - T_{i_n}| \le \delta \qquad \text{for all } T_i$$

where δ is some selected constant and n is the number of iterations. Alternatively, a nondimensional test may be selected such that

$$\epsilon \ge \left| \frac{T_{i_{n+1}} - T_{i_n}}{T_{i_n}} \right|$$

Obviously, the smaller the value of δ, the greater the calculation time required to obtain the desired result. The reader should note, however, that the *accuracy* of the solution to the physical problem is not dependent on the value of δ alone. This constant governs the accuracy of the solution to the set of difference equations. The solution to the physical problem also depends on the selection of the increment Δx.

As we noted in the discussion of solution techniques, the matrices encountered in the numerical formulations are very sparse; i.e., they contain a large number of zeros. In solving a problem with a large number of nodes it may be quite time-consuming to enter all these zeros and the simple form of the Gauss-Seidel equation may be preferable.

For nodes with $\Delta x = \Delta y$ and no heat generation, the form of Eq. (3-32) has been listed as the second equation in segments of Table 3-2. The nondimensional group

$$\frac{h\,\Delta x}{k} = \mathrm{Bi}$$

is called the *Biot number*.

3-8 ACCURACY CONSIDERATIONS

We have already noted that the finite-difference approximation to a physical problem improves as smaller and smaller and smaller increments of Δx and Δy are used. But, we have not said how to estimate the accuracy of this approximation. Two basic approaches are available.

1. Compare the numerical solution with an analytical solution for the problem, if available, or an analytical solution for a similar problem.
2. Choose progressively smaller values of Δx and observe the behavior of the solution. If the problem has been correctly formulated and solved, the nodal temperatures should converge as Δx becomes smaller. It should be noted that computational round-off errors increase with an increase in the number of nodes because of the increased number of machine calculations. This is why one needs to observe the convergence of the solution.

It can be shown that the error of the finite-difference approximation to $\partial T/\partial x$ is of the order of $(\Delta x/L)^2$ where L is some characteristic body dimension.

Analytical solutions are of limited utility in checking the accuracy of a numerical model because most problems which will need to be solved by numerical methods either do not have an analytical solution at all, or if one is available, it may be too cumbersome to compute.

In discussing solution techniques for nodal equations, we stated that an accurate solution of these equations does not ensure an accurate solution to the physical problem. In many cases the final solution is in serious error simply because the problem was not formulated correctly at the start. No computer or convergence criterion can correct this kind of error. One way to check for formulation errors is to perform some sort of energy balance using the final solution. The nature of the balance varies from problem to problem but for steady state it always takes the form of energy in equals energy out. If the energy balance does not check within reasonable limits, there is a likelihood that the problem has not been formulated correctly. Perhaps a constant is wrong here or there, or an input data point is incorrect, a faulty computer statement employed, or one or more nodal equations are incorrectly written. If the energy balance does check, one may then address the issue of using smaller values of Δx to improve accuracy.

In the examples we present energy balances as a check on problem formulation.

□ Accuracy of Properties and Boundary Conditions

From time to time we have mentioned that thermal conductivities of materials vary with temperature; however, over a temperature range of 100 to 200°C the variation is not great (on the order of 5 to 10 percent) and we are justified in assuming constant values to simplify problem solutions. Convection and radiation boundary conditions are particularly notorious for their nonconstant behavior. Even worse is the fact that for many practical problems the basic uncertainty in our knowledge of convection heat-transfer coefficients may not be better than ±20 percent. Uncertainties of surface-radiation properties of ±10 percent are not unusual at all. For example, a highly polished aluminum plate, if allowed to oxidize heavily, will absorb as much as *300 percent* more radiation than when it was polished.

The above remarks are not made to alarm the reader, but rather to show that selection of a large number of nodes for a numerical formulation does not necessarily produce an accurate solution to the physical problem; we must also examine uncertainties in the boundary conditions. At this point the reader is ill-equipped to estimate these uncertainties. Later chapters on convection and radiation will clarify the matter.

□ Some Remarks on Computer Solutions

It should be apparent by now that numerical methods and computers give the engineer tools for solving very complex heat-transfer problems. The advent of the microcomputer has made desktop computer power available to everyone at very economical prices. How should one choose between micro, mini, or mainframe computers for solution of heat-transfer problems? For modest-size problems, including many heat-exchanger design problems (Chap. 10), we may expect that they will be solved more and more with microcomputers. Large problems, particularly those involving many repetitive calculations with varying boundary conditions will probably remain a task for high-speed mainframe machines. Networks and communication links between micros and large machines will offer other opportunities.

Many software packages are available for solving heat-transfer problems on microcomputers, but their availability changes so rapidly that it would be futile to try to mention specific ones in a textbook. One characteristic common to almost *all* heat-transfer software is a requirement that the user *understand* something about the subject of heat transfer. Without such understanding it can become very easy to make gross mistakes and never detect them at all. Of course, our objective in this book is to give the reader such an understanding of the subject.

■ **EXAMPLE 3-4** Gauss-Seidel calculation

Apply the Gauss-Seidel technique to obtain the nodal temperatures for the four nodes in Fig. 3-6.

Solution

It is useful to think in terms of a resistance formulation for this problem because all the connecting resistances between the nodes in Fig. 3-6 are equal; that is,

$$R = \frac{\Delta y}{k\,\Delta y} = \frac{\Delta x}{k\,\Delta y} = \frac{1}{k} \tag{a}$$

Therefore, when we apply Eq. (3-32) to each node, we obtain ($q_i = 0$)

$$T_i = \frac{\sum_j k_j T_j}{\sum_j k_j} \tag{b}$$

Because each node has four resistances connected to it and k is assumed constant,

$$\sum_j k_j = 4k$$

and

$$T_i = \frac{1}{4}\sum_j T_j \tag{c}$$

We now set up an iteration table as shown and use initial temperature assumptions of 300 and 200°C as before. Equation (c) is then applied repeatedly until satisfactory convergence is achieved. In the table, five iterations produce convergence with 0.13 degree. To illustrate the calculation, we can note the two specific cases below:

$$(T_2)_{n=1} = \tfrac{1}{4}(500 + 100 + T_4 + T_1) = \tfrac{1}{4}(500 + 100 + 200 + 275) = 268.75$$

$$(T_3)_{n=4} = \tfrac{1}{4}(100 + T_1 + T_4 + 100) = \tfrac{1}{4}(100 + 250.52 + 150.52 + 100) = 150.26$$

Number of iterations n	T_1	T_2	T_3	T_4
0	300	300	200	200
1	275	268.75	168.75	159.38
2	259.38	254.69	154.69	152.35
3	251.76	251.03	151.03	150.52
4	250.52	250.26	150.26	150.13
5	250.13	250.07	150.07	150.03

Note that in computing $(T_3)_{n=4}$ we have used the most recent information available to us for T_1 and T_4.

■ **EXAMPLE 3-5** Numerical formulation with heat generation

We illustrate the resistance formulation in cylindrical coordinates by considering a 4.0-mm-diameter wire with uniform heat generation of 500 MW/m³. The outside surface temperature of the wire is 200°C, and the thermal conductivity is 19 W/m · °C. We wish to calculate the temperature distribution in the wire. For this purpose we select four nodes as shown in the accompanying figure. We shall make the calculations per unit length, so we let $\Delta z = 1.0$. Because the system is one-dimensional, we take $\Delta\phi = 2\pi$. For all the elements Δr is chosen as 0.5 mm. We then compute the resistances and volume elements using the relations from Table 3-6, and the values are given below. The computation of R_{m+} for node 4 is different from the others because the heat-flow path is shorter. For node 4, r_m is 1.75 mm, so the positive resistance extending to the known surface temperature is

$$R_{m+} = \frac{\Delta r/2}{(r_m + \Delta r/4)\,\Delta\phi\,\Delta z\,k} = \frac{1}{15\pi k}$$

The temperature equation for node 4 is written as

$$T_4 = \frac{2749 + 6\pi k T_3 + 15\pi k(200)}{21\pi k}$$

where the 200 is the known outer surface temperature.

Node	r_m, mm	R_{m+}, °C/W	R_{m-}, °C/W	$\Delta V = r_m\,\Delta r\,\Delta\phi\,\Delta z$, μm³	$q_i = \dot{q}\,\Delta V$, W
1	0.25	$\frac{1}{2\pi k}$	∞	0.785	392.5
2	0.75	$\frac{1}{4\pi k}$	$\frac{1}{2\pi k}$	2.356	1178
3	1.25	$\frac{1}{6\pi k}$	$\frac{1}{4\pi k}$	3.927	1964
4	1.75	$\frac{1}{15\pi k}$	$\frac{1}{6\pi k}$	5.498	2749

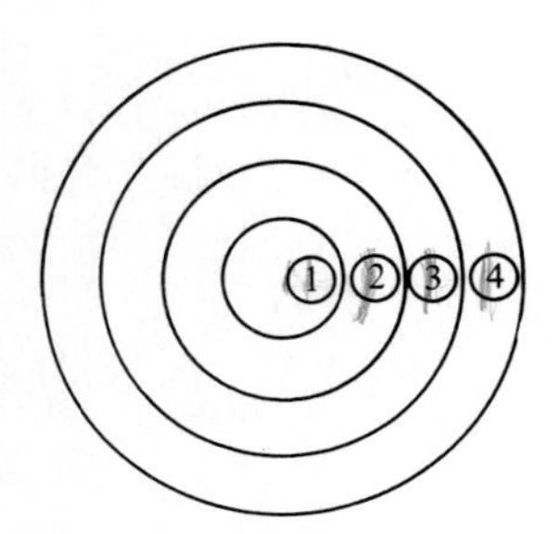

Fig. Ex. 3-5 Example schematic.

A summary of the values of $\Sigma(1/R_{ij})$ and T_i according to Eq. (3-32) is now given to be used in a Gauss-Seidel iteration scheme.

Node	$\sum \frac{1}{R_{ij}}$, W/°C	$T_i = \frac{q_i + \Sigma(T_j/R_{ij})}{\Sigma(1/R_{ij})}$
1	$2\pi k = 119.38$	$T_1 = 3.288 + T_2$
2	$6\pi k = 358.14$	$T_2 = 3.289 + \frac{1}{3}T_1 + \frac{2}{3}T_3$
3	$10\pi k = 596.90$	$T_3 = 3.290 + 0.4T_2 + 0.6T_4$
4	$21\pi k = 1253.50$	$T_4 = 2.193 + \frac{2}{7}T_3 + 142.857$

Thirteen iterations are now tabulated:

	Node temperature, °C			
Iteration n	T_1	T_2	T_3	T_4
0	240	230	220	210
1	233.29	227.72	220.38	208.02
2	231.01	227.21	218.99	207.62
3	230.50	226.12	218.31	207.42
4	229.41	225.30	217.86	207.30
5	228.59	224.73	217.56	207.21
6	228.02	224.34	217.35	207.15
7	227.63	224.07	217.21	207.11
8	227.36	223.88	217.11	207.08
9	227.17	223.75	217.04	207.06
10	227.04	223.66	216.99	207.04
11	226.95	223.60	216.95	207.04
12	226.89	223.55	216.93	207.03
13	226.84	223.52	216.92	207.03
Analytical	225.904	222.615	216.036	206.168
Gauss-Seidel check	225.903	222.614	216.037	206.775
Exact solution of nodal equations	226.75	223.462	216.884	207.017

We may compare the iterative solution with an exact calculation which makes use of Eq. (2-25*a*):

$$T - T_w = \frac{\dot{q}}{4k}(R^2 - r^2)$$

where T_w is the 200°C surface temperature, $R = 2.0$ mm, and r is the value of r_m for each node. The analytical values are shown below the last iteration, and then a Gauss-Seidel check is made on the analytical values. There is excellent agreement on the first three nodes and somewhat less on node 4. Finally, the exact solutions to the nodal equations are shown for comparison. These are the values the iterative scheme would

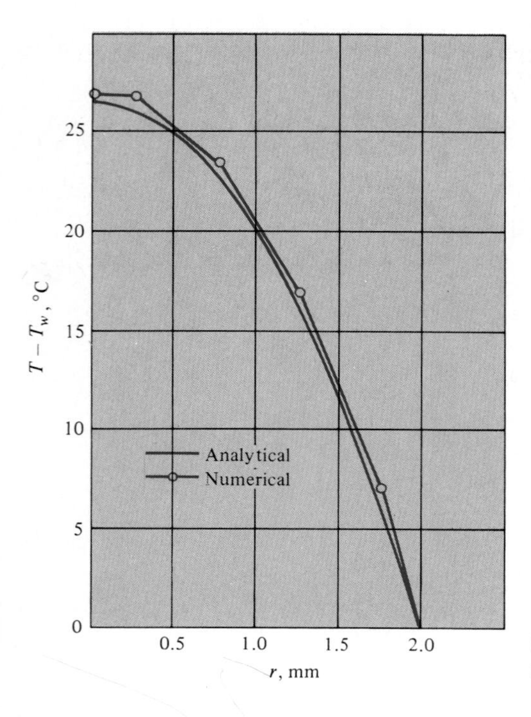

Fig. Ex. 3-5 Comparison of analytical and numerical solutions.

converge to if carried far enough. In this limit the analytical and numerical calculations differ by a constant factor of about 0.85°C, and this difference results mainly from the way in which the surface resistance and boundary condition are handled. A smaller value of Δr near the surface would produce better agreement. A graphical comparison of the analytical and numerical solutions is shown in an accompanying figure.

The total heat loss from the wire may be calculated as the conduction through R_{m+} at node 4. Then

$$q = \frac{T_4 - T_W}{R_{m+}} = 15\pi k(207.03 - 200) = 6.294 \text{ kW/m} \quad [6548 \text{ Btu/h} \cdot \text{ft}]$$

This must equal the total heat generated in the wire, or

$$q = \dot{q}V = (500 \times 10^6)\pi(2 \times 10^{-3})^2 = 6.283 \text{ kW/m} \quad [6536 \text{ Btu/h} \cdot \text{ft}]$$

The difference between the two values results from the inaccuracy in determination of T_4. Using the exact solution value of 207.017°C would give a heat loss of 6.2827 kW. For this problem the exact value of heat flow is 6.283 kW because the heat-generation calculation is independent of the finite-difference formulation.

■ **EXAMPLE 3-6** Heat generation with nonuniform nodal elements

A layer of glass [$k = 0.8$ W/m · °C] 3 mm thick has thin 1-mm electric conducting strips attached to the upper surface, as shown in the figure. The bottom surface of the glass is insulated, and the top surface is exposed to a convection environment at 30°C with $h = 100$ W/m² · °C. The strips generate heat at the rate of 40 or 20 W per meter of length. Determine the steady-state temperature distribution in a typical glass section, using the numerical method for both heat-generation rates.

Solution

The nodal network for a typical section of the glass is shown in the figure. In this example we have *not* chosen $\Delta x = \Delta y$. Because of symmetry, $T_1 = T_7$, $T_2 = T_6$, etc., and we only need to solve for the temperatures of 16 nodes. We employ the resistance formulation. As shown, we have chosen $\Delta x = 5$ mm and $\Delta y = 1$ mm. The various resistances may now be calculated:

Nodes 1, 2, 3, 4:

$$\frac{1}{R_{m+}} = \frac{1}{R_{m-}} = \frac{k(\Delta y/2)}{\Delta x} = \frac{(0.8)(0.001/2)}{0.005} = 0.08$$

$$\frac{1}{R_{n+}} = hA = (100)(0.005) = 0.5$$

$$\frac{1}{R_{n-}} = \frac{k\,\Delta x}{\Delta y} = \frac{(0.8)(0.005)}{0.001} = 4.0$$

Nodes 8, 9, 10, 11, 15, 16, 17, 18:

$$\frac{1}{R_{m+}} = \frac{1}{R_{m-}} = \frac{k\,\Delta y}{\Delta x} = \frac{(0.8)(0.001)}{0.005} = 0.16$$

$$\frac{1}{R_{n+}} = \frac{1}{R_{n-}} = \frac{k\,\Delta x}{\Delta y} = 4.0$$

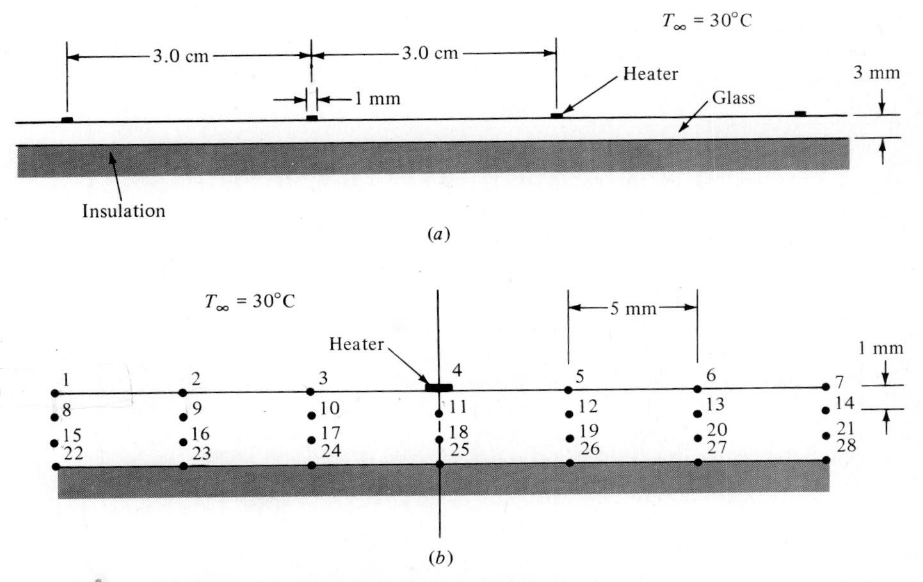

Fig. Ex. 3-6 (*a*) Physical system, (*b*) nodal arrangement.

Nodes 22, 23, 24, 25:

$$\frac{1}{R_{m+}} = \frac{1}{R_{m-}} = \frac{k(\Delta y/2)}{\Delta x} = 0.08$$

$$\frac{1}{R_{n+}} = \frac{k\,\Delta x}{\Delta y} = 4.0$$

$$\frac{1}{R_{n-}} = 0 \qquad \text{(insulated surface)}$$

The nodal equations are obtained from Eq. (3-31) in the general form

$$\Sigma(T_j/R_{ij}) + q_i - T_i\,\Sigma(1/R_{ij}) = 0$$

Only node 4 has a heat-generation term, and $q_i = 0$ for all other nodes. From the above resistances we may calculate the $\Sigma(1/R_{ij})$ as

Node	$\Sigma(1/R_{ij})$
1, 2, 3, 4	4.66
8, . . . , 18	8.32
22, 23, 24, 25	4.16

For node 4 the equation is

$$(2)(0.08)T_3 + 4.0T_5 + (0.5)(30) + q_4 - 4.66T_4 = 0$$

The factor of 2 on T_3 occurs because $T_3 = T_5$ from symmetry. When all equations are evaluated and the matrix is solved, the following temperatures are obtained:

Node temperature, °C	*q/L*, W/m	
	20	*40*
1	31.90309	33.80617
2	32.78716	35.57433
3	36.35496	42.70993
4	49.81266	69.62532
8	32.10561	34.21122
9	33.08189	36.16377
10	36.95154	43.90307
11	47.82755	65.65510
15	32.23003	34.46006
16	33.26087	36.52174
17	37.26785	44.53571
18	46.71252	63.42504
22	32.27198	34.54397
23	33.32081	36.64162
24	37.36667	44.73333
25	46.35306	62.70613

The results of the model and calculations may be checked by calculating the convection heat lost by the top surface. Because all the energy generated in the small heater strip must eventually be lost by convection (the bottom surface of the glass is insulated and thus loses no heat) we know the numerical value that the convection should have. The convection loss at the top surface is given by

$$q_c = \Sigma h_i A_i (T_i - T_\infty)$$
$$= (2)(100)\left[\frac{\Delta x}{2}(T_1 - T_\infty) + \Delta x(T_2 + T_3 - 2T_\infty) + \frac{\Delta x}{2}(T_4 - T_\infty)\right]$$

The factor of 2 accounts for both sides of the section. With $T_\infty = 30°\text{C}$ this calculation yields

$$q_c = 19.999995 \qquad \text{for } q/L = 20 \text{ W/m}$$
$$q_c = 40.000005 \qquad \text{for } q/L = 40 \text{ W/m}$$

Obviously, the agreement is excellent.

■ **EXAMPLE 3-7** Composite material with nonuniform nodal elements

A composite material is embedded in a high-thermal-conductivity material maintained at 400°C as shown. The upper surface is exposed to a convection environment at 30°C with $h = 25 \text{ W/m}^2 \cdot °\text{C}$. Determine the temperature distribution and heat loss from the upper surface for steady state.

Solution

For this example we choose nonsquare nodes as shown. Note also that nodes 1, 4, 7, 10, 13, 14, and 15 consist of *two* materials. We again employ the resistance formulation.

For node 1:

$$\frac{1}{R_{m+}} = \frac{kA}{\Delta x} = \frac{(2.0)(0.005)}{0.015} = 0.6667$$

$$\frac{1}{R_{m-}} = \frac{kA}{\Delta x} = \frac{(0.3)(0.005)}{0.01} = 0.15$$

$$\frac{1}{R_{n+}} = hA = (25)(0.005 + 0.0075) = 0.3125$$

$$\frac{1}{R_{n-}} = \left(\frac{kA}{\Delta y}\right)_L + \left(\frac{kA}{\Delta y}\right)_R = \frac{(0.3)(0.005) + (2.0)(0.0075)}{0.01} = 1.65$$

For nodes 4, 7, 10:

$$\frac{1}{R_{m+}} = \frac{(2.0)(0.01)}{0.015} = 1.3333$$

$$\frac{1}{R_{m-}} = \frac{(0.3)(0.01)}{0.01} = 0.3$$

$$\frac{1}{R_{n+}} = \frac{1}{R_{n-}} = 1.65$$

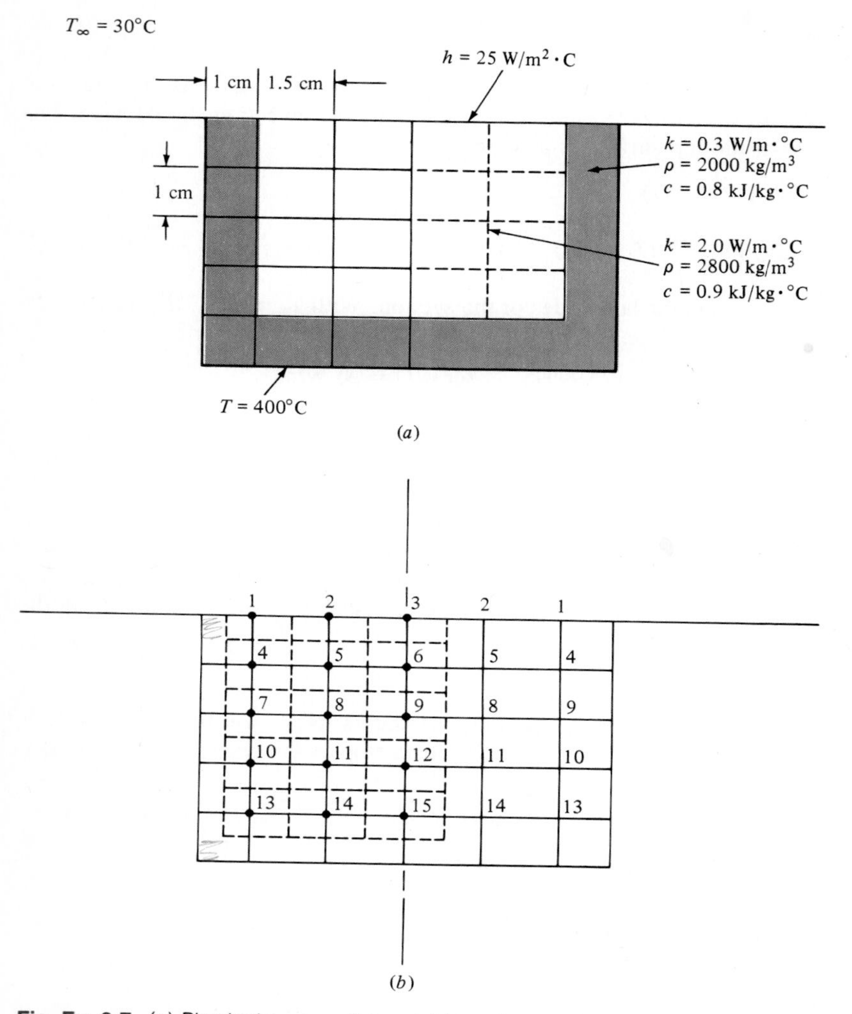

Fig. Ex. 3-7 (*a*) Physical system, (*b*) nodal boundaries.

For node 13:

$$\frac{1}{R_{m+}} = \frac{(2.0)(0.005) + (0.3)(0.005)}{0.015} = 0.76667$$

$$\frac{1}{R_{m-}} = \frac{(0.3)(0.01)}{0.01} = 0.3$$

$$\frac{1}{R_{n+}} = 1.65$$

$$\frac{1}{R_{n-}} = \frac{(0.3)(0.0075) + (0.3)(0.005)}{0.01} = 0.375$$

For nodes 5, 6, 8, 9, 11, 12:

$$\frac{1}{R_{m+}} = \frac{1}{R_{m-}} = \frac{(2.0)(0.01)}{0.015} = 1.3333$$

$$\frac{1}{R_{n+}} = \frac{1}{R_{n-}} = \frac{(2.0)(0.015)}{0.01} = 3.0$$

For nodes 2, 3:

$$\frac{1}{R_{m+}} = \frac{1}{R_{m-}} = \frac{(2.0)(0.005)}{0.015} = 0.6667$$

$$\frac{1}{R_{n+}} = hA = (25)(0.015) = 0.375$$

$$\frac{1}{R_{n-}} = 3.0$$

For nodes 14, 15:

$$\frac{1}{R_{m+}} = \frac{1}{R_{m-}} = \frac{(2.0)(0.005) + (0.3)(0.005)}{0.015} = 0.76667$$

$$\frac{1}{R_{n+}} = 3.0$$

$$\frac{1}{R_{n-}} = \frac{(0.3)(0.015)}{0.01} = 0.45$$

We shall use Eq. (3-32) for formulating the nodal equations. For node 1, $\Sigma(1/R_{ij}) = 2.7792$, and we obtain

$$T_1 = \frac{1}{2.7792}[(400)(0.15) + (30)(0.3125) + T_2(0.6667) + 1.65T_4]$$

For node 3, $\Sigma(1/R_{ij}) = 4.7083$, and the nodal equation is

$$T_3 = \frac{1}{4.7083}[T_2(0.6667)(2) + 3.0T_6 + (0.375)(30)]$$

The factor of 2 on T_2 occurs because of the mirror image of T_2 to the right of T_3.

A similar procedure is followed for the other nodes to obtain 15 nodal equations with the 15 unknown temperatures. These equations may then be solved by whatever computation method is most convenient. The resulting temperatures are:

$$T_1 = 254.956 \qquad T_2 = 247.637 \qquad T_3 = 244.454$$

$$T_4 = 287.334 \qquad T_5 = 273.921 \qquad T_6 = 269.844$$

$$T_7 = 310.067 \qquad T_8 = 296.057 \qquad T_9 = 291.610$$

$$T_{10} = 327.770 \qquad T_{11} = 313.941 \qquad T_{12} = 309.423$$

$$T_{13} = 343.516 \qquad T_{14} = 327.688 \qquad T_{15} = 323.220$$

The heat flow out the top face is obtained by summing the convection loss from the nodes:

$$\begin{aligned} q_{\text{conv}} &= \Sigma h A_i (T_i - T_\infty) \\ &= (2)(25)[(0.0125)(254.96 - 30) + (0.015)(247.64 - 30) \\ &\quad + (0.0075)(244.45 - 30)] \\ &= 384.24 \text{ W per meter of depth} \end{aligned}$$

As a check on this value, we can calculate the heat conducted in from the 400°C surface to nodes 1, 4, 7, 10, 13, 14, 15:

$$q_{\text{cond}} = \Sigma k A_i \frac{\Delta T}{\Delta x}$$

$$\begin{aligned} q_{\text{cond}} &= 2\frac{0.3}{0.01}\,[(0.005)(400 - 254.96) + (0.01)(400 - 287.33) \\ &\quad + (0.01)(400 - 310.07) + (0.01)(400 - 327.77) \\ &\quad + (0.0225)(400 - 343.52) \\ &\quad + (0.015)(400 - 327.69) \\ &\quad + (0.0075)(400 - 323.22)] \\ &= 384.29 \text{ W per meter of depth} \end{aligned}$$

The agreement is excellent.

EXAMPLE 3-8 Radiation boundary condition

A 1-by-2-cm ceramic strip [$k = 3.0$ W/m · °C, $\rho = 1600$ kg/m³, and $c = 0.8$ kJ/kg · °C] is embedded in a high-thermal-conductivity material, as shown, so that the sides are maintained at a constant temperature of 900°C. The bottom surface of the ceramic is insulated, and the top surface is exposed to a convection and radiation environment at $T_\infty = 50$°C; $h = 50$ W/m² · °C, and the radiation heat loss is calculated from

$$q = \sigma A \epsilon (T^4 - T_\infty^4)$$

where

A = surface area

$\sigma = 5.669 \times 10^{-8}$ W/m² · °K⁴

$\epsilon = 0.7$

Solve for the steady-state temperature distribution of the nodes shown and the rate of heat loss.

Solution

We shall employ the resistance formulation and note that the radiation can be written as

$$q = \sigma \epsilon A (T^4 - T_\infty^4) = \frac{T - T_\infty}{R_{\text{rad}}} \tag{a}$$

$$\frac{1}{R_{\text{rad}}} = \sigma \epsilon A (T^2 + T_\infty^2)(T + T_\infty) \tag{b}$$

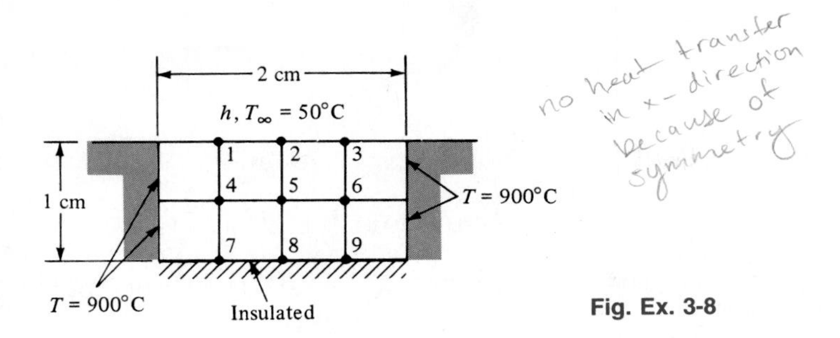

Fig. Ex. 3-8

From symmetry $T_1 = T_3$, $T_4 = T_6$, $T_7 = T_9$, so we have only six unknown nodes. The resistances are now computed:

Nodes 1, 2:

$$\frac{1}{R_{m+}} = \frac{1}{R_{m-}} = \frac{kA}{\Delta x} = \frac{(3.0)(0.0025)}{0.005} = 1.5 \qquad \frac{1}{R_{n-}} = \frac{(3.0)(0.005)}{0.005} = 3.0$$

$$\frac{1}{R_{n+,\text{conv}}} = hA = (50)(0.005) = 0.25 \tag{c}$$

$$\frac{1}{R_{n+,\text{rad}}} = \sigma\epsilon A(T^2 + T_\infty^2)(T + T_\infty)$$

The radiation term introduces nonlinearities and will force us to employ an iterative solution.

Nodes 4, 5:

$$\text{All } \frac{1}{R} = \frac{kA}{\Delta x} = \frac{(3.0)(0.005)}{0.005} = 3.0$$

Nodes 7, 8:

$$\frac{1}{R_{m+}} = \frac{1}{R_{m-}} = 1.5 \qquad \frac{1}{R_{n+}} = 3.0$$

Because the bottom surface is insulated, $1/R_{n-} = 0$. We now use Eq. (3-32)

$$T_i = \frac{\Sigma(T_j/T_{ij})}{\Sigma(1/R_{ij})} \tag{3-32}$$

and tabulate:

Node	$\Sigma(1/R_{ij})$
1	$6.25 + 1/R_{\text{rad}}$
2	$6.25 + 1/R_{\text{rad}}$
4	12
5	12
7	6
8	6

Our nodal equations are thus expressed in degrees Kelvin because of the radiation terms and become

$$T_1 = \frac{1}{\Sigma(1/R_{ij})}[1.5T_2 + 3T_4 + (1.5)(1173) + (323)(0.25) + \sigma\epsilon(0.005)(T_1^2 + 323^2)(T_1 + 323)(323)]$$

$$T_2 = \frac{1}{\Sigma(1/R_{ij})}[1.5T_1(2) + 3T_5 + (323)(0.25) + \sigma\epsilon(0.005)(T_2^2 + 323^2)(T_2 + 323)(323)]$$

$$T_4 = \tfrac{1}{12}[(1173)(3.0) + 3T_1 + 3T_7 + 3T_5] \qquad T_5 = \tfrac{1}{12}[2T_4(3.0) + 3T_2 + 3T_8]$$

$$T_7 = \tfrac{1}{6}[(1173)(1.5) + 3T_4 + 1.5T_8] \qquad T_8 = \tfrac{1}{6}[2T_7(1.5) + 3T_5]$$

The radiation terms create a very nonlinear set of equations. The computational algorithm we shall use is outlined as follows:

1. Assume $T_1 = T_2 = 1173$ K.
2. Compute $1/R_{\text{rad}}$ and $\Sigma(1/R_{ij})$ for nodes 1 and 2 on the basis of this assumption.
3. Solve the set of equations for T_1 through T_8.
4. Using new values of T_1 and T_2, recalculate $1/R_{\text{rad}}$ values.
5. Solve equations again, using new values.
6. Repeat the procedure until answers are sufficiently convergent.

The results of six iterations are shown in the table below. As can be seen, the convergence is quite rapid. The temperatures are in kelvins.

Iteration	T_1	T_2	T_4	T_5	T_7	T_8
1	990.840	944.929	1076.181	1041.934	1098.951	1070.442
2	1026.263	991.446	1095.279	1068.233	1113.622	1090.927
3	1019.879	982.979	1091.827	1063.462	1110.967	1087.215
4	1021.056	984.548	1092.464	1064.344	1111.457	1087.901
5	1020.840	984.260	1092.347	1064.182	1111.367	1087.775
6	1020.879	984.313	1092.369	1064.212	1111.384	1087.798

At this point we may note that in a practical problem the value of ϵ will only be known within a tolerance of several percent, and thus there is nothing to be gained by carrying the solution to unreasonable limits of accuracy.

The heat loss is determined by calculating the radiation and convection from the top surface (nodes 1, 2, 3):

$$\begin{aligned} q_{\text{rad}} &= \Sigma\sigma\epsilon A_i(T_i^4 - 323^4) \\ &= (5.669 \times 10^{-8})(0.7)(0.005)[(2)(1020.88^4 - 323^4) + 984.313^4 - 323^4] \\ &= 610.8 \text{ W/m depth} \end{aligned}$$

$$q_{\text{conv}} = \Sigma hA_i(T_i - 323)$$
$$= (50)(0.005)[(2)(1020.88 - 323) + 984.313 - 323] = 514.27\text{ W}$$
$$q_{\text{total}} = 610.8 + 514.27 = 1125.07\text{ W/m depth}$$

This can be checked by calculating the conduction input from the 900°C surfaces:

$$q_{\text{cond}} = \sum kA_i \frac{\Delta T}{\Delta x}$$
$$= \frac{(2)(3.0)}{0.005}[(0.0025)(1173 - 1020.879) + (0.005)(1173 - 1092.369) + (0.0025)(1173 - 1111.384)]$$
$$= 1124.99\text{ W/m depth}$$

The agreement is excellent.

■ **EXAMPLE 3-9** Use of variable mesh size

One may use a variable mesh size in a problem with a finer mesh to help in regions of large temperature gradients. This is illustrated in the accompanying figure, in which Fig. 3-6 is redrawn with a fine mesh in the corner. The boundary temperatures are the same as in Fig. 3-6. We wish to calculate the nodal temperatures and compare with the previous solution. Note the symmetry of the problem; $T_1 = T_2$, $T_3 = T_4$, etc.

Solution

Nodes 5, 6, 8, and 9 are internal nodes with $\Delta x = \Delta y$ and have nodal equations in the form of Eq. (3-24). Thus,

$$600 + T_6 + T_8 - 4T_5 = 0$$
$$500 + T_5 + T_7 + T_9 - 4T_6 = 0$$
$$100 + T_5 + T_9 + T_{11} - 4T_8 = 0$$
$$T_8 + T_6 + T_{10} + T_{12} - 4T_9 = 0$$

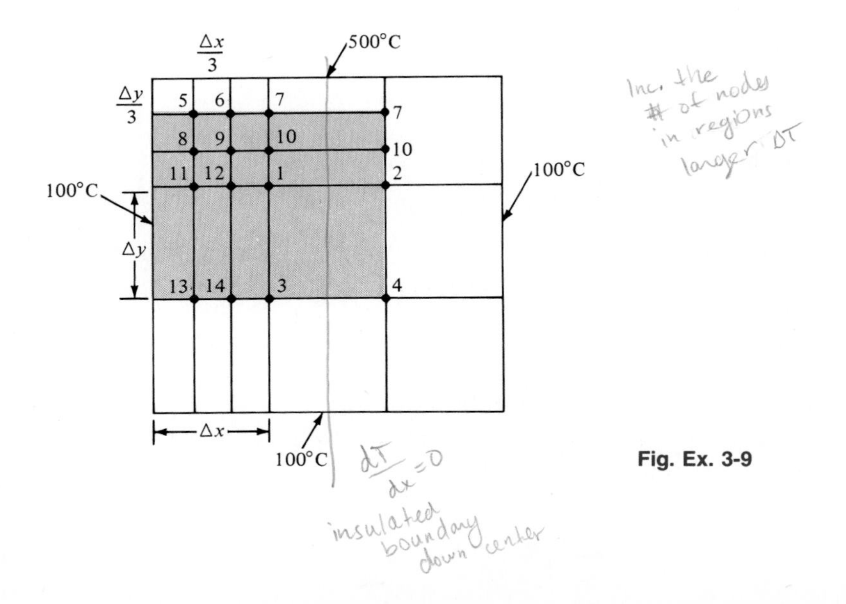

Fig. Ex. 3-9

For node 7 we can use a resistance formulation and obtain

$$1/R_{7-6} = k$$

$$1/R_{7-500^\circ} = \frac{k(\Delta x/6 + \Delta x/2)}{\Delta y/3} = 2k$$

$$1/R_{7-10} = 2k$$

and we find

$$1000 + T_6 + 2T_{10} - 5T_7 = 0$$

Similar resistors are obtained for node 10.

$$1/R_{10-9} = k$$

$$1/R_{10-7} = 2k = 1/R_{10-1}$$

so that

$$2T_7 + T_9 + 2T_1 - 5T_{10} = 0$$

For node 1,

$$1/R_{1-12} = \frac{k(\Delta y/6 + \Delta y/2)}{\Delta x/3} = 2k$$

$$1/R_{1-3} = \frac{k(\Delta x/6 + \Delta x/2)}{\Delta y} = 2k/3$$

$$1/R_{1-10} = 2k$$

and the nodal equation becomes

$$3T_{12} + 3T_{10} + T_3 - 7T_1 = 0$$

For node 11,

$$1/R_{11-100^\circ} = 1/R_{11-12} = \frac{k(\Delta y/6 + \Delta y/2)}{\Delta x/3} = 2k$$

$$1/R_{11-8} = k$$

$$1/R_{11-13} = \frac{k(\Delta x/3)}{\Delta y} = k/3$$

and the nodal equation becomes

$$600 + 6T_{12} + 3T_8 + T_{13} - 16T_{11} = 0$$

Similarly, the equation for node 12 is

$$3T_9 + 6T_{11} + 6T_1 + T_{14} - 16T_{12} = 0$$

For node 13,

$$1/R_{13-100^\circ} = \frac{k\Delta y}{\Delta x/3} = 3k = 1/R_{13-14}$$

$$1/R_{13-11} = 1/R_{13-100} = k/3$$

and we obtain

$$1000 + 9T_{14} + T_{11} - 20T_{13} = 0$$

Similarly for node 14,

$$100 + 9T_{13} + 9T_3 + T_{12} - 20T_{14} = 0$$

Finally, from resistances already found, the nodal equation for node 3 is

$$200 + 9T_{14} + 2T_1 - 13T_3 = 0$$

We choose to solve the set of equations by the Gauss-Seidel iteration technique and thus write them in the form $T_i = f(T_j)$. The solution was set up on a computer with all initial values for the T_i's taken as zero. The results of the computations are shown in the following table.

	Number of iterations				
Node	*2*	*10*	*20*	*30*	*50*
1	59.30662	232.6668	247.1479	247.7605	247.7875
2	59.30662	232.6668	247.1479	247.7605	247.7875
3	50.11073	139.5081	147.2352	147.5629	147.5773
4	50.11073	139.5081	147.2352	147.5629	147.5773
5	206.25	288.358	293.7838	294.0129	294.023
6	248.75	359.025	366.9878	367.3243	367.3391
7	291.45	390.989	398.7243	399.0513	399.0657
8	102.9297	200.5608	208.4068	208.7384	208.753
9	121.2334	264.2423	275.7592	276.2462	276.2677
10	164.5493	302.3108	313.5007	313.974	313.9948
11	70.95459	156.9976	164.3947	164.7076	164.7215
12	73.89051	203.6437	214.5039	214.9634	214.9836
13	70.18905	115.2635	119.2079	119.3752	119.3826
14	62.82942	129.8294	135.6246	135.8703	135.8811

Note that these solutions for $T_1 = T_2 = 247.79°\text{C}$ and $T_3 = T_4 = 147.58°\text{C}$ are somewhat below the values of 250°C and 150°C obtained when only four nodes were employed, but only modestly so.

3-9 ELECTRICAL ANALOGY FOR TWO-DIMENSIONAL CONDUCTION

Steady-state electric conduction in a homogeneous material of constant resistivity is analogous to steady-state heat conduction in a body of similar geometric shape. For two-dimensional electric conduction the Laplace equation applies:

$$\frac{\partial^2 E}{\partial x^2} + \frac{\partial^2 E}{\partial y^2} = 0$$

where E is the electric potential. A very simple way of solving a two-dimensional heat-conduction problem is to construct an electrical analog and experimentally determine the geometric shape factors for use in Eq. (3-23). One way to accomplish this is to use a commercially available paper which is coated with a thin conductive film. This paper may be cut to an exact geometric model of the two-dimensional heat-conduction system. At the appropriate edges of the paper, good electrical conductors are attached to simulate the temperature boundary conditions on the problem. An electric-potential difference is then impressed on the model. It may be noted that the paper has a very high resistance in comparison with the conductors attached to the edges, so that a constant-potential condition can be maintained at the region of contact.

Once the electric potential is impressed on the paper, an ordinary voltmeter may be used to plot lines of constant electric potential. With these constant-potential lines available, the flux lines may be easily constructed since they are orthogonal to the potential lines. These equipotential and flux lines have precisely the same arrangement as the isotherms and heat-flux lines in the corresponding heat-conduction problem. The shape factor is calculated immediately using the method which was applied to the curvilinear squares.

It may be noted that the conducting-sheet analogy is not applicable to problems where heat generation is present; however, by addition of appropriate resistances, convection boundary conditions may be handled with little trouble. Schneider [2] and Ozisik [12] discuss the conducting-sheet method, as well as other analogies for treating conduction heat-transfer problems, and Kayan [4, 5] gives a detailed discussion of the conducting-sheet method.

3-10 SUMMARY

There is a myriad of analytical solutions for steady-state conduction heat-transfer problems available in the literature. In this day of computers most of these solutions are of small utility, despite their exercise in mathematical facilities. This is not to say that we cannot use the results of past experience to anticipate answers to new problems. But, most of the time, the problem a person wants to solve can be attacked directly by numerical techniques, *except* when there is an easier way to do the job. As a summary, the following suggestions are offered:

1. When tackling a two- or three-dimensional heat-transfer problem *first* try to reduce it to a one-dimensional problem. An example is a cylinder with length much larger than its diameter.
2. If possible, select a simple shape factor model which may either exactly or approximately represent the physical situation. See comments under items 4 and 5.
3. Seek some simple analytical solutions but, if solutions are too complicated, go directly to the numerical techniques.

4. In practical problems recognize that convection and radiation boundary conditions are subject to large uncertainties. This means that, in most practical situations, undue concern over accuracy of solution to numerical nodal equations is unjustified.

5. In general, approach the solution in the direction of simple to complex, and make use of checkpoints along the way.

REVIEW QUESTIONS

1 What is the main assumption in the separation-of-variables method for solving Laplace's equation?

2 Define the conduction shape factor.

3 What is the basic procedure in setting up a numerical solution to a two-dimensional conduction problem?

4 Once finite-difference equations are obtained for a conduction problem, what methods are available to effect a solution? What are the advantages and disadvantages of each method, and when would each technique be applied?

5 Investigate the computer routines that are available at your computer center for solution of conduction heat-transfer problems.

PROBLEMS

3-1 Beginning with the separation-of-variables solutions for $\lambda^2 = 0$ and $\lambda^2 < 0$ [Eqs. (3-9) and (3-10)], show that it is not possible to satisfy the boundary conditions for the constant temperature at $y = H$ with either of these two forms of solution. That is, show that, in order to satisfy the boundary conditions

$$T = T_1 \qquad \text{at } y = 0$$

$$T = T_1 \qquad \text{at } x = 0$$

$$T = T_1 \qquad \text{at } x = W$$

$$T = T_2 \qquad \text{at } y = H$$

either a trivial or physically unreasonable solution results when either Eq. (3-9) or (3-10) is used.

3-2 Write out the first four nonzero terms of the series solutions given in Eq. (3-20). What percentage error results from using only these four terms at $y = H$ and $x = W/2$?

3-3 A 6.0-cm-diameter pipe whose surface temperature is maintained at 210°C passes through the center of a concrete slab 45 cm thick. The outer surface temperatures of the slab are maintained at 15°C. Using the flux plot, estimate the heat loss from the pipe per unit length.

3-4 A heavy-wall tube of Monel, 2.5-cm ID and 5-cm OD, is covered with a 2.5-cm layer of glass wool. The inside tube temperature is 300°C, and the temperature at the outside of the insulation is 40°C. How much heat is lost per foot of length? Take $k = 11$ Btu/h · ft · °F for Monel.

3-5 A symmetrical furnace wall has the dimensions shown. Using the flux plot, obtain the shape factor for this wall.

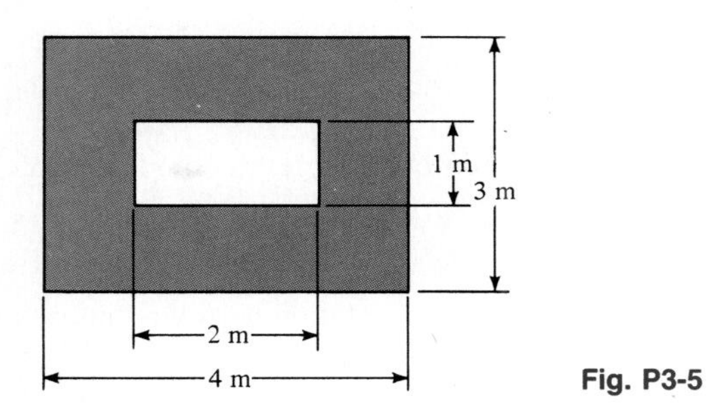

Fig. P3-5

3-6 A furnace of 1 by 2 by 3 ft inside dimensions is constructed of a material having a thermal conductivity of 0.5 Btu/h · ft · °F. The wall thickness is 6 in. The inner and outer surface temperatures are 1000 and 200°F, respectively. Calculate the heat loss through the furnace wall.

3-7 A cube 35 cm on each external side is constructed of fireclay brick. The wall thickness is 5.0 cm. The inner surface temperature is 500°C, and the outer surface temperature is 80°C. Compute the heat flow in watts.

3-8 Two long cylinders 8.0 and 3.0 cm in diameter are completely surrounded by a medium with $k = 1.4$ W/m · °C. The distance between centers is 10 cm, and the cylinders are maintained at 200 and 35°C. Calculate the heat-transfer rate per unit length.

3-9 A 1-m-diameter sphere maintained at 30°C is buried in the earth at a place where $k = 1.7$ W/m · °C. The depth to the centerline is 2.4 m, and the earth surface temperature is 0°C. Calculate the heat lost by the sphere.

3-10 A 20-cm-diameter sphere is totally enclosed by a large mass of glass wool. A heater inside the sphere maintains its outer surface temperature at 170°C while the temperature at the outer edge of the glass wool is 20°C. How much power must be supplied to the heater to maintain equilibrium conditions?

3-11 A large spherical storage tank, 2 m in diameter, is buried in the earth at a location where the thermal conductivity is 1.5 W/m · °C. The tank is used for the storage of an ice mixture at 0°C, and the ambient temperature of the earth is 20°C. Calculate the heat loss from the tank.

3-12 The solid shown in the figure has the upper surface, including the half-cylinder cutout, maintained at 100°C. At a large depth in the solid the temperature is 300

K; $k = 1$ W/m · °C. What is the heat transfer at the surface for the region where $L = 30$ cm and $D = 10$ cm?

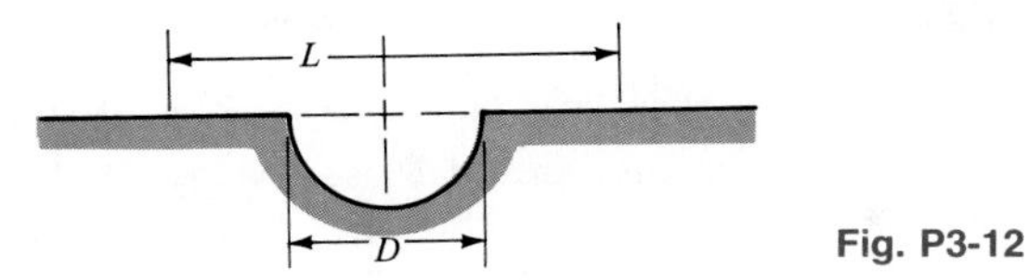

Fig. P3-12

3-13 In certain locales, power transmission is made by means of underground cables. In one example an 8.0-cm-diameter cable is buried at a depth of 1.3 m, and the resistance of the cable is 1.1×10^{-4} Ω/m. The surface temperature of the ground is 25°C, and $k = 1.2$ W/m · °C for earth. Calculate the maximum allowable current if the outside temperature of the cable cannot exceed 110°C.

3-14 A copper sphere 4.0 cm in diameter is maintained at 70°C and submerged in a large earth region where $k = 1.3$ W/m · °C. The temperature at a large distance from the sphere is 12°C. Calculate the heat lost by the sphere.

3-15 Two long, eccentric cylinders having diameters of 15 and 4 cm respectively are maintained at 100 and 20°C and separated by a material with $k = 3.0$ W/m · °C. The distance between centers is 4.5 cm. Calculate the heat transfer per unit length between the cylinders.

3-16 Two pipes are buried in the earth and maintained at temperatures of 300 and 125°C. The diameters are 8 and 16 cm, and the distance between centers is 40 cm. Calculate the heat-transfer rate per unit length if the thermal conductivity of earth at this location is 0.7 W/m · °C.

3-17 A hot sphere having a diameter of 1.5 m is maintained at 300°C and buried in a material with $k = 1.2$ W/m · °C and outside surface temperature of 30°C. The depth of the centerline of the sphere is 3.75 m. Calculate the heat loss.

3-18 A scheme is devised to measure the thermal conductivity of soil by immersing a long electrically heated rod in the ground in a vertical position. For design purposes, the rod is taken as 2.5 cm in diameter with a length of 1 m. To avoid improper alteration of the soil, the maximum surface temperature of the rod is 55°C while the soil temperature is 10°C. Assuming a soil conductivity of 1.7 W/m · °C, what are the power requirements of the electric heater in watts?

3-19 Two pipes are buried in an insulating material having $k = 0.8$ W/m · °C. One pipe is 10 cm in diameter and carries a hot fluid at 300°C while the other pipe is 2.8 cm in diameter and carries a cool fluid at 15°C. The pipes are parallel and separated by a distance of 12 cm on centers. Calculate the heat-transfer rate between the pipes per meter of length.

3-20 At a certain location the thermal conductivity of the earth is 1.5 W/m · °C. At this location an isothermal sphere having a temperature of 5°C and a diameter of 2.0 m is buried at a centerline depth of 5.0 m. The earth temperature is 25°C. Calculate the heat lost from the sphere.

3-21 People are sometimes careless at universities and bury steam pipes in the earth without insulation. Consider a 4-in pipe carrying steam at 300°F buried at a depth of 9 in to centerline. The buried length is 100 yards. Assuming that the earth thermal conductivity is 1.2 W/m^2 · °C and the surface temperature is 60°F, estimate the heat lost from the pipe.

3-22 Two parallel pipes 5 cm and 10 cm in diameter are totally surrounded by loosely packed asbestos. The distance between centers for the pipes is 20 cm. One pipe carries steam at 110°C while the other carries chilled water at 3°C. Calculate the heat lost by the hot pipe per unit length.

3-23 A long cylinder has its surface maintained at 135°C and is buried in a material having a thermal conductivity of 15.5 W/m · °C. The diameter of the cylinder is 3 cm and the depth to its centerline is 5 cm. The surface temperature of the material is 46°C. Calculate the heat lost by the cylinder per meter of length.

3-24 A 3-m-diameter sphere contains a mixture of ice and water at 0°C and is buried in a semi-infinite medium having a thermal conductivity of 0.2 W/m · °C. The top surface of the medium is isothermal at 30°C and the sphere centerline is at a depth of 8.5 m. Calculate the heat lost by the sphere.

3-25 An electric heater in the form of a 50-by-100-cm plate is laid on top of a semi-infinite insulating material having a thermal conductivity of 0.74 W/m · °C. The heater plate is maintained at a constant temperature of 120°C over all its surface, and the temperature of the insulating material a large distance from the heater is 15°C. Calculate the heat conducted into the insulating material.

3-26 A small furnace has inside dimensions of 60 by 70 by 80 cm with a wall thickness of 5 cm. Calculate the shape factor for this geometry.

3-27 A 15-cm-diameter steam pipe at 150°C is buried in the earth near a 5-cm pipe carrying chilled water at 5°C. The distance between centers is 15 cm and the thermal conductivity of the earth at this location may be taken as 0.7 W/m · °C. Calculate the heat lost by the steam pipe per unit length.

3-28 Derive an equation equivalent to Eq. (3-24) for an interior node in a three-dimensional heat-flow problem.

3-29 Derive an equation equivalent to Eq. (3-24) for an interior node in a one-dimensional heat-flow problem.

3-30 Derive an equation equivalent to Eq. (3-25) for a one-dimensional convection boundary condition.

3-31 Considering the one-dimensional fin problems of Chap. 2, show that a nodal equation for nodes along the fin in the accompanying figure may be expressed as

$$T_m\left[\frac{hP(\Delta x)^2}{kA} + 2\right] - \frac{hP(\Delta x)^2}{kA}T_\infty - (T_{m-1} + T_{m+1}) = 0$$

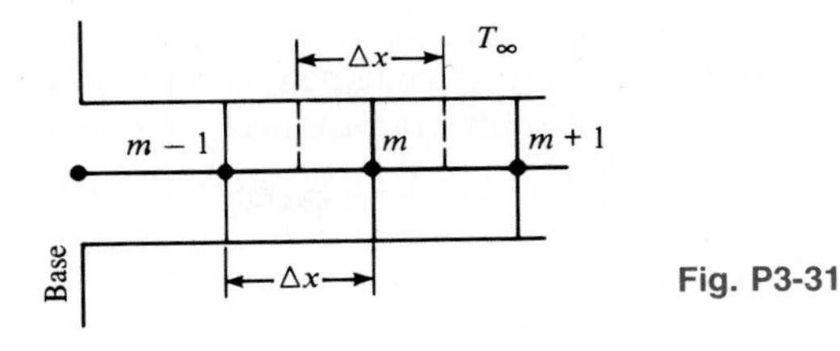

Fig. P3-31

3-32 Show that the nodal equation corresponding to an insulated wall shown in the accompanying figure is

$$T_{m,n+1} + T_{m,n-1} + 2T_{m-1,n} - 4T_{m,n} = 0$$

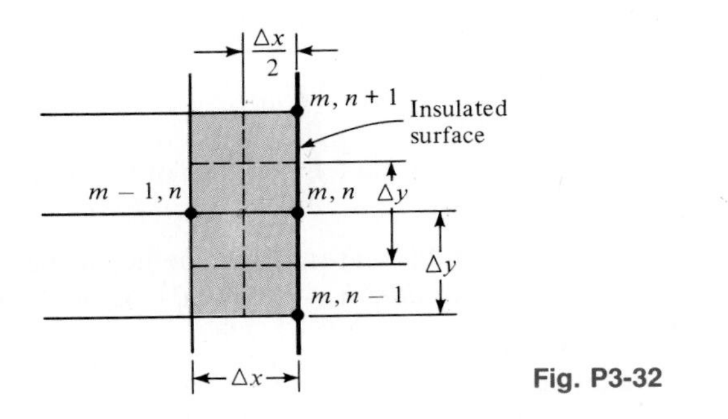

Fig. P3-32

3-33 For the insulated corner section shown, derive an expression for the nodal equation of node (m,n) under steady-state conditions.

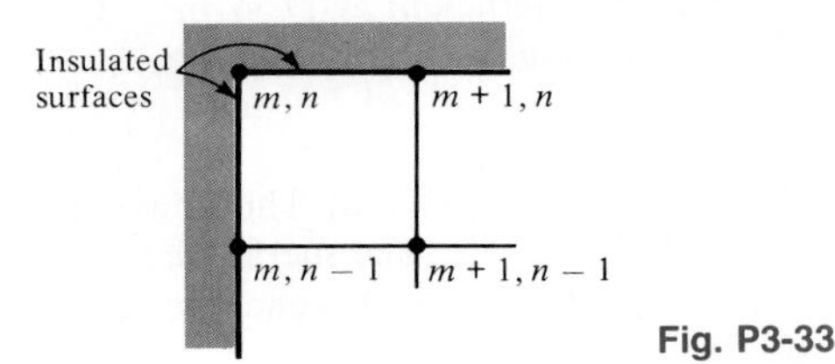

Fig. P3-33

3-34 Derive the equation in Table 3-2*f*.

3-35 Derive an expression for the equation of a boundary node subjected to a constant heat flux from the environment. Use the nomenclature of Fig. 3-7.

3-36 Set up the nodal equations for a modification of Example 3-5 in which the left half of the wire is insulated and the right half is exposed to a convection environment with $h = 200$ W/m² · °C and $T = 20$°C.

3-37 In a proposed solar-energy application, the solar flux is concentrated on a 5-cm-OD stainless-steel tube [$k = 16$ W/m · °C] 2 m long. The energy flux on the tube surface is 20,000 W/m², and the tube wall thickness is 2 mm. Boiling water flows inside the tube with a convection coefficient of 5000 W/m² · °C and a temperature of 250°C. Both ends of the tube are mounted in an appropriate supporting bracket, which maintains them at 100°C. For thermal-stress considerations the temperature gradient near the supports is important. Assuming a one-dimensional system, set up a numerical solution to obtain the temperature gradient near the supports.

3-38 An aluminum rod 2.5 cm in diameter and 15 cm long protrudes from a wall maintained at 300°C. The environment temperature is 38°C. The heat-transfer coefficient is 17 W/m² · °C. Using a numerical technique in accordance with the result of Prob. 3-31, obtain values for the temperature along the rod. Subsequently obtain the heat flow from the wall at $x = 0$. *Hint:* The boundary condition at the end of the rod may be expressed by

$$T_m \left[\frac{h\,\Delta x}{k} + \frac{hP(\Delta x)^2}{2kA} + 1 \right] - T_\infty \left[\frac{h\,\Delta x}{k} + \frac{hP(\Delta x)^2}{2kA} \right] - T_{m-1} = 0$$

where m denotes the node at the tip of the fin. The heat flow at the base is

$$q_{x=0} = \frac{-kA}{\Delta x}(T_{m+1} - T_m)$$

where T_m is the base temperature and T_{m+1} is the temperature of the first increment.

3-39 Repeat Prob. 3-38, using a linear variation of heat-transfer coefficient between base temperature and the tip of the fin. Assume $h = 28$ W/m² · °C at the base and $h = 11$ W/m² · °C at the tip.

3-40 For the wall in Prob. 3-5 a material with $k = 1.4$ W/m · °C is used. The inner and outer wall temperatures are 650 and 150°C, respectively. Using a numerical technique, calculate the heat flow through the wall.

3-41 Repeat Prob. 3-40, assuming that the outer wall is exposed to an environment at 38°C and that the convection heat-transfer coefficient is 17 W/m² · °C. Assume that the inner surface temperature is maintained at 650°C.

3-42 Repeat Prob. 3-3, using the numerical technique.

3-43 In the section illustrated, the surface 1-4-7 is insulated. The convection heat-transfer coefficient at surface 1-2-3 is 28 W/m² · °C. The thermal conductivity of the solid material is 5.2 W/m · °C. Using the numerical technique, compute the temperatures at nodes 1, 2, 4, and 5.

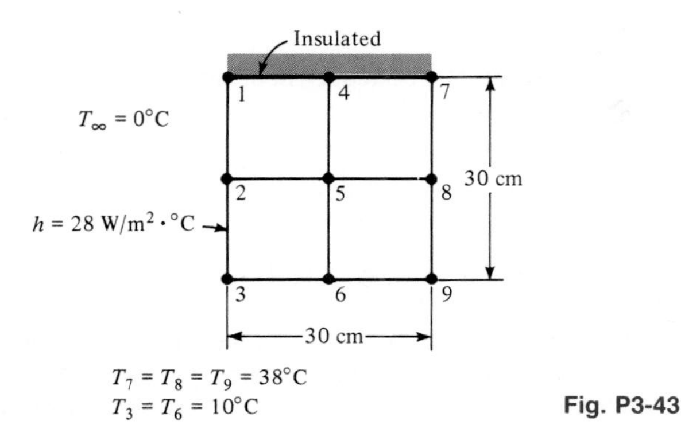

Fig. P3-43

3-44 A glass plate 3 by 12 by 12 in [$k = 0.7$ W/m · °C] is oriented with the 12 by 12 face in a vertical position. One face loses heat by convection to the surroundings at 70°F. The other vertical face is placed in contact with a constant-temperature block at 400°F. The other four faces are insulated. The convection heat-transfer coefficient varies approximately as

$$h_x = 0.22(T_S - T_\infty)^{1/4}x^{-1/4} \qquad \text{Btu/h} \cdot \text{ft}^2 \cdot {}^\circ\text{F}$$

where T_S and T_∞ are in degrees Fahrenheit, T_S is the local surface temperature, and x is the vertical distance from the bottom of the plate, measured in feet. Determine the convection heat loss from the plate, using an appropriate numerical analysis.

3-45 Calculate the temperatures at points 1, 2, 3, and 4 using the numerical method.

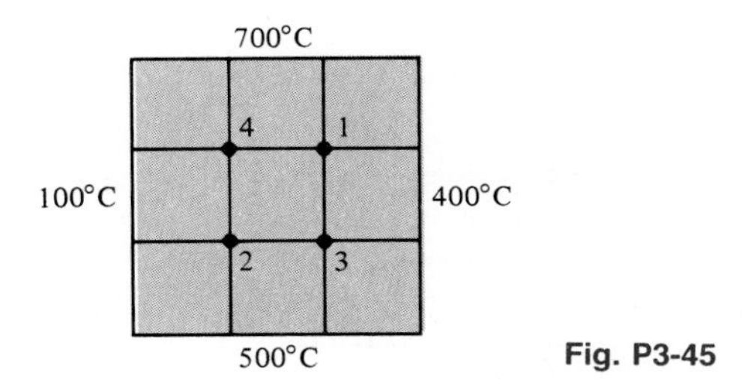

Fig. P3-45

3-46 For the block shown, calculate the steady-state temperature distribution at appropriate nodal locations using the numerical method. Employ a digital computer for the solution if possible, and take advantage of library subroutines available in the computer center; $k = 2.0$ Btu/h · ft · °F.

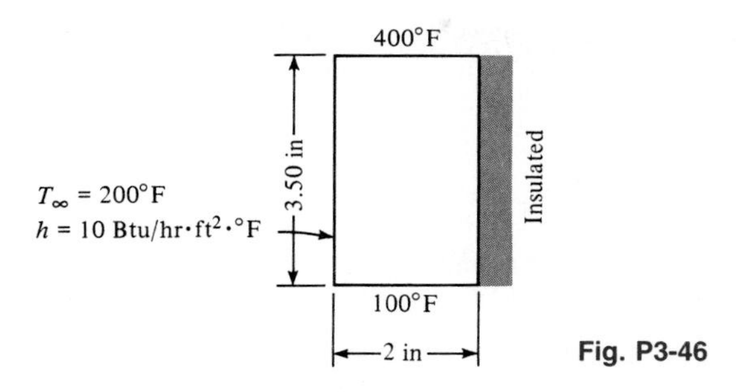

Fig. P3-46

3-47 Rework Prob. 3-43, using the Gauss-Seidel iteration method.

3-48 The composite strip in the accompanying figure is exposed to the convection environment at 300°C and $h = 40$ W/m² · °C. The material properties are $k_A = 20$ W/m · °C, $k_B = 1.2$ W/m · °C, and $k_C = 0.5$ W/m · °C. The strip is mounted on a plate maintained at the constant temperature of 50°C. Calculate the heat transfer from the strip to plate per unit length of strip. Assume two-dimensional heat flow.

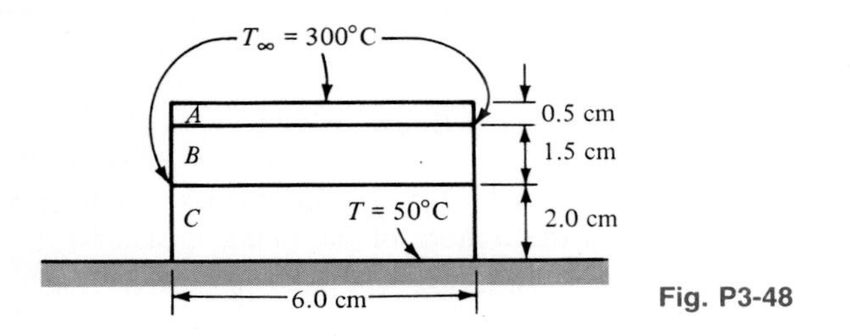

Fig. P3-48

3-49 The fin shown has a base maintained at 300°C and is exposed to the convection environment indicated. Calculate the steady-state temperatures of the nodes shown and the heat loss if $k = 1.0$ W/m · °C.

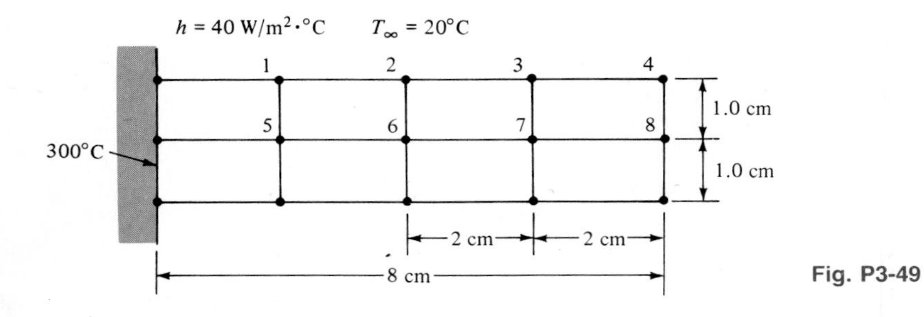

Fig. P3-49

3-50 Calculate the steady-state temperatures for nodes 1 to 16 in the figure.

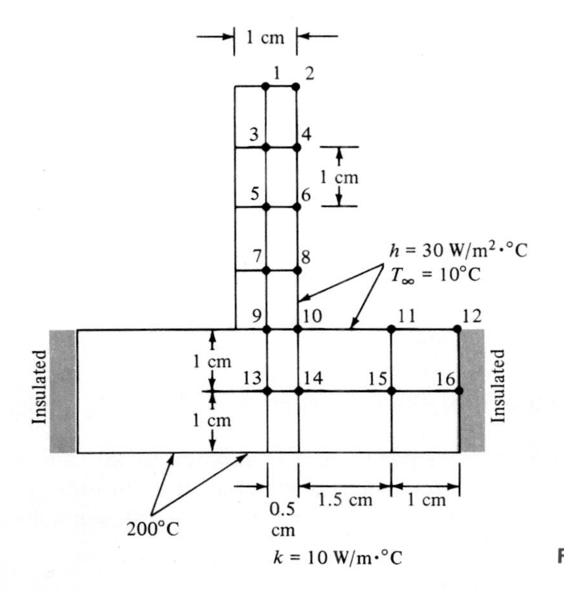

Fig. P3-50

3-51 Calculate the steady-state temperatures for nodes 1 to 9 in the figure.

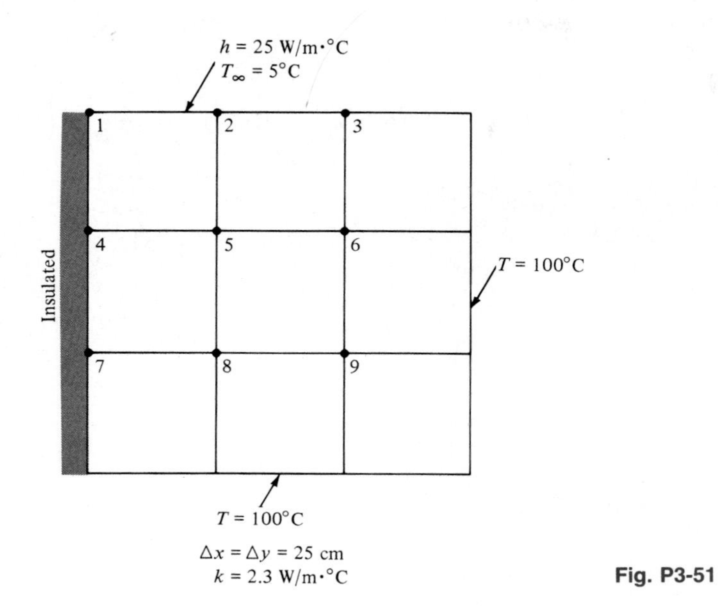

Fig. P3-51

3-52 Calculate the steady-state temperatures for nodes 1 to 6 in the figure.

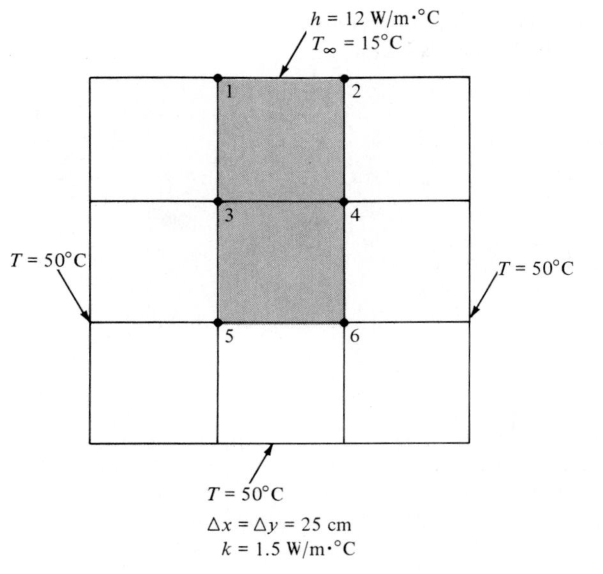

Fig. P3-52

3-53 Calculate the temperatures for the nodes indicated in the accompanying figure. The entire outer surface is exposed to the convection environment and the entire inner surface is at a constant temperature of 300°C. Properties for materials A and B are given in the figure.

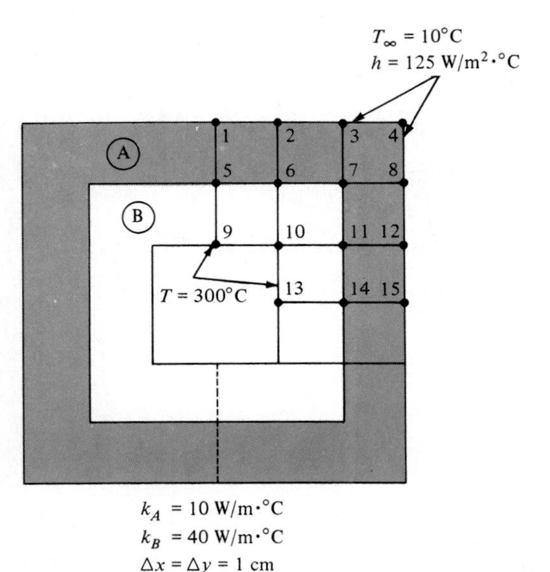

Fig. P3-53

3-54 A rod having a diameter of 2 cm and a length of 10 cm has one end maintained at 200°C and is exposed to a convection environment at 25°C with $h = 40\ W/m^2 \cdot °C$. The rod generates heat internally at the rate of 50 MW/m^3 and the thermal conductivity is 35 $W/m \cdot °C$. Calculate the temperatures of the nodes shown in the figure assuming one-dimensional heat flow.

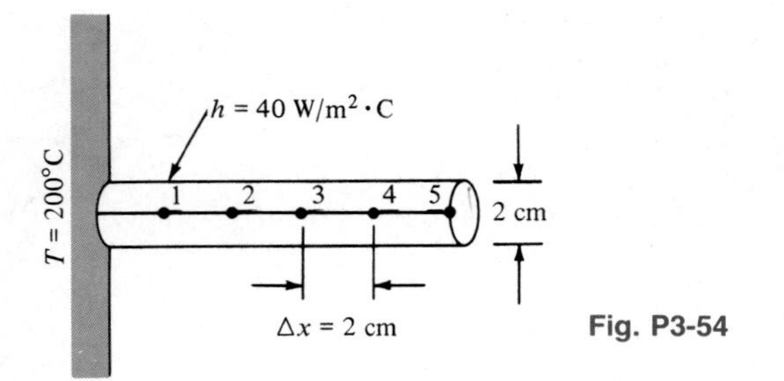

Fig. P3-54

3-55 Calculate the steady-state temperatures of the nodes in the accompanying figure. The entire outer surface is exposed to the convection environment at 20°C and the entire inner surface is constant at 500°C. Assume $k = 20\ W/m \cdot °C$.

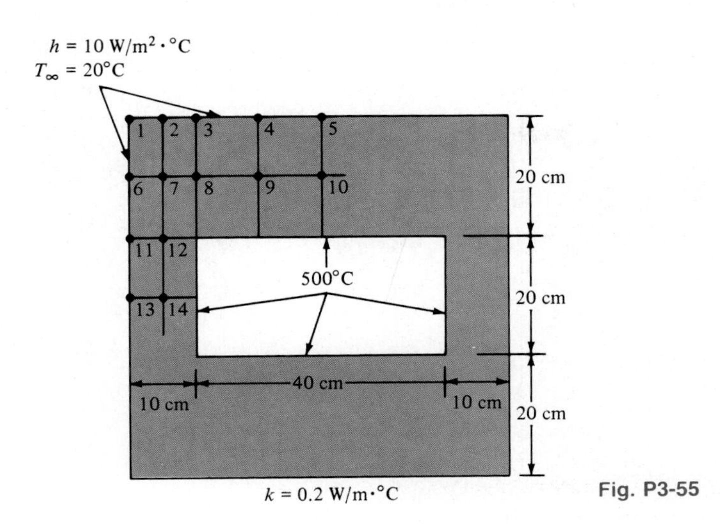

Fig. P3-55

3-56 A liner of stainless steel ($k = 20$ W/m · °C), having a thickness of 3 mm, is placed on the inside surface of the solid in Prob. 3-55. Assuming now that the inside surface of the stainless steel is at 500°C, calculate new values for the nodal temperatures in the low-conductivity material. Set up your nodes in the stainless steel as necessary.

3-57 Calculate the steady-state temperatures for the nodes indicated in the accompanying figure.

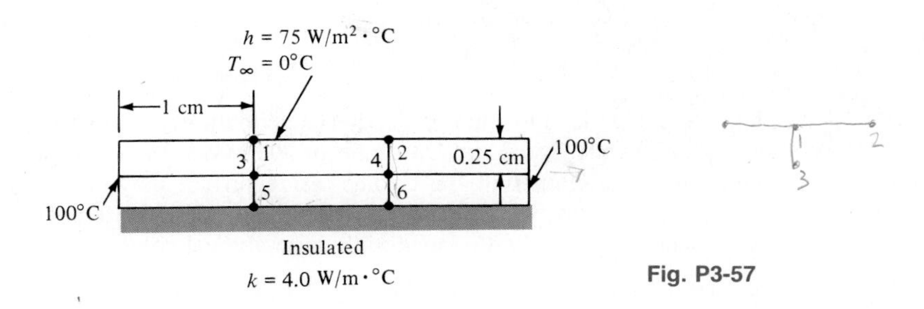

Fig. P3-57

3-58 The two-dimensional solid shown in the accompanying figure generates heat internally at the rate of 90 MW/m². Using the numerical method calculate the steady-state nodal temperatures for $k = 20$ W/m · °C.

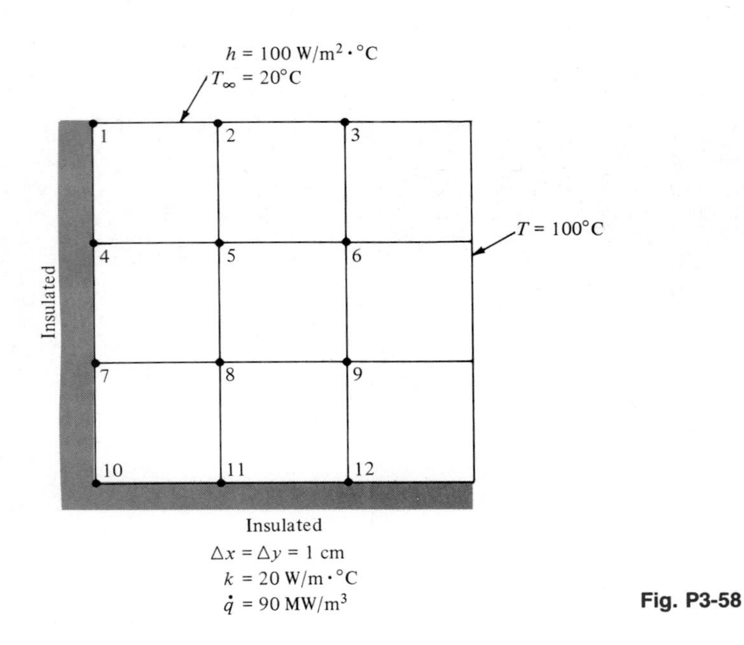

Fig. P3-58

3-59 The half-cylinder has $k = 20$ W/m · °C and is exposed to the convection environment at 20°C. The lower surface is maintained at 300°C. Compute the temperatures for the nodes shown and the heat loss for steady state.

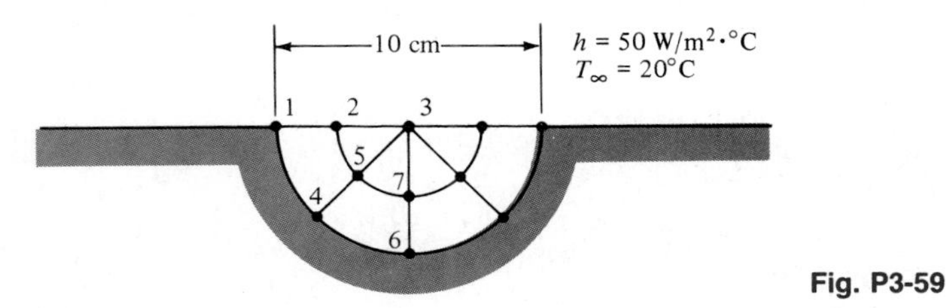

Fig. P3-59

3-60 A tube has diameters of 4 mm and 5 mm and a thermal conductivity 20 W/m^2 · °C. Heat is generated uniformly in the tube at a rate of 500 MW/m^3 and the outside surface temperature is maintained at 100°C. The inside surface may be assumed to be insulated. Divide the tube wall into four nodes and calculate the temperature at each using the numerical method. Check with an analytical solution.

3-61 Repeat Prob. 3-60 with the inside of the tube exposed to a convection condition with $h = 40$ W/m^2 · °C. Check with an analytical calculation.

3-62 Rework Prob. 3-50 with the surface absorbing a constant heat flux of 300 W/m^2 instead of the convection boundary condition. The bottom surface still remains at 200°C.

3-63 Rework Prob. 3-53 with the inner surface absorbing a constant heat flux of 300 W/m^2 instead of being maintained at a constant temperature of 300°C.

3-64 Rework Prob. 3-58 with the surface marked at a constant 100°C now absorbing a constant heat flux of 500 W/m². Add nodes as necessary.

3-65 The tapered aluminum pin fin shown in the figure is circular in cross section with a base diameter of 1 cm and a tip diameter of 0.5 cm. The base is maintained at 200°C and loses heat by convection to the surroundings at $T_\infty = 10$°C, $h = 200$ W/m² · °C. The tip is insulated. Assume one-dimensional heat flow and use the finite-difference method to obtain the temperature at nodes 1 through 4 and the heat lost by the fin. The length of the fin is 6 cm.

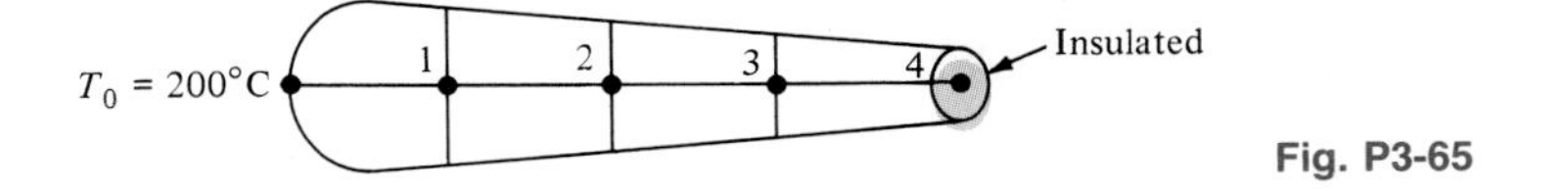

Fig. P3-65

3-66 Write the nodal equations 1 through 7 for the symmetrical solid shown. $\Delta x = \Delta y = 1$ cm.

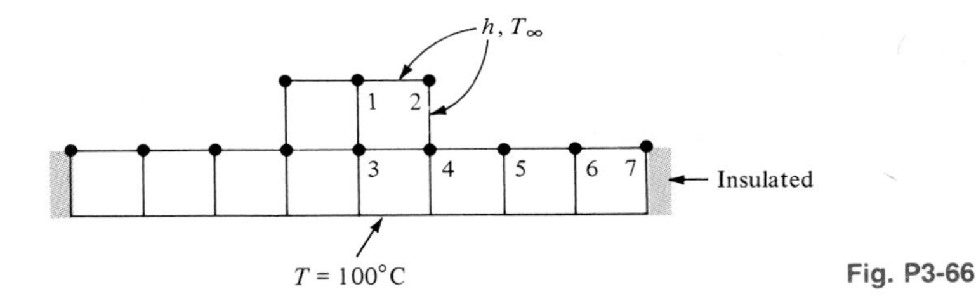

Fig. P3-66

3-67 Obtain the temperature for nodes 1 through 6 in the accompanying figure; $\Delta x = \Delta y = 1$ cm.

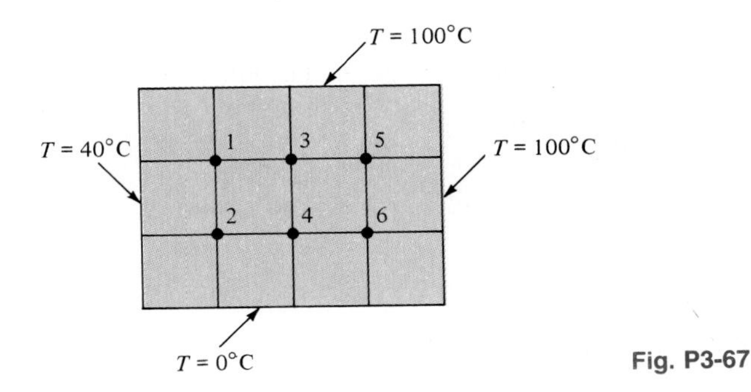

Fig. P3-67

3-68 Write the nodal equations for nodes 1 through 9 in the accompanying figure; $\Delta x = \Delta y = 1$ cm.

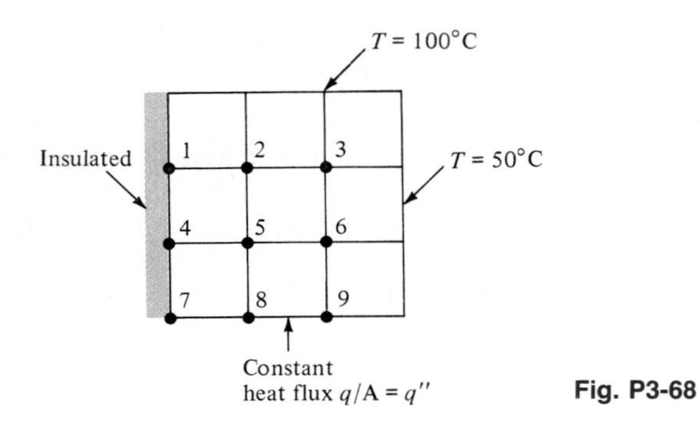

Fig. P3-68

3-69 Write the nodal equation for nodes 1 through 12 in the accompanying figure. Express the equations in a format for Gauss-Seidel iteration.

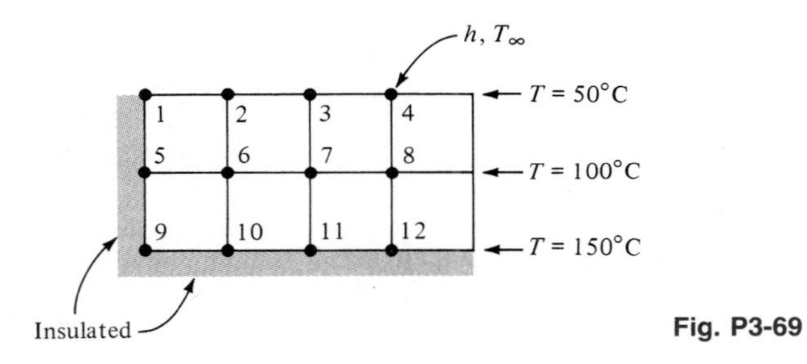

Fig. P3-69

3-70 Sometimes a square grid is desired even for a circular system. Consider the quadrant of a circle shown with $r = 10$ cm. $\Delta x = \Delta y = 3$ cm and $k = 10$ W/m · °C. Write the steady-state nodal equations for nodes 3 and 4. Make use of Tables 3–2 and 3–4.

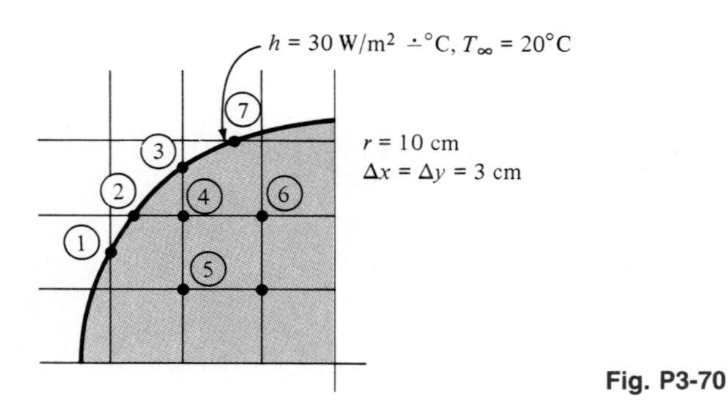

Fig. P3-70

3-71 Taking the accompanying figure as a special case of Table 3–2(f), write the nodal equations for nodes (m,n) and 2 for the case of $\Delta x = \Delta y$.

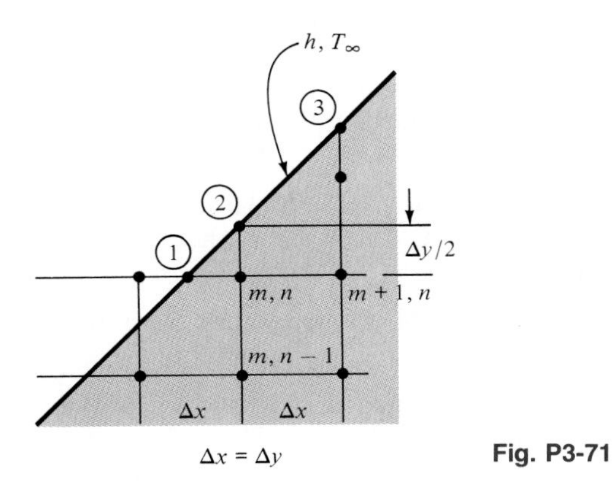

Fig. P3-71

3-72 Repeat Prob. 3-71 for a slanted surface which is insulated.

3-73 If the slanted surface of Prob. 3-71 is isothermal at T_∞, what is the nodal equation for node (m,n)?

3-74 The slanted intersection shown involves materials A and B. Write steady-state nodal equations for nodes 3, 4, 5, and 6 using Table 3–2(f and g) as a guide.

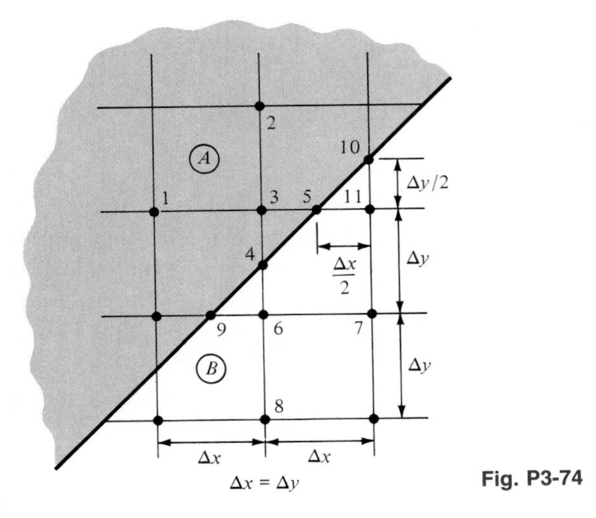

Fig. P3-74

3-75 A cube 20 cm on a side is maintained at 80°C and buried in a large medium at 10°C with a thermal conductivity of 2.3 W/m · °C. Calculate the heat lost by the cube. How does this compare with the heat which would be lost by a 20-cm-diameter sphere? Compare these heat transfers on a unit-volume basis.

3-76 A long horizontal cylinder having a diameter of 10 cm is maintained at a temperature of 100°C and centered in a 30-cm-thick slab of material for which k = 10 W/m · °C. The outside of the slab is at 20°C. Calculate the heat lost by the cylinder per unit length.

3-77 Work Prob. 3-76 using the flux plot.

3-78 A horizontal plate 10 by 100 cm is buried in a large medium at a depth of 2.0 m and maintained at 50°C. The surface of the medium is at 10°C and has $k = 1.5$ W/m · °C. Calculate the heat lost by the plate.

3-79 A thin disk 5 cm in diameter is maintained at 75°C and placed on the surface of a large medium at 15°C with $k = 3$ W/m · °C. Calculate the heat conducted into the medium.

3-80 Repeat Prob. 3-79 for a square 5 cm on a side. Compare the heat transfers on a per unit area basis.

3.81 A hot steam pipe 10 cm in diameter is maintained at 200°C and centered in a square mineral-fiber insulation 20 cm on a side. The outside surface temperature of the insulation is 35°C. Calculate the heat lost by a 20-m length of pipe if the thermal conductivity of the insulation can be taken as 50 mW/m · °C.

REFERENCES

1. Carslaw, H. S., and J. C. Jaeger: "Conduction of Heat in Solids," 2d ed., Oxford University Press, Fair Lawn, N.J., 1959.
2. Schneider, P. J.: "Conduction Heat Transfer," Addison-Wesley Publishing Company, Inc., Reading, Mass., 1955.
3. Dusinberre, G. M.: "Heat Transfer Calculations by Finite Differences," International Textbook Company, Scranton, Pa., 1961.
4. Kayan, C. F.: An Electrical Geometrical Analogue for Complex Heat Flow, *Trans. ASME,* vol. 67, p. 713, 1945.
5. Kayan, C. F.: "Heat Transfer Temperature Patterns of a Multicomponent Structure by Comparative Methods," *Trans. ASME,* vol. 71, p. 9, 1949.
6. Rudenberg, R.: Die Ausbreitung der Luft- und Erdfelder und Hochspannungsleitungen, besonders bei Erd- und Kurzschlussen, *Elecktrotech. Z.,* vol. 46, p. 1342, 1925.
7. Andrews, R. V.: Solving Conductive Heat Transfer Problems with Electrical-analogue Shape Factors, *Chem. Eng. Prog.,* vol. 51, no. 2, p. 67, 1955.
8. Sunderland, J. E., and K. R. Johnson: Shape Factors for Heat Conduction through Bodies with Isothermal or Convective Boundary Conditions, *Trans. ASHAE,* vol. 70, pp. 237–241, 1964.
9. Richtmeyer, R. D.: "Difference Methods for Initial Value Problems," Interscience Publishers, Inc., New York, 1957.
10. Cranck, J., and P. Nicolson: A Practical Method for Numerical Evaluation of Solutions of P.D.E. of Heat Conduction Type, *Proc. Camb. Phil. Soc.,* vol. 43, p. 50, 1947.
11. Barakat, H. Z., and J. A. Clark: On the Solution of Diffusion Equation by Numerical Methods, *J. Heat Transfer,* p. 421, November 1966.
12. Ozisik, M. N.: "Boundary Value Problems of Heat Conduction," International Textbook Company, Scranton, Pa., 1968.

13 Arpaci, V. S.: "Conduction Heat Transfer," Addison-Wesley Publishing Company, Inc., Reading, Mass., 1966.

14 Ames, W. F.: "Nonlinear Partial Differential Equations in Engineering," Academic Press, Inc., New York, 1965.

15 Myers, R. F.: "Conduction Heat Transfer," McGraw-Hill Book Company, New York, 1972.

16 Adams, J. A., and D. F. Rogers: "Computer Aided Analysis in Heat Transfer," McGraw-Hill Book Company, New York, 1973.

17 Bowers, J. C., and S. R. Sedore: "SCEPTRE: A Computer Program for Circuit and Systems Analysis," Prentice-Hall, Inc., Englewood Cliffs, N.J., 1971.

18 Rohsenow, W. M., and J. P. Hartnett (eds.): "Handbook of Heat Transfer," McGraw-Hill Book Company, New York, 1973.

19 Kern, D. Q., and A. D. Kraus: "Extended Surface Heat Transfer," McGraw-Hill Book Company, New York, 1972.

20 Strong, P. F., and A. G. Emslie: "The Method of Zones for the Calculation of Temperature Distribution," *NASA* CR-56800, July 1963.

21 Sepetoski, W. K., C. H. Sox, and P. F. Strong: "Description of a Transient Thermal Analysis Program for Use with the Method of Zones," *NASA* CR-56722, August 1963.

22 Schultz, H. D.: "Thermal Analyzer Computer Program for the Solution of General Heat Transfer Problems," *NASA* CR-65581, July 1965.

23 Hahne, E., and U. Grigull: Formfaktor und Formweiderstand der stationaren mehrdimensionalen Warmeleitung, *Int. J. Heat Mass Transfer,* vol. 18, p. 751, 1975.

UNSTEADY-STATE CONDUCTION

■ 4-1 INTRODUCTION

If a solid body is suddenly subjected to a change in environment, some time must elapse before an equilibrium temperature condition will prevail in the body. We refer to the equilibrium condition as the steady state and calculate the temperature distribution and heat transfer by methods described in Chaps. 2 and 3. In the transient heating or cooling process which takes place in the interim period before equilibrium is established, the analysis must be modified to take into account the change in internal energy of the body with time, and the boundary conditions must be adjusted to match the physical situation which is apparent in the unsteady-state heat-transfer problem. Unsteady-state heat-transfer analysis is obviously of significant practical interest because of the large number of heating and cooling processes which must be calculated in industrial applications.

To analyze a transient heat-transfer problem, we could proceed by solving the general heat-conduction equation by the separation-of-variables method, similar to the analytical treatment used for the two-dimensional steady-state problem discussed in Sec. 3-2. We give one illustration of this method of solution for a case of simple geometry and then refer the reader to the references for analysis of more complicated cases. Consider the infinite plate of thickness $2L$ shown in Fig. 4-1. Initially the plate is at a uniform temperature T_i, and at time zero the surfaces are suddenly lowered to $T = T_1$. The differential equation is

$$\frac{\partial^2 T}{\partial x^2} = \frac{1}{\alpha}\frac{\partial T}{\partial \tau} \tag{4-1}$$

The equation may be arranged in a more convenient form by introduction of the variable $\theta = T = T_1$. Then

$$\frac{\partial^2 \theta}{\partial x^2} = \frac{1}{\alpha}\frac{\partial \theta}{\partial \tau} \tag{4-2}$$

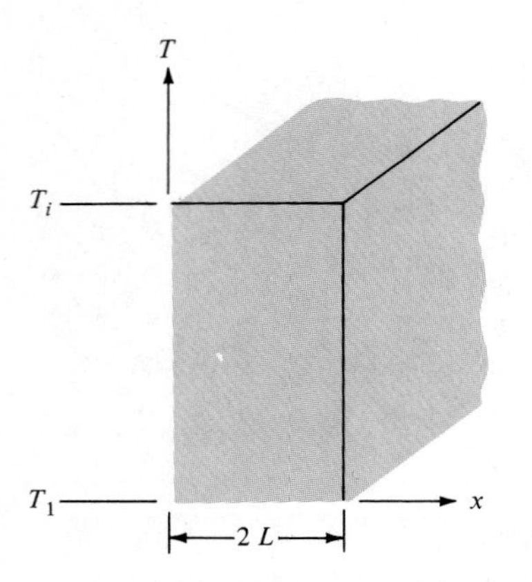

Fig. 4-1 Infinite plate subjected to sudden cooling of surfaces.

with the initial and boundary conditions

$$\theta = \theta_i = T_i - T_1 \qquad \text{at } \tau = 0,\ 0 \le x \le 2L \qquad (a)$$

$$\theta = 0 \qquad \text{at } x = 0,\ \tau > 0 \qquad (b)$$

$$\theta = 0 \qquad \text{at } x = 2L,\ \tau > 0 \qquad (c)$$

Assuming a product solution $\theta(x, \tau) = X(x)\mathcal{H}(\tau)$ produces the two ordinary differential equations

$$\frac{d^2X}{dx^2} + \lambda^2 X = 0$$

$$\frac{d\mathcal{H}}{d\tau} + \alpha\lambda^2\mathcal{H} = 0$$

where λ^2 is the separation constant. In order to satisfy the boundary conditions it is necessary that $\lambda^2 > 0$ so that the form of the solution becomes

$$\theta = (C_1 \cos \lambda x + C_2 \sin \lambda x)e^{-\lambda^2\alpha\tau}$$

From boundary condition (b), $C_1 = 0$ for $\tau > 0$. Because C_2 cannot also be zero, we find from boundary condition (c) that $\sin 2L\lambda = 0$, or

$$\lambda = \frac{n\pi}{2L} \qquad n = 1, 2, 3, \ldots$$

The final series form of the solution is therefore

$$\theta = \sum_{n=1}^{\infty} C_n e^{-[n\pi/2L]^2\alpha\tau} \sin \frac{n\pi x}{2L}$$

This equation may be recognized as a Fourier sine expansion with the constants C_n determined from the initial condition (a) and the following equation:

$$C_n = \frac{1}{L}\int_0^{2L} \theta_i \sin \frac{n\pi x}{2L}\, dx = \frac{4}{n\pi}\theta_i \qquad n = 1, 3, 5, \ldots$$

The final series solution is therefore

$$\frac{\theta}{\theta_i} = \frac{T - T_1}{T_i - T_1} = \frac{4}{\pi} \sum_{n=1}^{\infty} \frac{1}{n} e^{-[n\pi/2L]^2 \alpha\tau} \sin \frac{n\pi x}{2L} \qquad n = 1, 3, 5 \ldots \tag{4-3}$$

In Sec. 4-4, this solution will be presented in graphical form for calculation purposes. For now, our purpose has been to show how the unsteady-heat-conduction equation can be solved, for at least one case, with the separation-of-variables method. Further information on analytical methods in unsteady-state problems is given in the references

■ 4-2 LUMPED-HEAT-CAPACITY SYSTEM

We continue our discussion of transient heat conduction by analyzing systems which may be considered uniform in temperature. This type of analysis is called the *lumped-heat-capacity* method. Such systems are obviously idealized because a temperature gradient must exist in a material if heat is to be conducted into or out of the material. In general, the smaller the physical size of the body, the more realistic the assumption of a uniform temperature throughout; in the limit a differential volume could be employed as in the derivation of the general heat-conduction equation.

If a hot steel ball were immersed in a cool pan of water, the lumped-heat-capacity method of analysis might be used if we could justify an assumption of uniform ball temperature during the cooling process. Clearly, the temperature distribution in the ball would depend on the thermal conductivity of the ball material and the heat-transfer conditions from the surface of the ball to the surrounding fluid, i.e., the surface-convection heat-transfer coefficient. We should obtain a reasonably uniform temperature distribution in the ball if the resistance to heat transfer by conduction were small compared with the convection resistance at the surface, so that the major temperature gradient would occur through the fluid layer at the surface. The lumped-heat-capacity analysis, then, is one which assumes that the internal resistance of the body is negligible in comparison with the external resistance.

The convection heat loss from the body is evidenced as a decrease in the internal energy of the body, as shown in Fig. 4-2. Thus

$$q = hA\,(T - T_\infty) = -c\rho V \frac{dT}{d\tau} \tag{4-4}$$

where A is the surface area for convection and V is the volume. The initial condition is written

$$T = T_0 \qquad \text{at } \tau = 0$$

so that the solution to Eq. (4-4) is

$$\frac{T - T_\infty}{T_0 - T_\infty} = e^{-[hA/\rho c V]\tau} \tag{4-5}$$

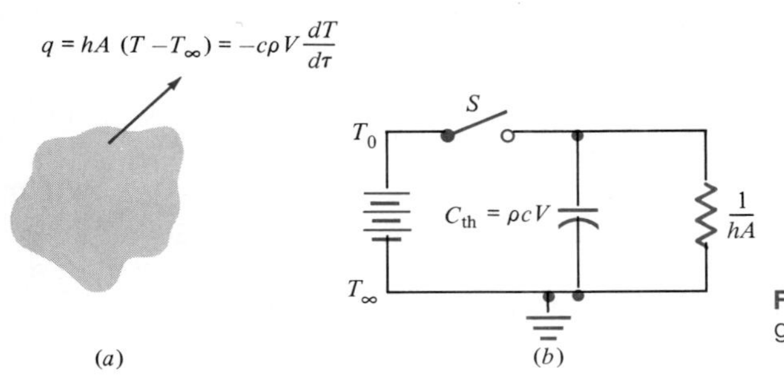

Fig. 4-2 Nomenclature for single-lump heat-capacity analysis.

The thermal network for the single-capacity system is shown in Fig. 4-2*b*. In this network we notice that the thermal capacity of the system is "charged" initially at the potential T_0 by closing the switch S. Then, when the switch is opened, the energy stored in the thermal capacitance is dissipated through the resistance $1/hA$. The analogy between this thermal system and an electric system is apparent, and we could easily construct an electric system which would behave exactly like the thermal system as long as we made the ratio

$$\frac{hA}{\rho c V} = \frac{1}{R_{\text{th}} C_{\text{th}}} \qquad R_{\text{th}} = \frac{1}{hA} \qquad C_{\text{th}} = \rho c V$$

equal to $1/R_e C_e$, where R_e and C_e are the electric resistance and capacitance, respectively. In the thermal system we store energy, while in the electric system we store electric charge. The flow of energy in the thermal system is called heat, and the flow of charge is called electric current. The quantity $c\rho V/hA$ is called the *time constant* of the system because it has the dimensions of time. When

$$\tau = \frac{c\rho V}{hA}$$

it is noted that the temperature difference $T - T_\infty$ has a value of 36.8 percent of the initial difference $T_0 - T_\infty$.

□ Applicability of Lumped-Capacity Analysis

We have already noted that the lumped-capacity type of analysis assumes a uniform temperature distribution throughout the solid body and that the assumption is equivalent to saying that the surface-convection resistance is large compared with the internal-conduction resistance. Such an analysis may be expected to yield reasonable estimates when the following condition is met:

$$\frac{h(V/A)}{k} < 0.1 \tag{4-6}$$

where k is the thermal conductivity of the solid. In sections which follow we examine those situations for which this condition does *not* apply. We shall see that the lumped-capacity analysis has a direct relationship to the numerical

Table 4-1 Examples of Lumped-Capacity Systems

Physical situation	k, W/m · °C	*Approximate value of h,* W/m² · °C	$\frac{h(V/A)}{k}$
1. 3.0-cm steel cube cooling in room air	40	7.0	8.75×10^{-4}
2. 5.0-cm-glass cylinder cooled by a 50-m/s airstream	0.8	180	2.81
3. Same as situation 2 but a copper cylinder	380	180	0.006
4. 3.0-cm hot copper cube submerged in water such that boiling occurs	380	10,000	0.132

methods discussed in Sec. 4-7. If one considers the ratio $V/A = s$ as a characteristic dimension of the solid, the dimensionless group is called the *Biot number:*

$$\frac{hs}{k} = \text{Biot number} = \text{Bi}$$

The reader should recognize that there are many practical cases where the lumped-capacity method may yield good results. In Table 4-1 we give some examples which illustrate the relative validity of such cases.

Do not dismiss lumped-capacity analysis because of its simplicity. In many cases one will not know the convection coefficient better than ±25 percent, so it is not necessary to use more elaborate analysis techniques.

■ EXAMPLE 4-1

A steel ball [$c = 0.46$ kJ/kg · °C, $k = 35$ W/m · °C] 5.0 cm in diameter and initially at a uniform temperature of 450°C is suddenly placed in a controlled environment in which the temperature is maintained at 100°C. The convection heat-transfer coefficient is 10 W/m² · °C. Calculate the time required for the ball to attain a temperature of 150°C.

Solution

We anticipate that the lumped-capacity method will apply because of the low value of h and high value of k. We can check by using Eq. (4-6):

$$\frac{h(V/A)}{k} = \frac{(10)[(4/3)\pi(0.025)^3]}{4\pi(0.025)^2(35)} = 0.0023 < 0.1$$

so we may use Eq. (4-5). We have

$$T = 150^\circ\text{C} \qquad \rho = 7800 \text{ kg/m}^3 \quad [486 \text{ lb}_m/\text{ft}^3]$$

$$T_\infty = 100^\circ\text{C} \qquad h = 10 \text{ W/m}^2 \cdot {}^\circ\text{C} \quad [1.76 \text{ Btu/h} \cdot \text{ft}^2 \cdot {}^\circ\text{F}]$$

$$T_0 = 450^\circ\text{C} \qquad c = 460 \text{ J/kg} \cdot {}^\circ\text{C} \quad [0.11 \text{ Btu/lb}_m \cdot {}^\circ\text{F}]$$

$$\frac{hA}{\rho c V} = \frac{(10)4\pi(0.025)^2}{(7800)(460)(4\pi/3)(0.025)^3} = 3.344 \times 10^{-4}\ \text{s}^{-1}$$

$$\frac{T - T_\infty}{T_0 - T_\infty} = e^{-[hA/\rho c V]\tau}$$

$$\frac{150 - 100}{450 - 100} = e^{-3.344 \times 10^{-4}\tau}$$

$$\tau = 5819\ \text{s} = 1.62\ \text{h}$$

4-3 TRANSIENT HEAT FLOW IN A SEMI-INFINITE SOLID

Consider the semi-infinite solid shown in Fig. 4-3 maintained at some initial temperature T_i. The surface temperature is suddenly lowered and maintained at a temperature T_0, and we seek an expression for the temperature distribution in the solid as a function of time. This temperature distribution may subsequently be used to calculate heat flow at any x position in the solid as a function of time. For constant properties, the differential equation for the temperature distribution $T(x, \tau)$ is

$$\frac{\partial^2 T}{\partial x^2} = \frac{1}{\alpha}\frac{\partial T}{\partial \tau} \tag{4-7}$$

The boundary and initial conditions are

$$T(x, 0) = T_i$$

$$T(0, \tau) = T_0 \qquad \text{for } \tau > 0$$

This is a problem which may be solved by the Laplace-transform technique. The solution is given in Ref. 1 as

$$\frac{T(x, \tau) - T_0}{T_i - T_0} = \text{erf}\,\frac{x}{2\sqrt{\alpha\tau}} \tag{4-8}$$

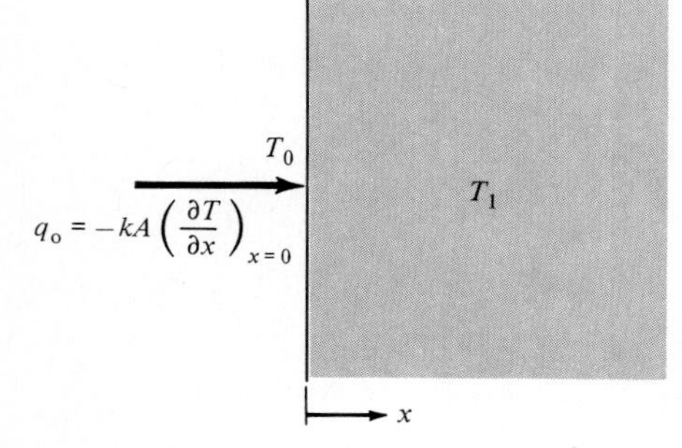

Fig. 4-3 Nomenclature for transient heat flow in a semi-infinite solid.

where the Gauss error function is defined as

$$\operatorname{erf} \frac{x}{2\sqrt{\alpha\tau}} = \frac{2}{\sqrt{\pi}} \int_0^{x/2\sqrt{\alpha\tau}} e^{-\eta^2}\, d\eta \tag{4-9}$$

It will be noted that in this definition η is a dummy variable and the integral is a function of its upper limit. When the definition of the error function is inserted in Eq. (4-8), the expression for the temperature distribution becomes

$$\frac{T(x,\tau) - T_0}{T_i - T_0} = \frac{2}{\sqrt{\pi}} \int_0^{x/2\sqrt{\alpha\tau}} e^{-\eta^2}\, d\eta \tag{4-10}$$

The heat flow at any x position may be obtained from

$$q_x = -kA \frac{\partial T}{\partial x}$$

Performing the partial differentiation of Eq. (4-10) gives

$$\begin{aligned} \frac{\partial T}{\partial x} &= (T_i - T_0) \frac{2}{\sqrt{\pi}} e^{-x^2/4\alpha\tau} \frac{\partial}{\partial x}\left(\frac{x}{2\sqrt{\alpha\tau}}\right) \\ &= \frac{T_i - T_0}{\sqrt{\pi\alpha\tau}} e^{-x^2/4\alpha\tau} \end{aligned} \tag{4-11}$$

At the surface the heat flow is

$$q_0 = \frac{kA(T_0 - T_i)}{\sqrt{\pi\alpha\tau}} \tag{4-12}$$

The surface heat flux is determined by evaluating the temperature gradient at $x = 0$ from Eq. (4-11). A plot of the temperature distribution for the semi-infinite solid is given in Fig. 4-4. Values of the error function are tabulated in Ref. 3, and an abbreviated tabulation is given in Appendix A.

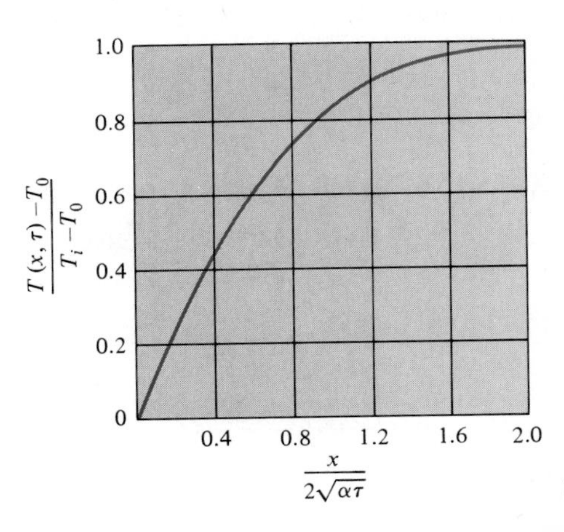

Fig. 4-4 Temperature distribution in the semi-infinite solid.

Constant Heat Flux on Semi-Infinite Solid

For the same uniform initial temperature distribution, we could suddenly expose the surface to a constant surface heat flux q_0/A. The initial and boundary conditions on Eq. (4-7) would then become

$$T(x, 0) = T_i$$

$$\frac{q_0}{A} = -k \frac{\partial T}{\partial x}\bigg]_{x=0} \qquad \text{for } \tau > 0$$

The solution for this case is

$$T - T_i = \frac{2q_0\sqrt{\alpha\tau/\pi}}{kA} \exp\left(\frac{-x^2}{4\alpha\tau}\right) - \frac{q_0 x}{kA}\left(1 - \text{erf}\frac{x}{2\sqrt{\alpha\tau}}\right) \tag{4-13}$$

EXAMPLE 4-2

A large block of steel [$k = 45$ W/m · °C, $\alpha = 1.4 \times 10^{-5}$ m²/s] is initially at a uniform temperature of 35°C. The surface is exposed to a heat flux (*a*) by suddenly raising the surface temperature to 250°C and (*b*) through a constant surface heat flux of 3.2×10^5 W/m². Calculate the temperature at a depth of 2.5 cm after a time of 0.5 min for both these cases.

Solution

We can make use of the solutions for the semi-infinite solid given as Eqs. (4-8) and (4-13). For case *a*,

$$\frac{x}{2\sqrt{\alpha\tau}} = \frac{0.025}{(2)[(1.4 \times 10^{-5})(30)]^{1/2}} = 0.61$$

The error function is determined from Appendix A as

$$\text{erf}\frac{x}{2\sqrt{\alpha\tau}} = \text{erf } 0.61 = 0.61164$$

We have $T_i = 35$°C and $T_0 = 250$°C, so the temperature at $x = 2.5$ cm is determined from Eq. (4-8) as

$$T(x, \tau) = T_0 + (T_i - T_0)\,\text{erf}\frac{x}{2\sqrt{\alpha\tau}}$$

$$= 250 + (35 - 250)(0.61164) = 118.5°\text{C}$$

For the constant-heat-flux case *b*, we make use of Eq. (4-13). Since q_0/A is given as 3.2×10^5 W/m², we can insert the numerical values to give

$$T(x, \tau) = 35 + \frac{(2)(3.2 \times 10^5)[(1.4 \times 10^{-5})(30)/\pi]^{1/2}}{45} e^{-(0.61)^2} - \frac{(0.025)(3.2 \times 10^5)}{45}(1 - 0.61164)$$

$$= 79.3°\text{C} \qquad x = 2.5 \text{ cm}, \ \tau = 30 \text{ s}$$

For the constant-heat-flux case the *surface* temperature after 30 s would be evaluated with $x = 0$ in Eq. (4-13). Thus

$$T(x = 0) = 35 + \frac{(2)(3.2 \times 10^5)[(1.4 \times 10^{-5})(30)/\pi]^{1/2}}{45} = 199.4°\text{C}$$

EXAMPLE 4-3

A large slab of aluminum at a uniform temperature of 200°C suddenly has its surface temperature lowered to 70°C. What is the total heat removed from the slab per unit surface area when the temperature at a depth 4.0 cm has dropped to 120°C?

Solution

We first find the time required to attain the 120°C temperature and then integrate Eq. (4-12) to find the total heat removed during this time interval. For aluminum,

$$\alpha = 8.4 \times 10^{-5} \text{ m}^2/\text{s} \qquad k = 215 \text{ W/m} \cdot °\text{C} \quad [124 \text{ Btu/h} \cdot \text{ft} \cdot °\text{F}]$$

We also have

$$T_i = 200°\text{C} \qquad T_0 = 70°\text{C} \qquad T(x, \tau) = 120°\text{C}$$

Using Eq. (4-8) gives

$$\frac{120 - 70}{200 - 70} = \text{erf}\,\frac{x}{2\sqrt{\alpha\tau}} = 0.3847$$

From Fig. 4-4 or Appendix A,

$$\frac{x}{2\sqrt{\alpha\tau}} = 0.3553$$

and

$$\tau = \frac{(0.04)^2}{(4)(0.3553)^2(8.4 \times 10^{-5})} = 37.72 \text{ s}$$

The total heat removed at the surface is obtained by integrating Eq. (4-12):

$$\frac{Q_0}{A} = \int_0^\tau \frac{q_0}{A}\,d\tau = \int_0^\tau \frac{k(T_0 - T_i)}{\sqrt{\pi\alpha\tau}}\,d\tau = 2k(T_0 - T_i)\sqrt{\frac{\tau}{\pi\alpha}}$$

$$= (2)(215)(70 - 200)\left[\frac{37.72}{\pi(8.4 \times 10^{-5})}\right]^{1/2} = -21.13 \times 10^6 \text{ J/m}^2 \quad [-1861 \text{ Btu/ft}^2]$$

4-4 CONVECTION BOUNDARY CONDITIONS

In most practical situations the transient heat-conduction problem is connected with a convection boundary condition at the surface of the solid. Naturally, the boundary conditions for the differential equation must be modified to take into account this convection heat transfer at the surface. For the semi-infinite-solid problem above, this would be expressed by

Heat convected into surface = heat conducted into surface

or

$$hA(T_\infty - T)_{x=0} = -kA\left.\frac{\partial T}{\partial x}\right]_{x=0} \tag{4-14}$$

The solution for this problem is rather involved, and is worked out in detail by Schneider [1]. The result is

$$\frac{T - T_i}{T_\infty - T_i} = 1 - \operatorname{erf} X - \left[\exp\left(\frac{hx}{k} + \frac{h^2\alpha\tau}{k^2}\right)\right] \times \left[1 - \operatorname{erf}\left(X + \frac{h\sqrt{\alpha\tau}}{k}\right)\right] \tag{4-15}$$

where $X = x/(2\sqrt{\alpha\tau})$

T_i = initial temperature of solid

T_∞ = environment temperature

This solution is presented in graphical form in Fig. 4-5.

Solutions have been worked out for other geometries. The most important

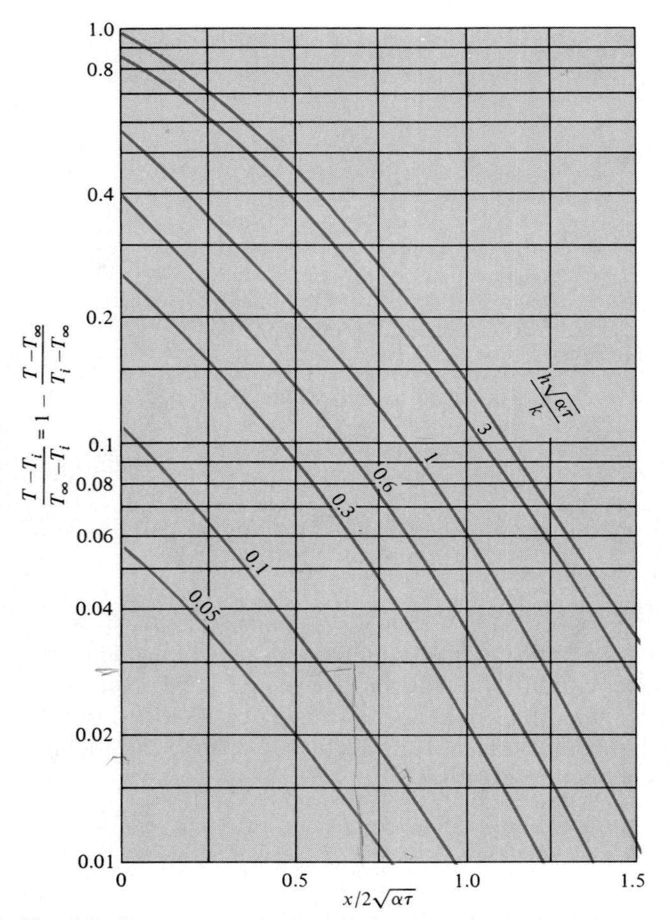

Fig. 4-5 Temperature distribution in the semi-infinite solid with convection boundary condition.

cases are those dealing with (1) plates whose thickness is small in relation to the other dimensions, (2) cylinders where the diameter is small compared to the length, and (3) spheres. Results of analyses for these geometries have been presented in graphical form by Heisler [2], and nomenclature for the three cases is illustrated in Fig. 4-6. In all cases the convection environment temperature is designated as T_∞ and the center temperature for $x = 0$ or $r = 0$ is T_0. At time zero, each solid is assumed to have a uniform initial temperature T_i. Temperatures in the solids are given in Figs. 4-7 to 4-13 as functions of time and spatial position. In these charts we note the definitions

$$\theta = T(x, \tau) - T_\infty \qquad \text{or} \qquad T(r, \tau) - T_\infty$$

$$\theta_i = T_i - T_\infty$$

$$\theta_0 = T_0 - T_\infty$$

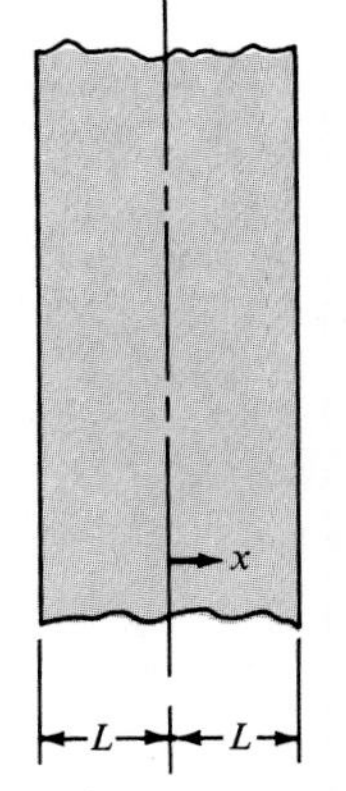

T_0 = centerline temperature

(a)

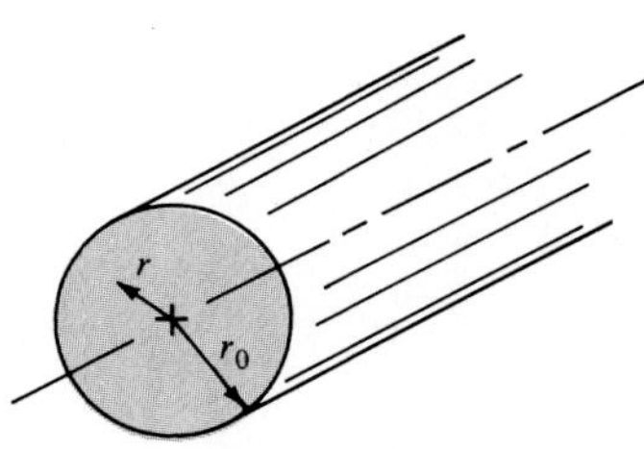

T_0 = centerline temperature

(b)

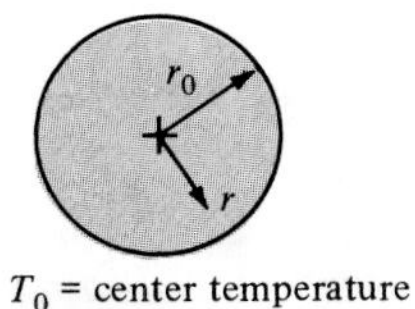

T_0 = center temperature

(c)

Fig. 4-6 Nomenclature for one-dimensional solids suddenly subjected to convection environment at T_∞: (a) infinite plate of thickness $2L$; (b) infinite cylinder of radius r_0; (c) sphere of radius r_0.

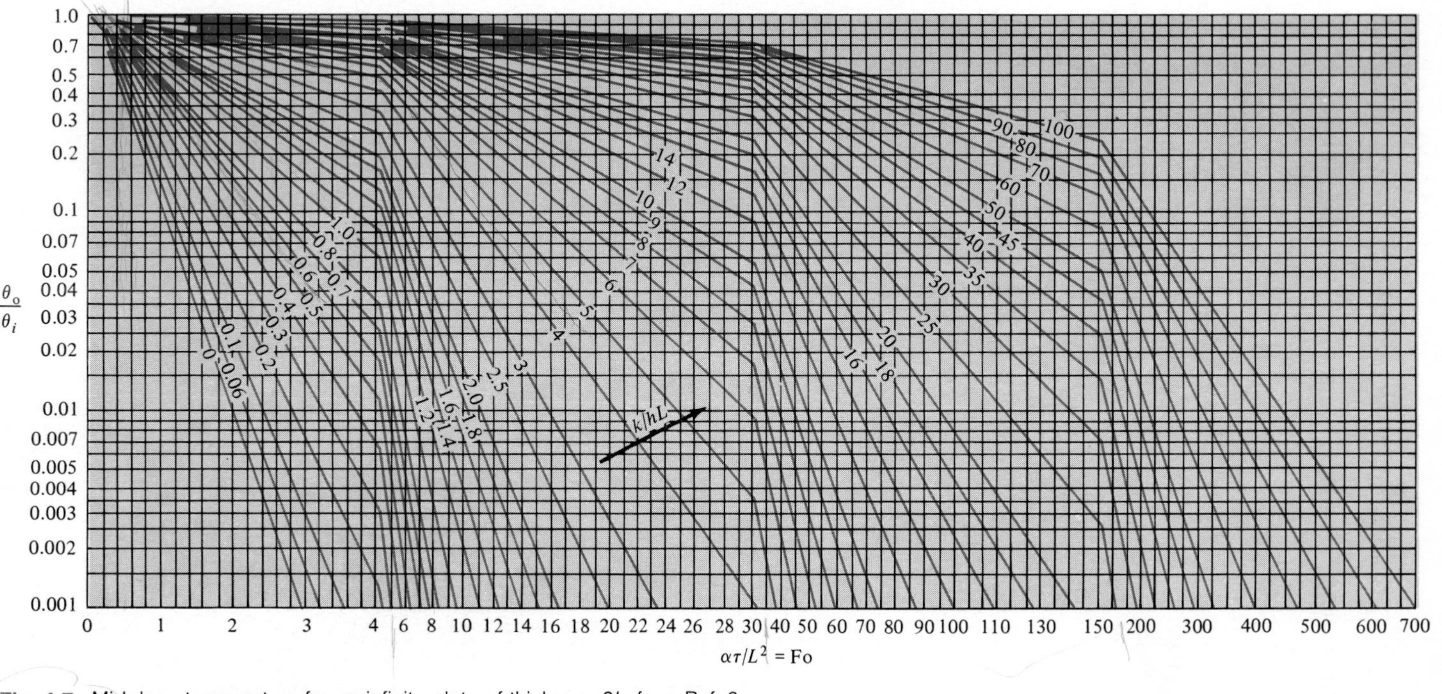

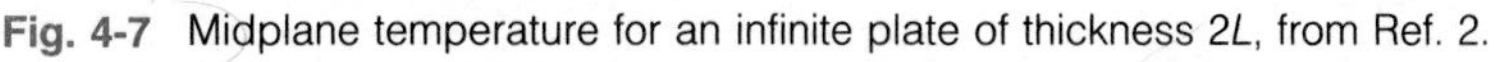

Fig. 4-7 Midplane temperature for an infinite plate of thickness $2L$, from Ref. 2.

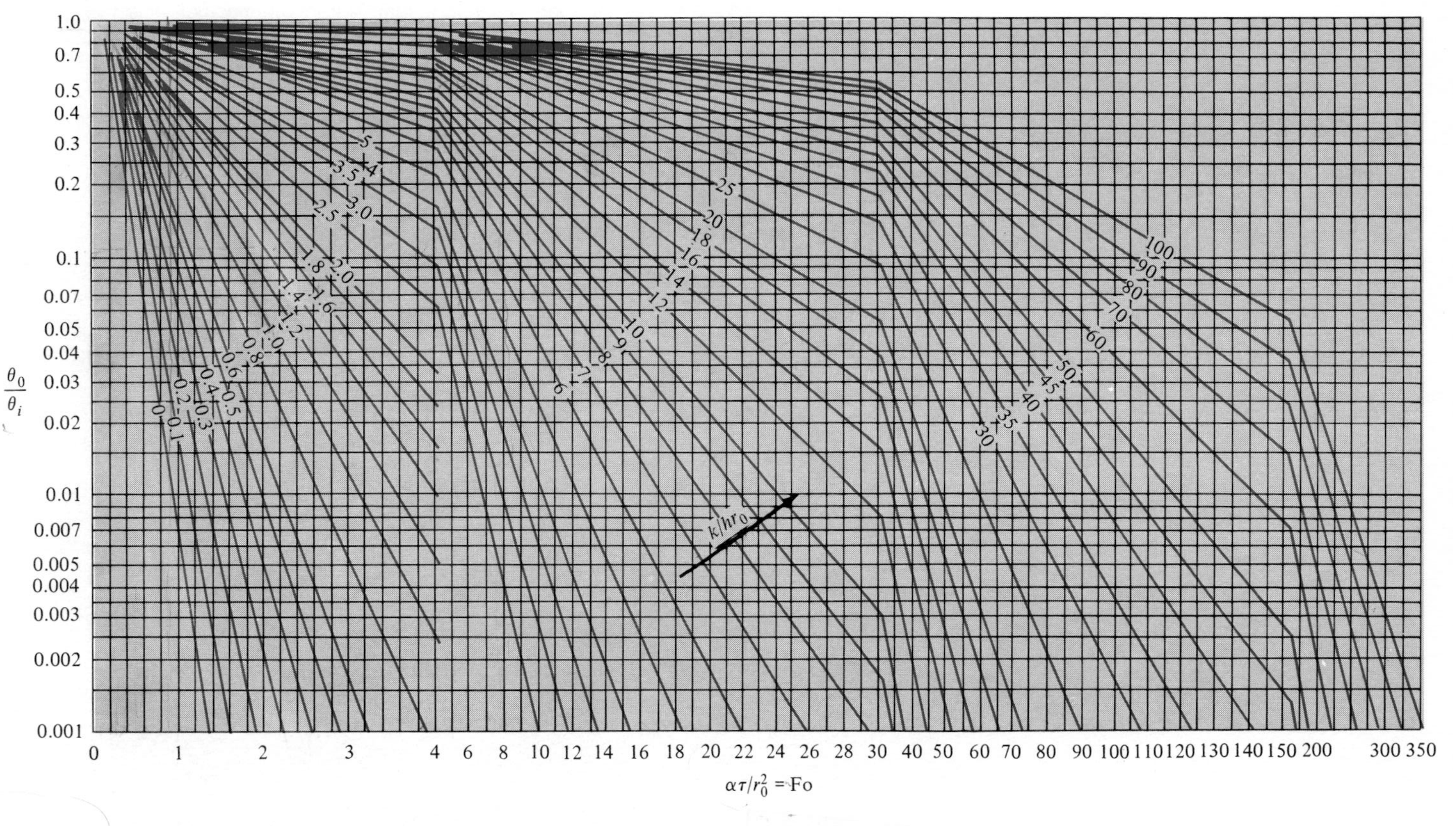

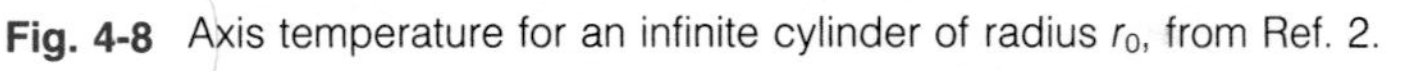

Fig. 4-8 Axis temperature for an infinite cylinder of radius r_0, from Ref. 2.

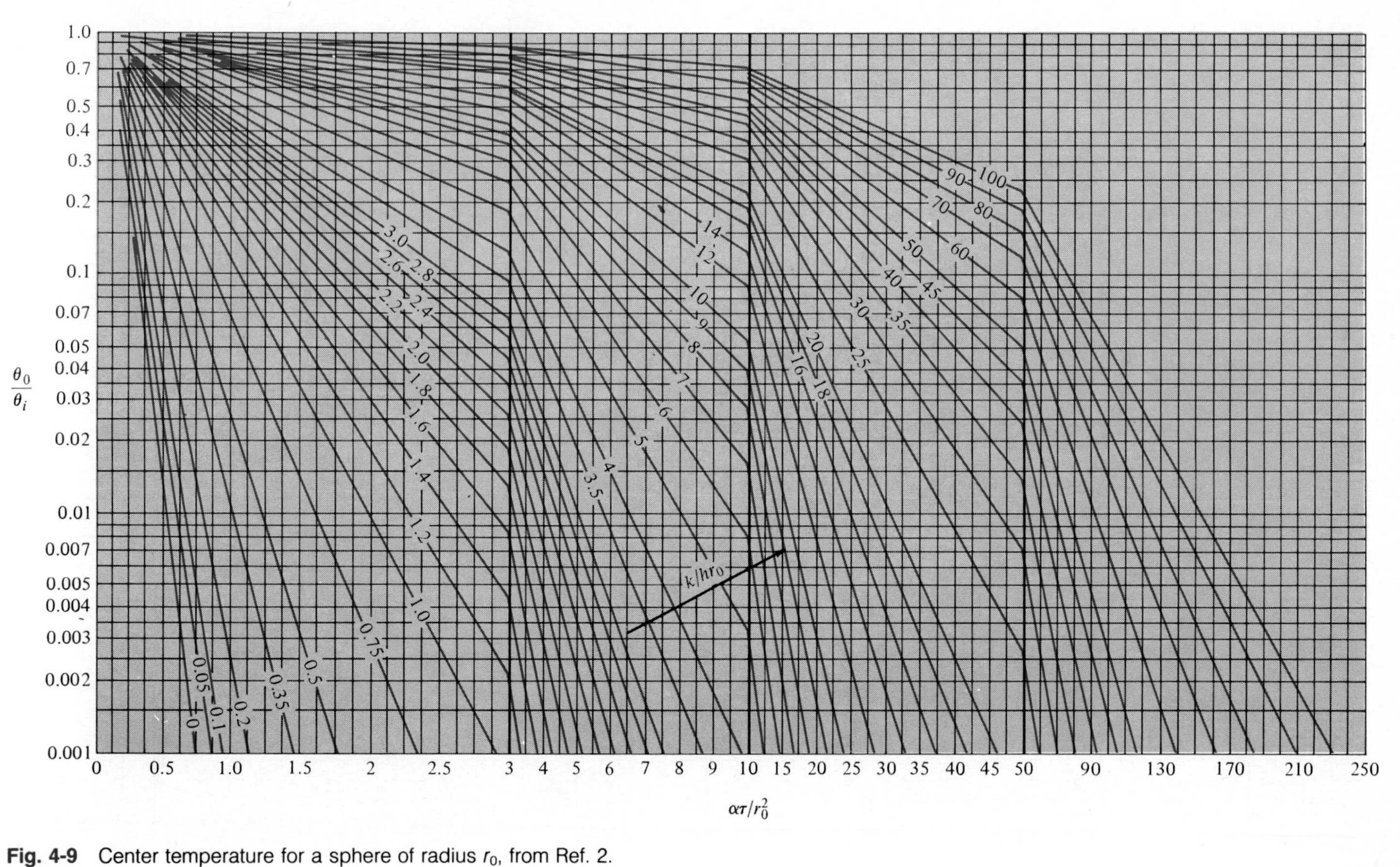

Fig. 4-9 Center temperature for a sphere of radius r_0, from Ref. 2.

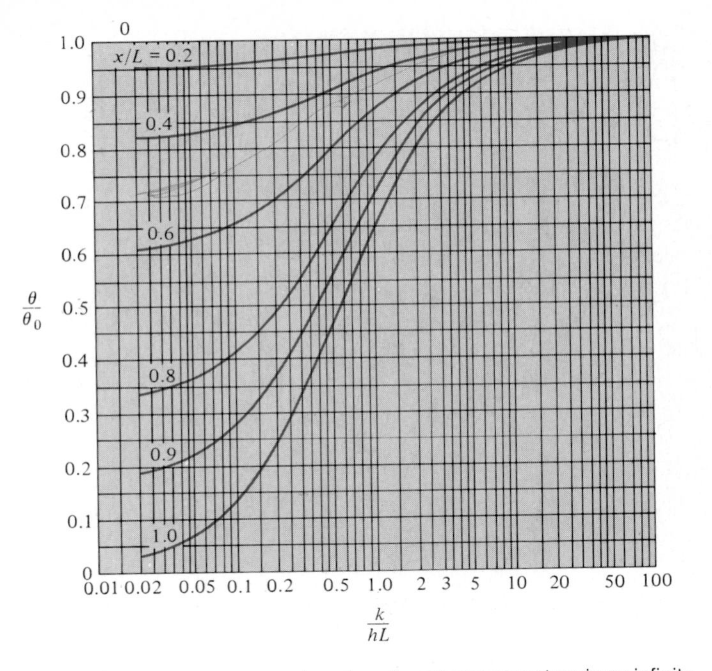

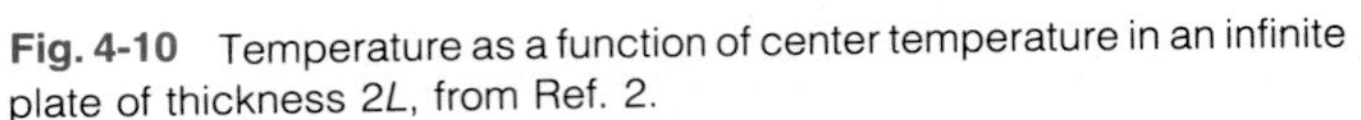

Fig. 4-10 Temperature as a function of center temperature in an infinite plate of thickness $2L$, from Ref. 2.

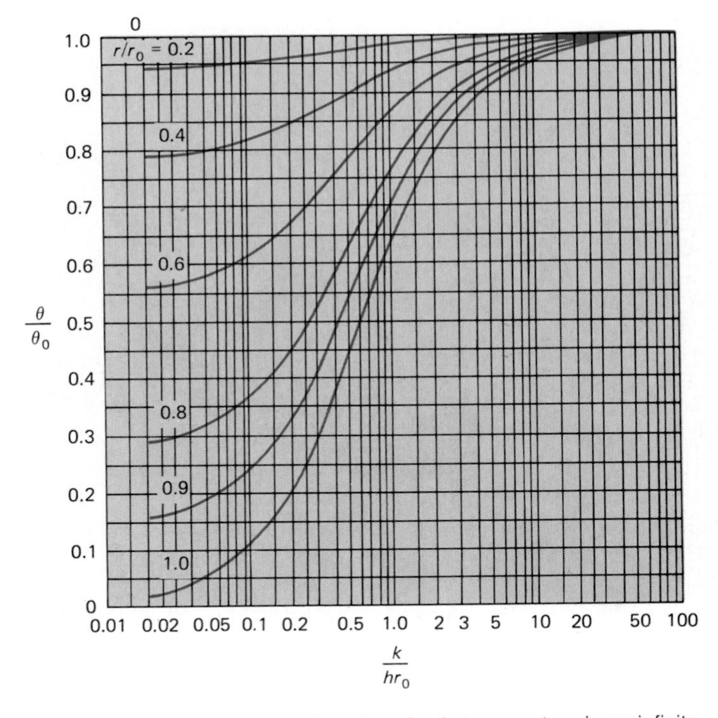

Fig. 4-11 Temperature as a function of axis temperature in an infinite cylinder of radius r_0, from Ref. 2.

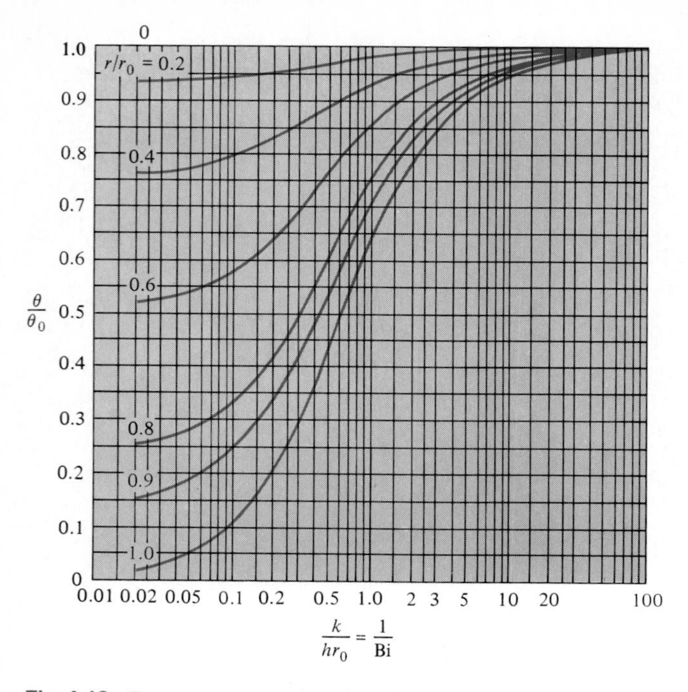

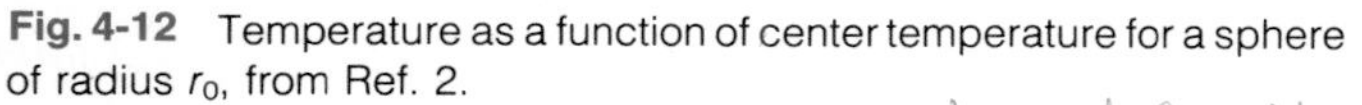

Fig. 4-12 Temperature as a function of center temperature for a sphere of radius r_0, from Ref. 2.

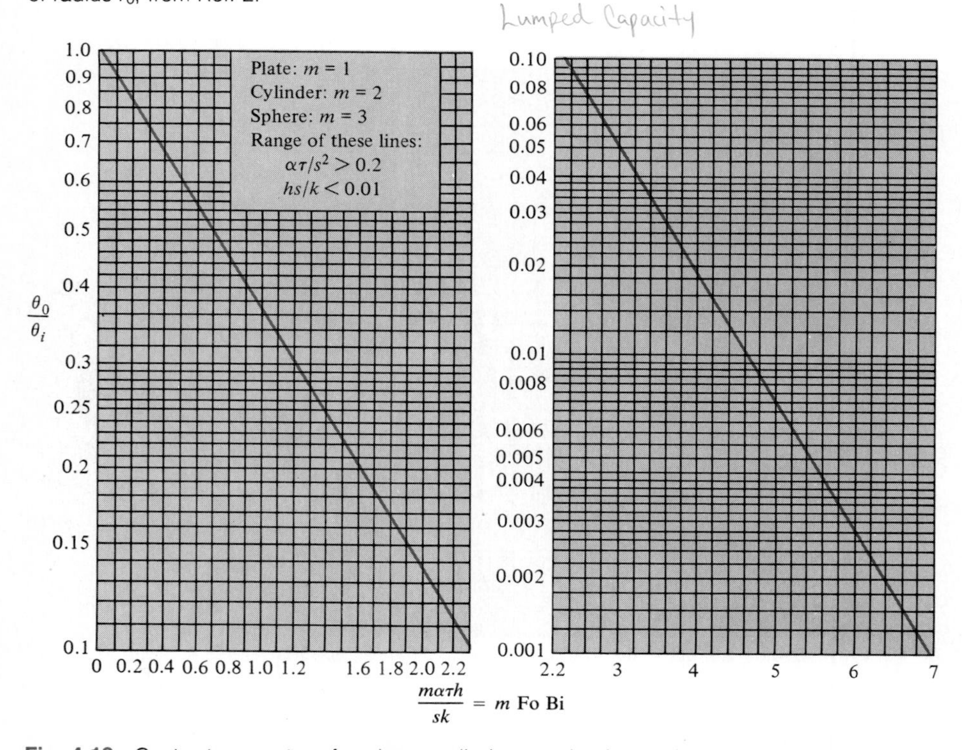

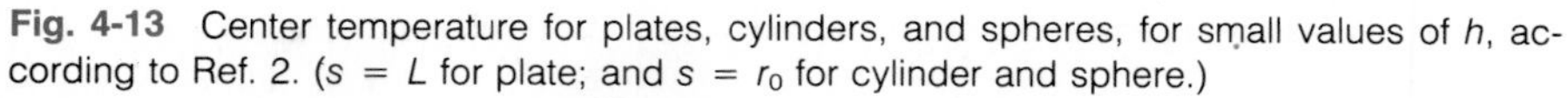

Fig. 4-13 Center temperature for plates, cylinders, and spheres, for small values of h, according to Ref. 2. ($s = L$ for plate; and $s = r_0$ for cylinder and sphere.)

$Fo = \frac{\alpha\tau}{s^2} = \frac{k\tau}{\rho c s^2}$

If a centerline temperature is desired, only one chart is required to obtain a value for θ_0 and then T_0. To determine an off-center temperature two charts are required to calculate the product

$$\frac{\theta}{\theta_i} = \frac{\theta_0}{\theta_i}\frac{\theta}{\theta_0}$$

For example, Figs. 4-7 and 4-10 would be employed to calculate an off-center temperature for an infinite plate.

The heat losses for the infinite plate, infinite cylinder, and sphere are given in Figs. 4-14 to 4-16, where Q_0 represents the initial internal energy content of the body in reference to the environment temperature

$$Q_0 = \rho c V (T_i - T_\infty) = \rho c V \theta_i \tag{4-16}$$

In these figures Q is the actual heat lost by the body in time τ.

Obviously, there are many other practical heating and cooling problems of interest. The solutions for a large number of cases are presented in graphical form by Schneider [7], and readers interested in such calculations will find this reference to be of great utility.

Figure 4-13 gives the center temperatures of the three types of solids for small values of h, or for conditions where the solids behave as a lumped capacity. In this figure the characteristic dimensions s is L for the plate and r_0 for the cylinder and sphere. Simplified charts are available in Ref. [17].

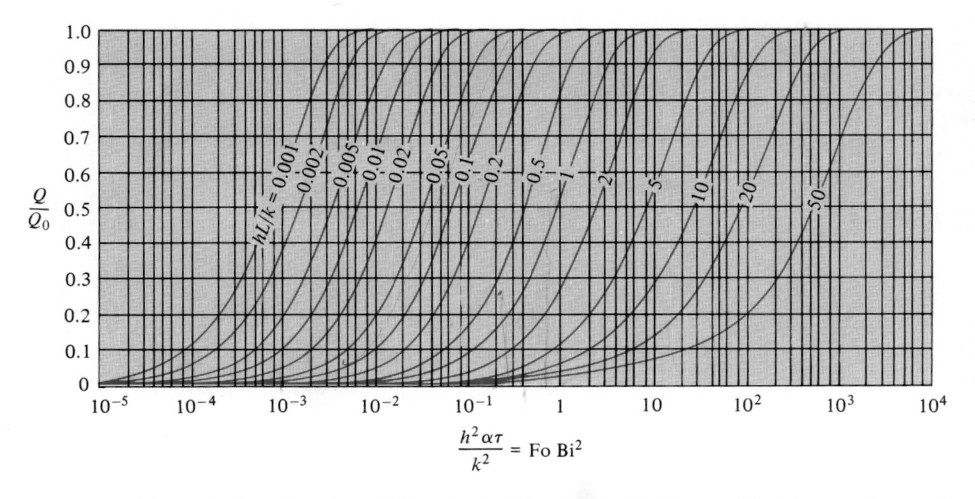

Fig. 4-14 Dimensionless heat loss Q/Q_0 of an infinite plane of thickness $2L$ with time, from Ref. 6.

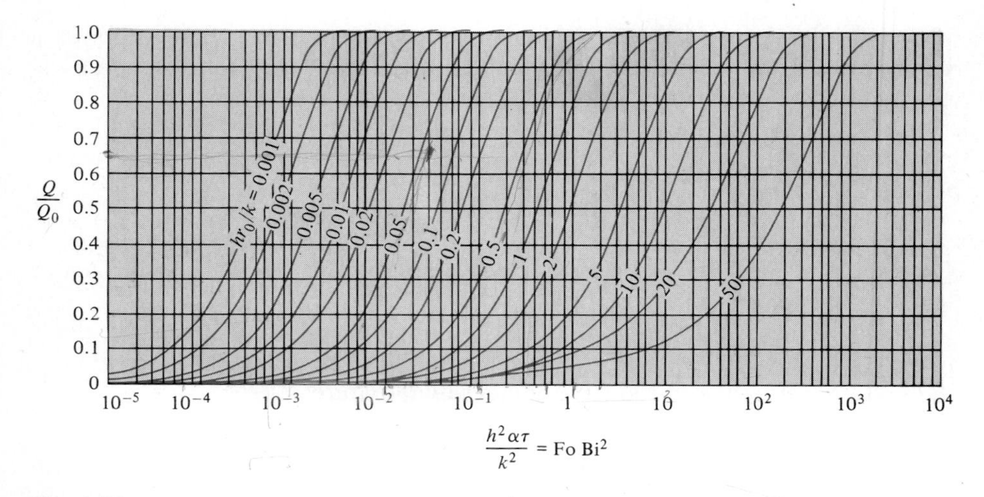

Fig. 4-15 Dimensionless heat loss Q/Q_0 of an infinite cylinder of radius r_0 with time, from Ref. 6.

The Biot and Fourier Numbers

A quick inspection of Figs. 4-5 to 4-16 indicates that the dimensionless temperature profiles and heat flows may all be expressed in terms of two dimensionless parameters called the Biot and Fourier numbers:

$$\text{Biot number} = \text{Bi} = \frac{hs}{k}$$

$$\text{Fourier number} = \text{Fo} = \frac{\alpha\tau}{s^2} = \frac{k\tau}{\rho c s^2}$$

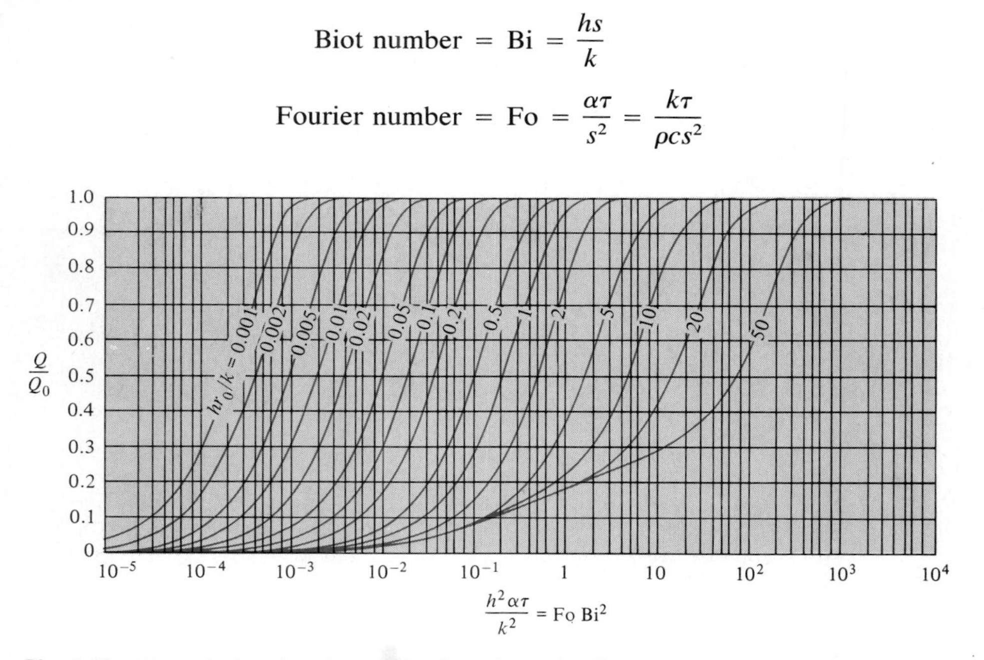

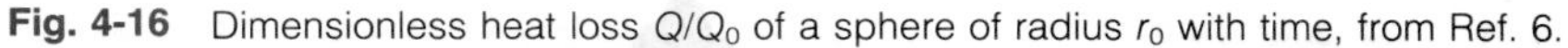

Fig. 4-16 Dimensionless heat loss Q/Q_0 of a sphere of radius r_0 with time, from Ref. 6.

In these parameters s designates some characteristic dimensions of the body; for the plate it is the half-thickness, whereas for the cylinders and sphere it is the radius. The Biot number compares the relative magnitudes of surface-convection and internal-conduction resistances to heat transfer. The Fourier modulus compares a characteristic body dimension with an approximate temperature-wave penetration depth for a given time τ.

A very low value of the Biot modulus means that internal-conduction resistance is negligible in comparison with surface-convection resistance. This in turn implies that the temperature will be nearly uniform throughout the solid, and its behavior may be approximated by the lumped-capacity method of analysis. It is interesting to note that the exponent of Eq. (4-5) may be expressed in terms of the Biot and Fourier numbers if one takes the ratio V/A as the characteristic dimension s. Then,

$$\frac{hA}{\rho c V}\tau = \frac{h\tau}{\rho c s} = \frac{hs}{k}\frac{k\tau}{\rho c s^2} = \text{Bi Fo}$$

Applicability of the Heisler Charts

The calculations for the Heisler charts were performed by truncating the infinite series solutions for the problems into a few terms. This restricts the applicability of the charts to values of the Fourier number greater than 0.2.

$$\text{Fo} = \frac{\alpha\tau}{s^2} > 0.2$$

For smaller values of this parameter the reader should consult the solutions and charts given in the references at the end of the chapter. Calculations using the truncated series solutions directly are discussed in Appendix C.

EXAMPLE 4-4

The slab of Example 4-3 is suddenly exposed to a convection-surface environment of 70°C with a heat-transfer coefficient of 525 W/m² · °C. Calculate the time required for the temperature to reach 120°C at the depth of 4.0 cm for this circumstance.

Solution

We may use either Eq. (4-15) or Fig. 4-5 for solution of this problem, but Fig. 4-5 is easier to apply because the time appears in two terms. Even when the figure is used, an iterative procedure is required because the time appears in both of the variables $h\sqrt{\alpha\tau}/k$ and $x/(2\sqrt{\alpha\tau})$. We seek the value of τ such that

$$\frac{T - T_i}{T_\infty - T_i} = \frac{120 - 200}{70 - 200} = 0.615 \qquad (a)$$

We therefore try values of τ and obtain readings of the temperature ratio from Fig. 4-5 until agreement with Eq. (*a*) is reached. The iterations are listed below. Values of k and α are obtained from Example 4-3.

τ, s	$\dfrac{h\sqrt{\alpha\tau}}{k}$	$\dfrac{x}{2\sqrt{\alpha\tau}}$	$\dfrac{T - T_i}{T_\infty - T_i}$ *from Fig. 4-5*
1000	0.708	0.069	0.41
3000	1.226	0.040	0.61
4000	1.416	0.035	0.68

Consequently, the time required is approximately 3000 s.

EXAMPLE 4-5

A large plate of aluminum 5.0 cm thick and initially at 200°C is suddenly exposed to the convection environment of Example 4-4. Calculate the temperature at a depth of 1.25 cm from one of the faces 1 min after the plate has been exposed to the environment. How much energy has been removed per unit area from the plate in this time?

Solution

The Heisler charts of Figs. 4-7 and 4-10 may be used for solution of this problem. We first calculate the center temperature of the plate, using Fig. 4-7, and then use Fig. 4-10 to calculate the temperature at the specified x position. From the conditions of the problem we have

$$\theta_i = T_i - T_\infty = 200 - 70 = 130 \qquad \alpha = 8.4 \times 10^{-5}\ \text{m}^2/\text{s} \quad [3.26\ \text{ft}^2/\text{h}]$$

$$2L = 5.0\ \text{cm} \qquad L = 2.5\ \text{cm} \qquad \tau = 1\ \text{min} = 60\ \text{s}$$

$$k = 215\ \text{W/m} \cdot {}^\circ\text{C} \quad [124\ \text{Btu/h} \cdot \text{ft} \cdot {}^\circ\text{F}]$$

$$h = 525\ \text{W/m}^2 \cdot {}^\circ\text{C} \quad [92.5\ \text{Btu/h} \cdot \text{ft}^2 \cdot {}^\circ\text{F}]$$

$$x = 2.5 - 1.25 = 1.25\ \text{cm}$$

Then

$$\frac{\alpha\tau}{L^2} = \frac{(8.4 \times 10^{-5})(60)}{(0.025)^2} = 8.064 \qquad \frac{k}{hL} = \frac{215}{(525)(0.025)} = 16.38$$

$$\frac{x}{L} = \frac{1.25}{2.5} = 0.5$$

From Fig. 4-7

$$\frac{\theta_0}{\theta_i} = 0.61$$

$$\theta_0 = T_0 - T_\infty = (0.61)(130) = 79.3$$

From Fig. 4-10 at $x/L = 0.5$,

$$\frac{\theta}{\theta_0} = 0.98$$

and

$$\theta = T - T_\infty = (0.98)(79.3) = 77.7$$
$$T = 77.7 + 70 = 147.7°\text{C}$$

We compute the energy lost by the slab by using Fig. 4-14. For this calculation we require the following properties of aluminum:

$$\rho = 2700 \text{ kg/m}^3 \qquad c = 0.9 \text{ kJ/kg} \cdot °\text{C}$$

For Fig. 4-14 we need

$$\frac{h^2\alpha\tau}{k^2} = \frac{(525)^2(8.4 \times 10^{-5})(60)}{(215)^2} = 0.03 \qquad \frac{hL}{k} = \frac{(525)(0.025)}{215} = 0.061$$

From Fig. 4-14

$$\frac{Q}{Q_0} = 0.41$$

For unit area

$$\frac{Q_0}{A} = \frac{\rho c V \theta_i}{A} = \rho c (2L)\theta_i$$
$$= (2700)(900)(0.05)(130)$$
$$= 15.8 \times 10^6 \text{ J/m}^2$$

so that the heat removed per unit surface area is

$$\frac{Q}{A} = (15.8 \times 10^6)(0.41) = 6.48 \times 10^6 \text{ J/m}^2 \quad [571 \text{ Btu/ft}^2]$$

EXAMPLE 4-6

A long aluminum cylinder 5.0 cm in diameter and initially at 200°C is suddenly exposed to a convection environment at 70°C and $h = 525$ W/m² · °C. Calculate the temperature at a radius of 1.25 cm and the heat lost per unit length 1 min after the cylinder is exposed to the environment.

Solution

This problem is like Example 4-5 except that Figs. 4-8 and 4-11 are employed for the solution. We have

$$\theta_i = T_i - T_\infty = 200 - 70 = 130 \qquad \alpha = 8.4 \times 10^5 \text{ m}^2/\text{s}$$
$$r_0 = 2.5 \text{ cm} \qquad \tau = 1 \text{ min} = 60 \text{ s}$$
$$k = 215 \text{ W/m} \cdot °\text{C} \qquad h = 525 \text{ W/m}^2 \cdot °\text{C} \qquad r = 1.25 \text{ cm}$$
$$\rho = 2700 \text{ kg/m}^3 \qquad c = 0.9 \text{ kJ/kg} \cdot °\text{C}$$

We compute

$$\frac{\alpha\tau}{r_0^2} = \frac{(8.4 \times 10^{-5})(60)}{(0.025)^2} = 8.064 \qquad \frac{k}{hr_0} = \frac{215}{(525)(0.025)} = 16.38$$

$$\frac{r}{r_0} = \frac{1.25}{2.5} = 0.5$$

From Fig. 4-8

$$\frac{\theta_0}{\theta_i} = 0.38$$

and from Fig. 4-11 at $r/r_0 = 0.5$

$$\frac{\theta}{\theta_0} = 0.98$$

so that

$$\frac{\theta}{\theta_i} = \frac{\theta_0}{\theta_i}\frac{\theta}{\theta_0} = (0.38)(0.98) = 0.372$$

and

$$\theta = T - T_\infty = (0.372)(130) = 48.4$$

$$T = 70 + 48.4 = 118.4°\text{C}$$

To compute the heat lost, we determine

$$\frac{h^2\alpha\tau}{k^2} = \frac{(525)^2(8.4 \times 10^{-5})(60)}{(215)^2} = 0.03 \qquad \frac{hr_0}{k} = \frac{(525)(0.025)}{215} = 0.061$$

Then from Fig. 4-15

$$\frac{Q}{Q_0} = 0.65$$

For unit length

$$\frac{Q_0}{L} = \frac{\rho c V \theta_i}{L} = \rho c \pi r_0^2 \theta_i = (2700)(900)\pi(0.025)^2(130) = 6.203 \times 10^5 \text{ J/m}$$

and the actual heat lost per unit length is

$$\frac{Q}{L} = (6.203 \times 10^5)(0.65) = 4.032 \times 10^5 \text{ J/m} \quad [116.5 \text{ Btu/ft}]$$

4-5 MULTIDIMENSIONAL SYSTEMS

The Heisler charts discussed above may be used to obtain the temperature distribution in the infinite plate of thickness $2L$, in the long cylinder, or in the sphere. When a wall whose height and depth dimensions are not large compared with the thickness or a cylinder whose length is not large compared with its diameter is encountered, additional space coordinates are necessary to specify the temperature, the above charts no longer apply, and we are forced to seek another method of solution. Fortunately, it is possible to combine the solutions for the one-dimensional systems in a very straightforward way to obtain solutions for the multidimensional problems.

It is clear that the infinite rectangular bar in Fig. 4-17 can be formed from two infinite plates of thickness $2L_1$ and $2L_2$, respectively. The differential equation governing this situation would be

$$\frac{\partial^2 T}{\partial x^2} + \frac{\partial^2 T}{\partial z^2} = \frac{1}{\alpha}\frac{\partial T}{\partial \tau} \tag{4-17}$$

and to use the separation-of-variables method to effect a solution, we should assume a product solution of the form

$$T(x, z, \tau) = X(x)Z(z)\Theta(\tau)$$

It can be shown that the dimensionless temperature distribution may be expressed as a product of the solutions for two plate problems of thickness $2L_1$ and $2L_2$, respectively:

$$\left(\frac{T - T_\infty}{T_i - T_\infty}\right)_{\text{bar}} = \left(\frac{T - T_\infty}{T_i - T_\infty}\right)_{2L1\ plate} \left(\frac{T - T_\infty}{T_i - T_\infty}\right)_{2L2\ plate} \tag{4-18}$$

where T_i is the initial temperature of the bar and T_∞ is the environment temperature.

For two infinite plates the respective differential equations would be

$$\frac{\partial^2 T_1}{\partial x^2} = \frac{1}{\alpha}\frac{\partial T_1}{\partial \tau} \qquad \frac{\partial^2 T_2}{\partial z^2} = \frac{1}{\alpha}\frac{\partial T_2}{\partial \tau} \tag{4-19}$$

and the product solutions assumed would be

$$T_1 = T_1(x, \tau) \qquad T_2 = T_2(z, \tau) \tag{4-20}$$

We shall now show that the product solution to Eq. (4-17) can be formed from a simple product of the functions (T_1, T_2), that is,

$$T(x, z, \tau) = T_1(x, \tau)T_2(z, \tau) \tag{4-21}$$

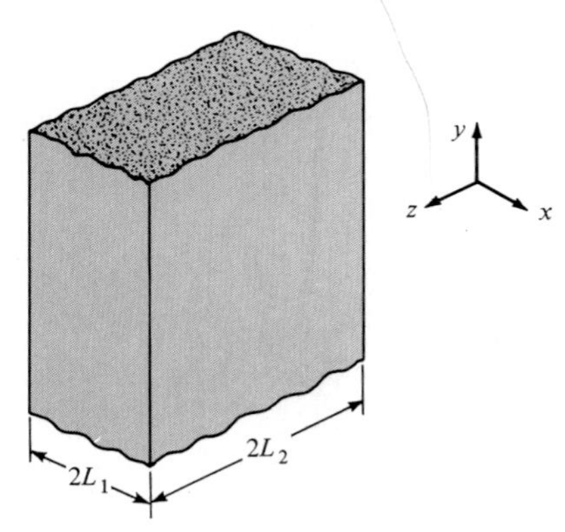

Fig. 4-17 Infinite rectangular bar.

The appropriate derivatives for substitution in Eq. (4-23) are obtained from Eq. (4-21) as

$$\frac{\partial^2 T}{\partial x^2} = T_2 \frac{\partial^2 T_1}{\partial x^2} \qquad \frac{\partial^2 T}{\partial z^2} = T_1 \frac{\partial^2 T_2}{\partial z^2}$$

$$\frac{\partial T}{\partial \tau} = T_1 \frac{\partial T_2}{\partial \tau} + T_2 \frac{\partial T_1}{\partial \tau}$$

Using Eqs. (4-19), we have

$$\frac{\partial T}{\partial \tau} = \alpha T_1 \frac{\partial^2 T_2}{\partial z^2} + \alpha T_2 \frac{\partial^2 T_1}{\partial x^2}$$

Substituting these relations in Eq. (4-17) gives

$$T_2 \frac{\partial^2 T_1}{\partial x^2} + T_1 \frac{\partial^2 T_2}{\partial z^2} = \frac{1}{\alpha}\left(\alpha T_1 \frac{\partial^2 T_2}{\partial z^2} + \alpha T_2 \frac{\partial^2 T_1}{\partial x^2}\right)$$

or the assumed product solution of Eq. (4-21) does indeed satisfy the original differential equation (4-17). This means that the dimensionless temperature distribution for the infinite rectangular bar may be expressed as a product of the solutions for two plate problems of thickness $2L_1$ and $2L_2$, respectively, as indicated by Eq. (4-18).

In a manner similar to that described above, the solution for a three-dimensional block may be expressed as a product of three infinite-plate solutions for plates having the thickness of the three sides of the block. Similarly, a solution for a cylinder of finite length could be expressed as a product of solutions of the infinite cylinder and an infinite plate having a thickness equal to the length of the cylinder. Combinations could also be made with the infinite-cylinder and infinite-plate solutions to obtain temperature distributions in semi-infinite bars and cylinders. Some of the combinations are summarized in Fig. 4-18, where

$$C(\Theta) = \text{solution for infinite cylinder}$$
$$P(X) = \text{solution for infinite plate}$$
$$S(X) = \text{solution for semi-infinite solid}$$

The general idea is then

$$\left(\frac{\theta}{\theta_i}\right)_{\substack{\text{combined}\\ \text{solid}}} = \left(\frac{\theta}{\theta_i}\right)_{\substack{\text{intersection}\\ \text{solid 1}}} \left(\frac{\theta}{\theta_i}\right)_{\substack{\text{intersection}\\ \text{solid 2}}} \left(\frac{\theta}{\theta_i}\right)_{\substack{\text{intersection}\\ \text{solid 3}}}$$

□ Heat Transfer in Multidimensional Systems

Langston [16] has shown that it is possible to superimpose the heat-loss solutions for one-dimensional bodies, as shown in Figs. 4-14, 4-15, and 4-16 to

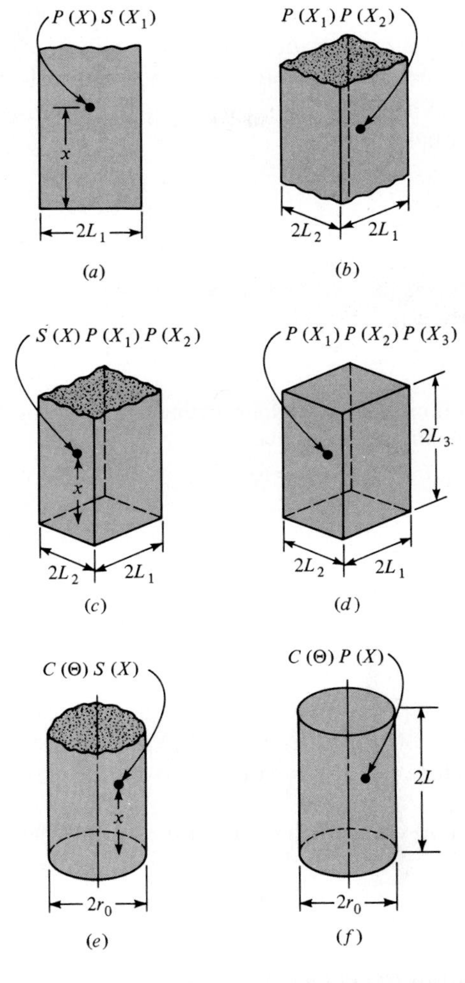

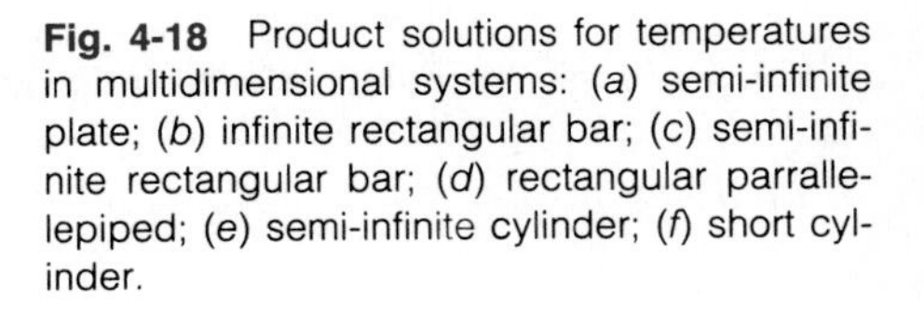
Fig. 4-18 Product solutions for temperatures in multidimensional systems: (*a*) semi-infinite plate; (*b*) infinite rectangular bar; (*c*) semi-infinite rectangular bar; (*d*) rectangular parrallelepiped; (*e*) semi-infinite cylinder; (*f*) short cylinder.

obtain the heat for a multidimensional body. The results of this analysis for intersection of two bodies is

$$\left(\frac{Q}{Q_0}\right)_{\text{total}} = \left(\frac{Q}{Q_0}\right)_1 + \left(\frac{Q}{Q_0}\right)_2 \left[1 - \left(\frac{Q}{Q_0}\right)_1\right] \tag{4-22}$$

where the subscripts refer to the two intersecting bodies. For a multidimensional body formed by intersection of three one-dimensional systems, the heat loss is given by

$$\left(\frac{Q}{Q_0}\right)_{\text{total}} = \left(\frac{Q}{Q_0}\right)_1 + \left(\frac{Q}{Q_0}\right)_2 \left[1 - \left(\frac{Q}{Q_0}\right)_1\right] + \left(\frac{Q}{Q_0}\right)_3 \left[1 - \left(\frac{Q}{Q_0}\right)_1\right]\left[1 - \left(\frac{Q}{Q_0}\right)_2\right] \tag{4-23}$$

If the heat loss is desired after a given time, the calculation is straightforward. On the other hand, if the *time to achieve a certain heat loss* is the desired quantity, a trial-and-error or iterative procedure must be employed. The following examples illustrate the use of the various charts for calculating temperatures and heat flows in multidimensional systems.

EXAMPLE 4-7

A semi-infinite aluminum cylinder 5 cm in diameter is initially at a uniform temperature of 200°C. It is suddenly subjected to a convection boundary condition at 70°C with $h = 525\ \text{W/m}^2 \cdot {}^\circ\text{C}$. Calculate the temperatures at the axis and surface of the cylinder 10 cm from the end 1 min after exposure to the environment.

Solution

This problem requires a combination of solutions for the infinite cylinder and semi-infinite slab in accordance with Fig. 4-18*e*. For the slab we have

$$x = 10\ \text{cm} \qquad \alpha = 8.4 \times 10^{-5}\ \text{m}^2/\text{s} \qquad k = 215\ \text{W/m} \cdot {}^\circ\text{C}$$

so that the parameters for use with Fig. 4-5 are

$$\frac{h\sqrt{\alpha\tau}}{k} = \frac{(525)[(8.4 \times 10^{-5})(60)]^{1/2}}{215} = 0.173$$

$$\frac{x}{2\sqrt{\alpha\tau}} = \frac{0.1}{(2)[(8.4 \times 10^{-5})(60)]^{1/2}} = 0.704$$

From Fig. 4-5

$$\left(\frac{\theta}{\theta_i}\right)_{\text{semi-infinite slab}} = 1 - 0.036 = 0.964 = S(X)$$

For the infinite cylinder we seek both the axis- and surface-temperature ratios. The parameters for use with Fig. 4-8 are

$$r_0 = 2.5\ \text{cm} \qquad \frac{k}{hr_0} = 16.38 \qquad \frac{\alpha\tau}{r_0^2} = 8.064 \qquad \frac{\theta_0}{\theta_i} = 0.38$$

This is the axis-temperature ratio. To find the surface-temperature ratio, we enter Fig. 4-11, using

$$\frac{r}{r_0} = 1.0 \qquad \frac{\theta}{\theta_0} = 0.97$$

Thus

$$C(\Theta) = \left(\frac{\theta}{\theta_i}\right)_{\text{inf cyl}} = \begin{cases} 0.38 & \text{at } r = 0 \\ (0.38)(0.97) = 0.369 & \text{at } r = r_0 \end{cases}$$

Combining the solutions for the semi-infinite slab and infinite cylinder, we have

$$\left(\frac{\theta}{\theta_i}\right)_{\text{semi-infinite cylinder}} = C(\Theta)S(X)$$

$$= (0.38)(0.964) = 0.366 \qquad \text{at } r = 0$$

$$= (0.369)(0.964) = 0.356 \qquad \text{at } r = r_0$$

The corresponding temperatures are

$$T = 70 + (0.366)(200 - 70) = 117.6 \qquad \text{at } r = 0$$
$$T = 70 + (0.356)(200 - 70) = 116.3 \qquad \text{at } r = r_0$$

■ EXAMPLE 4-8

A short aluminum cylinder 5.0 cm in diameter and 10.0 cm long is initially at a uniform temperature of 200°C. It is suddenly subjected to a convection environment at 70°C, and $h = 525$ W/m$^2 \cdot$ °C. Calculate the temperature at a radial position of 1.25 cm and a distance of 0.625 cm from one end of the cylinder 1 min after exposure to the environment.

Solution

To solve this problem we combine the solutions from the Heisler charts for an infinite cylinder and an infinite plate in accordance with the combination shown in Fig. 4-18*f*. For the infinite-plate problem

$$L = 5 \text{ cm}$$

The x position is measured from the center of the plate so that

$$x = 5 - 0.625 = 4.375 \text{ cm} \qquad \frac{x}{L} = \frac{4.375}{5} = 0.875$$

For aluminum

$$\alpha = 8.4 \times 10^{-5} \text{ m}^2\text{/s} \qquad k = 215 \text{ W/m} \cdot \text{°C}$$

so

$$\frac{k}{hL} = \frac{215}{(525)(0.05)} = 8.19 \qquad \frac{\alpha\tau}{L^2} = \frac{(8.4 \times 10^{-5})(60)}{(0.05)^2} = 2.016$$

From Figs. 4-7 and 4-10, respectively,

$$\frac{\theta_0}{\theta_i} = 0.75 \qquad \frac{\theta}{\theta_0} = 0.95$$

so that

$$\left(\frac{\theta}{\theta_i}\right)_{\text{plate}} = (0.75)(0.95) = 0.7125$$

For the cylinder $r_0 = 2.5$ cm

$$\frac{r}{r_0} = \frac{1.25}{2.5} = 0.5 \qquad \frac{k}{hr_0} = \frac{215}{(525)(0.025)} = 16.38$$

$$\frac{\alpha\tau}{r_0^2} = \frac{(8.4 \times 10^{-5})(60)}{(0.0025)^2} = 8.064$$

and from Figs. 4-8 and 4-11, respectively,

$$\frac{\theta_0}{\theta_i} = 0.38 \qquad \frac{\theta}{\theta_0} = 0.98$$

so that

$$\left(\frac{\theta}{\theta_i}\right)_{\text{cyl}} = (0.38)(0.98) = 0.3724$$

Combining the solutions for the plate and cylinder gives

$$\left(\frac{\theta}{\theta_i}\right)_{\text{short cylinder}} = (0.7125)(0.3724) = 0.265$$

Thus $\quad T = T_\infty + (0.265)(T_i - T_\infty) = 70 + (0.265)(200 - 70) = 104.5°\text{C}$

EXAMPLE 4-9

Calculate the heat loss for the short cylinder in Example 4-8.

Solution

We first calculate the dimensionless heat-loss ratio for the infinite plate and infinite cylinder which make up the multidimensional body. For the plate we have $L = 5$ cm $= 0.05$ m. Using the properties of aluminum from Example 4-8, we calculate

$$\frac{hL}{k} = \frac{(525)(0.05)}{215} = 0.122$$

$$\frac{h^2\alpha\tau}{k^2} = \frac{(525)^2\,(8.4 \times 10^{-5})(60)}{(215)^2} = 0.03$$

From Fig. 4-14, for the plate, we read

$$\left(\frac{Q}{Q_0}\right)_p = 0.22$$

For the cylinder $r_0 = 2.5$ cm $= 0.025$ m, so we calculate

$$\frac{hr_0}{k} = \frac{(525)(0.025)}{215} = 0.061$$

and from Fig. 4-15 we can read

$$\left(\frac{Q}{Q_0}\right)_c = 0.55$$

The two heat ratios may be inserted in Eq. (4-22) to give

$$\left(\frac{Q}{Q_0}\right)_{\text{tot}} = 0.22 + (0.55)(1 - 0.22) = 0.649$$

The specific heat of aluminum is 0.896 kJ/kg · °C and the density is 2707 kg/m^3, so we calculate Q_0 as

$$Q_0 = \rho c V \theta_i = (2707)(0.896)\pi(0.025)^2(0.1)(200 - 70)$$
$$= 61.9 \text{ kJ}$$

The actual heat loss in the 1-min time is thus

$$Q = (61.9 \text{ kJ})(0.649) = 40.2 \text{ kJ}$$

4-6 TRANSIENT NUMERICAL METHOD

The charts described above are very useful for calculating temperatures in certain regular-shaped solids under transient heat-flow conditions. Unfortunately, many geometric shapes of practical interest do not fall into these categories; in addition, one is frequently faced with problems in which the boundary conditions vary with time. These transient boundary conditions as well as the geometric shape of the body can be such that a mathematical solution is not possible. In these cases, the problems are best handled by a numerical technique with computers. It is the setup for such calculations which we now describe. For ease in discussion we limit the analysis to two-dimensional systems. An extension to three dimensions can then be made very easily.

Consider a two-dimensional body divided into increments as shown in Fig. 4-19. The subscript m denotes the x position, and the subscript n denotes the y position. Within the solid body the differential equation which governs the heat flow is

$$k\left(\frac{\partial^2 T}{\partial x^2} + \frac{\partial^2 T}{\partial y^2}\right) = \rho c \frac{\partial T}{\partial \tau} \tag{4-24}$$

assuming constant properties. We recall from Chap. 3 that the second partial derivatives may be approximated by

$$\frac{\partial^2 T}{\partial x^2} \approx \frac{1}{(\Delta x)^2}(T_{m+1,n} + T_{m-1,n} - 2T_{m,n}) \tag{4-25}$$

$$\frac{\partial^2 T}{\partial y^2} \approx \frac{1}{(\Delta y^2)}(T_{m,n+1} + T_{m,n-1} - 2T_{m,n}) \tag{4-26}$$

The time derivative in Eq. (4-24) is approximated by

$$\frac{\partial T}{\partial \tau} \approx \frac{T^{p+1}_{m,n} - T^{p}_{m,n}}{\Delta \tau} \tag{4-27}$$

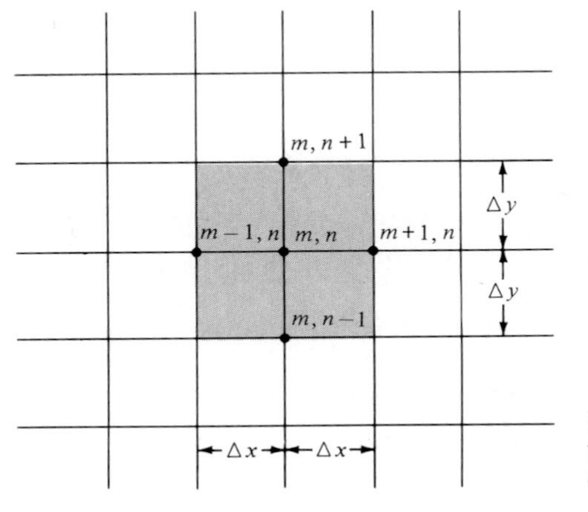

Fig. 4-19 Nomenclature for numerical solution of two-dimensional unsteady-state conduction problem.

In this relation the superscripts designate the time increment. Combining the relations above gives the difference equation equivalent to Eq. (4-24)

$$\frac{T^p_{m+1,n} + T^p_{m-1,n} - 2T^p_{m,n}}{(\Delta x)^2} + \frac{T^p_{m,n+1} + T^p_{m,n-1} - 2T^p_{m,n}}{(\Delta y)^2} = \frac{1}{\alpha}\frac{T^{p+1}_{m,n} - T^p_{m,n}}{\Delta\tau} \tag{4-28}$$

Thus, if the temperatures of the various nodes are known at any particular time, the temperatures after a time increment $\Delta\tau$ may be calculated by writing an equation like Eq. (4-28) for each node and obtaining the values of $T^{p+1}_{m,n}$. The procedure may be repeated to obtain the distribution after any desired number of time increments. If the increments of space coordinates are chosen such that

$$\Delta x = \Delta y$$

the resulting equation for $T^{p+1}_{m,n}$ becomes

$$T^{p+1}_{m,n} = \frac{\alpha\,\Delta\tau}{(\Delta x)^2}(T^p_{m+1,n} + T^p_{m-1,n} + T^p_{m,n+1} + T^p_{m,n-1}) + \left[1 - \frac{4\alpha\,\Delta\tau}{(\Delta x)^2}\right]T^p_{m,n} \tag{4-29}$$

If the time and distance increments are conveniently chosen so that

$$\frac{(\Delta x)^2}{\alpha\,\Delta\tau} = 4 \tag{4-30}$$

it is seen that the temperature of node (m, n) after a time increment is simply the arithmetic average of the four surrounding nodal temperatures at the beginning of the time increment.

When a one-dimensional system is involved, the equation becomes

$$T^{p+1}_{m} = \frac{\alpha\,\Delta\tau}{(\Delta x)^2}(T^p_{m+1} + T^p_{m-1}) + \left[1 - \frac{2\alpha\,\Delta\tau}{(\Delta x)^2}\right]T^p_{m} \tag{4-31}$$

and if the time and distance increments are chosen so that

$$\frac{(\Delta x)^2}{\alpha\,\Delta\tau} = 2 \tag{4-32}$$

the temperature of node m after the time increment is given as the arithmetic average of the two adjacent nodal temperatures at the beginning of the time increment.

Some general remarks concerning the use of numerical methods for solution of transient conduction problems are in order at this point. We have already noted that the selection of the value of the parameter

$$M = \frac{(\Delta x)^2}{\alpha\,\Delta\tau}$$

governs the ease with which we may proceed to effect the numerical solution; the choice of a value of 4 for a two-dimensional system or a value of 2 for a one-dimensional system makes the calculation particularly easy.

Once the distance increments and the value of M are established, the time increment is fixed; and we may not alter it without changing the value of either Δx or M, or both. Clearly, the larger the values of Δx and $\Delta\tau$, the more rapidly our solution will proceed. On the other hand, the smaller the value of these increments in the independent variables, the more accuracy will be obtained. At first glance one might assume that small distance increments could be used for greater accuracy in combination with large time increments to speed the solution. This is not the case, however, because the finite-difference equations limit the values of $\Delta\tau$ which may be used once Δx is chosen. Note that if $M < 2$ in Eq. (4-31), the coefficient of T_m^p becomes negative, and we generate a condition which will violate the second law of thermodynamics. Suppose, for example, that the adjoining nodes are equal in temperature but less than T_m^p. After the time increment $\Delta\tau$, T_m^p may not be lower than these adjoining temperatures; otherwise heat would have to flow uphill on the temperature scale, and this is impossible. A value of $M < 2$ would produce just such an effect; so we must restrict the values of M to

$$\frac{(\Delta x)^2}{\alpha\,\Delta\tau} = \begin{cases} M \geq 2 & \text{one-dimensional systems} \\ M \geq 4 & \text{two-dimensional systems} \end{cases}$$

This restriction automatically limits our choice of $\Delta\tau$, once Δx is established.

It so happens that the above restrictions, which are imposed in a physical sense, may also be derived on mathematical grounds. It may be shown that the finite-difference solutions will not converge unless these conditions are fulfilled. The problems of stability and convergence of numerical solutions are discussed in Refs. 7, 13, and 15 in detail.

The difference equations given above are useful for determining the internal temperature in a solid as a function of space and time. At the boundary of the solid, a convection resistance to heat flow is usually involved, so that the above relations no longer apply. In general, each convection boundary condition must be handled separately, depending on the particular geometric shape under consideration. The case of the flat wall will be considered as an example.

For the one-dimensional system shown in Fig. 4-20 we may make an energy balance at the convection boundary such that

$$-kA\left.\frac{\partial T}{\partial x}\right]_{\text{wall}} = hA(T_w - T_\infty) \tag{4-33}$$

The finite-difference approximation would be given by

$$-k\frac{\Delta y}{\Delta x}(T_{m+1} - T_m) = h\,\Delta y(T_{m+1} - T_\infty)$$

or

$$T_{m+1} = \frac{T_m + (h\,\Delta x/k)T_\infty}{1 + h\,\Delta x/k}$$

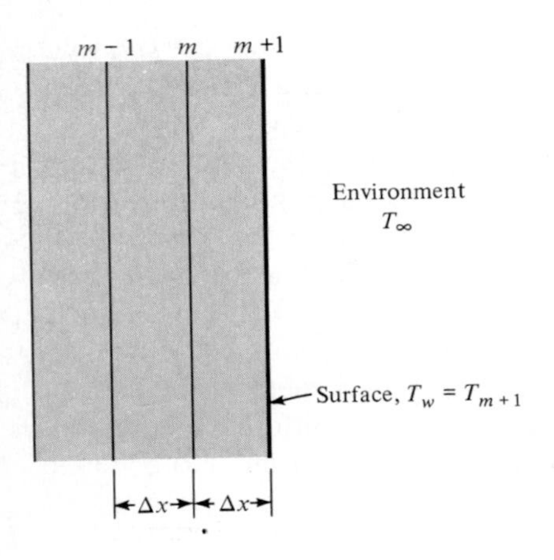

Fig. 4-20 Nomenclature for numerical solution of unsteady-state conduction problem with convection boundary condition.

To apply this condition, we should calculate the surface temperature T_{m+1} at each time increment and then use this temperature in the nodal equations for the interior points of the solid. This is only an approximation because we have neglected the heat capacity of the element of the wall at the boundary. This approximation will work fairly well when a large number of increments in x are used because the portion of the heat capacity which is neglected is then small in comparison with the total. We may take the heat capacity into account in a general way by considering the two-dimensional wall of Fig. 3-7 exposed to a convection boundary condition, which we duplicate here for convenience as Fig. 4-21. We make a transient energy balance on the node (m, n) by setting the sum of the energy conducted and convected into the node equal to the increase in the internal energy of the node. Thus

$$k\,\Delta y\,\frac{T^p_{m-1,n} - T^p_{m,n}}{\Delta x} + k\,\frac{\Delta x}{2}\,\frac{T^p_{m,n+1} - T^p_{m,n}}{\Delta y} + k\,\frac{\Delta x}{2}\,\frac{T^p_{m,n-1} - T^p_{m,n}}{\Delta y} + h\,\Delta y(T_\infty - T^p_{m,n}) = \rho c\,\frac{\Delta x}{2}\,\Delta y\,\frac{T^{p+1}_{m,n} - T^p_{m,n}}{\Delta \tau}$$

If $\Delta x = \Delta y$, the relation for $T^{p+1}_{m,n}$ becomes

$$T^{p+1}_{m,n} = \frac{\alpha\,\Delta\tau}{(\Delta x)^2}\left\{2\,\frac{h\,\Delta x}{k}\,T_\infty + 2T^p_{m-1,n} + T^p_{m,n+1} + T^p_{m,n-1} + \left[\frac{(\Delta x)^2}{\alpha\,\Delta\tau} - 2\,\frac{h\,\Delta x}{k} - 4\right]T^p_{m,n}\right\} \tag{4-34}$$

The corresponding one-dimensional relation is

$$T^{p+1}_{m} = \frac{\alpha\,\Delta\tau}{(\Delta x)^2}\left\{2\,\frac{h\,\Delta x}{k}\,T_\infty + 2T^p_{m-1} + \left[\frac{(\Delta x)^2}{\alpha\,\Delta\tau} - 2\,\frac{h\,\Delta x}{k} - 2\right]T^p_{m}\right\} \tag{4-35}$$

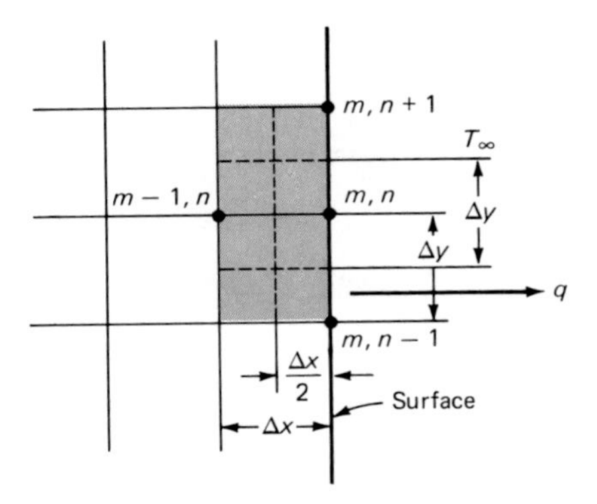

Fig. 4-21 Nomenclature for nodal equation with convective boundary condition.

Notice now that the selection of the parameter $(\Delta x)^2/\alpha\ \Delta\tau$ is not as simple as it is for the interior nodal points because the heat-transfer coefficient influences the choice. It is still possible to choose the value of this parameter so that the coefficient of T_m^p or $T_{m,n}^p$ will be zero. These values would then be

$$\frac{(\Delta x)^2}{\alpha\ \Delta\tau} = \begin{cases} 2\left(\dfrac{h\ \Delta x}{k} + 1\right) & \text{for the one-dimensional case} \\ 2\left(\dfrac{h\ \Delta x}{k} + 2\right) & \text{for the two-dimensional case} \end{cases}$$

To ensure convergence of the numerical solution, all selections of the parameter $(\Delta x)^2/\alpha\ \Delta\tau$ must be restricted according to

$$\frac{(\Delta x)^2}{\alpha \Delta\tau} \geq \begin{cases} 2\left(\dfrac{h\ \Delta x}{k} + 1\right) & \text{for the one-dimensional case} \\ 2\left(\dfrac{h\ \Delta x}{k} + 2\right) & \text{for the two-dimensional case} \end{cases}$$

□ Forward and Backward Differences

The equations above have been developed on the basis of a *forward-difference* technique in that the temperature of a node at a future time increment is expressed in terms of the surrounding nodal temperatures at the beginning of the time increment. The expressions are called *explicit* formulations because it is possible to write the nodal temperatures $T_{m,n}^{p+1}$ explicitly in terms of the previous nodal temperatures $T_{m,n}^p$. In this formulation, the calculation proceeds directly from one time increment to the next until the temperature distribution is calculated at the desired final state.

The difference equation may also be formulated by computing the space derivatives in terms of the temperatures at the $p + 1$ time increment. Such an arrangement is called a *backward-difference* formulation because the time derivative moves backward from the times for heat conduction into the node. The equation equivalent to Eq. (4-28) would then be

$$\frac{T_{m+1,n}^{p+1} + T_{m-1,n}^{p+1} - 2T_{m,n}^{p+1}}{(\Delta x)^2} + \frac{T_{m,n+1}^{p+1} + T_{m,n-1}^{p+1} - 2T_{m,n}^{p+1}}{(\Delta y)^2} = \frac{1}{\alpha}\frac{T_{m,n}^{p+1} - T_{m,n}^{p}}{\Delta\tau} \tag{4-36}$$

The equivalence to Eq. (4-29) is

$$T_{m,n}^{p} = \frac{-\alpha\,\Delta\tau}{(\Delta x)^2}\left(T_{m+1,n}^{p+1} + T_{m-1,n}^{p+1} + T_{m,n+1}^{p+1} + T_{m,n-1}^{p+1}\right) + \left[1 + \frac{4\alpha\,\Delta\tau}{(\Delta x)^2}\right]T_{m,n}^{p+1} \tag{4-37}$$

We may now note that this backward-difference formulation does not permit the explicit calculation of the T^{p+1} in terms of T^{p}. Rather, a whole set of equations must be written for the entire nodal system and solved simultaneously to determine the temperatures T^{p+1}. Thus we say that the backward-difference method produces an *implicit formulation* for the future temperatures in the transient analysis. The solution to the set of equations can be performed with the methods discussed in Chap. 3.

The Biot and Fourier numbers may also be defined in the following way for problems in the numerical format:

$$\text{Bi} = \frac{h\,\Delta x}{k} \tag{4-38}$$

$$\text{Fo} = \frac{\alpha\,\Delta\tau}{(\Delta x)^2} \tag{4-39}$$

By using this notation Tables 4-2 and 4-3 have been constructed to summarize some typical nodal equations in both the explicit and implicit formulations.

The advantage of an explicit forward-difference procedure is the direct calculation of future nodal temperatures; however, the stability of this calculation is governed by the selection of the values of Δx and $\Delta\tau$. A selection of a small value of Δx automatically forces the selection of some maximum value of $\Delta\tau$. On the other hand, no such restriction is imposed on the solution of the equations which are obtained from the implicit formulation. This means that larger time increments can be selected to speed the calculation. The obvious disadvantage of the implicit method is the larger number of calculations for each time step. For problems involving a large number of nodes, however, the implicit method may result in less total computer time expended for the final solution because very small time increments may be imposed in the explicit method from stability requirements. Much larger increments in $\Delta\tau$ can be employed with the implicit method to speed the solution.

Table 4-2 Explicit Nodal Equations
(Dashed lines indicate element volume)†

Physical situation	*Nodal equation for* $\Delta x = \Delta y$	*Stability requirement*
(*a*) Interior node 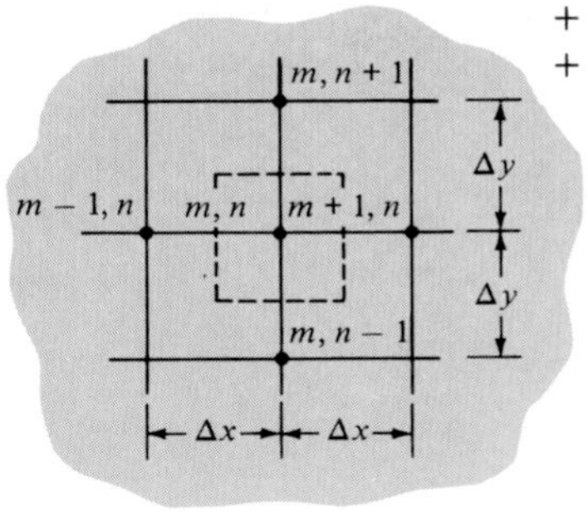	$T^{p+1}_{m,n} = \text{Fo}\,(T^p_{m-1,n} + T^p_{m,n+1} + T^p_{m+1,n} + T^p_{m,n-1}) + [1 - 4(\text{Fo})]\,T^p_{m,n}$	$\text{Fo} \le \frac{1}{4}$
(*b*) Convection boundary node 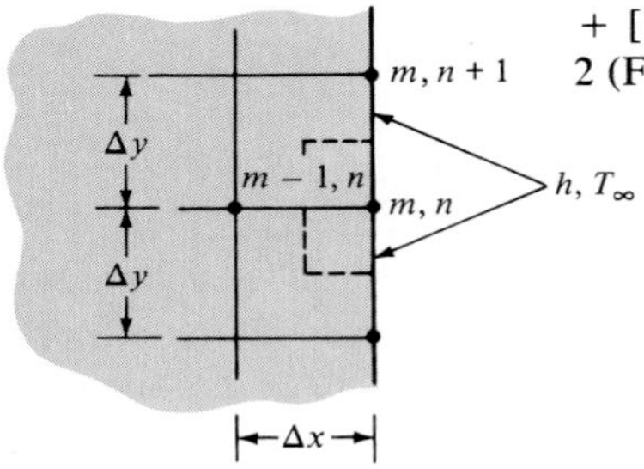	$T^{p+1}_{m,n} = \text{Fo}\,[2T^p_{m-1,n} + T^p_{m,n+1} + T^p_{m,n-1} + 2\,(\text{Bi})\,T^p_\infty] + [1 - 4\,(\text{Fo}) - 2\,(\text{Fo})\,(\text{Bi})]\,T^p_{m,n}$	$\text{Fo}\,(2 + \text{Bi}) \le \frac{1}{2}$
(*c*) Exterior corner with convection boundary 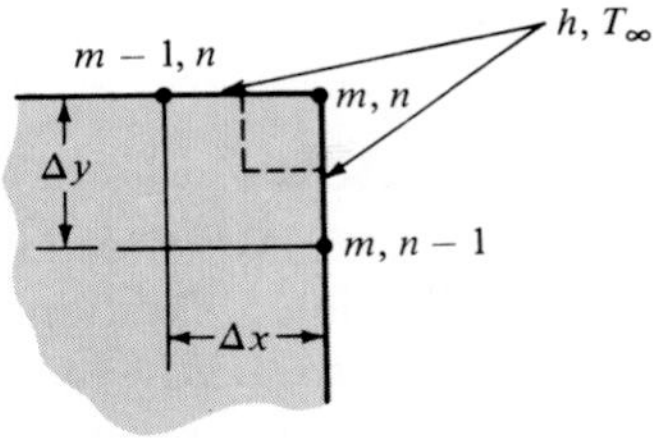	$T^{p+1}_{m,n} = 2\,(\text{Fo})\,[T^p_{m-1,n} + T^p_{m,n-1} + 2\,(\text{Bi})\,T^p_\infty] + [1 - 4\,(\text{Fo}) - 4\,(\text{Fo})\,(\text{Bi})]\,T^p_{m,n}$	$\text{Fo}\,(1 + \text{Bi}) \le \frac{1}{4}$
(*d*) Interior corner with convection boundary 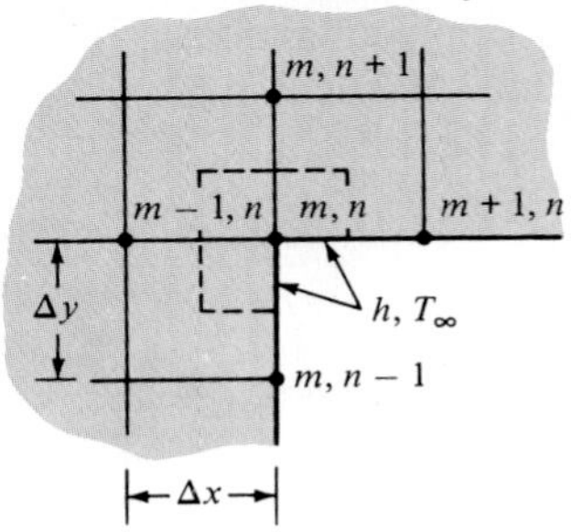	$T^{p+1}_{m,n} = \frac{2}{3}\,(\text{Fo})\,[2T^p_{m,n+1} + 2T^p_{m+1,n} + T^p_{m-1,n} + T^p_{m,n-1} + 2\,(\text{Bi})\,T^p_\infty] + [1 - 4\,(\text{Fo}) - \frac{4}{3}\,(\text{Fo})\,(\text{Bi})]\,T^p_{m,n}$	$\text{Fo}\,(3 + \text{Bi}) \le \frac{3}{4}$

Table 4-2 *(Continued)*
(Dashed lines indicate element volume)†

Physical situation	*Nodal equation for* $\Delta x = \Delta y$	*Stability requirement*
(*e*) Insulated boundary 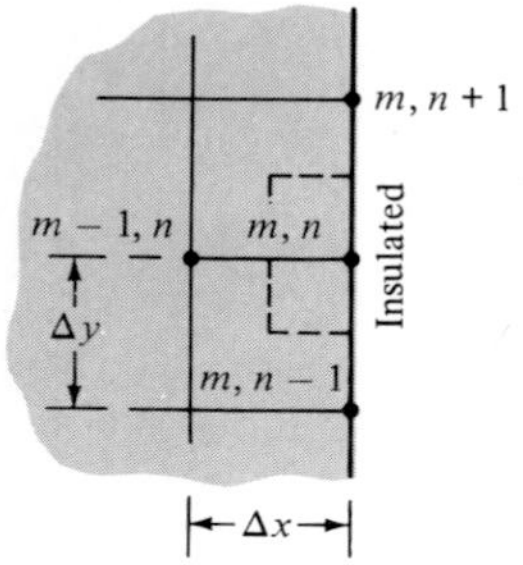	$T^{p+1}_{m,n} = \text{Fo}\,[2T^p_{m-1,n} + T^p_{m,n+1} + T^p_{m,n-1}] + [1 - 4\,(\text{Fo})]T^p_{m,n}$	$\text{Fo} \le \frac{1}{4}$

†Convection surfaces may be made insulated by setting $h = 0$ (Bi = 0).

For a discussion of many applications of numerical analysis to transient heat-conduction problems, the reader is referred to Refs. 4, 8, 13, 14, and 15.

It should be obvious to the reader by now that finite-difference techniques may be applied to almost any situation with just a little patience and care. Very complicated problems then become quite easy to solve when large digital computer facilities are available. Computer programs for several heat-transfer problems of interest are given in Refs. 17, 19, and 21 of Chap. 3. Finite-element methods for use in conduction heat-transfer problems are discussed in Refs. 9 to 13. A number of software packages are available commercially for use with microcomputers.

4-7 THERMAL RESISTANCE AND CAPACITY FORMULATION

As in Chap. 3, we can view each volume element as a node which is connected by thermal resistances to its adjoining neighbors. For steady-state conditions the net energy transfer into the node is zero, while for the unsteady-state problems of interest in this chapter the net energy transfer into the node must be evidenced as an increase in internal energy of the element. Each volume element behaves like a small "lumped capacity," and the interaction of all the elements determines the behavior of the solid during a transient process. If the internal energy of a node i can be expressed in terms of specific heat and temperature, then its rate of change with time is approximated by

$$\frac{\Delta E}{\Delta \tau} = \rho c\,\Delta V\,\frac{T_i^{p+1} - T_i^p}{\Delta \tau}$$

where ΔV is the volume element. If we define the thermal capacity as

$$C_i = \rho_i c_i\,\Delta V_i \tag{4-40}$$

Table 4-3 Implicit Nodal Equations
(Dashed lines indicate volume element)

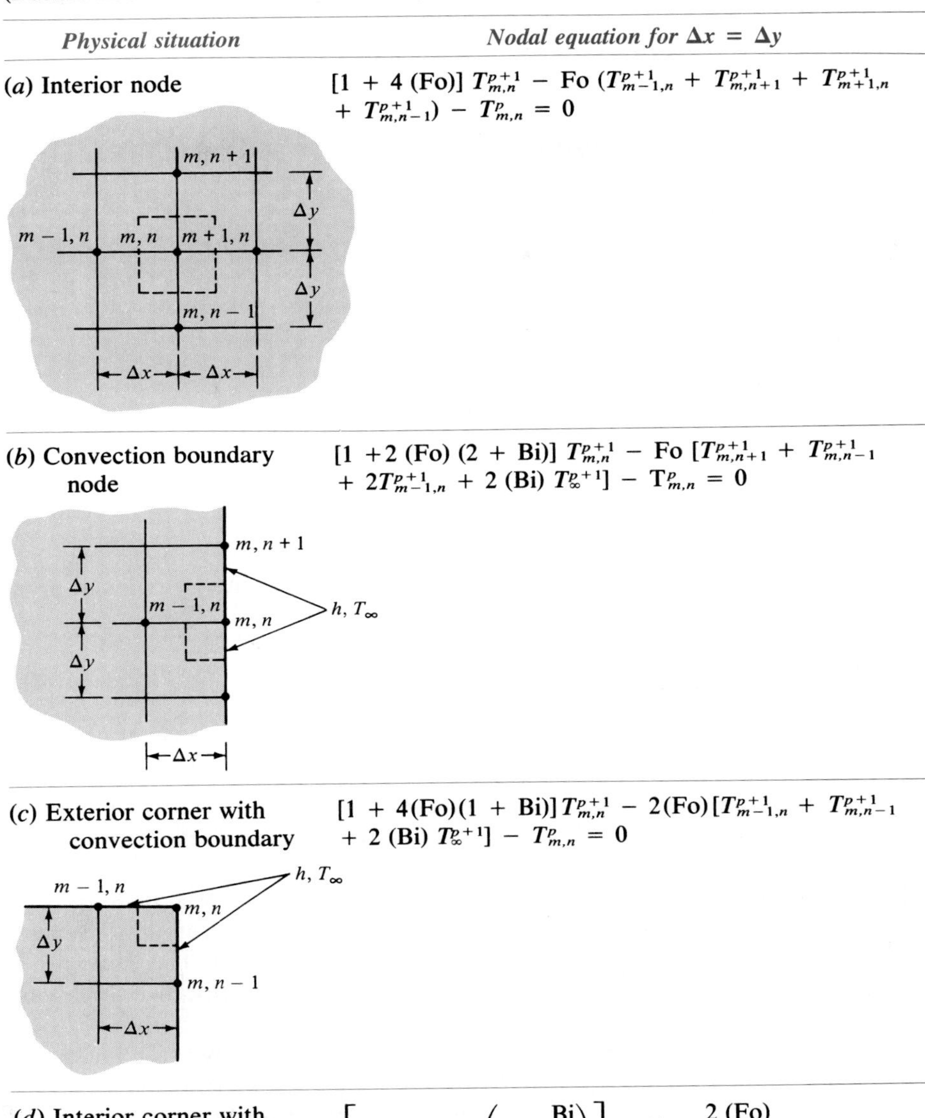

Physical situation	*Nodal equation for* $\Delta x = \Delta y$
(*a*) Interior node	$[1 + 4(\text{Fo})]\,T^{p+1}_{m,n} - \text{Fo}\,(T^{p+1}_{m-1,n} + T^{p+1}_{m,n+1} + T^{p+1}_{m+1,n} + T^{p+1}_{m,n-1}) - T^{p}_{m,n} = 0$
(*b*) Convection boundary node	$[1 + 2(\text{Fo})(2 + \text{Bi})]\,T^{p+1}_{m,n} - \text{Fo}\,[T^{p+1}_{m,n+1} + T^{p+1}_{m,n-1} + 2T^{p+1}_{m-1,n} + 2(\text{Bi})\,T^{p+1}_{\infty}] - T^{p}_{m,n} = 0$
(*c*) Exterior corner with convection boundary	$[1 + 4(\text{Fo})(1 + \text{Bi})]\,T^{p+1}_{m,n} - 2(\text{Fo})\,[T^{p+1}_{m-1,n} + T^{p+1}_{m,n-1} + 2(\text{Bi})\,T^{p+1}_{\infty}] - T^{p}_{m,n} = 0$
(*d*) Interior corner with convection boundary 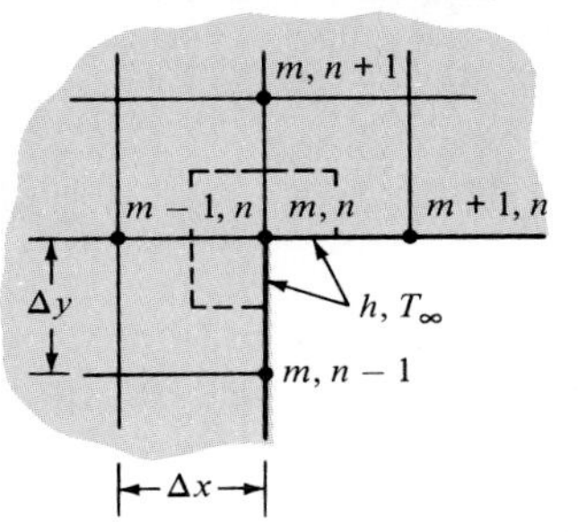	$\left[1 + 4(\text{Fo})\left(1 + \frac{\text{Bi}}{3}\right)\right] T^{p+1}_{m,n} - \frac{2(\text{Fo})}{3} \times [T^{p+1}_{m-1,n} + T^{p+1}_{m,n-1} + 2T^{p+1}_{m,n+1} + 2T^{p+1}_{m+1,n} + 2(\text{Bi})\,T^{p+1}_{\infty}] - T^{p}_{m,n} = 0$

Table 4-3 (*Continued*)
(Dashed lines indicate volume element)

Physical situation	*Nodal equation for* $\Delta x = \Delta y$
(*e*) Insulated boundary 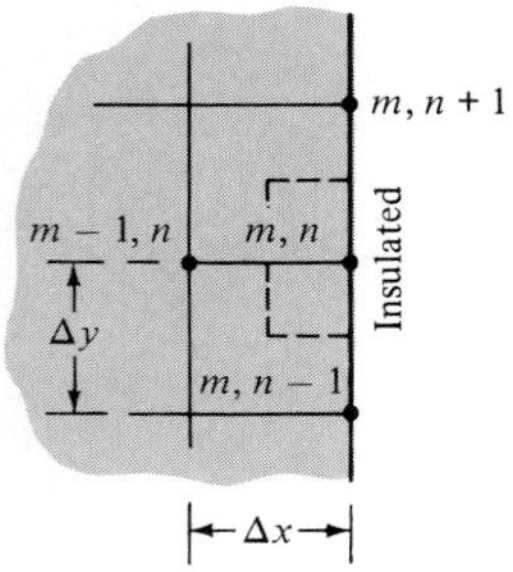	$[1 + 4\,(\text{Fo})]\,T_{m,n}^{p+1} - \text{Fo}(2T_{m-1,n}^{p+1} + T_{m,n+1}^{p+1} + T_{m,n-1}^{p+1}) - T_{m,n}^{p} = 0$

then the general resistance-capacity formulation for the energy balance on a node is

$$q_i + \sum_j \frac{T_j^p - T_i^p}{R_{ij}} = C_i \frac{T_i^{p+1} - T_i^p}{\Delta\tau} \tag{4-41}$$

where all the terms on the left are the same as in Eq. (3-31). The resistance and volume elements for a variety of geometries and boundary conditions were given in Tables 3-4 and 3-5. Physical systems where the internal energy E involves phase changes can also be accommodated in the above formulation but are beyond the scope of our discussion.

The central point is that use of the concepts of thermal resistance and capacitance enables us to write the forward-difference equation for all nodes and boundary conditions in the single compact form of Eq. (4-41). The setup for a numerical solution then becomes a much more organized process which can be adapted quickly to the computational methods at hand.

Equation (4-41) is developed by using the forward-difference concept to produce an explicit relation for each T_i^{p+1}. As in our previous discussion, we could also write the energy balance using backward differences, with the heat transfers into each ith node calculated in terms of the temperatures at the $p + 1$ time increment. Thus,

$$q_i + \sum_j \frac{T_j^{p+1} - T_i^{p+1}}{R_{ij}} = C_i \frac{T_i^{p+1} - T_i^p}{\Delta\tau} \tag{4-42}$$

Now, as before, the set of equations produces an *implicit* set which must be solved simultaneously for the T_i^{p+1}, etc. The solution can be carried out by a number of methods, as discussed in Chap. 3. If the solution is to be performed with a Gauss-Seidel iteration technique, then Eq. (4-42) should be solved for T_i^{p+1} and expressed as

$$T_i^{p+1} = \frac{q_i + \sum_i (T_j^{p+1}/R_{ij}) + (C_i/\Delta\tau)T_i^p}{\sum_j (1/R_{ij}) + C_i/\Delta\tau} \tag{4-43}$$

It is interesting to note that in the steady-state limit of $\Delta\tau \to \infty$ this equation becomes identical with Eq. (3-32), the formulation we employed for the iterative solution in Chap. 3.

The stability requirement in the explicit formulation may be examined by solving Eq. (4-41) for T_i^{p+1}:

$$T_i^{p+1} = \left(q_i + \sum_j \frac{T_j^p}{R_{ij}}\right)\frac{\Delta\tau}{C_i} + \left(1 - \frac{\Delta\tau}{C_i}\sum_j \frac{1}{R_{ij}}\right)T_i^p \tag{4-44}$$

The value of q_i can influence the stability, but we can choose a safe limit by observing the behavior of the equation for $q_i = 0$. Using the same type of thermodynamic argument as with Eq. (4-31), we find that the coefficient of T_i^p cannot be negative. Our minimum stability requirement is therefore

$$1 - \frac{\Delta\tau}{C_i}\sum_j \frac{1}{R_{ij}} \geq 0 \tag{4-45}$$

Suppose we have a complicated numerical problem to solve with a variety of boundary conditions, perhaps nonuniform values of the space increments, etc. Once we have all the nodal resistances and capacities formulated, we then have the task of choosing the time increment $\Delta\tau$ to use for the calculation. To ensure stability we must keep $\Delta\tau$ equal to or less than a value obtained from the most restrictive nodal relation like Eq. (4-45). Solving for $\Delta\tau$ gives

$$\Delta\tau \leq \left[\frac{C_i}{\sum_j (1/R_{ij})}\right]_{\min} \quad \text{for stability} \tag{4-46}$$

While Eq. (4-44) is very useful in establishing the maximum allowable time increment, it may involve problems of round-off errors in computer solutions when small thermal resistances are employed. The difficulty may be alleviated by expressing T_i^{p+1} in the following form for calculation purposes:

$$T_i^{p+1} = \frac{\Delta\tau}{C_i}\left[q_i + \sum_j \frac{T_j^p - T_i^p}{R_{ij}}\right] + T_i^p \tag{4-47}$$

We should remark that the resistance-capacity formulation is easily adapted to take into account thermal-property variations with temperature. One need only calculate the proper values of ρ, c, and k for inclusion in the C_i and R_{ij}. Depending on the nature of the problem and accuracy required, it may be necessary to calculate new values of C_i and R_{ij} for each iteration. Example 4-16 illustrates the effects of variable conductivity.

■ **EXAMPLE 4-10** Sudden cooling of a rod

A steel rod [k = 50 W/m · °C] 3 mm in diameter and 10 cm long is initially at a uniform temperature of 200°C. At time zero it is suddenly immersed in a fluid having h = 50 W/m² · °C and T_∞ = 40°C while one end is maintained at 200°C. Determine the temperature distribution in the rod after 100 s. The properties of steel are ρ = 7800 kg/m³ and c = 0.47 kJ/kg · °C.

Solution

The selection of increments on the rod is as shown in the figure. The cross-sectional area of the rod is $A = \pi(1.5)^2 = 7.069$ mm². The volume element for nodes 1, 2, and 3 is

$$\Delta V = A\,\Delta x = (7.069)(25) = 176.725 \text{ mm}^3$$

Node 4 has a ΔV of half this value, or 88.36 mm³. We can now tabulate the various resistances and capacities for use in an explicit formulation. For nodes 1, 2, and 3 we have

$$R_{m+} = R_{m-} = \frac{\Delta x}{kA} = \frac{0.025}{(50)(7.069 \times 10^{-6})} = 70.731\text{°C/W}$$

and

$$R_\infty = \frac{1}{h(\pi d\,\Delta x)} = \frac{1}{(50)\pi(3 \times 10^{-3})(0.025)} = 84.883\text{°C/W}$$

$$C = \rho c\,\Delta V = (7800)(470)(1.7673 \times 10^{-7}) = 0.6479 \text{ J/°C}$$

For node 4 we have

$$R_{m+} = \frac{1}{hA} = 2829\text{°C/W} \qquad R_{m-} = \frac{\Delta x}{kA} = 70.731\text{°C/W}$$

$$C = \frac{\rho c\,\Delta V}{2} = 0.3240 \text{ J/°C} \qquad R_\infty = \frac{2}{h\pi d\,\Delta x} = 169.77\text{°C/W}$$

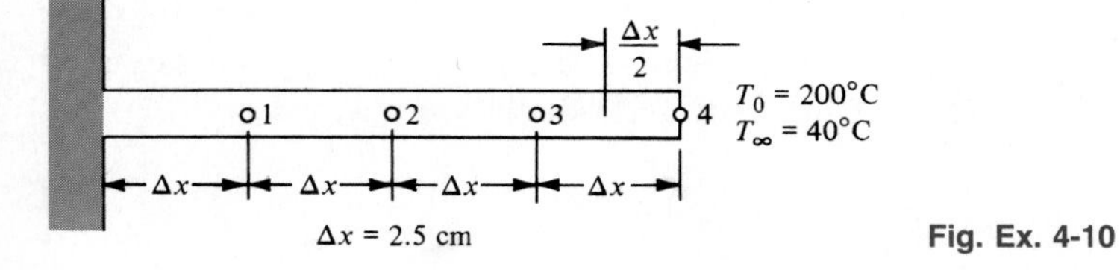

Fig. Ex. 4-10

To determine the stability requirement we form the following table:

Node	$\Sigma(1/R_{ij})$	C_i	$\dfrac{C_i}{\Sigma(1/R_{ij})}$, s
1	0.04006	0.6479	16.173
2	0.04006	0.6479	16.173
3	0.04006	0.6479	16.173
4	0.02038	0.3240	15.897

Thus node 4 is the most restrictive, and we must select $\Delta\tau < 15.9$ s. Since we wish to find the temperature distribution at 100 s, let us use $\Delta\tau = 10$ s and make the calculation for 10 time increments using Eq. (4-47) for the computation. We note, of course, that $q_i = 0$ because there is no heat generation. The calculations are shown in the following table.

Time increment	*Node temperature*			
	T_1	T_2	T_3	T_4
0	200	200	200	200
1	170.87	170.87	170.87	169.19
2	153.40	147.04	146.68	145.05
3	141.54	128.86	126.98	125.54
4	133.04	115.04	111.24	109.70
5	126.79	104.48	98.76	96.96
6	122.10	96.36	88.92	86.78
7	118.53	90.09	81.17	78.71
8	115.80	85.23	75.08	72.34
9	113.70	81.45	70.31	67.31
10	112.08	78.51	66.57	63.37

We can calculate the heat-transfer *rate* at the end of 100 s by summing the convection heat losses on the surface of the rod. Thus

$$q = \sum_i \frac{T_i - T_\infty}{R_{i\infty}}$$

and

$$q = \frac{200 - 40}{(2)(84.883)} + \frac{112.08 + 78.51 + 66.57 - (3)(40)}{84.883} + \left(\frac{1}{169.77} + \frac{1}{2829}\right)(63.37 - 40) = 2.704 \text{ W}$$

■ **EXAMPLE 4-11**

We can illustrate the calculation scheme for the implicit formulation by reworking Example 4-10 using only two time increments, that is, $\Delta\tau = 50$ s.

For this problem we employ the formulation indicated by Eq. (4-43), with $\Delta\tau = 50$ s. The following quantities are needed.

Node	$\dfrac{C_i}{\Delta\tau}$	$\displaystyle\sum_i \frac{1}{R_{ij}} + \frac{C_i}{\Delta\tau}$
1	0.01296	0.05302
2	0.01296	0.05302
3	0.01296	0.05302
4	0.00648	0.02686

We have already determined the R_{ij} in Example 4-10 and thus can insert them into Eq. (4-43) to write the nodal equations for the end of the first time increment, taking all $T_i^p = 200°$C. We use the prime to designate temperatures at the end of the time increment. For node 1,

$$0.05302T_1' = \frac{200}{70.731} + \frac{T_2'}{70.731} + \frac{40}{84.833} + (0.01296)(200)$$

For node 2,

$$0.05302T_2' = \frac{T_1'}{70.731} + \frac{T_3'}{70.731} + \frac{40}{84.833} + (0.01296)(200)$$

For nodes 3 and 4,

$$0.05302T_3' = \frac{T_2'}{70.731} + \frac{T_4'}{70.731} + \frac{40}{84.833} + (0.01296)(200)$$

$$0.02686T_4' = \frac{T_3'}{70.731} + \frac{40}{2829} + \frac{40}{169.77} + (0.00648)(200)$$

These equations can then be reduced to

$$\begin{aligned}
0.05302T_1' - 0.01414T_2' \qquad\qquad\qquad\qquad\qquad &= 5.8911 \\
-0.01414T_1' + 0.05302T_2' - 0.01414T_3' \qquad\qquad &= 3.0635 \\
-0.01414T_2' + 0.05302T_3' - 0.01414T_4' &= 3.0635 \\
-0.01414T_3' + 0.02686T_4' &= 1.5457
\end{aligned}$$

which have the solution

$$T_1' = 145.81°\text{C} \qquad T_2' = 130.12°\text{C}$$

$$T_3' = 125.43°\text{C} \qquad T_4' = 123.56°\text{C}$$

We can now apply the backward-difference formulation a second time using the double prime to designate the temperatures at the end of the second time increment:

$$0.05302T_1'' = \frac{200}{70.731} + \frac{T_2''}{70.731} + \frac{40}{84.833} + (0.01296)(145.81)$$

$$0.05302T_2'' = \frac{T_1''}{70.731} + \frac{T_3''}{70.731} + \frac{40}{84.833} + (0.01296)(130.12)$$

$$0.05302T_3'' = \frac{T_2''}{70.731} + \frac{T_4''}{70.731} + \frac{40}{84.833} + (0.01296)(125.43)$$

$$0.02686T_4'' = \frac{T_3''}{70.731} + \frac{40}{2829} + \frac{40}{169.77} + (0.00648)(123.56)$$

and this equation set has the solution

$$T_1'' = 123.81°\text{C} \qquad T_2'' = 97.27°\text{C}$$

$$T_3'' = 88.32°\text{C} \qquad T_4'' = 85.59°\text{C}$$

We find this calculation in substantial disagreement with the results of Example 4-10. With a larger number of time increments better agreement would be achieved. In a problem involving a large number of nodes, the implicit formulation might involve less computer time than the explicit method, and the purpose of this example has been to show how the calculation is performed.

■ **EXAMPLE 4-12** Cooling of a ceramic

A 1 by 2 cm ceramic strip [$k = 3.0$ W/m · °C] is embedded in a high-thermal-conductivity material, as shown, so that the sides are maintained at a constant temperature of 300°C. The bottom surface of the ceramic is insulated, and the top surface is exposed to a convection environment with $h = 200$ W/m² · °C and $T_\infty = 50$°C. At time zero the ceramic is uniform in temperature at 300°C. Calculate the temperatures at nodes 1 to 9 after a time of 12 s. For the ceramic $\rho = 1600$ kg/m³ and $c = 0.8$ kJ/kg · °C. Also calculate the total heat loss in this time.

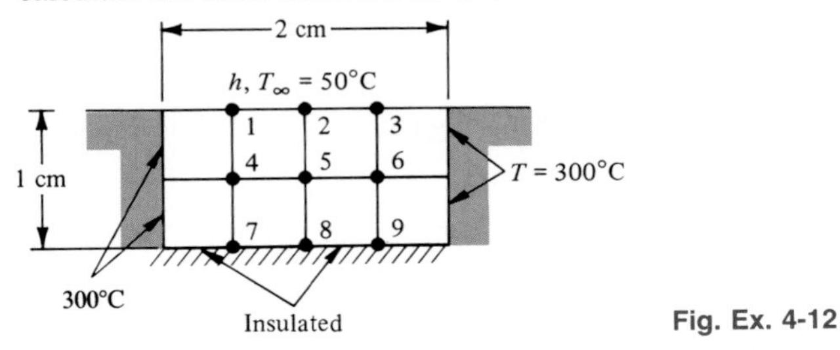

Fig. Ex. 4-12

Solution

We treat this as a two-dimensional problem with $\Delta x = \Delta y = 0.5$ cm. From symmetry $T_1 = T_3$, $T_4 = T_6$, and $T_7 = T_9$, so we have six unknown nodal temperatures. We now tabulate the various nodal resistances and capacities. For nodes 4 and 5

$$R_{m+} = R_{m-} = R_{n+} = R_{n-} = \frac{\Delta x}{kA} = \frac{0.005}{(3.0)(0.005)} = 0.3333$$

For nodes 1 and 2

$$R_{m+} = R_{m-} = \frac{\Delta x}{kA} = \frac{(0.005)(2)}{(3.0)(0.005)} = 0.6667°\text{C/W} \qquad R_{n-} = 0.3333°\text{C/W}$$

$$R_{n+} = \frac{1}{h\,\Delta x} = \frac{1}{(200)(0.005)} = 1.0°\text{C/W}$$

For nodes 7 and 8

$$R_{m+} = R_{m-} = 0.6667°\text{C/W} \qquad R_{n+} = 0.3333°\text{C/W} \qquad R_{n-} = \infty$$

For nodes 1, 2, 7, and 8 the capacities are

$$C = \frac{\rho c(\Delta x)^2}{2} = \frac{(1600)(800)(0.005)^2}{2} = 16\ \text{J/°C}$$

For nodes 4 and 5

$$C = \rho c(\Delta x)^2 = 32\ \text{J/°C}$$

The stability requirement for an explicit solution is now determined by tabulating the following quantities:

Node	$\sum \frac{1}{R_{ij}}$	C_i	$\frac{C_i}{\Sigma(1/R_{ij})}$, s
1	7	16	2.286
2	7	16	2.286
4	12	32	2.667
5	12	32	2.667
7	6	16	2.667
8	6	16	2.667

Thus the two convection nodes control the stability requirement, and we must choose $\Delta\tau \le 2.286$ s. Let us choose $\Delta\tau = 2.0$ s and make the calculations for six time increments with Eq. (4-47). We note once again the symmetry considerations when calculating the temperatures of nodes 2, 5, and 8, that is, $T_1 = T_3$, etc. The calculations are shown in the following table.

Time increment	*Node temperature*					
	T_1	T_2	T_4	T_5	T_7	T_8
0	300	300	300	300	300	300
1	268.75	268.75	300	300	300	300
2	258.98	253.13	294.14	294.14	300	300
3	252.64	245.31	289.75	287.55	297.80	297.80
4	284.73	239.48	285.81	282.38	295.19	293.96
5	246.67	235.35	282.63	277.79	292.34	290.08
6	243.32	231.97	279.87	273.95	289.71	286.32

The total heat loss during the 12-s time interval is calculated by summing the heat loss of each node relative to the initial temperature of 300°C. Thus

$$q = \Sigma C_i(300 - T_i)$$

where q is the heat *loss*. For this summation, since the constant-temperature boundary nodes experience no change in temperature, they can be left out. Recalling that $T_1 = T_3$, $T_4 = T_6$, and $T_7 = T_9$, we have

$$\begin{aligned}\Sigma C_i(300 - T_i) &= \text{nodes } (1, 2, 3, 7, 8, 9) + \text{nodes } (4, 5, 6)\\ &= 16[(6)(300) - (2)(243.32) - 231.97 - (2)(289.71)\\ &\quad - 286.32] + 32[(3)(300) - (2)(279.87) - 273.95]\\ &= 5572.3 \text{ J/m length of strip}\end{aligned}$$

The *average rate* of heat loss for the 12-s time interval is

$$\frac{q}{\Delta\tau} = \frac{5572.3}{12} = 464.4 \text{ W} \quad [1585 \text{ Btu/h}]$$

■ **EXAMPLE 4-13** Cooling of a steel rod, nonuniform *h*

A nickel-steel rod having a diameter of 2.0 cm is 10 cm long and initially at a uniform temperature of 200°C. It is suddenly exposed to atmospheric air at 30°C while one end

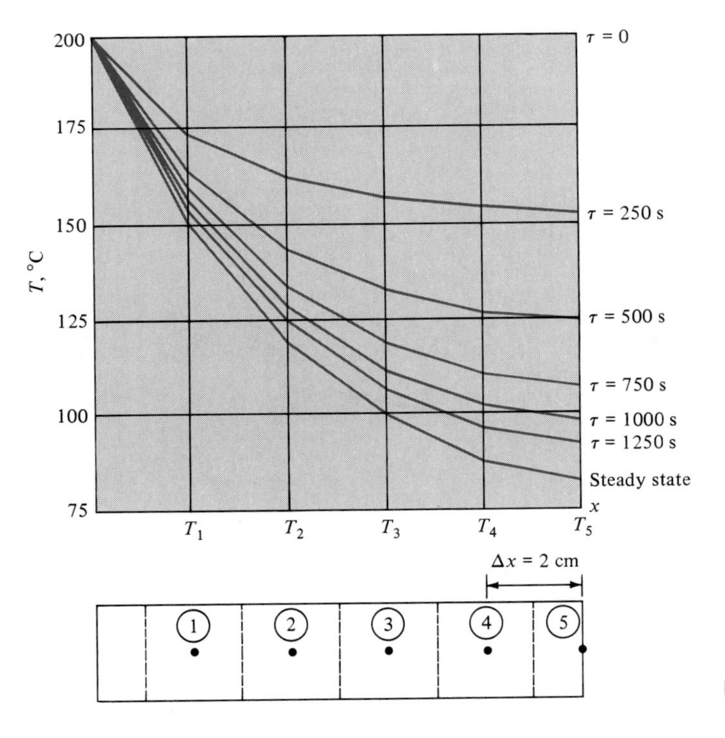

Fig. Ex. 4-13

of the rod is maintained at 200°C. The convection heat-transfer coefficient can be computed from

$$h = 9.0\ \Delta T^{0.175}\ \text{W/m}^2 \cdot {}^\circ\text{C}$$

where ΔT is the temperature difference between the rod and air surroundings. The properties of nickel steel may be taken as $k = 12$ W/m · °C, $c = 0.48$ kJ/kg · °C, and $\rho = 7800$ kg/m³. Using the numerical method, (*a*) determine the temperature distribution in the rod after 250, 500, 750, 1000, 1250 s, and for steady state; (*b*) determine the steady-state temperature distribution for a constant $h = 22.11$ W/m² · °C and compare with an analytical solution.

Solution

Five nodes are chosen as shown in the accompanying figure with $\Delta x = 2.0$ cm. The capacitances are then

$$C_1 = C_2 = C_3 = C_4 = \frac{(7800)(480)\pi(0.02)^2(0.02)}{4} = 23.524\ \text{J/°C}$$

$$C_5 = \tfrac{1}{2}C_1 = 11.762\ \text{J/°C}$$

The resistances for nodes 1, 2, 3, and 4 are

$$\frac{1}{R_{m+}} = \frac{1}{R_{m-}} = \frac{kA}{\Delta x} = \frac{(12)\pi(0.02)^2}{(4)(0.02)} = 0.188496$$

$$\frac{1}{R_\infty} = hP\,\Delta x = (9.0)\pi(0.02)(0.02)(T - 30)^{0.175} = (1.131 \times 10^{-2})(T - 30)^{0.175}$$

For node 5

$$\frac{1}{R_{m-}} = 0.188496$$

$$\frac{1}{R_{m+}} = hA = 9.0\,\frac{\pi(0.02)^2}{4}(T - 30)^{0.175} = (2.827 \times 10^{-3})(T - 30)^{0.175}$$

$$\frac{1}{R_{5\infty}} = \frac{1}{2R_{1\infty}} = (5.655 \times 10^{-3})(T - 30)^{0.175}$$

where $T_\infty = 30$°C for all nodes. We can compute the following table for worst-case conditions of $T = 200$°C throughout the rod. The stability requirement so established will then work for all other temperatures.

Node	$\Sigma(1/R_{ij})\vert_{\min}$	$\dfrac{C_i}{\Sigma(1/R_{ij})}$, s
1	0.4048	58.11
2	0.4048	58.11
3	0.4048	58.11
4	0.4048	58.11
5	0.2093	56.197

Thus, time steps below 56 s will ensure stability. The computational procedure is complicated by the fact that the convection-resistance elements must be recalculated for each time step. Selecting $\Delta\tau = 50$ s, we have:

Node	$\Delta\tau/C_i$
1	2.1255
2	2.1255
3	2.1255
4	2.1255
5	4.251

We then use the explicit formulation of Eq. (4-47) with no heat generation. The computational algorithm is thus:

1. Compute R_∞ values for the initial condition.
2. Compute temperatures at next time increment using Eq. (4-47).
3. Recalculate R_∞ values based on new temperatures.
4. Repeat temperature calculations and continue until the temperature distributions are obtained at the desired times.

Results of these calculations are shown in the accompanying figure.

To determine the steady-state distribution we could carry the unsteady method forward a large number of time increments or use the steady-state method and an iterative approach. The iterative approach is required because the equations are nonlinear as a result of the variations in the convection coefficient.

We still use a resistance formulation, which is now given as Eq. (3-31):

$$\sum \frac{T_j - T_i}{R_{ij}} = 0$$

The computational procedure is:

1. Calculate R_∞ values for all nodes assuming all $T_i = 200°\text{C}$.
2. Formulate nodal equations for the T_i's.
3. Solve the equations by an appropriate method.
4. Recalculate R_∞ values based on T_i values obtained in step 3.
5. Repeat the procedure until there are only small changes in T_i's.

The results of this iteration are shown in the following table:

Iteration	T_1, °C	T_2, °C	T_3, °C	T_4, °C	T_5, °C
1	148.462	114.381	92.726	80.310	75.302
2	151.381	119.557	99.409	87.853	83.188
3	151.105	119.038	98.702	87.024	82.306
4	151.132	119.090	98.774	87.109	82.396

This steady-state temperature distribution is also plotted with the transient profiles.

The value of h for $T_i = 200°\text{C}$ is 22.11 W/m² · °C, so the results of the first iteration correspond to a solution for a constant h of this value. The exact analytical solution is given in Eq. (2-34) as

$$\frac{\theta}{\theta_0} = \frac{T - T_\infty}{T_0 - T_\infty} = \frac{\cosh m(L - x) + [h/km] \sinh m(L - x)}{\cosh mL + [h/km] \sinh mL}$$

The required quantities are

$$m = \left(\frac{hP}{kA}\right)^{1/2} = \left[\frac{(22.11)\pi(0.02)}{(12)\pi(0.01)^2}\right]^{1/2} = 19.1964$$

$$mL = (19.1964)(0.1) = 1.91964$$

$$h/km = \frac{22.22}{(12)(19.1964)} = 0.09598$$

The temperatures at the nodal points can then be calculated and compared with the numerical results in the following table. As can be seen, the agreement is excellent.

Node	x, *m*	$(\theta/\theta_0)_{\text{num}}$	$(\theta/\theta_0)_{\text{anal}}$	*Percent deviation*
1	0.02	0.6968	0.6949	0.27
2	0.04	0.4964	0.4935	0.59
3	0.06	0.3690	0.3657	0.9
4	0.08	0.2959	0.2925	1.16
5	0.1	0.2665	0.2630	1.33

We may also check the heat loss with that predicted by the analytical relation in Eq. (2-34). When numerical values are inserted we obtain

$$q_{\text{anal}} = 11.874 \text{ W}$$

The heat loss for the numerical model is computed by summing the convection loss from the six nodes (including base node at 200°C). Using the temperatures for the first iteration corresponding to $h = 22.11$ W/m²·°C,

$$\begin{aligned} q &= (22.11)\pi(0.02)(0.02)[(200 - 30)(\tfrac{1}{2}) + (148.462 - 30) \\ &\quad + (114.381 - 30) + (92.726 - 30) + (80.31 - 30) \\ &\quad + (75.302 - 30)(\tfrac{1}{2})] + (22.11)\pi(0.01)^2(75.302 - 30) \\ &= 12.082 \text{ W} \end{aligned}$$

We may make a further check by calculating the energy conducted in the base. This must be the energy conducted to node 1 plus the convection lost by the base node or

$$q = (12)\pi(0.01)^2 \frac{(200 - 148.462)}{0.02} + (22.11)\pi(0.02)(0.01)(200 - 30)$$

$$= 12.076 \text{ W}$$

This agrees very well with the convection calculation and both are within 1.8 percent of the analytical value.

The results of this example illustrate the power of the numerical method in solving problems which could not be solved in any other way. Furthermore, only a modest number of nodes, and thus modest computation facilities, may be required to obtain a sufficiently accurate solution. For example, the accuracy with which h will be known is typically $\pm$ 10 to 15 percent. This would overshadow any inaccuracies introduced by using relatively large nodes, as was done here.

■ EXAMPLE 4-14 Radiation heating and cooling

The ceramic wall shown is initially uniform in temperature at 20°C and has a thickness of 3.0 cm. It is suddenly exposed to a radiation source on the right side at 1000°C. The left side is exposed to room air at 20°C with a radiation surrounding temperature of 20°C. Properties of the ceramic are $k = 3.0$ W/m · °C, $\rho = 1600$ kg/m³, and $c = 0.8$ kJ/kg · °C. Radiation heat transfer with the surroundings at T_r may be calculated from

$$q_r = \sigma\epsilon A(T^4 - T_r^4) \qquad \text{W} \tag{a}$$

where $\sigma = 5.669 \times 10^{-8}$, $\epsilon = 0.8$, and T is in degrees Kelvin. The convection heat-transfer coefficient from the left side of the plate is given by

$$h = 1.92\Delta T^{1/4} \qquad \text{W/m}^2 \cdot {}^\circ\text{C} \tag{b}$$

Convection on the right side is negligible. Determine the temperature distribution in the plate after 15, 30, 45, 60, 90, 120, and 150 s. Also determine the steady-state temperature distribution. Calculate the total heat gained by the plate for these times.

Solution

We divide the wall into five nodes as shown and must express temperatures in degrees Kelvin because of the radiation boundary condition. For node 1 the transient energy equation is

$$\sigma\epsilon(293^4 - T_1^{p4}) - 1.92(T_1^p - 293)^{5/4} + \frac{k}{\Delta x}(T_2^p - T_1^p) = \rho c \frac{\Delta x}{2} \frac{T_1^{p+1} - T_1^p}{\Delta\tau} \tag{c}$$

Similarly, for node 5

$$\sigma\epsilon(1273^4 - T_5^{p4}) + \frac{k}{\Delta x}(T_4^p - T_5^p) = \rho c \frac{\Delta x}{2} \frac{T_5^{p+1} - T_5^p}{\Delta\tau} \tag{d}$$

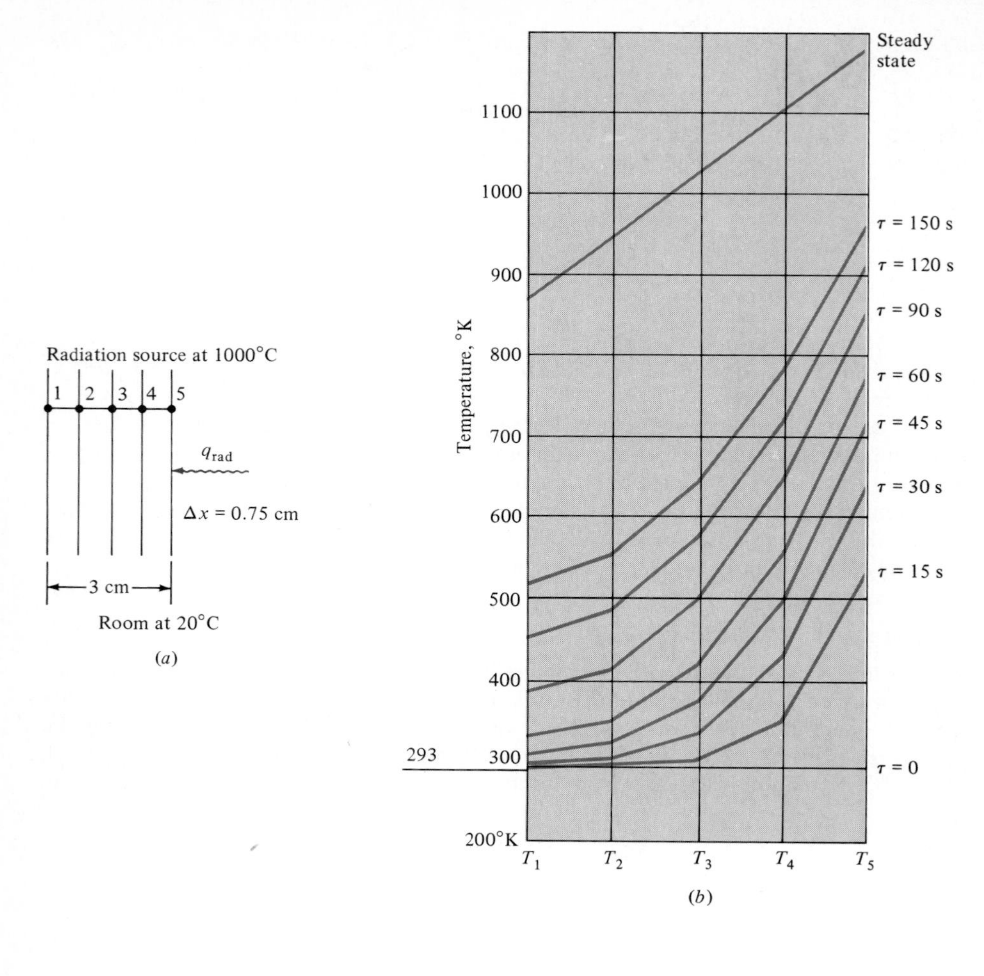

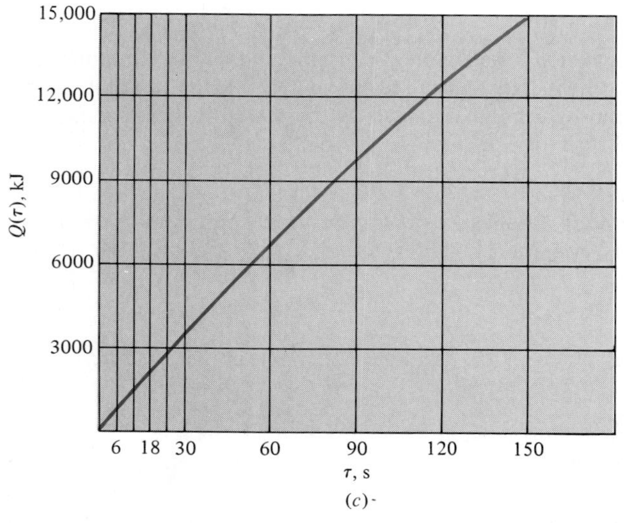

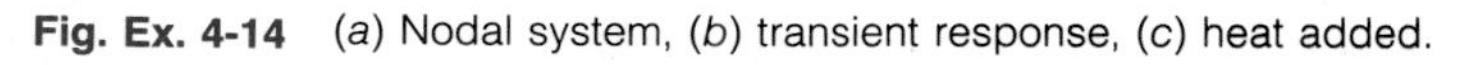

Fig. Ex. 4-14 (*a*) Nodal system, (*b*) transient response, (*c*) heat added.

Equations (c) and (d) may be subsequently written

$$T_1^{p+1} = \frac{\Delta\tau}{C_1}\left[\sigma\epsilon(293^2 + T_1^{p2})(293 + T_1^p)(293) - 1.92(T_1^p - 293)^{1/4}(293) + \frac{k}{\Delta x}T_2^p\right]$$
$$+ \left\{1 - \frac{\Delta\tau}{C_1}\left[\sigma\epsilon(293^2 + T_1^{p2})(293 + T_1^p) - 1.92(T_1^p - 293)^{1/4} + \frac{k}{\Delta x}\right]\right\}T_1^p \quad (e)$$

$$T_5^{p+1} = \frac{\Delta\tau}{C_5}\left[\sigma\epsilon(1273^2 + T_5^{p2})(1273 + T_5^p)(1273) + \frac{k}{\Delta x}T_4^p\right]$$
$$+ \left\{1 - \frac{\Delta\tau}{C_5}\left[\sigma\epsilon(1273^2 + T_5^{p2})(1273 + T_5^p) + \frac{k}{\Delta x}\right]\right\}T_5^p \quad (f)$$

where $C_1 = C_5 = \rho c\,\Delta x/2$. For the other three nodes the expressions are much simpler:

$$T_2^{p+1} = \frac{\Delta\tau}{C_2}\frac{k}{\Delta x}(T_1^p + T_3^p) + \left(1 - \frac{2k\,\Delta\tau}{C_2\,\Delta x}\right)T_2^p \quad (g)$$

$$T_3^{p+1} = \frac{\Delta\tau}{C_3}\frac{k}{\Delta x}(T_2^p + T_4^p) + \left(1 - \frac{2k\,\Delta\tau}{C_3\,\Delta x}\right)T_3^p \quad (h)$$

$$T_4^{p+1} = \frac{\Delta\tau}{C_4}\frac{k}{\Delta x}(T_3^p + T_5^p) + \left(1 - \frac{2k\,\Delta\tau}{C_3\,\Delta x}\right)T_4^p \quad (i)$$

where $C_2 = C_3 = C_4 = \rho c\ \Delta x$. So, to determine the transient response, we simply choose a suitable value of $\Delta\tau$ and march through the calculations. The stability criterion is such that the coefficients of the last term in each equation cannot be negative. For Eqs. (g), (h), and (i) the maximum allowable time increment is

$$\Delta\tau_{\text{max}} = \frac{C_3\,\Delta x}{2k} = \frac{(1600)(800)(0.0075)^2}{(2)(3)} = 12\text{ s}$$

For Eq. (f), the worst case is at the start when $T_5^p = 20°\text{C} = 293$ K. We have

$$C_5 = \frac{(1600)(800)(0.0075)}{2} = 4800$$

so that

$$\Delta\tau_{\text{max}} = \frac{4800}{(5.669 \times 10^{-8})(0.8)(1273^2 + 293^2)(1273 + 293) + 3.0/0.0075}$$
$$= 9.43\text{ s}$$

For node 1 [Eq. (e)] the most restrictive condition occurs when $T_1^p = 293$. We have

$$C_1 = C_5 = 4800$$

so that

$$\Delta\tau_{\text{max}} = \frac{4800}{(5.669 \times 10^{-8})(0.8)(293^2 + 293^2)(293 + 293) - 0 + 3.0/0.0075}$$
$$= 11.86\text{ s}$$

So, from these calculations we see that node 5 is most restrictive and we must choose $\Delta\tau < 9.43$ s.

The calculations were performed with $\Delta\tau = 3.0$ s, and the results are shown in the accompanying figures. To compute the heat added at any instant of time we perform the sum:

$$Q(\tau) = \Sigma C_i(T_i - 293) \tag{j}$$

and plot the results in Fig. Ex. 4-14*c*.

■ **EXAMPLE 4-15** Transient conduction with heat generation

The plane wall shown has internal heat generation of 50 MW/m³ and thermal properties of $k = 19$ W/m · °C, $\rho = 7800$ kg/m³, and $c = 460$ J/kg · °C. It is initially at a uniform temperature of 100°C and is suddenly subjected to the heat generation and the convective boundary conditions indicated in the figure. Calculate the temperature distribution after several time increments.

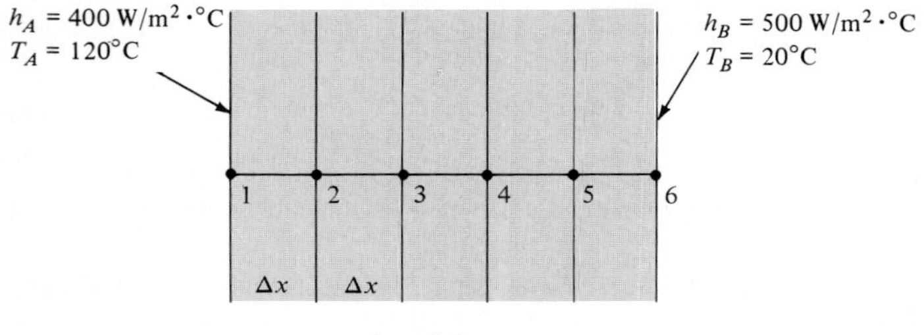

Fig. Ex. 4-15(*a*)

Solution

We use this resistance and capacity formulation and write, for unit area,

$$1/R_{12} = kA/\Delta x = (19)(1)/0.001 = 19{,}000 \text{ W/°C}$$

All the conduction resistances have this value. Also,

$$1/R_{1A} = hA = (400)(1) = 400 \text{ W/°C}$$

$$1/R_{1B} = hA = (500)(1) = 500 \text{ W/°C}$$

The capacities are

$$C_1 = C_6 = \rho(\Delta x/2) = (7800)(0.001/2)(460) = 1794 \text{ J/°C}$$

$$C_2 = C_3 = C_4 = C_5 = \rho(\Delta x)c = 3588 \text{ J/°C}$$

We next tabulate values.

Node	$\Sigma(1/R_{ij})$	C_i	$\dfrac{C_i}{\Sigma(1/R_{ij})}$
1	19,400	1794	0.092
2	38,000	3588	0.094
3	38,000	3588	0.094
4	38,000	3588	0.094
5	38,000	3588	0.094
6	19,500	1794	0.092

Any time increment $\Delta\tau$ less than 0.09 s will be satisfactory. The nodal equations are now written in the form of Eq. (4-47) and the calculation marched forward on a computer.

The heat-generation terms are

$$q_i = \dot{q}\,\Delta V_i$$

so that

$$q_1 = q_6 = (50 \times 10^6)(1)(0.001/2) = 25{,}000 \text{ W}$$

$$q_2 = q_3 = q_4 = q_5 = (50 \times 10^6)(1)(0.001) = 50{,}000 \text{ W}$$

The computer results for several time increments of 0.09 s are shown in the following table. Because the solid stays nearly uniform in temperature at any instant of time it behaves almost like a lumped capacity. The temperature of node 3 is plotted versus time to illustrate this behavior.

	Number of time increments ($\Delta\tau$ = *0.09 s*)			
Node	5	20	100	200
1	106.8826	123.0504	190.0725	246.3855
2	106.478	122.8867	190.9618	248.1988
3	106.1888	122.1404	190.7033	248.3325
4	105.3772	120.9763	189.3072	246.7933
5	104.4622	119.2217	186.7698	243.5786
6	102.4416	117.0056	183.0735	238.6773
Node	500	800	1200	3000
1	320.5766	340.1745	346.0174	347.2085
2	323.6071	343.5267	349.4654	350.676
3	324.2577	344.3137	350.2931	351.512
4	322.5298	342.536	348.5006	349.7165
5	318.4229	338.1934	344.0877	345.2893
6	311.9341	331.2853	337.0545	338.2306

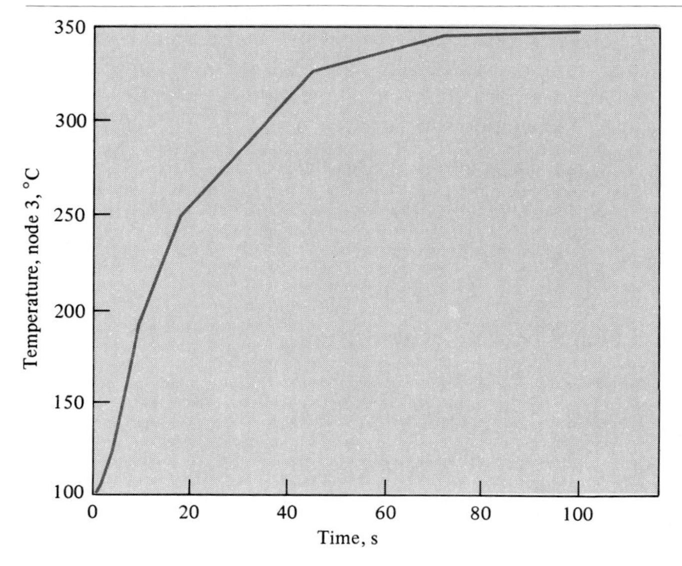

Fig. Ex. 4-15(*b*)

EXAMPLE 4-16 Numerical solution for variable conductivity

A 4.0-cm-thick slab of stainless steel (18% Cr, 8% Ni) is initially at a uniform temperature of 0°C with the left face perfectly insulated as shown in the accompanying figure. The right face is suddenly raised to a constant 1000°C by an intense radiation source. Calculate the temperature distribution after (*a*) 25 s, (*b*) 50 s, (*c*) 100 s, (*d*) an interval long enough for the slab to reach a steady state, taking into account variation in thermal conductivity. Approximate the conductivity data in Appendix A with a linear relation. Repeat the calculation for the left face maintained at 0°C.

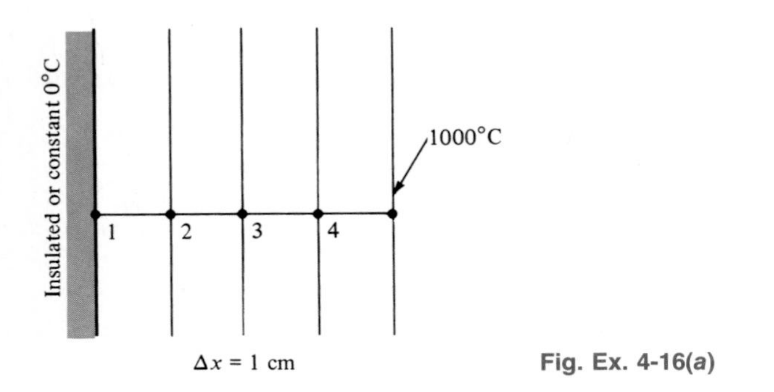

Fig. Ex. 4-16(*a*)

Solution

From Table A-2 we have $k = 16.3$ W/m · °C at 0°C and $k = 31$ W/m · °C at 1000°C. A linear relation for k is assumed so that

$$k = k_0(1 + \beta T)$$

where T is in degrees Celsius. Inserting the data gives

$$k = 16.3(1 + 9.02 \times 10^{-4}T) \text{ W/m} \cdot {}^\circ\text{C}$$

We also have $\rho = 7817$ kg/m³ and $c = 460$ J/kg · °C, and use the thermal resistance-capacitance formula assuming that the resistances are evaluated at the arithmetic mean of their connecting nodal temperatures; i.e., R_{3-4} is evaluated at $(T_3 + T_4)/2$.

First, the thermal capacities are evaluated for unit area:

$$C_1 = \rho(\Delta x/2)c = (7817)(0.01/2)(460) = 17980 \text{ J/m}^2 \cdot {}^\circ\text{C}$$

$$C_2 = C_3 = C_4 = \rho(\Delta x)c = (7817)(0.01)(460) = 35960 \text{ J/m}^2 \cdot {}^\circ\text{C}$$

For the resistances we have the form, for unit area,

$$1/R = k/\Delta x = k_0(1 + \beta T)/\Delta x$$

Evaluating at the mean temperatures between nodes gives

$$1/R_{1-2} = (16.3)[1 + 4.51 \times 10^{-4}(T_1 + T_2)]/0.01 = 1/R_{2-1}$$

$$1/R_{2-3} = (16.3)[1 + 4.51 \times 10^{-4}(T_2 + T_3)]/0.01 = 1/R_{3-2}$$

$$1/R_{3-4} = (16.3)[1 + 4.51 \times 10^{-4}(T_3 + T_4)]/0.01 = 1/R_{4-3}$$

$$1/R_{4-1000} = (16.3)[1 + 4.51 \times 10^{-4}(T_4 + T_{1000})]/0.01 = 1/R_{1000-4}$$

The stability requirement is most severe on node 1 because it has the lowest capacity. To be on the safe side we can choose a large k of about 31 W/m · °C and calculate

$$\Delta\tau_{max} = \frac{(17{,}980)(0.01)}{31} = 5.8 \text{ s}$$

The nodal equations are now written in the form of Eq. (4-47); viz., the equation for node 2 would be

$$T_2^{p+1} = \frac{\Delta\tau}{C_2}\{1630[1 + 4.51 \times 10^{-4}(T_1^p + T_2^p)](T_1^p - T_2^p) + 1630[1 + 4.51 \times 10^{-4}(T_3^p + T_2^p)](T_3^p - T_2^p)\} + T_2^p$$

A computer solution has been performed with $\Delta\tau = 5$ s and the results are shown in the tables. The steady-state solution for the insulated left face is, of course, a constant 1000°C. The steady-state distribution for the left face at 0°C corresponds to Eq. (2-2) of Chap. 2. Note that, because of the nonconstant thermal conductivity, the steady-state temperature profile is not a straight line.

Temperatures for Left Face at Constant 0°C, $\Delta\tau = 5$ s

Node	25 s	50 s	100 s	*Steady state*
1	0	0	0	0
2	94.57888	236.9619	308.2699	317.3339
3	318.7637	486.5802	565.7786	575.9138
4	653.5105	748.1359	793.7976	799.7735

Temperatures for Left Face Insulated, $\Delta\tau = 5$ s

Node	25 s	50 s	100 s	*Steady state*
1	30.55758	232.8187	587.021	1000
2	96.67601	310.1737	623.5018	1000
3	318.7637	505.7613	721.5908	1000
4	653.5105	752.3268	855.6965	1000

These temperatures are plotted in the accompanying figure.

The purpose of this example has been to show how the resistance-capacity formulation can be used to take into account property variations in a rather straightforward way. These variations may or may not be important when one considers uncertainties in boundary conditions.

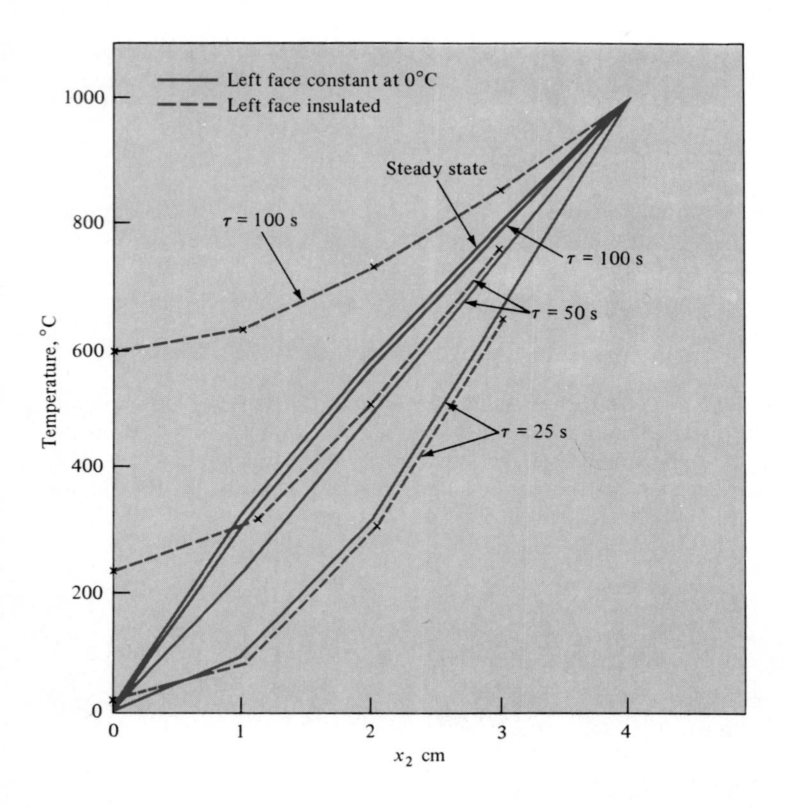

Fig. Ex. 4-16(*b*)

■ 4-8 GRAPHICAL ANALYSIS—THE SCHMIDT PLOT

A graphical technique of historical note may be employed to establish transient temperature distributions when the problem is one-dimensional. The method is based on the choice of the parameter

$$\frac{(\Delta x)^2}{\alpha\,\Delta\tau} = 2 \tag{4-48}$$

so that the temperature at any node after the time increment $\Delta\tau$ is the arithmetic average of the temperatures of the adjacent nodes at the beginning of the time increment. Such an arithmetic average is very easy to construct graphically, as shown in Fig. 4-22. The value of T_m^{p+1} is obtained by drawing a line between T_{m-1}^p and T_{m+1}^p. Thus, to find the temperature distribution in a solid after some specified time, the solid is divided into increments of Δx. Then, by using Eq. (4-48), the value of $\Delta\tau$ is obtained. This value of $\Delta\tau$, when divided into the total time, gives the number of time increments necessary to establish the desired temperature distribution. The graphical construction is repeated until the final temperature distribution is obtained. A nonintegral number of time increments is usually required, and it will probably be necessary to interpolate between the last two increments to obtain the final temperature distribution.

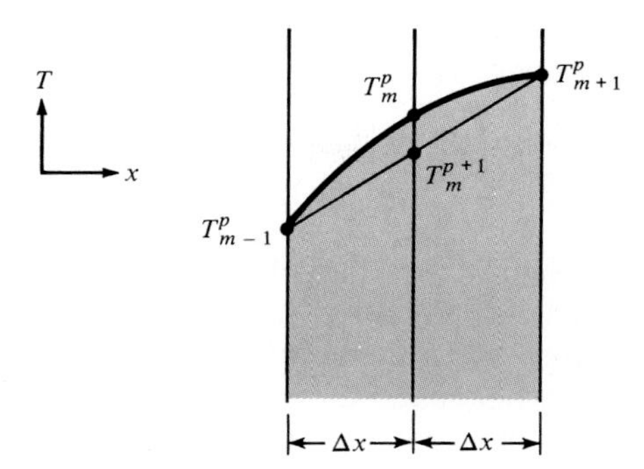

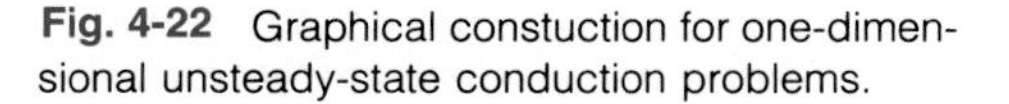

Fig. 4-22 Graphical constuction for one-dimensional unsteady-state conduction problems.

An example of the method is shown in Fig. 4-23, where an initial temperature distribution is given and the construction is carried out for four time increments. The boundary temperatures are maintained at constant values throughout the cooling process indicated in this example. Notice that the construction approaches the steady-state straight-line temperature distribution with increasing time.

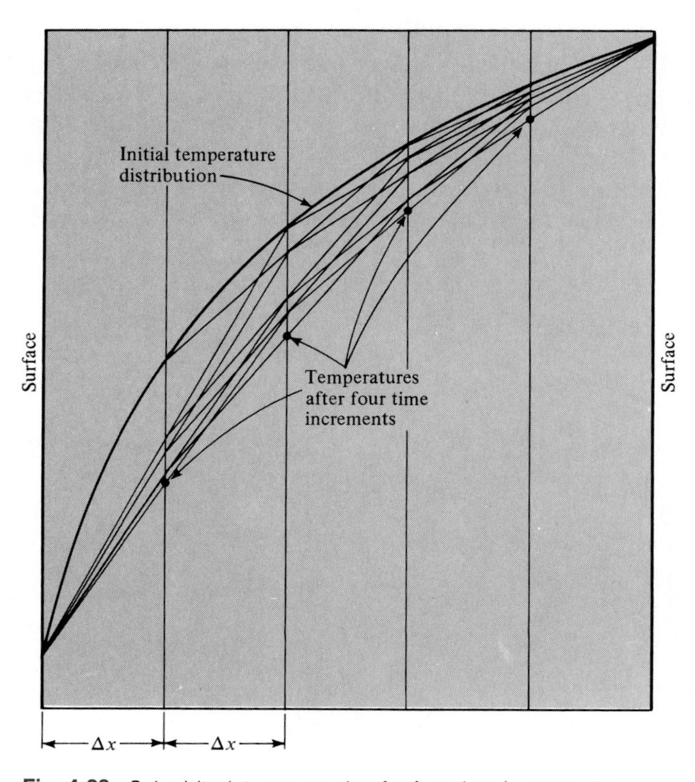

Fig. 4-23 Schmidt-plot construction for four time increments.

When a convection boundary condition is involved, the construction at the boundary must be modified. Rewriting Eq. (4-39), we have

$$\left.\frac{\partial T}{\partial x}\right]_{\text{wall}} = \frac{T_w - T_\infty}{k/h} \tag{4-49}$$

and the temperature gradient at the surface is approximated by the construction shown in Fig. 4-24. A line is drawn between the temperature T_{m+1} and the environment temperature T_∞. The intersection of this line with the surface determines the surface temperature at that particular time. This type of construction is used at each time increment to establish the surface temperature. Once this temperature is established, the construction to determine the internal temperatures in the solid proceeds as described above. An example of the construction for the convection boundary-condition problem with four time increments is shown in Fig. 4-25. In this example the temperature of the right face and the environment temperature T_∞ are maintained constant. If the environment temperature changes with time, according to some known variation, this could easily be incorporated in the construction by moving the T_∞ point up or down as required. In a similar fashion a variable heat-transfer coefficient could be considered by changing the value of k/h, according to some specified variation, and moving the environment point in or out a corresponding distance.

Refinements in the Schmidt graphical method are discussed by Jakob [5], particularly the techniques for improving accuracy at the boundary for either convection or other boundary conditions. The accuracy of the method is improved when smaller Δx increments are taken, but this requires a larger number of time increments to obtain a temperature distribution after a given time.

Graphical techniques are seldom used anymore because of the ready avail-

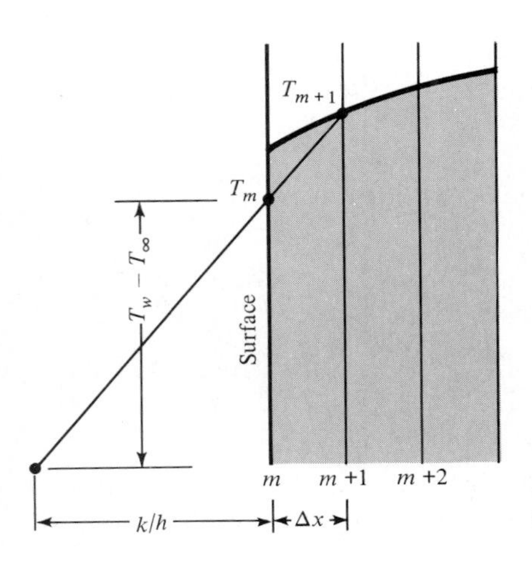

Fig. 4-24 Graphical technique of representing convection boundary condition with the Schmidt plot.

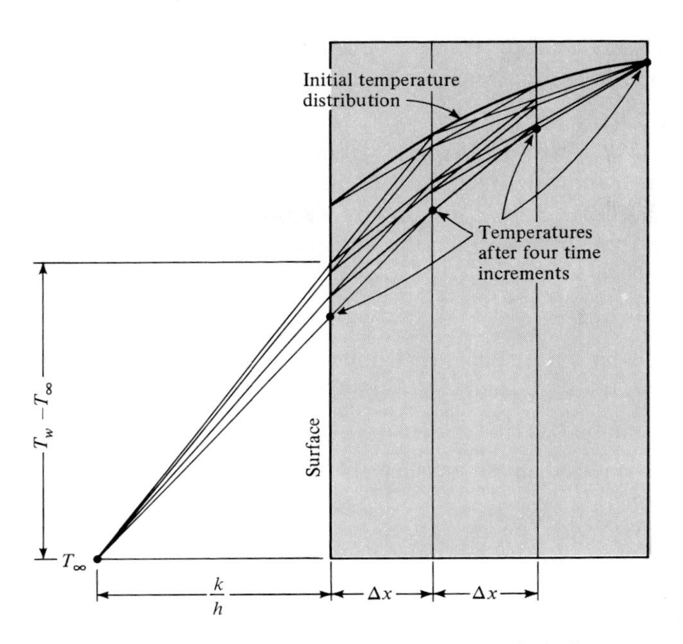

Fig. 4-25 Schmidt plot for four time increments, including convection boundary condition.

ability of computing power. We mention them here mainly to indicate the resourcefulness of heat-transfer workers before the time of computers.

4-9 SUMMARY

In progressing through this chapter the reader will have noted analysis techniques of varying complexity, ranging from simple lumped-capacity systems to numerical computer solutions. At this point some suggestions are offered for a general approach to follow in the solution of transient heat-transfer problems.

1. First, determine if a lumped capacity analysis can apply. If so, you may be led to a much easier calculation.
2. Check to see if an analytical solution is available with such aids as the Heisler charts and approximations.
3. If analytical solutions are very complicated, *even when already available,* move directly to numerical techniques. This is particularly true where repetitive calculations must be performed.
4. When approaching a numerical solution recognize the large uncertainties present in convection and radiation boundary conditions. Do not insist upon

a large number of nodes and computer time (and chances for error) which cannot possibly improve upon the basic uncertainty in the boundary conditions.

5. Finally, recognize that it is a rare occurrence when one has a "pure" conduction problem; there is almost always a coupling with convection and radiation. The reader should keep this in mind as we progress through subsequent chapters which treat heat convection and radiation in detail.

REVIEW QUESTIONS

1 What is meant by a lumped capacity? What are the physical assumptions necessary for a lumped-capacity unsteady-state analysis to apply?

2 What is meant by a semi-infinite solid?

3 What initial conditions are imposed on the transient solutions presented in graphical form in this chapter?

4 What boundary conditions are applied to problems in this chapter?

5 Define the error function.

6 Define the Biot and Fourier numbers.

7 Describe how one-dimensional transient solutions may be used for solution of two- and three-dimensional problems.

8 What are the advantages and disadvantages of the forward- and backward-difference formulations in the unsteady-state numerical method? Under what conditions would you choose one method over the other?

PROBLEMS

4-1 A copper sphere initially at a uniform temperature T_0 is immersed in a fluid. Electric heaters are placed in the fluid and controlled so that the temperature of the fluid follows a periodic variation given by

$$T_\infty - T_m = A \sin \omega\tau$$

where T_m = time-average mean fluid temperature
A = amplitude of temperature wave
ω = frequency

Derive an expression for the temperature of the sphere as a function of time and the heat-transfer coefficient from the fluid to the sphere. Assume that the temperatures of the sphere and fluid are uniform at any instant so that the lumped-capacity method of analysis may be used.

4-2 An infinite plate having a thickness of 2.5 cm is initially at a temperature of 150°C, and the surface temperature is suddenly lowered to 30°C. The thermal diffusivity of the material is 1.8×10^{-6} m²/s. Calculate the center-plate temperature after 1 min by summing the first four nonzero terms of Eq. (4-3). Check the answer using the Heisler charts.

4-3 What error would result from using the first four terms of Eq. (4-3) to compute the temperature at $\tau = 0$ and $x = L$?

4-4 A solid body at some initial temperature T_0 is suddenly placed in a room where the air temperature is T_∞ and the walls of the room are very large. The heat-transfer coefficient for the convection heat loss is h, and the surface of the solid may be assumed black. Assuming that the temperature in the solid is uniform at any instant, write the differential equation for the variation in temperature with time, considering both radiation and convection.

4-5 A 30 by 30 cm slab of copper 5 cm thick at a uniform temperature of 260°C suddenly has its surface temperature lowered to 35°C. Using the concepts of thermal resistance and capacitance and the lumped-capacity analysis, find the time at which the center temperature becomes 90°C; $\rho = 8900$ kg/m³, $c_p = 0.38$ kJ/kg · °C, and $k = 370$ W/m · °C.

4-6 A piece of aluminum weighing 5.5 kg and initially at a temperature of 290°C is suddenly immersed in a fluid at 15°C. The convection heat-transfer coefficient is 58 W/m² · °C. Taking the aluminum as a sphere having the same weight as that given, estimate the time required to cool the aluminum to 90°C, using the lumped-capacity method of analysis.

4-7 Two identical 7.5-cm cubes of copper at 425 and 90°C are brought into contact. Assuming that the blocks exchange heat only with each other and that there is no resistance to heat flow as a result of the contact of the blocks, plot the temperature of each block as a function of time, using the lumped-capacity method of analysis. That is, assume the resistance to heat transfer is the conduction resistance of the two blocks. Assume that all surfaces are insulated except those in contact.

4-8 Repeat Prob. 4-7 for a 7.5-cm copper cube at 425°C in contact with a 7.5-cm steel cube at 90°C. Sketch the thermal circuit.

4-9 An infinite plate of thickness $2L$ is suddenly exposed to a constant-temperature radiation heat source or sink of temperature T_s. The plate has a uniform initial temperature of T_i. The radiation heat loss from each side of the plate is given by $q = \sigma\epsilon A(T^4 - T_s^4)$, where σ and ϵ are constants and A is the surface area. Assuming that the plate behaves as a lumped capacity, that is, $k \rightarrow \infty$, derive an expression for the temperature of the plate as a function of time.

4-10 A stainless-steel rod (18% Cr, 8% Ni) 6.4 mm in diameter is initially at a uniform temperature of 50°C and is suddenly immersed in a liquid at 200°C with $h = 120$ W/m² · °C. Using the lumped-capacity method of analysis, calculate the time necessary for the rod temperature to reach 120°C.

4-11 A 5-cm-diameter copper sphere is initially at a uniform temperature of 250°C. It is suddenly exposed to an environment at 30°C having a heat-transfer coefficient $h = 28$ W/m² · °C. Using the lumped-capacity method of analysis, calculate the time necessary for the sphere temperature to reach 90°C.

4-12 A stack of common building brick 1 m high, 3 m long, and 0.5 m thick leaves an oven, where it has been heated to a uniform temperature of 300°C. The stack is allowed to cool in a room at 35°C with an air-convection coefficient of 15 W/m² · °C.

The bottom surface of the brick is on an insulated stand. How much heat will have been lost when the bricks cool to room temperature? How long will it take to lose half this amount, and what will the temperature at the geometric center of the stack be at this time?

4-13 A copper sphere having a diameter of 3.0 cm is initially at a uniform temperature of 50°C. It is suddenly exposed to an airstream of 10°C with $h = 15 \text{ W/m}^2 \cdot {}^\circ\text{C}$. How long does it take the sphere temperature to drop to 25°C?

4-14 An aluminum sphere, 5.0 cm in diameter, is initially at a uniform temperature of 50°C. It is suddenly exposed to an outer-space radiation environment at 0 K (no convection). Assuming the surface of aluminum is blackened and lumped-capacity analysis applies, calculate the time required for the temperature of the sphere to drop to −110°C.

4-15 An aluminum can having a volume of about 350 cm^3 contains beer at 1°C. Using a lumped-capacity analysis, estimate the time required for the contents to warm to 15°C when the can is placed in a room at 22°C with a convection coefficient of 15 $\text{W/m}^2 \cdot {}^\circ\text{C}$. Assume beer has the same properties as water.

4-16 A 12-mm-diameter aluminum sphere is heated to a uniform temperature of 400°C and then suddenly subjected to room air at 20°C with a convection heat-transfer coefficient of 10 $\text{W/m}^2 \cdot {}^\circ\text{C}$. Calculate the time for the center temperature of the sphere to reach 200°C.

4-17 A 4-cm-diameter copper sphere is initially at a uniform temperature of 200°C. It is suddenly exposed to a convection environment at 30°C with $h = 20 \text{ W/m}^2 \cdot {}^\circ\text{C}$. Calculate the time necessary for the center of the sphere to reach a temperature of 80°C.

4-18 When a sine-wave temperature distribution is impressed on the surface of a semi-infinite solid, the temperature distribution is given by

$$T_{x,\tau} - T_m = A \exp\left(-x\sqrt{\frac{\pi n}{\alpha}}\right) \sin\left(2\pi n\tau - x\sqrt{\frac{\pi n}{\alpha}}\right)$$

where $T_{x,\tau}$ = temperature at depth x and time τ after start of temperature wave at surface
T_m = mean surface temperature
n = frequency of wave, cycles per unit time
A = amplitude of temperature wave at surface

If a sine-wave temperature distribution is impressed on the surface of a large slab of concrete such that the temperature varies from 35 to 90°C and a complete cycle is accomplished in 15 min, find the heat flow through a plane 5 cm from the surface 2 h after the start of the initial wave.

4-19 Using the temperature distribution of Prob. 4-18, show that the time lag between maximum points in the temperature wave at the surface and at a depth x is given by

$$\Delta\tau = \frac{x}{2}\sqrt{\frac{1}{\alpha\pi n}}$$

4-20 A thick concrete wall having a uniform temperature of 54°C is suddenly subjected to an airstream at 10°C. The heat-transfer coefficient is 2.6 W/m^2 · °C. Calculate the temperature in the concrete slab at a depth of 7 cm after 30 min.

4-21 A very large slab of copper is initially at a temperature of 300°C. The surface temperature is suddenly lowered to 35°C. What is the temperature at a depth of 7.5 cm 4 min after the surface temperature is changed?

4-22 On a hot summer day a concrete driveway may reach a temperature of 50°C. Suppose that a stream of water is directed on the driveway so that the surface temperature is suddenly lowered to 10°C. How long will it take to cool the concrete to 25°C at a depth of 5 cm from the surface?

4-23 A semi-infinite slab of copper is exposed to a constant heat flux at the surface of 0.32 MW/m^2. Assume that the slab is in a vacuum, so that there is no convection at the surface. What is the surface temperature after 5 min if the initial temperature of the slab is 30°C? What is the temperature at a distance of 15 cm from the surface after 5 min?

4-24 A large slab of copper is initially at a uniform temperature of 90°C. Its surface temperature is suddenly lowered to 30°C. Calculate the heat-transfer rate through a plane 7.5 cm from the surface 5 s after the surface temperature is lowered.

4-25 A large slab of aluminum at a uniform temperature of 30°C is suddenly exposed to a constant surface heat flux of 15 kW/m^2. What is the temperature at a depth of 2.5 cm after 2 min?

4-26 For the slab in Prob. 4-25, how long would it take for the temperature to reach 150°C at the depth of 2.5 cm?

4-27 A piece of ceramic material [k = 0.8 W/m · °C, ρ = 2700 kg/m^3, c = 0.8 kJ/kg · °C] is quite thick and initially at a uniform temperature of 30°C. The surface of the material is suddenly exposed to a constant heat flux of 650 W/m^2 · °C. Plot the temperature at a depth of 1 cm as a function of time.

4-28 A large thick layer of ice is initially at a uniform temperature of −20°C. If the surface temperature is suddenly raised to −1°C, calculate the time required for the temperature at a depth of 1.5 cm to reach −11°C. The properties of ice are ρ = 57 lb$_m$/ft^3, c_p = 0.46 Btu/lb$_m$, k = 1.28 Btu/h · ft · °F, α = 0.048 ft^2/h.

4-29 A large slab of concrete (stone 1-2-4 mix) is suddenly exposed to a constant radiant heat flux of 900 W/m^2. The slab is initially uniform in temperature at 20°C. Calculate the temperature at a depth of 10 cm in the slab after a time of 9 h.

4-30 A very thick plate of stainless steel (18% Cr, 8% Ni) at a uniform temperature of 300°C has its surface temperature suddenly lowered to 100°C. Calculate the time required for the temperature at a depth of 3 cm to attain a value of 200°C.

4-31 A large slab has properties of common building brick and is heated to a uniform temperature of 40°C. The surface is suddenly exposed to a convection environment at 2°C with h = 25 W/m^2 · °C. Calculate the time for the temperature to reach 20°C at a depth of 8 cm.

4-32 A large block having the properties of chrome brick at 200°C is at a uniform temperature of 30°C when it is suddenly exposed to a surface heat flux of 3 × 10^4

W/m^2. Calculate the temperature at a depth of 3 cm after a time of 10 min. What is the surface temperature at this time?

4-33 A slab of copper having a thickness of 3.0 cm is initially at 300°C. It is suddenly exposed to a convection environment on the top surface at 80°C while the bottom surface is insulated. In 6 min the surface temperature drops to 140°C. Calculate the value of the convection heat-transfer coefficient.

4-34 A large slab of aluminum has a thickness of 10 cm and is initially uniform in temperature at 400°C. Suddenly it is exposed to a convection environment at 90°C with $h = 1400$ W/m$^2 \cdot$ °C. How long does it take the centerline temperature to drop to 180°C?

4-35 A horizontal copper plate 10 cm thick is initially uniform in temperature at 250°C. The bottom surface of the plate is insulated. The top surface is suddenly exposed to a fluid stream at 100°C. After 6 min the surface temperature has dropped to 150°C. Calculate the convection heat-transfer coefficient which causes this drop.

4-36 A large slab of aluminum has a thickness of 10 cm and is initially uniform in temperature at 400°C. It is then suddenly exposed to a convection environment at 90°C with $h = 1400$ W/m$^2 \cdot$ °C. How long does it take the center to cool to 180°C?

4-37 A plate of stainless steel (18% Cr, 8% Ni) has a thickness of 3.0 cm and is initially uniform in temperature at 500°C. The plate is suddenly exposed to a convection environment on both sides at 40°C with $h = 150$ W/m$^2 \cdot$ °C. Calculate the times for the center and face temperatures to reach 100°C.

4-38 A steel cylinder 10 cm in diameter and 10 cm long is initially at 300°C. It is suddenly immersed in an oil bath which is maintained at 40°C, with $h = 280$ W/m$^2 \cdot$ °C. Find (*a*) the temperature at the center of the solid after 2 min and (*b*) the temperature at the center of one of the circular faces after 2 min.

4-39 An aluminum bar has a diameter of 11 cm and is initially uniform in temperature at 300°C. If it is suddenly exposed to a convection environment at 50°C with $h = 1200$ W/m$^2 \cdot$ °C, how long does it take the center temperature to cool to 80°C? Also calculate the heat loss per unit length.

4-40 A 5-lb roast initially at 70°F is placed in an oven at 350°F. Assuming that the heat-transfer coefficient is 2.5 Btu/h $\cdot$ ft$^2 \cdot$ °F and that the thermal properties of the roast may be approximated by those of water, estimate the time required for the center of the roast to attain a temperature of 200°F.

4-41 A fused-quartz sphere has a thermal diffusivity of 9.5×10^{-7} m^2/s, a diameter of 2.5 cm, and a thermal conductivity of 1.52 W/m $\cdot$ °C. The sphere is initially at a uniform temperature of 25°C and is suddenly subjected to a convection environment at 200°C. The convection heat-transfer coefficient is 110 W/m$^2 \cdot$ °C. Calculate the temperatures at the center and at a radius of 6.4 mm after a time of 4 min.

4-42 Lead shot may be manufactured by dropping molten-lead droplets into water. Assuming that the droplets have the properties of solid lead at 300°C, calculate the time for the center temperature to reach 120°C when the water is at 100°C with $h = 5000$ W/m$^2 \cdot$ °C, $d = 1.5$ mm.

4-43 A steel sphere 10 cm in diameter is suddenly immersed in a tank of oil at 10°C. The initial temperature of the sphere is 220°C; $h = 5000$ W/m² · °C. How long will it take the center of the sphere to cool to 120°C?

4-44 A boy decides to place his glass marbles in an oven at 200°C. The diameter of the marbles is 15 mm. After a while he takes them from the oven and places them in room air at 20°C to cool. The convection heat-transfer coefficient is approximately 14 W/m² · °C. Calculate the time the boy must wait until the center temperature of the marbles reaches 35°C.

4-45 A lead sphere with $d = 1.5$ mm and initial temperature of 200°C is suddenly exposed to a convection environment at 100°C and $h = 5000$ W/m² · °C. Calculate the time for the center temperature to reach 120°C.

4-46 A long steel bar 5 by 10 cm is initially maintained at a uniform temperature of 250°C. It is suddenly subjected to a change such that the environment temperature is lowered to 35°C. Assuming a heat-transfer coefficient of 23 W/m² · °C, use a numerical method to estimate the time required for the center temperature to reach 90°C. Check this result with a calculation, using the Heisler charts.

4-47 A steel bar 2.5 cm square and 7.5 cm long is initially at a temperature of 250°C. It is immersed in a tank of oil maintained at 30°C. The heat-transfer coefficient is 570 W/m² · °C. Calculate the temperature in the center of the bar after 2 min.

4-48 A cube of aluminum 10 cm on each side is initially at a temperature of 300°C and is immersed in a fluid at 100°C. The heat-transfer coefficient is 900 W/m² · °C. Calculate the temperature at the center of one face after 1 min.

4-49 A short concrete cylinder 15 cm in diameter and 30 cm long is initially at 25°C. It is allowed to cool in an atmospheric environment in which the temperature is 0°C. Calculate the time required for the center temperature to reach 6°C if the heat-transfer coefficient is 17 W/m² · °C.

4-50 A 4.0-cm cube of aluminum is initially at 450°C and is suddenly exposed to a convection environment at 100°C with $h = 120$ W/m² · °C. How long does it take the cube to cool to 200°C?

4-51 A cube of aluminum 11 cm on each side is initially at a temperature of 400°C. It is suddenly immersed in a tank of oil maintained at 85°C. The convection coefficient is 1100 W/m² · °C. Calculate the temperature at the center of one face after a time of 1 min.

4-52 An aluminum cube 5 cm on a side is initially at a uniform temperature of 100°C and is suddenly exposed to room air at 25°C. The convection heat-transfer coefficient is 20 W/m² · °C. Calculate the time required for the geometric center temperature to reach 50°C.

4-53 A stainless steel cylinder (18% Cr, 8% Ni) is heated to a uniform temperature of 200°C and then allowed to cool in an environment where the air temperature is maintained constant at 30°C. The convection heat-transfer coefficient may be taken as 200 W/m² · C. The cylinder has a diameter of 10 cm and a length of 15 cm. Calculate the temperature of the geometric center of the cylinder after a time of 10 min. Also calculate the heat loss.

4-54 A cylinder having a diameter of 15 cm and a length of 30 cm is initially uniform in temperature at 300°C. It is suddenly exposed to a convection environment at 20°C with $h = 35\ \text{W/m}^2 \cdot {}^\circ\text{C}$. Properties of the solid are $k = 2.3\ \text{W/m} \cdot {}^\circ\text{C}$, $\rho = 3000\ \text{kg/m}^3$, and $c = 840\ \text{J/kg} \cdot {}^\circ\text{C}$. Calculate the time for (*a*) the center and (*b*) the center of one face to reach a temperature of 150°C. Also calculate the heat loss for each case.

4-55 A rectangular solid is 15 by 10 by 20 cm and has the properties of fireclay brick. It is initially uniform in temperature at 300°C and then suddenly exposed to a convection environment at 80°C and $h = 110\ \text{W/m}^2 \cdot {}^\circ\text{C}$. Calculate the time for (*a*) the geometric center and (*b*) the center of each face to reach a temperature of 200°C. Also calculate the heat loss for each of these times.

4-56 Calculate the heat loss for both cases in Prob. 4-38.

4-57 Calculate the heat loss for the bar in Prob. 4-47 per unit length.

4-58 Calculate the heat loss for the cube in Prob. 4-48.

4-59 Develop a backward-difference formulation for a boundary node subjected to a convection environment. Check with Table 4-3.

4-60 The stainless-steel plate is surrounded by an insulating block as shown and is initially at a uniform temperature of 50°C with a convection environment at 50°C. The plate is suddenly exposed to a radiant heat flux of $20\ \text{kW/m}^2$. Calculate the temperatures at the indicated nodes after 10 s, 1 min, and 10 min. Take the properties of stainless steel as $k = 16\ \text{W/m} \cdot {}^\circ\text{C}$, $\rho = 7800\ \text{kg/m}^3$, and $c = 0.46\ \text{kJ/kg} \cdot {}^\circ\text{C}$, $h = 30\ \text{W/m}^2 \cdot {}^\circ\text{C}$. Assume all the radiation is absorbed.

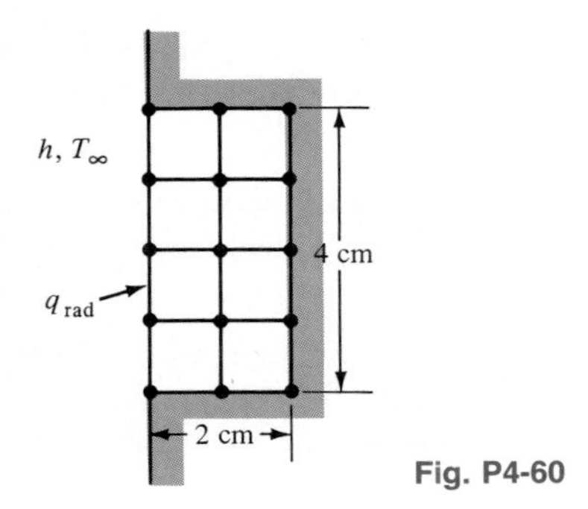

Fig. P4-60

4-61 The composite plate shown has one face insulated and is initially at a uniform temperature of 100°C. At time zero the face is suddenly exposed to a convection environment at 10°C and $h = 70\ \text{W/m}^2 \cdot {}^\circ\text{C}$. Determine the temperatures at the indicated nodes after 1 s, 10 s, 1 min, and 10 min.

Material	k, $\text{W/m} \cdot {}^\circ\text{C}$	ρ, kg/m^3	c, $\text{kJ/kg} \cdot {}^\circ\text{C}$
A	20	7800	0.46
B	1.2	1600	0.85
C	0.5	2500	0.8

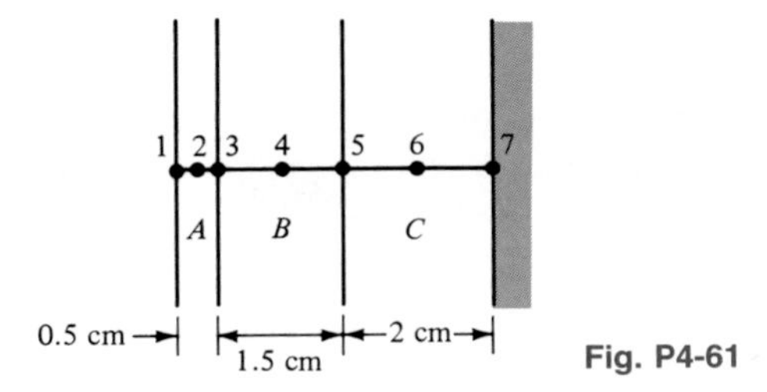

Fig. P4-61

4-62 The 4.0-mm-diameter stainless-steel wire shown is initially at 20°C and is exposed to a convection environment at 20°C where h may be taken as 200 W/m^2 · °C. An electric current is applied to the wire such that there is a uniform internal heat generation of 500 MW/m^3. The left side of the wire is insulated as shown. Set up the nodal equations and stability requirement for calculating the temperature in the wire as a function of time, using increments of $\Delta r = 0.5$ mm and $\Delta\phi = \pi/4$. Take the properties of stainless steel as $k = 16$ W/m · °C, $\rho = 7800$ kg/m^3, and $c = 0.46$ kJ/kg · °C.

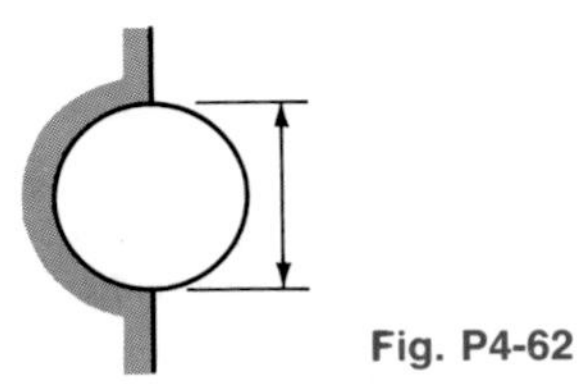

Fig. P4-62

4-63 The corner shown is initially uniform at 200°C and then suddenly exposed to convection around the edge with $h = 50$ W/m^2 · °C and $T = 30$°C. Assume the solid has the properties of fireclay brick. Examine nodes 1, 2, 3, and 4 and determine the maximum time increment which may be used for a transient numerical calculation.

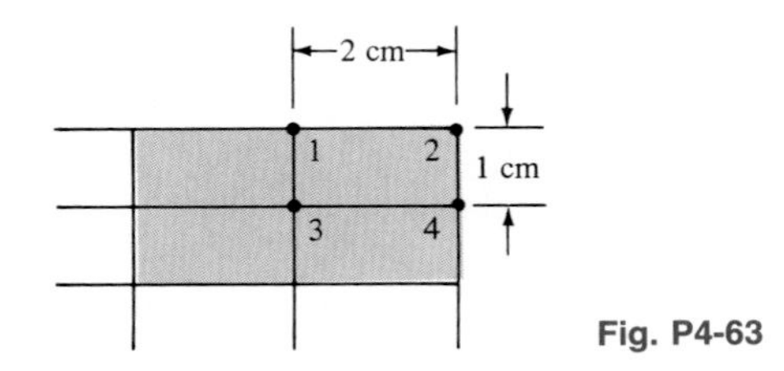

Fig. P4-63

4-64 Write the nodal equation for node 3 in the figure for use in a transient analysis. Determine the stability criterion for this node.

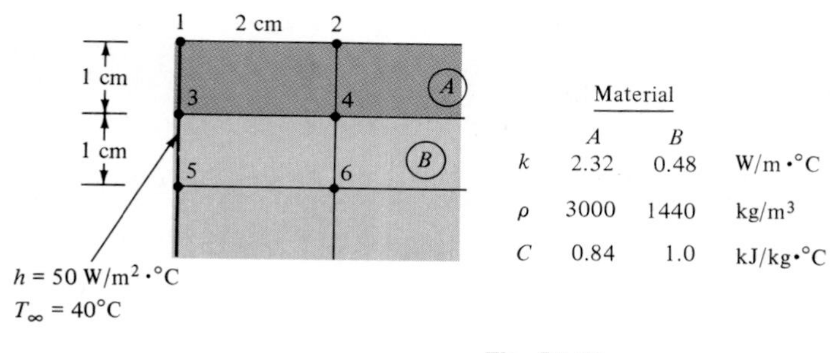

	Material		
	A	B	
k	2.32	0.48	W/m·°C
ρ	3000	1440	kg/m^3
C	0.84	1.0	kJ/kg·°C

Fig. P4-64

4-65 Write a nodal equation for analysis of node (m,n) in the figure to be used in a transient analysis of the solid.

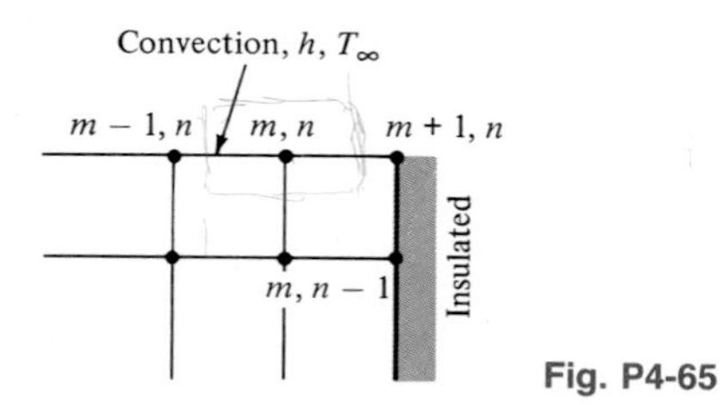

Fig. P4-65

4-66 Write the nodal equation and establish the stability criteria for node 1 in the figure (transient analysis). Materials A and B have the properties given in Prob. 4-61.

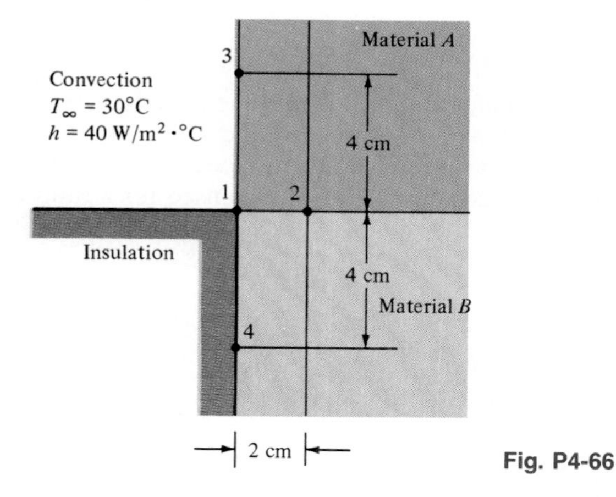

Fig. P4-66

4-67 Write a computer program which will solve Example 4-15 for different input properties. For nomenclature take T(N) = temperature of node N at beginning of time increment, TP(N) = temperature of node at end of time increment, X = number of nodes, W = width of plate, TA = temperature of left fluid, HA = convection coefficient of left fluid, TB = temperature of right fluid, HB

= convection coefficient of right fluid, DT = time increment, C = specific heat, D = density, K = thermal conductivity, Q = heat-generation rate per unit volume, TI = total time. Write the program so that the user can easily rerun the program for new times and print out the results for each.

4-68 Calculate the maximum time increment that can be used for node 5 in the accompanying figure for a transient numerical analysis. Also write the nodal equation for this node.

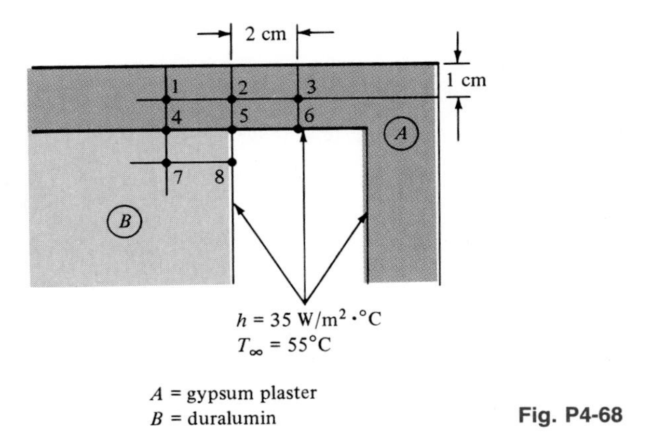

Fig. P4-68

4-69 The corner shown is initially uniform at 300°C and then suddenly exposed to a convection environment at 50°C with h = 60 W/m² · °C. Assume the solid has the properties of fireclay brick. Examine nodes 1, 2, 3, 4, and 5 and determine the maximum time increment which may be used for a transient numerical calculation.

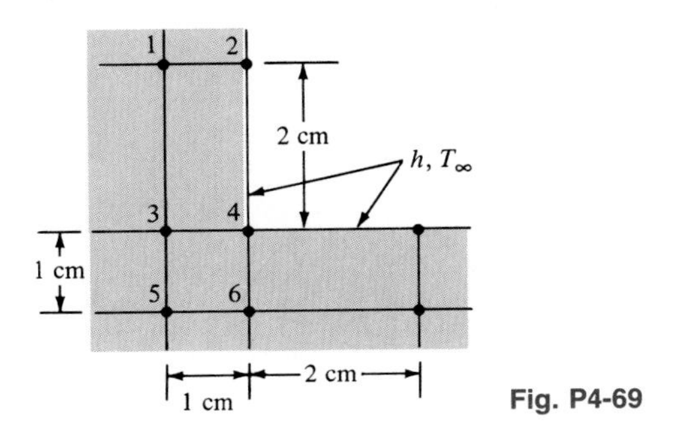

Fig. P4-69

4-70 Write a steady-state nodal equation for node 3 in the figure assuming unit depth perpendicular to the page and using the node spacing shown. The thermal conductivity of the solid is 15 W/m · °C and the convection heat-transfer coefficient on the side surface is 25 W/m² · °C.

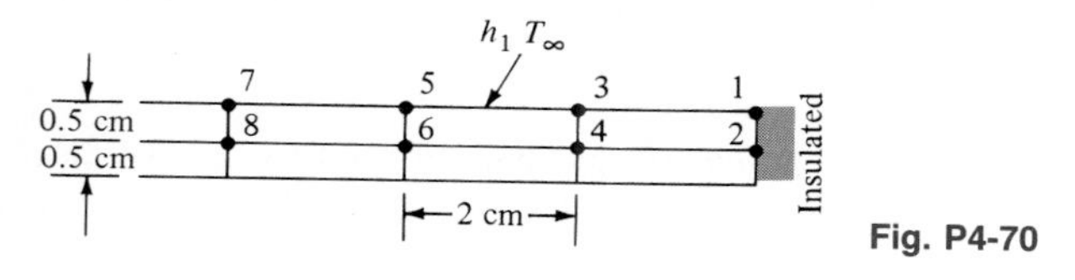

Fig. P4-70

4-71 For the section shown, calculate the maximum time increment allowed for node 2 in a transient numerical analysis. Also write the entire nodal equation for this node.

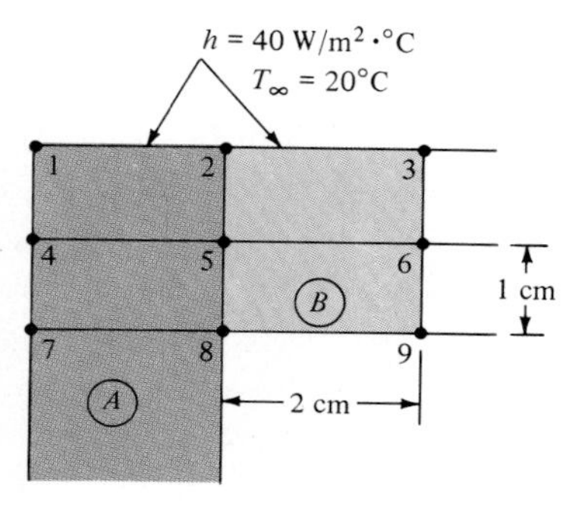

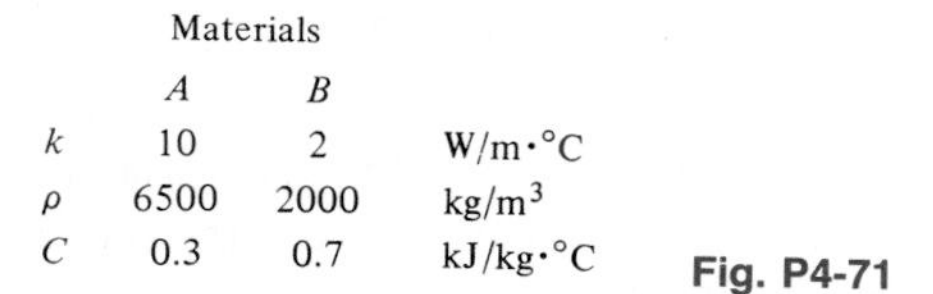

Materials

	A	B	
k	10	2	W/m·°C
ρ	6500	2000	kg/m³
C	0.3	0.7	kJ/kg·°C

Fig. P4-71

4-72 A transient numerical analysis is to be performed on the composite material section shown. Calculate the maximum time increment that can be used for node 5 to ensure convergence.

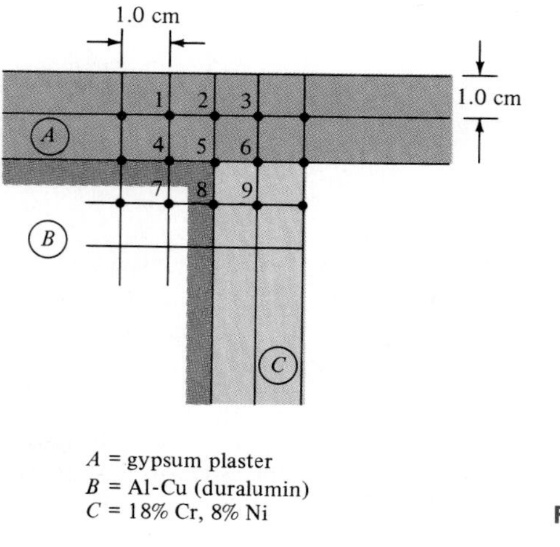

Fig. P4-72

4-73 For the section shown, calculate the maximum time increment allowed for node 4 in a transient numerical environment. Also write the complete nodal equation for node 4.

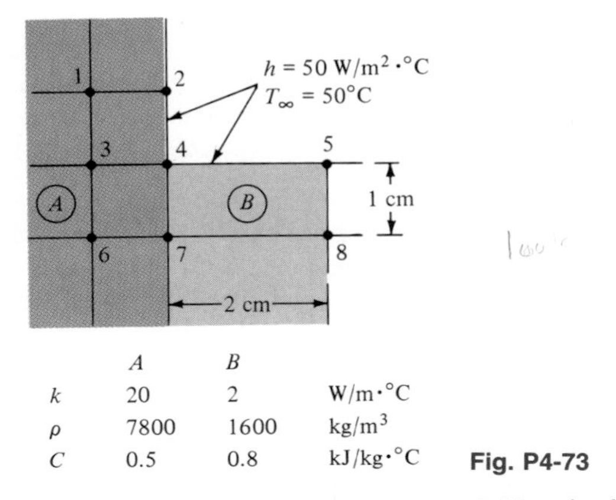

	A	B	
k	20	2	W/m·°C
ρ	7800	1600	kg/m^3
C	0.5	0.8	kJ/kg·°C

Fig. P4-73

4-74 A node like that shown in Table 3-2d has both x and y increments equal to 1.0 cm. The convection boundary condition is at 50°C and $h = 60\ \text{W/m}^2 \cdot {}^\circ\text{C}$. The solid material is stainless steel (18% Cr, 8% Ni). Using the thermal resistance and capacitance formulation for a transient analysis write the nodal equation for this node and determine the maximum allowable time increment.

4-75 The solid in Prob. 3-43 is initially uniform in temperature at 10°C. At time zero the right face is suddenly changed to 38°C and the left face exposed to the convection environment. Nodes 3 and 6 remain at 10°C. Select an appropriate value for $\Delta\tau$ and calculate the temperatures of nodes 1, 2, 4, and 5 after 10 time increments. Carry the calculation forward to verify the steady-state distribution. Take $\rho = 3000\ \text{kg/m}^3$ and $c = 840\ \text{J/kg} \cdot {}^\circ\text{C}$.

4-76 The solid in Prob. 3-45 has $k = 11\ \text{W/m} \cdot {}^\circ\text{C}$ and is initially uniform in temperature at 1000°C. At time zero the four surfaces are changed to the values shown. Select an appropriate $\Delta\tau$ and calculate the temperatures of nodes 1, 2, 3, and 4 after 10 time increments. Also obtain the limiting steady-state temperatures. Take $\rho = 2800\ \text{kg/m}^3$ and $c = 940\ \text{J/kg} \cdot {}^\circ\text{C}$.

4-77 The fin in Prob. 3-49 is initially uniform in temperature at 300°C and then suddenly exposed to the convection environment. Select an appropriate $\Delta\tau$ and calculate the nodal temperatures after 10 time increments. Take $\rho = 2200\ \text{kg/m}^3$ and $c = 820\ \text{J/kg} \cdot {}^\circ\text{C}$.

4-78 The fin in Prob. 3-50 is initially uniform in temperature at 200°C and then is suddenly exposed to the convection environment shown while maintaining the bottom face at 200°C. Select an appropriate $\Delta\tau$ and calculate the nodal temperatures after 10 time increments. Repeat for 100 $\Delta\tau$. Take $\rho = 7800\ \text{kg/m}^3$ and $c = 460\ \text{J/kg} \cdot {}^\circ\text{C}$.

4-79 The solid in Prob. 3-51 is initially uniform in temperature at 100°C and then suddenly exposed to the convection condition while the right and bottom faces are held constant at 100°C. Select a value for $\Delta\tau$ and calculate the nodal temperatures after 10 time increments. Take $\rho = 3000\ \text{kg/m}^3$ and $c = 800\ \text{J/kg} \cdot {}^\circ\text{C}$.

4-80 The solid in Prob. 3-52 is initially uniform in temperature at 50°C and suddenly is exposed to the convection condition. Select a value for $\Delta\tau$ and calculate the nodal temperatures after 10 time increments. Take ρ = 2500 kg/m^3 and c = 900 J/kg · °C.

4-81 The solids in Prob. 3-53 are initially uniform in temperature at 300°C and suddenly are exposed to the convection boundary, while the inner temperature is kept constant at 300°C. Select a value for $\Delta\tau$ and calculate the nodal temperatures after 10 time increments. Take ρ_A = 2900 kg/m^3, c_A = 810 J/kg · °C, ρ_B = 7800 kg/m^3, and c_B = 470 J/kg · °C.

4-82 The fin in Prob. 3-54 is initially uniform in temperature at 200°C, and then suddenly exposed to the convection boundary and heat generation. Select a value for $\Delta\tau$ and calculate the nodal temperatures for 10 time increments. Take ρ = 7600 kg/m^3 and c = 450 J/kg · °C. The base stays constant at 200°C.

4-83 The solid in Prob. 3-55 is initially uniform in temperature at 500°C and suddenly exposed to the convection boundary while the inner surface is kept constant at 500°C. Select a value for $\Delta\tau$ and calculate the nodal temperatures after 10 time increments. Take ρ = 500 kg/m^3 and c = 810 J/kg · °C.

4-84 Repeat Prob. 4-83 for the steel liner of Prob. 3-56. Take ρ = 7800 kg/m^3 and c = 460 J/kg · °C for the steel.

4-85 The plate in Prob. 3-57 is initially uniform in temperature at 100°C and suddenly exposed to the convection boundary. Select a value for $\Delta\tau$ and calculate the nodal temperatures after 10 time increments. Take ρ = 7500 kg/m^3 and c = 440 J/kg · °C.

4-86 The solid shown in Prob. 3-58 is initially uniform in temperature at 100°C and suddenly exposed to the convection boundary and heat generation while the right face is kept at 100°C. Select a value for $\Delta\tau$ and calculate the nodal temperatures after 10 time increments. Take ρ = 7600 kg/m^3 and c = 460 J/kg · °C.

4-87 A steel rod 12.5 mm in diameter and 20 cm long has one end attached to a heat reservoir at 250°C. The bar is initially maintained at this temperature throughout. It is then subjected to an airstream at 30°C such that the convection heat-transfer coefficient is 35 W/m^2 · °C. Estimate the time required for the temperature midway along the length of the rod to attain a value of 190°C.

4-88 A concrete slab 15 cm thick has a thermal conductivity of 0.87 W/m · °C and has one face insulated and the other face exposed to an environment. The slab is initially uniform in temperature at 300°C, and the environment temperature is suddenly lowered to 90°C. The heat-transfer coefficient is proportional to the fourth root of the temperature difference between the surface and environment and has a value of 11 W/m^2 · °C at time zero. The environment temperature increases linearly with time and has a value of 200°C after 20 min. Using the numerical method, obtain the temperature distribution in the slab after 5, 10, 15, and 20 min.

4-89 The two-dimensional body of Fig. 3-6 has the initial surface and internal temperature as shown and calculated in Table 3-3. At time zero the 500°C face is suddenly lowered to 30°C. Taking $\Delta x = \Delta y$ = 15 cm and $\alpha = 1.29 \times 10^{-5}$ m^2/s, calculate the temperatures at nodes 1, 2, 3, and 4 after 30 min. Perform the calculation using both a forward- and backward-difference method. For the backward-difference method use only two time increments. Take $k = 45$ W/m · °C.

4-90 The strip of material shown has a thermal conductivity of 20 W/m · °C and is placed firmly on the isothermal surface maintained at 50°C. At time zero the strip is suddenly exposed to an airstream with $T_\infty = 300$°C and $h = 40$ W/m² · °C. Using a numerical technique, calculate the temperatures at nodes 1 to 8 after 1 s, 10 s, 1 min, and steady state; $\rho = 7000$ kg/m³ and $c = 0.5$ kJ/kg · °C.

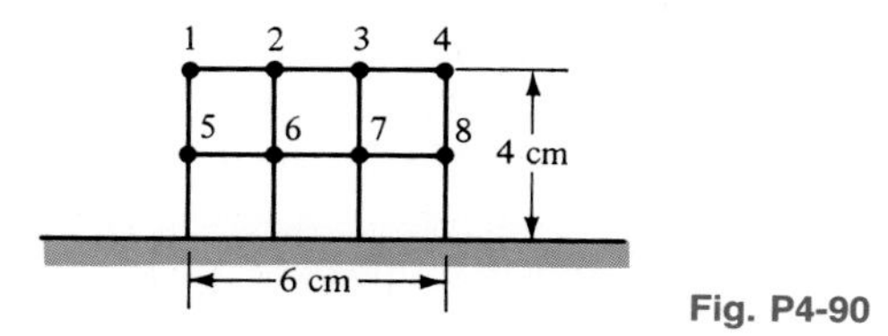

Fig. P4-90

4-91 Rework Probs. 4-7 and 4-8 using the numerical technique.

4-92 Rework Prob. 4-87 using the numerical technique.

4-93 A blackened ceramic sphere of 10 cm diameter is initially uniform in temperature at 1000°K and is suddenly placed in outer space where it loses heat by radiation (no convection) according to

$$q_{\text{rad}} = \sigma A T^4 \qquad T \text{ in degrees Kelvin}$$

$$\sigma = 5.669 \times 10^{-8}\ \text{W/m}^2 \cdot \text{K}^4$$

Calculate the temperatures of the nodes shown for several increments of time and the corresponding heat losses. Use the values of k, ρ, and c from Prob. 4-60.

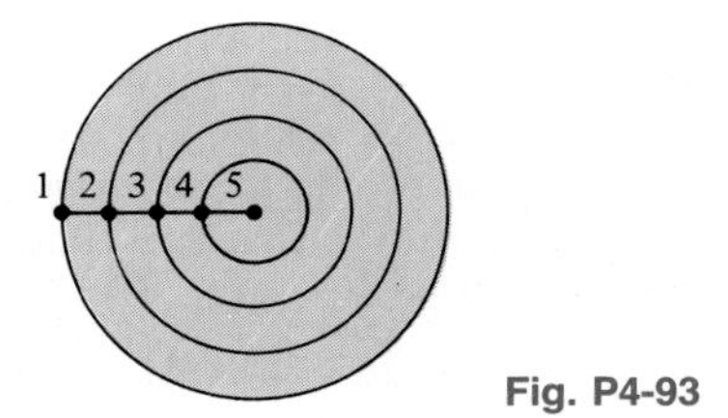

Fig. P4-93

4-94 A hollow concrete sphere [$k = 1.3$ W/m · °C, $\alpha = 7 \times 10^{-7}$ m²/s] has inside and outside diameters of 0.5 and 1.0 m and is initially uniform in temperature at 200°C. The outside surface is suddenly lowered to 20°C. Calculate the nodal temperatures shown for several increments of time. Assume the inside surface acts as though it were insulated.

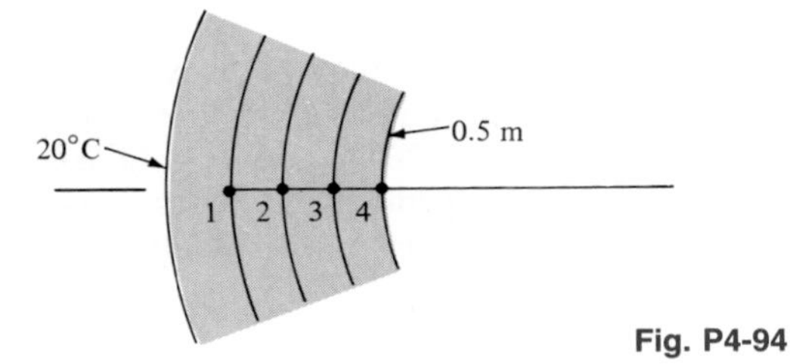

Fig. P4-94

4-95 The stainless-steel plate shown is initially at a uniform temperature of 150°C and is suddenly exposed to a convection environment at 30°C with $h = 17$ W/m² · °C. Using numerical techniques, calculate the time necessary for the temperature at a depth of 6.4 mm to reach 65°C.

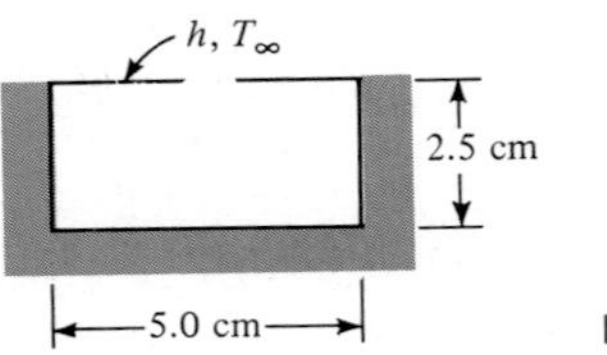

Fig. P4-95

4-96 Repeat Prob. 4-50 with the top surface also losing heat by radiation according to

$$q_{rad} = \sigma A \epsilon (T^4 - T_\infty^4) \qquad T \text{ in degrees Kelvin}$$

$$\sigma = 5.669 \times 10^{-8} \text{ W/m}^2 \cdot \text{K}^4$$

$$\epsilon = 0.7$$

4-97 A fireproof safe is constructed of loosely packed asbestos contained between thin sheets of stainless steel. The safe is built in the form of a cube with inside and outside dimensions of 0.5 and 1.0 m. If the safe is initially uniform in temperature at 30°C and the outside is suddenly exposed to a convection environment at 600°C, $h = 100$ W/m² · °C, calculate the time required for the inside temperature to reach 150°C. Assume the inside surface is insulated, and neglect the resistance and capacitance of the stainless steel. Take the properties of asbestos as $k = 0.16$ W/m · °C, $\alpha = 3.5 \times 10^{-7}$ m²/s.

4-98 The half-cylinder in Prob. 3-59 is initially uniform in temperature at 300°C and then suddenly exposed to the convection boundary while the bottom side is maintained at 300°C. Calculate the nodal temperatures for several time increments, and compute the heat loss in each period. Take $\alpha = 0.5 \times 10^{-5}$ m²/s.

4-99 A large slab of brick [$k = 1.07$ W/m · °C, $\alpha = 5.4 \times 10^{-7}$ m²/s] is initially at a uniform temperature of 20°C. One surface is suddenly exposed to a uniform heat flux of 4500 W/m². Calculate and plot the surface temperature as a function of time. Also calculate the heat flux through the plane 2.0 cm deep when the surface temperature reaches 150°C.

4-100 Repeat Prob. 4-95 with the top surface also losing heat by radiation according to

$$q_{rad} = \sigma A \epsilon (T^4 - T_\infty^4) \qquad T \text{ in degrees Kelvin}$$
$$\sigma = 5.669 \times 10^{-8} \text{ W/m}^2 \cdot \text{K}^4$$
$$\epsilon = 0.7$$

Repeat the calculation for 10 and 20 min.

4-101 Work Prob. 4-22 using the numerical method.

4-102 Work Prob. 4-20 using the numerical method.

4-103 Work Prob. 4-22 using the Schmidt plot.

4-104 Work Prob. 4-20 using the Schmidt plot.

4-105 A ceramic plate having a thickness of 2.0 cm is heated to a uniform temperature of 1000°K and suddenly exposed to radiation on both sides at 300°K. The properties of the solid are $k = 1.2$ W/m · °C, $\rho = 2500$ kg/m³, $c = 0.9$ kJ/kg · °C, and $\epsilon = 0.85$. Divide the plate into eight segments ($\Delta x = 0.25$ cm) and, using a numerical technique, obtain information to plot the center and surface temperatures as a function of time.

4-106 Suppose the ceramic of Prob. 4-105 is in the form of a long cylinder having a diameter of 2.0 cm. Divide the cylinder into four increments ($\Delta r = 0.25$ cm) and obtain information to plot the center and surface temperatures as a function of time.

4-107 A granite sphere having a diameter of 15 cm and initially at a uniform temperature of 120°C is suddenly exposed to a convection environment with $h = 350$ W/m² · °C and $T_\infty = 30$°C. Calculate the temperature at a radius of 4.5 cm after 21 min and the energy removed from the sphere in this time. Take the properties of granite as $k = 3.2$ W/m · °C and $\alpha = 13 \times 10^{-7}$ m²/s.

4-108 Oranges with a diameter of about 3 in are to be cooled from room temperature of 25°C to 3°C using an air-convection environment with $h = 45$ W/m² · °C and $T_\infty = 0$°C. Assuming that the oranges have the properties of water at 10°C, calculate the time required for the cooling and the total cooling required for 100 oranges.

4-109 A 10-cm-thick brick wall having the properties of common building brick initially at a uniform temperature of 80°C is suddenly exposed to a convection environment of $T_\infty = 20$°C and $h = 100$ W/m² · °C. Using $\Delta x = 2.5$ cm, calculate the time for the center temperature to reach 50°C using the numerical method. Also determine the maximum time increment for these calculations.

4-110 A chrome steel plate (1% Cr) is heated in an oven to a uniform temperature of 200°C and then subjected to a convection environment having $T_\infty = 20$°C and $h = 300$ W/m² · °C on both sides. The plate thickness is 10 cm. Taking $\Delta x = 1$ cm, calculate the center temperature after 5 and 10 min using the numerical method. Also solve using the Heisler charts.

4-111 A long slab of oak 4.1 by 9.2 cm is initially at 20°C and is placed in an oven with $T_\infty = 200$°C and $h = 40$ W/m² · °C. Calculate the time required for the surface to reach 150°C. Repeat for the geometric center.

4-112 Consider two solids initially at uniform temperatures of 200°C with $k = 1.4$ W/m · °C and $\alpha = 7 \times 10^{-7}$ m²/s: (*a*) a semi-infinite solid and (*b*) an infinite plate 10 cm thick. Both solids are suddenly exposed to a convection environment at 25°C with $h = 40$ W/m² · °C. Calculate the temperatures at the center of the plate and for $x = 5$ cm in the semi-infinite solid for 5, 10, 20, and 30 min. What do you conclude from these calculations?

4-113 Make the calculations for Prob. 4-112 based on a lumped-capacity analysis and comment on the results.

4-114 The rate at which cooling can be accomplished is of considerable importance in the food-processing industry. In a pizza-cooking application hot-air jets at 200°C can achieve heat-transfer coefficients of $h = 75$ W/m² · °C. Suppose the jets impinge on both sides of a pizza layer having a thickness of 1.2 cm at an initial uniform temperature of 25°C. How long does it take to reach a center temperature of 100°C? Take the properties of pizza as those of water ($k = 0.6$ W/m · °C, $\alpha = 1.5 \times 10^{-7}$ m²/s).

4-115 For the square grid imposed on the circular quadrant shown, write the transient explicit nodal equations for nodes 3 and 4. Take $k = 10$ W/m · °C, $\rho = 2000$ kg/m³, and $c = 840$ J/kg · °C. Use information from Tables 3-2 and 3-4. What is the maximum allowable time increment for each node?

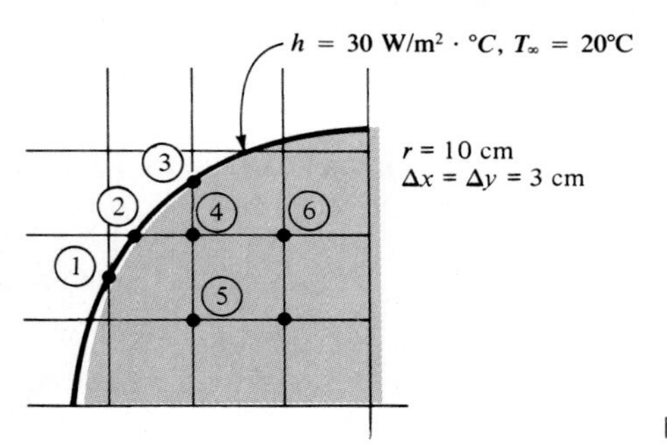

Fig. P4-115

4-116 Taking the accompanying figure as a special case of Table 3-2(*f*), write an explicit formulation for nodes (*m, n*) and 2 using the resistance-capacity formulation and the information of Table 3-4.

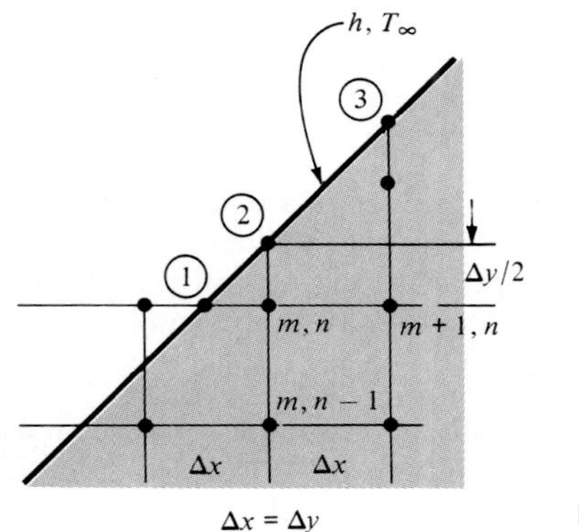

Fig. P4-116

4-117 Repeat Prob. 4-116 for a slanted surface that is (*a*) insulated and (*b*) isothermal at T_∞.

4-118 The slanted intersection shown in the figure is an intersection of materials A and B. Write the transient nodal equations for nodes 3, 4, and 6 using information from Tables 3-2(*f*) and 3-4(*f* and *g*).

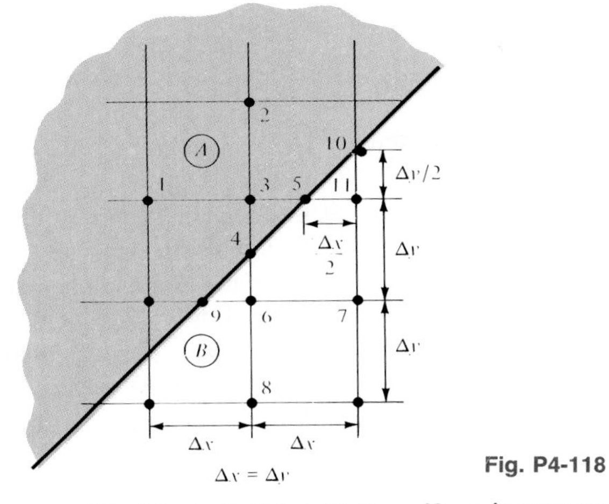

Fig. P4-118

4-119 The solid of Prob. 3-67 is initially uniform in temperature at 100°C, but suddenly the two surfaces are lowered to 0 and 40°C. If the solid has $k = 20$ W/m · °C and $\alpha = 5 \times 10^{-6}$ m²/s, find the steady-state temperature of each node and the nodal temperatures after 1 min.

4-120 The solid in Prob. 3-69 is initially at a uniform temperature of 150°C and then suddenly exposed to the given boundary conditions with $h = 50$ W/m² · °C and $T_\infty = 20$°C. Taking the properties as $k = 61$ W/m · °C and $\alpha = 1.7 \times 10^{-5}$ m²/s, determine the steady-state values of the 12 nodes and the nodal temperature after 10 min.

4-121 The pin fin of Prob. 3-65, initially uniform in temperature at 200°C, is suddenly exposed to the convection environment. Determine (*a*) the steady-state temperature distribution by a transient analysis taken to a long time and (*b*) the distribution for a time approximately equal to half the "long" time.

4-122 The solid in Prob. 3-66 is initially uniform in temperature at 100°C before suddenly being exposed to $h = 100$ W/m² · °C and $T_\infty = 0$°C. Take the properties as $k = 2$ W/m · °C and $\alpha = 7 \times 10^{-7}$ m²/s. Determine (*a*) the steady-state temperature distribution by taking a transient analysis to a long time and (*b*) the temperature distribution at a time approximately half the "long" time.

■ REFERENCES

1 Schneider, P. J.: "Conduction Heat Transfer," Addison-Wesley Publishing Company, Inc., Reading, Mass., 1955.

2 Heisler, M. P.: Temperature Charts for Induction and Constant Temperature Heating, *Trans. ASME,* vol. 69, pp. 227–236, 1947.

3 Abramowitz, M., and I. Stegun (eds.): Handbook of Mathematical Functions, NBS AMS 55, U.S. Government Printing Office, 1964.

4 Dusinberre, G. M.: "Heat Transfer Calculations by Finite Differences," International Textbook Company, Scranton, Pa., 1961.

5 Jakob, M.: "Heat Transfer," vol. 1, John Wiley & Sons, Inc., New York, 1949.

6 Gröber, H., S. Erk, and U. Grigull: "Fundamentals of Heat Transfer," McGraw-Hill Book Company, New York, 1961.

7 Schneider, P. J.: "Temperature Response Charts," John Wiley & Sons, Inc., New York, 1963.

8 Schenck, H.: "Fortran Methods in Heat Flow," The Ronald Press Company, New York, 1963.

9 Richardson, P. D., and Y. M. Shum: Use of Finite-Element Methods in Solution of Transient Heat Conduction Problems, *ASME Pap.* 69-WA/HT-36.

10 Emery, A. F., and W. W. Carson: Evaluation of Use of the Finite Element Method in Computation of Temperature, *ASME Pap.* 69-WA/HT-38.

11 Wilson, E. L., and R. E. Nickell: Application of the Finite Element Method to Heat Conduction Analysis, *Nucl. Eng. Des.,* vol. 4, pp. 276–286, 1966.

12 Zienkiewicz, O. C.: "The Finite Element Method in Structural and Continuum Mechanics," McGraw-Hill Book Company, New York, 1967.

13 Myers, G. E.: "Conduction Heat Transfer," McGraw-Hill Book Company, New York, 1972.

14 Arpaci, V. S.: "Conduction Heat Transfer," Addison-Wesley Publishing Company, Inc., Reading, Mass., 1966.

15 Ozisik, M. N.: "Boundary Value Problems of Heat Conduction," International Textbook Company, Scranton, Pa., 1968.

16 Langston, L. S.: Heat Transfer from Multidimensional Objects Using One-Dimensional Solutions for Heat Loss, *Int. J. Heat Mass Transfer,* vol. 25, p. 149, 1982.

17 Colakyan M., R. Turton, and O. Levenspiel: Unsteady State Heat Transfer to Variously Shaped Objects, *Heat Transfer Engr.,* vol. 5, p. 82, 1984.

PRINCIPLES OF CONVECTION

5-1 INTRODUCTION

The preceding chapters have considered the mechanism and calculation of conduction heat transfer. Convection was considered only insofar as it related to the boundary conditions imposed on a conduction problem. We now wish to examine the methods of calculating convection heat transfer and, in particular, the ways of predicting the value of the convection heat-transfer coefficient h. The subject of convection heat transfer requires an energy balance along with an analysis of the fluid dynamics of the problems concerned. Our discussion in this chapter will first consider some of the simple relations of fluid dynamics and boundary-layer analysis which are important for a basic understanding of convection heat transfer. Next, we shall impose an energy balance on the flow system and determine the influence of the flow on the temperature gradients in the fluid. Finally, having obtained a knowledge of the temperature distribution, the heat-transfer rate from a heated surface to a fluid which is forced over it may be determined.

Our development in this chapter is primarily analytical in character and is concerned only with forced-convection flow systems. Subsequent chapters will present empirical relations for calculating forced-convection heat transfer and will also treat the subjects of natural convection and boiling and condensation heat transfer.

5-2 VISCOUS FLOW

Consider the flow over a flat plate as shown in Figs. 5-1 and 5-2. Beginning at the leading edge of the plate, a region develops where the influence of viscous forces is felt. These viscous forces are described in terms of a shear stress τ between the fluid layers. If this stress is assumed to be proportional to the normal velocity gradient, we have the defining equation for the viscosity,

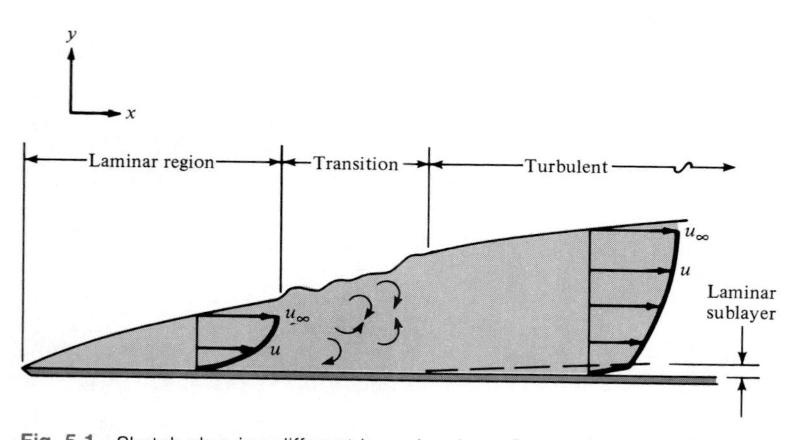

Fig. 5-1 Sketch showing different boundary-layer flow regimes on a flat plate.

$$\tau = \mu \frac{du}{dy} \tag{5-1}$$

The constant of proportionality μ is called the *dynamic viscosity*. A typical set of units is newton-seconds per square meter; however, many sets of units are used for the viscosity, and care must be taken to select the proper group which will be consistent with the formulation at hand.

The region of flow which develops from the leading edge of the plate in which the effects of viscosity are observed is called the boundary layer. Some arbitrary point is used to designate the y position where the boundary layer ends; this point is usually chosen as the y coordinate where the velocity becomes 99 percent of the free-stream value.

Initially, the boundary-layer development is laminar, but at some critical distance from the leading edge, depending on the flow field and fluid properties, small disturbances in the flow begin to become amplified, and a transition process takes place until the flow becomes turbulent. The turbulent-flow region may be pictured as a random churning action with chunks of fluid moving to and fro in all directions. The transition from laminar to turbulent flow occurs when

$$\frac{u_\infty x}{\nu} = \frac{\rho u_\infty x}{\mu} > 5 \times 10^5$$

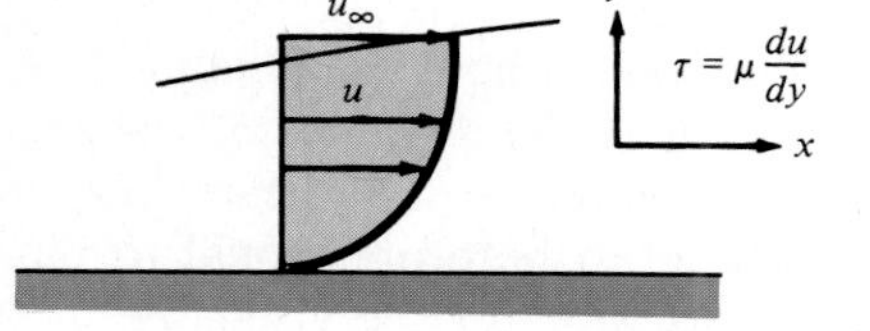

Fig. 5-2 Laminar velocity profile on a flat plate.

where u_∞ = free-stream velocity
x = distance from leading edge
$\nu = \mu/\rho$ = kinematic viscosity

This particular grouping of terms is called the Reynolds number, and is dimensionless if a consistent set of units is used for all the properties:

$$\mathrm{Re}_x = \frac{u_\infty x}{\nu} \tag{5-2}$$

Although the critical Reynolds number for transition on a flat plate is usually taken as 5×10^5 for most analytical purposes, the critical value in a practical situation is strongly dependent on the surface-roughness conditions and the "turbulence level" of the free stream. The normal range for the beginning of transition is between 5×10^5 and 10^6. With very large disturbances present in the flow, transition may begin with Reynolds numbers as low as 10^5, and for flows which are very free from fluctuations, it may not start until $\mathrm{Re} = 2 \times 10^6$ or more. In reality, the transition process is one which covers a range of Reynolds numbers, with transition being complete and with developed turbulent flow usually observed at Reynolds numbers twice the value at which transition began.

The relative shapes for the velocity profiles in laminar and turbulent flow are indicated in Fig. 5-1. The laminar profile is approximately parabolic, while the turbulent profile has a portion near the wall which is very nearly linear. This linear portion is said to be due to a laminar sublayer which hugs the surface very closely. Outside this sublayer the velocity profile is relatively flat in comparison with the laminar profile.

The physical mechanism of viscosity is one of momentum exchange. Consider the laminar-flow situation. Molecules may move from one lamina to another, carrying with them a momentum corresponding to the velocity of the flow. There is a net momentum transport from regions of high velocity to regions of low velocity, thus creating a force in the direction of the flow. This force is the viscous-shear stress which is calculated with Eq. (5-1).

The rate at which the momentum transfer takes place is dependent on the rate at which the molecules move across the fluid layers. In a gas, the molecules would move about with some average speed proportional to the square root of the absolute temperature since, in the kinetic theory of gases, we identify temperature with the mean kinetic energy of a molecule. The faster the molecules move, the more momentum they will transport. Hence we should expect the viscosity of a gas to be approximately proportional to the square root of temperature, and this expectation is corroborated fairly well by experiment. The viscosities of some typical fluids are given in Appendix A.

In the turbulent-flow region distinct fluid layers are no longer observed, and we are forced to seek a somewhat different concept for viscous action. A qualitative picture of the turbulent-flow process may be obtained by imagining macroscopic chunks of fluid transporting energy and momentum instead of microscopic transport on the basis of individual molecules. Naturally, we should expect the larger mass of the macroscopic elements of fluid to transport more energy and momentum than the individual molecules, and we should also expect a larger viscous-shear force in turbulent flow than in laminar flow (and a larger thermal conductivity as well). This expectation is verified by experiment, and it is this larger viscous action in turbulent flow which causes the flat velocity profile indicated in Fig. 5-1.

Consider the flow in a tube as shown in Fig. 5-3. A boundary layer develops at the entrance, as shown. Eventually the boundary layer fills the entire tube, and the flow is said to be fully developed. If the flow is laminar, a parabolic velocity profile is experienced, as shown in Fig. 5-3*a*. When the flow is turbulent, a somewhat blunter profile is observed, as in Fig. 5-3*b*. In a tube, the Reynolds number is again used as a criterion for laminar and turbulent flow. For

$$\mathrm{Re}_d = \frac{u_m d}{\nu} > 2300 \tag{5-3}$$

the flow is usually observed to be turbulent.

Again, a range of Reynolds numbers for transition may be observed, depending on the pipe roughness and smoothness of the flow. The generally accepted range for transition is

$$2000 < \mathrm{Re}_d < 4000$$

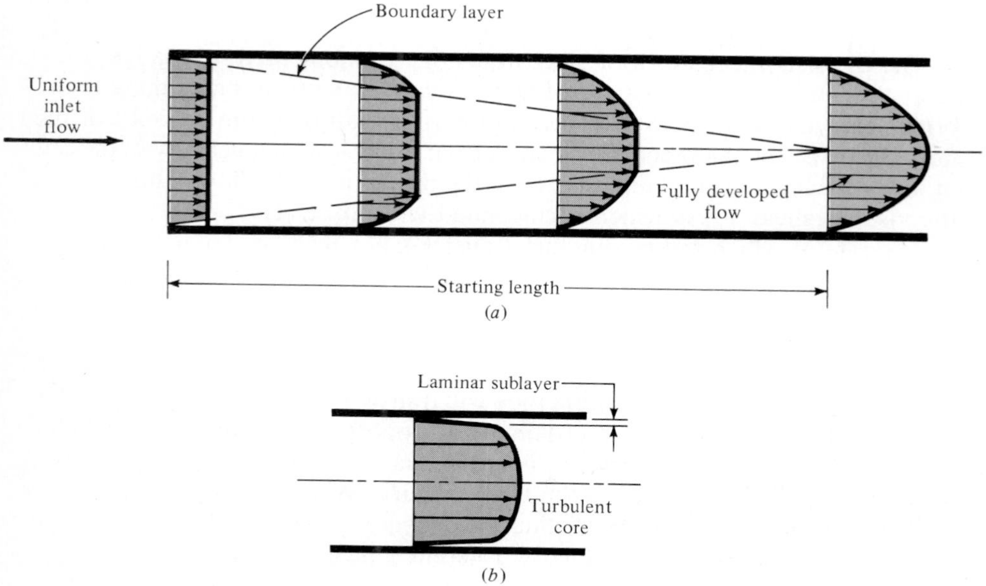

Fig. 5-3 Velocity profile for (*a*) laminar flow in a tube and (*b*) turbulent tube flow.

although laminar flow has been maintained up to Reynolds numbers of 25,000 in carefully controlled laboratory conditions.

The continuity relation for one-dimensional flow in a tube is

$$\dot{m} = \rho u_m A \tag{5-4}$$

where $\dot{m}$ = mass rate of flow
u_m = mean velocity
A = cross-sectional area

We define the mass velocity as

$$\text{Mass velocity} = G = \frac{\dot{m}}{A} = \rho u_m \tag{5-5}$$

so that the Reynolds number may also be written

$$\text{Re}_d = \frac{Gd}{\mu} \tag{5-6}$$

Equation (5-6) is sometimes more convenient to use than Eq. (5-3).

5-3 INVISCID FLOW

Although no real fluid is inviscid, in some instances the fluid may be treated as such, and it is worthwhile to present some of the equations which apply in these circumstances. For example, in the flat-plate problem discussed above, the flow at a sufficiently large distance from the plate will behave as a nonviscous flow system. The reason for this behavior is that the velocity gradients normal to the flow direction are very small, and hence the viscous-shear forces are small.

If a balance of forces is made on an element of incompressible fluid and these forces are set equal to the change in momentum of the fluid element, the Bernoulli equation for flow along a streamline results:

$$\frac{p}{\rho} + \frac{1}{2}\frac{V^2}{g_c} = \text{const} \tag{5-7a}$$

or, in differential form,

$$\frac{dp}{\rho} + \frac{V\,dV}{g_c} = 0 \tag{5-7b}$$

where ρ = fluid density
p = pressure at particular point in flow
V = velocity of flow at that point

The Bernoulli equation is sometimes considered an energy equation because the $V^2/2g_c$ term represents kinetic energy and the pressure represents potential energy; however, it must be remembered that these terms are derived on the basis of a dynamic analysis, so that the equation is fundamentally a dynamic equation. In fact, the concept of kinetic energy is based on a dynamic analysis.

When the fluid is compressible, an energy equation must be written which will take into account changes in internal thermal energy of the system and the corresponding changes in temperature. For a one-dimensional flow system this equation is the steady-flow energy equation for a control volume,

$$i_1 + \frac{1}{2g_c} V_1^2 + Q = i_2 + \frac{1}{2g_c} V_2^2 + Wk \tag{5-8}$$

where i is the enthalpy defined by

$$i = e + pv \tag{5-9}$$

and where e = internal energy
Q = heat added to control volume
Wk = net external work done in the process
v = specific volume of fluid

(The symbol i is used to denote the enthalpy instead of the customary h to avoid confusion with the heat-transfer coefficient.) The subscripts 1 and 2 refer to entrance and exit conditions to the control volume. To calculate pressure drop in compressible flow, it is necessary to specify the equation of state of the fluid, viz., for an ideal gas,

$$p = \rho RT \qquad \Delta e = c_v \, \Delta T \qquad \Delta i = c_p \, \Delta T$$

The gas constant for a particular gas is given in terms of the universal gas constant $\mathcal{R}$ as

$$R = \frac{\mathcal{R}}{M}$$

where M is the molecular weight and $\mathcal{R}$ = 8314.5 J/kg · mol · K. For air, the appropriate ideal-gas properties are

$$R_{\text{air}} = 287 \text{ J/kg} \cdot \text{K} \qquad c_{p,\text{air}} = 1.005 \text{ kJ/kg} \cdot {}^\circ\text{C} \qquad c_{v,\text{ air}} = 0.718 \text{ kJ/kg} \cdot {}^\circ\text{C}$$

To solve a particular problem, we must also specify the process. For example, reversible adiabatic flow through a nozzle yields the following familiar expressions relating the properties at some point in the flow to the Mach number and the stagnation properties, i.e., the properties where the velocity is zero:

$$\frac{T_0}{T} = 1 + \frac{\gamma - 1}{2} M^2$$

$$\frac{p_0}{p} = \left(1 + \frac{\gamma - 1}{2} M^2\right)^{\gamma/(\gamma - 1)}$$

$$\frac{\rho_0}{\rho} = \left(1 + \frac{\gamma - 1}{2} M^2\right)^{1/(\gamma - 1)}$$

where T_0, p_0, ρ_0 = stagnation properties
γ = ratio of specific heats c_p/c_v
M = Mach number

$$M = \frac{V}{a}$$

where a is the local velocity of sound, which may be calculated from

$$a = \sqrt{\gamma g_c R T} \tag{5-10}$$

for an ideal gas.† For air behaving as an ideal gas this equation reduces to

$$a = 20.045\sqrt{T} \quad \text{m/s} \tag{5-11}$$

where T is in degrees Kelvin.

■ EXAMPLE 5-1

Water at 20°C flows at 8 kg/s through the diffuser arrangement shown in the accompanying figure. The diameter at section 1 is 3.0 cm, and the diameter at section 2 is 7.0 cm. Determine the increase in static pressure between sections 1 and 2. Assume frictionless flow.

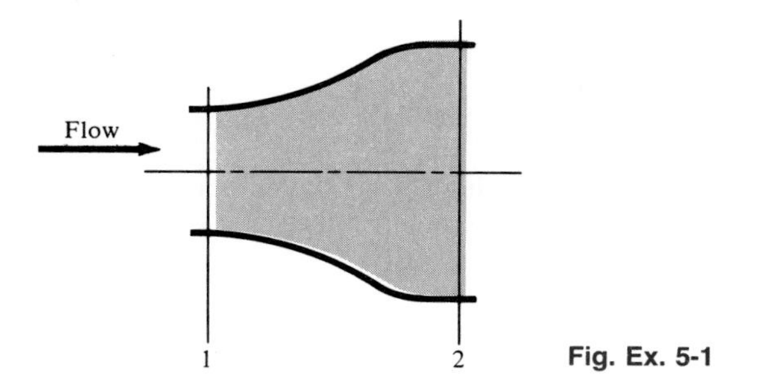

Fig. Ex. 5-1

Solution

The flow cross-sectional areas are

$$A_1 = \frac{\pi d_1^2}{4} = \frac{\pi(0.03)^2}{4} = 7.069 \times 10^{-4} \text{ m}^2$$

$$A_2 = \frac{\pi d_2^2}{4} = \frac{\pi(0.07)^2}{4} = 3.848 \times 10^{-3} \text{ m}^2$$

The density of water at 20°C is 1000 kg/m³, and so we may calculate the velocities from the mass-continuity relation

$$u = \frac{\dot{m}}{\rho A}$$

†The isentropic flow formulas are derived in Ref. 7, p. 629.

$$u_1 = \frac{8.0}{(1000)(7.069 \times 10^{-4})} = 11.32 \text{ m/s} \quad [37.1 \text{ ft/s}]$$

$$u_2 = \frac{8.0}{(1000)(3.848 \times 10^{-3})} = 2.079 \text{ m/s} \quad [6.82 \text{ ft/s}]$$

The pressure difference is obtained from the Bernoulli equation (5-7*a*):

$$\frac{p_2 - p_1}{\rho} = \frac{1}{2g_c}(u_1^2 - u_2^2)$$

$$p_2 - p_1 = \frac{1000}{2}[(11.32)^2 - (2.079)^2]$$

$$= 61.91 \text{ kPa} \quad [8.98 \text{ lb/in}^2 \text{ abs}]$$

EXAMPLE 5-2

Air at 300°C and 0.7 MPa pressure is expanded isentropically from a tank until the velocity is 300 m/s. Determine the static temperature, pressure, and Mach number of the air at the high-velocity condition. $\gamma = 1.4$ for air.

Solution

We may write the steady-flow energy equation as

$$i_1 = i_2 + \frac{u_2^2}{2g_c}$$

because the initial velocity is small and the process is adiabatic. In terms of temperature,

$$c_p(T_1 - T_2) = \frac{u_2^2}{2g_c}$$

$$(1005)(300 - T_2) = \frac{(300)^2}{(2)(1.0)}$$

$$T_2 = 255.2°\text{C} = 528.2 \text{ K} \quad [491.4°\text{F}]$$

We may calculate the pressure from the isentropic relation

$$\frac{p_2}{p_1} = \left(\frac{T_2}{T_1}\right)^{\gamma/(\gamma-1)}$$

$$p_2 = (0.7)\left(\frac{528.2}{573}\right)^{3.5} = 0.526 \text{ MPa} \quad [76.3 \text{ lb/in}^2 \text{ abs}]$$

The velocity of sound at condition 2 is

$$a_2 = (20.045)(528.2)^{1/2} = 460.7 \text{ m/s} \quad [1511 \text{ ft/s}]$$

so that the Mach number is

$$M_2 = \frac{u_2}{a_2} = \frac{300}{460.7} = 0.651$$

5-4 LAMINAR BOUNDARY LAYER ON A FLAT PLATE

Consider the elemental control volume shown in Fig. 5-4. We derive the equation of motion for the boundary layer by making a force-and-momentum balance on this element. To simplify the analysis we assume:

1. The fluid is incompressible and the flow is steady.
2. There are no pressure variations in the direction perpendicular to the plate.
3. The viscosity is constant.
4. Viscous-shear forces in the y direction are negligible.

We apply Newton's second law of motion,

$$\sum F_x = \frac{d(mV)_x}{d\tau}$$

The above form of Newton's second law of motion applies to a system of constant mass. In fluid dynamics it is not usually convenient to work with elements of mass; rather, we deal with elemental control volumes such as that shown in Fig. 5-4, where mass may flow in or out of the different sides of the

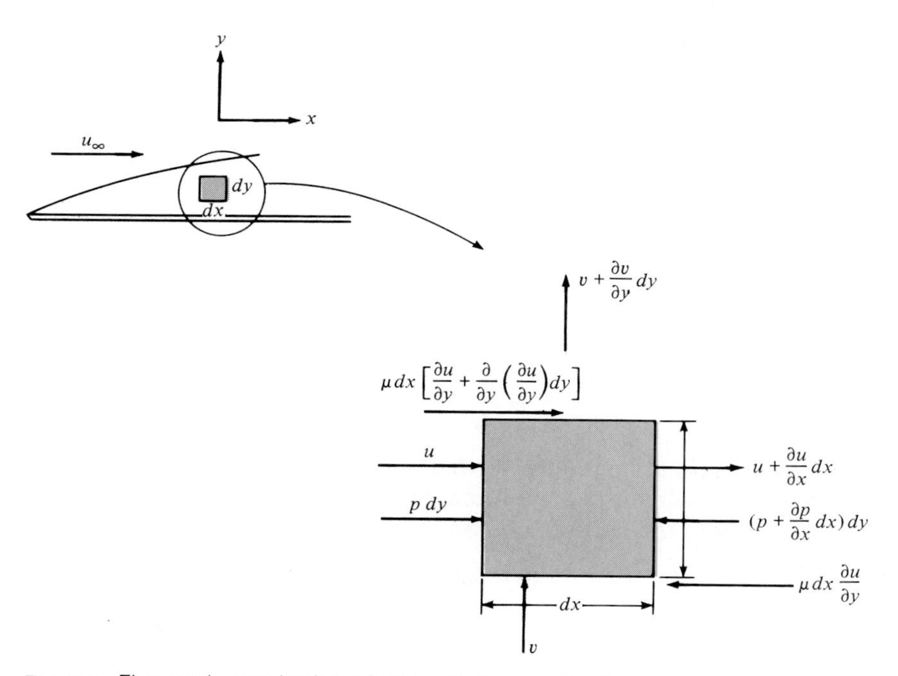

Fig. 5-4 Elemental control volume for force balance on laminar boundary layer.

volume, which is fixed in space. For this system the force balance is then written

$$\Sigma F_x = \text{increase in momentum flux in } x \text{ direction}$$

The momentum flux in the x direction is the product of the mass flow through a particular side of the control volume and the x component of velocity at that point.

The mass entering the left face of the element per unit time is

$$\rho u \, dy$$

if we assume unit depth in the z direction. Thus the momentum entering the left face per unit time is

$$\rho u \, dy \, u = \rho u^2 \, dy$$

The mass flow leaving the right face is

$$\rho \left(u + \frac{\partial u}{\partial x} dx \right) dy$$

and the momentum leaving the right face is

$$\rho \left(u + \frac{\partial u}{\partial x dx} \right)^2 dy$$

The mass flow entering the bottom face is

$$\rho v \, dx$$

and the mass flow leaving the top face is

$$\rho \left(v + \frac{\partial v}{\partial y} dy \right) dx$$

A mass balance on the element yields

$$\rho u \, dy + \rho v \, dx = \rho \left(u + \frac{\partial u}{\partial x} dx \right) dy + \rho \left(v + \frac{\partial u}{\partial y} dy \right) dx$$

or

$$\frac{\partial u}{\partial x} + \frac{\partial v}{\partial y} = 0 \qquad (5\text{-}12)$$

This is the mass continuity equation for the boundary layer.

Returning to the momentum-and-force analysis, the momentum *in the x direction* which enters the bottom face is

$$\rho v u \, dx$$

and the momentum *in the x direction* which leaves the top face is

$$\rho \left(v + \frac{\partial v}{\partial y} dy \right) \left(u + \frac{\partial u}{\partial y} dy \right) dx$$

We are interested only in the momentum in the x direction because the forces considered in the analysis are those in the x direction. These forces are those due to viscous shear and the pressure forces on the element. The pressure force on the left face is $p\,dy$, and that on the right is $-[p + (\partial p/\partial x)\,dx]\,dy$, so that the net pressure force in the direction of motion is

$$-\frac{\partial p}{\partial x}\,dx\,dy$$

The viscous-shear force on the bottom face is

$$-\mu\frac{\partial u}{\partial y}\,dx$$

and the shear force on the top is

$$\mu\,dx\left[\frac{\partial u}{\partial y} + \frac{\partial}{\partial y}\left(\frac{\partial u}{\partial y}\right)dy\right]$$

The net viscous-shear force in the direction of motion is the sum of the above:

$$\text{Net viscous-shear force} = \mu\frac{\partial^2 u}{\partial y^2}\,dx\,dy$$

Equating the sum of the viscous-shear and pressure forces to the net momentum transfer in the x direction, we have

$$\mu\frac{\partial^2 u}{\partial y^2}\,dx\,dy - \frac{\partial p}{\partial x}\,dx\,dy = \rho\left(u + \frac{\partial u}{\partial x}\,dx\right)^2 dy - \rho u^2\,dy$$
$$+ \rho\left(v + \frac{\partial v}{\partial y}\,dy\right)\left(u + \frac{\partial u}{\partial y}\,dy\right)dx - \rho v u\,dx$$

Clearing terms, making use of the continuity relation (5-12) and neglecting second-order differentials, gives

$$\rho\left(u\frac{\partial u}{\partial x} + v\frac{\partial u}{\partial y}\right) = \mu\frac{\partial^2 u}{\partial y^2} - \frac{\partial p}{\partial x} \tag{5-13}$$

This is the momentum equation of the laminar boundary layer with constant properties. The equation may be solved exactly for many boundary conditions, and the reader is referred to the treatise by Schlichting [1] for details of the various methods employed in the solutions. In Appendix B we have included the classical method for obtaining an exact solution to Eq. (5-13) for laminar flow over a flat plate. For the development in this chapter we shall be satisfied with an approximate analysis which furnishes an easier solution without a loss in physical understanding of the processes involved. The approximate method is due to von Kármán [2].

Consider the boundary-layer flow system shown in Fig. 5-5. The free-stream velocity outside the boundary layer is u_∞, and the boundary-layer thickness is

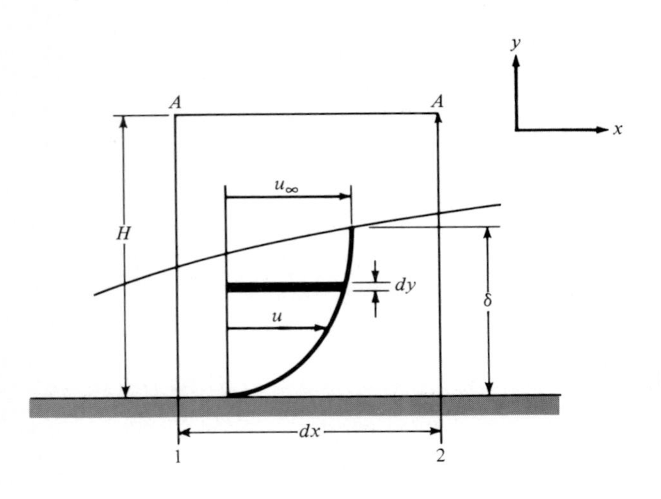

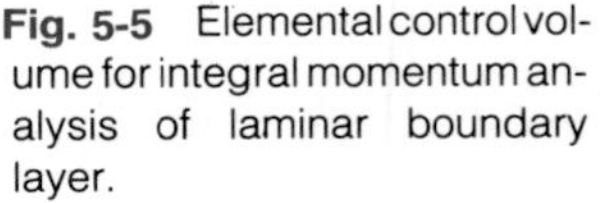

Fig. 5-5 Elemental control volume for integral momentum analysis of laminar boundary layer.

δ. We wish to make a momentum-and-force balance on the control volume bounded by the planes 1, 2, A-A, and the solid wall. The velocity components normal to the wall are neglected, and only those in the x direction are considered. We assume that the control volume is sufficiently high that it always encloses the boundary layer; that is, $H > \delta$.

The mass flow through plane 1 is

$$\int_0^H \rho u \, dy \tag{a}$$

and the momentum flow through plane 1 is

$$\int_0^H \rho u^2 \, dy \tag{b}$$

The momentum flow through plane 2 is

$$\int_0^H \rho u^2 \, dy + \frac{d}{dx}\left(\int_0^H \rho u^2 \, dy\right) dx \tag{c}$$

and the mass flow through plane 2 is

$$\int_0^H \rho u \, dy + \frac{d}{dx}\left(\int_0^H \rho u \, dy\right) dx \tag{d}$$

Considering the conservation of mass and the fact that no mass can enter the control volume through the solid wall, the additional mass flow in expression (d) over that in (a) must enter through plane A-A. This mass flow carries with it a momentum in the x direction equal to

$$u_\infty \frac{d}{dx}\left(\int_0^H \rho u \, dy\right) dx$$

The net momentum flow out of the control volume is therefore

$$\frac{d}{dx}\left(\int_0^H \rho u^2\, dy\right) dx \;-\; u_\infty \frac{d}{dx}\left(\int_0^H \rho u\, dy\right) dx$$

This expression may be put in a somewhat more useful form by recalling the product formula from the differential calculus:

$$d(\eta\phi) = \eta d\phi + \phi d\eta$$

or

$$\eta d\phi = d(\eta\phi) - \phi d\eta$$

In the momentum expression given above, the integral

$$\int_0^H \rho u\, dy$$

is the ϕ function and u_∞ is the η function. Thus

$$\begin{aligned} u_\infty \frac{d}{dx}\left(\int_0^H \rho u\, dy\right) dx &= \frac{d}{dx}\left(u_\infty \int_0^H \rho u\, dy\right) dx - \frac{du_\infty}{dx}\left(\int_0^H \rho u\, dy\right) dx \\ &= \frac{d}{dx}\left(\int_0^H \rho u u_\infty\, dy\right) dx - \frac{du_\infty}{dx}\left(\int_0^H \rho u\, dy\right) dx \end{aligned} \tag{5-14}$$

The u_∞ may be placed inside the integral since it is not a function of y and thus may be treated as a constant insofar as an integral with respect to y is concerned.

Returning to the analysis, the force on plane 1 is the pressure force pH and that on plane 2 is $[p + (dp/dx)dx]H$. The shear force at the wall is

$$-\tau_w\, dx = -\mu\, dx \left.\frac{\partial u}{\partial y}\right]_{y=0}$$

There is no shear force at plane A-A since the velocity gradient is zero outside the boundary layer. Setting the forces on the element equal to the net increase in momentum and collecting terms gives

$$-\tau_w - \frac{dp}{dx}H = -\rho \frac{d}{dx}\int_0^H (u_\infty - u)u\, dy + \frac{du_\infty}{dx}\int_0^H \rho u\, dy \tag{5-15}$$

This is the integral momentum equation of the boundary layer. If the pressure is constant throughout the flow,

$$\frac{dp}{dx} = 0 = -\rho u_\infty \frac{du_\infty}{dx} \tag{5-16}$$

since the pressure and free-stream velocity are related by the Bernoulli equation. For the constant-pressure condition the integral boundary-layer equation becomes

$$\rho \frac{d}{dx}\int_0^\delta (u_\infty - u)u\, dy = \tau_w = \mu \left.\frac{\partial u}{\partial y}\right]_{y=0} \tag{5-17}$$

The upper limit on the integral has been changed to δ because the integrand is zero for $y > \delta$ since $u = u_\infty$ for $y > \delta$.

If the velocity profile were known, the appropriate function could be inserted in Eq. (5-17) to obtain an expression for the boundary-layer thickness. For our approximate analysis we first write down some conditions which the velocity function must satisfy:

$$u = 0 \qquad \text{at } y = 0 \tag{a}$$

$$u = u_\infty \qquad \text{at } y = \delta \tag{b}$$

$$\frac{\partial u}{\partial y} = 0 \qquad \text{at } y = \delta \tag{c}$$

For a constant-pressure condition Eq. (5-13) yields

$$\frac{\partial^2 u}{\partial y^2} = 0 \qquad \text{at } y = 0 \tag{d}$$

since the velocities u and v are zero at $y = 0$. We assume that the velocity profiles at various x positions are similar; i.e., they have the same functional dependence on the y coordinate. There are four conditions to satisfy. The simplest function which we can choose to satisfy these conditions is a polynomial with four arbitrary constants. Thus

$$u = C_1 + C_2 y + C_3 y^2 + C_4 y^3 \tag{5-18}$$

Applying the four conditions (a) to (d),

$$\frac{u}{u_\infty} = \frac{3}{2}\frac{y}{\delta} - \frac{1}{2}\left(\frac{y}{\delta}\right)^3 \tag{5-19}$$

Inserting the expression for the velocity into Eq. (5-17) gives

$$\frac{d}{dx}\left\{\rho u_\infty^2 \int_0^\delta \left[\frac{3}{2}\frac{y}{\delta} - \frac{1}{2}\left(\frac{y}{\delta}\right)^3\right]\left[1 - \frac{3}{2}\frac{y}{\delta} + \frac{1}{2}\left(\frac{y}{\delta}\right)^3\right] dy\right\} = \mu \left.\frac{\partial u}{\partial y}\right]_{y=0}$$

$$= \frac{3}{2}\frac{\mu u_\infty}{\delta}$$

Carrying out the integration leads to

$$\frac{d}{dx}\left(\frac{39}{280}\rho u_\infty^2 \delta\right) = \frac{3}{2}\frac{\mu u_\infty}{\delta}$$

Since ρ and u_∞ are constants, the variables may be separated to give

$$\delta\, d\delta = \frac{140}{13}\frac{\mu}{\rho u_\infty}dx = \frac{140}{13}\frac{\nu}{u_\infty}dx$$

and

$$\frac{\delta^2}{2} = \frac{140}{13}\frac{\nu x}{u_\infty} + \text{const}$$

At $x = 0$, $\delta = 0$, so that

$$\delta = 4.64\sqrt{\frac{\nu x}{u_\infty}} \tag{5-20}$$

This may be written in terms of the Reynolds number as

$$\frac{\delta}{x} = \frac{4.64}{\text{Re}_x^{1/2}} \tag{5-21}$$

where $$\text{Re}_x = \frac{u_\infty x}{\nu}$$

$\nu = \frac{\mu}{\rho}$

The exact solution of the boundary-layer equations as given in Appendix B yields

$$\frac{\delta}{x} = \frac{5.0}{\text{Re}_x^{1/2}} \tag{5-21a}$$

EXAMPLE 5-3

Air at 27°C and 1 atm flows over a flat plate at a speed of 2 m/s. Calculate the boundary-layer thickness at distances of 20 and 40 cm from the leading edge of the plate. Calculate the mass flow which enters the boundary layer between $x = 20$ cm and $x = 40$ cm. The viscosity of air at 27°C is 1.85×10^{-5} kg/m · s. Assume unit depth in the z direction.

Solution

The density of air is calculated from

$$\rho = \frac{p}{RT} = \frac{1.0132 \times 10^5}{(287)(300)} = 1.177 \text{ kg/m}^3 \quad [0.073 \text{ lb}_m/\text{ft}^3]$$

The Reynolds number is calculated as

At $x = 20$ cm: $$\text{Re} = \frac{(1.177)(2.0)(0.2)}{1.85 \times 10^{-5}} = 27{,}580$$

$\frac{\rho}{\mu}$

At $x = 40$ cm: $$\text{Re} = \frac{(1.177)(2.0)(0.4)}{1.85 \times 10^{-5}} = 55{,}160$$

The boundary-layer thickness is calculated from Eq. (5-21):

At $x = 20$ cm: $$\delta = \frac{(4.64)(0.2)}{(27{,}580)^{1/2}} = 0.00559 \text{ m} \quad [0.24 \text{ in}]$$

At $x = 40$ cm: $$\delta = \frac{(4.64)(0.4)}{(55{,}160)^{1/2}} = 0.0079 \text{ m} \quad [0.4 \text{ in}]$$

To calculate the mass flow which enters the boundary layer from the free stream between $x = 20$ cm and $x = 40$ cm, we simply take the difference between the mass flow in the boundary layer at these two x positions. At any x position the mass flow in the boundary layer is given by the integral

$$\int_0^\delta \rho u \, dy$$

where the velocity is given by Eq. (5-19),

$$u = u_\infty \left[\frac{3}{2}\frac{y}{\delta} - \frac{1}{2}\left(\frac{y}{\delta}\right)^3\right]$$

Evaluating the integral with this velocity distribution, we have

$$\int_0^\delta \rho u_\infty \left[\frac{3}{2}\frac{y}{\delta} - \frac{1}{2}\left(\frac{y}{\delta}\right)^3\right] dy = \frac{5}{8}\rho u_\infty \delta$$

Thus the mass flow entering the boundary layer is

$$\begin{aligned}\Delta m &= \tfrac{5}{8}\rho u_\infty(\delta_{40} - \delta_{20})\\ &= (\tfrac{5}{8})(1.177)(2.0)(0.0079 - 0.00559)\\ &= 3.399 \times 10^{-3}\ \text{kg/s} \quad [7.48 \times 10^{-3}\ \text{lb}_m/\text{s}]\end{aligned}$$

5-5 ENERGY EQUATION OF THE BOUNDARY LAYER

The foregoing analysis considered the fluid dynamics of a laminar-boundary-layer flow system. We shall now develop the energy equation for this system and then proceed to an integral method of solution.

Consider the elemental control volume shown in Fig. 5-6. To simplify the analysis we assume

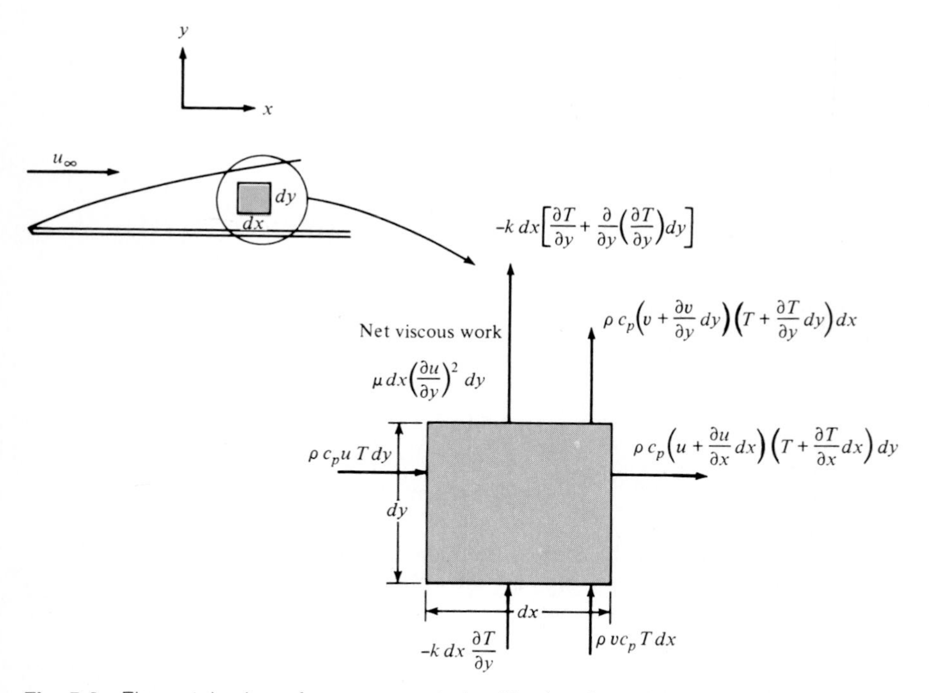

Fig. 5-6 Elemental volume for energy analysis of laminar boundary layer.

1. Incompressible steady flow

2. Constant viscosity, thermal conductivity, and specific heat

3. Negligible heat conduction in the direction of flow (x direction)

Then, for the element shown, the energy balance may be written

Energy convected in left face + energy convected in bottom face
+ heat conducted in bottom face + net viscous work done on element
= energy convected out right face + energy convected out top face
+ heat conducted out top face

The convective and conduction energy quantities are indicated in Fig. 5-6, and the energy term for the viscous work may be derived as follows. The viscous work may be computed as a product of the net viscous-shear force and the distance this force moves in unit time. The viscous-shear force is the product of the shear-stress and the area dx,

$$\mu \frac{\partial u}{\partial y} dx$$

and the distance through which it moves per unit time in respect to the elemental control volume $dx\ dy$ is

$$\frac{\partial u}{\partial y} dy$$

so that the net viscous energy delivered to the element is

$$\mu \left(\frac{\partial u}{\partial y}\right)^2 dx\ dy$$

Writing the energy balance corresponding to the quantities shown in Fig. 5-6, assuming unit depth in the z direction, and neglecting second-order differentials yields

$$\rho c_p \left[u \frac{\partial T}{\partial x} + v \frac{\partial T}{\partial y} + T \left(\frac{\partial u}{\partial x} + \frac{\partial v}{\partial y} \right) \right] dx\ dy = k \frac{\partial^2 T}{\partial y^2} dx\ dy + \mu \left(\frac{\partial u}{\partial y}\right)^2 dx\ dy$$

Using the continuity relation

$$\frac{\partial u}{\partial x} + \frac{\partial v}{\partial y} = 0 \tag{5-12}$$

and dividing by ρc_p gives

$$u \frac{\partial T}{\partial x} + v \frac{\partial T}{\partial y} = \alpha \frac{\partial^2 T}{\partial y^2} + \frac{\mu}{\rho c_p} \left(\frac{\partial u}{\partial y}\right)^2 \tag{5-22}$$

This is the energy equation of the laminar boundary layer. The left side represents the net transport of energy into the control volume, and the right side represents the sum of the net heat conducted out of the control volume and the net viscous work done on the element. The viscous-work term is of importance only at high velocities since its magnitude will be small compared with the other terms when low-velocity flow is studied. This may be shown with an order-of-magnitude analysis of the two terms on the right side of Eq. (5-22). For this order-of-magnitude analysis we might consider the velocity as having the order of the free-stream velocity u_∞ and the y dimension of the order of δ. Thus

$$u \sim u_\infty \qquad \text{and} \qquad y \sim \delta$$

so that

$$\alpha \frac{\partial^2 T}{\partial y^2} \sim \alpha \frac{T}{\delta^2}$$

$$\frac{\mu}{\rho c_p}\left(\frac{\partial u}{\partial y}\right)^2 \sim \frac{\mu}{\rho c_p}\frac{u_\infty^2}{\delta^2}$$

If the ratio of these quantities is small, that is,

$$\frac{\mu}{\rho c_p \alpha}\frac{u_\infty^2}{T} \ll 1 \tag{5-23}$$

then the viscous dissipation is small in comparison with the conduction term. Let us rearrange Eq. (5-23) by introducing

$$\text{Pr} = \frac{\nu}{\alpha} = \frac{c_p \mu}{k}$$

where Pr is called the Prandtl number, which we shall discuss later. Equation (5-23) becomes

$$\text{Pr}\frac{u_\infty{}^2}{c_p T} \ll 1 \tag{5-24}$$

As an example, consider the flow of air at

$$u_\infty = 70 \text{ m/s} \qquad T = 20°\text{C} = 293 \text{ K} \qquad p = 1 \text{ atm}$$

For these conditions $c_p = 1005$ J/kg · °C and Pr = 0.7 so that

$$\text{Pr}\frac{u_\infty{}^2}{c_p T} = \frac{(0.7)(70)^2}{(1005)(293)} = 0.012 \ll 1.0$$

indicating that the viscous dissipation is small. Thus, for low-velocity incompressible flow, we have

$$u\frac{\partial T}{\partial x} + v\frac{\partial T}{\partial y} = \alpha\frac{\partial^2 T}{\partial y^2} \tag{5-25}$$

In reality, our derivation of the energy equation has been a simplified one, and several terms have been left out of the analysis because they are small in comparison with others. In this way we arrive at the boundary-layer approximation immediately, without resorting to a cumbersome elimination process to obtain the final simplified relation. The general derivation of the boundary-layer energy equation is very involved and quite beyond the scope of our discussion. The interested reader should consult the books by Schlichting [1] and White [5] for more information.

There is a striking similarity between Eq. (5-25) and the momentum equation for constant pressure,

$$u \frac{\partial u}{\partial x} + v \frac{\partial u}{\partial y} = \nu \frac{\partial^2 u}{\partial y^2} \tag{5-26}$$

The solution to the two equations will have exactly the same form when $\alpha = \nu$. Thus we should expect that the relative magnitudes of the thermal diffusivity and kinematic viscosity would have an important influence on convection heat transfer since these magnitudes relate the velocity distribution to the temperature distribution. This is exactly the case, and we shall see the role which these parameters play in the subsequent discussion.

5-6 THE THERMAL BOUNDARY LAYER

Just as the hydrodynamic boundary layer was defined as that region of the flow where viscous forces are felt, a thermal boundary layer may be defined as that region where temperature gradients are present in the flow. These temperature gradients would result from a heat-exchange process between the fluid and the wall.

Consider the system shown in Fig. 5-7. The temperature of the wall is T_w, the temperature of the fluid outside the thermal boundary layer is T_∞, and the thickness of the thermal boundary layer is designated as δ_t. At the wall, the velocity is zero, and the heat transfer into the fluid takes place by conduction. Thus the local heat flux per unit area, q'', is

$$\frac{q}{A} = q'' = -k \left. \frac{\partial T}{\partial y} \right]_{\text{wall}} \tag{5-27}$$

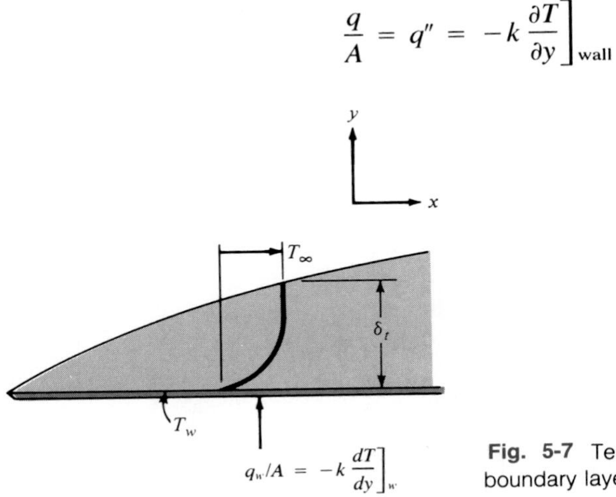

Fig. 5-7 Temperature profile in the thermal boundary layer.

From Newton's law of cooling [Eq. (1-8)],

$$q'' = h(T_w - T_\infty) \tag{5-28}$$

where h is the convection heat-transfer coefficient. Combining these equations, we have

$$h = \frac{-k(\partial T/\partial y)_{\text{wall}}}{T_w - T_\infty} \tag{5-29}$$

so that we need only find the temperature gradient at the wall in order to evaluate the heat-transfer coefficient. This means that we must obtain an expression for the temperature distribution. To do this, an approach similar to that used in the momentum analysis of the boundary layer is followed.

The conditions which the temperature distribution must satisfy are

$$T = T_w \qquad \text{at } y = 0 \tag{a}$$

$$\frac{\partial T}{\partial y} = 0 \qquad \text{at } y = \delta_t \tag{b}$$

$$T = T_\infty \qquad \text{at } y = \delta_t \tag{c}$$

and by writing Eq. (5-25) at $y = 0$ with no viscous heating we find

$$\frac{\partial^2 T}{\partial y^2} = 0 \qquad \text{at } y = 0 \tag{d}$$

since the velocities must be zero at the wall.

Conditions (a) to (d) may be fitted to a cubic polynomial as in the case of the velocity profile, so that

$$\frac{\theta}{\theta_\infty} = \frac{T - T_w}{T_\infty - T_w} = \frac{3}{2}\frac{y}{\delta_t} - \frac{1}{2}\left(\frac{y}{\delta_t}\right)^3 \tag{5-30}$$

where $\theta = T - T_w$. There now remains the problem of finding an expression for δ_t, the thermal-boundary-layer thickness. This may be obtained by an integral analysis of the energy equation for the boundary layer.

Consider the control volume bounded by the planes 1, 2, A-A, and the wall as shown in Fig. 5-8. It is assumed that the thermal boundary layer is thinner than the hydrodynamic boundary layer, as shown. The wall temperature is T_w, the free-stream temperature is T_∞, and the heat given up to the fluid over the length dx is dq_w. We wish to make the energy balance

Energy convected in + viscous work within element
+ heat transfer at wall = energy convected out (5-31)

The energy convected in through plane 1 is

$$\rho c_p \int_0^H uT\,dy$$

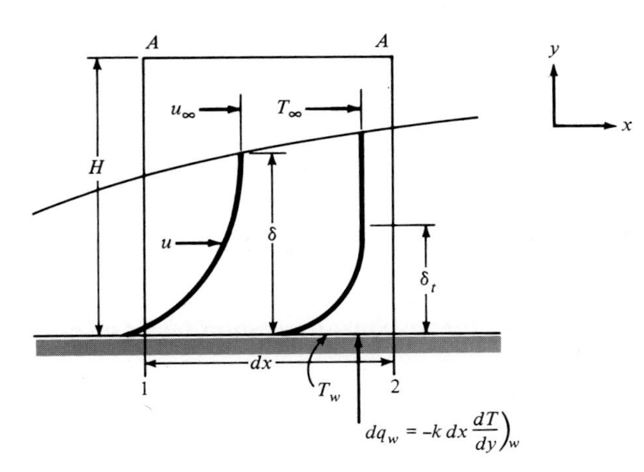

Fig. 5-8 Control volume for integral energy analysis of laminar boundary flow.

and the energy convected out through plane 2 is

$$\rho c_p \left(\int_0^H uT\,dy \right) + \frac{d}{dx}\left(\rho c_p \int_0^H uT\,dy \right) dx$$

The mass flow through plane *A-A* is

$$\frac{d}{dx}\left(\int_0^H \rho u\,dy \right) dx$$

and this carries with it an energy equal to

$$c_p T_\infty \frac{d}{dx}\left(\int_0^H \rho u\,dy \right) dx$$

The net viscous work done within the element is

$$\mu \left[\int_0^H \left(\frac{du}{dy} \right)^2 dy \right] dx$$

and the heat transfer at the wall is

$$dq_w = -k\,dx \left.\frac{\partial T}{\partial y}\right]_w$$

Combining these energy quantities according to Eq. (5-31) and collecting terms gives

$$\frac{d}{dx}\left[\int_0^H (T_\infty - T)u\,dy \right] + \frac{\mu}{\rho c_p}\left[\int_0^H \left(\frac{du}{dy} \right)^2 dy \right] = \alpha \left.\frac{\partial T}{dy}\right]_w \tag{5-32}$$

This is the integral energy equation of the boundary layer for constant properties and constant free-stream temperature T_∞.

To calculate the heat transfer at the wall, we need to derive an expression for the thermal-boundary-layer thickness which may be used in conjunction with Eqs. (5-29) and (5-30) to determine the heat-transfer coefficient. For now, we neglect the viscous-dissipation term; this term is very small unless the velocity of the flow field becomes very large, and the calculation of high-velocity heat transfer will be considered later.

The plate under consideration need not be heated over its entire length. The situation which we shall analyze is shown in Fig. 5-9, where the hydrodynamic boundary layer develops from the leading edge of the plate, while heating does not begin until $x = x_0$.

Inserting the temperature distribution Eq. (5-30) and the velocity distribution Eq. (5-19) into Eq. (5-32) and neglecting the viscous-dissipation term, gives

$$\frac{d}{dx}\left[\int_0^H (T_\infty - T)u\,dy\right] = \frac{d}{dx}\left[\int_0^H (\theta_\infty - \theta)u\,dy\right]$$

$$= \theta_\infty u_\infty \frac{d}{dx}\left\{\int_0^H \left[1 - \frac{3}{2}\frac{y}{\delta_t} + \frac{1}{2}\left(\frac{y}{\delta_t}\right)^3\right]\left[\frac{3}{2}\frac{y}{\delta} - \frac{1}{2}\left(\frac{y}{\delta}\right)^3\right] dy\right\}$$

$$= \alpha \left.\frac{\partial T}{\partial y}\right]_{y=0} = \frac{3\alpha\theta_\infty}{2\delta_t}$$

Let us assume that the thermal boundary layer is thinner than the hydrodynamic boundary layer. Then we only need to carry out the integration to $y = \delta_t$ since the integrand is zero for $y > \delta_t$. Performing the necessary algebraic manipulation, carrying out the integration, and making the substitution $\zeta = \delta_t/\delta$ yields

$$\theta_\infty u_\infty \frac{d}{dx}\left[\delta\left(\frac{3}{20}\zeta^2 - \frac{3}{280}\zeta^4\right)\right] = \frac{3}{2}\frac{\alpha\theta_\infty}{\delta\zeta} \tag{5-33}$$

Because $\delta_t < \delta$, $\zeta < 1$, and the term involving ζ^4 is small compared with the ζ^2 term, we neglect the ζ^4 term and write

$$\frac{3}{20}\theta_\infty u_\infty \frac{d}{dx}(\delta\zeta^2) = \frac{3}{2}\frac{\alpha\theta_\infty}{\zeta\delta} \tag{5-34}$$

Performing the differentiation gives

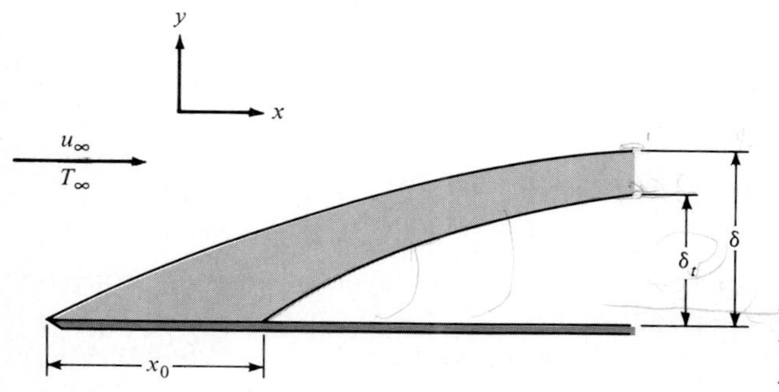

Fig. 5-9 Hydrodynamic and thermal boundary layers on a flat plate. Heating starts at $x = x_0$.

$$\frac{1}{10} u_\infty \left(2\delta\zeta \frac{d\zeta}{dx} + \zeta^2 \frac{d\delta}{dx}\right) = \frac{\alpha}{\delta\zeta}$$

or

$$\frac{1}{10} u_\infty \left(2\delta^2\zeta^2 \frac{d\zeta}{dx} + \zeta^3\delta \frac{d\delta}{dx}\right) = \alpha$$

But

$$\delta\, d\delta = \frac{140}{13} \frac{\nu}{u_\infty} dx$$

and

$$\delta^2 = \frac{280}{13} \frac{\nu x}{u_\infty}$$

so that we have

$$\zeta^3 + 4x\zeta^2 \frac{d\zeta}{dx} = \frac{13}{14} \frac{\alpha}{\nu} \tag{5-35}$$

Noting that

$$\zeta^2 \frac{d\zeta}{dx} = \frac{1}{3} \frac{d}{dx} \zeta^3$$

we see that Eq. (5-35) is a linear differential equation of the first order in ζ^3, and the solution is

$$\zeta^3 = Cx^{-3/4} + \frac{13}{14} \frac{\alpha}{\nu}$$

When the boundary condition

$$\delta_t = 0 \qquad \text{at } x = x_0$$

$$\zeta = 0 \qquad \text{at } x = x_0$$

is applied, the final solution becomes

$$\zeta = \frac{\delta_t}{\delta} = \frac{1}{1.026} \mathrm{Pr}^{-1/3} \left[1 - \left(\frac{x_0}{x}\right)^{3/4}\right]^{1/3} \tag{5-36}$$

where

$$\mathrm{Pr} = \frac{\nu}{\alpha} \tag{5-37}$$

has been introduced. The ratio ν/α is called the Prandtl number after Ludwig Prandtl, the German scientist who introduced the concepts of boundary-layer theory.

When the plate is heated over the entire length, $x_0 = 0$, and

$$\frac{\delta_t}{\delta} = \zeta = \frac{1}{1.026} \mathrm{Pr}^{-1/3} \tag{5-38}$$

In the foregoing analysis the assumption was made that $\zeta < 1$. This assumption is satisfactory for fluids having Prandtl numbers greater than about

0.7. Fortunately, most gases and liquids fall within this category. Liquid metals are a notable exception, however, since they have Prandtl numbers of the order of 0.01.

The Prandtl number ν/α has been found to be the parameter which relates the relative thicknesses of the hydrodynamic and thermal boundary layers. The kinematic viscosity of a fluid conveys information about the rate at which momentum may diffuse through the fluid because of molecular motion. The thermal diffusivity tells us the same thing in regard to the diffusion of heat in the fluid. Thus the ratio of these two quantities should express the relative magnitudes of diffusion of momentum and heat in the fluid. But these diffusion rates are precisely the quantities that determine how thick the boundary layers will be for a given external flow field; large diffusivities mean that the viscous or temperature influence is felt farther out in the flow field. The Prandtl number is thus the connecting link between the velocity field and the temperature field.

The Prandtl number is dimensionless when a consistent set of units is used:

$$\mathrm{Pr} = \frac{\nu}{\alpha} = \frac{\mu/\rho}{k/\rho c_p} = \frac{c_p\mu}{k} \tag{5-39}$$

In the SI system a typical set of units for the parameters would be μ in kilograms per second per meter, c_p in kilojoules per kilogram per Celsius degree, and k in kilowatts per meter per Celsius degree. In the English system one would typically employ μ in pound mass per hour per foot, c_p in Btu per pound mass per Fahrenheit degree, and k in Btu per hour per foot per Fahrenheit degree.

Returning now to the analysis, we have

$$h = \frac{-k(\partial T/\partial y)_w}{T_w - T_\infty} = \frac{3}{2}\frac{k}{\delta_t} = \frac{3}{2}\frac{k}{\zeta\delta} \tag{5-40}$$

Substituting for the hydrodynamic-boundary-layer thickness from Eq. (5-21) and using Eq. (5-36) gives

$$h_x = 0.332k\,\mathrm{Pr}^{1/3}\left(\frac{u_\infty}{\nu x}\right)^{1/2}\left[1 - \left(\frac{x_0}{x}\right)^{3/4}\right]^{-1/3} \tag{5-41}$$

The equation may be nondimensionalized by multiplying both sides by x/k, producing the dimensionless group on the left side,

$$\mathrm{Nu}_x = \frac{h_x x}{k} \tag{5-42}$$

called the Nusselt number after Wilhelm Nusselt, who made significant contributions to the theory of convection heat transfer. Finally,

$$\mathrm{Nu}_x = 0.332\,\mathrm{Pr}^{1/3}\,\mathrm{Re}_x^{1/2}\left[1 - \left(\frac{x_0}{x}\right)^{3/4}\right]^{-1/3} \tag{5-43}$$

or, for the plate heated over its entire length, $x_0 = 0$ and

$$\mathrm{Nu}_x = 0.332\ \mathrm{Pr}^{1/3}\ \mathrm{Re}_x^{1/2} \tag{5-44}$$

Equations (5-41), (5-43), and (5-44) express the local values of the heat-transfer coefficient in terms of the distance from the leading edge of the plate and the fluid properties. For the case where $x_0 = 0$ the average heat-transfer coefficient and Nusselt number may be obtained by integrating over the length of the plate:

$$\bar{h} = \frac{\int_0^L h_x\, dx}{\int_0^L dx} = 2h_{x=L} \tag{5-45}$$

$$\overline{\mathrm{Nu}_L} = \frac{\bar{h}L}{k} = 2\ \mathrm{Nu}_{x=L} \tag{5-46a}$$

or

$$\overline{\mathrm{Nu}_L} = \frac{\bar{h}L}{k} = 0.664\ \mathrm{Re}_L{}^{1/2}\ \mathrm{Pr}^{1/3} \tag{5-46b}$$

where

$$\mathrm{Re}_L = \frac{\rho u_\infty L}{\mu}$$

The reader should carry out the integrations to verify these results.

The foregoing analysis was based on the assumption that the fluid properties were constant throughout the flow. When there is an appreciable variation between wall and free-stream conditions, it is recommended that the properties be evaluated at the so-called *film temperature* T_f, defined as the arithmetic mean between the wall and free-stream temperature,

$$T_f = \frac{T_w + T_\infty}{2} \tag{5-47}$$

An exact solution to the energy equation is given in Appendix B. The results of the exact analysis are the same as those of the approximate analysis given above.

Constant Heat Flux

The above analysis has considered the laminar heat transfer from an isothermal surface. In many practical problems the surface *heat flux* is essentially constant, and the objective is to find the distribution of the plate-surface temperature for given fluid-flow conditions. For the constant-heat-flux case it can be shown that the local Nusselt number is given by

$$\mathrm{Nu}_x = \frac{hx}{k} = 0.453\ \mathrm{Re}_x^{1/2}\ \mathrm{Pr}^{1/3} \tag{5-48}$$

which may be expressed in terms of the wall heat flux and temperature difference as

$$\mathrm{Nu}_x = \frac{q_w x}{k(T_w - T_\infty)} \tag{5-49}$$

The average temperature difference along the plate, for the constant heat flux condition, may be obtained by performing the integration

$$\overline{T_w - T_\infty} = \frac{1}{L}\int_0^L (T_w - T_\infty)\, dx = \frac{1}{L}\int_0^L \frac{q_w x}{k\,\mathrm{Nu}_x}\, dx$$

$$= \frac{q_w L/k}{0.6795\ \mathrm{Re}_L^{1/2}\ \mathrm{Pr}^{1/3}} \tag{5-50}$$

or

$$q_w = \tfrac{3}{2} h_{x=L}(\overline{T_w - T_\infty})$$

In these equations q_w is the heat flux per unit area and will have the units of watts per square meter (W/m²) in SI units or British thermal units per hour per square foot (Btu/h · ft²) in the English system.

Other Relations

Equation (5-44) is applicable to fluids having Prandtl numbers between about 0.6 and 50. It would not apply to fluids with very low Prandtl numbers like liquid metals or to high-Prandtl-number fluids like heavy oils or silicones. For a very wide range of Prandtl numbers, Churchill and Ozoe [9] have correlated a large amount of data to give the following relation for laminar flow on an isothermal flat plate:

$$\mathrm{Nu}_x = \frac{0.3387\ \mathrm{Re}_x^{1/2}\ \mathrm{Pr}^{1/3}}{\left[1 + \left(\dfrac{0.0468}{\mathrm{Pr}}\right)^{2/3}\right]^{1/4}} \qquad \text{for } \mathrm{Re}_x\,\mathrm{Pr} > 100 \tag{5-51}$$

For the constant-heat-flux case, 0.3387 is changed to 0.4637 and 0.0468 is changed to 0.0207. Properties are still evaluated at the film temperature.

EXAMPLE 5-4

For the flow system in Example 5-3 assume that the plate is heated over its entire length to a temperature of 60°C. Calculate the heat transferred in (*a*) the first 20 cm of the plate and (*b*) the first 40 cm of the plate.

Solution

The total heat transfer over a certain length of the plate is desired; so we wish to calculate average heat-transfer coefficients. For this purpose we use Eqs. (5-44) and (5-45), evaluating the properties at the film temperature:

$$T_f = \frac{27 + 60}{2} = 43.5°\mathrm{C} = 316.5\ \mathrm{K} \quad [110.3°\mathrm{F}]$$

From Appendix A the properties are

$$\nu = 17.36 \times 10^{-6}\ \mathrm{m^2/s} \quad [1.87 \times 10^{-4}\ \mathrm{ft^2/s}]$$

$$k = 0.02749\ \mathrm{W/m \cdot °C} \quad [0.0159\ \mathrm{Btu/h \cdot ft \cdot °F}]$$

$$\text{Pr} = 0.7$$

$$c_p = 1.006 \text{ kJ/kg} \cdot °\text{C} \quad [0.24 \text{ Btu/lb}_m \cdot °\text{F}]$$

At $x = 20$ cm

$$\text{Re}_x = \frac{u_\infty x}{\nu} = \frac{(2)(0.2)}{17.36 \times 10^{-6}} = 23{,}041$$

$$\text{Nu}_x = \frac{h_x x}{k} = 0.332 \text{ Re}_x^{1/2} \text{ Pr}^{1/3}$$

$$= (0.332)(23{,}041)^{1/2} (0.7)^{1/3} = 44.74$$

$$h_x = \text{Nu}_x \left(\frac{k}{x}\right) = \frac{(44.74)(0.02749)}{0.2}$$

$$= 6.15 \text{ W/m}^2 \cdot °\text{C} \quad [1.083 \text{ Btu/h} \cdot \text{ft}^2 \cdot °\text{F}]$$

The average value of the heat-transfer coefficient is twice this value, or

$$\bar{h} = (2)(6.15) = 12.3 \text{ W/m}^2 \cdot °\text{C} \quad [2.17 \text{ Btu/h} \cdot \text{ft}^2 \cdot °\text{F}]$$

The heat flow is

$$q = \bar{h}A(T_w - T_\infty)$$

If we assume unit depth in the z direction,

$$q = (12.3)(0.2)(60 - 27) = 81.18 \text{ W} \quad [277 \text{ Btu/h}]$$

At $x = 40$ cm

$$\text{Re}_x = \frac{u_\infty x}{\nu} = \frac{(2)(0.4)}{17.36 \times 10^{-6}} = 46{,}082$$

$$\text{Nu}_x = (0.332)(46{,}082)^{1/2} (0.7)^{1/3} = 63.28$$

$$h_x = \frac{(63.28)(0.02749)}{0.4} = 4.349 \text{ W/m}^2 \cdot °\text{C}$$

$$\bar{h} = (2)(4.349) = 8.698 \text{ W/m}^2 \cdot °\text{C} \quad [1.53 \text{ Btu/h} \cdot \text{ft}^2 \cdot °\text{F}]$$

$$q = (8.698)(0.4)(60 - 27) = 114.8 \text{ W} \quad [392 \text{ Btu/h}]$$

■ **EXAMPLE 5-5**

A 1.0-kW heater is constructed of a glass plate with an electrically conducting film which produces a constant heat flux. The plate is 60 by 60 cm and placed in an airstream at 27°C, 1 atm with $u_\infty = 5$ m/s. Calculate the average temperature difference along the plate and the temperature difference at the trailing edge.

Solution

Properties should be evaluated at the film temperature, but we do not know the plate temperature so for an initial calculation we take the properties at the free-stream conditions of

$$T_\infty = 27°\text{C} = 300 \text{ K}$$

$$\nu = 15.69 \times 10^{-6} \text{ m}^2/\text{s} \qquad \text{Pr} = 0.708 \qquad k = 0.02624 \text{ W/m} \cdot °\text{C}$$

$$\text{Re}_L = \frac{(0.6)(5)}{15.69 \times 10^{-6}} = 1.91 \times 10^5$$

From Eq. (5-50) the average temperature difference is

$$\overline{T_w - T_\infty} = \frac{[1000/(0.6)^2](0.6)/0.02624}{0.6795(1.91 \times 10^5)^{1/2}(0.708)^{1/3}} = 240°\text{C}$$

Now, we go back and evaluate properties at

$$T_f = \frac{240 + 27}{2} = 133.5°\text{C} = 406.5 \text{ K}$$

and obtain

$$\nu = 26.66 \times 10^{-6} \text{ m}^2/\text{s} \qquad \text{Pr} = 0.687 \qquad k = 0.0344 \text{ W/m} \cdot °\text{C}$$

$$\text{Re}_L = \frac{(0.6)(5)}{26.66 \times 10^{-6}} = 1.13 \times 10^5$$

$$\overline{T_w - T_\infty} = \frac{[1000/(0.6)^2](0.6)/0.0344}{0.6795(1.13 \times 10^5)^{1/2}(0.687)^{1/3}} = 240.3°\text{C}$$

At the end of the plate ($x = L = 0.6$ m) the temperature difference is obtained from Eqs. (5-48) and (5-50) with the constant 0.453 to give

$$(T_w - T_\infty)_{x=L} = \frac{(240.3)(0.6795)}{0.453} = 360.5°\text{C}$$

An alternate solution would be to base the Nusselt number on Eq. (5-51).

■ EXAMPLE 5-6

Engine oil at 20°C is forced over a 20-cm-square plate at a velocity of 1.2 m/s. The plate is heated to a uniform temperature of 60°C. Calculate the heat lost by the plate.

Solution

We first evaluate the film temperature:

$$T = \frac{20 + 60}{2} = 40°\text{C}$$

The properties of engine oil are

$$\rho = 876 \text{ kg/m}^3 \qquad \nu = 0.00024 \text{ m}^2/\text{s}$$

$$k = 0.144 \text{ W/m} \cdot °\text{C} \qquad \text{Pr} = 2870$$

The Reynolds number is

$$\text{Re} = \frac{u_\infty L}{\nu} = \frac{(1.2)(0.2)}{0.00024} = 1000$$

Because the Prandtl number is so large we will employ Eq. (5-51) for the solution. We see that h_x varies with x in the same fashion as in Eq. (5-44), i.e., $h_x \propto x^{-1/2}$, so that we get the same solution as in Eq. (5-45) for the average heat-transfer coefficient. Evaluating Eq. (5-51) at $x = 0.2$ gives

$$\text{Nu}_x = \frac{(0.3387)(1000)^{1/2}(2870)^{1/3}}{\left[1 + \left(\frac{0.0468}{2870}\right)^{2/3}\right]^{1/4}} = 152.2$$

and
$$h_x = \frac{(152.2)(0.144)}{0.2} = 109.6 \text{ W/m}^2 \cdot {}^\circ\text{C}$$

The average value of the convection coefficient is

$$h = (2)(109.6) = 219.2 \text{ W/m}^2 \cdot {}^\circ\text{C}$$

so that the total heat transfer is

$$q = hA(T_w - T_\infty) = (219.2)(0.2)^2(60 - 20) = 350.6 \text{ W}$$

5-7 THE RELATION BETWEEN FLUID FRICTION AND HEAT TRANSFER

We have already seen that the temperature and flow fields are related. Now we seek an expression whereby the frictional resistance may be directly related to heat transfer.

The shear stress at the wall may be expressed in terms of a friction coefficient C_f:

$$\tau_w = C_f \frac{\rho u_\infty^2}{2} \tag{5-52}$$

Equation (5-52) is the defining equation for the friction coefficient. The shear stress may also be calculated from the relation

$$\tau_w = \mu \left.\frac{\partial u}{\partial y}\right]_w$$

Using the velocity distribution given by Eq. (5-19), we have

$$\tau_w = \frac{3}{2}\frac{\mu u_\infty}{\delta}$$

and making use of the relation for the boundary-layer thickness gives

$$\tau_w = \frac{3}{2}\frac{\mu u_\infty}{4.64}\left(\frac{u_\infty}{\nu x}\right)^{1/2} \tag{5-53}$$

Combining Eqs. (5-52) and (5-53) leads to

$$\frac{C_{fx}}{2} = \frac{3}{2}\frac{\mu u_\infty}{4.64}\left(\frac{u_\infty}{\nu x}\right)^{1/2}\frac{1}{\rho u_\infty^2} = 0.323\ \text{Re}_x^{-1/2} \tag{5-54}$$

Equation (5-44) may be rewritten in the following form:

$$\frac{\text{Nu}_x}{\text{Re}_x\ \text{Pr}} = \frac{h_x}{\rho c_p u_\infty} = 0.332\ \text{Pr}^{-2/3}\ \text{Re}_x^{-1/2}$$

The group on the left is called the Stanton number,

$$\mathrm{St}_x = \frac{h_x}{\rho c_p u_\infty}$$

so that
$$\mathrm{St}_x \, \mathrm{Pr}^{2/3} = 0.332 \, \mathrm{Re}_x^{-1/2} \tag{5-55}$$

Upon comparing Eqs. (5-54) and (5-55), we note that the right sides are alike except for a difference of about 3 percent in the constant, which is the result of the approximate nature of the integral boundary-layer analysis. We recognize this approximation and write

$$\mathrm{St}_x \, \mathrm{Pr}^{2/3} = \frac{C_{fx}}{2} \tag{5-56}$$

Equation (5-56), called the *Reynolds-Colburn analogy,* expresses the relation between fluid friction and heat transfer for laminar flow on a flat plate. The heat-transfer coefficient thus could be determined by making measurements of the frictional drag on a plate under conditions in which no heat transfer is involved.

It turns out that Eq. (5-56) can also be applied to turbulent flow over a flat plate and in a modified way to turbulent flow in a tube. It does not apply to laminar tube flow. In general, a more rigorous treatment of the governing equations is necessary when embarking on new applications of the heat-transfer-fluid-friction analogy, and the results do not always take the simple form of Eq. (5-56). The interested reader may consult the references at the end of the chapter for more information on this important subject. At this point, the simple analogy developed above has served to amplify our understanding of the physical processes in convection and to reinforce the notion that heat-transfer and viscous-transport processes are related at both the microscopic and macroscopic levels.

■ EXAMPLE 5-7

For the flow system in Example 5-4 compute the drag force exerted on the first 40 cm of the plate using the analogy between fluid friction and heat transfer.

Solution

We use Eq. (5-56) to compute the friction coefficient and then calculate the drag force. An average friction coefficient is desired, so

$$\overline{\mathrm{St}} \, \mathrm{Pr}^{2/3} = \frac{\overline{C}_f}{2} \tag{a}$$

The density at 316.5 K is

$$\rho = \frac{p}{RT} = \frac{1.0132 \times 10^5}{(287)(316.5)} = 1.115 \text{ kg/m}^3$$

For the 40-cm length

$$\overline{\mathrm{St}} = \frac{\bar{h}}{\rho c_p u_\infty} = \frac{8.698}{(1.115)(1006)(2)} = 3.88 \times 10^{-3}$$

Then from Eq. (*a*)

$$\frac{\overline{C}_f}{2} = (3.88 \times 10^{-3})(0.7)^{2/3} = 3.06 \times 10^{-3}$$

The average shear stress at the wall is computed from Eq. (5-52):

$$\begin{aligned}\overline{\tau}_w &= \overline{C}_f \rho \frac{u_\infty^2}{2} \\ &= (3.06 \times 10^{-3})(1.115)(2)^2 \\ &= 0.0136 \text{ N/m}^2\end{aligned}$$

The drag force is the product of this shear stress and the area,

$$D = (0.0136)(0.4) = 5.44 \text{ mN} \quad [1.23 \times 10^{-3} \text{ lb}_f]$$

5-8 TURBULENT-BOUNDARY-LAYER HEAT TRANSFER

Consider a portion of a turbulent boundary layer as shown in Fig. 5-10. A very thin region near the plate surface has a laminar character, and the viscous action and heat transfer take place under circumstances like those in laminar flow. Farther out, at larger y distances from the plate, some turbulent action is experienced, but the molecular viscous action and heat conduction are still important. This region is called the *buffer layer*. Still farther out, the flow is fully turbulent, and the main momentum- and heat-exchange mechanism is one involving macroscopic lumps of fluid moving about in the flow. In this fully turbulent region we speak of *eddy viscosity* and *eddy thermal conductivity*. These eddy properties may be 10 times as large as the molecular values.

The physical mechanism of heat transfer in turbulent flow is quite similar to that in laminar flow; the primary difference is that one must deal with the eddy properties instead of the ordinary thermal conductivity and viscosity. The main difficulty in an analytical treatment is that these eddy properties vary across the boundary layer, and the specific variation can be determined only from experimental data. This is an important point. All analyses of turbulent

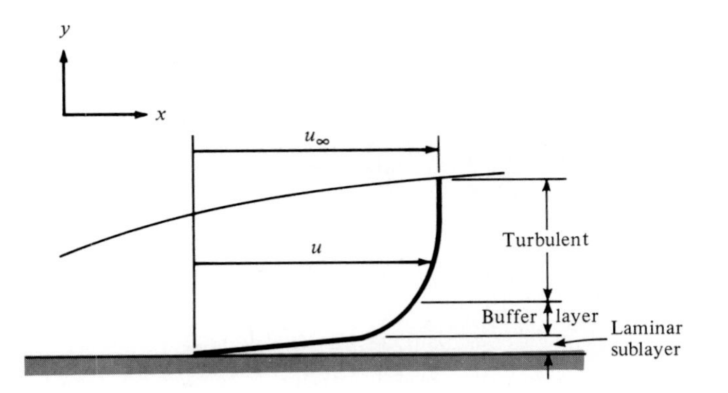

Fig. 5-10 Velocity profile in turbulent boundary layer on a flat plate.

flow must eventually rely on experimental data because there is no completely adequate theory to predict turbulent-flow behavior.

If one observes the instantaneous macroscopic velocity in a turbulent-flow system, as measured with a laser anemometer or other sensitive device, significant fluctuations about the mean flow velocity are observed as indicated in Fig. 5-11, where $\bar{u}$ is designated as the mean velocity and u' is the *fluctuation* from the mean. The instantaneous velocity is therefore

$$u = \bar{u} + u' \tag{5-57}$$

The mean value of the fluctuation u' must be zero over an extended period for steady flow conditions. There are also fluctuations in the y component of velocity, so we would write

$$v = \bar{v} + v' \tag{5-58}$$

The fluctuations give rise to a turbulent-shear stress which may be analyzed by referring to Fig. 5-12.

For a unit area of the plane *P-P*, the instantaneous turbulent mass-transport rate across the plane is $\rho v'$. Associated with this mass transport is a change in the x component of velocity u'. The net momentum flux per unit area, in the x direction, represents the turbulent-shear stress at the plane *P-P*, or $\rho v'u'$. When a turbulent lump moves upward ($v' > 0$), it enters a region of higher $\bar{u}$ and is therefore likely to effect a slowing-down fluctuation in u', that is, $u' < 0$. A similar argument can be made for $v' < 0$, so that the average turbulent-shear stress will be given as

$$\tau_t = -\overline{\rho v'u'} \tag{5-59}$$

We must note that even though $\overline{v'} = \overline{u'} = 0$, the average of the *fluctuation product* $\overline{u'v'}$ is *not* zero.

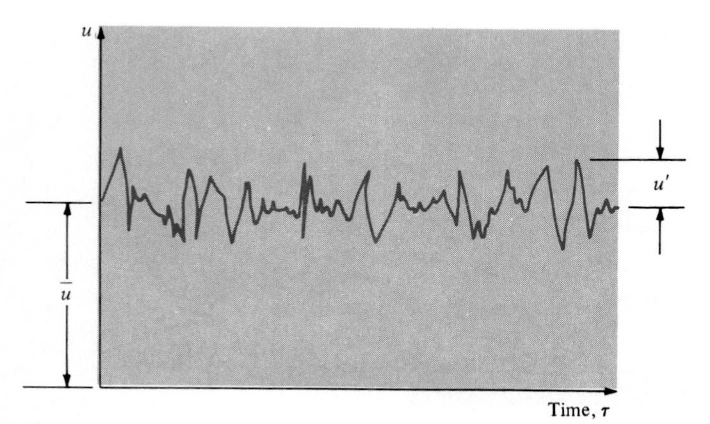

Fig. 5-11 Turbulent fluctuations with time.

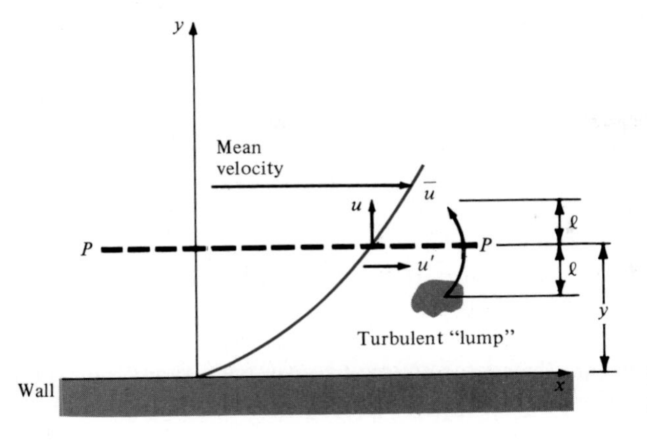

Fig. 5-12 Turbulent shear stress and mixing length.

Eddy Viscosity and the Mixing Length

Let us define an eddy viscosity or eddy diffusivity for momentum ϵ_M such that

$$\tau_t = -\overline{\rho v' u'} = \rho \epsilon_M \frac{du}{dy} \tag{5-60}$$

We have already likened the macroscopic transport of heat and momentum in turbulent flow to their molecular counterparts in laminar flow, so the definition in Eq. (5-60) is a natural consequence of this analogy. To analyze molecular-transport problems (see, for example, Ref. 7, p. 369) one normally introduces the concept of *mean free path,* or the average distance a particle travels between collisions. Prandtl introduced a similar concept for describing turbulent-flow phenomena. The *Prandtl mixing length* is the distance traveled, on the average, by the turbulent lumps of fluid in a direction normal to the mean flow.

Let us imagine a turbulent lump which is located a distance ℓ above or below the plane *P-P*, as shown in Fig. 5-12. These lumps of fluid move back and forth across the plane and give rise to the eddy or turbulent-shear-stress effect. At $y + \ell$ the velocity would be approximately

$$u(y + \ell) \approx u(y) + \ell \frac{\partial u}{\partial y}$$

while at $y - \ell$,

$$u(y - \ell) \approx u(y) - \ell \frac{\partial u}{\partial y}$$

Prandtl postulated that the turbulent fluctuation u' is proportional to the mean of the above two quantities, or

$$u' \approx \ell \frac{\partial u}{\partial y} \tag{5-61}$$

The distance ℓ is called the Prandtl mixing length. Prandtl also postulated that v' would be of the same order of magnitude as u' so that the turbulent-shear stress of Eq. (5-60) could be written

$$\tau_t = -\overline{\rho u'v'} = \rho \ell^2 \left(\frac{\partial u}{\partial y}\right)^2 = \rho \epsilon_M \frac{\partial u}{\partial y} \tag{5-62}$$

The eddy viscosity ϵ_M thus becomes

$$\epsilon_M = \ell^2 \frac{\partial u}{\partial y} \tag{5-63}$$

We have already noted that the eddy properties, and hence the mixing length, vary markedly through the boundary layer. Many analysis techniques have been applied over the years to take this variation into account. Prandtl's hypothesis was that the mixing length is proportional to distance from the wall, or

$$\ell = Ky \tag{5-64}$$

where K is the proportionality constant. The additional assumption was made that in the near-wall region the shear stress is approximately constant so that $\tau_t \approx \tau_w$. When this assumption is used along with Eq. (5-64), Eq. (5-62) yields

$$\tau_w = \rho K^2 y^2 \left(\frac{\partial u}{\partial y}\right)^2$$

Taking the square root and integrating with respect to y gives

$$u = \frac{1}{K} \sqrt{\frac{\tau_w}{\rho}} \ln y + C \tag{5-65}$$

where C is the constant of integration. Equation (5-65) matches very well with experimental data except in the region very close to the wall, where the laminar sublayer is present. In the sublayer the velocity distribution is essentially linear.

Let us now quantify our earlier qualitative description of a turbulent boundary layer by expressing the shear stress as the sum of a molecular and turbulent part:

$$\frac{\tau}{\rho} = (\nu + \epsilon_M) \frac{\partial u}{\partial y} \tag{5-66}$$

The so-called universal velocity profile is obtained by introducing two nondimensional coordinates

$$u^+ = \frac{u}{\sqrt{\tau_w/\rho}} \tag{5-67}$$

$$y^+ = \frac{\sqrt{\tau_w/\rho}\, y}{\nu} \tag{5-68}$$

Using these parameters and assuming $\tau \approx$ constant, we can rewrite Eq. (5-66) as

$$du^+ = \frac{dy^+}{1 + \epsilon_M/\nu} \tag{5-69}$$

In terms of our previous qualitative discussion, the laminar sublayer is the region where $\epsilon_M \sim 0$, the buffer layer has $\epsilon_M \sim \nu$, and the turbulent layer has $\epsilon_M \gg \nu$. Therefore, taking $\epsilon_M = 0$ in Eq. (5-69) and integrating yields

$$u^+ = y^+ + c$$

At the wall, $u^+ = 0$ for $y^+ = 0$ so that $c = 0$ and

$$u^+ = y^+ \tag{5-70}$$

is the velocity relation (a linear one) for the laminar sublayer. In the fully turbulent region $\epsilon_M/\nu \gg 1$. From Eq. (5-65)

$$\frac{\partial u}{\partial y} = \frac{1}{K}\sqrt{\frac{\tau_w}{\rho}}\frac{1}{y}$$

Substituting this relation along with Eq. (5-64) into Eq. (5-63) gives

$$\epsilon_M = K\sqrt{\frac{\tau_w}{\rho}}\,y$$

or

$$\frac{\epsilon_m}{\nu} = Ky^+ \tag{5-71}$$

Substituting this relation in Eq. (5-69) for $\epsilon_M/\nu \gg 1$ and integrating gives

$$u^+ = \frac{1}{K}\ln y^+ + c \tag{5-72}$$

This same *form* of equation will also be obtained for the buffer region. The limits of each region are obtained by comparing the above equations with experimental velocity measurements, with the following generally accepted constants:

$$\begin{aligned} &\text{Laminar sublayer: } 0 < y^+ < 5 && u^+ = y^+ \\ &\text{Buffer layer: } 5 < y^+ < 30 && u^+ = 5.0 \ln y^+ - 3.05 \\ &\text{Turbulent layer: } 30 < y^+ < 400 && u^+ = 2.5 \ln y^+ + 5.5 \end{aligned} \tag{5-73}$$

The equation set (5-73) is called the *universal velocity profile* and matches very well with experimental data; however, we should note once again that the constants in the equations must be determined from experimental velocity measurements. The satisfying point is that the simple Prandtl mixing-length model yields an equation form which fits the data so well.

Turbulent heat transfer is analogous to turbulent momentum transfer. The turbulent momentum flux postulated by Eq. (5-59) carries with it a turbulent energy fluctuation proportional to the temperature gradient. We thus have, in analogy to Eq. (5-62),

$$\left(\frac{q}{A}\right)_{\text{turb}} = -\rho c_p \epsilon_H \frac{\partial T}{\partial y} \tag{5-74}$$

or, for regions where both molecular and turbulent energy transport are important,

$$\frac{q}{A} = -\rho c_p(\alpha + \epsilon_H)\frac{\partial T}{\partial y} \tag{5-75}$$

Turbulent Heat Transfer Based on Fluid-Friction Analogy

Various analyses, similar to the one for the universal velocity profile above, have been performed to predict turbulent-boundary-layer heat transfer. The analyses have met with good success, but for our purposes the Colburn analogy between fluid friction and heat transfer is easier to apply and yields results which are in agreement with experiment and of simpler form.

In the turbulent-flow region, where $\epsilon_M \gg \nu$ and $\epsilon_H \gg \alpha$, we define the turbulent Prandtl number as

$$\mathrm{Pr}_t = \frac{\epsilon_M}{\epsilon_H} \tag{5-76}$$

If we can expect that the eddy momentum and energy transport will both be increased in the same proportion compared with their molecular values, we might anticipate that heat-transfer coefficients can be calculated by Eq. (5-56) with the ordinary molecular Prandtl number used in the computation. It turns out that the assumption that $\mathrm{Pr}_t \approx \mathrm{Pr}$ is a good one because heat-transfer calculations based on the fluid-friction analogy match experimental data very well. For this calculation we need experimental values of C_f for turbulent flow.

Schlichting [1] has surveyed experimental measurements of friction coefficients for turbulent flow on flat plates. We present the results of that survey so that they may be employed in the calculation of turbulent heat transfer with the fluid-friction–heat-transfer analogy. The *local* skin-friction coefficient is given by

$$C_{fx} = 0.0592\,\mathrm{Re}_x^{-1/5} \tag{5-77}$$

for Reynolds numbers between 5×10^5 and 10^7. At higher Reynolds numbers from 10^7 to 10^9 the formula of Schultz-Grunow [8] is recommended:

$$C_{fx} = 0.370\,(\log \mathrm{Re}_x)^{-2.584} \tag{5-78}$$

The *average-friction coefficient* for a flat plate with a laminar boundary layer up to $\mathrm{Re}_{\mathrm{crit}}$ and turbulent thereafter can be calculated from

$$\overline{C}_f = \frac{0.455}{(\log \mathrm{Re}_L)^{2.584}} - \frac{A}{\mathrm{Re}_L} \qquad \mathrm{Re}_L < 10^9 \tag{5-79}$$

where the constant A depends on $\mathrm{Re}_{\mathrm{crit}}$ in accordance with Table 5-1. A somewhat simpler formula can be obtained for lower Reynolds numbers as

$$\overline{C}_f = \frac{0.074}{\mathrm{Re}_l^{1/5}} - \frac{A}{\mathrm{Re}_l} \qquad \mathrm{Re}_L < 10^7 \tag{5-80}$$

Equations (5-79) and (5-80) are in agreement within their common range of applicability, and the one to be used in practice will depend on computational convenience.

Applying the fluid-friction analogy $\mathrm{St}\,\mathrm{Pr}^{2/3} = C_f/2$, we obtain the local turbulent heat transfer as:

$$\text{St}_x\,\text{Pr}^{2/3} = 0.0296\,\text{Re}_x^{-1/5} \qquad 5 \times 10^5 < \text{Re}_x < 10^7 \tag{5-81}$$

or

$$\text{St}_x\,\text{Pr}^{2/3} = 0.185(\log \text{Re}_x)^{-2.584} \qquad 10^7 < \text{Re}_x < 10^9 \tag{5-82}$$

The average heat transfer over the entire laminar-turbulent boundary layer is

$$\overline{\text{St}}\,\text{Pr}^{2/3} = \frac{\overline{C}_f}{2} \tag{5-83}$$

For $\text{Re}_{\text{crit}} = 5 \times 10^5$ and $\text{Re}_L < 10^7$, Eq. (5-80) can be used to obtain

$$\overline{\text{St}}\,\text{Pr}^{2/3} = 0.037\,\text{Re}_L^{-1/5} - 871\,\text{Re}_L^{-1} \tag{5-84}$$

Recalling that $\overline{\text{St}} = \overline{\text{Nu}}/(\text{Re}_L\,\text{Pr})$, we can rewrite Eq. (5-84) as

$$\overline{\text{Nu}}_L = \frac{\bar{h}L}{k} = \text{Pr}^{1/3}(0.037\,\text{Re}_L^{0.8} - 871) \tag{5-85}$$

The average heat-transfer coefficient can also be obtained by integrating the local values over the entire length of the plate. Thus,

$$h = \frac{1}{L}\left(\int_0^{x_{\text{crit}}} h_{\text{lam}}\,dx + \int_{x_{\text{crit}}}^{L} h_{\text{turb}}\,dx\right)$$

Using Eq. (5-55) for the laminar portion, $\text{Re}_{\text{crit}} = 5 \times 10^5$, and Eq. (5-81) for the turbulent portion gives the same result as Eq. (5-85). For higher Reynolds numbers the friction coefficient from Eq. (5-79) may be used, so that

$$\text{Nu}_L = \frac{\bar{h}l}{k} = [0.228\,\text{Re}_L(\log \text{Re}_L)^{-2.584} - 871]\text{Pr}^{1/3} \tag{5-85a}$$

for $10^7 < \text{Re}_L < 10^9$ and $\text{Re}_{\text{crit}} = 5 \times 10^5$.

The reader should note that if a transition Reynolds number different from 500,000 is chosen, then Eqs. (5-84) and (5-85) must be changed accordingly. An alternative equation is suggested by Whitaker [10] which may give better results with some liquids because of the viscosity-ratio term:

$$\overline{\text{Nu}}_L = 0.036\,\text{Pr}^{0.43}\,(\text{Re}_L^{0.8} - 9200)\left(\frac{\mu_\infty}{\mu_w}\right)^{1/4} \tag{5-86}$$

for

$$0.7 < \text{Pr} < 380$$
$$2 \times 10^5 < \text{Re}_L < 5.5 \times 10^6$$
$$0.26 < \frac{\mu_\infty}{\mu_W} < 3.5$$

All properties except μ_w are evaluated at the free stream temperature. For gases the viscosity ratio is dropped and the properties are evaluated at the film temperature.

Table 5-1

Re_{crit}	3×10^5	5×10^5	10^6	3×10^6
A	1055	1742	3340	8940

Constant Heat Flux

For constant wall heat flux in turbulent flow it is shown in Ref. 12 that the local Nusselt number is only about 4 percent higher than for the isothermal surface; i.e.,

$$\mathrm{Nu}_x = 1.04\ \mathrm{Nu}_x \Big]_{T_w\,=\,\mathrm{const}} \tag{5-87}$$

Some more comprehensive methods of correlating turbulent-boundary-layer heat transfer are given by Churchill [11].

EXAMPLE 5-8

Air at 20°C and 1 atm flows over a flat plate at 35 m/s. The plate is 75 cm long and is maintained at 60°C. Assuming unit depth in the z direction, calculate the heat transfer from the plate.

Solution

We evaluate properties at the film temperature:

$$T_f = \frac{20 + 60}{2} = 40°\mathrm{C} = 313\ \mathrm{K}$$

$$\rho = \frac{p}{RT} = \frac{1.0132 \times 10^5}{(287)(313)} = 1.128\ \mathrm{kg/m^3}$$

$$\mu = 1.906 \times 10^{-5}\ \mathrm{kg/m \cdot s}$$

$$\mathrm{Pr} = 0.7 \qquad k = 0.02723\ \mathrm{W/m \cdot °C} \qquad c_p = 1.007\ \mathrm{kJ/kg \cdot °C}$$

The Reynolds number is

$$\mathrm{Re}_L = \frac{\rho u_\infty L}{\mu} = \frac{(1.128)(35)(0.75)}{1.906 \times 10^{-5}} = 1.553 \times 10^6$$

and the boundary layer is turbulent because the Reynolds number is greater than 5×10^5. Therefore, we use Eq. (5-85) to calculate the average heat transfer over the plate:

$$\overline{\mathrm{Nu}}_L = \frac{\bar{h}L}{k} = \mathrm{Pr}^{1/3}\,(0.037\ \mathrm{Re}_L^{0.8} - 871)$$

$$= (0.7)^{1/3}\,[(0.037)(1.553 \times 10^6)^{0.8} - 871] = 2180$$

$$\bar{h} = \overline{\mathrm{Nu}}_L \frac{k}{L} = \frac{(2180)(0.02723)}{0.75} = 79.1\ \mathrm{W/m^2 \cdot °C} \quad [13.9\ \mathrm{Btu/h \cdot ft^2 \cdot °F}]$$

$$q = \bar{h}A\,(T_w - T_\infty) = (79.1)(0.75)(60 - 20) = 2373\ \mathrm{W} \quad [8150\ \mathrm{Btu/h}]$$

5-9 TURBULENT-BOUNDARY-LAYER THICKNESS

A number of experimental investigations have shown that the velocity profile in a turbulent boundary layer, outside the laminar sublayer, can be described by a one-seventh-power relation

$$\frac{u}{u_\infty} = \left(\frac{y}{\delta}\right)^{1/7} \tag{5-88}$$

where δ is the boundary-layer thickness as before. For purposes of an integral analysis the momentum integral can be evaluated with Eq. (5-88) because the laminar sublayer is so thin. However, the wall shear stress cannot be calculated from Eq. (5-88) because it yields an infinite value at $y = 0$.

To determine the turbulent-boundary-layer thickness we employ Eq. (5-17) for the integral momentum relation and evaluate the wall shear stress from the empirical relations for skin friction presented previously. According to Eq. (5-52),

$$\tau_w = \frac{C_f \rho u_\infty^2}{2}$$

and so for $\mathrm{Re}_x < 10^7$ we obtain from Eq. (5-77)

$$\tau_w = 0.0296 \left(\frac{\nu}{u_\infty x}\right)^{1/5} \rho u_\infty^2 \tag{5-89}$$

Now, using the integral momentum equation for zero pressure gradient [Eq. (5-17)] along with the velocity profile and wall shear stress, we obtain

$$\frac{d}{dx}\int_0^\delta \left[1 - \left(\frac{y}{\delta}\right)^{1/7}\right]\left(\frac{y}{\delta}\right)^{1/7} dy = 0.0296 \left(\frac{\nu}{u_\infty x}\right)^{1/5}$$

Integrating and clearing terms gives

$$\frac{d\delta}{dx} = \frac{72}{7}(0.0296)\left(\frac{\nu}{u_\infty}\right)^{1/5} x^{-1/5} \tag{5-90}$$

We shall integrate this equation for two physical situations:

1. The boundary layer is fully turbulent from the leading edge of the plate.
2. The boundary layer follows a laminar growth pattern up to $\mathrm{Re}_{\mathrm{crit}} = 5 \times 10^5$ and a turbulent growth thereafter.

For the first case we integrate Eq. (5-89) with the condition that $\delta = 0$ at $x = 0$, to obtain

$$\frac{\delta}{x} = 0.381\ \mathrm{Re}_x^{-1/5} \tag{5-91}$$

For case 2 we have the condition

$$\delta = \delta_{\mathrm{lam}} \qquad \text{at } x_{\mathrm{crit}} = 5 \times 10^5 \frac{\nu}{u_\infty} \tag{5-92}$$

Now, δ_{lam} is calculated from the exact relation of Eq. (5-21a):

$$\delta_{\mathrm{lam}} = 5.0 x_{\mathrm{crit}} (5 \times 10^5)^{-1/2} \tag{5-93}$$

Integrating Eq. (5-89) gives

$$\delta - \delta_{lam} = \frac{72}{7}(0.0296)\left(\frac{\nu}{u_\infty}\right)^{1/5}\frac{5}{4}(x^{4/5} - x_{crit}{}^{4/5}) \tag{5-94}$$

Combining the various relations above gives

$$\frac{\delta}{x} = 0.381\ \mathrm{Re}_x{}^{-1/5} - 10{,}256\ \mathrm{Re}_x{}^{-1} \tag{5-95}$$

This relation applies only for the region $5 \times 10^5 < \mathrm{Re}_x < 10^7$.

EXAMPLE 5-9

Calculate the turbulent-boundary-layer thickness at the end of the plate for Example 5-7, assuming that it develops (*a*) from the leading edge of the plate and (*b*) from the transition point at $\mathrm{Re}_{crit} = 5 \times 10^5$.

Solution

Since we have already calculated the Reynolds number as $\mathrm{Re}_L = 1.553 \times 10^6$, it is a simple matter to insert this value in Eqs. (5-91) and (5-95) along with $x = L = 0.75$ m to give

(*a*) $\delta = (0.75)(0.381)(1.553 \times 10^6)^{-0.2} = 0.0165\ \text{m} = 16.5\ \text{mm}$ [0.65 in]

(*b*) $\delta = (0.75)[(0.381)(1.553 \times 10^6)^{-0.2} - 10{,}256(1.553 \times 10^6)^{-1}]$
$= 0.0099\ \text{m} = 9.9\ \text{mm}$ [0.39 in]

The two values differ by 40 percent.

5-10 HEAT TRANSFER IN LAMINAR TUBE FLOW

Consider the tube-flow system in Fig. 5-13. We wish to calculate the heat transfer under developed flow conditions when the flow remains laminar. The wall temperature is T_w, the radius of the tube is r_o, and the velocity at the center of the tube is u_0. It is assumed that the pressure is uniform at any cross section. The velocity distribution may be derived by considering the fluid element shown in Fig. 5-14. The pressure forces are balanced by the viscous-

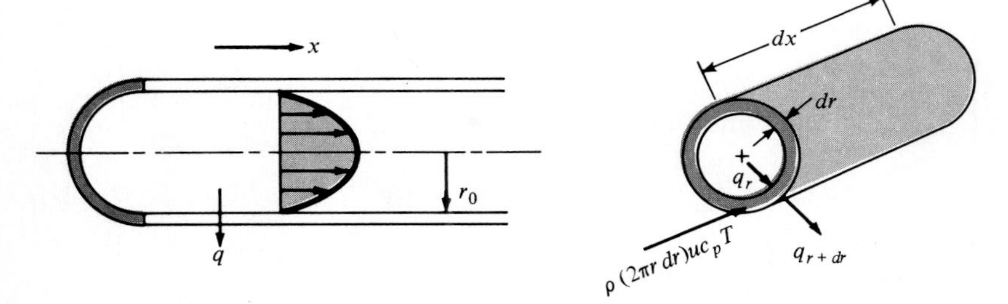

Fig. 5-13 Control volume for energy analysis in tube flow.

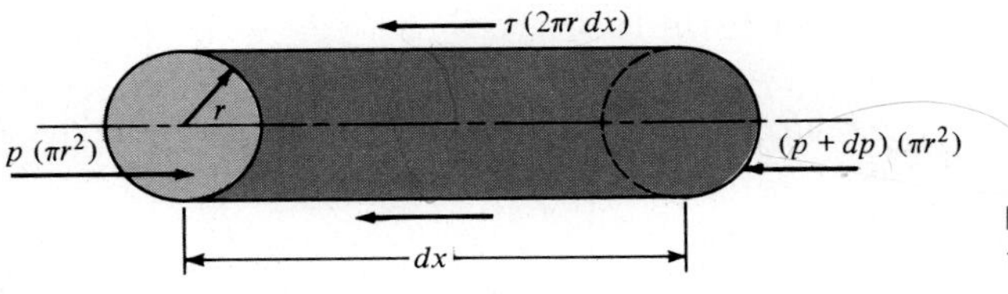

Fig. 5-14 Force balance on fluid element in tube flow.

shear forces so that

$$\pi r^2\,dp = \tau 2\pi r\,dx = 2\pi r\mu\,dx\,\frac{du}{dr}$$

or
$$du = \frac{1}{2\mu}\,r\,\frac{dp}{dx}\,dr$$

and
$$u = \frac{1}{4\mu}\frac{dp}{dx}\,r^2 + \text{const} \tag{5-96}$$

With the boundary condition

$$u = 0 \qquad \text{at } r = r_o$$

$$u = \frac{1}{4\mu}\frac{dp}{dx}\,(r^2 - r_o^2)$$

the velocity at the center of the tube is given by

$$u_0 = -\frac{r_o^2}{4\mu}\frac{dp}{dx} \tag{5-97}$$

so that the velocity distribution may be written

$$\frac{u}{u_0} = 1 - \frac{r^2}{r_o^2} \tag{5-98}$$

which is the familiar parabolic distribution for laminar tube flow. Now consider the heat-transfer process for such a flow system. To simplify the analysis, we assume that there is a constant heat flux at the tube wall; that is,

$$\frac{dq_w}{dx} = 0$$

The heat flow conducted into the annular element is

$$dq_r = -k2\pi r\,dx\,\frac{\partial T}{\partial r}$$

and the heat conducted out is

$$dq_{r+dr} = -k2\pi(r + dr)\,dx\left(\frac{\partial T}{\partial r} + \frac{\partial^2 T}{\partial r^2}\,dr\right)$$

The net heat convected out of the element is

$$2\pi r\, dr\, \rho c_p u \frac{\partial T}{\partial x} dx$$

The energy balance is

Net energy convected out = net heat conducted in

or, neglecting second-order differentials,

$$r\rho c_p u \frac{\partial T}{\partial x} dx\, dr = k\left(\frac{\partial T}{\partial r} + r\frac{\partial^2 T}{\partial r^2}\right) dx\, dr$$

which may be rewritten

$$\frac{1}{ur}\frac{\partial}{\partial r}\left(r\frac{\partial T}{\partial r}\right) = \frac{1}{\alpha}\frac{\partial T}{\partial x} \tag{5-99}$$

We assume that the heat flux at the wall is constant, so that the average fluid temperature must increase linearly with x, or

$$\frac{\partial T}{\partial x} = \text{const}$$

This means that the temperature profiles will be similar at various x distances along the tube. The boundary conditions on Eq. (5-99) are

$$\frac{\partial T}{\partial r} = 0 \qquad \text{at } r = 0$$

$$k\frac{\partial T}{\partial r}\bigg]_{r=r_o} = q_w = \text{const}$$

To obtain the solution to Eq. (5-99), the velocity distribution given by Eq. (5-99) must be inserted. It is assumed that the temperature and velocity fields are independent; i.e., a temperature gradient does not affect the calculation of the velocity profile. This is equivalent to specifying that the properties remain constant in the flow. With the substitution of the velocity profile, Eq. (5-99) becomes

$$\frac{\partial}{\partial r}\left(r\frac{\partial T}{\partial r}\right) = \frac{1}{\alpha}\frac{\partial T}{\partial x} u_0\left(1 - \frac{r^2}{r_o^2}\right) r$$

Integration yields

$$r\frac{\partial T}{\partial r} = \frac{1}{\alpha}\frac{\partial T}{\partial x} u_0\left(\frac{r^2}{2} - \frac{r^4}{4r_o^2}\right) + C_1$$

and a second integration gives

$$T = \frac{1}{\alpha}\frac{\partial T}{\partial x} u_0\left(\frac{r^2}{4} - \frac{r^4}{16r_o^2}\right) + C_1 \ln r + C_2$$

Applying the first boundary condition, we find that

$$C_1 = 0$$

The second boundary condition has been satisfied by noting that the axial temperature gradient $\partial T/\partial x$ is constant. The temperature distribution may finally be written in terms of the temperature at the center of the tube:

$$T = T_c \qquad \text{at } r = 0 \qquad \text{so that} \qquad C_2 = T_c$$

$$T - T_c = \frac{1}{\alpha}\frac{\partial T}{\partial x}\frac{u_0 r_o^2}{4}\left[\left(\frac{r}{r_o}\right)^2 - \frac{1}{4}\left(\frac{r}{r_o}\right)^4\right] \tag{5-100}$$

The Bulk Temperature

In tube flow the convection heat-transfer coefficient is usually defined by

$$\text{Local heat flux} = q'' = h(T_w - T_b) \tag{5-101}$$

where T_w is the wall temperature and T_b is the so-called *bulk temperature,* or energy-average fluid temperature across the tube, which may be calculated from

$$T_b = \overline{T} = \frac{\int_0^{r_o} \rho 2\pi r \, dr \, u c_p T}{\int_0^{r_o} \rho 2\pi r \, dr \, u c_p} \tag{5-102}$$

The reason for using the bulk temperature in the definition of heat-transfer coefficients for tube flow may be explained as follows. In a tube flow there is no easily discernible free-stream condition as is present in the flow over a flat plate. Even the centerline temperature T_c is not easily expressed in terms of the inlet flow variables and the heat transfer. For most tube- or channel-flow heat-transfer problems the topic of central interest is the total energy transferred to the fluid in either an elemental length of the tube or over the entire length of the channel. At any x position, the temperature that is indicative of the total energy of the flow is an integrated mass-energy average temperature over the entire flow area. The numerator of Eq. (5-102) represents the total energy flow through the tube, and the denominator represents the product of mass flow and specific heat integrated over the flow area. The bulk temperature is thus representative of the total energy of the flow at the particular location. For this reason, the bulk temperature is sometimes referred to as the "mixing cup" temperature, since it is the temperature the fluid would assume if placed in a mixing chamber and allowed to come to equilibrium. For the temperature distribution given in Eq. (5-100), the bulk temperature is a linear function of x because the heat flux at the tube wall is constant. Calculating the bulk temperature from Eq. (5-102), we have

$$T_b = T_c + \frac{7}{96}\frac{u_0 r_o^2}{\alpha}\frac{\partial T}{\delta x} \tag{5-103}$$

and for the wall temperature

$$T_w = T_c + \frac{3}{16}\frac{u_0 r_o^2}{\alpha}\frac{\partial T}{\partial x} \tag{5-104}$$

The heat-transfer coefficient is calculated from

$$q = hA\,(T_w - T_b) = kA\left(\frac{\partial T}{\partial r}\right)_{r=r_o} \tag{5-105}$$

$$y = \frac{k(\partial T/\partial r)_{r=r_o}}{T_w - T_b}$$

The temperature gradient is given by

$$\left.\frac{\partial T}{\partial r}\right]_{r=r_o} = \frac{u_0}{\alpha}\frac{\partial T}{\partial x}\left(\frac{r}{2} - \frac{r^3}{4r_o^2}\right)_{r=r_o} = \frac{u_0 r_o}{4\alpha}\frac{\partial T}{\partial x} \tag{5-106}$$

Substituting Eqs. (5-103), (5-104), and (5-106) in Eq. (5-105) gives

$$h = \frac{24}{11}\frac{k}{r_o} = \frac{48}{11}\frac{k}{d_o}$$

Expressed in terms of the Nusselt number, the result is

$$\mathrm{Nu}_d = \frac{nd_o}{k} = 4.364 \tag{5-107}$$

which is in agreement with an exact calculation by Sellars, Tribus, and Klein [3], which considers the temperature profile as it develops. Some empirical relations for calculating heat transfer in laminar tube flow will be presented in Chap. 6.

We may remark at this time that when the statement is made that a fluid enters a tube at a certain temperature, it is the bulk temperature to which we refer. The bulk temperature is used for overall energy balances on systems.

5-11 TURBULENT FLOW IN A TUBE

The developed velocity profile for turbulent flow in a tube will appear as shown in Fig. 5-15. A laminar sublayer, or "film," occupies the space near the surface, while the central core of the flow is turbulent. To determine the heat transfer analytically for this situation, we require, as usual, a knowledge of the temperature distribution in the flow. To obtain this temperature distribution, the

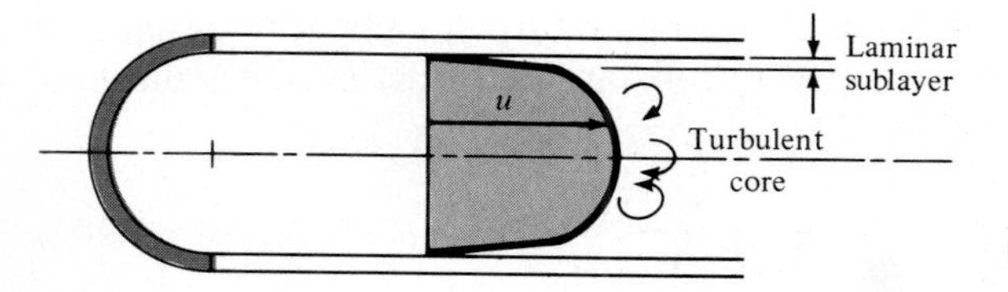

Fig. 5-15 Velocity profile in turbulent tube flow.

analysis must take into consideration the effect of the turbulent eddies in the transfer of heat and momentum. We shall use an approximate analysis which relates the conduction and transport of heat to the transport of momentum in the flow, i.e., viscous effects.

The heat flow across a fluid element in laminar flow may be expressed by

$$\frac{q}{A} = -k\frac{dT}{dy}$$

Dividing both sides of the equation by ρc_p,

$$\frac{q}{\rho c_p A} = -\alpha\frac{dT}{dy}$$

It will be recalled that α is the molecular diffusivity of heat. In turbulent flow one might assume that the heat transport could be represented by

$$\frac{q}{\rho c_p A} = -(\alpha + \epsilon_H)\frac{dT}{dy} \tag{5-108}$$

where ϵ_H is an eddy diffusivity of heat.

Equation (5-108) expresses the total heat conduction as a sum of the molecular conduction and the macroscopic eddy conduction. In a similar fashion, the shear stress in turbulent flow could be written

$$\frac{\tau}{\rho} = \left(\frac{\mu}{\rho} + \epsilon_M\right)\frac{du}{dy} = (\nu + \epsilon_M)\frac{du}{dy} \tag{5-109}$$

where ϵ_M is the eddy diffusivity for momentum. We now assume that the heat and momentum are transported at the same rate; that is, $\epsilon_M = \epsilon_H$ and $\nu = \alpha$, or $\mathrm{Pr} = 1$.

Dividing Eq. (5-108) by Eq. (5-109) gives

$$\frac{q}{c_p A\tau}\,du = -dT$$

An additional assumption is that the ratio of the heat transfer per unit area to the shear stress is constant across the flow field. This is consistent with the assumption that heat and momentum are transported at the same rate. Thus

$$\frac{q}{A\tau} = \text{const} = \frac{q_w}{A_w\tau_w} \tag{5-110}$$

Then, integrating Eq. (5-109) between wall conditions and mean bulk conditions gives

$$\frac{q_w}{A_w\tau_w c_p}\int_{u=0}^{u=u_m} du = \int_{T_w}^{T_b} -\,dT$$

$$\frac{q_w u_m}{A_w\tau_w c_p} = T_w - T_b \tag{5-111}$$

But the heat transfer at the wall may be expressed by

$$q_w = hA_w (T_w - T_b)$$

and the shear stress may be calculated from

$$\tau_w = \frac{\Delta p\ (\pi d_o^2)}{4\pi\ d_o L} = \frac{\Delta p}{4}\frac{d_o}{L}$$

The pressure drop may be expressed in terms of a friction factor f by

$$\Delta p = f \frac{L}{d} \rho \frac{u_m^2}{2} \qquad \text{(5-112)}$$

so that (5-113)

$$\tau_w = \frac{f}{8} \rho u_m^2$$

Substituting the expressions for τ_w and q_w in Eq. (5-111) gives

$$\text{St} = \frac{h}{\rho c_p u_m} = \frac{\text{Nu}_d}{\text{Re}_d \text{ Pr}} = \frac{f}{8} \qquad \text{(5-114)}$$

Equation (5-114) is called the Reynolds analogy for tube flow. It relates the heat-transfer rate to the frictional loss in tube flow and is in fair agreement with experiments when used with gases whose Prandtl numbers are close to unity. (Recall that Pr = 1 was one of the assumptions in the analysis.)

An empirical formula for the turbulent-friction factor up to Reynolds numbers of about 2×10^5 for the flow in smooth tubes is

$$f = \frac{0.316}{\text{Re}_d^{1/4}} \qquad \text{(5-115)}$$

Inserting this expression in Eq. (5-113) gives

$$\frac{\text{Nu}_d}{\text{Re}_d \text{ Pr}} = 0.0395 \text{ Re}_d^{-1/4}$$

or

$$\text{Nu}_d = 0.0395 \text{ Re}_d^{3/4} \qquad \text{(5-116)}$$

since we assumed the Prandtl number to be unity. This derivation of the relation for turbulent heat transfer in smooth tubes is highly restrictive because of the $\text{Pr} \approx 1.0$ assumption. The heat-transfer–fluid-friction analogy of Sec. 5-7 indicated a Prandtl-number dependence of $\text{Pr}^{2/3}$ for the flat-plate problem and, as it turns out, this dependence works fairly well for turbulent tube flow. Equations (5-114) and (5-116) may be modified by this factor to yield

$$\text{St Pr}^{2/3} = \frac{f}{8} \qquad \text{(5-114a)}$$

$$\text{Nu}_d = 0.0395 \text{ Re}_d^{3/4} \text{ Pr}^{1/3} \qquad \text{(5-116a)}$$

As we shall see in Chap. 6, Eq. (5-116*a*) predicts heat-transfer coefficients that are somewhat higher than those observed in experiments. The purpose of the discussion at this point has been to show that one may arrive at a relation for turbulent heat transfer in a fairly simple analytical fashion. As we have indicated earlier, a rigorous development of the Reynolds analogy between heat transfer and fluid friction involves considerations beyond the scope of our discussion and the simple path of reasoning chosen here is offered for the purpose of indicating the general nature of the physical processes.

For calculation purposes, a more correct relation to use for turbulent flow in a smooth tube is Eq. (6-4), which we list here for comparison:

$$\mathrm{Nu}_d = 0.023\ \mathrm{Re}_d^{0.8}\ \mathrm{Pr}^{0.4} \tag{6-4}$$

All properties in Eq. (6-4) are evaluated at the bulk temperature.

5-12 HEAT TRANSFER IN HIGH-SPEED FLOW

Our previous analysis of boundary-layer heat transfer (Sec. 5-6) neglected the effects of viscous dissipation within the boundary layer. When the free-stream velocity is very high, as in high-speed aircraft, these dissipation effects must be considered. We begin our analysis by considering the adiabatic case, i.e., a perfectly insulated wall. In this case the wall temperature may be considerably higher than the free-stream temperature even though no heat transfer takes place. This high temperature results from two situations: (1) the increase in temperature of the fluid as it is brought to rest at the plate surface while the kinetic energy of the flow is converted to internal thermal energy and (2) the heating effect due to viscous dissipation. Consider the first situation. The kinetic energy of the gas is converted to thermal energy as the gas is brought to rest, and this process is described by the steady-flow energy equation for an adiabatic process:

$$i_0 = i_\infty + \frac{1}{2g_c} u_\infty^2 \tag{5-117}$$

where i_0 is the stagnation enthalpy of the gas. This equation may be written in terms of temperature as

$$c_p (T_0 - T_\infty) = \frac{1}{2g_c} u_\infty^2$$

where T_0 is the stagnation temperature and T_∞ is the static free-stream temperature. Expressed in terms of the free-stream Mach number, it is

$$\frac{T_0}{T_\infty} = 1 + \frac{\gamma - 1}{2} M_\infty^2 \tag{5-118}$$

where M_∞ is the Mach number, defined as $M_\infty = u_\infty/a$, and a is the acoustic velocity, which for a perfect gas may be calculated with

$$a = \sqrt{\gamma g_c R T} \tag{5-119}$$

where R is the gas constant.

In the actual case of a boundary-layer flow problem, the fluid is not brought to rest reversibly because the viscous action is basically an irreversible process in a thermodynamic sense. In addition, not all the free-stream kinetic energy is converted to thermal energy—part is lost as heat, and part is dissipated in the form of viscous work. To take into account the irreversibilities in the boundary-layer flow system, a *recovery factor* is defined by

$$r = \frac{T_{aw} - T_\infty}{T_0 - T_\infty} \tag{5-120}$$

where T_{aw} is the actual adiabatic wall temperature and T_∞ is the static temperature of the free stream. The recovery factor may be determined experimentally, or, for some flow systems, analytical calculations may be made.

The boundary-layer energy equation

$$u \frac{\partial T}{\partial x} + v \frac{\partial T}{\partial y} = \alpha \frac{\partial^2 T}{\partial y^2} + \frac{\mu}{\rho c_p} \left(\frac{\partial u}{\partial y} \right)^2$$

has been solved for the high-speed-flow situation, taking into account the viscous-heating term. Although the complete solution is somewhat tedious, the final results are remarkably simple. For our purposes we present only the results and indicate how they may be applied. The reader is referred to Appendix B for an exact solution to Eq. (5-22). An excellent synopsis of the high-speed heat-transfer problem is given in a report by Eckert [4]. Some typical boundary-layer temperature profiles for an adiabatic wall in high-speed flow are given in Fig. B-3.

The essential result of the high-speed heat-transfer analysis is that heat-transfer rates may generally be calculated with the same relations used for low-speed incompressible flow when the average heat-transfer coefficient is redefined with the relation

$$q = \bar{h} A \,(T_w - T_{aw}) \tag{5-121}$$

Notice that the difference between the adiabatic wall temperature and the actual wall temperature is used in the definition so that the expression will yield a value of zero heat flow when the wall is at the adiabatic wall temperature. For gases with Prandtl numbers near unity the following relations for the recovery factor have been derived:

Laminar flow: $$r = \mathrm{Pr}^{1/2} \tag{5-122}$$

Turbulent flow: $$r = \mathrm{Pr}^{1/3} \tag{5-123}$$

These recovery factors may be used in conjunction with Eq. (5-119) to obtain the adiabatic wall temperature.

In high-velocity boundary layers substantial temperature gradients may occur, and there will be correspondingly large property variations across the boundary layer. The constant-property heat-transfer equations may still be used if the properties are introduced at a reference temperature T^* as recommended by Eckert:

$$T^* = T_\infty + 0.50(T_w - T_\infty) + 0.22(T_{aw} - T_\infty) \tag{5-124}$$

The analogy between heat transfer and fluid friction [Eq. (5-56)] may also be used when the friction coefficient is known. Summarizing the relations used for high-speed heat-transfer calculations:

Laminar boundary layer ($\mathrm{Re}_x < 5 \times 10^5$):

$$\mathrm{St}_x^* \, \mathrm{Pr}^{*2/3} = 0.332 \, \mathrm{Re}_x^{*\,-1/2} \tag{5-125}$$

Turbulent boundary layer ($5 \times 10^5 < \mathrm{Re}_x < 10^7$):

$$\mathrm{St}_x^* \, \mathrm{Pr}^{*2/3} = 0.0296 \, \mathrm{Re}_x^{*\,-1/5} \tag{5-126}$$

Turbulent boundary layer ($10^7 < \mathrm{Re}_x < 10^9$):

$$\mathrm{St}_x^* \, \mathrm{Pr}^{*2/3} = 0.185 \, (\log \mathrm{Re}_x^*)^{-2.584} \tag{5-127}$$

The superscript * in the above equations indicates that the properties are evaluated at the reference temperature given by Eq. (5-124).

To obtain an average heat-transfer coefficient, the above expressions must be integrated over the length of the plate. If the Reynolds number falls in a range such that Eq. (5-127) must be used, the integration cannot be expressed in closed form, and a numerical integration must be used. Care must be taken in performing the integration for the high-speed heat-transfer problem since the reference temperature is different for the laminar and turbulent portions of the boundary layer. This results from the different value of the recovery factor used for laminar and turbulent flow as given by Eqs. (5-122) and (5-123).

When very high flow velocities are encountered, the adiabatic wall temperature may become so high that dissociation of the gas will take place and there will be a very wide variation of the properties in the boundary layer. Eckert [4] recommends that these problems be treated on the basis of a heat-transfer coefficient defined in terms of *enthalpy* difference:

$$q = h_i A \, (i_w - i_{aw}) \tag{5-128}$$

The enthalpy recovery factor is then defined as

$$r_i = \frac{i_{aw} - i_\infty}{i_0 - i_\infty} \tag{5-129}$$

where i_{aw} is the enthalpy at the adiabatic wall conditions. The same relations as before are used to calculate the recovery factor and heat transfer except that all properties are evaluated at a reference enthalpy i^* given by

$$i^* = i_\infty + 0.5(i_w - i_\infty) + 0.22(i_{aw} - i_\infty) \qquad \text{(5-130)}$$

The Stanton number is redefined as

$$\text{St}_i = \frac{h_i}{\rho u_\infty} \qquad \text{(5-131)}$$

This Stanton number is then used in Eq. (5-125), (5-126), or (5-127) to calculate the heat-transfer coefficient. When calculating the enthalpies for use in the above relations, the *total* enthalpy must be used; i.e., chemical energy of dissociation as well as internal thermal energy must be included. The reference-enthalpy method has proved successful for calculating high-speed heat transfer with an accuracy of better than 10 percent.

■ **EXAMPLE 5-10**

A flat plate 70 cm long and 1.0 m wide is placed in a wind tunnel where the flow conditions are $M = 3$, $p = \frac{1}{20}$ atm, and $T = -40°\text{C}$. How much cooling must be used to maintain the plate temperature at 35°C?

Solution

We must consider the laminar and turbulent portions of the boundary layer separately because the recovery factors, and hence the adiabatic wall temperatures, used to establish the heat flow will be different for each flow regime. It turns out that the difference is rather small in this problem, but we shall follow a procedure which would be used if the difference were appreciable, so that the general method of solution may be indicated. The free-stream acoustic velocity is calculated from

$$a = \sqrt{\gamma g_c R T_\infty} = [(1.4)(1.0)(287)(233)]^{1/2} = 306 \text{ m/s} \quad [1003 \text{ ft/s}]$$

so that the free-stream velocity is

$$u_\infty = (3)(306) = 918 \text{ m/s} \quad [3012 \text{ ft/s}]$$

The maximum Reynolds number is estimated by making a computation based on properties evaluated at free-stream conditions:

$$\rho_\infty = \frac{(1.0132 \times 10^5)(\frac{1}{20})}{(287)(233)} = 0.0758 \text{ kg/m}^3 \quad [4.73 \times 10^{-3} \text{ lb}_m/\text{ft}^3]$$

$$\mu_\infty = 1.434 \times 10^{-5} \text{ kg/m} \cdot \text{s} \quad [0.0347 \text{ lb}_m/\text{h} \cdot \text{ft}]$$

$$\text{Re}_{L,\infty} = \frac{(0.0758)(918)(0.70)}{1.434 \times 10^{-5}} = 3.395 \times 10^6$$

Thus we conclude that both laminar and turbulent boundary-layer heat transfer must be considered. We first determine the reference temperatures for the two regimes and then evaluate properties at these temperatures.

Laminar Portion

$$T_0 = T_\infty \left(1 + \frac{\gamma - 1}{2} M_\infty^2\right) = (233)[1 + (0.2)(3)^2] = 652 \text{ K}$$

Assuming a Prandtl number of about 0.7, we have

$$r = \text{Pr}^{1/2} = (0.7)^{1/2} = 0.837$$

$$r = \frac{T_{aw} - T_\infty}{T_0 - T_\infty} = \frac{T_{aw} - 233}{652 - 233}$$

and $T_{aw} = 584$ K $= 311$°C [592°F]. Then the reference temperature from Eq. (5-123) is

$$T^* = 233 + (0.5)(35 + 40) + (0.22)(584 - 233) = 347.8 \text{ K}$$

Checking the Prandtl number at this temperature, we have

$$\text{Pr}^* = 0.697$$

so that the calculation is valid. If there were an appreciable difference between the value of Pr* and the value used to determine the recovery factor, the calculation would have to be repeated until agreement was reached.

The other properties to be used in the laminar heat-transfer analysis are

$$\rho^* = \frac{(1.0132 \times 10^5)(1/20)}{(287)(347.8)} = 0.0508 \text{ kg/m}^3$$

$$\mu^* = 2.07 \times 10^{-5} \text{ kg/m} \cdot \text{s}$$

$$k^* = 0.03 \text{ W/m} \cdot {}^\circ\text{C} \quad [0.0173 \text{ Btu/h} \cdot \text{ft} \cdot {}^\circ\text{F}]$$

$$c_p^* = 1.009 \text{ kJ/kg} \cdot {}^\circ\text{C}$$

Turbulent Portion

Assuming Pr = 0.7 gives

$$r = \text{Pr}^{1/3} = 0.888 = \frac{T_{aw} - T_\infty}{T_0 - T_\infty} = \frac{T_{aw} - 233}{652 - 233}$$

$$T_{aw} = 605 \text{ K} = 332^\circ\text{C}$$

$$T^* = 233 + (0.5)(35 + 40) + (0.22)(605 - 233) = 352.3 \text{ K}$$

$$\text{Pr}^* = 0.695$$

The agreement between Pr* and the assumed value is sufficiently close. The other properties to be used in the turbulent heat-transfer analysis are

$$\rho^* = \frac{(1.0132 \times 10^5)(1/20)}{(287)(352.3)} = 0.0501 \text{ kg/m}^3$$

$$\mu^* = 2.09 \times 10^{-5} \text{ kg/m} \cdot \text{s}$$

$$k^* = 0.0302 \text{ W/m} \cdot {}^\circ\text{C} \qquad c_p^* = 1.009 \text{ kJ/kg} \cdot {}^\circ\text{C}$$

Laminar Heat Transfer

We assume

$$\mathrm{Re}^*_{\mathrm{crit}} = 5 \times 10^5 = \frac{\rho^* u_\infty x_c}{\mu^*}$$

$$x_c = \frac{(5 \times 10^5)(2.07 \times 10^{-5})}{(0.0508)(918)} = 0.222 \text{ m}$$

$$\overline{\mathrm{Nu}}^* = \frac{\bar{h} x_c}{k^*} = 0.664\,(\mathrm{Re}^*_{\mathrm{crit}})^{1/2}\,\mathrm{Pr}^{*1/3}$$

$$= (0.664)(5 \times 10^5)^{1/2}(0.697)^{1/3} = 416.3$$

$$\bar{h} = \frac{(416.3)(0.03)}{0.222} = 56.25 \text{ W/m}^2 \cdot {}^\circ\mathrm{C} \quad [9.91 \text{ Btu/h} \cdot \text{ft}^2 \cdot {}^\circ\mathrm{F}]$$

This is the average heat-transfer coefficient for the laminar portion of the boundary layer, and the heat transfer is calculated from

$$q = \bar{h}A(T_w - T_{aw})$$

$$= (56.26)(0.222)(35 - 311)$$

$$= -3445 \text{ W} \quad [-11{,}750 \text{ Btu/h}]$$

so that 3445 W of cooling is required in the laminar region of the plate per meter of depth in the z direction.

Turbulent Heat Transfer

To determine the turbulent heat transfer we must obtain an expression for the local heat-transfer coefficient from

$$\mathrm{St}^*_x\,\mathrm{Pr}^{*2/3} = 0.0296\,\mathrm{Re}^{*\,-1/5}_x$$

and then integrate from $x = 0.222$ m to $x = 0.7$ m to determine the total heat transfer:

$$h_x = \mathrm{Pr}^{*\,-2/3}\rho^* u_\infty c_p(0.0296)\left(\frac{\rho^* u_\infty x}{\mu^*}\right)^{-1/5}$$

Inserting the numerical values for the properties gives

$$h_x = 94.34x^{-1/5}$$

The average heat-transfer coefficient in the turbulent region is determined from

$$\bar{h} = \frac{\int_{0.222}^{0.7} h_x\,dx}{\int_{0.222}^{0.7} dx} = 111.46 \text{ W/m}^2 \cdot {}^\circ\mathrm{C} \quad [19.6 \text{ Btu/h} \cdot \text{ft}^2 \cdot {}^\circ\mathrm{F}]$$

Using this value we may calculate the heat transfer in the turbulent region of the flat plate:

$$q = \bar{h}A\,(T_w - T_{aw})$$

$$= (111.46)(0.7 - 0.222)(35 - 332)$$

$$= -15{,}823 \text{ W} \quad [-54{,}006 \text{ Btu/h}]$$

The total amount of cooling required is the sum of the heat transfers for the laminar and turbulent portions:

$$\text{Total cooling} = 3445 + 15{,}823 = 19{,}268 \text{ W} \quad [65{,}761 \text{ Btu/h}]$$

These calculations assume unit depth of 1 m in the z direction.

5-13 SUMMARY

Most of this chapter has been concerned with flow over flat plates and the associated heat transfer. For convenience to the reader we have summarized the equations in Table 5-2 along with restrictions which apply. The general procedure then is to:

1. Evaluate the fluid properties; this will usually be at the film temperature.
2. Establish the the boundary conditions, i.e., constant temperature or constant heat flux.
3. Establish the flow regime as determined by the Reynolds number.
4. Select the appropriate equation, taking into account the flow regime and any fluid property restrictions which may apply.
5. Calculate the value(s) of the convection heat-transfer coefficient and/or heat transfer.

REVIEW QUESTIONS

1. What is meant by a hydrodynamic boundary level?
2. Define the Reynolds number. Why is it important?
3. What is the physical mechanism of viscous action?
4. Distinguish between laminar and turbulent flow in a physical sense.
5. What is the momentum equation for the laminar boundary layer on a flat plate? What assumptions are involved in the derivation of this equation?
6. How is the boundary-layer thickness defined?
7. What is the energy equation for the laminar boundary layer on a flat plate? What assumptions are involved in the derivation of this equation?
8. What is meant by a thermal boundary layer?
9. Define the Prandtl number. Why is it important?
10. Describe the physical mechanism of convection. How is the convection heat-transfer coefficient related to this mechanism?
11. Describe the relation between fluid friction and heat transfer.
12. Define the bulk temperature. How is it used?
13. How is the heat-transfer coefficient defined for high-speed heat-transfer calculations?

Table 5-2 Summary of Equations for Flow over Flat Plates
Properties evaluated at $T_f = (T_w + T_\infty)/2$ unless otherwise noted.

Flow regime	*Restrictions*	*Equation*	*Equation number*
		Heat transfer	
Laminar, local	T_w = const, $\mathrm{Re}_x < 5 \times 10^5$ $0.6 < \mathrm{Pr} < 50$	$\mathrm{Nu}_x = 0.332\,\mathrm{Re}_x^{1/2}\,\mathrm{Pr}^{1/3}$	(5-44)
Laminar, local	T_w = const, $\mathrm{Re}_x < 5 \times 10^5$, $\mathrm{Re}_x\,\mathrm{Pr} > 100$	$\mathrm{Nu}_x = \dfrac{0.3387\,\mathrm{Re}_x^{1/2}\,\mathrm{Pr}^{1/3}}{\left[1 + \left(\dfrac{0.0468}{\mathrm{Pr}}\right)^{2/3}\right]^{1/4}}$	(5-51)
Laminar, local	q_w = const, $\mathrm{Re}_x < 5 \times 10^5$, $0.6 < \mathrm{Pr} < 50$	$\mathrm{Nu}_x = 0.453\,\mathrm{Re}_x^{1/2}\,\mathrm{Pr}^{1/3}$	(5-48)
Laminar, local	q_w = const, $\mathrm{Re}_x < 5 \times 10^5$	$\mathrm{Nu}_x = \dfrac{0.4637\,\mathrm{Re}_x^{1/2}\,\mathrm{Pr}^{1/3}}{\left[1 + \left(\dfrac{0.0207}{\mathrm{Pr}}\right)^{2/3}\right]^{1/4}}$	(5-51)
Laminar, average	$\mathrm{Re}_L < 5 \times 10^5$, T_w = const	$\overline{\mathrm{Nu}}_L = 2\,\mathrm{Nu}_{x=L} = 0.664\,\mathrm{Re}_L^{1/2}\,\mathrm{Pr}^{1/3}$	(5-46)
Laminar, local	T_w = const, $\mathrm{Re}_x < 5 \times 10^5$, $\mathrm{Pr} \ll 1$ (liquid metals)	$\mathrm{Nu}_x = 0.564(\mathrm{Re}_x\,\mathrm{Pr})^{1/3}$	
Laminar, local	T_w = const, starting at $x = x_0$, $\mathrm{Re}_x < 5 \times 10^5$, $0.6 < \mathrm{Pr} < 50$	$\mathrm{Nu}_x = 0.332\,\mathrm{Re}_x^{1/2}\,\mathrm{Pr}^{1/3}\left[1 - \left(\dfrac{x_0}{x}\right)^{3/4}\right]^{-1/3}$	(5-43)
Turbulent, local	T_w = const, $5 \times 10^5 < \mathrm{Re}_x < 10^7$	$\mathrm{St}_x\,\mathrm{Pr}^{2/3} = 0.0296\,\mathrm{Re}_x^{-0.2}$	(5-81)
Turbulent, local	T_w = const, $10^7 < \mathrm{Re}_x < 10^9$	$\mathrm{St}_x\,\mathrm{Pr}^{2/3} = 0.185(\log \mathrm{Re}_x)^{-2.584}$	(5-82)
Turbulent, local	q_w = const, $5 \times 10^5 < \mathrm{Re}_x < 10^7$	$\mathrm{Nu}_x = 1.04\,\mathrm{Nu}_{x T_w = \mathrm{const}}$	(5-87)
Laminar-turbulent, average	T_w = const, $\mathrm{Re}_x < 10^7$, $\mathrm{Re}_{\mathrm{crit}} = 5 \times 10^5$	$\overline{\mathrm{St}}\,\mathrm{Pr}^{2/3} = 0.037\,\mathrm{Re}_L^{-0.2} - 871\,\mathrm{Re}_L^{-1}$ $\overline{\mathrm{Nu}}_L = \mathrm{Pr}^{1/3}(0.037\,\mathrm{Re}_L^{0.8} - 871)$	(5-84) (5-85)

Laminar-turbulent, average	T_w = const, $Re_x < 10^7$, liquids, μ at T_∞, μ_w at T_w	$\overline{Nu_L} = 0.036\ Pr^{0.43}(Re_L^{0.8} - 9200)\left(\frac{\mu_\infty}{\mu_w}\right)^{1/4}$	(5-86)
High-speed flow	T_w = const, $q = hA(T_w - T_{aw})$	Same as for low-speed flow with properties evaluated at $T^* = T_\infty + 0.5(T_w - T_\infty) + 0.22(T_{aw} - T_\infty)$	
		Boundary layer thickness	
Laminar	$Re_x < 5 \times 10^5$	$\frac{\delta}{x} = 5.0\ Re_x^{-1/2}$	(5-21*a*)
Turbulent	$Re_x < 10^7$, $\delta = 0$ at $x = 0$	$\frac{\delta}{x} = 0.381\ Re_x^{-1/5}$	(5-91)
Turbulent	$5 \times 10^5 < Re_x < 10^7$, $Re_{crit} = 5 \times 10^5$, $\delta = \delta_{lam}$ at Re_{crit}	$\frac{\delta}{x} = 0.381\ Re_x^{-1/5} - 10256\ Re_x^{-1}$	(5-95)
		Friction coefficients	
Laminar, local	$Re_x < 5 \times 10^5$	$C_{fx} = 0.664\ Re_x^{1/2}$	(5-54)
Turbulent, local	$5 \times 10^5 < Re_x < 10^7$	$C_{fx} = 0.0592\ Re_x^{-1/5}$	(5-77)
Turbulent, local	$10^7 < Re_x < 10^9$	$C_{fx} = 0.37(\log Re_x)^{-2.584}$	(5-78)
Turbulent, average	$Re_{crit} < Re_x < 10^9$	$\overline{C_f} = \frac{0.455}{(\log Re_L)^{2.584}} - \frac{A}{Re_L}$, *A* from Table 5-1	(5-79)

PROBLEMS

5-1 A certain nozzle is designed to expand air from stagnation conditions of 1.38 MPa and 200°C to 0.138 MPa. The mass rate of flow is designed to be 4.5 kg/s. Suppose this nozzle is used in conjunction with a blowdown wind-tunnel facility so that the nozzle is suddenly allowed to discharge into a perfectly evacuated tank. What will the temperature of the air in the tank be when the pressure in the tank equals 0.138 MPa? Assume that the tank is perfectly insulated and that air behaves as a perfect gas. Assume that the expansion in the nozzle is isentropic.

5-2 Using a linear velocity profile

$$\frac{u}{u_\infty} = \frac{y}{\delta}$$

for a flow over a flat plate, obtain an expression for the boundary-layer thickness as a function of x.

5-3 Using the continuity relation

$$\frac{\partial u}{\partial x} + \frac{\partial v}{\partial y} = 0$$

along with the velocity distribution

$$\frac{u}{u_\infty} = \frac{3}{2}\frac{y}{\delta} - \frac{1}{2}\left(\frac{y}{\delta}\right)^3$$

and the expression for the boundary-layer thickness

$$\frac{\delta}{x} = \frac{4.64}{\sqrt{\mathrm{Re}_x}}$$

derive an expression for the y component of velocity v as a function of x and y. Calculate the value of v at the outer edge of the boundary layer at distances of 6 and 12 in from the leading edge for the conditions of Example 5-3.

5-4 Repeat Prob. 5-3 for the linear velocity profile of Prob. 5-2.

5-5 Using the linear-velocity profile in Prob. 5-2 and a cubic-parabola temperature distribution [Eq. (5-30)], obtain an expression for heat-transfer coefficient as a function of the Reynolds number for a laminar boundary layer on a flat plate.

5-6 Air at 20 kPa and 5°C enters a 2.5-cm-diameter tube at a velocity of 1.5 m/s. Using a flat-plate analysis, estimate the distance from the entrance at which the flow becomes fully developed.

5-7 A fluid flows between two large parallel plates. Develop an expression for the velocity distribution as a function of distance from the centerline between the two plates under developed flow conditions.

5-8 Using the energy equation given by Eq. (5-32), determine an expression for heat-transfer coefficient under the conditions

$$u = u_\infty = \text{const} \qquad \frac{T - T_w}{T_\infty - T_w} = \frac{y}{\delta_t}$$

where δ_t is the thermal-boundary-layer thickness.

5-9 Derive an expression for the heat transfer in a laminar boundary layer on a flat plate under the condition $u = u_\infty =$ constant. Assume that the temperature distribution is given by the cubic-parabola relation in Eq. (5-30). This solution approximates the condition observed in the flow of a liquid metal over a flat plate.

5-10 Show that $\partial^3 u/\partial y^3 = 0$ at $y = 0$ for an incompressible laminar boundary layer on a flat plate with zero-pressure gradient.

5-11 Review the analytical developments of this chapter and list the restrictions which apply to the following equations: (5-25), (5-26), (5-44), (5-46), (5-85), and (5-107).

5-12 Calculate the ratio of thermal-boundary-layer thickness to hydrodynamic-boundary-layer thickness for the following fluids: air at 1 atm and 20°C, water at 20°C, helium at 1 atm and 20°C, liquid ammonia at 20°C, glycerine at 20°C.

5-13 For water flowing over a flat plate at 15°C and 3 m/s, calculate the mass flow through the boundary layer at a distance of 5 cm from the leading edge of the plate.

5-14 Air at 90°C and 1 atm flows over a flat plate at a velocity of 30 m/s. How thick is the boundary layer at a distance of 2.5 cm from the leading edge of the plate?

5-15 Air flows over a flat plate at a constant velocity of 20 m/s and ambient conditions of 20 kPa and 20°C. The plate is heated to a constant temperature of 75°C, starting at a distance of 7.5 cm from the leading edge. What is the total heat transfer from the leading edge to a point 35 cm from the leading edge?

5-16 Water at 15°C flows between two large parallel plates at a velocity of 1.5 m/s. The plates are separated by a distance of 15 mm. Estimate the distance from the leading edge where the flow becomes fully developed.

5-17 Air at standard conditions of 1 atm and 30°C flows over a flat plate at 20 m/s. The plate is 60 cm square and is maintained at 90°C. Calculate the heat transfer from the plate.

5-18 Air at 7 kPa and 35°C flows across a 30-cm-square flat plate at 7.5 m/s. The plate is maintained at 65°C. Estimate the heat lost from the plate.

5-19 Air at 90°C and atmospheric pressure flows over a horizontal flat plate at 60 m/s. The plate is 60 cm square and is maintained at a uniform temperature of 10°C. What is the total heat transfer?

5-20 Plot the heat-transfer coefficient versus length for flow over a 1-m-long flat plate under the following conditions: (*a*) helium at 1 lb/in^2 abs, 80°F, $u_\infty = 10$ ft/s [3.048 m/s]; (*b*) hydrogen at 1 lb/in^2 abs, 80°F, $u_\infty = 10$ ft/s; (*c*) air at 1 lb/in^2 abs, 80°F, $u_\infty = 10$ ft/s; (*d*) water at 80°F, $u_\infty = 10$ ft/s; (*e*) helium at 20 lb/in^2 abs, 80°F, $u_\infty = 10$ ft/s.

5-21 Calculate the heat transfer from a 30-cm-square plate over which air flows at 35°C and 14 kPa. The plate temperature is 250°C, and the free-stream velocity is 6 m/s.

5-22 Air at 20 kPa and 20°C flows across a flat plate 60 cm long. The free-stream velocity is 30 m/s, and the plate is heated over its total length to a temperature of 55°C. For $x = 30$ cm calculate the value of y for which u will equal 22.5 m/s.

5-23 For the flow system in Prob. 5-22 calculate the value of the friction coefficient at a distance of 15 cm from the leading edge.

5-24 Air at 5°C and 70 kPa flows over a flat plate at 6 m/s. A heater strip 2.5 cm long is placed on the plate at a distance of 15 cm from the leading edge. Calculate the heat lost from the strip per unit depth of plate for a heater surface temperature of 65°C.

5-25 Air at 1 atm and 27°C blows across a large concrete surface 15 m wide maintained at 55°C. The flow velocity is 4.5 m/s. Calculate the convection heat loss from the surface.

5-26 Air at 300 K and 75 kPa flows over a 1-m-square plate at a velocity of 45 m/s. The plate is maintained at a constant temperature of 400 K. Calculate the heat lost by the plate.

5-27 A horizontal flat plate is maintained at 50°C and has dimensions of 50 by 50 cm. Air at 50 kPa and 10°C is blown across the plate at 20 m/s. Calculate the heat lost from the plate.

5-28 Air flows across a 20-cm-square plate with a velocity of 5 m/s. Free-stream conditions are 10°C and 0.2 atm. A heater in the plate surface furnishes a constant heat-flux condition at the wall so that the average wall temperature is 100°C. Calculate the surface heat flux and the value of h at an x position of 10 cm.

5-29 Calculate the flow velocity necessary to produce a Reynolds number of 10^7 for flow across a 1-m-square plate with the following fluids: (*a*) water at 20°C, (*b*) air at 1 atm and 20°C, (*c*) Freon 12 at 20°C, (*d*) ammonia at 20°C, and (*e*) helium at 20°C.

5-30 Calculate the average heat-transfer coefficient for each of the cases in Prob. 5-29 assuming all properties are evaluated at 20°C.

5-31 Calculate the boundary-layer thickness at the end of the plate for each case in Prob. 5-29.

5-32 A blackened plate is exposed to the sun so that a constant heat flux of 800 W/m^2 is absorbed. The back side of the plate is insulated so that all the energy absorbed is dissipated to an airstream which blows across the plate at conditions of 25°C, 1 atm, and 3 m/s. The plate is 25 cm square. Estimate the average temperature of the plate. What is the plate temperature at the trailing edge?

5-33 Air at 1 atm and 300 K blows across a 50-cm-square flat plate at a velocity such that the Reynolds number at the downstream edge of the plate is 1.1×10^5. Heating does not begin until halfway along the plate and then the surface temperature is 400 K. Calculate the heat transfer from the plate.

5-34 Air at 20°C and 14 kPa flows at a velocity of 150 m/s past a flat plate 1 m long which is maintained at a constant temperature of 150°C. What is the average heat-transfer rate per unit area of plate?

5-35 Derive equations equivalent to Eq. (5-85) for critical Reynolds numbers of 3×10^5, 10^6, and 3×10^6.

5-36 Assuming that the local heat-transfer coefficient for flow on a flat plate can be represented by Eq. (5-81) and that the boundary layer starts at the leading edge of the plate, determine an expression for the average heat-transfer coefficient.

5-37 A 10-cm-square plate has an electric heater installed which produces a constant heat flux. Water at 10°C flows across the plate at a velocity of 3 m/s. What is the total heat which can be dissipated if the plate temperature is not to exceed 80°C?

5-38 Repeat Prob. 5-37 for air at 1 atm and 300 K.

5-39 Helium at 1 atm and 300 K is used to cool a 1-m-square plate maintained at 500 K. The flow velocity is 50 m/s. Calculate the total heat loss from the plate. What is the boundary-layer thickness as the flow leaves the plate?

5-40 For the flow system in Prob. 5-40 calculate the y position in the boundary layer at the trailing edge where $u = 25$ m/s.

5-41 A low-speed wind tunnel is to be designed to study boundary layers up to $\text{Re}_x = 10^7$ with air at 1 atm and 25°C. The maximum flow velocity which can be expected from an existing fan system is 30 m/s. How long must the flat-plate test-section be to produce the required Reynolds numbers? What will the maximum boundary-layer thickness be under these conditions? What would the maximum boundary-layer thicknesses be for flow velocities at 7 and 12 m/s?

5-42 A light breeze at 10 mi/h blows across a metal building in the summer. The height of the building wall is 12 ft, and the width is 20 ft. A net energy flux of 110 Btu/h · ft² from the sun is absorbed in the wall and subsequently dissipated to the surrounding air by convection. Assuming that the air is 80°F and 1 atm and blows across the wall as on a flat plate, estimate the average temperature the wall will attain for equilibrium conditions.

5-43 The bottom of a corn-chip fryer is 10 ft long by 3 ft wide and is maintained at a temperature of 420°F. Cooking oil flows across this surface at a velocity of 1 ft/s and has a free-stream temperature of 400°F. Calculate the heat transfer to the oil and estimate the maximum boundary-layer thickness. Properties of the oil may be taken as $\nu = 2 \times 10^{-6}$ m²/s, $k = 0.12$ W/m · °C, and Pr = 40.

5-44 Air at 27°C and 1 atm blows over a 4.0-m-square flat plate at a velocity of 40 m/s. The plate temperature is 77°C. Calculate the total heat transfer.

5-45 The roof of a building is 30 m by 60 m, and because of heat loading by the sun it attains a temperature of 300 K when the ambient air temperature is 0°C. Calculate the heat loss from the roof for a mild breeze blowing at 5 mi/h across the roof ($L = 30$ m).

5-46 Air at 1 atm and 30°C flows over a 15-cm-square plate at a velocity of 10 m/s. Calculate the maximum boundary layer thickness.

5-47 Air at 0.2 MPa and 25°C flows over a square flat plate at a velocity of 60 m/s. The plate is 0.5 m on a side and is maintained at a constant temperature of 150°C. Calculate the heat lost from the plate.

5-48 Helium at a pressure of 150 kPa and a temperature of 20°C flows across a 1-m-square plate at a velocity of 50 m/s. The plate is maintained at a constant temperature of 100°C. Calculate the heat lost by the plate.

5-49 Air at 50 kPa and 250 K flows across a 2-m-square plate at a velocity of 20 m/s. The plate is maintained at a constant temperature of 350 K. Calculate the heat lost by the plate.

5-50 Using Eqs. (5-55), (5-81), and (5-82) for the local heat transfer in their respective ranges, obtain an expression for the average heat transfer coefficient, or Nusselt number, over the range $5 \times 10^5 < \mathrm{Re}_L < 10^9$ with $\mathrm{Re}_{crit} = 5 \times 10^5$. Use a numerical technique to perform the necessary integration and a curve fit to simplify the results.

5-51 An experiment is to be designed to demonstrate measurement of heat loss for water flowing over a flat plate. The plate is 30 cm square and it will be maintained nearly constant in temperature at 50°C while the water temperature will be about 10°C. (*a*) Calculate the flow velocities necessary to study a range of Reynolds numbers from 10^4 to 10^7. (*b*) Estimate the heat-transfer coefficients and heat-transfer rates for several points in the specified range.

5-52 Nitrogen at 50 kPa and 300 K flows over a flat plate at a velocity of 100 m/s. The length of the plate is 1.2 m and the plate is maintained at a constant temperature of 400 K. Calculate the heat lost by the plate.

5-53 Hydrogen at 2 atm and 15°C flows across a 1-m-square flat plate at a velocity of 6 m/s. The plate is maintained at a constant temperature of 139°C. Calculate the heat lost by the plate.

5-54 Liquid ammonia at 10°C is forced across a square plate 45 cm on a side at a velocity of 5 m/s. The plate is maintained at 50°C. Calculate the heat lost by the plate.

5-55 Helium flows across a 1.0-m-square plate at a velocity of 50 m/s. The helium is at a pressure of 45 kPa and a temperature of 50°C. The plate is maintained at a constant temperature of 136°C. Calculate the heat lost by the plate.

5-56 Air at 0.1 atm flows over a flat plate at a velocity of 300 m/s. The plate temperature is maintained constant at 100°C and the free-stream air temperature is 10°C. Calculate the heat transfer for a plate which is 80 cm square.

5-57 Water at 70°F flows across a 1-ft-square flat plate at a velocity of 20 ft/s. The plate is maintained at a constant temperature of 130°F. Calculate the heat lost by the plate.

5-58 Plot h_x versus x for air at 1 atm and 300 K flowing at a velocity of 30 m/s across a flat plate. Take $\mathrm{Re}_{crit} = 5 \times 10^5$ and use semilog plotting paper. Extend the plot to an x value equivalent to $\mathrm{Re} = 10^9$. Also plot the average heat-transfer coefficient over this same range.

5-59 Air flows over a flat plate at 1 atm and 350 K with a velocity of 30 m/s. Calculate the mass flow through the boundary layer at x locations where $\mathrm{Re}_x = 10^6$ and 10^7.

5-60 Air flows with a velocity of 6 m/s across a 20-cm-square plate at 50 kPa and 300 K. An electrical heater is installed in the plate such that it produces a constant heat flux. What is the total heat which can be dissipated if the plate temperature cannot exceed 600 K?

5-61 "Slug" flow in a tube may be described as that flow in which the velocity is constant across the entire flow area of the tube. Obtain an expression for the heat-transfer coefficient in this type of flow with a constant-heat-flux condition maintained at the wall. Compare the results with those of Sec. 5-10. Explain the reason for the difference in answers on a physical basis.

5-62 Assume that the velocity distribution in the turbulent core for tube flow may be represented by

$$\frac{u}{u_c} = \left(1 - \frac{r}{r_o}\right)^{1/7}$$

where u_c is the velocity at the center of the tube and r_o is the tube radius. The velocity in the laminar sublayer may be assumed to vary linearly with the radius. Using the friction factor given by Eq. (5-115), derive an equation for the thickness of the laminar sublayer. For this problem the average flow velocity may be calculated using only the turbulent velocity distribution.

5-63 Using the velocity profile in Prob. 5-62, obtain an expression for the eddy diffusivity of momentum as a function of radius.

5-64 In heat-exchanger applications, it is frequently important to match heat-transfer requirements with pressure-drop limitations. Assuming a fixed total heat-transfer requirement and a fixed temperature difference between wall and bulk conditions as well as a fixed pressure drop through the tube, derive expressions for the length and diameter of the tube, assuming turbulent flow of a gas with the Prandtl number near unity.

5-65 Water flows in a 2.5-cm-diameter pipe so that the Reynolds number based on diameter is 1500 (laminar flow is assumed). The average bulk temperature is 35°C. Calculate the maximum water velocity in the tube. (Recall that $u_m = 0.5u_0$.) What would the heat-transfer coefficient be for such a system if the tube wall was subjected to a constant heat flux and the velocity and temperature profiles were completely developed? Evaluate properties at bulk temperature.

5-66 A slug flow is encountered in an annular-flow system which is subjected to a constant heat flux at both the inner and outer surfaces. The temperature is the same at both inner and outer surfaces at identical x locations. Derive an expression for the temperature distribution in such a flow system, assuming constant properties and laminar flow.

5-67 Air at Mach 4 and 3 lb/in^2 abs, 0°F, flows past a flat plate. The plate is to be maintained at a constant temperature of 200°F. If the plate is 18 in long, how much cooling will be required to maintain this temperature?

5-68 Air flows over an isothermal flat plate maintained at a constant temperature of 65°C. The velocity of the air is 600 m/s at static properties of 15°C and 7 kPa. Calculate the average heat-transfer coefficient for a plate 1 m long.

5-69 Air at 7 kPa and −40°C flows over a flat plate at Mach 4. The plate temperature is 35°C, and the plate length is 60 cm. Calculate the adiabatic wall temperature for the laminar portion of the boundary layer.

5-70 A wind tunnel is to be constructed to produce flow conditions of Mach 2.8 at $T_\infty = -40$°C and $p = 0.05$ atm. What is the stagnation temperature for these conditions? What would be the adiabatic wall temperature for the laminar and turbulent portions of a boundary layer on a flat plate? If a flat plate were installed in the tunnel such that $Re_L = 10^7$, what would the heat transfer be for a constant wall temperature of 0°C?

5-71 Compute the drag force exerted on the plate by each of the systems in Prob. 5-20.

5-72 Glycerin at 30°C flows past a 30-cm-square flat plate at a velocity of 1.5 m/s. The drag force is measured as 8.9 N (both sides of the plate). Calculate the heat-transfer coefficient for such a flow system.

5-73 Calculate the drag (viscous-friction) force on the plate in Prob. 5-21 under the conditions of no heat transfer. Do not use the analogy between fluid friction and heat transfer for this calculation; i.e., calculate the drag directly by evaluating the viscous-shear stress at the wall.

5-74 Nitrogen at 1 atm and 20°C is blown across a 130-cm-square flat plate at a velocity of 3.0 m/s. The plate is maintained at a constant temperature of 100°C. Calculate the average-friction coefficient and the heat transfer from the plate.

5-75 Using the velocity distribution for developed laminar flow in a tube, derive an expression for the friction factor as defined by Eq. 5-112.

5-76 Engine oil at 10°C flows across a 15-cm-square plate upon which is imposed a constant heat flux of 10 kW/m². Determine (*a*) the average temperature difference, (*b*) the temperature difference at the trailing edge, and (*c*) the average heat-transfer coefficient. Use the Churchill relation [Eq. (5-51)]. $u_\infty = 0.5$ m/s.

5-77 Work Prob. 5-76 for a constant plate surface temperature equal to that at the trailing edge, and determine the total heat transfer.

5-78 For air at 25°C and 1 atm, with a free-stream velocity of 45 m/s, calculate the length of a flat plate to produce Reynolds numbers of 5×10^5 and 10^8. What are the boundary-layer thicknesses at these Reynolds numbers?

5-79 Determine the boundary-layer thickness at $Re = 5 \times 10^5$ for the following fluids flowing over a flat plate at 20 m/s: (*a*) air at 1 atm and 10°C, (*b*) saturated liquid water at 10°C, (*c*) hydrogen at 1 atm and 10°C, (*d*) saturated liquid ammonia at 10°C, and (*e*) saturated liquid Freon 12 at 10°C.

5-80 Many of the heat-transfer relations for flow over a flat plate are of the form

$$Nu_x = \frac{h_x x}{k} = C\, Re_x^n f(Pr)$$

Obtain an expression for $\bar{h}_L/h_{x=L}$ in terms of the constants C and n.

5-81 Compare Eqs. (5-51) and (5-44) for engine oil at 20°C and a Reynolds number of 10,000.

5-82 Air at 1 atm and 300 K blows across a square plate 75 cm on a side which is maintained at 350 K. The free-stream velocity is 45 m/s. Calculate the heat transfer and drag force on one side of the plate. Also calculate the heat transfer for just the laminar portion of the boundary layer.

5-83 Taking the critical Reynolds number as 5×10^5 for Prob. 5-82, calculate the boundary-layer thickness at this point and at the trailing edge of the plate assuming

(a) laminar flow to Re_{crit} and turbulent thereafter and (b) turbulent flow from the leading edge.

5-84 If the plate temperature in Prob. 5-82 is raised to 500 K while keeping the free-stream conditions the same, calculate the total heat transfer evaluating properties at (a) free-stream conditions, (b) film temperature, and (c) wall temperature. Comment on the results.

5-85 Air at 250 K and 1 atm blows across a 30-cm-square plate at a velocity of 10 m/s. The plate maintains a constant heat flux of 700 W/m^2. Determine the plate temperatures at x locations of 1, 5, 10, 20, and 30 cm.

5-86 Engine oil at 20°C is forced across a 20-cm-square plate at 10 m/s. The plate surface is maintained at 40°C. Calculate the heat lost by the plate and the drag force for one side of an unheated plate.

5-87 A large flat plate 4.0 m long and 1.0 m wide is exposed to an atmospheric air at 27°C with a velocity of 30 mi/h in a direction parallel to the 4.0-m dimension. If the plate is maintained at 77°C, calculate the total heat loss. Also calculate the heat flux in watts per square meter at x locations of 3 cm, 50 cm, 1.0 m, and 4.0 m.

5-88 Air at 1 atm and 300 K blows across a 10-cm-square plate at 30 m/s. Heating does not begin until x = 5.0 cm, after which the plate surface is maintained at 400 K. Calculate the total heat lost by the plate.

5-89 For the plate and flow conditions of Prob. 5-88 only a 0.5-cm strip centered at x = 5.0 cm is heated to 400 K. Calculate the heat lost by this strip.

5-90 Two 20-cm-square plates are separated by a distance of 3.0 cm. Air at 1 atm, 300 K, and 15 m/s enters the space separating the plates. Will there be interference between the two boundary layers?

REFERENCES

1 Schlichting, H.: "Boundary Layer Theory," 7th ed., McGraw-Hill Book Company, New York, 1979.

2 von Kármán, T.: Über laminaire und turbulente Reibung, *Angew. Math. Mech.*, vol. 1, pp. 233–252, 1921; also *NACA Tech. Mem.* 1092, 1946.

3 Sellars, J. R., M. Tribus, and J. S. Klein: Heat Transfer to Laminar Flows in a Round Tube or Flat Conduit: The Graetz Problem Extended, *Trans. ASME*, vol. 78, p. 441, 1956.

4 Eckert, E. R. G.: Survey of Boundary Layer Heat Transfer at High Velocities and High Temperatures, *WADC Tech. Rep.*, pp. 59–624, April 1960.

5 White, F. M.: "Viscous Fluid Flow," McGraw-Hill Book Company, New York, 1974.

6 Knudsen, J. D., and D. L. Katz: "Fluid Dynamics and Heat Transfer," McGraw-Hill Book Company, New York, 1958.

7 Holman, J. P.: "Thermodynamics," 3d ed., McGraw-Hill Book Company, New York, 1980.

8 Schultz-Grunow, F.: Neues Widerstandsgesetz für glatte Platten, *Luftfahrtforschung,* vol. 17, p. 239, 1940; also *NACA Tech. Mem.* 986, 1941.

9 Churchill, S. W., and H. Ozoe: Correlations for Laminar Forced Convection in Flow over an Isothermal Flat Plate and in Developing and Fully Developed Flow in an Isothermal Tube, *J. Heat Transfer,* vol. 95, p. 46, 1973.

10 Whitaker, S.: Forced Convection Heat Transfer Correlation for Flow in Pipes, Past Flat Plates, Single Cylinders, Single Spheres, and for Flow in Packed Beds and Tube Bundles, *AIChE J.,* vol. 18, p. 361, 1972.

11 Churchill, S. W.: A. Comprehensive Correlating Equation for Forced Convection from Flat Plates, *AIChE J.,* vol. 22, p. 264, 1976.

EMPIRICAL AND PRACTICAL RELATIONS FOR FORCED-CONVECTION HEAT TRANSFER

■ 6-1 INTRODUCTION

The discussion and analyses of Chap. 5 have shown how forced-convection heat transfer may be calculated for several cases of practical interest; the problems considered, however, were those which could be solved in an analytical fashion. In this way, the principles of the convection process and their relation to fluid dynamics were demonstrated, with primary emphasis being devoted to a clear understanding of physical mechanism. Regrettably, it is not always possible to obtain analytical solutions to convection problems, and the individual is forced to resort to experimental methods to obtain design information, as well as to secure the more elusive data which increase the physical understanding of the heat-transfer processes.

Results of experimental data are usually expressed in the form of either empirical formulas or graphical charts so that they may be utilized with a maximum of generality. It is in the process of trying to generalize the results of one's experiments, in the form of some empirical correlation, that difficulty is encountered. If an analytical solution is available for a similar problem, the correlation of data is much easier, since one may guess at the functional form of the results, and hence use the experimental data to obtain values of constants or exponents for certain significant parameters such as the Reynolds or Prandtl numbers. If an analytical solution for a similar problem is not available, the individual must resort to intuition based on physical understanding of the problem, or shrewd inferences which one may be able to draw from the differential equations of the flow processes based upon dimensional or order-of-magnitude estimates. In any event, there is no substitute for physical insight and understanding.

To show how one might proceed to analyze a new problem to obtain an important functional relationship from the differential equations, consider the problem of determining the hydrodynamic-boundary-layer thickness for flow over a flat plate. This problem was solved in Chap. 5, but we now wish to make an order-of-magnitude analysis of the differential equations to obtain the functional form of the solution. The momentum equation

$$u \frac{\partial u}{\partial x} + v \frac{\partial u}{\partial y} = \nu \frac{\partial^2 u}{\partial y^2}$$

must be solved in conjunction with the continuity equation

$$\frac{\partial u}{\partial x} + \frac{\partial v}{\partial y} = 0$$

Within the boundary layer we may say that the velocity u is of the order of the free-stream velocity u_∞. Similarly, the y dimension is of the order of the boundary-layer thickness δ. Thus

$$u \sim u_\infty$$

$$y \sim \delta$$

and we might write the continuity equation in an approximate form as

$$\frac{\partial u}{\partial x} + \frac{\partial v}{\partial y} = 0$$

$$\frac{u_\infty}{x} + \frac{v}{\delta} \approx 0$$

or

$$v \sim \frac{u_\infty \delta}{x}$$

Then, by using this order of magnitude for v, the analysis of the momentum equation would yield

$$u \frac{\partial u}{\partial x} + v \frac{\partial u}{\partial y} = \nu \frac{\partial^2 u}{\partial y^2}.$$

$$u_\infty \frac{u_\infty}{x} + \frac{u_\infty \delta}{x} \frac{u_\infty}{\delta} \approx \nu \frac{u_\infty}{\delta^2}$$

or

$$\delta^2 \sim \frac{\nu x}{u_\infty}$$

$$\delta \sim \sqrt{\frac{\nu x}{u_\infty}}$$

Dividing by x to express the result in dimensionless form gives

$$\frac{\delta}{x} \sim \sqrt{\frac{\nu}{u_\infty x}} = \frac{1}{\sqrt{\text{Re}_x}}$$

This functional variation of the boundary-layer thickness with the Reynolds number and x position is precisely that which was obtained in Sec. 5-4. Although this analysis is rather straightforward and does indeed yield correct results, the order-of-magnitude analysis may not always be so fortunate when applied to more complex problems, particularly those involving turbulent- or separated-flow regions. Nevertheless, one may often obtain valuable information and physical insight by examining the order of magnitude of various terms in a governing differential equation for the particular problem at hand.

A conventional technique used in correlation of experimental data is that of dimensional analysis, in which appropriate dimensionless groups such as the Reynolds and Prandtl numbers are derived from purely dimensional and functional considerations. There is, of course, the assumption of flow-field and temperature-profile similarity for geometrically similar heating surfaces. Generally speaking, the application of dimensional analysis to any new problem is extremely difficult when a previous analytical solution of some sort is not available. It is usually best to attempt an order-of-magnitude analysis such as the one above if the governing differential equations are known. In this way it may be possible to determine the significant dimensionless variables for correlating experimental data. In some complex flow and heat-transfer problems a clear physical model of the processes may not be available, and the engineer must first try to establish this model before the experimental data can be correlated.

Schlichting [6], Giedt [7], and Kline [28] discuss similarity considerations and their use in boundary-layer and heat-transfer problems.

The purpose of the foregoing discussion has not been to emphasize or even to imply any new method for solving problems, but rather to indicate the necessity of applying intuitive physical reasoning to a difficult problem and to point out the obvious advantage of using any and all information which may be available. When the problem of correlation of experimental data for a previously unsolved situation is encountered, one must frequently adopt devious methods to accomplish the task.

6-2 EMPIRICAL RELATIONS FOR PIPE AND TUBE FLOW

The analysis of Sec. 5-10 has shown how one might analytically attack the problem of heat transfer in fully developed laminar tube flow. The cases of undeveloped laminar flow, flow systems where the fluid properties vary widely with temperature, and turbulent-flow systems are considerably more complicated but are of very important practical interest in the design of heat exchangers and associated heat-transfer equipment. These more complicated problems may sometimes be solved analytically, but the solutions, when possible, are very tedious. For design and engineering purposes, empirical correlations are usually of greatest practical utility. In this section we present some of the more important and useful empirical relations and point out their limitations.

☐ The Bulk Temperature

First let us give some further consideration to the bulk-temperature concept which is important in all heat-transfer problems involving flow inside closed channels. In Chap. 5 we noted that the bulk temperature represents energy average or "mixing cup" conditions. Thus, for the tube flow depicted in Fig. 6-1 the total energy added can be expressed in terms of a bulk-temperature difference by

$$q = \dot{m}c_p(T_{b_2} - T_{b_1}) \tag{6-1}$$

provided c_p is reasonably constant over the length. In some differential length dx the heat added dq can be expressed either in terms of a bulk-temperature difference or in terms of the heat-transfer coefficient

$$dq = \dot{m}c_p dT_b = h(2\pi r)\, dx\, (T_w - T_b) \tag{6-2}$$

where T_w and T_b are the wall and bulk temperatures at the particular x location. The total heat transfer can also be expressed as

$$q = hA(T_w - T_b)_{\text{av}} \tag{6-3}$$

where A is the total surface area for heat transfer. Because both T_w and T_b can vary along the length of the tube, a suitable averaging process must be adopted for use with Eq. (6-3). In this chapter most of our attention will be focused on methods for determining h, the convection heat-transfer coefficient. Chapter 10 will discuss different methods for taking proper account of temperature variations in heat exchangers.

For fully developed turbulent flow in smooth tubes the following relation is recommended by Dittus and Boelter [1]:

$$\text{Nu}_d = 0.023\ \text{Re}_d^{0.8}\ \text{Pr}^n \tag{6-4}$$

The properties in this equation are evaluated at the fluid bulk temperature, and the exponent n has the following values:

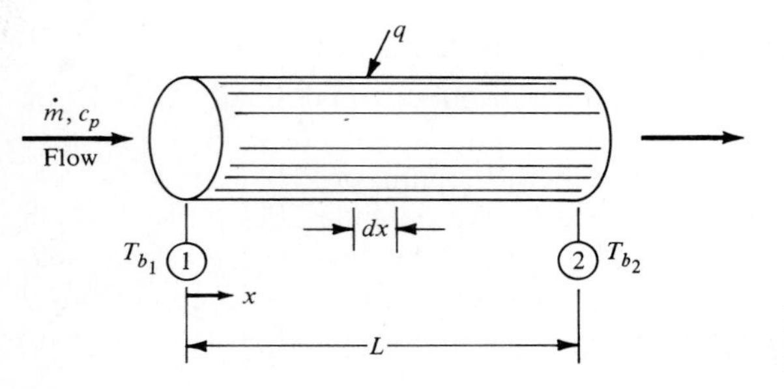

Fig. 6-1 Total heat transfer in terms of bulk-temperature difference.

$$n = \begin{cases} 0.4 & \text{for heating of the fluid} \\ 0.3 & \text{for cooling of the fluid} \end{cases}$$

Equation (6-4) is valid for fully developed turbulent flow in smooth tubes for fluids with Prandtl numbers ranging from about 0.6 to 100 and with moderate temperature differences between wall and fluid conditions.

One may ask the reason for the functional form of Eq. (6-4). Physical reasoning, based on the experience gained with the analyses of Chap. 5, would certainly indicate a dependence of the heat-transfer process on the flow field, and hence on the Reynolds number. The relative rates of diffusion of heat and momentum are related by the Prandtl number, so that the Prandtl number is expected to be a significant parameter in the final solution. We can be rather confident of the dependence of the heat transfer on the Reynolds and Prandtl numbers. But the question arises as to the correct functional form of the relation; i.e., would one necessarily expect a product of two exponential functions of the Reynolds and Prandtl numbers? The answer is that one might expect this functional form since it appears in the flat-plate analytical solutions of Chap. 5, as well as the Reynolds analogy for turbulent flow. In addition, this type of functional relation is convenient to use in correlating experimental data, as described below.

Suppose a number of experiments are conducted with measurements taken of heat-transfer rates of various fluids in turbulent flow inside smooth tubes under different temperature conditions. Different-diameter tubes may be used to vary the range of the Reynolds number in addition to variations in the mass-flow rate. We wish to generalize the results of these experiments by arriving at one empirical equation which represents all the data. As described above, we may anticipate that the heat-transfer data will be dependent on the Reynolds and Prandtl numbers. An exponential function for each of these parameters is perhaps the simplest type of relation to use, so we assume

$$\text{Nu}_d = C\ \text{Re}_d^{\,m}\ \text{Pr}^n$$

where C, m, and n are constants to be determined from the experimental data.

A log-log plot of Nu_d versus Re_d is first made for one fluid to estimate the dependence of the heat transfer on the Reynolds number, i.e., to find an approximate value of the exponent m. This plot is made for one fluid at a constant temperature, so that the influence of the Prandtl number will be small, since the Prandtl number will be approximately constant for the one fluid. By using this first estimate for the exponent m, the data for all fluids are plotted as $\log(\text{Nu}_d/\text{Re}_d^{\,m})$ versus log Pr, and a value for the exponent n is determined. Then, by using this value of n, all the data are plotted again as $\log(\text{Nu}_d/\text{Pr}^n)$ versus $\log \text{Re}_d$, and a final value of the exponent m is determined as well as a value for the constant C. An example of this final type of data plot is shown in Fig. 6-2. The final correlation equation usually represents the data within ± 25 percent.

If wide temperature differences are present in the flow, there may be an

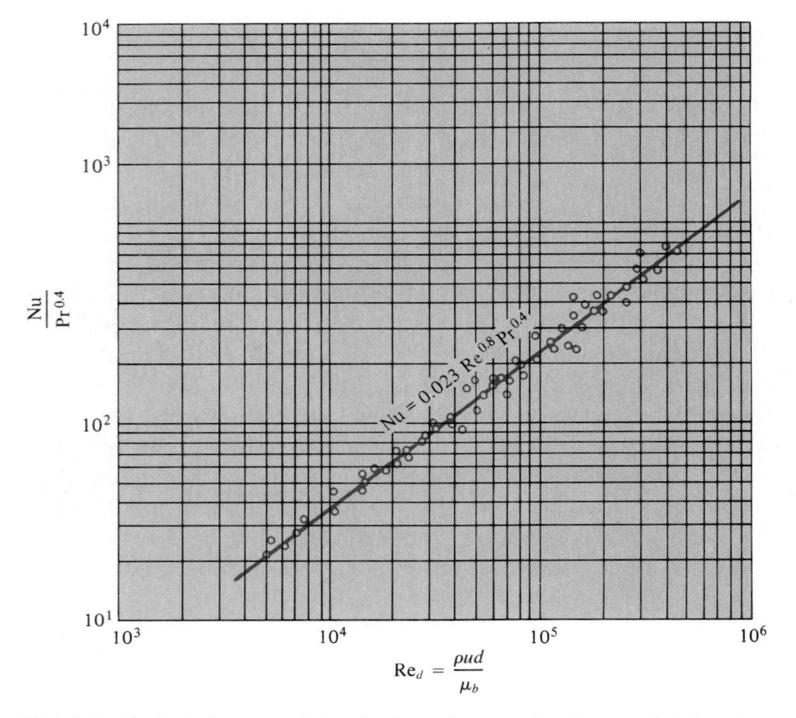

Fig. 6-2 Typical data correlation for forced convection in smooth tubes, turbulent flow.

appreciable change in the fluid properties between the wall of the tube and the central flow. These property variations may be evidenced by a change in the velocity profile as indicated in Fig. 6-3. The deviations from the velocity profile for isothermal flow as shown in this figure are a result of the fact that the viscosity of gases increases with an increase in temperature, while the viscosities of liquids decrease with an increase in temperature.

To take into account the property variations, Sieder and Tate [2] recommend the following relation:

$$\mathrm{Nu}_d = 0.027\ \mathrm{Re}_d^{0.8}\ \mathrm{Pr}^{1/3}\left(\frac{\mu}{\mu_w}\right)^{0.14} \tag{6-5}$$

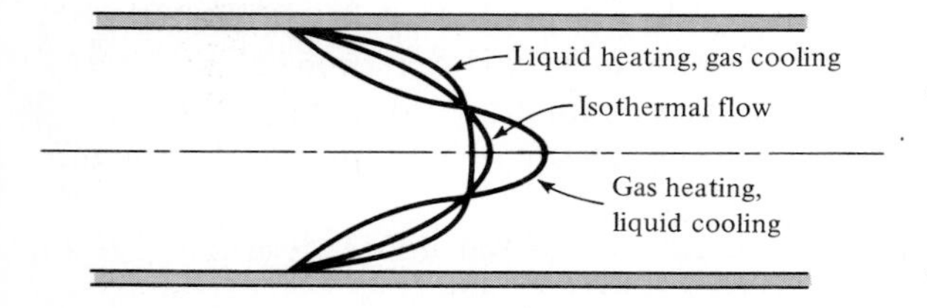

Fig. 6-3 Influence of heating on velocity profile in laminar tube flow.

All properties are evaluated at bulk-temperature conditions, except μ_w, which is evaluated at the wall temperature.

Equations (6-4) and (6-5) apply to fully developed turbulent flow in tubes. In the entrance region the flow is not developed, and Nusselt [3] recommended the following equation:

$$\mathrm{Nu}_d = 0.036\ \mathrm{Re}_d^{0.8}\ \mathrm{Pr}^{1/3}\left(\frac{d}{L}\right)^{0.055} \qquad \text{for } 10 < \frac{L}{d} < 400 \tag{6-6}$$

where L is the length of the tube and d is the tube diameter. The properties in Eq. (6-6) are evaluated at the mean bulk temperature. Hartnett [24] has given experimental data on the thermal entrance region for water and oils. Definitive studies of turbulent transfer with water in smooth tubes and at uniform heat flux have been presented by Allen and Eckert [25].

The above equations offer simplicity in computation, but errors on the order of ±25 percent are not uncommon. Petukhov [42] has developed a more accurate, although more complicated, expression for fully developed turbulent flow in smooth tubes:

$$\mathrm{Nu}_d = \frac{(f/8)\ \mathrm{Re}_d\ \mathrm{Pr}}{1.07 + 12.7(f/8)^{1/2}(\mathrm{Pr}^{2/3} - 1)}\left(\frac{\mu_b}{\mu_w}\right)^n \tag{6-7}$$

where $n = 0.11$ for $T_w > T_b$, $n = 0.25$ for $T_w < T_b$, and $n = 0$ for constant heat flux or for gases. All properties are evaluated at $T_f = (T_w + T_b)/2$ except for μ_b and μ_w. The friction factor may be obtained either from Fig. 6-4 or from the following for smooth tubes:

$$f = (1.82 \log_{10} \mathrm{Re}_d - 1.64)^{-2} \tag{6-8}$$

Equation (6-7) is applicable for the following ranges:

$0.5 < \mathrm{Pr} < 200$ for 6 percent accuracy

$0.5 < \mathrm{Pr} < 2000$ for 10 percent accuracy

$10^4 < \mathrm{Re}_d < 5 \times 10^6$

$0.08 < \mu_b/\mu_w < 40$

Hausen [4] presents the following empirical relation for fully developed laminar flow in tubes at constant wall temperature:

$$\mathrm{Nu}_d = 3.66 + \frac{0.0668(d/L)\ \mathrm{Re}_d\ \mathrm{Pr}}{1 + 0.04[(d/L)\ \mathrm{Re}_d\ \mathrm{Pr}]^{2/3}} \tag{6-9}$$

The heat-transfer coefficient calculated from this relation is the average value over the entire length of tube. Note that the Nusselt number approaches a constant value of 3.66 when the tube is sufficiently long. This situation is similar to that encountered in the constant-heat-flux problem analyzed in Chap. 5 [Eq. (5-107)], except that in this case we have a constant wall temperature instead of a linear variation with length. The temperature profile is fully developed when the Nusselt number approaches a constant value.

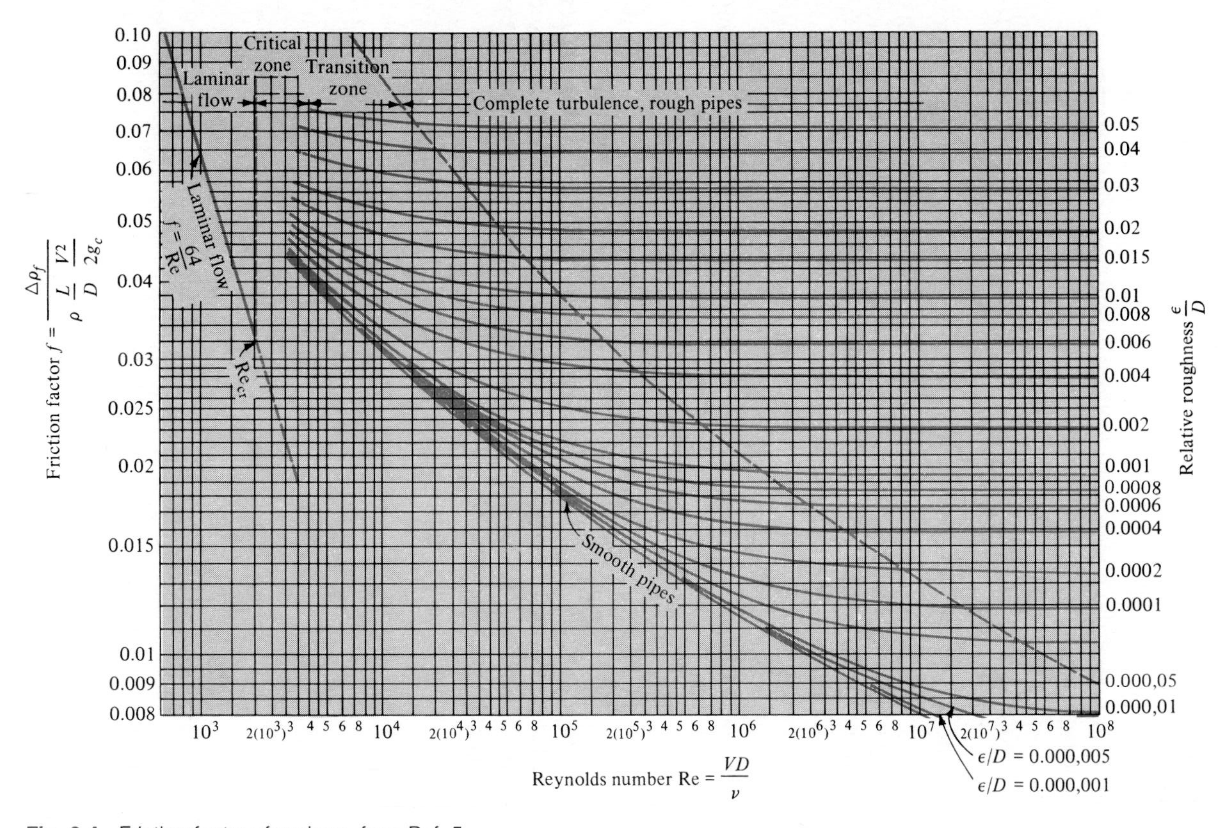

Fig. 6-4 Friction factors for pipes, from Ref. 5.

A somewhat simpler empirical relation was proposed by Sieder and Tate [2] for laminar heat transfer in tubes:

$$\mathrm{Nu}_d = 1.86\,(\mathrm{Re}_d\,\mathrm{Pr})^{1/3}\left(\frac{d}{L}\right)^{1/3}\left(\frac{\mu}{\mu_w}\right)^{0.14} \tag{6-10}$$

In this formula the average heat-transfer coefficient is based on the arithmetic average of the inlet and outlet temperature differences, and all fluid properties are evaluated at the mean bulk temperature of the fluid, except μ_w, which is evaluated at the wall temperature. Equation (6-10) obviously cannot be used for extremely long tubes since it would yield a zero heat-transfer coefficient. A comparison by Knudsen and Katz [9, p. 377] of Eq. (6-10) with other relationships indicates that it is valid for

$$\mathrm{Re}_d\,\mathrm{Pr}\,\frac{d}{L} > 10$$

The product of the Reynold and Prandtl numbers which occurs in the laminar-flow correlations is called the Peclet number.

$$\mathrm{Pe} = \frac{du\rho c_p}{k} = \mathrm{Re}_d\,\mathrm{Pr} \tag{6-11}$$

The calculation of laminar heat-transfer coefficients is frequently complicated by the presence of natural-convection effects which are superimposed on the forced-convection effects. The treatment of combined forced- and free-convection problems is discussed in Chap. 7.

The empirical correlations presented above, with the exception of Eq. (6-7), apply to smooth tubes. Correlations are, in general, rather sparse where rough tubes are concerned, and it is sometimes appropriate that the Reynolds analogy between fluid friction and heat transfer be used to effect a solution under these circumstances. Expressed in terms of the Stanton number,

$$\mathrm{St}_b\,\mathrm{Pr}_f^{2/3} = \frac{f}{8} \tag{6-12}$$

The friction coefficient f is defined by

$$\Delta p = f\,\frac{L}{d}\,\rho\,\frac{u_m^2}{2g_c} \tag{6-13}$$

where u_m is the mean flow velocity. Values of the friction coefficient for different roughness conditions are shown in Fig. 6-4.

Note that the relation in Eq. (6-12) is the same as Eq. (5-114), except that the Stanton number has been multiplied by $\mathrm{Pr}^{2/3}$ to take into account the variation of the thermal properties of different fluids. This correction follows the recommendation of Colburn [15], and is based on the reasoning that fluid friction and heat transfer in tube flow are related to the Prandtl number in the same way as they are related in flat-plate flow [Eq. (5-56)]. In Eq. (6-12) the Stanton number is based on bulk temperature, while the Prandtl number and friction factor are based on properties evaluated at the film temperature. Further information on the effects of tube roughness on heat transfer is given in Refs. 27, 29, 30, and 31.

If the channel through which the fluid flows is not of circular cross section, it is recommended that the heat-transfer correlations be based on the hydraulic diameter D_H, defined by

$$D_H = \frac{4A}{P} \tag{6-14}$$

where A is the cross-sectional area of the flow and P is the wetted perimeter. This particular grouping of terms is used because it yields the value of the physical diameter when applied to a circular cross section. The hydraulic diameter should be used in calculating the Nusselt and Reynolds numbers, and in establishing the friction coefficient for use with the Reynolds analogy.

Although the hydraulic-diameter concept frequently yields satisfactory relations for fluid friction and heat transfer in many practical problems, there are some notable exceptions where the method does not work. Some of the problems involved in heat transfer in noncircular channels have been summarized by Irvine [20] and Knudsen and Katz [9]. The interested reader should consult these discussions for additional information.

Shah and London [40] have compiled the heat-transfer and fluid-friction information for fully developed laminar flow in ducts with a variety of flow cross sections as shown in Table 6-1. In this table the following nomenclature applies:

Nu_{H1} = average Nusselt number for uniform heat flux in flow direction and uniform wall temperature at particular flow cross section

Nu_{H2} = average Nusselt number for uniform heat flux both in flow direction and around periphery

Nu_T = average Nusselt number for uniform wall temperature

$f\,\text{Re}$ = product of friction factor and Reynolds number

Kays [36] and Sellars, Tribus, and Klein (Ref. 3, Chap. 5) have calculated the local and average Nusselt numbers for laminar entrance regions of circular tubes for the case of a fully developed velocity profile. Results of these analyses are shown in Fig. 6-5 in terms of the inverse Graetz number, where

$$\text{Graetz number} = \text{Gz} = \text{Re Pr}\,\frac{d}{x} \tag{6-15}$$

Entrance effects for turbulent flow in tubes are more complicated than for laminar flow and cannot be expressed in terms of a simple function of the Graetz number. Kays [36] has computed the influence for several values of Re and Pr with the results summarized in Fig. 6-6. The ordinate is the ratio of the

Table 6-1 Heat Transfer and Fluid Friction for Fully Developed Laminar Flow in Ducts of Various Cross Sections.

Geometry ($L/D_h > 100$)	Nu_{H1}	Nu_{H2}	Nu_T	$f\,Re/4$
60°, 2a, 2b; $\frac{2b}{2a} = \frac{\sqrt{3}}{2}$	3.111	1.892	2.47	13.333
2b, 2a; $\frac{2b}{2a} = 1$	3.608	3.091	2.976	14.227
	4.002	3.862	3.34	15.054
2b, 2a; $\frac{2b}{2a} = \frac{1}{2}$	4.123	3.017	3.391	15.548
	4.364	4.364	3.657	16.000
2b, 2a; $\frac{2b}{2a} = \frac{1}{4}$	5.099	4.35	3.66	18.700
2b, 2a; $\frac{2b}{2a} = 0.9$	5.331	2.930	4.439	18.233
2b, 2a; $\frac{2b}{2a} = \frac{1}{8}$	6.490	2.904	5.597	20.585
$\frac{2b}{2a} = 0$	8.235	8.235	7.541	24.000
Insulated; $\frac{b}{a} = 0$	5.385	—	4.861	24.000

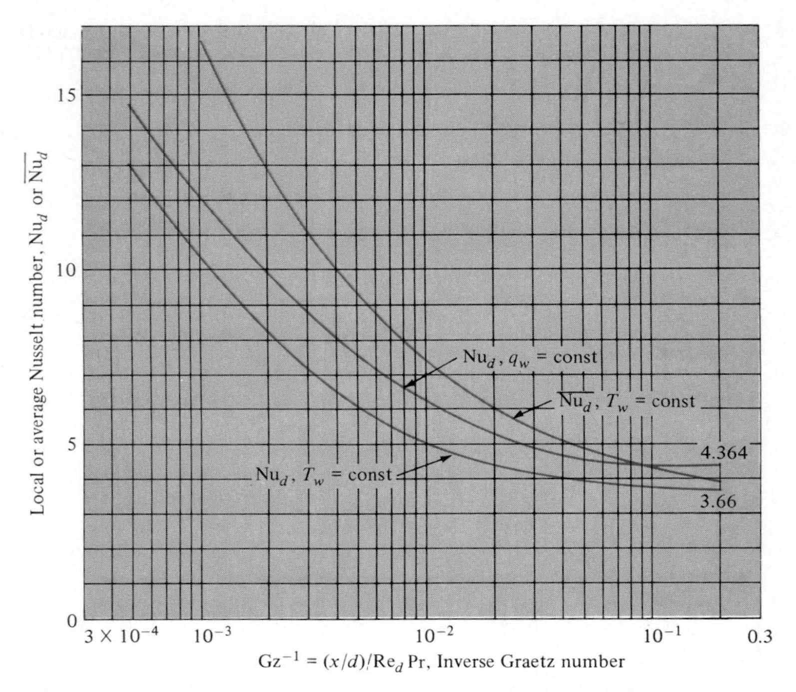

Fig. 6-5 Local and average Nusselt numbers for circular tube thermal entrance regions in fully developed laminar flow.

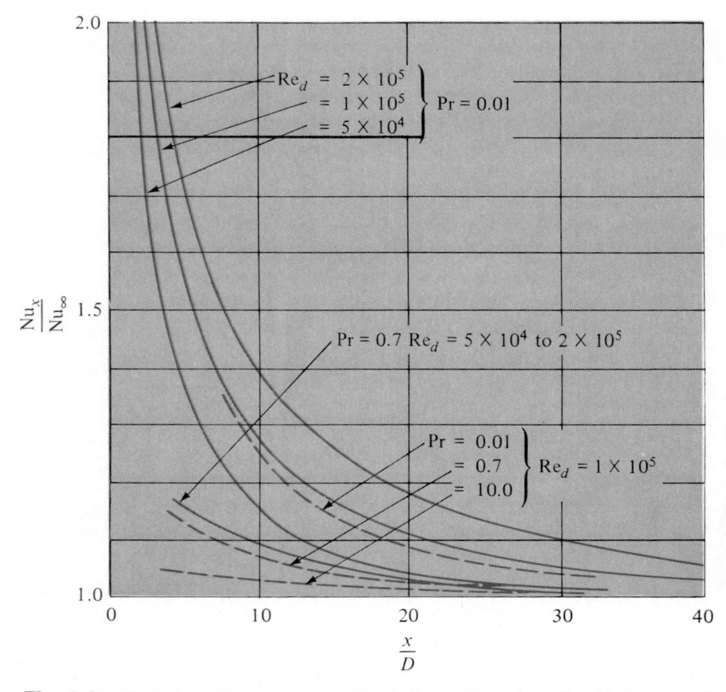

Fig. 6-6 Turbulent thermal entry Nusselt numbers for circular tubes with q_w = constant.

local Nusselt number to that a long distance from the inlet, or for fully developed thermal conditions. In general, the higher the Prandtl number, the shorter the entry length. We can see that the thermal entry lengths are much shorter for turbulent flow than for the laminar counterpart.

■ **EXAMPLE 6-1** Turbulent flow in a tube

Air at 2 atm and 200°C is heated as it flows through a tube with a diameter of 1 in (2.54 cm) at a velocity of 10 m/s. Calculate the heat transfer per unit length of tube if a constant-heat-flux condition is maintained at the wall and the wall temperature is 20°C above the air temperature, all along the length of the tube. How much would the bulk temperature increase over a 3-m length of the tube?

Solution

We first calculate the Reynolds number to determine if the flow is laminar or turbulent, and then select the appropriate empirical correlation to calculate the heat transfer. The properties of air at a bulk temperature of 200°C are

$$\rho = \frac{p}{RT} = \frac{(2)(1.0132 \times 10^5)}{(287)(473)} = 1.493 \text{ kg/m}^3 \qquad [0.0932 \text{ lb}_m/\text{ft}^3]$$

$$\text{Pr} = 0.681$$

$$\mu = 2.57 \times 10^{-5} \text{ kg/m} \cdot \text{s} \qquad [0.0622 \text{ lb}_m/\text{h} \cdot \text{ft}]$$

$$k = 0.0386 \text{ W/m} \cdot {}^\circ\text{C} \qquad [0.0223 \text{ Btu/h} \cdot \text{ft} \cdot {}^\circ\text{F}]$$

$$c_p = 1.025 \text{ kJ/kg} \cdot {}^\circ\text{C}$$

$$\text{Re}_d = \frac{\rho u_m d}{\mu} = \frac{(1.493)(10)(0.0254)}{2.57 \times 10^{-5}} = 14{,}756$$

so that the flow is turbulent. We therefore use Eq. (6-4) to calculate the heat-transfer coefficient.

$$\text{Nu}_d = \frac{hd}{k} = 0.023\,\text{Re}_d^{0.8}\,\text{Pr}^{0.4} = (0.023)(14{,}756)^{0.8}(0.681)^{0.4} = 42.67$$

$$h = \frac{k}{d}\text{Nu}_d = \frac{(0.0386)(42.67)}{0.0254} = 64.85 \text{ W/m}^2 \cdot {}^\circ\text{C} \qquad [11.42 \text{ Btu/h} \cdot \text{ft}^2 \cdot {}^\circ\text{F}]$$

The heat flow per unit length is then

$$\frac{q}{L} = h\,\pi d\,(T_w - T_b) = (64.85)\pi(0.0254)(20) = 103.5 \text{ W/m} \qquad [107.7 \text{ Btu/ft}]$$

We can now make an energy balance to calculate the increase in bulk temperature in a 3.0-m length of tube:

$$q = \dot{m} c_p\,\Delta T_b = L\left(\frac{q}{L}\right)$$

We also have

$$\dot{m} = \rho u_m \frac{\pi d^2}{4} = (1.493)(10)\pi\frac{(0.0254)^2}{4}$$
$$= 7.565 \times 10^{-3} \text{ kg/s} \qquad [0.0167 \text{ lb}_m/\text{s}]$$

so that we insert the numerical values in the energy balance to obtain

$$(7.565 \times 10^{-3})(1025)\ \Delta T_b = (3.0)(103.5)$$

and

$$\Delta T_b = 40.04°\text{C} \qquad [104.07°\text{F}]$$

■ **EXAMPLE 6-2** Laminar tube flow

Water at 60°C enters a tube of 1-in (2.54-cm) diameter at a mean flow velocity of 2 cm/s. Calculate the exit water temperature if the tube is 3.0 m long and the wall temperature is constant at 80°C.

Solution

We first evaluate the Reynolds number at the inlet bulk temperature to determine the flow regime. The properties of water at 60°C are

$$\rho = 985\ \text{kg/m}^3 \qquad c_p = 4.18\ \text{kJ/kg} \cdot °\text{C}$$

$$\mu = 4.71 \times 10^{-4}\ \text{kg/m} \cdot \text{s} \qquad [1.139\ \text{lb}_m/\text{h} \cdot \text{ft}]$$

$$k = 0.651\ \text{W/m} \cdot °\text{C} \qquad \text{Pr} = 3.02$$

$$\text{Re}_d = \frac{\rho u_m d}{\mu} = \frac{(985)(0.02)(0.0254)}{4.71 \times 10^{-4}} = 1062$$

so the flow is laminar. Calculating the additional parameter, we have

$$\text{Re}_d \, \text{Pr} \frac{d}{L} = \frac{(1062)(3.02)(0.0254)}{3} = 27.15 > 10$$

so Eq. (6-10) is applicable. We do not yet know the mean bulk temperature to evaluate properties so we first make the calculation on the basis of 60°C, determine an exit bulk temperature, and then make a second iteration to obtain a more precise value. When inlet and outlet conditions are designated with the subscripts 1 and 2, respectively, the energy balance becomes

$$q = h\pi\, dL \left(T_w - \frac{T_{b_1} + T_{b_2}}{2} \right) = \dot{m} c_p (T_{b_2} - T_{b_1}) \tag{a}$$

At the wall temperature of 80°C we have

$$\mu_w = 3.55 \times 10^{-4}\ \text{kg/m} \cdot \text{s}$$

From Eq. (6-10)

$$\text{Nu}_d = (1.86) \left[\frac{(1062)(3.02)(0.0254)}{3} \right]^{1/3} \left(\frac{4.71}{3.55} \right)^{0.14} = 5.816$$

$$h = \frac{k\,\text{Nu}_d}{d} = \frac{(0.651)(5.816)}{0.0254} = 149.1\ \text{W/m}^2 \cdot °\text{C} \qquad [26.26\ \text{Btu/h} \cdot \text{ft}^2 \cdot °\text{F}]$$

The mass flow rate is

$$\dot{m} = \rho \frac{\pi d^2}{4} u_m = \frac{(985)\pi(0.0254)^2(0.02)}{4} = 9.982 \times 10^{-3}\ \text{kg/s}$$

Inserting the value for h into Eq. (a) along with $\dot{m}$ and $T_{b_1} = 60°\text{C}$ and $T_w = 80°\text{C}$ gives

$$(149.1)\pi(0.0254)(3.0)\left(80 - \frac{T_{b2} + 60}{2}\right) = (9.982 \times 10^{-3})(4180)(T_{b2} - 60) \qquad (b)$$

This equation can be solved to give

$$T_{b2} = 71.98°\text{C}$$

Thus, we should go back and evaluate properties at

$$T_{b,\text{mean}} = \frac{71.98 + 60}{2} = 66°\text{C}$$

We obtain

$$\rho = 982 \text{ kg/m}^3 \qquad c_p = 4185 \text{ J/kg} \cdot °\text{C} \qquad \mu = 4.36 \times 10^{-4} \text{ kg/m} \cdot \text{s}$$

$$k = 0.656 \text{ W/m} \cdot °\text{C} \qquad \text{Pr} = 2.78$$

$$\text{Re}_d = \frac{(1062)(4.71)}{4.36} = 1147$$

$$\text{Re Pr}\,\frac{d}{L} = \frac{(1147)(2.78)(0.0254)}{3} = 27.00$$

$$\text{Nu}_d = (1.86)(27.00)^{1/3}\left(\frac{4.36}{3.55}\right)^{0.14} = 5.743$$

$$h = \frac{(0.656)(5.743)}{0.0254} = 148.3 \text{ W/m}^3 \cdot °\text{C}$$

We insert this value of h back into Eq. (a) to obtain

$$T_{b2} = 71.88°\text{C} \qquad [161.4°\text{F}]$$

The iteration makes very little difference in this problem. If a large bulk-temperature difference had been encountered, the change in properties could have had a larger effect.

■ **EXAMPLE 6-3**

Air at 1 atm and 27°C enters a 5.0-mm-diameter smooth tube with a velocity of 3.0 m/s. The length of the tube is 10 cm. A constant heat flux is imposed on the tube wall. Calculate the heat transfer if the exit bulk temperature is 77°C. Also calculate the exit wall temperature and the value of h at exit.

Solution

We first must evaluate the flow regime and do so by taking properties at the average bulk temperature

$$\overline{T_b} = \frac{27 + 77}{2} = 52°\text{C} = 325 \text{ K}$$

$$\nu = 18.22 \times 10^{-6} \text{ m}^2\text{/s} \qquad \text{Pr} = 0.703 \qquad k = 0.02814 \text{ W/m} \cdot °\text{C}$$

$$\text{Re}_d = \frac{ud}{\nu} = \frac{(3)(0.005)}{18.22 \times 10^{-6}} = 823 \qquad (a)$$

so that the flow is laminar. The tube length is rather short, so we expect a thermal entrance effect and shall consult Fig. 6-5. The inverse Graetz number is computed as

$$\mathrm{Gz}^{-1} = \frac{1}{\mathrm{Re}_d \, \mathrm{Pr}} \frac{x}{d} = \frac{0.1}{(823)(0.703)(0.005)} = 0.0346$$

Therefore, for q_w = constant, we obtain the Nusselt number at exit from Fig. 6-5 as

$$\mathrm{Nu} = \frac{hd}{k} = 4.7 = \frac{q_w d}{(T_w - T_b)k} \tag{b}$$

The total heat transfer is obtained in terms of the overall energy balance:

$$q = \dot{m} c_p (T_{b2} - T_{b1})$$

At entrance $\rho = 1.1774 \text{ kg/m}^3$, so the mass flow is

$$\dot{m} = (1.1774)\pi(0.0025)^2(3.0) = 6.94 \times 10^{-5} \text{ kg/s}$$

and

$$q = (6.94 \times 10^{-5})(1006)(77 - 27) = 3.49 \text{ W}$$

Thus we may find the heat transfer without actually determining wall temperatures or values of h. However, to determine T_w we must compute q_w for insertion in Eq. (b). We have

$$q = q_w \pi \, dL = 3.49 \text{ W}$$

and

$$q_w = 2222 \text{ W/m}^2$$

Now, from Eq. (b)

$$(T_w - T_b)_{x=L} = \frac{(2222)(0.005)}{(4.7)(0.02814)} = 84°\text{C}$$

The wall temperature at exit is thus

$$T_w]_{x=L} = 84 + 77 = 161°\text{C}$$

and the heat-transfer coefficient is

$$h_{x=L} = \frac{q_w}{(T_w - T_b)_{x=L}} = \frac{2222}{84} = 26.45 \text{ W/m}^2 \cdot °\text{C}$$

EXAMPLE 6-4

Repeat Example 6-3 for the case of constant wall temperature.

Solution

We evaluate properties as before and now enter Fig. 6-5 to determine $\overline{\mathrm{Nu}}_d$ for T_w = constant. For $\mathrm{Gz}^{-1} = 0.0346$ we read

$$\mathrm{Nu}_d = 5.15$$

We thus calculate the average heat-transfer coefficient as

$$\bar{h} = (5.15)\left(\frac{k}{d}\right) = \frac{(5.15)(0.02814)}{0.005} = 29.98 \text{ W/m}^2 \cdot °\text{C}$$

We base the heat transfer on mean bulk temperature of 52°C, so that

$$q = \bar{h}\pi\, dL\, (T_w - \overline{T_b}) = 3.49 \text{ W}$$

and

$$T_w = 76.67 + 52 = 128.67°\text{C}$$

■ **EXAMPLE 6-5** Heat transfer in a rough tube

A 2.0-cm-diameter tube having a relative roughness of 0.001 is maintained at constant wall temperature of 90°C. Water enters the tube at 40°C and leaves at 60°C. If the entering velocity is 3 m/s, calculate the length of tube necessary to accomplish the heating.

Solution

We first calculate the heat transfer from

$$q = \dot{m}c_p\, \Delta T_b = (989)(3.0)\pi(0.01)^2(4174)(60 - 40) = 77{,}812 \text{ W}$$

For the rough-tube condition, we may employ the Petukhov relation, Eq. (6-7). The mean film temperature is

$$T_f = \frac{90 + 50}{2} = 70°\text{C}$$

and the fluid properties are

$$\rho = 978 \text{ kg/m}^3 \qquad \mu = 4.0 \times 10^{-4} \text{ kg/m} \cdot \text{s}$$

$$k = 0.664 \text{ W/m} \cdot °\text{C} \qquad \text{Pr} = 2.54$$

Also,

$$\mu_b = 5.55 \times 10^{-4} \text{ kg/m} \cdot \text{s}$$

$$\mu_w = 2.81 \times 10^{-4} \text{ kg/m} \cdot \text{s}$$

The Reynolds number is thus

$$\text{Re}_d = \frac{(978)(3)(0.02)}{4 \times 10^{-4}} = 146{,}700$$

Consulting Fig. 6-4, we find the friction factor as

$$f = 0.0218 \qquad f/8 = 0.002725$$

Because $T_w > T_b$, we take $n = 0.11$ and obtain

$$\text{Nu}_d = \frac{(0.002725)(146{,}700)(2.54)}{1.07 + (12.7)(0.002725)^{1/2}(2.54^{2/3} - 1)}\left(\frac{5.55}{2.81}\right)^{0.11}$$

$$= 666.8$$

$$h = \frac{(666.8)(0.664)}{0.02} = 22138 \text{ W/m}^2 \cdot °\text{C}$$

The tube length is then obtained from the energy balance

$$q = \bar{h}\pi\, dL(T_w - \overline{T_b}) = 77{,}812 \text{ W}$$

$$L = 1.40 \text{ m}$$

6-3 FLOW ACROSS CYLINDERS AND SPHERES

While the engineer may frequently be interested in the heat-transfer characteristics of flow systems inside tubes or over flat plates, equal importance must be placed on the heat transfer which may be achieved by a cylinder in cross flow, as shown in Fig. 6-7. As would be expected, the boundary-layer development on the cylinder determines the heat-transfer characteristics. As long as the boundary layer remains laminar and well behaved, it is possible to compute the heat transfer by a method similar to the boundary-layer analysis of Chap. 5. It is necessary, however, to include the pressure gradient in the analysis because this influences the boundary-layer velocity profile to an appreciable extent. In fact, it is this pressure gradient which causes a separated-flow region to develop on the back side of the cylinder when the free-stream velocity is sufficiently large.

The phenomenon of boundary-layer separation is indicated in Fig. 6-8. The physical reasoning which explains the phenomenon in a qualitative way is as follows. Consistent with boundary-layer theory, the pressure through the boundary layer is essentially constant at any x position on the body. In the case of the cylinder, one might measure x distance from the front stagnation point of the cylinder. Thus the pressure in the boundary layer should follow that of the free stream for potential flow around a cylinder, provided this behavior would not contradict some basic principle which must apply in the boundary layer. As the flow progresses along the front side of the cylinder, the pressure would decrease and then increase along the back side of the cylinder, resulting in an increase in free-stream velocity on the front side of the cylinder and a decrease on the back side. The transverse velocity (that velocity parallel to the surface) would decrease from a value of u_∞ at the outer edge of the boundary layer to zero at the surface. As the flow proceeds to the back side of the cylinder, the pressure increase causes a reduction in velocity in the free stream and throughout the boundary layer. The pressure increase and reduction in velocity are related through the Bernoulli equation written along a streamline:

$$\frac{dp}{\rho} = -d\left(\frac{u^2}{2g_c}\right)$$

Since the pressure is assumed constant throughout the boundary layer, we note that reverse flow may begin in the boundary layer near the surface; i.e., the momentum of the fluid layers near the surface is not sufficiently high to over-

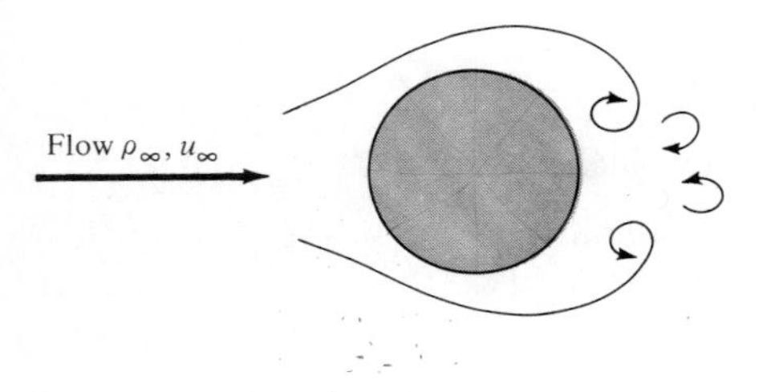

Fig. 6-7 Cylinder in cross flow.

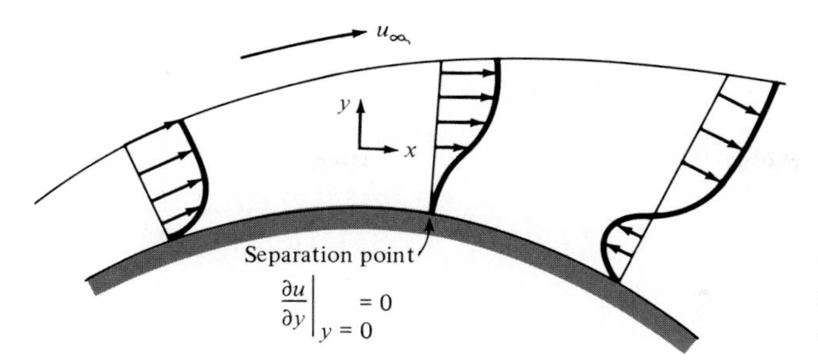

Fig. 6-8 Velocity distributions indicating flow separation on a cylinder in cross flow.

come the increase in pressure. When the velocity gradient at the surface becomes zero, the flow is said to have reached a separation point:

$$\text{Separation point at } \left.\frac{\partial u}{\partial y}\right]_{y=0} = 0$$

This separation point is indicated in Fig. 6-8. As the flow proceeds past the separation point, reverse-flow phenomena may occur, as also shown in Fig. 6-8. Eventually, the separated-flow region on the back side of the cylinder becomes turbulent and random in motion.

The drag coefficient for bluff bodies is defined by

$$\text{Drag force} = F_D = C_D A \frac{\rho u_\infty^2}{2g_c} \tag{6-16}$$

where C_D is the drag coefficient and A is the *frontal area* of the body exposed to the flow, which, for a cylinder, is the product of diameter and length. The values of the drag coefficient for cylinders and spheres are given as a function of the Reynolds number in Figs. 6-9 and 6-10.

The drag force on the cylinder is a result of a combination of frictional resistance and so-called form, or pressure drag, resulting from a low-pressure region on the rear of the cylinder created by the flow-separation process. At low Reynolds numbers of the order of unity, there is no flow separation, and all the drag results from viscous friction. At Reynolds numbers of the order of 10, the friction and form drag are of the same order, while the form drag resulting from the turbulent separated-flow region predominates at Reynolds numbers greater than 1000. At Reynolds numbers of approximately 10^5, based on diameter, the boundary-layer flow may become turbulent, resulting in a steeper velocity profile and extremely late flow separation. Consequently, the form drag is reduced, and this is represented by the break in the drag-coefficient curve at about Re $= 3 \times 10^5$. The same reasoning applies to the sphere as to the circular cylinder. Similar behavior is observed with other bluff bodies, such as elliptic cylinders and airfoils.

The flow processes discussed above obviously influence the heat transfer from a heated cylinder to a fluid stream. The detailed behavior of the heat transfer from a heated cylinder to air has been investigated by Giedt [7], and

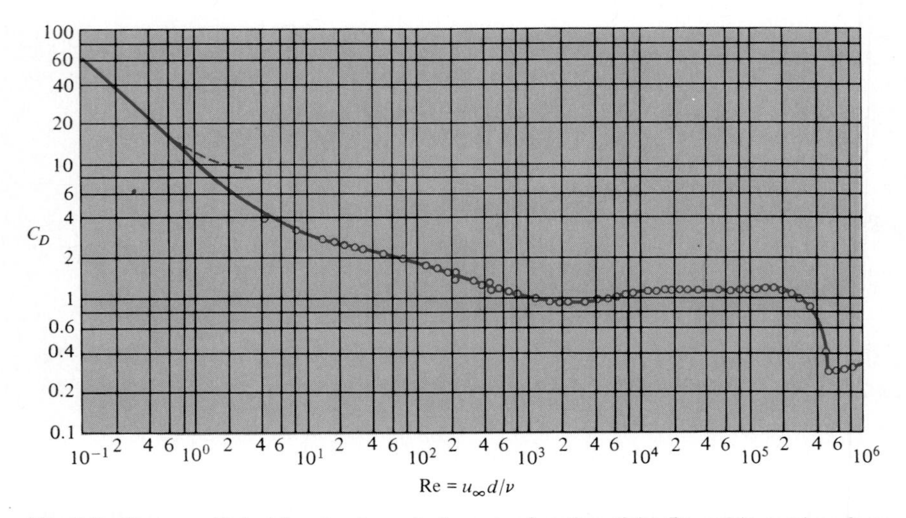

Fig. 6-9 Drag coefficient for circular cylinders as a function of the Reynolds number, from Ref. 6.

the results are summarized in Fig. 6-11. At the lower Reynolds numbers (70,800 and 101,300) a minimum point in the heat-transfer coefficient occurs at approximately the point of separation. There is a subsequent increase in the heat-transfer coefficient on the rear side of the cylinder, resulting from the turbulent eddy motion in the separated flow. At the higher Reynolds numbers two minimum points are observed. The first occurs at the point of transition from laminar to turbulent boundary layer, and the second minimum occurs when the turbulent boundary layer separates. There is a rapid increase in heat transfer

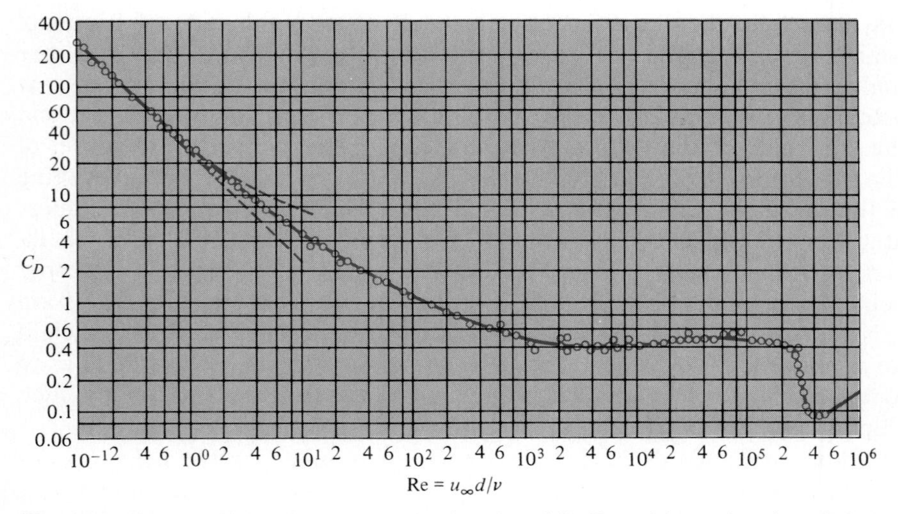

Fig. 6-10 Drag coefficient for spheres as a function of the Reynolds number, from Ref. 6.

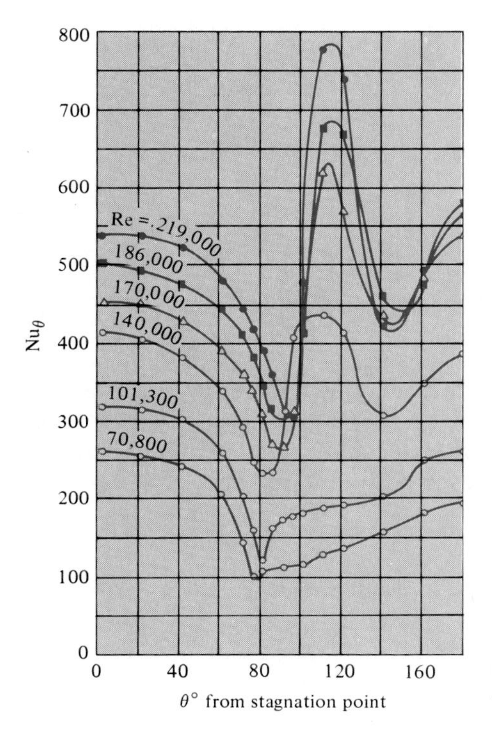

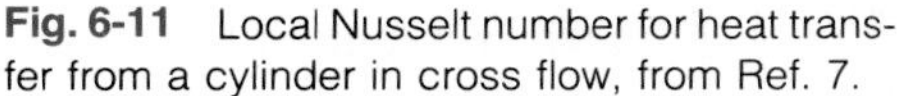
Fig. 6-11 Local Nusselt number for heat transfer from a cylinder in cross flow, from Ref. 7.

when the boundary layer becomes turbulent and another when the increased eddy motion at separation is encountered.

Because of the complicated nature of the flow-separation processes, it is not possible to calculate analytically the average heat-transfer coefficients in cross flow; however, correlations of the experimental data of Hilpert [8] for gases and Knudsen and Katz [9] for liquids indicate that the average heat-transfer coefficients may be calculated with

$$\frac{hd}{k_f} = C\left(\frac{u_\infty d}{\nu_f}\right)^n \mathrm{Pr}_f^{1/3} \tag{6-17}$$

where the constants C and n are tabulated in Table 6-2. The heat-transfer data for air are plotted in Fig. 6-12. Properties for use with Eq. (6-17) are evaluated at the film temperature as indicated by the subscript f.

Figure 6-13 shows the temperature field around heated cylinders placed in a transverse airstream. The dark lines are lines of constant temperature, made visible through the use of an interferometer. Note the separated-flow region which develops on the back side of the cylinder at the higher Reynolds numbers and the turbulent field which is present in that region.

We may note that the original correlation for gases omitted the Prandtl number term in Eq. (6-17) with little error because most diatomic gases have

Table 6-2 Constants for Use with Eq. (6-17), Based on Refs. 8 and 9.

Re_{df}	C	n
0.4–4	0.989	0.330
4–40	0.911	0.385
40–4000	0.683	0.466
4000–40,000	0.193	0.618
40,000–400,000	0.0266	0.805

$\mathrm{Pr} \sim 0.7$. The introduction of the $\mathrm{Pr}^{1/3}$ factor follows from the previous reasoning in Chap. 5.

Fand [21] has shown that the heat-transfer coefficients from liquids to cylinders in cross flow may be better represented by the relation

$$\mathrm{Nu}_f = (0.35 + 0.56\,\mathrm{Re}_f^{0.52})\,\mathrm{Pr}_f^{0.3} \tag{6-18}$$

This relation is valid for $10^{-1} < \mathrm{Re}_f < 10^5$ provided excessive free-stream turbulence is not encountered.

In some instances, particularly those involving calculations on a computer, it may be more convenient to utilize a more complicated expression than Eq. (6-17) if it can be applied over a wider range of Reynolds numbers. Eckert and Drake [34] recommend the following relations for heat transfer from tubes in cross flow, based on the extensive study of Refs. 33 and 39:

$$\mathrm{Nu} = (0.43 + 0.50\,\mathrm{Re}^{0.5})\,\mathrm{Pr}^{0.38}\left(\frac{\mathrm{Pr}_f}{\mathrm{Pr}_w}\right)^{0.25} \quad \text{for } 1 < \mathrm{Re} < 10^3 \tag{6-19}$$

$$\mathrm{Nu} = 0.25\,\mathrm{Re}^{0.6}\,\mathrm{Pr}^{0.38}\left(\frac{\mathrm{Pr}_f}{\mathrm{Pr}_w}\right)^{0.25} \quad \text{for } 10^3 < \mathrm{Re} < 2 \times 10^5 \tag{6-20}$$

For gases the Prandtl number ratio may be dropped, and fluid properties are evaluated at the film temperature. For liquids the ratio is retained, and fluid properties are evaluated at the free-stream temperature. Equations (6-19) and (6-20) are in agreement with results obtained using Eq. (6-17) within 5 to 10 percent.

Still a more comprehensive relation is given by Churchill and Bernstein [37] which is applicable over the complete range of available data:

$$\mathrm{Nu}_d = 0.3 + \frac{0.62\,\mathrm{Re}^{1/2}\,\mathrm{Pr}^{1/3}}{\left[1 + \left(\dfrac{0.4}{\mathrm{Pr}}\right)^{2/3}\right]^{1/4}}\left[1 + \left(\frac{\mathrm{Re}}{282{,}000}\right)^{5/8}\right]^{4/5}$$

$$\text{for } 10^2 < \mathrm{Re}_d < 10^7;\ \mathrm{Pe}_d > 0.2 \tag{6-21}$$

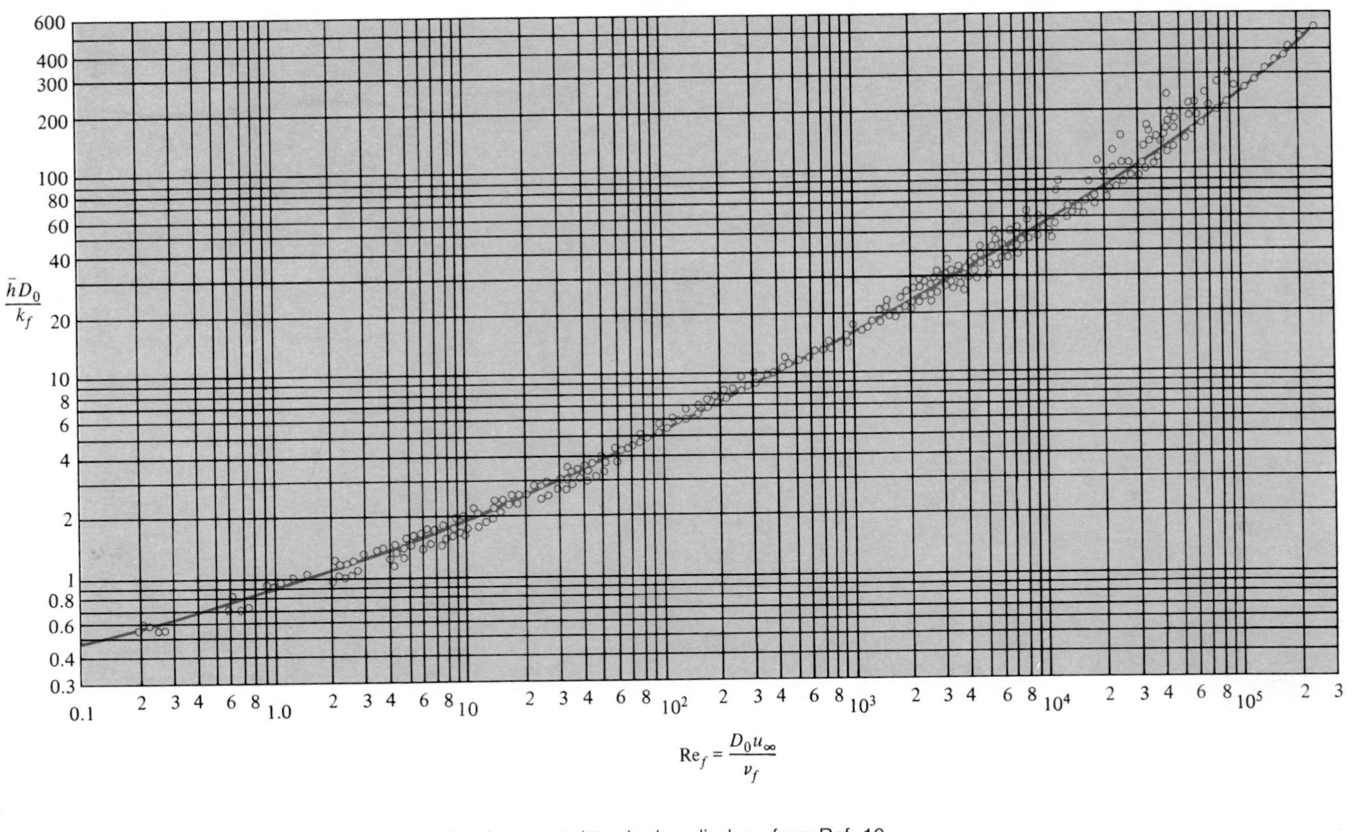

Fig. 6-12 Data for heating and cooling of air flowing normal to single cylinders, from Ref. 10.

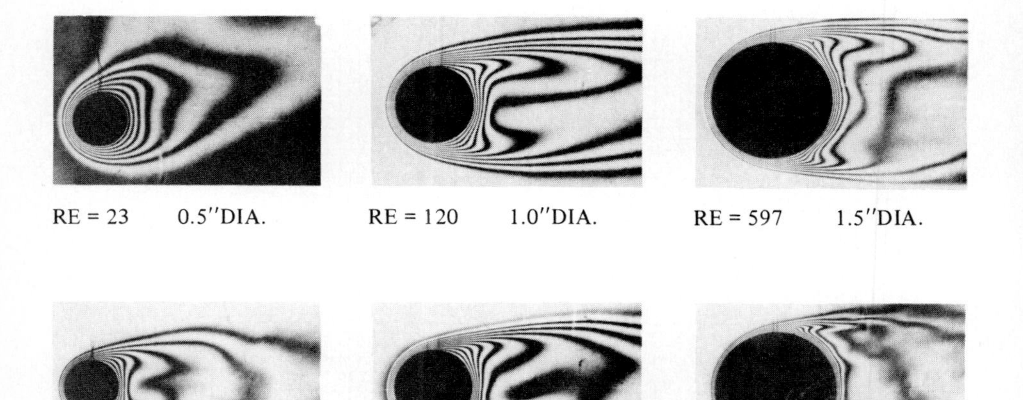

Fig. 6-13 Interferometer photograph showing isotherms around heated horizontal cylinders placed in a transverse airstream. (*Photograph courtesy E. Soehngen.*)

This relation underpredicts the data somewhat in the midrange of Reynolds numbers between 20,000 and 400,000, and it is suggested that the following be employed for this range:

$$\mathrm{Nu}_d = 0.3 + \frac{0.62\,\mathrm{Re}_d^{1/2}\,\mathrm{Pr}^{1/3}}{\left[1 + \left(\frac{0.4}{\mathrm{Pr}}\right)^{2/3}\right]^{1/4}}\left[1 + \left(\frac{\mathrm{Re}_d}{282{,}000}\right)^{1/2}\right]$$

$$\text{for } 20{,}000 < \mathrm{Re}_d < 400{,}000;\ \mathrm{Pe}_d > 0.2 \qquad (6\text{-}22)$$

The heat-transfer data which were used to arrive at Eqs. (6-21) and (6-22) include fluids of air, water, and liquid sodium. Still another correlation equation is given by Whitaker [35] as

$$\mathrm{Nu} = \frac{\bar{h}d}{k} = (0.4\,\mathrm{Re}^{0.5} + 0.06\,\mathrm{Re}^{2/3})\,\mathrm{Pr}^{0.4}\left(\frac{\mu_\infty}{\mu_w}\right)^{0.25} \qquad (6\text{-}23)$$

for $40 < \mathrm{Re} < 10^5$, $0.65 < \mathrm{Pr} < 300$, and $0.25 < \mu_\infty/\mu_w < 5.2$. All properties are evaluated at the free-stream temperature except that μ_w is at the wall temperature.

Below $\mathrm{Pe}_d = 0.2$, Nakai and Okazaki [38] present the following relation:

$$\mathrm{Nu}_d = [0.8237 - \ln(\mathrm{Pe}_d^{1/2})]^{-1} \quad \text{for } \mathrm{Pe}_d < 0.2 \qquad (6\text{-}24)$$

Properties in Eqs. (6-21), (6-22), and (6-24) are evaluated at the film temperature.

Choice of Equation for Cross Flow Over Cylinders

The choice of equation to use for cross flow over cylinders is subject to some conjecture. Clearly, Eq. (6-17) is easiest to use from a computational standpoint, and Eq. (6-21) is the most comprehensive. The more comprehensive relations are preferable for computer setups because of the wide range of fluids and Reynolds numbers covered. For example, Eq. (6-21) has been successful in correlating data for fluids ranging from air to liquid sodium. Equation (6-17) could not be used for liquid metals. If one were making calculations for air, either relation would be satisfactory.

Noncircular Cylinders

Jakob [22] has summarized the results of experiments with heat transfer from noncircular cylinders. Equation (6-17) is employed in order to obtain an empirical correlation for gases, and the constants for use with this equation are summarized in Table 6-3.

Spheres

McAdams [10] recommends the following relation for heat transfer from spheres to a flowing gas:

$$\frac{hd}{k_f} = 0.37\left(\frac{u_\infty d}{\nu_f}\right)^{0.6} \qquad \text{for } 17 < \text{Re}_d < 70{,}000 \tag{6-25}$$

Achenbach [43] has obtained relations applicable over a still wider range of Reynolds numbers for air with $\text{Pr} = 0.71$:

$$\text{Nu} = 2 + (0.25 + 3 \times 10^{-4}\,\text{Re}^{1.6})^{1/2} \qquad \text{for } 100 < \text{Re} < 3 \times 10^5 \tag{6-26}$$

$$\text{Nu} = 430 + a\text{Re} + b\text{Re}^2 + c\text{Re}^3 \qquad \text{for } 3 \times 10^5 < \text{Re} < 5 \times 10^6 \tag{6-27}$$

with $a = 5 \times 10^{-3}$ $\quad b = 0.25 \times 10^{-9}$ $\quad c = -3.1 \times 10^{-17}$

For flow of liquids past spheres, the data of Kramers [11] may be used to obtain the correlation

$$\frac{hd}{k_f}\text{Pr}_f^{-0.3} = 0.97 + 0.68\left(\frac{u_\infty d}{\nu_f}\right)^{0.5} \qquad \text{for } 1 < \text{Re}_d < 2000 \tag{6-28}$$

Vliet and Leppert [19] recommend the following expression for heat transfer from spheres to oil and water over a more extended range of Reynolds numbers from 1 to 200,000:

$$\text{Nu}\,\text{Pr}^{-0.3}\left(\frac{\mu_w}{\mu}\right)^{0.25} = 1.2 + 0.53\,\text{Re}_d^{0.54} \tag{6-29}$$

Table 6-3 Constants for Heat Transfer from Noncircular Cylinders According to Ref. 22, for Use with Eq. (6-17).

Geometry	Re_{df}	C	n
u_∞ → square (diagonal to flow), d	$5 \times 10^3 - 10^5$	0.246	0.588
u_∞ → square, d	$5 \times 10^3 - 10^5$	0.102	0.675
u_∞ → hexagon, d	$5 \times 10^3 - 1.95 \times 10^4$	0.160	0.638
	$1.95 \times 10^4 - 10^5$	0.0385	0.782
u_∞ → hexagon, d	$5 \times 10^3 - 10^5$	0.153	0.638
u_∞ → vertical plate, d	$4 \times 10^3 - 1.5 \times 10^4$	0.228	0.731

where all properties are evaluated at free-stream conditions, except μ_w, which is evaluated at the surface temperature of the sphere. Equation (6-26) represents the data of Ref. 11, as well as the more recent data of Ref. 19.

All the above data have been brought together by Whitaker [35] to develop a single equation for gases and liquids flowing past spheres:

$$\text{Nu} = 2 + (0.4\,\text{Re}_d^{1/2} + 0.06\,\text{Re}_d^{2/3})\,\text{Pr}^{0.4}\,(\mu_\infty/\mu_w)^{1/4} \qquad (6\text{-}30)$$

which is valid for the range $3.5 < \text{Re}_d < 8 \times 10^4$ and $0.7 < \text{Pr} < 380$. Properties in Eq. (6-30) are evaluated at the free-stream temperature.

■ EXAMPLE 6-6

Air at 1 atm and 35°C flows across a 5.0-cm-diameter cylinder at a velocity of 50 m/s. The cylinder surface is maintained at a temperature of 150°C. Calculate the heat loss per unit length of the cylinder.

Solution

We first determine the Reynolds number and then find the applicable constants from Table 6-2 for use with Eq. (6-17). The properties of air are evaluated at the film temperature:

$$T_f = \frac{T_w + T_\infty}{2} = \frac{150 + 35}{2} = 92.5°\text{C} = 365.5 \text{ K}$$

$$\rho_f = \frac{p}{RT} = \frac{1.0132 \times 10^5}{(287)(365.5)} = 0.966 \text{ kg/m}^3 \quad [0.0603 \text{ lb}_m/\text{ft}^3]$$

$$\mu_f = 2.14 \times 10^{-5} \text{ kg/m} \cdot \text{s} \quad [0.0486 \text{ lb}_m/\text{h} \cdot \text{ft}]$$

$$k_f = 0.0312 \text{ W/m} \cdot °\text{C} \quad [0.018 \text{ Btu/h} \cdot \text{ft} \cdot °\text{F}]$$

$$\text{Pr}_f = 0.695$$

$$\text{Re}_f = \frac{\rho u_\infty d}{\mu} = \frac{(0.966)(50)(0.05)}{2.14 \times 10^{-5}} = 1.129 \times 10^5$$

From Table 6-2

$$C = 0.0266 \qquad n = 0.805$$

so from Eq. (6-17)

$$\frac{hd}{k_f} = (0.0266)(1.129 \times 10^5)^{0.805}(0.695)^{1/3} = 275.1$$

$$h = \frac{(275.1)(0.0312)}{0.05} = 171.7 \text{ W/m}^2 \cdot °\text{C} \quad [30.2 \text{ Btu/h} \cdot \text{ft}^2 \cdot °\text{F}]$$

The heat transfer per unit length is therefore

$$\frac{q}{L} = h\,\pi d\,(T_w - T_\infty)$$

$$= (171.7)\pi(0.05)(150 - 35)$$

$$= 3100 \text{ W/m} \quad [3226 \text{ Btu/ft}]$$

■ **EXAMPLE 6-7**

A fine wire having a diameter of 3.94×10^{-5} m is placed in a 1-atm airstream at 25°C having a flow velocity of 50 m/s perpendicular to the wire. An electric current is passed through the wire, raising its surface temperature to 50°C. Calculate the heat loss per unit length.

Solution

We first obtain the properties at the film temperature:

$$T_f = (25 + 50)/2 = 37.5°\text{C} = 310 \text{ K}$$

$$\nu_f = 16.7 \times 10^{-6} \text{ m}^2/\text{s} \qquad k = 0.02704 \text{ W/m} \cdot °\text{C}$$

$$\text{Pr}_f = 0.706$$

The Reynolds number is

$$\text{Re}_d = \frac{u_\infty d}{\nu_f} = \frac{(50)(3.94 \times 10^{-5})}{16.7 \times 10^{-6}} = 118$$

The Peclet number is Pe = Re Pr = 83.3, and we find that Eqs. (6-17), (6-21), or (6-19) apply. Let us make the calculation with both the simplest expression, (6-17), and the most complex, (6-21), and compare results.

Using Eq. (6-17) with $C = 0.683$ and $n = 0.466$, we have

$$\mathrm{Nu}_d = (0.683)(118)^{0.466}(0.705)^{1/3} = 5.615$$

and the value of the heat-transfer coefficient is

$$h = \mathrm{Nu}_d \left(\frac{k}{d}\right) = 5.615 \frac{0.02704}{3.94 \times 10^{-5}} = 3854 \text{ W/m}^2 \cdot {}^\circ\mathrm{C}$$

The heat transfer per unit length is then

$$q/L = \pi d h(T_w - T_\infty) = \pi(3.94 \times 10^{-5})(3854)(50 - 25)$$
$$= 11.93 \text{ W/m}$$

Using Eq. (6-21), we calculate the Nusselt number as

$$\mathrm{Nu}_d = 0.3 + \frac{(0.62)(118)^{1/2}(0.705)^{1/3}}{[1 + (0.4/0.705)^{2/3}]^{1/4}} [1 + (118/282{,}000)^{5/8}]^{4/5}$$
$$= 5.593$$

and

$$h = \frac{(5.593)(0.02704)}{3.94 \times 10^{-5}} = 3838 \text{ W/m}^2 \cdot {}^\circ\mathrm{C}$$

and $$q/L = (3838)\pi(3.94 \times 10^{-5})(50 - 25) = 11.88 \text{ W/m}$$

Here, we find the two correlations differing by 0.4 percent if the value from Eq. (6-21) is taken as correct, or 0.2 percent from the mean value. Examining Fig. 6-12, we see that data scatter of ±15 percent is not unusual.

■ **EXAMPLE 6-8**

Air at 1 atm and 27°C blows across a 12-mm-diameter sphere at a free-stream velocity of 4 m/s. A small heater inside the sphere maintains the surface temperature at 77°C. Calculate the heat lost by the sphere.

Solution

Consulting Eq. (6-30) we find that the Reynolds number is evaluated at the free-stream temperature. We therefore need the following properties: at $T_\infty = 27°\mathrm{C} = 300$ K,

$$\nu = 15.69 \times 10^{-6} \text{ m}^2/\text{s} \qquad k = 0.02624 \text{ W/m}^2 \cdot {}^\circ\mathrm{C},$$
$$\mathrm{Pr} = 0.708 \qquad \mu_\infty = 1.8462 \times 10^{-5} \text{ kg/m} \cdot \text{s}$$

At $T_w = 77°\mathrm{C} = 350$ K,

$$\mu_w = 2.075 \times 10^{-5}$$

The Reynolds number is thus

$$\mathrm{Re}_d = \frac{(4)(0.012)}{15.69 \times 10^{-6}} = 3059$$

From Eq. (6-30),

$$\overline{\text{Nu}} = 2 + [(0.4)(3059)^{1/2} + (0.06)(3059)^{2/3}](0.708)^{0.4}\left(\frac{1.8462}{2.075}\right)^{1/4}$$

$$= 31.40$$

and

$$\bar{h} = \overline{\text{Nu}}\left(\frac{k}{d}\right) = \frac{(31.4)(0.02624)}{0.012} = 68.66 \text{ W/m}^2 \cdot {}^\circ\text{C}$$

The heat transfer is then

$$q = \bar{h}A(T_w - T_\infty) = (68.66)(4\pi)(0.006)^2(77 - 27) = 1.553 \text{ W}$$

For comparison purposes let us also calculate the heat-transfer coefficient using Eq. (6-25). The film temperature is $T_f = (350 + 300)/2 = 325$ K so that

$$\nu_f = 18.23 \times 10^{-6} \text{ m}^2/\text{s} \qquad k_f = 0.02814 \text{ W/m} \cdot {}^\circ\text{C}$$

and the Reynolds number is

$$\text{Re}_d = \frac{(4)(0.012)}{18.23 \times 10^{-6}} = 2633$$

From Eq. (6-25)

$$\text{Nu}_f = (0.37)(2633)^{0.6} = 41.73$$

and $\bar{h}$ is calculated as

$$\bar{h} = \text{Nu}\left(\frac{k_f}{d}\right) = \frac{(41.73)(0.02814)}{0.012} = 97.9 \text{ W/m}^2 \cdot {}^\circ\text{C}$$

or about 42 percent higher than the value calculated before.

6-4 FLOW ACROSS TUBE BANKS

Because many heat-exchanger arrangements involve multiple rows of tubes, the heat-transfer characteristics for tube banks are of important practical interest. The heat-transfer characteristics of staggered and in-line tube banks were studied by Grimson [12], and on the basis of a correlation of the results of various investigators, he was able to represent data in the form of Eq. (6-17). The values of the constant C and the exponent n are given in Table 6-4 in terms of the geometric parameters used to describe the tube-bundle arrangement. The Reynolds number is based on the maximum velocity occurring in the tube bank, i.e., the velocity through the minimum-flow area. This area will depend on the geometric tube arrangement. The nomenclature for use with Table 6-4 is shown in Fig. 6-14. The data of Table 6-4 pertain to tube banks having 10 or more rows of tubes in the direction of flow. For fewer rows the ratio of h for N rows deep to that for 10 rows is given in Table 6-5.

Table 6-4 Correlation of Grimson for Heat Transfer in Tube Banks of 10 Rows or More, From Ref. 12, for Use with Eq. (6-17).

	$\frac{S_n}{d}$							
$\frac{S_p}{d}$	1.25		1.5		2.0		3.0	
	C	n	C	n	C	n	C	n
In line								
1.25	0.386	0.592	0.305	0.608	0.111	0.704	0.0703	0.752
1.5	0.407	0.586	0.278	0.620	0.112	0.702	0.0753	0.744
2.0	0.464	0.570	0.332	0.602	0.254	0.632	0.220	0.648
3.0	0.322	0.601	0.396	0.584	0.415	0.581	0.317	0.608
Staggered								
0.6	—	—	—	—	—	—	0.236	0.636
0.9	—	—	—	—	0.495	0.571	0.445	0.581
1.0	—	—	0.552	0.558	—	—	—	—
1.125	—	—	—	—	0.531	0.565	0.575	0.560
1.25	0.575	0.556	0.561	0.554	0.576	0.556	0.579	0.562
1.5	0.501	0.568	0.511	0.562	0.502	0.568	0.542	0.568
2.0	0.448	0.572	0.462	0.568	0.535	0.556	0.498	0.570
3.0	0.344	0.592	0.395	0.580	0.488	0.562	0.467	0.574

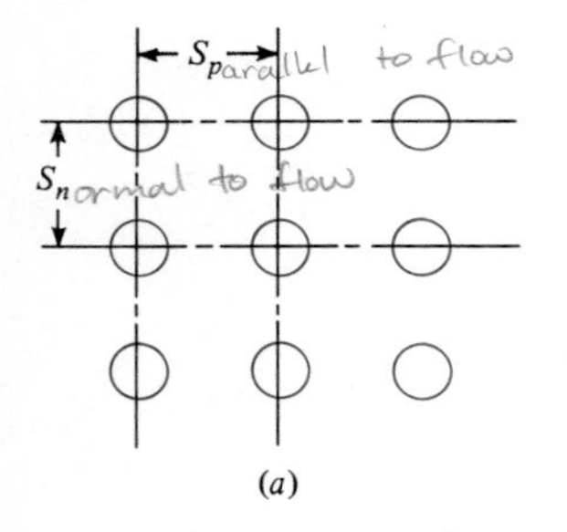

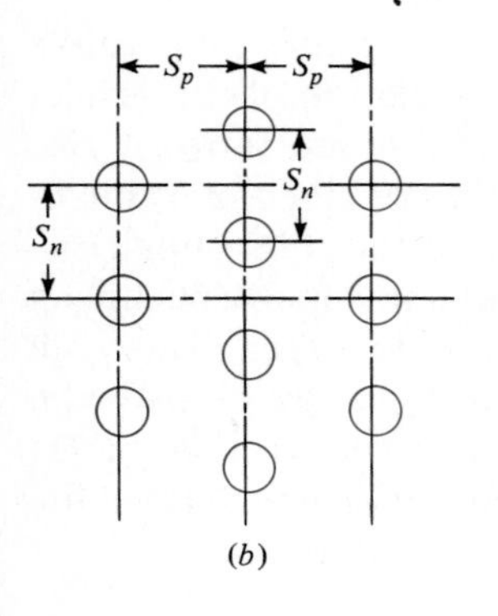

Fig. 6-14 Nomenclature for use with Table 6-4: (*a*) in-line tube rows; (*b*) staggered tube rows.

Table 6-5 Ratio of h for N Rows Deep to that for 10 Rows Deep. From Ref. 17.

N	1	2	3	4	5	6	7	8	9	10
Ratio for staggered tubes	0.68	0.75	0.83	0.89	0.92	0.95	0.97	0.98	0.99	1.0
Ratio for in-line tubes	0.64	0.80	0.87	0.90	0.92	0.94	0.96	0.98	0.99	1.0

Pressure drop for flow of gases over a bank of tubes may be calculated with Eq. (6-31), expressed in pascals:

$$\Delta p = \frac{2f' G_{max}^2 N}{\rho} \left(\frac{\mu_w}{\mu_b}\right)^{0.14} \tag{6-31}$$

where G_{max} = mass velocity at minimum flow area, kg/m$^2 \cdot$ s
ρ = density evaluated at free-stream conditions, kg/m^3
N = number of transverse rows
μ_b = average free-stream viscosity

The empirical friction factor f' is given by Jakob [18] as

$$f' = \left\{0.25 + \frac{0.118}{[(S_n - d)/d]^{1.08}}\right\} \text{Re}_{max}^{-0.16} \tag{6-32}$$

for staggered tube arrangements, and

$$f' = \left\{0.044 + \frac{0.08 S_p/d}{[(S_n - d)/d]^{0.43 + 1.13d/S_p}}\right\} \text{Re}_{max}^{-0.15} \tag{6-33}$$

for in-line arrangements.

Zukauskas [39] has presented additional information for tube bundles which takes into account wide ranges of Reynolds numbers and property variations. The correlating equation takes the form

$$\text{Nu} = \frac{\bar{h}d}{k} = C\,\text{Re}_{d,max}^n\,\text{Pr}^{0.36}\left(\frac{\text{Pr}}{\text{Pr}_w}\right)^{1/4} \tag{6-34}$$

where all properties except Pr_w are evaluated at T_∞ and the values of the constants are given in Table 6-6 for greater than 20 rows of tubes. This equation is applicable for $0.7 < \text{Pr} < 500$ and $10 < \text{Re}_{d,max} < 10^6$. For gases the Prandtl number ratio has little influence and is dropped. Once again, note that the Reynolds number is based on the maximum velocity in the tube bundle. For less than 20 rows in the direction of flow the correction factor in Table 6-7 should be applied. It is essentially the same as for the Grimson correlation. Additional information is given by Morgan [44]. Further information on pressure drop is given in Ref. 39.

Table 6-6 Constant for Zukauskas Correlation [Eq. (6-34)] for Heat Transfer in Tube Banks of 20 Rows or More. From Ref. 39.

Geometry	$Re_{d,max}$	C	n
In-line	10–100	0.8	0.4
	$100–10^3$	Treat as individual tubes	
	$10^3 - 2 \times 10^5$	0.27	0.63
	$>2 \times 10^5$	0.21	0.84
Staggered	10–100	0.9	0.4
	$100–10^3$	Treat as individual tubes	
	$10^3 - 2 \times 10^5$	$0.35 \left(\frac{S_n}{S_L}\right)^{0.2}$ for $\frac{S_n}{S_L} < 2$	0.60
	$10^3 - 2 \times 10^5$	0.40 for $\frac{S_n}{S_L} > 2$	0.60
	$>2 \times 10^5$	0.022	0.84

The reader should keep in mind that these relations correlate experimental data with an accuracy of about ±25 percent.

EXAMPLE 6-9

Air at 1 atm and 10°C flows across a bank of tubes 15 rows high and 5 rows deep at a velocity of 7 m/s measured at a point in the flow before the air enters the tube bank. The surfaces of the tubes are maintained at 65°C. The diameter of the tubes is 1 in [2.54 cm]; they are arranged in an in-line manner so that the spacing in both the normal and parallel directions to the flow is 1.5 in [3.81 cm]. Calculate the total heat transfer per unit length for the tube bank and the exit air temperature.

Solution

The constants for use with Eq. (6-17) may be obtained from Table 6-4, using

$$\frac{S_p}{d} = \frac{3.81}{2.54} = 1.5 \qquad \frac{S_n}{d} = \frac{3.81}{2.54} = 1.5$$

so that

$$C = 0.278 \qquad n = 0.620$$

Table 6-7 Ratio of h for N Rows Deep to that for 20 Rows Deep According to Ref. 39 and for Use with Eq. (6-34).

N	2	3	4	5	6	8	10	16	20
Staggered	0.77	0.84	0.89	0.92	0.94	0.97	0.98	0.99	1.0
In-line	0.70	0.80	0.90	0.92	0.94	0.97	0.98	0.99	1.0

The properties of air are evaluated at the film temperature, which at entrance to the tube bank is

$$T_{f1} = \frac{T_w + T_\infty}{2} = \frac{65 + 10}{2} = 37.5°\text{C} = 310.5 \text{ K} \quad [558.9°\text{R}]$$

Then
$$\rho_f = \frac{p}{RT} = \frac{1.0132 \times 10^5}{(287)(310.5)} = 1.137 \text{ kg/m}^3$$

$$\mu_f = 1.894 \times 10^{-5} \text{ kg/m} \cdot \text{s}$$

$$k_f = 0.027 \text{ W/m} \cdot °\text{C} \quad [0.0156 \text{ Btu/h} \cdot \text{ft} \cdot °\text{F}]$$

$$c_p = 1007 \text{ J/kg} \cdot °\text{C} \quad [0.24 \text{ Btu/lb}_m \cdot °\text{F}]$$

$$\text{Pr} = 0.706$$

To calculate the maximum velocity, we must determine the minimum flow area. From Fig. 6-14 we find that the ratio of the minimum flow area to the total frontal area is $(S_n - d)/S_n$. The maximum velocity is thus

$$u_{\max} = u_\infty \frac{S_n}{S_n - d} = \frac{(7)(3.81)}{3.81 - 2.54} = 21 \text{ m/s} \quad [68.9 \text{ ft/s}] \tag{a}$$

where u_∞ is the incoming velocity before entrance to the tube bank. The Reynolds number is computed by using the maximum velocity.

$$\text{Re} = \frac{\rho u_{\max} d}{\mu} = \frac{(1.137)(21)(0.0254)}{1.894 \times 10^{-5}} = 32{,}020 \tag{b}$$

The heat-transfer coefficient is then calculated with Eq. (6-17):

$$\frac{hd}{k_f} = (0.278)(32{,}020)^{0.62}(0.706)^{1/3} = 153.8 \tag{c}$$

$$h = \frac{(153.8)(0.027)}{0.0254} = 164 \text{ W/m}^2 \cdot °\text{C} \quad [28.8 \text{ Btu/h} \cdot \text{ft}^2 \cdot °\text{F}] \tag{d}$$

This is the heat-transfer coefficient which would be obtained if there were 10 rows of tubes in the direction of the flow. Because there are only 5 rows, this value must be multiplied by the factor 0.92, as determined from Table 6-5.

The total surface area for heat transfer, considering unit length of tubes, is

$$A = N\,\pi d(1) = (15)(5)\pi(0.0254) = 5.985 \text{ m}^2/\text{m}$$

where N is the total number of tubes.

Before calculating the heat transfer, we must recognize that the air temperature increases as the air flows through the tube bank. Therefore, this must be taken into account when using

$$q = hA(T_w - T_\infty) \tag{e}$$

As a good approximation, we can use an arithmetic average value of T_∞ and write for the energy balance

$$q = hA\left(T_w - \frac{T_{\infty,1} + T_{\infty,2}}{2}\right) = \dot{m}c_p(T_{\infty,2} - T_{\infty,1}) \tag{f}$$

where now the subscripts 1 and 2 designate entrance and exit to the tube bank. The mass flow at entrance to the 15 tubes is

$$\dot{m} = \rho_\infty u_\infty (15) S_n$$

$$\rho_\infty = \frac{p}{RT_\infty} = \frac{1.0132 \times 10^5}{(287)(283)} = 1.246 \text{ kg/m}^3 \qquad (g)$$

$$\dot{m} = (1.246)(7)(15)(0.0381) = 4.99 \text{ kg/s} \quad [11.0 \text{ lb}_m\text{/s}]$$

so that Eq. (f) becomes

$$(0.92)(164)(5.985)\left(65 - \frac{10 + T_{\infty,2}}{2}\right) = (4.99)(1006)(T_{\infty,2} - 10)$$

which may be solved to give

$$T_{\infty,2} = 19.08°\text{C}$$

The heat transfer is then obtained from the right side of Eq. (f):

$$q = (4.99)(1005)(19.08 - 10) = 45.6 \text{ kW/m}$$

This answer could be improved somewhat by recalculating the air properties based on a mean value of T_∞, but the improvement would be small and well within the accuracy of the empirical heat-transfer correlation of Eq. (6-17).

■ **EXAMPLE 6-10**

Compare the heat-transfer coefficient calculated with Eq. (6-34) with the value obtained in Example 6-9.

Solution

Properties for use in Eq. (6-34) are evaluated at free-stream conditions of 10°C, so we have

$$\nu = 14.86 \times 10^{-6} \qquad \text{Pr} = 0.71 \qquad k = 0.0249$$

The Reynolds number is

$$\text{Re}_{d,\max} = \frac{(21)(0.0254)}{14.86 \times 10^{-6}} = 35{,}895$$

so that the constants for Eq. (6-34) are $C = 0.27$ and $n = 0.63$.
Inserting values, we obtain

$$\frac{\bar{h}d}{k} = (0.27)(35{,}895)^{0.63}(0.71)^{0.36} = 176.8$$

and

$$h = \frac{(176.8)(0.0249)}{0.254} = 173.3 \text{ W/m}^2 \cdot °\text{C}$$

or a value about 9 percent higher than in Example 6-9. Both values are within the accuracies of the correlations.

6-5 LIQUID-METAL HEAT TRANSFER

In recent years concentrated interest has been placed on liquid-metal heat transfer because of the high heat-transfer rates which may be achieved with these media. These high heat-transfer rates result from the high thermal conductivities of liquid metals as compared with other fluids; as a consequence, they are particularly applicable to situations where large energy quantities must be removed from a relatively small space, as in a nuclear reactor. In addition, the liquid metals remain in the liquid state at higher temperatures than conventional fluids like water and various organic coolants. This also makes more compact heat-exchanger design possible. Liquid metals are difficult to handle because of their corrosive nature and the violent action which may result when they come into contact with water or air; even so, their advantages in certain heat-transfer applications have overshadowed their shortcomings, and suitable techniques for handling them have been developed.

Let us first consider the simple flat plate with a liquid metal flowing across it. The Prandtl number for liquid metals is very low, of the order of 0.01, so that the thermal-boundary-layer thickness should be substantially larger than the hydrodynamic-boundary-layer-thickness. The situation results from the high values of thermal conductivity for liquid metals and is depicted in Fig. 6-15. Since the ratio of δ/δ_t is small, the velocity profile has a very blunt shape over most of the thermal boundary layer. As a first approximation, then, we might assume a slug-flow model for calculation of the heat transfer; i.e., we take

$$u = u_\infty \tag{6-35}$$

throughout the thermal boundary layer for purposes of computing the energy-transport term in the integral energy equation (Sec. 5-6):

$$\frac{d}{dx}\left[\int_0^{\delta_t} (T_\infty - T)u\,dy\right] = \alpha \left.\frac{dT}{dy}\right]_w \tag{6-36}$$

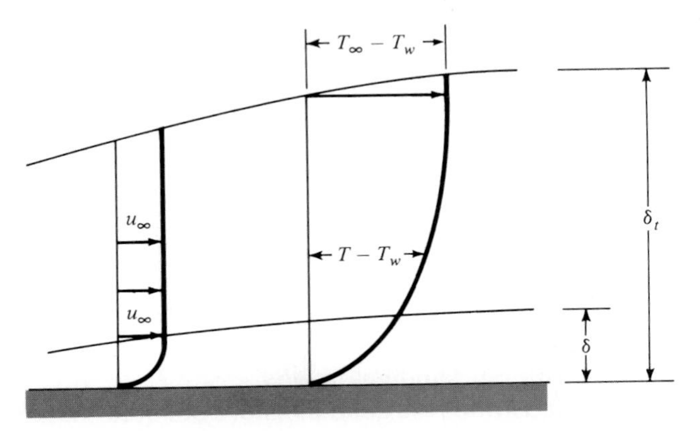

Fig. 6-15 Boundary-layer regimes for analysis of liquid-metal heat transfer.

The conditions on the temperature profile are the same as those in Sec. 5-6, so that we use the cubic parabola as before:

$$\frac{\theta}{\theta_\infty} = \frac{T - T_w}{T_\infty - T_w} = \frac{3}{2}\frac{y}{\delta_t} - \frac{1}{2}\left(\frac{y}{\delta_t}\right)^3 \tag{6-37}$$

Inserting Eqs. (6-31) and (6-33) in (6-32) gives

$$\theta_\infty u_\infty \frac{d}{dx}\left\{\int_0^{\delta_t}\left[1 - \frac{3}{2}\frac{y}{\delta_t} + \frac{1}{2}\left(\frac{u}{\delta_t}\right)^3\right] dy\right\} = \frac{3\alpha\theta_\infty}{2\delta_t} \tag{6-38}$$

which may be integrated to give

$$2\delta_t\, d\delta_t = \frac{8\alpha}{u_\infty}\, dx \tag{6-39}$$

The solution to this differential equation is

$$\delta_t = \sqrt{\frac{8\alpha x}{u_\infty}} \tag{6-40}$$

for a plate heated over its entire length.

The heat-transfer coefficient may be expressed by

$$h_x = \frac{-k(\partial T/\partial y)_w}{T_w - T_\infty} = \frac{3k}{2\delta_t} = \frac{3\sqrt{2}}{8}\, k\sqrt{\frac{u_\infty}{\alpha x}} \tag{6-41}$$

This relationship may be put in dimensionless form as

$$\text{Nu}_x = \frac{h_x x}{k} = 0.530\,(\text{Re}_x\,\text{Pr})^{1/2} = 0.530\,\text{Pe}^{1/2} \tag{6-42}$$

Using Eq. (5-21) for the hydrodynamic-boundary-layer thickness,

$$\frac{\delta}{x} = \frac{4.64}{\text{Re}_x^{1/2}} \tag{6-43}$$

we may compute the ratio δ/δ_t:

$$\frac{\delta}{\delta_t} = \frac{4.64}{\sqrt{8}}\sqrt{\text{Pr}} = 1.64\sqrt{\text{Pr}} \tag{6-44}$$

Using Pr $\sim$ 0.01, we obtain

$$\frac{\delta}{\delta_t} \sim 0.16$$

which is in reasonable agreement with our slug-flow model.

The flow model discussed above illustrates the general nature of liquid-metal heat transfer, and it is important to note that the heat transfer is dependent on the Peclet number. Empirical correlations are usually expressed in terms of this parameter, four of which we present below.

Extensive data on liquid metals are given in Ref. 13, and the heat-transfer characteristics are summarized in Ref. 23. Lubarsky and Kaufman [14] recommended the following relation for calculation of heat-transfer coefficients in fully developed turbulent flow of liquid metals in smooth tubes with uniform heat flux at the wall:

$$\mathrm{Nu}_d = \frac{hd}{k} = 0.625\,(\mathrm{Re}_d\,\mathrm{Pr})^{0.4} \tag{6-45}$$

All properties for use in Eq. (6-45) are evaluated at the bulk temperature. Equation (6-45) is valid for $10^2 < \mathrm{Pe} < 10^4$ and for $L/d > 60$. Seban and Shimazaki [16] propose the following relation for calculation of heat transfer to liquid metals in tubes with constant wall temperature:

$$\mathrm{Nu}_d = 5.0 + 0.025\,(\mathrm{Re}_d\,\mathrm{Pr})^{0.8} \tag{6-46}$$

where all properties are evaluated at the bulk temperature. Equation (6-42) is valid for $\mathrm{Pe} > 10^2$ and $L/d > 60$.

More recent data by Skupinshi, Tortel, and Vautrey [26] with sodium-potassium mixtures indicate that the following relation may be preferable to that of Eq. (6-45) for constant-heat-flux conditions:

$$\mathrm{Nu} = 4.82 + 0.0185\,\mathrm{Pe}^{0.827} \tag{6-47}$$

This relation is valid for $3.6 \times 10^3 < \mathrm{Re} < 9.05 \times 10^5$ and $10^2 < \mathrm{Pe} < 10^4$.

Witte [32] has measured the heat transfer from a sphere to liquid sodium during forced convection, with the data being correlated by

$$\mathrm{Nu} = 2 + 0.386\,(\mathrm{Re}\,\mathrm{Pr})^{0.5} \tag{6-48}$$

for the Reynolds number range $3.56 \times 10^4 < \mathrm{Re} < 1.525 \times 10^5$.

Kalish and Dwyer [41] have presented information on liquid-metal heat transfer in tube bundles.

In general, there are many open questions concerning liquid-metal heat transfer, and the reader is referred to Refs. 13 and 23 for more information.

■ EXAMPLE 6-11

Liquid bismuth flows at a rate of 4.5 kg/s through a 5.0-cm-diameter stainless-steel tube. The bismuth enters at 415°C and is heated to 440°C as it passes through the tube. If a constant heat flux is maintained along the tube and the tube wall is at a temperature 20°C higher than the bismuth bulk temperature, calculate the length of tube required to effect the heat transfer.

Solution

Because a constant heat flux is maintained, we may use Eq. (6-47) to calculate the heat-transfer coefficient. The properties of bismuth are evaluated at the average bulk tem-

Table 6-8 Summary of Forced-Convection Relations (See text for property evaluation)

Geometry	*Equation*	*Restrictions*	*Equation number*
Tube flow	$\mathrm{Nu}_d = 0.023\ \mathrm{Re}_d^{0.8}\ \mathrm{Pr}^n$	Fully developed turbulent flow $n = 0.4$ for heating $n = 0.3$ for cooling $0.6 < \mathrm{Pr} < 100$	(6-4)
Tube flow	$\mathrm{Nu}_d = 0.027\ \mathrm{Re}_d^{0.8}\ \mathrm{Pr}^{1/3} \left(\frac{\mu}{\mu_w}\right)^{0.14}$	Fully developed turbulent flow	(6-5)
Tube flow, entrance region	$\mathrm{Nu}_d = 0.036\ \mathrm{Re}_d^{0.8}\ \mathrm{Pr}^{1/3} \left(\frac{d}{L}\right)^{0.055}$ See also Figs. 6-5 and 6-6	Turbulent flow, $10 < \frac{L}{d} < 400$	(6-6)
Tube flow	Petukov relation	Fully developed turbulent flow, $0.5 < \mathrm{Pr} < 2000$, $10^4 < \mathrm{Re}_d < 5 \times 10^6$, $0 < \frac{\mu_b}{\mu_w} < 40$	(6-7)
Tube flow	$\mathrm{Nu}_d = 3.66 + \frac{0.0668(d/L)\ \mathrm{Re}_d\ \mathrm{Pr}}{1 + 0.04[(d/L)\ \mathrm{Re}_d\ \mathrm{Pr}]^{2/3}}$	Laminar	(6-9)
Tube flow	$\mathrm{Nu}_d = 1.86\ (\mathrm{Re}_d\ \mathrm{Pr})^{1/3} \left(\frac{d}{L}\right)^{1/3} \left(\frac{\mu}{\mu_w}\right)^{0.14}$	Fully developed laminar flow, $\mathrm{Re}_d\ \mathrm{Pr}\ \frac{d}{L} > 10$	(6-10)
Rough tubes	$\mathrm{St}_b\ \mathrm{Pr}_f^{2/3} = \frac{f}{8}$ or Eq. (6-7)	Fully developed turbulent flow	(6-12)

Noncircular ducts	Reynolds number evaluated on basis of hydraulic diameter $D_H = \frac{4A}{P}$	Same as particular equation for tube flow	(6-14)
Flow across cylinders	$\text{Nu}_f = C\,\text{Re}_{df}^n\,\text{Pr}^{1/3}$ C and n from Table 6-2	$0.4 < \text{Re}_{df} < 400{,}000$	(6-17)
Flow across cylinders	$\text{Nu}_{df} = 0.3 + \frac{0.62\,\text{Re}_f^{1/2}\,\text{Pr}^{1/3}}{\left[1 + \left(\frac{0.4}{\text{Pr}}\right)^{2/3}\right]^{1/4}}\left[1 + \left(\frac{\text{Re}_f}{282{,}000}\right)^{5/8}\right]^{4/5}$	$10^2 < \text{Re}_f < 10^7$, $\text{Pe} > 0.2$	(6-21)
Flow across cylinders		See text	(6-18) to (6-20) (6-22) to (6-24)
Flow across noncircular cylinders	See Table 6-3		
Flow across spheres		See text	(6-25) to (6-30)
Flow across tube banks	$\text{Nu}_f = C\,\text{Re}_{f,\max}^n\,\text{Pr}_f^{1/3}$ C and n from Table 6-4	See text	(6-17)
Flow across tube banks	$\text{Nu} = C\,\text{Re}_{d,\max}^n\,\text{Pr}^{0.36}\left(\frac{\text{Pr}}{\text{Pr}_w}\right)^{1/4}$	$0.7 < \text{Pr} < 500$ $10 < \text{Re}_{d,\max} < 10^6$	(6-34)
Liquid metals		See text	(6-37) to (6-48)

perature of (415 + 440)/2 = 427.5°C

$$\mu = 1.34 \times 10^{-3}\ \text{kg/m} \cdot \text{s} \quad [3.242\ \text{lb}_m/\text{h} \cdot \text{ft}]$$
$$c_p = 0.149\ \text{kJ/kg} \cdot {}^\circ\text{C} \quad [0.0356\ \text{Btu/lb}_m \cdot {}^\circ\text{F}]$$
$$k = 15.6\ \text{W/m} \cdot {}^\circ\text{C} \quad [9.014\ \text{Btu/h} \cdot \text{ft} \cdot {}^\circ\text{F}]$$
$$\text{Pr} = 0.013$$

The total transfer is calculated from

$$q = \dot{m}c_p\,\Delta T_b = (4.5)(149)(440 - 415) = 16.76\ \text{kW} \quad [57{,}186\ \text{Btu/h}] \tag{a}$$

We calculate the Reynolds and Peclet numbers as

$$\text{Re}_d = \frac{dG}{\mu} = \frac{(0.05)(4.5)}{[\pi(0.05)^2/4](1.34 \times 10^{-3})} = 85{,}520 \tag{b}$$
$$\text{Pe} = \text{Re Pr} = (85{,}520)(0.013) = 1111$$

The heat-transfer coefficient is then calculated from Eq. 6-47

$$\text{Nu}_d = 4.82 + (0.0185)(1111)^{0.827} = 10.93 \tag{c}$$
$$h = \frac{(10.93)(15.6)}{0.05} = 3410\ \text{W/m}^2 \cdot {}^\circ\text{C} \quad [600\ \text{Btu/h} \cdot \text{ft}^2 \cdot {}^\circ\text{F}]$$

The total required surface area of the tube may now be computed from

$$q = hA(T_w - T_b) \tag{d}$$

where we use the temperature difference of 20°C;

$$A = \frac{16{,}760}{(3410)(20)} = 0.246\ \text{m}^2 \quad [2.65\ \text{ft}^2]$$

This area in turn can be expressed in terms of the tube length

$$A = \pi dL \quad \text{and} \quad L = \frac{0.246}{\pi(0.05)} = 1.57\ \text{m} \quad [5.15\ \text{ft}]$$

6-6 SUMMARY REMARKS

In contrast to Chap. 5, which was mainly analytical in character, this chapter has dealt almost entirely with empirical correlations which may be used to calculate convection heat transfer. The general calculation procedure is as follows:

1. Establish the geometry of the situation.
2. Make a preliminary determination of appropriate fluid properties.
3. Establish the flow regime by calculating the Reynolds or Peclet number.

4. Select an equation which fits the geometry and flow regime and reevaluate properties, if necessary, in accordance with stipulations and the equation.
5. Proceed to calculate the value of h and/or the heat-transfer rate.

We should note that the data upon which the empirical equations are based are most often taken under laboratory conditions where it is possible to exert careful control over temperature and flow variables. In a practical application such careful control may not be present and there may be deviations from heat-transfer rates calculated from the equations given here. Our purpose is not to discourage the reader by this remark, but rather to indicate that sometimes it will be quite satisfactory to use a simple correlation over a more elaborate expression even if the simple relation has a larger scatter in its data representation. Our purpose has been to present a variety of expressions (where available) so that some choices can be made.

Finally, the most important relations of this chapter are listed in Table 6-8 for quick reference purposes.

REVIEW QUESTIONS

1 What is the Dittus-Boelter equation? When does it apply?

2 How may heat-transfer coefficients be calculated for flow in rough pipes?

3 What is the hydraulic diameter? When is it used?

4 What is the form of equation used to calculate heat transfer for flow over cylinders and bluff bodies?

5 Why does a slug-flow model yield reasonable results when applied to liquid-metal heat transfer?

6 What is the Peclet number?

7 What is the Graetz number?

PROBLEMS

6-1 Engine oil enters a 5.0-mm-diameter tube at 120°C. The tube wall is maintained at 50°C, and the inlet Reynolds number is 1000. Calculate the heat transfer, average heat-transfer coefficient, and exit oil temperature for tube lengths of 10, 20, and 50 cm.

6-2 Water at an average bulk temperature of 80°F flows inside a horizontal smooth tube with wall temperature maintained at 180°F. The tube length is 6.0 ft, and diameter is 0.125 in. The flow velocity is 0.125 ft/s. Calculate the heat-transfer rate.

6-3 Liquid ammonia flows in a duct having a cross section of an equilateral triangle 1.0 cm on a side. The average bulk temperature is 20°C, and the duct wall temperature is 50°C. Fully developed laminar flow is experienced with a Reynolds number of 1000. Calculate the heat transfer per unit length of duct.

6-4 Water flows in a duct having a cross section 5 × 10 mm with a mean bulk temperature of 20°C. If the duct wall temperature is constant at 60°C and fully developed laminar flow is experienced, calculate the heat transfer per unit length.

6-5 Water at the rate of 3 kg/s is heated from 5 to 15°C by passing it through a 5-cm-ID copper tube. The tube wall temperature is maintained at 90°C. What is the length of the tube?

6-6 Water at the rate of 0.8 kg/s is heated from 35 to 40°C in a 2.5-cm-diameter tube whose surface is at 90°C. How long must the tube be to accomplish this heating?

6-7 Water flows through a 2.5-cm-ID pipe 1.5 m long at a rate of 1.0 kg/s. The pressure drop is 7 kPa through the 1.5-m length. The pipe wall temperature is maintained at a constant temperature of 50°C by a condensing vapor, and the inlet water temperature is 20°C. Estimate the exit water temperature.

6-8 Water at the rate of 1 kg/s is forced through a tube with a 2.5-cm ID. The inlet water temperature is 15°C, and the outlet water temperature is 50°C. The tube wall temperature is 14°C higher than the water temperature all along the length of the tube. What is the length of the tube?

6-9 Engine oil enters a 1.25-cm-diameter tube 3 m long at a temperature of 38°C. The tube wall temperature is maintained at 65°C, and the flow velocity is 30 cm/s. Estimate the total heat transfer to the oil and the exit temperature of the oil.

6-10 Air at 1 atm and 15°C flows through a long rectangular duct 7.5 by 15 cm. A 1.8-m section of the duct is maintained at 120°C, and the average air temperature at exit from this section is 65°C. Calculate the air flow rate and the total heat transfer.

6-11 A heat exchanger is constructed so that hot flue gases at 700 K flow inside a 2.5-cm-ID copper tube with 1.6-mm wall thickness. A 5.0-cm tube is placed around the 2.5-cm-diameter tube, and high-pressure water at 150°C flows in the annular space between the tubes. If the flow rate of water is 1.5 kg/s and the total heat transfer is 17.5 kW, estimate the length of the heat exchanger for a gas mass flow of 0.8 kg/s. Assume that the properties of the flue gas are the same as those of air at atmospheric pressure and 700 K.

6-12 Water at the rate of 0.5 kg/s is forced through a smooth 2.5-cm-ID tube 15 m long. The inlet water temperature is 10°C, and the tube wall temperature is 15°C higher than the water temperature all along the length of the tube. What is the exit water temperature?

6-13 Water at an average temperature of 300 K flows at 0.7 kg/s in a 2.5-cm-diameter tube 6 m long. The pressure drop is measured as 2 kPa. A constant heat flux is imposed, and the average wall temperature is 55°C. Estimate the exit temperature of the water.

6-14 An oil with Pr = 1960, ρ = 860 kg/m^3, $\nu = 1.6 \times 10^{-4}$ m^2/s, and k = 0.14 W/m · °C enters a 2.5-mm-diameter tube 60 cm long. The oil entrance temperature is 20°C, the mean flow velocity is 30 cm/s, and the tube wall temperature is 120°C. Calculate the heat-transfer rate.

6-15 Liquid ammonia flows through a 2.5-cm-diameter smooth tube 2.5 m long at a rate of 1 lb_m/s. The ammonia enters at 10°C and leaves at 38°C, and a constant heat flux is imposed on the tube wall. Calculate the average wall temperature necessary to effect the indicated heat transfer.

6-16 Liquid Freon 12 (CCl_2F_2) flows inside a 1.25-cm-diameter tube at a velocity of 3 m/s. Calculate the heat-transfer coefficient for a bulk temperature of 10°C. How does this compare with water at the same conditions?

6-17 Water at an average temperature of 10°C flows in a 2.5-cm-diameter tube 6 m long at a rate of 0.4 kg/s. The pressure drop is measured as 3 kPa. A constant heat flux is imposed, and the average wall temperature is 50°C. Estimate the exit temperature of the water.

6-18 Water at the rate of 0.5 kg/s is to be cooled from 71 to 32°C. Which would result in less pressure drop—to run the water through a 12.5-mm-diameter pipe at a constant temperature of 4°C or through a constant-temperature 25-mm-diameter pipe at 20°C?

6-19 Air at 1400 kPa enters a duct 7.5 cm in diameter and 6 m long at a rate of 0.5 kg/s. The duct wall is maintained at an average temperature of 500 K. The average air temperature in the duct is 550 K. Estimate the decrease in temperature of the air as it passes through the duct.

6-20 An annulus consists of the region between two concentric tubes having diameters of 4 cm and 5 cm. Ethylene glycol flows in this space at a velocity of 6.9 m/s. Entrance temperature is 20°C and the exit temperature is 40°C. Only the inner tube is a heating surface and it is maintained constant at 80°C. Calculate the length of annulus necessary to effect the heat transfer.

6-21 An air-conditioning duct has a cross section of 45 by 90 cm. Air flows in the duct at a velocity of 7.5 m/s at conditions of 1 atm and 300 K. Calculate the heat-transfer coefficient for this system and the pressure drop per unit length.

6-22 Water flows in a 3.0-cm-diameter tube having a relative roughness of 0.002 with a constant wall temperature of 80°C. If the water enters at 20°C, estimate the convection coefficient for a Reynolds number of 10^5.

6-23 Liquid Freon 12 (CCl_2F_2) enters a 3.5-mm-diameter tube at 0°C and with a flow rate such that the Reynolds number is 700 at entrance conditions. Calculate the length of tube necessary to raise the fluid temperature to 20°C if the tube wall temperature is constant at 40°C.

6-24 Air enters a small duct having a cross section of an equilateral triangle, 3.0 mm on a side. The entering temperature is 27°C and the exit temperature is 77°C. If the flow rate is 5×10^{-5} kg/s and the tube length is 30 cm, calculate the tube wall temperature necessary to effect the heat transfer. Also calculate the pressure drop. The pressure is 1 atm.

6-25 Air at 90 kPa and 27°C enters a 4.0-mm-diameter tube with a mass flow rate of 7×10^{-5} kg/s. A constant heat flux is imposed at the tube surface so that the tube wall temperature is 70°C above the fluid bulk temperature. Calculate the exit air temperature for a tube length of 12 cm.

6-26 Air at 110 kPa and 40°C enters a 6.0-mm-diameter tube with a mass flow rate of 8×10^{-5} kg/s. The tube wall temperature is maintained constant at 140°C. Calculate the exit air temperature for a tube length of 14 cm.

6-27 Engine oil at 40°C enters a 1-cm-diameter tube at a flow rate such that the Reynolds number at entrance is 50. Calculate the exit oil temperature for a tube length of 8 cm and a constant tube wall temperature of 80°C.

6-28 Water flows in a 2-cm-diameter tube at an average flow velocity of 8 m/s. If the water enters at 20°C and leaves at 30°C and the tube length is 10 m, estimate the average wall temperature necessary to effect the required heat transfer.

6-29 Engine oil at 20°C enters a 2.0-mm-diameter tube at a velocity of 1.2 m/s. The tube wall temperature is constant at 60°C and the tube is 1.0 m long. Calculate the exit oil temperature.

6-30 Water flows inside a smooth tube at a mean flow velocity of 10 ft/s. The tube diameter is 1.0 inch and a constant heat flux condition is maintained at the tube wall such that the tube temperature is always 20°C above the water temperature. The water enters the tube at 30°C and leaves at 50°C. Calculate the tube length necessary to accomplish the indicated heating.

6-31 Water enters a 3-mm-diameter tube at 21°C and leaves at 32°C. The flow rate is such that the Reynolds number is 600. The tube length is 10 cm and is maintained at a constant temperature of 60°C. Calculate the water flow rate.

6-32 Water enters a 3.0-cm-diameter tube at 60°F and leaves at 100°F. The flow rate is 1.0 kg/s and the tube wall temperature is 140°F. Calculate the length of the tube.

6-33 Glycerin flows in a 5-mm-diameter tube at such a rate that the Reynolds number is 10. The glycerine enters at 10°C and leaves at 30°C. The tube wall is maintained constant at 40°C. Calculate the length of the tube.

6-34 A 5-cm-diameter cylinder maintained at 100°C is placed in a nitrogen flow stream at 2 atm pressure and 10°C. The nitrogen flows across the cylinder with a velocity of 5 m/s. Calculate the heat lost by the cylinder per meter of length.

6-35 Air at 1 atm and 0°C blows across a 4-cm-diameter cylinder maintained at a surface temperature of 54°C. The air velocity is 25 m/s. Calculate the heat loss from the cylinder per unit length.

6-36 Air at 200 kPa blows across a 20-cm-diameter cylinder at a velocity of 25 m/s and temperature of 10°C. The cylinder is maintained at a constant temperature of 80°C. Calculate the heat transfer and drag force per unit length.

6-37 Water at 43°C enters a 5-cm-ID pipe having a relative roughness of 0.002 at a rate of 6 kg/s. If the pipe is 9 m long and is maintained at 71°C, calculate the exit water temperature and the total heat transfer.

6-38 A short tube is 6.4 mm in diameter and 15 cm long. Water enters the tube at 1.5 m/s and 38°C, and a constant-heat-flux condition is maintained such that the tube wall temperature remains 28°C above the water bulk temperature. Calculate the heat-transfer rate and exit water temperature.

6-39 Ethylene glycol is to be cooled from 60 to 40°C in a 3.0-cm-diameter tube. The tube wall temperature is maintained constant at 20°C. The glycol enters the tube with a velocity of 10 m/s. Calculate the length of tube necessary to accomplish this cooling.

6-40 Using the values of the local Nusselt number given in Fig. 6-11, obtain values for the average Nusselt number as a function of the Reynolds number. Plot the results as log Nu versus log Re, and obtain an equation which represents all the data. Compare this correlation with that given by Eq. (6-17) and Table 6-2.

6-41 Air at 70 kPa and 20°C flows across a 5-cm-diameter cylinder at a velocity of 20 m/s. Compute the drag force exerted on the cylinder.

6-42 A heated cylinder at 450 K and 2.5 cm in diameter is placed in an atmospheric airstream at 1 atm and 325 K. The air velocity is 30 m/s. Calculate the heat loss per meter of length for the cylinder.

6-43 Assuming that a human can be approximated by a cylinder 1 ft in diameter and 6 ft high with a surface temperature of 75°F, calculate the heat the person would lose while standing in a 30-mi/h wind whose temperature is 30°F.

6-44 Assume that one-half the heat transfer from a cylinder in cross flow occurs on the front half of the cylinder. On this assumption, compare the heat transfer from a cylinder in cross flow with the heat transfer from a flat plate having a length equal to the distance from the stagnation point on the cylinder. Discuss this comparison.

6-45 A 0.13-mm-diameter wire is exposed to an airstream at −30°C and 54 kPa. The flow velocity is 230 m/s. The wire is electrically heated and is 12.5 mm long. Calculate the electric power necessary to maintain the wire surface temperature at 175°C.

6-46 Air at 90°C and 1 atm flows past a heated $\frac{1}{16}$-in-diameter wire at a velocity of 6 m/s. The wire is heated to a temperature of 150°C. Calculate the heat transfer per unit length of wire.

6-47 A fine wire 0.025 mm in diameter and 15 cm long is to be used to sense flow velocity by measuring the electrical heat which can be dissipated from the wire when placed in an airflow stream. The resistivity of the wire is 70 $\mu\Omega \cdot$ cm. The temperature of the wire is determined by measuring its electric resistance relative to some reference temperature T_0 so that

$$R = R_0[1 + a(T - T_0)]$$

For this particular wire the value of the temperature coefficient a is 0.006°C^{-1}. The resistance can be determined from measurements of the current and voltage impressed on the wire, and

$$R = \frac{E}{I}$$

Suppose a measurement is made for air at 20°C with a flow velocity of 10 m/s and the wire temperature is 40°C. What values of voltage and current would be measured for these conditions if R_0 is evaluated at $T_0 = 20°C$? What values of voltage and current would be measured for the same wire temperature but flow velocities of 15 and 20 m/s?

6-48 Helium at 1 atm and 325 K flows across a 3-mm-diameter cylinder which is heated to 425 K. The flow velocity is 9 m/s. Calculate the heat transfer per unit length of wire. How does this compare with the heat transfer for air under the same conditions?

6-49 Calculate the heat-transfer rate per unit length for flow over a 0.025-mm-diameter cylinder maintained at 65°C. Perform the calculation for (*a*) air at 20°C and 1 atm and (*b*) water at 20°C; $u_\infty = 6$ m/s.

6-50 Compare the heat-transfer results of Eqs. (6-17) and (6-18) for water at Reynolds numbers of 10^3, 10^4, and 10^5 and a film temperature of 90°C.

6-51 Compare Eqs. (6-19), (6-20), and (6-21) with Eq. (6-17) for a gas with Pr = 0.7 at the following Reynolds numbers: (*a*) 500, (*b*) 1000, (*c*) 2000, (*d*) 10,000, (*e*) 100,000.

6-52 A pipeline in the Arctic carries hot oil at 50°C. A strong arctic wind blows across the 50-cm-diameter pipe at a velocity of 13 m/s and a temperature of −35°C. Estimate the heat loss per meter of pipe length.

6-53 Two tubes are available, a 4.0-cm-diameter tube and a 4.0-cm-square tube. Air at 1 atm and 27°C is blown across the tubes with a velocity of 20 m/s. Calculate the heat transfer in each case if the tube wall temperature is maintained at 50°C.

6-54 A 3.0-cm-diameter cylinder is subjected to a cross flow of carbon dioxide at 200°C and a pressure of 1 atm. The cylinder is maintained at a constant temperature of 50°C and the carbon dioxide velocity is 40 m/s. Calculate the heat transfer to the cylinder per meter of length.

6-55 Helium at 150 kPa and 20°C is forced at 50 m/s across a horizontal cylinder having a diameter of 30 cm and a length of 6 m. Calculate the heat lost by the cylinder if its surface temperature is maintained constant at 100°C.

6-56 A 0.25-inch-diameter cylinder is maintained at a constant temperature of 300°C and placed in a cross flow of CO_2 at $p = 100$ kPa and $T = 30°C$. Calculate the heat loss for a 4.5-m length of the cylinder if the CO_2 velocity is 35 m/s.

6-57 A 20-cm-diameter cylinder is placed in a cross flow CO_2 stream at 1 atm and 300 K. The cylinder is maintained at a constant temperature of 400 K and the CO_2 velocity is 50 m/s. Calculate the heat lost by the cylinder per meter of length.

6-58 Air flows across a 4-cm-square cylinder at a velocity of 10 m/s. The surface temperature is maintained at 85°C. Free-stream air conditions are 20°C and 0.6 atm. Calculate the heat loss from the cylinder per meter of length.

6-59 Water flows over a 3-mm-diameter sphere at 6 m/s. The free-stream temperature is 38°C, and the sphere is maintained at 93°C. Calculate the heat-transfer rate.

6-60 A spherical water droplet having a diameter of 1.3 mm is allowed to fall from rest in atmospheric air at 1 atm and 20°C. Estimate the velocities the droplet will attain after a drop of 30, 60, and 300 m.

6-61 A spherical tank having a diameter of 4.0 m is maintained at a surface temperature of 40°C. Air at 1 atm and 20°C blows across the tank at 6 m/s. Calculate the heat loss.

6-62 A heated sphere having a diameter of 3 cm is maintained at a constant temperature of 90°C and placed in a water flow stream at 20°C. The water flow velocity is 3.5 m/s. Calculate the heat lost by the sphere.

6-63 A small sphere having a diameter of 6 mm has an electric heating coil inside which maintains the outer surface temperature at 220°C. The sphere is exposed to an airstream at 1 atm and 20°C with a velocity of 20 m/s. Calculate the heating rate which must be supplied to the sphere.

6-64 Air at a pressure of 3 atm blows over a flat plate at a velocity of 100 m/s. The plate is maintained at 200°C and the free-stream temperature is 30°C. Calculate the heat loss for a plate which is 1 m square.

6-65 Air at 3.5 MPa and 38°C flows across a tube bank consisting of 400 tubes of 1.25-cm OD arranged in a staggered manner 20 rows high; $S_p = 3.75$ cm and $S_n = 2.5$ cm. The incoming-flow velocity is 9 m/s, and the tube wall temperatures are maintained constant at 200°C by a condensing vapor on the inside of the tubes. The length of the tubes is 1.5 m. Estimate the exit air temperature as it leaves the tube bank.

6-66 A tube bank uses an in-line arrangement with $S_n = S_p = 1.9$ cm and 6.33-mm-diameter tubes. Six rows of tubes are employed with a stack 50 tubes high. The surface temperature of the tubes is constant at 90°C, and atmospheric air at 20°C is forced across them at an inlet velocity of 4.5 m/s before the flow enters the tube bank. Calculate the total heat transfer per unit length for the tube bank. Estimate the pressure drop for this arrangement.

6-67 Repeat Prob. 6-66 for a staggered-tube arrangement with the same values of S_p and S_n.

6-68 A more compact version of the tube bank in Prob. 6-66 can be achieved by reducing the S_p and S_n dimensions while still retaining the same number of tubes. Investigate the effect of reducing S_p and S_n in half, that is, $S_p = S_n = 0.95$ cm. Calculate the heat transfer and pressure drop for this new arrangement.

6-69 Condensing steam at 150°C is used on the inside of a bank of tubes to heat a cross-flow stream of CO_2 which enters at 3 atm, 35°C, and 5 m/s. The tube bank consists of 100 tubes of 1.25-cm OD in a square in-line array with $S_n = S_p = 1.875$ cm. The tubes are 60 cm long. Assuming the outside tube wall temperature is constant at 150°C, calculate the overall heat transfer to the CO_2 and its exit temperature.

6-70 An in-line tube bank is constructed of 2.5-cm-diameter tubes with 15 rows high and 7 rows deep. The tubes are maintained at 90°C, and atmospheric air is blown across them at 20°C and $u_\infty = 12$ m/s. The arrangement has $S_p = 3.75$ and $S_n = 5.0$ cm. Calculate the heat transfer from the tube bank per meter of length. Also calculate the pressure drop.

6-71 Air at 300 K and 1 atm enters an in-line tube bank consisting of five rows of 10 tubes each. The tube diameter is 2.5 cm and $S_n = S_p = 5.0$ cm. The incoming velocity is 10 m/s and the tube wall temperatures are constant at 350 K. Calculate the exit air temperature.

6-72 Atmospheric air at 20°C flows across a 5-cm-square rod at a velocity of 15 m/s. The velocity is normal to one of the faces of the rod. Calculate the heat transfer per unit length for a surface temperature of 90°C.

6-73 A certain home electric heater uses thin metal strips to dissipate heat. The strips are 6 mm wide and are oriented normal to the airstream, which is produced by a small fan. The air velocity is 2 m/s, and seven 35-cm strips are employed. If the strips are heated to 870°C, estimate the total convection heat transfer to the room air at 20°C. (Note that in such a heater, much of the *total* transfer will be by thermal radiation.)

6-74 A square duct, 30 by 30 cm, is maintained at a constant temperature of 30°C and an airstream of 50°C and 1 atm is forced across it with a velocity of 6 m/s. Calculate the heat gained by the duct. How much would the heat flow be reduced if the flow velocity were reduced in half?

6-75 The drag coefficient for a sphere at Reynolds numbers less than 100 may be approximated by $C_D = b\,\mathrm{Re}^{-1}$, where b is a constant. Assuming that the Colburn analogy between heat transfer and fluid friction applies, derive an expression for the heat loss from a sphere of diameter d and temperature T_s, released from rest and allowed to fall in a fluid of temperature T_∞. (Obtain an expression for the heat lost between the time the sphere is released and the time it reaches some velocity v. Assume that the Reynolds number is less than 100 during this time and that the sphere remains at a constant temperature.)

6-76 Using the slug-flow model, show that the boundary-layer energy equation reduces to the same form as the transient-conduction equation for the semi-infinite solid of Sec. 4-3. Solve this equation and compare the solution with the integral analysis of Sec. 6-5.

6-77 Liquid bismuth enters a 2.5-cm-diameter stainless-steel pipe at 400°C at a rate of 1 kg/s. The tube wall temperature is maintained constant at 450°C. Calculate the bismuth exit temperature if the tube is 60 cm long.

6-78 Liquid sodium is to be heated from 120 to 149°C at a rate of 2.3 kg/s. A 2.5-cm-diameter electrically heated tube is available (constant heat flux). If the tube wall temperature is not to exceed 200°C, calculate the minimum length required.

6-79 Determine an expression for the average Nusselt number for liquid metals flowing over a flat plate. Use Eq. (6-42) as a starting point.

6-80 Water at the rate of 0.8 kg/s at 93°C is forced through a 5-cm-ID copper tube at a suitable velocity. The wall thickness is 0.8 mm. Air at 15°C and atmospheric pressure is forced over the outside of the tube at a velocity of 15 m/s in a direction normal to the axis of the tube. What is the heat loss per meter of length of the tube?

6-81 Air at 1 atm and 350 K enters a 1.25-cm-diameter tube with a flow rate of 35 g/s. The surface temperature of the tube is 300 K, and its length is 12 m. Calculate the heat lost by the air and the exit air temperature.

6-82 Air flows across a 5.0-cm-diameter smooth tube with free-stream conditions of 20°C, 1 atm, and $u_\infty = 25$ m/s. If the tube surface temperature is 77°C, calculate the heat loss per unit length.

6-83 Engine oil enters an 8-m-long tube at 20°C. The tube diameter is 1.0 in, and the flow rate is 0.4 kg/s. Calculate the outlet temperature of the oil if the tube surface temperature is maintained at 80°C.

6-84 Air at 1 atm and 300 K with a flow rate of 0.2 kg/s enters a rectangular 10 by 20 cm duct that is 250 cm long. If the duct surface temperature is maintained constant at 400 K, calculate the heat transfer to the air and the exit air temperature.

6-85 Air at 1 atm and 300 K flows inside a 1.5-mm-diameter smooth tube such that the Reynolds number is 1200. Calculate the heat-transfer coefficients for tube lengths of 1, 10, 20, and 100 cm.

6-86 Water at an average bulk temperature of 10°C flows inside a channel shaped like an equilateral triangle 2.5 cm on a side. The flow rate is such that a Reynolds number of 50,000 is obtained. If the tube-wall temperature is maintained 15°C higher than the water bulk temperature, calculate the length of tube needed to effect a 10°C increase in bulk temperature. What is the total heat transfer under this condition?

6-87 Air at 1 atm and 300 K flows normal to a square noncircular cylinder such that the Reynolds number is 10^4. Compare the heat transfer for this system with that for a circular cylinder having diameter equal to a side of the square. Repeat the calculation for the first, third, and fourth entries of Table 6-3.

6-88 Air at 1 atm and 300 K flows across a sphere such that the Reynolds number is 50,000. Compare Eqs. (6-25) and (6-26) for these conditions. Also compare with Eq. (6-30).

6-89 Water at 10°C flows across a 2.5-cm-diameter sphere at a free-stream velocity of 4 m/s. If the surface temperature of the sphere is 60°C, calculate the heat loss.

6-90 A tube bank consists of a square array of 144 tubes arranged in an in-line position. The tubes have a diameter of 1.5 cm and length of 1.0 m; the center-to-center tube spacing is 2.0 cm. If the surface temperature of the tubes is maintained at 350 K and air enters the tube bank at 1 atm, 300 K, and $u_\infty = 6$ m/s, calculate the total heat lost by the tubes.

6-91 Though it may be classified as a rather simple mistake, a frequent cause for substantial error in convection calculations is failure to select the correct geometry for the problem. Consider the following three geometries for flow of air at 1 atm, 300 K, and a Reynolds number of 50,000: (*a*) flow across a cylinder with diameter of 10 cm, (*b*) flow inside a tube with diameter of 10 cm, and (*c*) flow along a flat plate of length 10 cm. Calculate the average heat-transfer coefficient for each of these geometries and comment on the results.

6-92 It has been noted that convection heat transfer is dependent on fluid properties, which in turn are dependent on temperature. Consider flow of atmospheric air at 0.012 kg/s in a smooth 2.5-cm-diameter tube. Assuming that the Dittus-Boelter relation (6-4) applies, calculate the average heat-transfer coefficient for properties evaluated at 300, 400, 500, and 800 K. Comment on the results.

6-93 Repeat Prob. 6-92 for the same mass flow of atmospheric helium with properties evaluated at 255, 477, and 700 K and comment on the results.

■ REFERENCES

1 Dittus, F. W., and L. M. K. Boelter: *Univ. Calif. (Berkeley) Pub. Eng.*, vol. 2, p. 443, 1930.

2 Sieder, E. N., and C. E. Tate: Heat Transfer and Pressure Drop of Liquids in Tubes, *Ind. Eng. Chem.*, vol. 28, p. 1429, 1936.

3 Nusselt, W.: Der Wärmeaustausch zwischen Wand und Wasser im Rohr, *Forsch. Geb. Ingenieurwes.*, vol. 2, p. 309, 1931.

4 Hausen, H.: Darstellung des Wärmeuberganges in Rohren durch verallgemeinerte Potenzbeziehungen, *VDI Z.*, no. 4, p. 91, 1943.

5 Moody, F. F.: Friction Factors for Pipe Flow, *Trans. ASME*, vol. 66, p. 671, 1944.

6 Schlichting, H.: "Boundary Layer Theory," 7th ed., McGraw-Hill Book Company, New York, 1979.

7 Giedt, W. H.: Investigation of Variation of Point Unit-Heat-Transfer Coefficient around a Cylinder Normal to an Air Stream, *Trans. ASME*, vol. 71, pp. 375–381, 1949.

8 Hilpert, R.: Wärmeabgabe von geheizen Drahten und Rohren, *Forsch. Geb. Ingenieurwes.*, vol. 4, p. 220, 1933.

9 Knudsen, J. D., and D. L. Katz: "Fluid Dynamics and Heat Transfer," McGraw-Hill Book Company, New York, 1958.

10 McAdams, W. H.: "Heat Transmission," 3d ed., McGraw-Hill Book Company, New York, 1954.

11 Kramers, H.: Heat Transfer from Spheres to Flowing Media, *Physica*, vol. 12, p. 61, 1946.

12 Grimson, E. D.: Correlation and Utilization of New Data on Flow Resistance and Heat Transfer for Cross Flow of Gases over Tube Banks, *Trans. ASME*, vol. 59, pp. 583–594, 1937.

13 Lyon, R. D. (ed.): "Liquid Metals Handbook," 3d ed., Atomic Energy Commission and U.S. Navy Department, Washington, D.C., 1952.

14 Lubarsky, B., and S. J. Kaufman: Review of Experimental Investigations of Liquid-Metal Heat Transfer, *NACA Tech. Note* 3336, 1955.

15 Colburn, A. P.: A Method of Correlating Forced Convection Heat Transfer Data and a Comparison with Fluid Friction, *Trans. AIChE*, vol. 29, p.174, 1933.

16 Seban, R. A., and T. T. Shimazaki: Heat Transfer to a Fluid Flowing Turbulently in a Smooth Pipe with Walls at Constant Temperature, *Trans. ASME*, vol. 73, p. 803, 1951.

17 Kays, W. M., and R. K. Lo: Basic Heat Transfer and Flow Friction Data for Gas Flow Normal to Banks of Staggered Tubes: Use of a Transient Technique, *Stanford Univ. Tech. Rep.* 15, Navy Contract N6-ONR251 T.O. 6, 1952.

18 Jakob, M.: Heat Transfer and Flow Resistance in Cross Flow of Gases over Tube Banks, *Trans. ASME,* vol. 60, p. 384, 1938.

19 Vliet, G. C., and G. Leppert: Forced Convection Heat Transfer from an Isothermal Sphere to Water, *J. Heat Transfer,* serv. C, vol. 83, p. 163, 1961.

20 Irvine, T. R.: Noncircular Duct Convective Heat Transfer, in W. Ibele (ed.), "Modern Developments in Heat Transfer," Academic Press, Inc., New York, 1963.

21 Fand, R. M.: Heat Transfer by Forced Convection from a Cylinder to Water in Crossflow, *Int. J. Heat Mass Transfer,* vol. 8, p. 995, 1965.

22 Jakob, M.: "Heat Transfer," vol. 1, John Wiley & Sons, Inc., New York, 1949.

23 Stein, R.: Liquid Metal Heat Transfer, *Adv. Heat Transfer,* vol. 3, 1966.

24 Hartnett, J. P.: Experimental Determination of the Thermal Entrance Length for the Flow of Water and of Oil in Circular Pipes, *Trans. ASME,* vol. 77, p. 1211, 1955.

25 Allen, R. W., and E. R. G. Eckert: Friction and Heat Transfer Measurements to Turbulent Pipe Flow of Water (Pr = 7 and 8) at Uniform Wall Heat Flux, *J. Heat Transfer,* ser. C, vol. 86, p. 301, 1964.

26 Skupinshi, E., J. Tortel, and L. Vautrey: Détermination des coéfficients de convection a'un alliage sodium-potassium dans un tube circulaire, *Int. J. Heat Mass Transfer,* vol. 8, p. 937, 1965.

27 Dipprey, D. F., and R. H. Sabersky: Heat and Momentum Transfer in Smooth and Rough Tubes at Various Prandtl Numbers, *Int. J. Heat Mass Transfer,* vol. 6, p. 329, 1963.

28 Kline, S. J.: "Similitude and Approximation Theory," McGraw-Hill Book Company, New York, 1965.

29 Townes, H. W., and R. H. Sabersky: Experiments on the Flow over a Rough Surface, *Int. J. Heat Mass Transfer,* vol. 9, p. 729, 1966.

30 Gowen, R. A., and J. W. Smith: Turbulent Heat Transfer from Smooth and Rough Surfaces, *Int. J. Heat Mass Transfer,* vol. 11, p. 1657, 1968.

31 Sheriff, N., and P. Gumley: Heat Transfer and Friction Properties of Surfaces with Discrete Roughness, *Int. J. Heat Mass Transfer,* vol. 9, p. 1297, 1966.

32 Witte, L. C.: An Experimental Study of Forced-Convection Heat Transfer from a Sphere to Liquid Sodium, *J. Heat Transfer,* vol. 90, p. 9, 1968.

33 Zukauskas, A. A., V. Makarevicius, and A. Schlanciauskas: "Heat Transfer in Banks of Tubes in Crossflow of Fluid," Mintis, Vilnius, Lithuania, 1968.

34 Eckert, E. R. G., and R. M. Drake: "Analysis of Heat and Mass Transfer," McGraw-Hill Book Company, New York, 1972.

35 Whitaker, S.: "Forced Convection Heat-Transfer Correlations for Flow in Pipes, Past Flat Plates, Single Cylinders, Single Spheres, and Flow in Packed Bids and Tube Bundles," *AIChE J.,* vol. 18, p. 361, 1972.

36 Kays, W. M.: "Convective Heat and Mass Transfer," pp. 187–190, McGraw-Hill Book Company, New York, 1966.

37 Churchill, S. W., and M. Bernstein: A Correlating Equation for Forced Convection from Gases and Liquids to a Circular Cylinder in Crossflow, *J. Heat Transfer,* vol. 99, pp. 300–306, 1977.

38 Nakai, S., and T. Okazaki: Heat Transfer from a Horizontal Circular Wire at Small Reynolds and Grashof Numbers—1 Pure Convection, *Int. J. Heat Mass Transfer,* vol. 18, p. 387, 1975.

39 Zukauskas, A.: Heat Transfer from Tubes in Cross Flow, *Adv. Heat Transfer,* vol. 8, pp. 93–160, 1972.

40 Shah, R. K., and A. L. London: "Laminar Flow: Forced Convection in Ducts," Academic Press, New York, 1978.

41 Kalish, S., and O. E. Dwyer: Heat Transfer to NaK Flowing through Unbaffled Rod Bundles, *Int. J. Heat Mass Transfer,* vol. 10, p. 1533, 1967.

42 Petukhov, B. S.: Heat Transfer and Friction in Turbulent Pipe Flow with Variable Physical Properties, in J. P. Hartnett and T. F. Irvine, (eds.), "Advances in Heat Transfer" Academic Press, Inc., New York, pp. 504–564, 1970.

43 Achenbach, E.: Heat Transfer from Spheres up to Re $= 6 \times 10^6$, *Proc. Sixth Int. Heat Trans. Conf.,* vol. 5, Hemisphere Pub. Co., Washington, D.C., pp. 341–346, 1978.

44 Morgan, V. T.: The Overall Convective Heat Transfer from Smooth Circular Cylinders, in T. F. Irvine and J. P. Hartnett, (eds.), "Advances in Heat Transfer" vol. 11, Academic Press, Inc., 1975.

NATURAL-CONVECTION SYSTEMS

7-1 INTRODUCTION

Our previous discussions of convection heat transfer have considered only the calculation of forced-convection systems where the fluid is forced by or through the heat-transfer surface. Natural, or free, convection is observed as a result of the motion of the fluid due to density changes arising from the heating process. A hot radiator used for heating a room is one example of a practical device which transfers heat by free convection. The movement of the fluid in free convection, whether it is a gas or a liquid, results from the buoyancy forces imposed on the fluid when its density in the proximity of the heat-transfer surface is decreased as a result of the heating process. The buoyancy forces would not be present if the fluid were not acted upon by some external force field such as gravity, although gravity is not the only type of force field which can produce the free-convection currents; a fluid enclosed in a rotating machine is acted upon by a centrifugal force field, and thus could experience free-convection currents if one or more of the surfaces in contact with the fluid were heated. The buoyancy forces which give rise to the free-convection currents are called *body forces*.

7-2 FREE-CONVECTION HEAT TRANSFER ON A VERTICAL FLAT PLATE

Consider the vertical flat plate shown in Fig. 7-1. When the plate is heated, a free-convection boundary layer is formed, as shown. The velocity profile in this boundary layer is quite unlike the velocity profile in a forced-convection boundary layer. At the wall the velocity is zero because of the no-slip condition; it increases to some maximum value and then decreases to zero at the edge of the boundary layer since the "free-stream" conditions are at rest in the free-convection system. The initial boundary-layer development is laminar; but at

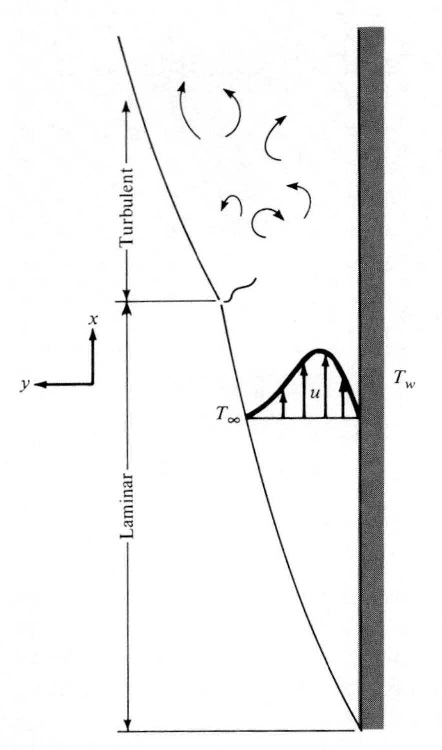

Fig. 7-1 Boundary layer on a vertical flat plate.

some distance from the leading edge, depending on the fluid properties and the temperature difference between wall and environment, turbulent eddies are formed, and transition to a turbulent boundary layer begins. Farther up the plate the boundary layer may become fully turbulent.

To analyze the heat-transfer problem, we must first obtain the differential equation of motion for the boundary layer. For this purpose we choose the x coordinate along the plate and the y coordinate perpendicular to the plate as in the analyses of Chap. 5. The only new force which must be considered in the derivation is the weight of the element of fluid. As before, we equate the sum of the external forces in the x direction to the change in momentum flux through the control volume $dx\ dy$. There results

$$\rho\left(u\frac{\partial u}{\partial x} + v\frac{\partial u}{\partial y}\right) = -\frac{\partial p}{\partial x} - \rho g + \mu\frac{\partial^2 u}{\partial y^2} \tag{7-1}$$

where the term $-\rho g$ represents the weight force exerted on the element. The pressure gradient in the x direction results from the change in elevation up the plate. Thus

$$\frac{\partial p}{\partial x} = -\rho_\infty g \tag{7-2}$$

In other words, the change in pressure over a height dx is equal to the weight per unit area of the fluid element. Substituting Eq. (7-2) into Eq. (7-1) gives

$$\rho\left(u\frac{\partial u}{\partial x}+v\frac{\partial u}{\partial y}\right)=g(\rho_\infty-\rho)+\mu\frac{\partial^2 u}{\partial y^2} \tag{7-3}$$

The density difference $\rho_\infty - \rho$ may be expressed in terms of the volume coefficient of expansion β, defined by

$$\beta=\frac{1}{V}\left(\frac{\partial V}{\partial T}\right)_p=\frac{1}{V_\infty}\frac{V-V_\infty}{T-T_\infty}=\frac{\rho_\infty-\rho}{\rho(T-T_\infty)}$$

so that

$$\rho\left(u\frac{\partial u}{\partial x}+v\frac{\partial u}{\partial y}\right)=g\rho\beta(T-T_\infty)+\mu\frac{\partial^2 u}{\partial y^2} \tag{7-4}$$

This is the equation of motion for the free-convection boundary layer. Notice that the solution for the velocity profile demands a knowledge of the temperature distribution. The energy equation for the free-convection system is the same as that for a forced-convection system at low velocity:

$$\rho c_p\left(u\frac{\partial T}{\partial x}+v\frac{\partial T}{\partial y}\right)=k\frac{\partial^2 T}{\partial y^2} \tag{7-5}$$

The volume coefficient of expansion β may be determined from tables of properties for the specific fluid. For ideal gases it may be calculated from (see Prob. 7-3)

$$\beta=\frac{1}{T}$$

where T is the absolute temperature of the gas.

Even though the fluid motion is the result of density variations, these variations are quite small, and a satisfactory solution to the problem may be obtained by assuming incompressible flow, that is, ρ = constant. To effect a solution of the equation of motion, we use the integral method of analysis similar to that used in the forced-convection problem of Chap. 5. Detailed boundary-layer analyses have been presented in Refs. 13, 27, and 32.

For the free-convection system, the integral momentum equation becomes

$$\frac{d}{dx}\left(\int_0^\delta \rho u^2\,dy\right)=-\tau_w+\int_0^\delta \rho g\beta(T-T_\infty)\,dy$$

$$=-\mu\left.\frac{\partial u}{\partial y}\right]_{y=0}+\int_0^\delta \rho g\beta(T-T_\infty)\,dy \tag{7-6}$$

and we observe that the functional form of both the velocity and the temperature distributions must be known in order to arrive at the solution. To obtain these functions, we proceed in much the same way as in Chap. 5. The following conditions apply for the temperature distribution:

$$T = T_w \qquad \text{at } y = 0$$

$$T = T_\infty \qquad \text{at } y = \delta$$

$$\frac{\partial T}{\partial y} = 0 \qquad \text{at } y = \delta$$

so that we obtain for the temperature distribution

$$\frac{T - T_\infty}{T_w - T_\infty} = \left(1 - \frac{y}{\delta}\right)^2 \tag{7-7}$$

Three conditions for the velocity profile are

$$u = 0 \qquad \text{at } y = 0$$

$$u = 0 \qquad \text{at } y = \delta$$

$$\frac{\partial u}{\partial y} = 0 \qquad \text{at } y = \delta$$

An additional condition may be obtained from Eq. (7-4) by noting that

$$\frac{\partial^2 u}{\partial y^2} = -g\beta \frac{T_w - T_\infty}{\nu} \qquad \text{at } y = 0$$

As in the integral analysis for forced-convection problems, we assume that the velocity profiles have geometrically similar shapes at various x distances along the plate. For the free-convection problem, we now assume that the velocity may be represented as a polynomial function of y multiplied by some arbitrary function of x. Thus,

$$\frac{u}{u_x} = a + by + cy^2 + dy^3$$

where u_x is a fictitious velocity which is a function of x. The cubic-polynomial form is chosen because there are four conditions to satisfy, and this is the simplest type of function which may be used. Applying the four conditions to the velocity profile listed above, we have

$$\frac{u}{u_x} = \frac{\beta \delta^2 g(T_w - T_\infty)}{4u_x \nu} \frac{y}{\delta}\left(1 - \frac{y}{\delta}\right)^2$$

The term involving the temperature difference, δ^2, and u_x may be incorporated into the function u_x so that the final relation to be assumed for the velocity profile is

$$\frac{u}{u_x} = \frac{y}{\delta}\left(1 - \frac{y}{\delta}\right)^2 \tag{7-8}$$

A plot of Eq. (7-8) is given in Fig. 7-2. Substituting Eqs. (7-7) and (7-8) into Eq. (7-6) and carrying out the integrations and differentiations yields

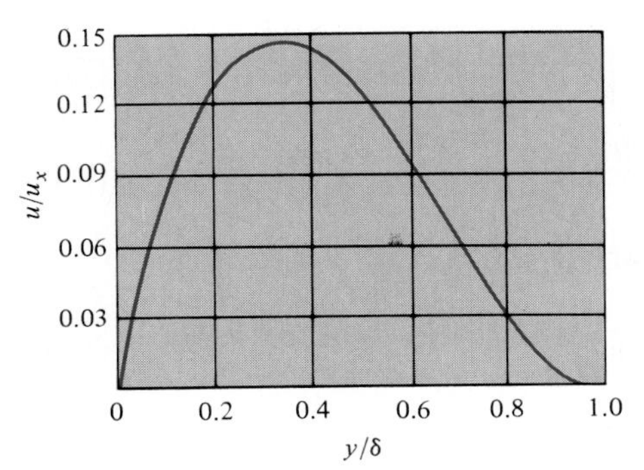

Fig. 7-2 Free-convection velocity profile given by Eq. (7-8).

$$\frac{1}{105}\frac{d}{dx}(u_x^2\,\delta) = \frac{1}{3}g\beta(T_w - T_\infty)\,\delta - \nu\frac{u_x}{\delta} \tag{7-9}$$

The integral form of the energy equation for the free-convection system is

$$\frac{d}{dx}\left[\int_0^\delta u(T - T_\infty)\,dy\right] = -\alpha\left.\frac{dT}{dy}\right]_{y=0} \tag{7-10}$$

and when the assumed velocity and temperature distributions are inserted into this equation and the operations are performed, there results

$$\frac{1}{30}(T_w - T_\infty)\frac{d}{dx}(u_x\,\delta) = 2\alpha\frac{T_w - T_\infty}{\delta} \tag{7-11}$$

It is clear from the reasoning which led to Eq. (7-8) that

$$u_x \sim \delta^2 \tag{7-12}$$

Inserting this type of relation in Eq. (7-9) yields the result that

$$\delta \sim x^{1/4} \tag{7-13}$$

We therefore assume the following exponential functional variations for u_x and δ:

$$u_x = C_1 x^{1/2} \tag{7-14}$$

$$\delta = C_2 x^{1/4} \tag{7-15}$$

Introducing these relations into Eqs. (7-9) and (7-11) gives

$$\frac{5}{420}C_1^2 C_2 x^{1/4} = g\beta(T_w - T_\infty)\frac{C_2}{3}x^{1/4} - \frac{C_1}{C_2}\nu x^{1/4} \tag{7-16}$$

and

$$\frac{1}{40}C_1 C_2 x^{-1/4} = \frac{2\alpha}{C_2}x^{-1/4} \tag{7-17}$$

These two equations may be solved for the constants C_1 and C_2 to give

$$C_1 = 5.17\,\nu\left(\frac{20}{21} + \frac{\nu}{\alpha}\right)^{-1/2}\left[\frac{g\beta(T_w - T_\infty)}{\nu^2}\right]^{1/2} \tag{7-18}$$

$$C_2 = 3.93\left(\frac{20}{21} + \frac{\nu}{\alpha}\right)^{1/4}\left[\frac{g\beta(T_w - T_\infty)}{\nu^2}\right]^{-1/4}\left(\frac{\nu}{\alpha}\right)^{-1/2} \tag{7-19}$$

The resultant expression for the boundary-layer thickness is

$$\frac{\delta}{x} = 3.93\ \mathrm{Pr}^{-1/2}\,(0.952 + \mathrm{Pr})^{1/4}\,\mathrm{Gr}_x^{-1/4} \tag{7-20}$$

where the Prandtl number $\mathrm{Pr} = \nu/\alpha$ has been introduced along with a new dimensionless group called the *Grashof number* Gr_x:

$$\mathrm{Gr}_x = \frac{g\beta(T_w - T_\infty)x^3}{\nu^2} \tag{7-21}$$

The heat-transfer coefficient may be evaluated from

$$q_w = -kA\left.\frac{dT}{dy}\right]_w = hA\,(T_w - T_\infty)$$

Using the temperature distribution of Eq. (7-7), one obtains

$$h = \frac{2k}{\delta} \qquad \text{or} \qquad \frac{hx}{k} = \mathrm{Nu}_x = 2\,\frac{x}{\delta}$$

so that the dimensionless equation for the heat-transfer coefficient becomes

$$\mathrm{Nu}_x = 0.508\ \mathrm{Pr}^{1/2}\,(0.952 + \mathrm{Pr})^{-1/4}\,\mathrm{Gr}_x^{1/4} \tag{7-22}$$

Equation (7-22) gives the variation of the local heat-transfer coefficient along the vertical plate. The average heat-transfer coefficient may then be obtained by performing the integration

$$\bar{h} = \frac{1}{L}\int_0^L h_x\,dx \tag{7-23}$$

For the variation given by Eq. (7-22), the average coefficient is

$$\bar{h} = \tfrac{4}{3}h_{x=L} \tag{7-24}$$

The Grashof number may be interpreted physically as a dimensionless group representing the ratio of the buoyancy forces to the viscous forces in the free-convection flow system. It has a role similar to that played by the Reynolds number in forced-convection systems and is the primary variable used as a criterion for transition from laminar to turbulent boundary-layer flow. For air in free convection on a vertical flat plate, the critical Grashof number has been observed by Eckert and Soehngen [1] to be approximately 4×10^8. Values ranging between 10^8 and 10^9 may be observed for different fluids and environment "turbulence levels."

A very complete survey of the stability and transition of free-convection boundary layers has been given by Gebhart et al. [13–15].

The foregoing analysis of free-convection heat transfer on a vertical flat plate is the simplest case that may be treated mathematically, and it has served to introduce the new dimensionless variable, the Grashof number,† which is important in all free-convection problems. But as in some forced-convection problems, experimental measurements must be relied upon to obtain relations for heat transfer in other circumstances. These circumstances are usually those in which it is difficult to predict temperature and velocity profiles analytically. Turbulent free convection is an important example, just as is turbulent forced convection, of a problem area in which experimental data are necessary; however, the problem is more acute with free-convection flow systems than with forced-convection systems because the velocities are usually so small that they are very difficult to measure. Despite the experimental difficulties, velocity measurements have been performed using hydrogen-bubble techniques [26], hot-wire anemometry [28], and quartz-fiber anemometers. Temperature field measurements have been obtained through the use of the Zehnder-Mach interferometer. The laser anemometer [29] is particularly useful for free-convection measurements because it does not disturb the flow field.

An interferometer indicates lines of constant density in a fluid flow field. For a gas in free convection at low pressure these lines of constant density are equivalent to lines of constant temperature. Once the temperature field is obtained, the heat transfer from a surface in free convection may be calculated by using the temperature gradient at the surface and the thermal conductivity of the gas. Several interferometric studies of free convection have been made [1–3], and some typical photographs of the flow fields are shown in Figs. 7-3 to 7-6. Figure 7-3 shows the lines of constant temperature around a heated vertical flat plate. Notice that the lines are closest together near the plate surface, indicating a higher temperature gradient in that region. Figure 7-4 shows the lines of constant temperature around a heated horizontal cylinder in free convection, and Fig. 7-5 shows the boundary-layer interaction between a group of four horizontal cylinders. A similar phenomenon would be observed for forced-convection flow across a heated tube bank. Interferometric studies have been conducted to determine the point at which eddies are formed in the free-convection boundary layer [1], and these studies have been used in predicting the start of transition to turbulent flow in free-convection systems.

It was mentioned earlier that the velocities in free convection are so small that for most systems they are difficult to measure without influencing the flow field by the insertion of a measuring device. A rough visual indication of the free-convection velocity is given in Fig. 7-6, where a free-convection boundary-

†History is not clear on the point, but it appears that the Grashof number was named for Franz Grashof, a professor of applied mechanics at Karlsruhe around 1863 and one of the founding directors of *Verein deutscher Ingenieure* in 1855. He developed some early steam-flow formulas, but made no significant contributions to free convection [36].

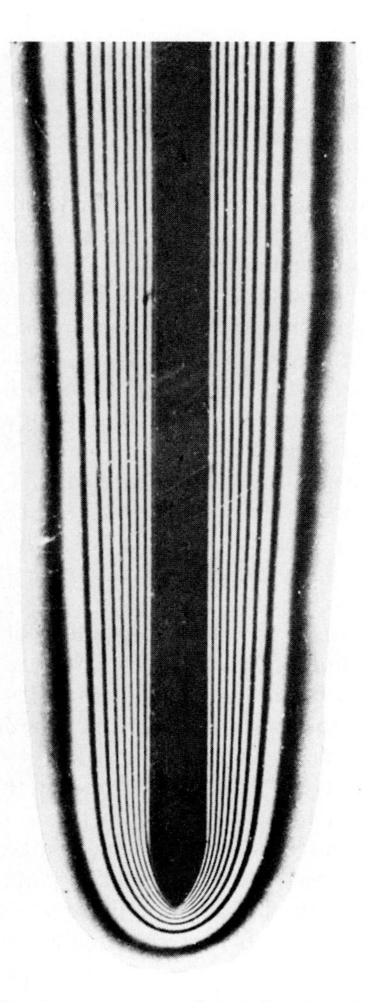

Fig. 7-3 Interferometer photograph showing lines of constant temperature around a heated vertical flat plate in free convection. (*Photograph courtesy E. Soehngen.*)

layer wave resulting from a heat pulse near the leading edge of the plate is presented. It may be noted that the maximum points in the isotherms experience a phase lag and that a line drawn through these maximum points has the approximate shape of the free-convection velocity profile.

A number of references treat the various theoretical and empirical aspects of free-convection problems. One of the most extensive discussions is given by Gebhart [13], and the interested reader may wish to consult this reference for additional information.

7-3 EMPIRICAL RELATIONS FOR FREE CONVECTION

Over the years it has been found that average free-convection heat-transfer coefficients can be represented in the following functional form for a variety of circumstances:

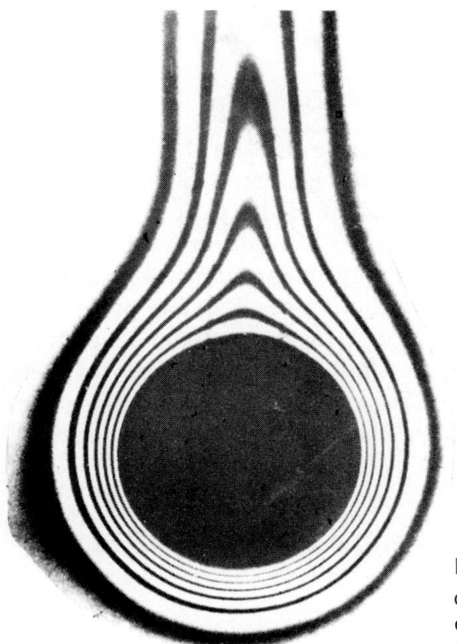

Fig. 7-4 Interferometer photograph showing lines of constant temperature around a heated horizontal cylinder in free convection. (*Photograph courtesy E. Soehngen.*)

$$\overline{\text{Nu}}_f = C\,(\text{Gr}_f\,\text{Pr}_f)^m \tag{7-25}$$

where the subscript f indicates that the properties in the dimensionless groups are evaluated at the film temperature

$$T_f = \frac{T_\infty + T_w}{2}$$

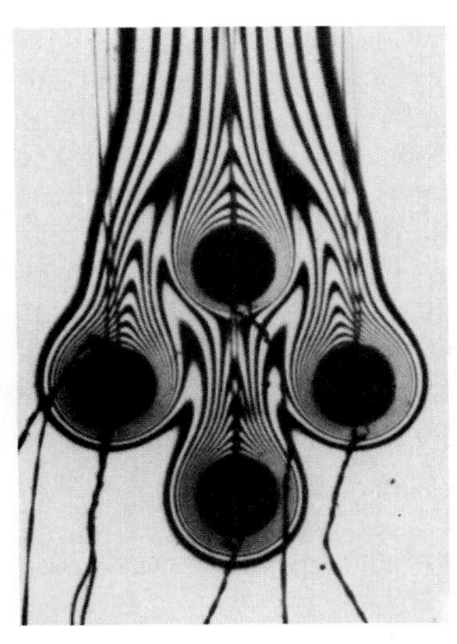

Fig. 7-5 Interferometer photograph showing the boundary-layer interaction between four heated horizontal cylinders in free convection. (*Photograph courtesy E. Soehngen.*)

The product of the Grashof and Prandtl numbers is called the Rayleigh number:

$$\text{Ra} = \text{Gr Pr} \tag{7-26}$$

The characteristic dimension to be used in the Nusselt and Grashof numbers depends on the geometry of the problem. For a vertical plate it is the height of the plate L; for a horizontal cylinder it is the diamter d; and so forth. Experimental data for free-convection problems appear in a number of references, with some conflicting results. The purpose of the sections that follow is to give these results in a summary form that may be easily used for calculation purposes. The functional form of Eq. (7-25) is used for many of these presentations, with the values of the constants C and m specified for each case.

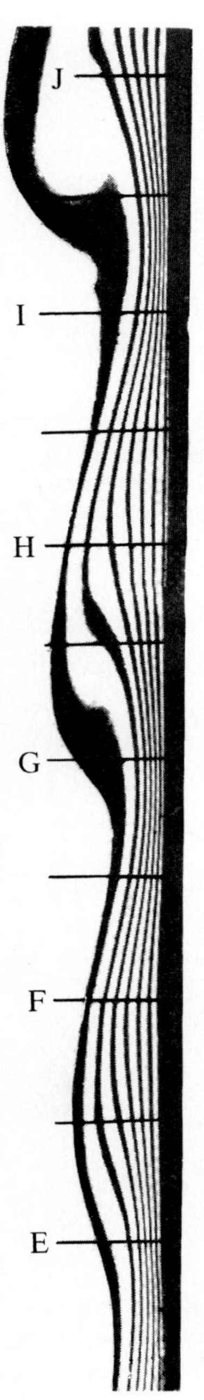

7-4 FREE CONVECTION FROM VERTICAL PLANES AND CYLINDERS

Isothermal Surfaces

For vertical surfaces, the Nusselt and Grashof numbers are formed with L, the height of the surface as the characteristic dimension. If the boundary-layer thickness is not large compared with the diameter of the cylinder, the heat transfer may be calculated with the same relations used for vertical plates. The general criterion is that a vertical cylinder may be treated as a vertical flat plate [13] when

$$\frac{D}{L} \geq \frac{35}{\text{Gr}_L^{1/4}} \tag{7-27}$$

where D is the diameter of the cylinder. For *isothermal* surfaces, the values of the constants are given in Table 7-1 with the appropriate references noted for further consultation. The reader's attention is directed to the two sets of constants given for the turbulent case ($\text{Gr}_f \text{ Pr}_f > 10^9$). Although there may appear to be a decided difference in these constants, a comparison by Warner and Arpaci [22] of the two relations with experimental data indicates that both sets of constants fit available data. There are some indications from the analytical work of Bayley [16], as well as heat flux measurements of Ref. 22, that the relation

Fig. 7-6 Interferometer photograph showing isotherms on a heated vertical flat plate resulting from a periodic disturbance of the boundary layer. Note phase shifts in maximum points of isotherms. (*From Holman, Gartrell, and Soehngen [3].*)

Table 7-1 Constants for Use with Eq. (7-25) for Isothermal Surfaces.

Geometry	$Gr_f\ Pr_f$	C	m	*Ref(s).*
Vertical planes and cylinders	10^{-1}–10^4	Use Fig. 7-7	Use Fig. 7-7	4
	10^4–10^9	0.59	$\frac{1}{4}$	4
	10^9–10^{13}	0.021	$\frac{2}{5}$	30
	10^9–10^{13}	0.10	$\frac{1}{3}$	22, 16†
Horizontal cylinders	0–10^{-5}	0.4	0	4
	10^{-5}–10^4	Use Fig. 7-8	Use Fig. 7-8	4
	10^4–10^9	0.53	$\frac{1}{4}$	4
	10^9–10^{12}	0.13	$\frac{1}{3}$	4
	10^{-10}–10^{-2}	0.675	0.058	76†
	10^{-2}–10^2	1.02	0.148	76†
	10^2–10^4	0.850	0.188	76
	10^4–10^7	0.480	$\frac{1}{4}$	76
	10^7–10^{12}	0.125	$\frac{1}{3}$	76
Upper surface of heated plates or lower surface of cooled plates	2×10^4–8×10^6	0.54	$\frac{1}{4}$	44, 52
Upper surface of heated plates or lower surface of cooled plates	8×10^6–10^{11}	0.15	$\frac{1}{3}$	44, 52
Lower surface of heated plates or upper surface of cooled plates	10^5–10^{11}	0.27	$\frac{1}{4}$	44, 37, 75
Vertical cylinder, height = diameter Characteristic length = diameter	10^4–10^6	0.775	0.21	77
Irregular solids, characteristic length = distance fluid particle travels in boundary layer	10^4–10^9	0.52	$\frac{1}{4}$	78

† Preferred.

$$\mathrm{Nu}_f = 0.10\,(\mathrm{Gr}_f\,\mathrm{Pr}_f)^{1/3}$$

may be preferable.

More complicated relations have been provided by Churchill and Chu [71] and are applicable over wider ranges of the Rayleigh number:

$$\overline{\mathrm{Nu}} = 0.68 + \frac{0.670\ \mathrm{Ra}^{1/4}}{[1 + (0.492/\mathrm{Pr})^{9/16}]^{4/9}} \qquad \text{for } \mathrm{Ra}_L < 10^9 \tag{7-28}$$

$$\overline{\mathrm{Nu}}^{1/2} = 0.825 + \frac{0.387\ \mathrm{Ra}^{1/6}}{[1 + (0.492/\mathrm{Pr})^{9/16}]^{8/27}} \qquad \text{for } 10^{-1} < \mathrm{Ra}_L < 10^{12} \tag{7-29}$$

Equation (7-28) is also a satisfactory representation for constant heat flux. Properties for these equations are evaluated at the film temperature.

□ Constant Heat Flux

Extensive experiments have been reported in Refs. 25, 26, and 39 for free convection from vertical and inclined surfaces to water under constant-heat-

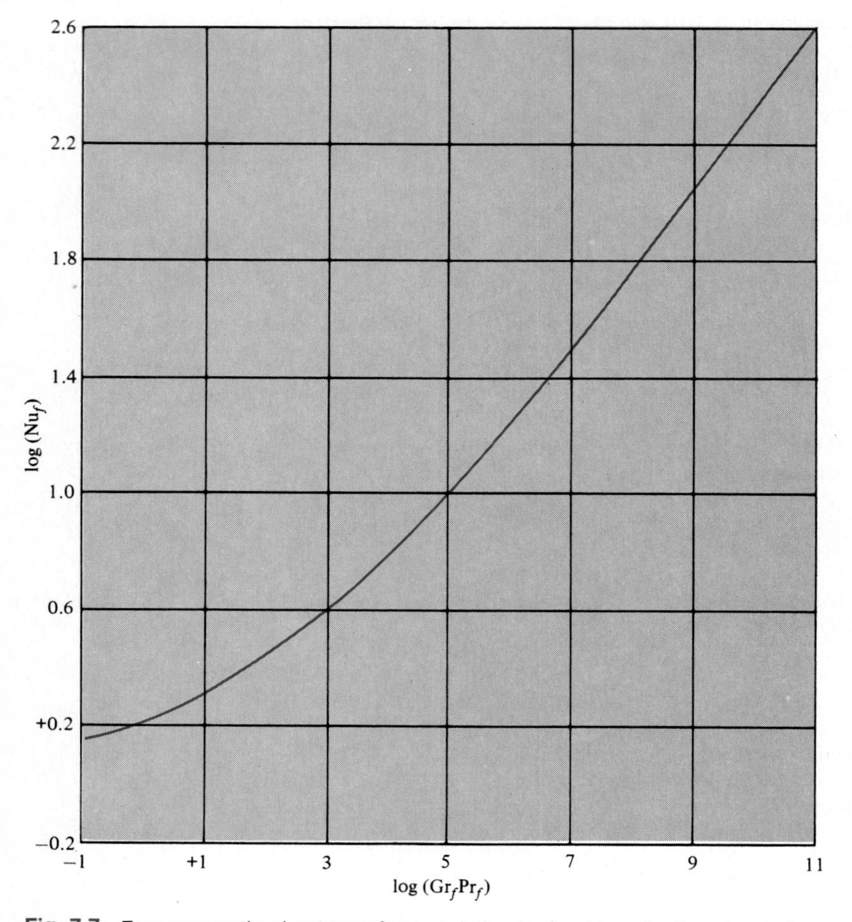

Fig. 7-7 Free-convection heat-transfer correlation for heat transfer from heated vertical plates, according to Ref. 4.

flux conditions. In such experiments, the results are presented in terms of a modified Grashof number, Gr*:

$$\mathrm{Gr}_x^* = \mathrm{Gr}_x\,\mathrm{Nu}_x = \frac{g\beta q_w x^4}{k\nu^2} \tag{7-30}$$

where q_w is the wall heat flux in watts per square meter. The *local* heat-transfer coefficients were correlated by the following relation for the laminar range:

$$\mathrm{Nu}_{xf} = \frac{hx}{k_f} = 0.60\,(\mathrm{Gr}_x^*\,\mathrm{Pr}_f)^{1/5} \qquad 10^5 < \mathrm{Gr}_x^* < 10^{11};\ q_w = \text{const} \tag{7-31}$$

It is to be noted that the criterion for laminar flow expressed in terms of Gr_x^* is not the same as that expressed in terms of Gr_x. Boundary-layer transition

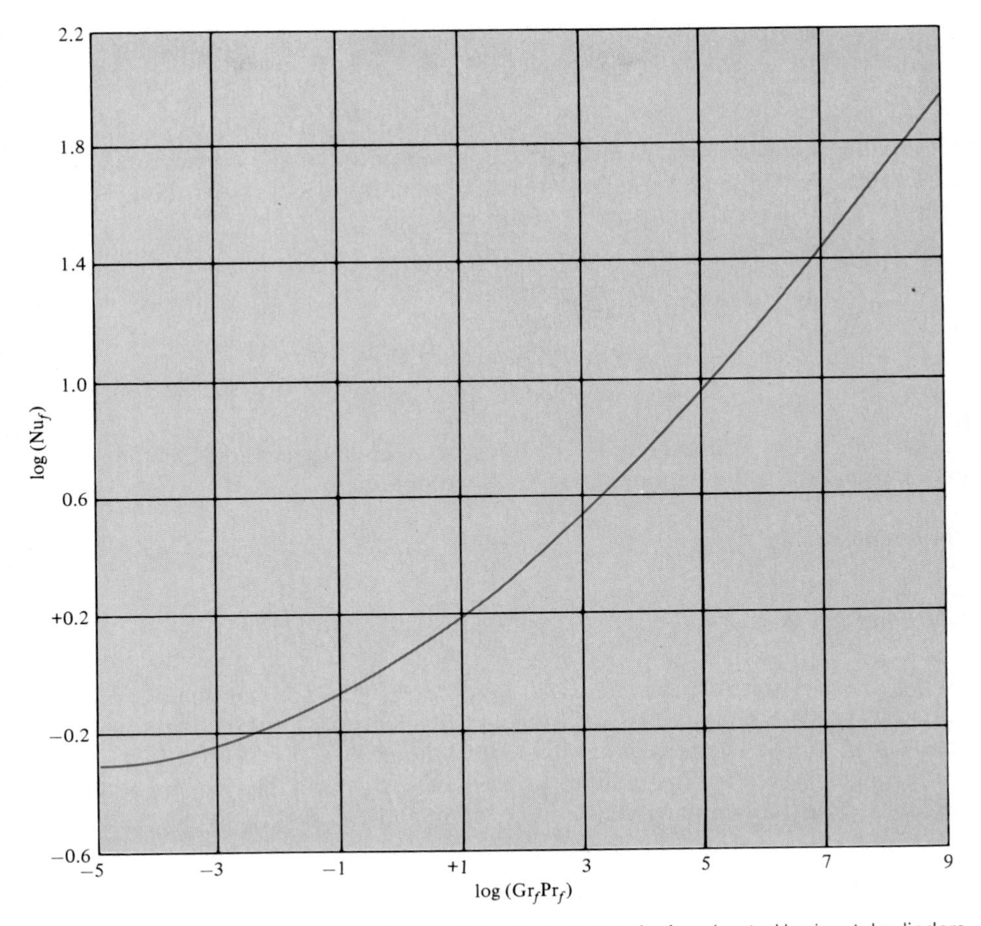

Fig. 7-8 Free-convection heat-transfer correlation for heat transfer from heated horizontal cylinders, according to Ref. 4.

was observed to begin between $Gr_x^* Pr = 3 \times 10^{12}$ and 4×10^{13} and to end between 2×10^{13} and 10^{14}. Fully developed turbulent flow was present by $Gr_x^* Pr = 10^{14}$, and the experiments were extended up to $Gr_x^* Pr = 10^{16}$. For the turbulent region, the local heat-transfer coefficients were correlated with

$$Nu_x = 0.17\,(Gr_x^*\,Pr)^{1/4} \qquad 2 \times 10^{13} < Gr_x^*\,Pr < 10^{16};\ q_w = \text{const} \qquad (7\text{-}32)$$

All properties in Eqs. (7-31) and (7-32) are evaluated at the local film temperature. Although these experiments were conducted for water, the resulting correlations are shown to work for air as well. The average heat-transfer coefficient for the constant-heat-flux case may not be evaluated from Eq. (7-24) but must be obtained through a separate application of Eq. (7-23). Thus, for the laminar region, using Eq. (7-31) to evaluate h_x,

$$\bar{h} = \frac{1}{L}\int_0^L h_x\,dx$$

$$\bar{h} = \tfrac{5}{4}h_{x=L} \qquad q_w = \text{const}$$

At this point we may note the relationship between the correlations in the form of Eq. (7-25) and those just presented in terms of $Gr_x^* = Gr_x Nu_x$. Writing Eq. (7-25) as a *local* heat-transfer form gives

$$Nu_x = C\,(Gr_x\,Pr)^m \tag{7-33}$$

Inserting $Gr_x = Gr_x^*/Nu_x$ gives

$$Nu^{1+m} = C\,(Gr_x^*\,Pr)^m$$

or

$$Nu_x = C^{1/(1+m)}\,(Gr_x^*\,Pr)^{m/(1+m)} \tag{7-34}$$

Thus, when the "characteristic" values of m for laminar and turbulent flow are compared to the exponents on Gr_x^*, we obtain

Laminar, $m = \frac{1}{4}$: $\qquad \dfrac{m}{1+m} = \dfrac{1}{5}$

Turbulent, $m = \frac{1}{3}$: $\qquad \dfrac{m}{1+m} = \dfrac{1}{4}$

While the Gr^* formulation is easier to employ for the constant-heat-flux case, we see that the characteristic exponents fit nicely into the scheme which is presented for the isothermal surface correlations.

It is also interesting to note the variation of h_x with x in the two characteristic regimes. For the laminar range $m = \frac{1}{4}$, and from Eq. (7-25)

$$h_x \sim \frac{1}{x}(x^3)^{1/4} = x^{-1/4}$$

In the turbulent regime $m = \frac{1}{3}$, and we obtain

$$h_x \sim \frac{1}{x}(x^3)^{1/3} = \text{const with } x$$

So when turbulent free convection is encountered, the local heat-transfer coefficient is essentially constant with x.

Churchill and Chu [71] show that Eq. (7-28) may be modified to apply to the constant-heat-flux case if the average Nusselt number is based on the wall heat flux and the temperature difference at the center of the plate ($x = L/2$). The result is

$$\overline{Nu}_L^{1/4}(\overline{Nu}_L - 0.68) = \frac{0.67\,(Gr_L^*\,Pr)^{1/4}}{[1 + (0.492/Pr)^{9/16}]^{4/9}} \tag{7-35}$$

where $\overline{Nu}_L = q_w L/(k\,\overline{\Delta T})$ and $\overline{\Delta T} = T_w$ at $L/2 - T_\infty$.

■ **EXAMPLE 7-1**

In a plant location near a furnace, a net radiant energy flux of 800 W/m² is incident on a vertical metal surface 3.5 m high and 2 m wide. The metal is insulated on the back side and painted black so that all the incoming radiation is lost by free convection to the surrounding air at 30°C. What average temperature will be attained by the plate?

Solution

We treat this problem as one with constant heat flux on the surface. Since we do not know the surface temperature, we must make an estimate for determining T_f and the air properties. An *approximate* value of h for free-convection problems is 10 W/m² · °C, and so, *approximately*,

$$\Delta T = \frac{q_w}{h} \approx \frac{800}{10} = 80°\text{C}$$

Then

$$T_f \approx \frac{80}{2} + 30 = 70°\text{C} = 343 \text{ K}$$

At 70°C the properties of air are

$$\nu = 2.043 \times 10^{-5} \text{ m}^2/\text{s} \qquad \beta = \frac{1}{T_f} = 2.79 \times 10^{-3} \text{ K}^{-1}$$

$$k = 0.0295 \text{ W/m} \cdot °\text{C} \qquad \text{Pr} = 0.7$$

From Eq. (7-30), with $x = 3.5$ m,

$$\text{Gr}_x^* = \frac{g\beta q_w x^4}{k\nu^2} = \frac{(9.8)(2.92 \times 10^{-3})(800)(3.5)^4}{(0.0295)(2.005 \times 10^{-5})^2} = 2.90 \times 10^{14}$$

We may therefore use Eq. (7-32) to evaluate h_x:

$$h_x = \frac{k}{x}(0.17)(\text{Gr}_x^* \text{ Pr})^{1/4}$$

$$= \frac{0.0295}{3.5}(0.17)(2.79 \times 10^{14} \times 0.7)^{1/4}$$

$$= 5.36 \text{ W/m}^2 \cdot °\text{C} \; [0.944 \text{ Btu/h} \cdot \text{ft}^2 \cdot °\text{F}]$$

In the turbulent heat transfer governed by Eq. (7-32), we note that

$$\text{Nu}_x = \frac{hx}{k} \sim (\text{Gr}_x^*)^{1/4} \sim (x^4)^{1/4}$$

or h_x does not vary with x, and we may take this as the average value. The value of $h = 5.41$ W/m² · °C is less than the approximate value we used to estimate T_f. Recalculating ΔT, we obtain

$$\Delta T = \frac{q_w}{h} = \frac{800}{5.36} = 149°\text{C}$$

Our new film temperature would be

$$T_f = 30 + \frac{149}{2} = 104.5°\text{C}$$

At 104.5°C the properties of air are

$$\nu = 2.354 \times 10^{-5}\ \text{m}^2/\text{s} \qquad \beta = \frac{1}{T_f} = 2.65 \times 10^{-3}/\text{K}$$

$$k = 0.0320\ \text{W/m} \cdot °\text{C} \qquad \text{Pr} = 0.695$$

Then

$$\text{Gr}_x^* = \frac{(9.8)(2.65 \times 10^{-3})(800)(3.5)^4}{(0.0320)(2.354 \times 10^{-5})^2} = 1.758 \times 10^{14}$$

and h_x is calculated from

$$h_x = \frac{k}{x}(0.17)(\text{Gr}_x^*\ \text{Pr})^{1/4}$$

$$= \frac{(0.0320)(0.17)}{3.5}[(1.758 \times 10^{14})(0.695)]^{1/4}$$

$$= 5.17\ \text{W/m}^2 \cdot °\text{C}\ [0.91\ \text{Btu/h} \cdot \text{ft}^2 \cdot °\text{F}]$$

Our new temperature difference is calculated as

$$\Delta T = (T_w - T_\infty)_{\text{av}} = \frac{q_w}{h} = \frac{800}{5.17} = 155°\text{C}$$

The average wall temperature is therefore

$$T_{w,\text{av}} = 155 + 30 = 185°\text{C}$$

Another iteration on the value of T_f is not warranted by the improved accuracy which would result.

■ EXAMPLE 7-2

A large vertical plate 4.0 m high is maintained at 60°C and exposed to atmospheric air at 10°C. Calculate the heat transfer if the plate is 10 m wide.

Solution

We first determine the film temperature as

$$T_f = \frac{60 + 10}{2} = 35°\text{C} = 308\ \text{K}$$

The properties of interest are thus

$$\beta = \frac{1}{308} = 3.25 \times 10^{-3} \qquad k = 0.02685$$

$$\nu = 16.5 \times 10^{-6} \qquad \text{Pr} = 0.7$$

and

$$\text{Gr Pr} = \frac{(9.8)(3.25 \times 10^{-3})(60 - 10)(4)^3}{(16.5 \times 10^{-6})^2} 0.7$$

$$= 3.743 \times 10^{11}$$

We then may use Eq. (7-29) to obtain

$$\overline{\text{Nu}}^{1/2} = 0.825 + \frac{(0.387)(3.743 \times 10^{11})^{1/6}}{[1 + (0.492/0.7)^{9/16}]^{8/27}}$$

$$= 28.34$$

$$\overline{\text{Nu}} = 803$$

The heat-transfer coefficient is then

$$\bar{h} = \frac{(803)(0.02685)}{4.0} = 5.39 \text{ W/m}^2 \cdot {}^\circ\text{C}$$

The heat transfer is

$$q = \bar{h}A(T_w - T_\infty)$$

$$= (5.39)(4)(10)(60 - 10) = 10{,}781 \text{ W}$$

As an alternative, we could employ the simpler relation

$$\text{Nu} = 0.10 \text{ (Gr Pr)}^{1/3}$$

$$= (0.10)(3.743 \times 10^{11})^{1/3} = 720.7$$

which gives a value about 10 percent lower than Eq. (7-29).

7-5 FREE CONVECTION FROM HORIZONTAL CYLINDERS

The values of the constants C and m are given in Table 7-1 according to Refs. 4 and 76. The predictions of Morgan (Ref. 76 in Table 7-1) are the most reliable for Gr Pr of approximately 10^{-5}. A more complicated expression for use over a wider range of Gr Pr is given by Churchill and Chu [70]:

$$\overline{\text{Nu}}^{1/2} = 0.60 + 0.387 \left\{ \frac{\text{Gr Pr}}{[1 + (0.559/\text{Pr})^{9/16}]^{16/9}} \right\}^{1/6} \quad \text{for } 10^{-5} < \text{Gr Pr} < 10^{12} \tag{7-36}$$

A simpler equation is available from Ref. 70, but is restricted to the laminar range of $10^{-6} < \text{Gr Pr} < 10^9$:

$$\text{Nu}_d = 0.36 + \frac{0.518 \text{ (Gr}_d \text{ Pr)}^{1/4}}{[1 + (0.559/\text{Pr})^{9/16}]^{4/9}} \tag{7-37}$$

Properties in Eqs. (7-36) and (7-37) are evaluated at the film temperature.

Heat transfer from horizontal cylinders to liquid metals may be calculated from Ref. 46:

$$\text{Nu}_d = 0.53 \text{ (Gr}_d \text{ Pr}^2)^{1/4} \tag{7-38}$$

■ **EXAMPLE 7-3**

A 2.0-cm-diameter horizontal heater is maintained at a surface temperature of 38°C and submerged in water at 27°C. Calculate the free-convection heat loss per unit length of the heater.

Solution

The film temperature is

$$T_f = \frac{38 + 27}{2} = 32.5°\text{C}$$

From Appendix A the properties of water are

$$k = 0.630 \text{ W/m} \cdot °\text{C},$$

and the following term is particularly useful in obtaining the Gr Pr product when it is multiplied by $d^3 \, \Delta T$:

$$\frac{g\beta\rho^2 c_p}{\mu k} = 2.48 \times 10^{10} \qquad [1/\text{m}^3 \cdot °\text{C}]$$

$$\text{Gr Pr} = (2.48 \times 10^{10})(38 - 27)(0.02)^3 = 2.18 \times 10^6$$

Using Table 7-1, we get $C = 0.53$ and $m = \frac{1}{4}$, so that

$$\text{Nu} = (0.53)(2.18 \times 10^6)^{1/4} = 38.425$$

$$h = \frac{(38.425)(0.63)}{0.02} = 1210 \text{ W/m}^2 \cdot °\text{C}$$

The heat transfer is thus

$$\frac{q}{L} = h\pi d(T_w - T_\infty)$$

$$= (1210)\pi(0.02)(38 - 27) = 836.3 \text{ W/m}$$

■ **EXAMPLE 7-4**

A fine wire having a diameter of 0.02 mm is maintained at a constant temperature of 54°C by an electric current. The wire is exposed to air at 1 atm and 0°C. Calculate the electric power necessary to maintain the wire temperature if the length is 50 cm.

Solution

The film temperature is $T_f = (54 + 0)/2 = 27°\text{C} = 300$ K, so the properties are

$$\beta = 1/300 = 0.00333 \qquad \nu = 15.69 \times 10^{-6} \text{ m}^2/\text{s}$$

$$k = 0.02624 \text{ W/m} \cdot °\text{C} \qquad \text{Pr} = 0.708$$

The Gr Pr product is then calculated as

$$\text{Gr Pr} = \frac{(9.8)(0.00333)(54 - 0)(0.02 \times 10^{-3})^3}{(15.69 \times 10^{-6})^2}(0.708) = 4.05 \times 10^{-5}$$

From Table 7-1 we find $C = 0.675$ and $m = 0.058$ so that

$$\overline{\text{Nu}} = (0.675)(4.05 \times 10^{-5})^{0.058} = 0.375$$

and

$$\bar{h} = \overline{\text{Nu}}\left(\frac{k}{d}\right) = \frac{(0.375)(0.02624)}{0.02 \times 10^{-3}} = 492.6 \text{ W/m}^2 \cdot {}^\circ\text{C}$$

The heat transfer or power required is then

$$q = \bar{h}A(T_w - T_\infty) = (492.6)\pi(0.02 \times 10^{-3})(0.5)(54 - 0) = 0.836 \text{ W}$$

■ EXAMPLE 7-5

A horizontal pipe 1 ft (0.3048 m) in diameter is maintained at a temperature of 250°C in a room where the ambient air is at 15°C. Calculate the free-convection heat loss per meter of length.

Solution

We first determine the Grashof-Prandtl number product and then select the appropriate constants from Table 7-1 for use with Eq. (7-25). The properties of air are evaluated at the film temperature:

$$T_f = \frac{T_w + T_\infty}{2} = \frac{250 + 15}{2} = 132.5^\circ\text{C} = 405.5 \text{ K}$$

$$k = 0.03406 \text{ W/m} \cdot {}^\circ\text{C} \qquad \beta = \frac{1}{T_f} = \frac{1}{405.5} = 2.47 \times 10^{-3} \text{ K}^{-1}$$

$$\nu = 26.54 \times 10^{-5} \text{ m}^2\text{/s} \qquad \text{Pr} = 0.687$$

$$\text{Gr}_d \text{Pr} = \frac{g\beta(T_w - T_\infty)d^3}{\nu^2}\text{Pr}$$

$$= \frac{(9.8)(2.47 \times 10^{-3})(250 - 15)(0.3048)^3(0.687)}{(26.54 \times 10^{-6})^2}$$

$$= 1.571 \times 10^8$$

From Table 7-1, $C = 0.53$ and $m = \frac{1}{4}$, so that

$$\text{Nu}_d = 0.53(\text{Gr}_d \text{Pr})^{1/4} = (0.53)(1.571 \times 10^8)^{1/4} = 59.4$$

$$h = \frac{k \text{Nu}_d}{d} = \frac{(0.03406)(59.4)}{0.3048} = 6.63 \text{ W/m}^2 \cdot {}^\circ\text{C} \text{ [1.175 Btu/h} \cdot \text{ft}^2 \cdot {}^\circ\text{F]}$$

The heat transfer per unit length is then calculated from

$$\frac{q}{L} = h\pi d(T_w - T_\infty) = (6.63)\pi(0.3048)(250 - 15) = 1.49 \text{ kW/m [1560 Btu/h} \cdot \text{ft]}$$

As an alternative, we could employ the more complicated expression, Eq. (7-36), for solution of the problem. The Nusselt number thus would be calculated as

$$\overline{\text{Nu}}^{1/2} = 0.60 + 0.387\left\{\frac{1.571 \times 10^8}{[1 + (0.559/0.687)^{9/16}]^{16/9}}\right\}^{1/6}$$

$$\text{Nu} = 64.7$$

or a value about 8 percent higher.

7-6 FREE CONVECTION FROM HORIZONTAL PLATES

Isothermal Surfaces

The average heat-transfer coefficient from horizontal flat plates is calculated with Eq. (7-25) and the constants given in Table 7-1. The characteristic dimension for use with these relations has traditionally [4] been taken as the length of a side for a square, the mean of the two dimensions for a rectangular surface, and $0.9d$ for a circular disk. References 52 and 53 indicate that better agreement with experimental data can be achieved by calculating the characteristic dimension with

$$L = \frac{A}{P} \tag{7-39}$$

where A is the area and P is the perimeter of the surface. This characteristic dimension is also applicable to unsymmetrical planforms.

Constant Heat Flux

The experiments of Ref. 44 have produced the following correlations for constant heat flux on a horizontal plate. For the heated surface facing upward,

$$\overline{\mathrm{Nu}}_L = 0.13\,(\mathrm{Gr}_L\,\mathrm{Pr})^{1/3} \qquad \text{for } \mathrm{Gr}_L\,\mathrm{Pr} < 2 \times 10^8 \tag{7-40}$$

and

$$\overline{\mathrm{Nu}}_L = 0.16\,(\mathrm{Gr}_L\,\mathrm{Pr})^{1/3} \qquad \text{for } 2 \times 10^8 < \mathrm{Gr}_L\,\mathrm{Pr} < 10^{11} \tag{7-41}$$

For the heated surface facing downward,

$$\overline{\mathrm{Nu}}_L = 0.58\,(\mathrm{Gr}_L\,\mathrm{Pr})^{1/5} \qquad \text{for } 10^6 < \mathrm{Gr}_L\,\mathrm{Pr} < 10^{11} \tag{7-42}$$

In these equations all properties except β are evaluated at a temperature T_e defined by

$$T_e = T_w - 0.25(T_w - T_\infty)$$

and T_w is the *average* wall temperature related, as before, to the heat flux by

$$\bar{h} = \frac{q_w}{T_w - T_\infty}$$

The Nusselt number is formed as before:

$$\overline{\mathrm{Nu}}_L = \frac{\bar{h}L}{k} = \frac{q_w L}{(T_w - T_\infty)k}$$

Section 7-7 discusses an extension of these equations to inclined surfaces.

Irregular Solids

There is no general correlation which can be applied to irregular solids. The results of Ref. 77 indicate that Eq. (7-25) may be used with $C = 0.775$ and $m = 0.208$ for a vertical cylinder with height equal to diameter. Nusselt and Grashof numbers are evaluated by using the diameter as characteristic length. Lienhard [78] offers a prescription that takes the characteristic length as the distance a fluid particle travels in the boundary layer and uses values of $C = 0.52$ and $m = \frac{1}{4}$ in Eq. (7-25) in the laminar range. This may serve as an estimate for calculating the heat-transfer coefficient in the absence of specific information on a particular geometric shape. Bodies of unity aspect ratio are studied extensively in Ref. 81.

EXAMPLE 7-6

A cube, 20 cm on a side, is maintained at 60°C and exposed to atmospheric air at 20°C. Calculate the heat transfer.

Solution

This is an irregular solid so we use the information in the last entry of Table 7-1 in the absence of a specific correlation for this geometry. The properties were evaluated in Ex. 7-2 as

$$\beta = 3.25 \times 10^{-3} \qquad k = 0.02685$$

$$\nu = 17.47 \times 10^{-6} \qquad \text{Pr} = 0.7$$

The characteristic length is the distance a particle travels in the boundary layer, which is $L/2$ along the bottom plus L along the side plus $L/2$ on the top, or $2L = 40$ cm. The Gr Pr product is thus:

$$\text{Gr Pr} = \frac{(9.8)(3.25 \times 10^{-3})(60 - 10)(0.4)^3}{(17.47 \times 10^{-6})^2}(0.7) = 3.34 \times 10^8$$

From the last entry in Table 7-1 we find $C = 0.52$ and $n = 1/4$ and calculate the Nusselt number as

$$\text{Nu} = (0.52)(3.34 \times 10^8)^{1/4} = 135.2$$

and

$$\bar{h} = \text{Nu}\,\frac{k}{L} = \frac{(135.2)(0.02685)}{(0.4)} = 9.07 \text{ W/m}^2 \cdot {}^\circ\text{C}$$

The cube has six sides so the area is $6(0.2)^2 = 0.24$ m² and the heat transfer is

$$q = \bar{h}A(T_w - T_\infty) = (9.07)(0.24)(60 - 10) = 108.8 \text{ W}$$

7-7 FREE CONVECTION FROM INCLINED SURFACES

Extensive experiments have been conducted by Fujii and Imura [44] for heated plates in water at various angels of inclination. The angle which the plate makes with the vertical is designated θ, with positive angles indicating that the heater surface faces downward, as shown in Fig. 7-9. For the inclined plate facing downward with approximately constant heat flux the following correlation was obtained for the average Nusselt number:

$$\overline{\mathrm{Nu}}_e = 0.56\,(\mathrm{Gr}_e\,\mathrm{Pr}_e \cos\theta)^{1/4} \qquad \theta < 88°;\ 10^5 < \mathrm{Gr}_e\,\mathrm{Pr}_e \cos\theta < 10^{11} \tag{7-43}$$

In Eq.(7-43) all properties except β are evaluated at a reference temperature T_e defined by

$$T_e = T_w - 0.25(T_w - T_\infty) \tag{7-44}$$

where T_w is the *mean* wall temperature and T_∞ is the free-stream temperature; β is evaluated at a temperature of $T_\infty + 0.50(T_w - T_\infty)$. For almost-horizontal plates facing downward, that is, $88° < \theta < 90°$, an additional relation was obtained as

$$\overline{\mathrm{Nu}}_e = 0.58\,(\mathrm{Gr}_e\,\mathrm{Pr}_e)^{1/5} \qquad 10^6 < \mathrm{Gr}_e\,\mathrm{Pr}_e < 10^{11} \tag{7-45}$$

For an inclined plate with heated surface facing upward the empirical correlations become more complicated. For angles between -15 and $-75°$ a suitable correlation is

$$\overline{\mathrm{Nu}}_e = 0.14\,[(\mathrm{Gr}_e\,\mathrm{Pr}_e)^{1/3} - (\mathrm{Gr}_c\,\mathrm{Pr}_e)^{1/3}] + 0.56\,(\mathrm{Gr}_e\,\mathrm{Pr}_e \cos\theta)^{1/4} \tag{7-46}$$

for the range $10^5 < \mathrm{Gr}_e\,\mathrm{Pr}_e \cos\theta < 10^{11}$. The quantity Gr_c is a critical Grashof relation indicating when the Nusselt number starts to separate from the laminar relation of Eq. (7-43) and is given in the following tabulation:

θ, degrees	Gr_c
−15	5×10^9
−30	2×10^9
−60	10^8
−75	10^6

For $\mathrm{Gr}_e < \mathrm{Gr}_c$ the first term of Eq. (7-46) is dropped out. Additional information is given by Vliet [39] and Pera and Gebhart [45]. There is some evidence to indicate that the above relations may also be applied to constant-temperature surfaces.

Experimental measurements with air on constant-heat-flux surfaces [51] have shown that Eq. (7-31) may be employed for the laminar region if we replace

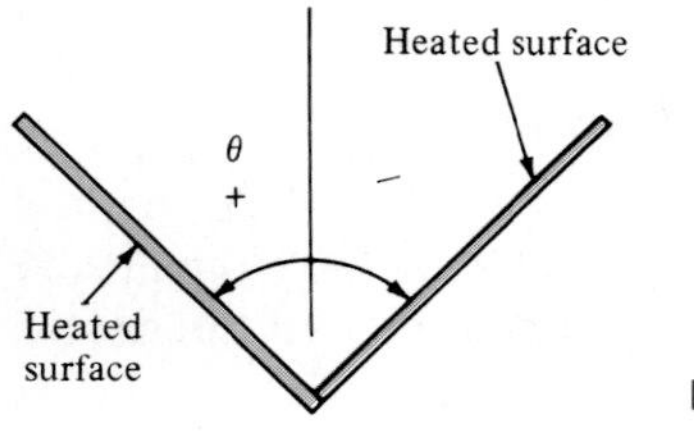

Fig. 7-9 Coordinate system for inclined plates.

Gr_x^* by $Gr_x^* \cos\theta$ for both upward- and downward-facing heated surfaces. In the turbulent region with air, the following empirical correlation was obtained:

$$Nu_x = 0.17\,(Gr_x^*\,Pr)^{1/4} \qquad 10^{10} < Gr_x^*\,Pr < 10^{15} \tag{7-47}$$

where the Gr_x^* is the same as for the vertical plate when the heated surface faces upward. When the heated surface faces downward, Gr_x^* is replaced by $Gr^* \cos^2\theta$. Equation (7-47) reduces approximately to the relation recommended in Table 7-1 for an isothermal vertical plate.

For inclined cylinders the data of Ref. 73 indicate that laminar heat transfer under constant-heat-flux conditions may be calculated with the following relation:

$$Nu_L = [0.60 - 0.488(\sin\theta)^{1.03}]\,(Gr_L\,Pr)^{\frac{1}{4}+\frac{1}{12}(\sin\theta)^{1.75}} \qquad \text{for } Gr_L\,Pr < 2\times 10^8 \tag{7-48}$$

where θ is the angle the cylinder makes with the vertical; that is, 0° corresponds to a vertical cylinder. Properties are evaluated at the film temperature except β, which is evaluated at ambient conditions.

Uncertainties still remain in the prediction of free convection from inclined surfaces, and an experimental-data scatter of ±20 percent is not unusual for the empirical relations presented above.

■ 7-8 NONNEWTONIAN FLUIDS

When the shear-stress viscosity relation of the fluid does not obey the simple newtonian expression of Eq. (5-1), the above equations for free-convection heat transfer do not apply. Extremely viscous polymers and lubricants are examples of fluids with nonnewtonian behavior. Successful analytical and experimental studies have been carried out with such fluids, but the results are very complicated. The interested reader should consult Refs. 48 to 50 for detailed information on this subject.

■ 7-9 SIMPLIFIED EQUATIONS FOR AIR

Simplified equations for the heat-transfer coefficient from various surfaces to air at atmospheric pressure and moderate temperatures are given in Table 7-2. These relations may be extended to higher or lower pressures by multiplying by the following factors:

$$\left(\frac{p}{101.32}\right)^{1/2} \qquad \text{for laminar cases}$$

$$\left(\frac{p}{101.32}\right)^{2/3} \qquad \text{for turbulent cases}$$

where p is the pressure in kilopascals. Due caution should be exercised in the use of these simplified relations because they are only approximations of the more precise equations stated earlier.

Table 7-2 Simplified Equations for Free Convection from Various Surfaces to Air at Atmospheric Pressure, Adapted from Table 7-1.

Surface	*Laminar,* $10^4 < \mathrm{Gr}_f \mathrm{Pr}_f < 10^9$	*Turbulent,* $\mathrm{Gr}_f \mathrm{Pr}_f > 10^9$
Vertical plane or cylinder	$h = 1.42 \left(\frac{\Delta T}{L}\right)^{1/4}$	$h = 1.31(\Delta T)^{1/3}$
Horizontal cylinder	$h = 1.32 \left(\frac{\Delta T}{d}\right)^{1/4}$	$h = 1.24(\Delta T)^{1/3}$
Horizontal plate:		
Heated plate facing upward or cooled plate facing downward	$h = 1.32 \left(\frac{\Delta T}{L}\right)^{1/4}$	$h = 1.52(\Delta T)^{1/3}$
Heated plate facing downward or cooled plate facing upward	$h = 0.59 \left(\frac{\Delta T}{L}\right)^{1/4}$	

where h = heat-transfer coefficient, W/m² · °C
$\Delta T = T_w - T_\infty$, °C
L = vertical or horizontal dimension, m
d = diameter, m

EXAMPLE 7-7

Compute the heat transfer for the conditions of Example 7-5 using the simplified relations of Table 7-2.

Solution

The heat-transfer coefficient is given by

$$h = 1.32 \left(\frac{\Delta T}{d}\right)^{1/4} = 1.32 \left(\frac{250 - 15}{0.3048}\right)^{1/4}$$
$$= 6.96 \text{ W/m}^2 \cdot {}^\circ\text{C}$$

The heat transfer is then

$$\frac{q}{L} = (6.96)\pi(0.3048)(250 - 15) = 1.57 \text{ kW/m}$$

Note that the simplified relation gives a value approximately 4 percent higher than Eq. (7-25).

7-10 FREE CONVECTION FROM SPHERES

Yuge [5] recommends the following empirical relation for free-convection heat transfer from spheres to air:

$$\text{Nu}_f = \frac{\bar{h}d}{k_f} = 2 + 0.392\ \text{Gr}_f^{1/4} \qquad \text{for } 1 < \text{Gr}_f < 10^5 \tag{7-49}$$

This equation may be modified by the introduction of the Prandtl number to give

$$\text{Nu}_f = 2 + 0.43\ (\text{Gr}_f\ \text{Pr}_f)^{1/4} \tag{7-50}$$

Properties are evaluated at the film temperature, and it is expected that this relation would be primarily applicable to calculations for free convection in gases. However, in the absence of more specific information it may also be used for liquids. We may note that for very low values of the Grashof-Prandtl number product the Nusselt number approaches a value of 2.0. This is the value which would be obtained for pure conduction through an infinite stagnant fluid surrounding the sphere.

For higher ranges of the Rayleigh numbers the experiments of Amato and Tien [79] with water suggest the following correlation:

$$\text{Nu}_f = 2 + 0.50\ (\text{Gr}_f\ \text{Pr}_f)^{1/4} \tag{7-51}$$

for $3 \times 10^5 < \text{Gr Pr} < 8 \times 10^8$.

7-11 FREE CONVECTION IN ENCLOSED SPACES

The free-convection flow phenomena inside an enclosed space are interesting examples of very complex fluid systems that may yield to analytical, empirical, and numerical solutions. Consider the system shown in Fig. 7-10, where a fluid is contained between two vertical plates separated by the distance δ. As a temperature difference $\Delta T_w = T_1 - T_2$ is impressed on the fluid, a heat transfer will be experienced with the approximate flow regions shown in Fig. 7-11, according to MacGregor and Emery [18]. In this figure, the Grashof number is calculated as

$$\text{Gr}_\delta = \frac{g\beta(T_1 - T_2)\delta^3}{\nu^2} \tag{7-52}$$

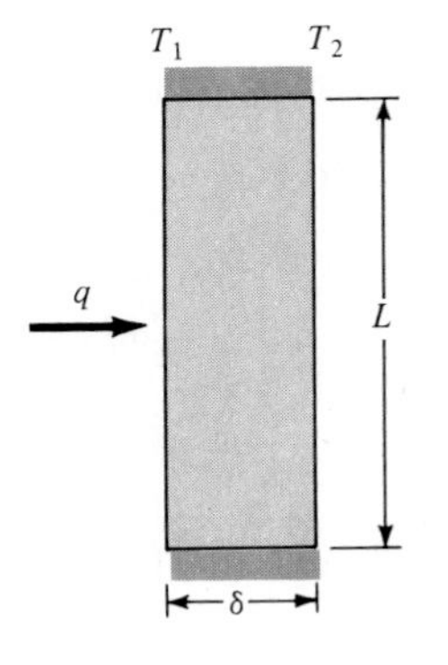

Fig. 7-10 Nomenclature for free convection in enclosed vertical spaces.

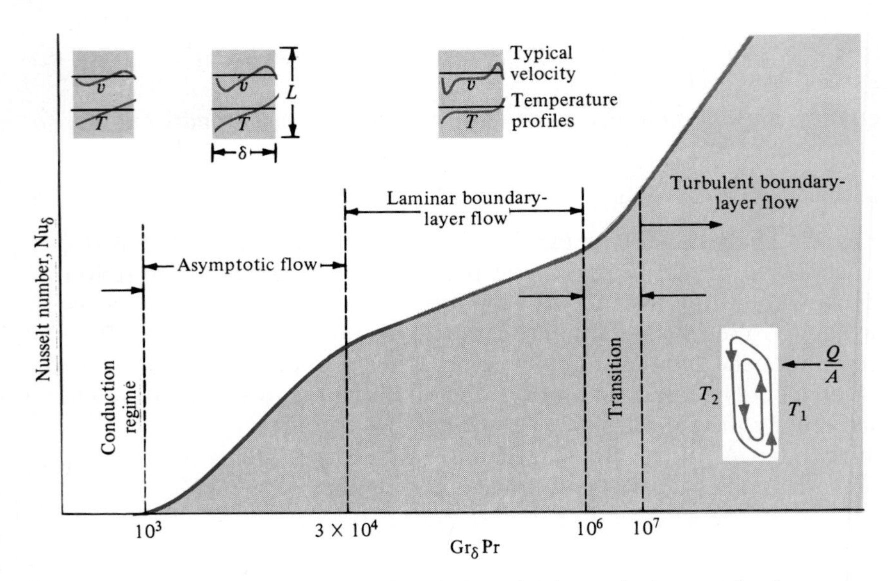

Fig. 7-11 Schematic diagram and flow regimes for the vertical convection layer, according to Ref. 18.

At very low Grashof numbers, there are very minute free-convection currents and the heat transfer occurs mainly by conduction across the fluid layer. As the Grashof number is increased, different flow regimes are encountered, as shown, with a progressively increasing heat transfer as expressed through the Nusselt number

$$\mathrm{Nu}_\delta = \frac{h\delta}{k}$$

Although some open questions still remain, the experiments of Ref. 18 may be used to predict the heat transfer to a number of liquids under constant-heat-flux conditions. The empirical correlations obtained were:

$$\mathrm{Nu}_\delta = 0.42\,(\mathrm{Gr}_\delta\,\mathrm{Pr})^{1/4}\,\mathrm{Pr}^{0.012}\left(\frac{L}{\delta}\right)^{-0.30} \qquad \begin{aligned} q_w &= \text{const} \\ 10^4 < \mathrm{Gr}_\delta\,&\mathrm{Pr} < 10^7 \\ 1 < \mathrm{Pr} &< 20{,}000 \\ 10 < L/\delta &< 40 \end{aligned} \tag{7-53}$$

$$\mathrm{Nu}_\delta = 0.046\,(\mathrm{Gr}_\delta\,\mathrm{Pr})^{1/3} \qquad \begin{aligned} q_w &= \text{const} \\ 10^6 < \mathrm{Gr}_\delta\,&\mathrm{Pr} < 10^9 \\ 1 < \mathrm{Pr} &< 20 \\ 1 < L/\delta &< 40 \end{aligned} \tag{7-54}$$

The heat flux is calculated as

$$\frac{q}{A} = q_w = h(T_1 - T_2) = \mathrm{Nu}_\delta \frac{k}{\delta}(T_1 - T_2) \tag{7-55}$$

The results are sometimes expressed in terms of an *effective* or *apparent thermal conductivity* k_e, defined by

$$\frac{q}{A} = k_e \frac{T_1 - T_2}{\delta} \tag{7-56}$$

By comparing Eqs. (7-55) and (7-56), we see that

$$\mathrm{Nu}_\delta \equiv \frac{k_e}{k} \tag{7-57}$$

In the building industry the heat transfer across an air gap is sometimes expressed in terms of the R values (see Sec. 2-3), so that

$$\frac{q}{A} = \frac{\Delta T}{R}$$

In terms of the above discussion, the R value would be

$$R = \frac{\delta}{k_e} \tag{7-58}$$

Heat transfer in horizontal enclosed spaces involves two distinct situations. If the upper plate is maintained at a higher temperature than the lower plate, the lower-density fluid is above the higher-density fluid and no convection currents will be experienced. In this case the heat transfer across the space will be by conduction alone and $\mathrm{Nu}_\delta = 1.0$, where δ is still the separation distance between the plates. The second, and more interesting, case is experienced when the lower plate has a higher temperature than the upper plate. For values of Gr_δ below about 1700, pure conduction is still observed and $\mathrm{Nu}_\delta = 1.0$. As convection begins, a pattern of hexagonal cells is formed as shown in Fig. 7-12. These patterns are called Benard cells [33]. Turbulence begins at about $\mathrm{Gr}_\delta = 50{,}000$ and destroys the cellular pattern.

Free convection in inclined enclosures is discussed by Dropkin and Somerscales [12]. Evans and Stefany [9] have shown that transient natural-convection heating or cooling in closed vertical or horizontal cylindrical enclosures may be calculated with

$$\mathrm{Nu}_f = 0.55\,(\mathrm{Gr}_f\,\mathrm{Pr}_f)^{1/4} \tag{7-59}$$

for the range $0.75 < L/d < 2.0$. The Grashof number is formed with the length of the cylinder L.

The analysis and experiments of Ref. 43 indicate that it is possible to represent

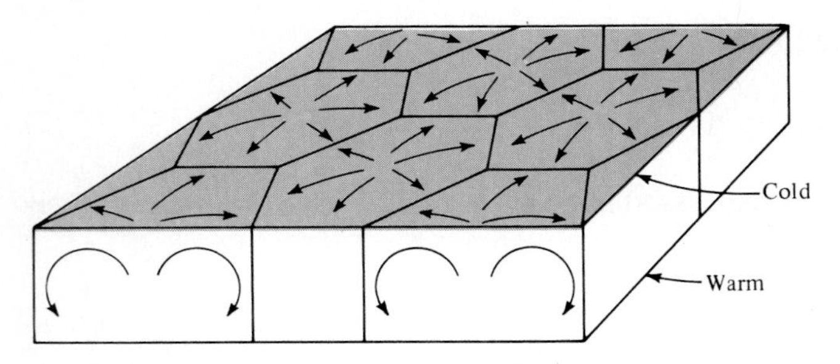

Fig. 7-12 Benard-cell pattern in enclosed fluid layer heated from below, from Ref. 33.

the effective thermal conductivity for fluids between concentric spheres with the relation

$$\frac{k_e}{k} = 0.228\ (\mathrm{Gr}_\delta\ \mathrm{Pr})^{0.226} \tag{7-60}$$

where now the gap spacing is $\delta = r_o - r_i$. The effective thermal conductivity given by Eq. (7-60) is to be used with the conventional relation for steady-state conduction in a spherical shell:

$$q = \frac{4\pi k r_i r_o\ \Delta T}{r_o - r_i} \tag{7-61}$$

Equation (7-60) is valid for $0.25 \leq \delta/r_i \leq 1.5$ and

$$1.2 \times 10^2 < \mathrm{Gr\ Pr} < 1.1 \times 10^9 \qquad 0.7 < \mathrm{Pr} < 4150$$

Properties are evaluated at a volume mean temperature T_m defined by

$$T_m = \frac{(r_m{}^3 - r_i^3)T_i + (r_o{}^3 - r_m{}^3)T_o}{r_o{}^3 - r_i^3} \tag{7-62}$$

where $r_m = (r_i + r_o)/2$. Equation (7-60) may also be used for eccentric spheres with a coordinate transformation as described in Ref. 43.

Experimental results for free convection in enclosures are not always in agreement, but we can express them in a general form as

$$\frac{k_e}{k} = C\ (\mathrm{Gr}_\delta\ \mathrm{Pr})^n \left(\frac{L}{\delta}\right)^m \tag{7-63}$$

Table 7-3 lists values of the constants C, n, and m for a number of physical circumstances. These values may be used for design purposes in the absence of specific data for the geometry or fluid being studied. We should remark that some of the data correlations represented by Table 7-3 have been artificially adjusted by Holman [74] to give the characteristic exponents of $\frac{1}{4}$ and $\frac{1}{3}$ for the laminar and turbulent regimes of free convection. However, it appears that the error introduced by this adjustment is not significantly greater than the disa-

Table 7-3 Summary of Empirical Relations for Free Convection in Enclosures in the Form of Eq. (7-60), Correlation Constants Adjusted by Holman [74].

Fluid	*Geometry*	Gr_δ Pr	Pr	$\frac{L}{\delta}$	*C*	*n*	*m*	*Ref(s).*
Gas	Vertical plate,	<2000	$k_e/k = 1.0$					6, 7, 55, 59
	isothermal	6000–200,000	0.5–2	11–42	0.197	$\frac{1}{4}$	$-\frac{1}{9}$	
		200,000–1.1×10^7	0.5–2	11–42	0.073	$\frac{1}{3}$	$-\frac{1}{9}$	
	Horizontal plate,	<1700	$k_e/k = 1.0$					
	isothermal	1700–7000	0.5–2	—	0.059	0.4	0	6, 7, 55, 59, 62, 63,
	heated from							
	below	7000–3.2×10^5	0.5–2	—	0.212	$\frac{1}{4}$	0	66
		$>3.2 \times 10^5$	0.5–2	—	0.061	$\frac{1}{3}$	0	
Liquid	Vertical plate,	<2000	$k_e/k = 1.0$					
	constant heat	10^4–10^7	1–20,000	10–40	Eq. 7-52	—	—	18, 61
	flux or	10^6–10^9	1–20	1–40	0.046	$\frac{1}{3}$	0	
	isothermal							
	Horizontal plate,	<1700	$k_e/k = 1.0$	—				7, 8, 58, 63, 66
	isothermal,	1700–6000	1–5000	—	0.012	0.6	0	
	heated from	6000–37,000	1–5000	—	0.375	0.2	0	
	below	37,000–10^8	1–20	—	0.13	0.3	0	
		$>10^8$	1–20		0.057	$\frac{1}{3}$	0	
Gas or liquid	Vertical annulus	Same as vertical plates						
	Horizontal annulus,	6000–10^6	1–5000	—	0.11	0.29	0	56, 57, 60
	isothermal	10^6–10^8	1–5000	—	0.40	0.20	0	
	Spherical annulus	120–1.1×10^9	0.7–4000		0.228	0.226	0	43
						—	—	

greement between different experimental investigations. The interested reader may wish to consult the specific references for more details.

For the annulus space the heat transfer is based on

$$q = \frac{2\pi k L\ \Delta T}{\ln (r_o/r_i)} \tag{7-64}$$

where L is the length of the annulus and the gap spacing is $\delta = r_o - r_i$.

Extensive correlations for free convection between cylindrical, cubical, and spherical bodies and various enclosure geometries are given by Warrington and Powe [80]. The correlations cover a wide range of fluids.

Free convection through vertical plane layers of nonnewtonian fluids is discussed in Ref. 38, but the results are too complicated to present here.

In the absence of more specific design information the heat transfer for inclined enclosures may be calculated by substituting g' for g in the Grashof number, where

$$g' = g \cos \theta \tag{7-65}$$

and θ is the angle which the heater surface makes with the horizontal. This transformation may be expected to hold up to inclination angles of 60° and applies *only* to those cases where the hotter surface is facing upward. Further information is available from Hollands et al. [66, 67, 69, 82].

■ **EXAMPLE 7-8**

Air at atmospheric pressure is contained between two 0.5-m-square vertical plates separated by a distance of 15 mm. The temperatures of the plates are 100 and 40°C, respectively. Calculate the free-convection heat transfer across the air space.

Solution

We evaluate the air properties at the mean temperature between the two plates:

$$T_f = \frac{100 + 40}{2} = 70°\text{C} = 343 \text{ K}$$

$$\rho = \frac{p}{RT} = \frac{1.0132 \times 10^5}{(287)(343)} = 1.029 \text{ kg/m}^3$$

$$\beta = \frac{1}{T_f} = \frac{1}{343} = 2.915 \times 10^{-3} \text{ K}^{-1}$$

$$\mu = 2.043 \times 10^{-5} \text{ kg/m} \cdot \text{s} \qquad k = 0.0295 \text{ W/m} \cdot °\text{C} \qquad \text{Pr} = 0.7$$

The Grashof-Prandtl number product is now calculated as

$$\text{Gr}_\delta \text{ Pr} = \frac{(9.8)(1.029)^2(2.915 \times 10^{-3})(100 - 40)(15 \times 10^{-3})^3}{(2.043 \times 10^{-5})^2}\, 0.7$$

$$= 1.027 \times 10^4$$

We may now use Eq. (7-63) to calculate the effective thermal conductivity, with $L = 0.5$ m, $\delta = 0.015$ m, and the constants taken from Table 7-3:

$$\frac{k_e}{k} = (0.197)(1.027 \times 10^4)^{1/4}\left(\frac{0.5}{0.015}\right)^{-1/9} = 1.343$$

The heat transfer may now be calculated with Eq. (7-53). The area is $(0.5)^2 = 0.25\ \text{m}^2$, so that

$$q = \frac{(1.343)(0.0295)(0.25)(100 - 40)}{0.015} = 39.62\ \text{W} \qquad [135.2\ \text{Btu/h}]$$

■ **EXAMPLE 7-9**

Two horizontal plates 20 cm on a side are separated by a distance of 1 cm with air at 1 atm in the space. The temperatures of the plates are 100°C for the lower and 40°C for the upper plate. Calculate the heat transfer across the air space.

Solution

The properties are the same as given in Example 7-8:

$$\rho = 1.029\ \text{kg/m}^3 \qquad \beta = 2.915 \times 10^{-3}\text{K}^{-1}$$

$$\mu = 2.043 \times 10^{-5}\ \text{kg/m·s} \qquad k = 0.0295\ \text{W/m·°C}$$

$$\text{Pr} = 0.7$$

The Gr Pr product is evaluated on the basis of the separating distance, so we have

$$\text{Gr Pr} = \frac{(9.8)(1.029)^2(2.915 \times 10^{-3})(100 - 40)(0.01)^3}{(2.043 \times 10^{-5})^2}(0.7) = 3043$$

Consulting Table 7-3, we find $C = 0.059$, $n = 0.4$, and $m = 0$ so that

$$\frac{k_e}{k} = (0.059)(3043)^{0.4}\left(\frac{0.2}{0.01}\right)^0 = 1.46$$

and

$$q = \frac{k_e A(T_1 - T_2)}{\delta} = \frac{(1.460)(0.0295)(0.2)^2(100 - 40)}{0.01} = 10.34\ \text{W}$$

■ **Example 7-10** Heat transfer across water layer

Two 50-cm horizontal square plates are separated by a distance of 1 cm. The lower plate is maintained at a constant temperature of 100°F and the upper plate is constant at 80°F. Water at atmospheric pressure occupies the space between the plates. Calculate the heat lost by the lower plate.

Solution

We evaluate properties at the mean temperature of 90°F and obtain, for water,

$$k = 0.623\ \text{W/m} \cdot \text{°C} \qquad \frac{g\beta\rho^2 c_p}{\mu k} = 2.48 \times 10^{10}$$

The Grashof-Prandtl number product is now evaluated using the plate spacing of 1 cm as the characteristic dimension.

$$\text{Gr Pr} = (2.48 \times 10^{10})(0.01)^3(100 - 80)(5/9) = 2.76 \times 10^5$$

Now, using Eq. (7-63) and consulting Table 7-3 we obtain

$$C = 0.13 \qquad n = 0.3 \qquad m = 0$$

Therefore, Eq. (7-63) becomes

$$\frac{k_e}{k} = (0.13)(2.76 \times 10^5)^{0.3} = 5.57$$

The effective thermal conductivity is thus

$$k_e = (0.623)(5.57) = 3.47 \text{ W/m} \cdot {}^\circ\text{C}$$

and the heat transfer is

$$q = k_e A \Delta T/\delta = \frac{(3.47)(0.5)^2(100 - 80)(5/9)}{0.01} = 964 \text{ W}$$

■ 7-12 COMBINED FREE AND FORCED CONVECTION

A number of practical situations involve convection heat transfer which is neither "forced" nor "free" in nature. The circumstances arise when a fluid is forced over a heated surface at a rather low velocity. Coupled with the forced-flow velocity is a convective velocity which is generated by the buoyancy forces resulting from a reduction in fluid density near the heated surface.

A summary of combined free- and forced-convection effects in tubes has been given by Metais and Eckert [10], and Fig. 7-13 presents the regimes for combined convection in vertical tubes. Two different combinations are indicated in this figure. *Aiding flow* means that the forced- and free-convection currents are in the same direction, while *opposing flow* means that they are in the opposite direction. The abbreviation UWT means uniform wall temperature, and the abbreviation UHF indicates data for uniform heat flux. It is fairly easy to anticipate the qualitative results of the figure. A large Reynolds number implies a large forced-flow velocity, and hence less influence of free-convection currents. The larger the value of the Grashof-Prandtl product, the more one would expect free-convection effects to prevail.

Figure 7-14 presents the regimes for combined convection in horizontal tubes. In this figure the Graetz number is defined as

$$\text{Gz} = \text{Re Pr} \frac{d}{L} \tag{7-66}$$

The applicable range of Figs. 7-13 and 7-14 is for

$$10^{-2} < \text{Pr}\left(\frac{d}{L}\right) < 1$$

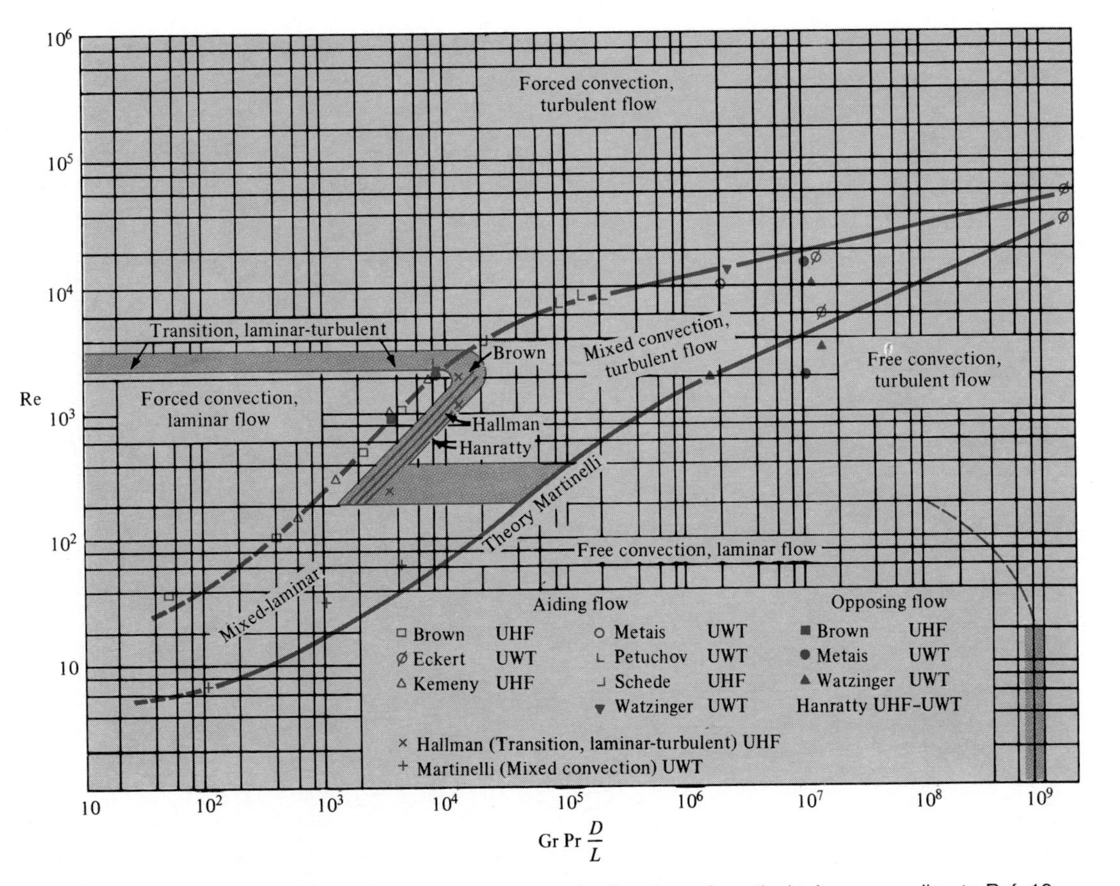

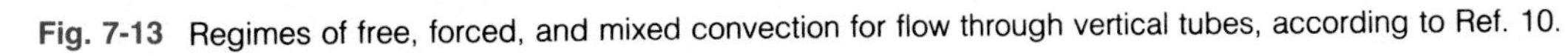
Fig. 7-13 Regimes of free, forced, and mixed convection for flow through vertical tubes, according to Ref. 10.

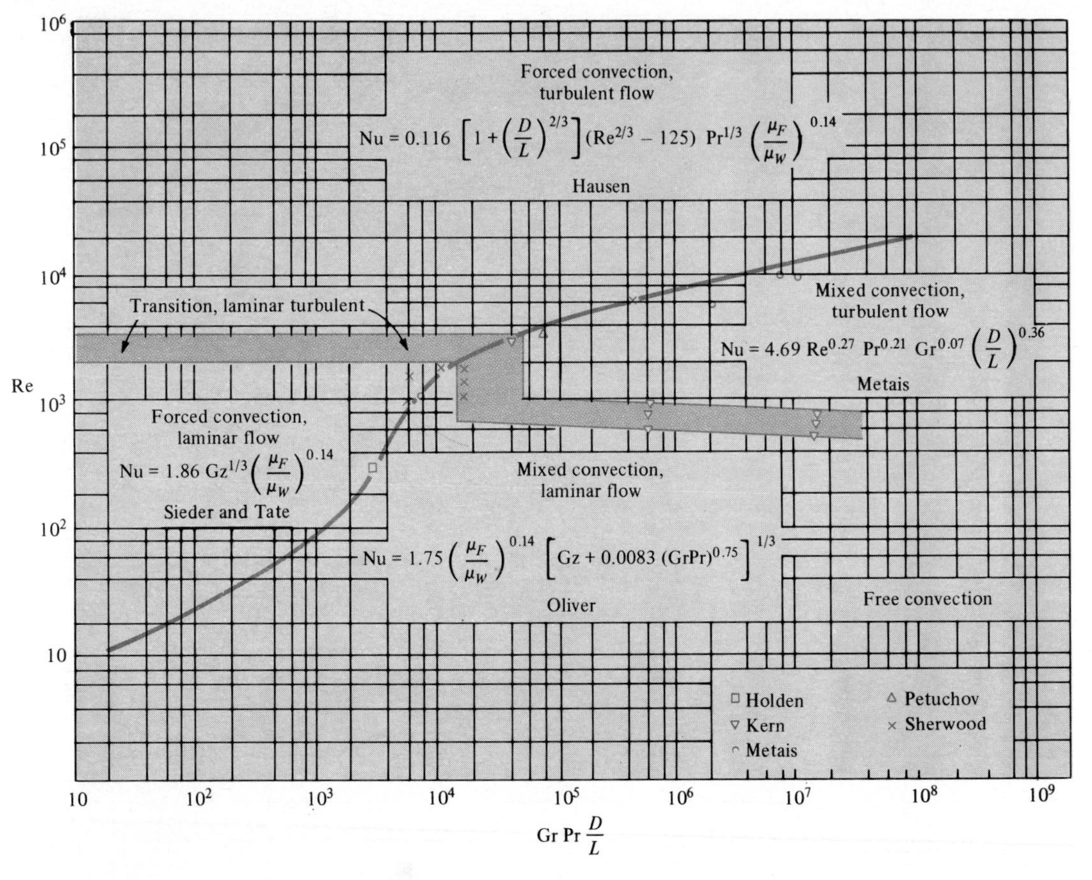

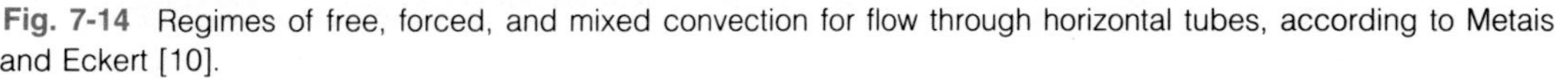
Fig. 7-14 Regimes of free, forced, and mixed convection for flow through horizontal tubes, according to Metais and Eckert [10].

The correlations presented in the figures are for constant wall temperature. All properties are evaluated at the film temperature.

Brown and Gauvin [17] have developed a better correlation for the mixed-convection, laminar flow region of Fig. 7-14:

$$\mathrm{Nu} = 1.75 \left(\frac{\mu_b}{\mu_w}\right)^{0.14} [\mathrm{Gz} + 0.012\,(\mathrm{Gz}\,\mathrm{Gr}^{1/3})^{4/3}]^{1/3} \tag{7-67}$$

where μ_b is evaluated at the bulk temperature. This relation is preferred over that shown in Fig. 7-14. Further information is available in Ref. 68. The problem of combined free and forced convection from horizontal cylinders is treated in detail by Fand and Keswani [47].

The general notion which is applied in combined-convection analysis is that the predominance of a heat-transfer mode is governed by the fluid velocity associated with that mode. A forced-convection situation involving a fluid velocity of 30 m/s, for example, would be expected to overshadow most free-convection effects encountered in ordinary gravitational fields because the velocities of the free-convection currents are small in comparison with 30 m/s. On the other hand, a forced-flow situation at very low velocities (~0.3 m/s) might be influenced appreciably by free-convection currents. An order-of-magnitude analysis of the free-convection boundary-layer equations will indicate a general criterion for determining whether free-convection effects dominate. The criterion is that when

$$\mathrm{Gr}/\mathrm{Re}^2 > 10 \tag{7-68}$$

free convection is of primary importance. This result is in agreement with Figs. 7-13 and 7-14.

■ EXAMPLE 7-11

Air at 1 atm and 27°C is forced through a horizontal 25-mm-diameter tube at an average velocity of 30 cm/s. The tube wall is maintained at a constant temperature of 140°C. Calculate the heat-transfer coefficient for this situation if the tube is 0.4 m long.

Solution

For this calculation we evaluate properties at the film temperature:

$$T_f = \frac{140 + 27}{2} = 83.5°\mathrm{C} = 356.5\ k$$

$$\rho_f = \frac{p}{RT} = \frac{1.0132 \times 10^5}{(287)(356.5)} = 0.99\ \mathrm{kg/m^3}$$

$$\beta = \frac{1}{T_f} = 2.805 \times 10^{-3}\ \mathrm{K}^{-1} \qquad \mu_w = 2.337 \times 10^{-5}\ \mathrm{kg/m \cdot s}$$

$$\mu_f = 2.102 \times 10^{-5}\ \mathrm{kg/m \cdot s} \qquad k_f = 0.0305\ \mathrm{W/m \cdot °C} \qquad \mathrm{Pr} = 0.695$$

Let us take the bulk temperature as 27°C for evaluating μ_b; then

$$\mu_b = 1.8462 \times 10^{-5}\ \mathrm{kg/m \cdot s}$$

The significant parameters are calculated as

$$\mathrm{Re}_f = \frac{\rho u d}{\mu} = \frac{(0.99)(0.3)(0.025)}{2.102 \times 10^{-5}} = 3.53$$

$$\mathrm{Gr} = \frac{\rho^2 g \beta (T_w - T_b) d^3}{\mu^2} = \frac{(0.99)^2(9.8)(2.805 \times 10^{-3})(140 - 27)(0.025)^3}{(2.102 \times 10^{-5})^2}$$

$$= 1.007 \times 10^5$$

$$\mathrm{Gr}\,\mathrm{Pr}\,\frac{d}{L} = (1.077 \times 10^5)(0.695)\,\frac{0.025}{0.4} = 4677$$

According to Fig. 7-14, the mixed-convection-flow regime is encountered. Thus we must use Eq. (7-67). The Graetz number is calculated as

$$\mathrm{Gz} = \mathrm{Re}\,\mathrm{Pr}\,\frac{d}{L} = \frac{(353)(0.695)(0.025)}{0.4} = 15.33$$

and the numerical calculation for Eq. (7-67) becomes

$$\mathrm{Nu} = 1.75\left(\frac{1.8462}{2.337}\right)^{0.14}\{15.33 + (0.012)[(15.33)(1.077 \times 10^5)^{1/3}]^{4/3}\}^{1/3}$$

$$= 7.70$$

The average heat-transfer coefficient is then calculated as

$$\bar{h} = \frac{k}{d}\,\mathrm{Nu} = \frac{(0.0305)(7.70)}{0.025} = 9.40\ \mathrm{W/m^2 \cdot {}^\circ C} \qquad [1.67\ \mathrm{Btu/h \cdot ft^2 \cdot {}^\circ F}]$$

It is interesting to compare this value with that which would be obtained for strictly laminar forced convection. The Sieder-Tate relation [Eq. (6-10)] applies, so that

$$\mathrm{Nu} = 1.86\,(\mathrm{Re}\,\mathrm{Pr})^{1/3}\left(\frac{\mu_f}{\mu_w}\right)^{0.14}\left(\frac{d}{L}\right)^{1/3}$$

$$= 1.86\,\mathrm{Gz}^{1/3}\left(\frac{\mu_f}{\mu_w}\right)^{0.14}$$

$$= (1.86)(15.33)^{1/3}\left(\frac{2.102}{2.337}\right)^{0.14}$$

$$= 4.55$$

and

$$\bar{h} = \frac{(4.55)(0.0305)}{0.025} = 5.55\ \mathrm{W/m^2 \cdot {}^\circ C} \quad [0.977\ \mathrm{Btu/h \cdot ft^2 \cdot {}^\circ F}]$$

Thus there would be an error of −41 percent if the calculation were made strictly on the basis of laminar forced convection.

7-13 SUMMARY

By now the reader will have sensed that there is an abundance of empirical relations for natural convection systems. Our purposes in this section are to

Table 7-4 Summary of Free-Convection Heat-Transfer Relations.

Geometry	*Equation*	*Restrictions*	*Equation number*
A variety of isothermal surfaces	$\mathrm{Nu}_f = C\,(\mathrm{Gr}_f\,\mathrm{Pr}_f)^m$ C and m from Table 7-1	See Table 7-1	(7-25)
Vertical isothermal surface	$\overline{\mathrm{Nu}}^{1/2} = 0.825 + \dfrac{0.387\,\mathrm{Ra}^{1/6}}{[1 + (0.492/\mathrm{Pr})^{9/16}]^{8/27}}$	$10^{-1} < \mathrm{Ra}_L < 10^{12}$	(7-29)
Vertical surface, constant heat flux, local h	$\mathrm{Nu}_{xf} = C\,(\mathrm{Gr}_x^*\,\mathrm{Pr}_f)^m$	$C = 0.60$, $m = \frac{1}{5}$ for $10^5 < \mathrm{Gr}_x^* < 10^{11}$ $C = 0.17$, $m = \frac{1}{4}$ for $2 \times 10^{13} < \mathrm{Gr}^* < 10^{16}$	(7-31) (7-32)
Isothermal horizontal cylinders	$\overline{\mathrm{Nu}}^{1/2} = 0.60 + 0.387\left\{\dfrac{\mathrm{Gr\,Pr}}{[1 + (0.559/\mathrm{Pr})^{9/16}]^{16/9}}\right\}^{1/6}$	$10^{-5} < \mathrm{Gr\,Pr} < 10^{12}$	(7-36)
Horizontal surface, constant heat flux		See text	(7-39) to (7-42)
Inclined surfaces	Sec. 7-7	See text	
Spheres		See text	(7-49) to (7-51)
Enclosed spaces	Sec. 7-11 and Table 7-3		

(1) issue a few words of caution and (2) provide a convenient table to summarize the relations.

Most free-convection data are collected under laboratory conditions in still air, still water, etc. A practical free-convection problem might not be so fortunate and the boundary layer could have a slightly added forced-convection effect. In addition, real surfaces in practice are *seldom* isothermal or constant heat flux so the correlations developed from laboratory data for these conditions may not strictly apply. The net result, of course, is that the engineer must realize that calculated values of the heat-transfer coefficient can vary ± 25 percent from what will actually be experienced.

For solution of free-convection problems one should follow a procedure similar to that given in Chap. 6 for forced-convection problems.

1. Decide that the problem is indeed a *free*-convection problem.
2. Establish the geometry of the situation; vertical plate, horizontal cylinder, etc.
3. Make a preliminary determination of appropriate fluid properties.
4. Establish the flow regime by calculating the Grashof-Prandtl number product. Be careful to employ the correct characteristic dimension for the particular geometry.
5. Select an equation which fits the geometry and flow regime and reevaluate properties, if necessary, in accordance with stipulations and the equation.
6. Proceed to calculate the value of h and/or the heat-transfer rate.

To aid the reader in selecting correlations for free convection we have given a summary in Table 7-4.

REVIEW QUESTIONS

1 Why is an analytical solution of a free-convection problem more involved than its forced-convection counterpart?

2 Define the Grashof number. What is its physical significance?

3 What is the approximate criterion for transition to turbulence in a free-convection boundary layer?

4 What functional form of equation is normally used for correlation of free-convection heat-transfer data?

5 Discuss the problem of combined free and forced convection.

6 What is the approximate criterion dividing pure conduction and free convection in an enclosed space between vertical walls?

7 How is a modified Grashof number defined for a constant-heat-flux condition on a vertical plate?

PROBLEMS

7-1 Suppose the heat-transfer coefficients for forced or free convection over vertical flat plates are to be compared. Develop an approximate relation between the Reynolds and Grashof numbers such that the heat-transfer coefficients for pure forced convection and pure free convection are equal. Assume laminar flow.

7-2 For a vertical isothermal flat plate at 93°C exposed to air at 20°C and 1 atm, plot the free-convection velocity profiles as a function of distance from the plate surface at x positions of 15, 30, and 45 cm.

7-3 Show that $\beta = 1/T$ for an ideal gas having the equation of state $p = \rho RT$.

7-4 A 1-ft-square vertical plate is maintained at 65°C and is exposed to atmospheric air at 15°C. Compare the free-convection heat transfer from this plate with that which would result from forcing air over the plate at a velocity equal to the maximum velocity which occurs in the free-convection boundary layer. Discuss this comparison.

7-5 Plot the free-convection boundary-layer thickness as a function of x for a vertical plate maintained at 80°C and exposed to air at atmospheric pressure and 15°C. Consider the laminar portion only.

7-6 Derive an expression for the maximum velocity in the free-convection boundary layer on a vertical flat plate. At what position in the boundary layer does this maximum velocity occur?

7-7 Two vertical flat plates at 65°C are placed in a tank of water at 25°C. If the plates are 30 cm high, what is the minimum spacing which will prevent interference of the free-convection boundary layers?

7-8 A vertical cylinder having a length of 30 cm is maintained at 100°C and exposed to room air at 15°C. Calculate the minimum diameter the cylinder can have in order to behave as a vertical flat plate.

7-9 A 1-m-square vertical plate is heated to 400°C and placed in room air at 25°C. Calculate the heat loss from one side of the plate.

7-10 A vertical flat plate is maintained at a constant temperature of 120°F and exposed to atmospheric air at 70°F. At a distance of 14 in from the leading edge of the plate the boundary layer thickness is 1.0 in. Estimate the thickness of the boundary layer at a distance of 24 in from the leading edge.

7-11 A vertical cylinder 1.8 m high and 7.5 cm in diameter is maintained at a temperature of 93°C in an atmospheric environment of 30°C. Calculate the heat lost by free convection from this cylinder. For this calculation the cylinder may be treated as a vertical flat plate.

7-12 The outside wall of a building 6 m high receives an average radiant heat flux from the sun of 1100 W/m². Assuming that 95 W/m² is conducted through the wall, estimate the outside wall temperature. Assume the atmospheric air on the outside of the building is at 20°C.

7-13 Assuming that a human may be approximated by a vertical cylinder 1 ft in diameter and 6 ft tall, estimate the free-convection heat loss for a surface temperature of 75°F in ambient air at 68°F.

7-14 A 30-cm-square vertical plate is heated electrically such that a constant-heat-flux condition is maintained with a total heat dissipation of 30 W. The ambient air is at 1 atm and 20°C. Calculate the value of the heat-transfer coefficient at heights of 15 and 30 cm. Also calculate the average heat-transfer coefficient for the plate.

7-15 A 1-ft-square vertical plate is maintained at 120°F and exposed to room air at 1 atm and 65°F. Calculate the heat lost from *both* sides of the plate.

7-16 Calculate the free-convection heat loss from a 2-ft-square vertical plate maintained at 100°C and exposed to helium at 20°C and a pressure of 2 atm.

7-17 A large vertical plate 20 ft high and 4 ft wide is maintained at a constant temperature of 135°F and exposed to atmospheric air at 40°F. Calculate the heat lost by the plate.

7-18 A 1-m-square vertical plate is maintained at 120°F and exposed to room air at 70°F. Calculate the heat lost by the plate.

7-19 What vertical distance is necessary to produce a Rayleigh number of 10^{12} in air at standard conditions and $\Delta T = 10°C$?

7-20 A 25 by 25 cm vertical plate is fitted with an electric heater which produces a constant heat flux of 1000 W/m^2. The plate is submerged in water at 15°C. Calculate the heat-transfer coefficient and the average temperature of the plate. How much heat would be lost by an isothermal surface at this average temperature?

7-21 Assume that one-half of the heat transfer by free convection from a horizontal cylinder occurs on each side of the cylinder because of symmetry considerations. Going by this assumption, compare the heat transfer on each side of the cylinder with that from a vertical flat plate having a height equal to the circumferential distance from the bottom stagnation point to the top stagnation point on the cylinder. Discuss this comparison.

7-22 A horizontal cylindrical heater with $d = 2$ cm is placed in a pool of sodium-potassium mixture with 22 percent sodium. The mixture is at 120°C, and the heater surface is constant at 200°C. Calculate the heat transfer for a heater 40 cm long.

7-23 A horizontal heating rod having a diameter of 3.0 cm and a length of 1 m is placed in a pool of saturated liquid ammonia at 20°C. The heater is maintained at a constant surface temperature of 70°C. Calculate the heat-transfer rate.

7-24 Condensing steam at 120°C is to be used inside a 7.5-cm-diameter horizontal pipe to provide heating for a certain work area where the ambient air temperature is 17°C. The total heating required is 100,000 Btu/h. What length pipe would be required to accomplish this heating?

7-25 A 10-cm length of platinum wire 0.4 mm in diameter is placed horizontally in a container of water at 38°C and is electrically heated so that the surface temperature is maintained at 93°C. Calculate the heat lost by the wire.

7-26 Water at the rate of 0.8 kg/s at 90°C flows through a steel pipe with 2.5-cm ID and 3-cm OD. The outside surface temperature of the pipe is 85°C, and the temperature of the surrounding air is 20°C. The room pressure is 1 atm, and the pipe is 15 m long. How much heat is lost by free convection to the room?

7-27 A horizontal pipe 8.0 cm in diameter is located in a room where atmospheric air is at 20°C. The surface temperature of the pipe is 140°C. Calculate the free-convection heat loss per meter of pipe.

7-28 A horizontal 1.25-cm-OD tube is heated to a surface temperature of 250°C and exposed to air at room temperature of 20°C and 1 atm. What is the free-convection heat transfer per unit length of tube?

7-29 A horizontal electric heater 2.5 cm in diameter is submerged in a light-oil bath at 93°C. The heater surface temperature is maintained at 150°C. Calculate the heat lost per foot of length of the heater.

7-30 A 1-ft-square air-conditioning duct carries air at a temperature such that the outside temperature of the duct is maintained at 60°F and is exposed to room air at 80°F. Estimate the heat gained by the duct per foot of length.

7-31 A fine wire having a diameter of 0.001 in (0.0254 mm) is heated by an electric current and placed horizontally in a chamber containing helium at 3 atm and 10°C. If the surface temperature of the wire is not to exceed 240°C, calculate the electric power to be supplied per unit length.

7-32 A large circular duct, 3.0 m in diameter, carries hot gases at 250°C. The outside of the duct is exposed to room air at 1 atm and 20°C. Estimate the heat loss per unit length of the duct.

7-33 A 2.0-cm diameter is placed in a tank of glycerine at 20°C. The surface temperature of the heater is 60°C, and its length is 60 cm. Calculate the heat transfer.

7-34 A 3.5-cm-diameter cylinder contains an electric heater which maintains a constant heat flux at the surface of 1500 W/m^2. If the cylinder is inclined at an angle of 35° with the horizontal and exposed to room air at 20°C, estimate the average surface temperature.

7-35 A 30-cm-diameter horizontal pipe is maintained at a constant temperature of 25°C and placed in room air at 20°C. Calculate the free-convection heat loss from the pipe per unit length.

7-36 A 5-in-diameter duct is maintained at a constant temperature of 500°F by hot combustion gases inside. The duct is located horizontally in a small warehouse area having an ambient temperature of 60°F. Calculate the length of the duct necessary to provide 125,000 Btu/h of convection heating.

7-37 A horizontal cylinder with diameter of 5 cm and length of 3 m is maintained at 180°F and submerged in water which is at 60°F. Calculate the heat lost by the cylinder.

7-38 A 2.0-m-diameter horizontal cylinder is maintained at a constant temperature of 77°C and exposed to a large warehouse space at 27°C. The cylinder is 20 m long. Calculate the heat lost by the cylinder.

7-39 Calculate the rate of free-convection heat loss from a 30-cm-diameter sphere maintained at 90°C and exposed to atmospheric air at 20°C.

7-40 A 2.5-cm-diameter sphere at 32°C is submerged in water at 10°C. Calculate the rate of free-convection heat loss.

7-41 A spherical balloon gondola 2.4 m in diameter rises to an altitude where the ambient pressure is 1.4 kPa and the ambient temperature is −50°C. The outside surface of the sphere is at approximately 0°C. Estimate the free-convection heat loss from the outside of the sphere. How does this compare with the forced-convection loss from such a sphere with a low free-stream velocity of approximately 30 cm/s?

7-42 A 2.5-cm diameter sphere is maintained at 38°C and submerged in water at 15°C. Calculate the heat-transfer rate under these conditions.

7-43 Apply the reasoning pertaining to the last entry of Table 7-1 to free convection from a sphere and compare with Eq. (7-50).

7-44 Using the information in Table 7-1 and the simplified relations of Table 7-2 devise a simplified relation which may be used as a substitute for Eq. (7-50) to calculate free convection from a sphere to air at 1 atm.

7-45 A circular hot plate, 15 cm in diameter, is maintained at 150°C in atmospheric air at 20°C. Calculate the free-convection heat loss when the plate is in a horizontal position.

7-46 An engine-oil heater consists of a large vessel with a square-plate electric-heater surface in the bottom of the vessel. The heater plate is 30 by 30 cm and is maintained at a constant temperature of 60°C. Calculate the heat-transfer rate for an oil temperature of 20°C.

7-47 Small electric strip heaters with a width of 6 mm are oriented in a horizontal position. The strips are maintained at 500°C and exposed to room air at 20°C. Assuming that the strips dissipate heat from both the top and the bottom surfaces, estimate the strip length required to dissipate 2 kW of heat by free convection.

7-48 The top surface of a 10 by 10 m horizontal plate is maintained at 25°C and exposed to room temperature at 28°C. Estimate the heat transfer.

7-49 A 4 by 4 m horizontal heater is placed in room air at 15°C. Both the top and the bottom surfaces are heated to 50°C. Estimate the total heat loss by free convection.

7-50 A horizontal plate, uniform in temperature at 400 K, has the shape of an equilateral triangle 45 cm on each side and is exposed to atmospheric air at 300 K. Calculate the heat lost by the plate.

7-51 A heated plate, 20 by 20 cm, is inclined at an angle of 60° with the horizontal and placed in water. Approximately constant-heat-flux conditions prevail with a mean plate temperature of 40°C and the heated surface facing downward. The water temperature is 20°C. Calculate the heat lost by the plate.

7-52 Repeat Prob. 7-51 for the heated plate facing upward.

7-53 A double plate-glass window is constructed with a 1.25-cm air space. The plate dimensions are 1.2 by 1.8 m. Calculate the free-convection heat-transfer rate through the air space for a temperature difference of 30°C and $T_1 = 20$°C.

7-54 A flat-plate solar collector is 1 m square and is inclined at an angle of 20° with the horizontal. The hot surface at 160°C is placed in an enclosure which is evacuated to a pressure of 0.1 atm. Above the hot surface, and parallel to it, is the transparent window which admits the radiant energy from the sun. The hot surface and window are separated by a distance of 8 cm. Because of convection to the surroundings, the window temperature is maintained at 40°C. Calculate the free-convection heat transfer between the hot surface and the transparent window.

7-55 A flat plate 1 by 1 m is inclined at 30° with the horizontal and exposed to atmospheric air at 30°C and 1 atm. The plate receives a net radiant-energy flux from the sun of 700 W/m^2, which then is dissipated to the surroundings by free convection. What average temperature will be attained by the plate?

7-56 A horizontal cylinder having a diameter of 5 cm and an emissivity of 0.5 is placed in a large room, the walls of which are maintained at 35°C. The cylinder loses heat by natural convection with an h of 6.5 $W/m^2 \cdot °C$. A sensitive thermocouple placed on the surface of the cylinder measures the temperature as 30°C. What is the temperature of the air in the room?

7-57 A 10 by 10 cm plate is maintained at 80°C and inclined at 45° with the horizontal. Calculate the heat loss from both sides of the plate to room air at 20°C.

7-58 A 5 by 5 cm plate is maintained at 50°C and inclined at 60° with the horizontal. Calculate the heat loss from both sides of the plate to water at 20°C.

7-59 Air at 1 atm and 38°C is forced through a horizontal 6.5-mm-diameter tube at an average velocity of 30 m/s. The tube wall is maintained at 540°C, and the tube is 30 cm long. Calculate the average heat-transfer coefficient. Repeat for a velocity of 30 m/s and a tube wall temperature of 800°C.

7-60 A small copper block having a square bottom 2.5 by 2.5 cm and a vertical height of 5 cm cools in room air at 1 atm and 20°C. The block is isothermal at 93°C. Calculate the heat-transfer rate.

7-61 A horizontal plate in the shape of an equilateral triangle 40 cm on a side is maintained at a constant temperature of 55°C and exposed to atmospheric air at 25°C. Calculate the heat lost by the top surface of the plate.

7-62 A small horizontal heater is in the shape of a circular disk with a diameter of 3 cm. The disk is maintained at 50°C and exposed to atmospheric air at 30°C. Calculate the heat loss.

7-63 A hot ceramic block at 400°C has dimensions of 15 by 15 by 8 cm high. It is exposed to room air at 27°C. Calculate the free-convection heat loss.

7-64 A magnetic amplifier is encased in a cubical box 6 in on a side and must dissipate 50 W to surrounding air at 70°F. Estimate the surface temperature of the box.

7-65 A glass thermometer is placed in a large room, the walls of which are maintained at 10°C. The convection coefficient between the thermometer and the room air is 5 $W/m^2 \cdot °C$, and the thermometer indicates a temperature of 30°C. Determine the temperature of the air in the room. Take $\epsilon = 1.0$.

7-66 A horizontal air-conditioning duct having a horizontal dimension of 30 cm and a vertical dimension of 15 cm is maintained at 120°F and exposed to atmospheric air at 70°F. Calculate the heat lost per unit length of duct.

7-67 A free-convection heater is to be designed which will dissipate 10,000 kJ/h to room air at 300 K. The heater surface temperature must not exceed 350 K. Consider four alternatives: (*a*) a group of vertical surfaces, (*b*) a single vertical surface, (*c*) a single horizontal surface, and (*d*) a group of horizontal cylindrical surfaces. Examine these alternatives and suggest a design.

7-68 Two 30-cm-square vertical plates are separated by a distance of 1.25 cm, and the space between them is filled with water. A constant-heat-flux condition is imposed on the plates such that the average temperature is 38°C for one and 60°C for the other. Calculate the heat-transfer rate under these conditions. Evaluate properties at the mean temperature.

7-69 An enclosure contains helium at a pressure of 1.3 atm and has two vertical heating surfaces, which are maintained at 80 and 20°C, respectively. The vertical surfaces are 40 by 40 cm and are separated by a gap of 2.0 cm. Calculate the free-convection heat transfer between the vertical surfaces.

7-70 A horizontal annulus with inside and outside diameters of 8 and 10 cm, respectively, contains liquid water. The inside and outside surfaces are maintained at 40 and 20°C, respectively. Calculate the heat transfer across the annulus space per meter of length.

7-71 Two concentric spheres are arranged to provide storage of brine inside the inner sphere at a temperature of −10°C. The inner-sphere diameter is 2 m, and the gap spacing is 5 cm. The outer sphere is maintained at 30°C, and the gap space is evacuated to a pressure of 0.05 atm. Estimate the free-convection heat transfer across the gap space.

7-72 A large vat used in food processing contains a hot oil at 400°F. Surrounding the vat on the vertical sides is a shell which is cooled to 140°F. The air space separating the vat and the shell is 35 cm high and 3 cm thick. Estimate the free-convection loss per square meter of surface area.

7-73 A special double-pane insulating window glass is to be constructed of two glass plates separated by an air gap. The plates are square, 60 by 60 cm, and are designed to be used with temperatures of −10 and +20°C on the respective plates. Assuming the air in the gap is at 1 atm, calculate and plot the free convection across the gap as a function of gap spacing for a vertical window. What conclusions can you draw from this plot from a design standpoint?

7-74 Repeat Prob. 7-73 for a horizontal window with the hot surface on the lower side.

7-75 Two 30-cm-square vertical plates are separated by a distance of 2.5 cm and air at 1 atm. The two plates are maintained at temperatures of 200 and 90°C, respectively. Calculate the heat-transfer rate across the air space.

7-76 A horizontal air space is separated by a distance of 1.6 mm. Estimate the heat-transfer rate per unit area for a temperature difference of 165°C, with one plate temperature at 90°C.

7-77 Repeat Prob. 7-76 for a horizontal space filled with water.

7-78 An atmospheric vertical air space 4.0 ft high has a temperature differential of 20°F at 300 K. Calculate and plot k_e/k and the R value for spacings of 0 to 10 in. At approximately what spacing is the R value a maximum?

7-79 Two vertical plates 50 by 50 cm are separated by a space of 4 cm which is filled with water. The plate temperatures are 50 and 20°C. Calculate the heat transfer across the space.

7-80 Repeat Prob. 7-79 for the plates oriented in a horizontal position with the 50°C surface as the lower plate.

7-81 Two vertical plates 1.1 by 1.1 m are separated by a 4.0-cm air space. The two surface temperatures are at 300 and 350 K. The heat transfer in the space can be reduced by decreasing the pressure of the air. Calculate the plot k_e/k and the R value as a function of pressure. To what value must the pressure be reduced to make $k_e/k = 1.0$?

7-82 Repeat Prob. 7-81 for two horizontal plates with the 350 K surface on the bottom.

7-83 An air space in a certain building wall is 10 cm thick and 2 m high. Estimate the free-convection heat transfer through this space for a temperature difference of 17°C.

7-84 Develop an expression for the optimum spacing for vertical plates in air in order to achieve minimum heat transfer, assuming that the heat transfer results from pure conduction at $Gr_\delta < 2000$. Plot this optimum spacing as a function of temperature difference for air at 1 atm.

7-85 Air at atmospheric pressure is contained between two vertical plates maintained at 100°C and 20°C, respectively. The plates are 1.0 m on a side and spaced 8 cm apart. Calculate the convection heat transfer across the air space.

7-86 A special section of insulating glass is constructed of two glass plates 30 cm square separated by an air space of 1 cm. Calculate the percent reduction in heat transfer of this arrangement compared to free convection from a vertical plate with a temperature difference of 30°C.

7-87 One way to reduce the free-convection heat loss in a horizontal solar collector is to reduce the pressure in the space separating the glass admitting the solar energy and the black absorber below. Assume the bottom surface is at 120°C and the top surface is at 20°C. Calculate the pressures that are necessary to eliminate convection for spacings of 1, 2, 5, and 10 cm.

7-88 Air at 20°C and 1 atm is forced upward through a vertical 2.5-cm-diameter tube 30 cm long. Calculate the total heat-transfer rate where the tube wall is maintained at 200°C and the flow velocity is 45 cm/s.

7-89 A horizontal tube is maintained at a surface temperature of 55°C and exposed to atmospheric air at 27°C. Heat is supplied to the tube by a suitable electric heater which produces an input of 175 W for each meter of length. Find the expected power input if the surface temperature is raised to 83°C.

7-90 A large vertical plate is maintained at a surface temperature of 140°F and exposed to air at 1 atm and 70°F. Estimate the vertical position on the plate where the boundary layer becomes turbulent. What is the average q/A for the portion of the plate preceding this location? What is the maximum velocity in the boundary layer at this location?

7-91 The horizontal air space over a solar collector has a spacing of 2.5 cm. The lower plate is maintained at 70°C while the upper plate is at 30°C. Calculate the free convection across the space for air at 1 atm. If the spacing is reduced to 1.0 cm, by how much is the heat transfer changed?

7-92 One concept of a solar collector reduces the pressure of the air gap to a value low enough to eliminate free-convection effects. For the air space in Prob. 7-91 determine the pressures to eliminate convection; that is, $\text{Gr Pr} < 1700$.

7-93 A 2.5-cm sphere is maintained at a surface temperature of 120°F and exposed to a fluid at 80°F. Compare the heat loss for (*a*) air and (*b*) water.

7-94 Air at 1 atm is contained between two concentric spheres having diameters of 10 and 8 cm and maintained at temperatures of 300 and 350 K. Calculate the free-convection heat transfer across the air gap.

7-95 Energy-conservation advocates claim that storm windows can substantially reduce energy losses (or gains). Consider a vertical 1.0-m-square window covered by a storm window with an air gap of 2.5 cm. The inside window is at 15°C and the outside storm window is at −10°C. Calculate the *R* value for the gap. What would the *R* value be for the same thickness of fiberglass blanket?

7-96 Some canned goods are to be cooled from room temperature of 300 K by placing them in a refrigerator maintained at 275 K. The cans have diameter and height of 8.0 cm. Calculate the cooling rate. Approximately how long will it take the temperature of the can to drop to 290 K if the contents have the properties of water?

7-97 A 5.0-cm-diameter horizontal disk is maintained at 120°F and submerged in water at 80°F. Calculate the heat lost from the top and bottom of the disk.

7-98 A 10-cm-square plate is maintained at 400 K on the bottom side, and exposed to air at 1 atm and 300 K. The plate is inclined at 45° with the vertical. Calculate the heat lost by the bottom surface of the plate.

7-99 Calculate the heat transfer for the plate of Prob. 7-98 if the heated surface faces upward.

7-100 A vertical cylinder 50 cm high is maintained at 400 K and exposed to air at 1 atm and 300 K. What is the minimum diameter for which the vertical-flat-plate relations may be used to calculate the heat transfer? What would the heat transfer be for this diameter?

REFERENCES

1. Eckert, E. R. G., and E. Soehngen: Interferometric Studies on the Stability and Transition to Turbulence of a Free Convection Boundary Layer, *Proc. Gen. Discuss. Heat Transfer ASME-IME, London, 1951.*
2. Eckert, E. R. G., and E. Soehngen: Studies on Heat Transfer in Laminar Free Convection with the Zehnder-Mach Interferometer, *USAF Tech. Rep.* 5747, December 1948.

3 Holman, J. P., H. E. Gartrell, and E. E. Soehngen: An Interferometric Method of Studying Boundary Layer Oscillations, *J. Heat Transfer,* ser. C, vol. 80, August 1960.

4 McAdams, W. H.: "Heat Transmission," 3d ed., McGraw-Hill Book Company, New York, 1954.

5 Yuge, T.: Experiments on Heat Transfer from Spheres Including Combined Natural and Forced Convection, *J. Heat Transfer,* ser. C, vol. 82, p. 214, 1960.

6 Jakob, M.: Free Convection through Enclosed Gas Layers, *Trans. ASME,* vol. 68, p. 189, 1946.

7 Jakob, M.: "Heat Transfer," vol. 1, John Wiley & Sons, Inc., New York, 1949.

8 Globe, S., and D. Dropkin: *J. Heat Transfer,* February 1959, pp. 24–28.

9 Evans, L. B., and N. E. Stefany: An Experimental Study of Transient Heat Transfer to Liquids in Cylindrical Enclosures, *AIChE Pap. 4, Heat Transfer Conf. Los Angeles, August 1965.*

10 Metais, B., and E. R. G. Eckert: Forced, Mixed, and Free Convection Regimes, *J. Heat Transfer,* ser. C, vol. 86, p. 295, 1964.

11 Bishop, E. N., L. R. Mack, and J. A. Scanlan: Heat Transfer by Natural Convection between Concentric Spheres, *Int. J. Heat Mass Transfer,* vol. 9, p. 649, 1966.

12 Dropkin, D., and E. Somerscales: Heat Transfer by Natural Convection in Liquids Confined by Two Parallel Plates Which Are Inclined at Various Angles with Respect to the Horizontal, *J. Heat Transfer,* vol. 87, p. 71, 1965.

13 Gebhart, B.: "Heat Transfer," 2d ed., chap. 8, McGraw-Hill Book Company, New York, 1970.

14 Gebhart, B.: Natural Convection Flow, Instability, and Transition, *ASME Pap.* 69-HT-29, August 1969.

15 Mollendorf, J. C., and B. Gebhart: An Experimental Study of Vigorous Transient Natural Convection, *ASME Pap.* 70-HT-2, May 1970.

16 Bayley, F. J.: An Analysis of Turbulent Free Convection Heat Transfer, *Proc. Inst. Mech. Eng.,* vol. 169, no. 20, p. 361, 1955.

17 Brown, C. K., and W. H. Gauvin: Combined Free and Forced Convection, I, II, *Can. J. Chem. Eng.,* vol. 43, no. 6, pp. 306, 313, 1965.

18 MacGregor, R. K., and A. P. Emery: Free Convection through Vertical Plane Layers: Moderate and High Prandtl Number Fluids, *J. Heat Transfer,* vol. 91, p. 391, 1969.

19 Newell, M. E., and F. W. Schmidt: Heat Transfer by Laminar Natural Convection within Rectangular Enclosures, *J. Heat Transfer,* vol. 92, pp. 159–168, 1970.

20 Husar, R. B., and E. M. Sparrow: Patterns of Free Convection Flow Adjacent to Horizontal Heated Surfaces, *Int. J. Heat Mass Trans.,* vol. 11, p. 1206, 1968.

21 Habne, E. W. P.: Heat Transfer and Natural Convection Patterns on a Horizontal Circular Plate, *Int. J. Heat Mass Transfer,* vol. 12, p. 651, 1969.

22 Warner, C. Y., and V. S. Arpaci: An Experimental Investigation of Turbulent Natural Convection in Air at Low Pressure along a Vertical Heated Flat Plate, *Int. J. Heat Mass Transfer,* vol. 11, p. 397, 1968.

23 Gunness, R. C., Jr., and B. Gebhart: Stability of Transient Convection, *Phys. Fluids,* vol. 12, p. 1968, 1969.

24 Rotern, Z., and L. Claassen: Natural Convection above Unconfined Horizontal Surfaces, *J. Fluid Mech.,* vol. 39, pt. 1, p. 173, 1969.

25 Vliet, G. C.: Natural Convection Local Heat Transfer on Constant Heat Flux Inclined Surfaces, *J. Heat Transfer,* vol. 91, p. 511, 1969.

26 Vliet, G. C., and C. K. Lin: An Experimental Study of Turbulent Natural Convection Boundary Layers, *J. Heat Transfer,* vol. 91, p. 517, 1969.

27 Ostrach, S.: An Analysis of Laminar-Free-Convection Flow and Heat Transfer about a Flat Plate Parallel to the Direction of the Generating Body Force, *NACA Tech. Rep.* 1111, 1953.

28 Cheesewright, R.: Turbulent Natural Convection from a Vertical Plane Surface, *J. Heat Transfer,* vol. 90, p. 1, February 1968.

29 Flack, R. D., and C. L. Witt: Velocity Measurements in Two Natural Convection Air Flows Using a Laser Velocimeter, *J. Heat Transfer,* vol. 101, p. 256, 1979.

30 Eckert, E. R. G., and T. W. Jackson: Analysis of Turbulent Free Convection Boundary Layer on a Flat Plate, *NACA Rep.* 1015, 1951.

31 King, W. J.: The Basic Laws and Data of Heat Transmission, *Mech. Eng.,* vol. 54, p. 347, 1932.

32 Sparrow, E. M., and J. L. Gregg: Laminar Free Convection from a Vertical Flat Plate, *Trans. ASME,* vol. 78, p. 435, 1956.

33 Benard, H.: Les Tourbillons cellulaires dans une nappe liquide transportant de la chaleur par convection en régime permanent, *Ann. Chim. Phys.,* vol. 23, pp. 62–144, 1901.

34 "Progress in Heat and Mass Transfer," vol. 2, Eckert Presentation Volume, Pergamon Press, New York, 1969.

35 Gebhart, B., T. Audunson, and L. Pera: *Fourth Int. Heat Transfer Conf., Paris, August 1970.*

36 Sanders, C. J., and J. P. Holman: Franz Grashof and the Grashof Number, *Int. J. Heat Mass Transfer,* vol. 15, p. 562, 1972.

37 Clifton, J. V., and A. J. Chapman: Natural Convection on a Finite-Size Horizontal Plate, *Int. J. Heat Mass Transfer,* vol. 12, p. 1573, 1969.

38 Emery, A. F., H. W. Chi, and J. D. Dale: Free Convection through Vertical Plane Layers of Non-Newtonian Power Law Fluids, *ASME Pap.* 70-WA/HT-1.

39 Vliet, G. C.: Natural Convection Local Heat Transfer on Constant Heat Flux Inclined Surfaces, *Trans. ASME,* vol. 91C, p. 511, 1969.

40 Bergles, A. E., and R. R. Simonds: Combined Forced and Free Convection for Laminar Flow in Horizontal Tubes with Uniform Heat Flux, *Int. J. Heat Mass Transfer,* vol. 14, p. 1989, 1971.

41 Aihara, T., Y. Yamada, and S. Endo: Free Convection along the Downward-facing Surface of a Heated Horizontal Plate, *Int. J. Heat Mass Transfer,* vol. 15, p. 2535, 1972.

42 Saunders, O. A., M. Fishenden, and H. D. Mansion: Some Measurement of Convection by an Optical Method, *Engineering,* p. 483, May 1935.

43 Weber, N., R. E. Rowe, E. H. Bishop, and J. A. Scanlan: Heat Transfer by Natural Convection between Vertically Eccentric Spheres, *ASME Pap*. 72-WA/HT-2.

44 Fujii, T., and H. Imura: Natural Convection Heat Transfer from a Plate with Arbitrary Inclination, *Int. J. Heat Mass Transfer,* vol. 15, p. 755, 1972.

45 Pera, L., and B. Gebhart: Natural Convection Boundary Layer Flow over Horizontal and Slightly Inclined Surfaces, *Int. J. Heat Mass Transfer,* vol. 16, p. 1131, 1973.

46 Hyman, S. C., C. F. Bonilla, and S. W. Ehrlich: Heat Transfer to Liquid Metals from Horizontal Cylinders, *AiChE Symp. Heat Transfer, Atlantic City, 1953,* p. 21.

47 Fand, R. M., and K. K. Keswani: Combined Natural and Forced Convection Heat Transfer from Horizontal Cylinders to Water, *Int. J. Heat Mass Transfer,* vol. 16, p. 175, 1973.

48 Dale, J. D., and A. F. Emery: The Free Convection of Heat from a Vertical Plate to Several Non-Newtonian Pseudoplastic Fluids, *ASME Pap*. 71-HT-S.

49 Fujii, T., O. Miyatake, M. Fujii, H. Tanaka, and K. Murakami: Natural Convective Heat Transfer from a Vertical Isothermal Surface to a Non-Newtonian Sutterby Fluid, *Int. J. Heat Mass Transfer,* vol. 16, p. 2177, 1973.

50 Soehngen, E. E.: Experimental Studies on Heat Transfer at Very High Prandtl Numbers, *Prog. Heat Mass Transfer,* vol. 2, p. 125, 1969.

51 Vliet, G. C., and D. C. Ross: Turbulent Natural Convection on Upward and Downward Facing Inclined Constant Heat Flux Surfaces, *ASME Pap*. 74-WA/HT-32.

52 Llyod, J. R., and W. R. Moran: Natural Convection Adjacent to Horizontal Surface of Various Planforms, *ASME Pap*. 74-WA/HT-66.

53 Goldstein, R. J., E. M. Sparrow, and D. C. Jones: Natural Convection Mass Transfer Adjacent to Horizontal Plates, *Int. J. Heat Mass Transfer,* vol. 16, p. 1025, 1973.

54 Holman, J. P., and J. H. Boggs: Heat Transfer to Freon 12 near the Critical State in a Natural Circulation Loop, *J. Heat Transfer,* vol. 80, p. 221, 1960.

55 Mull, W., and H. Reiher: Der Wärmeschutz von Luftschichten, *Beih. Gesund. Ing.*, ser. 1, no. 28, 1930.

56 Krasshold, H.: Wärmeabgabe von zylindrischen Flussigkeitsschichten bei natürlichen Konvektion, *Forsch, Geb. Ingenieurwes,* vol. 2, p. 165, 1931.

57 Beckmann, W.: Die Wärmeübertragung in zylindrischen Gasschichten bei natürlicher Konvektion, *Forsch. Geb. Ingenieurwes,* vol. 2, p. 186, 1931.

58 Schmidt, E.: Free Convection in Horizontal Fluid Spaces Heated from Below, *Proc. Int. Heat Transfer Conf., Boulder, Colo., ASME,* 1961.

59 Graff, J. G. A., and E. F. M. Van der Held: The Relation between the Heat Transfer and Convection Phenomena in Enclosed Plain Air Players, *Appl. Sci. Res.,* ser. A, vol. 3, p. 393, 1952.

60 Liu, C. Y., W. K. Mueller, and F. Landis: Natural Convection Heat Transfer in Long Horizontal Cylindrical Annuli, *Int. Dev. Heat Transfer,* pt. 5, pap. 117, p. 976, 1961.

61 Emery, A., and N. C. Chu: Heat Transfer across Vertical Layers, *J. Heat Transfer,* vol. 87, p. 110, 1965.

62 O'Toole, J., and P. L. Silveston: Correlation of Convective Heat Transfer in Confined Horizontal Layers, *Chem. Eng. Prog. Symp.,* vol. 57, no. 32, p. 81, 1961.

63 Goldstein, R. J., and T. Y. Chu: Thermal Convection in a Horizontal Layer of Air, *Prog. Heat Mass Transfer,* vol. 2, p. 55, 1969.

64 Singh, S. N., R. C. Birkebak, and R. M. Drake: Laminar Free Convection Heat Transfer from Downward-facing Horizontal Surfaces of Finite Dimensions, *Prog. Heat Mass Transfer,* vol. 2, p. 87, 1969.

65 McDonald, J. S., and T. J. Connally: Investigation of Natural Convection Heat Transfer in Liquid Sodium, *Nucl Sci. Eng.,* vol. 8, p. 369, 1960.

66 Hollands, K. G. T., G. D. Raithby, and L. Konicek: Correlation Equations for Free Convection Heat Transfer in Horizontal Layers of Air and Water, *Int. J. Heat Mass Transfer,* vol. 18, p. 879, 1975.

67 Hollands, K. G. T., T. E. Unny, and G. D. Raithby: Free Convective Heat Transfer across Inclined Air Layers, *ASME Pap.* 75-HT-55, August 1975.

68 Depew, C. A., J. L. Franklin, and C. H. Ito: Combined Free and Forced Convection in Horizontal, Uniformly Heated Tubes, *ASME Pap.* 75-HT-19, August 1975.

69 Raithby, G. D., and K. G. T. Hollands: A General Method of Obtaining Approximate Solutions to Laminar and Turbulent Free Convection Problems, *Advances in Heat Transfer,* Academic Press, New York, 1974.

70 Churchill, S. W., and H. H. S. Chu: Correlating Equations for Laminar and Turbulent Free Convection from a Horizontal Cylinder, *Int. J. Heat Mass Transfer,* vol. 18, p. 1049, 1975.

71 Churchill, S. W., and H. H. S. Chu: Correlating Equations for Laminar and Turbulent Free Convection from a Vertical Plate, *Int. J. Heat Mass Transfer,* vol. 18, p. 1323, 1975.

72 Churchill, S. W.: A Comprehensive Correlating Equation for Laminar, Assisting, Forced and Free Convection, *AiChE J.,* vol. 23, no. 1, p. 10, 1977.

73 Al-Arabi, M., and Y. K. Salman: Laminar Natural Convection Heat Transfer from an Inclined Cylinder, *Int. J. Heat Mass Transfer,* vol. 23, pp. 45–51, 1980.

74 Holman, J. P.: "Heat Transfer," 4th ed., McGraw-Hill Book Co., New York, 1976.

75 Hatfield, D. W., and D. K. Edwards: Edge and Aspect Ratio Effects on Natural Convection from the Horizontal Heated Plate Facing Downwards, *Int. J. Heat Mass Transfer,* vol. 24, p. 1019, 1981.

76 Morgan, V. T.: "The Overall Convective Heat Transfer from Smooth Circular Cylinders," "Advances in Heat Transfer" (T. F. Irvine and J. P. Hartnett, eds.), vol. 11, Academic Press, Inc., New York, 1975.

77 Sparrow, E. M., and M. A. Ansari: A Refutation of King's Rule for Multi-Dimensional External Natural Convection, *Int. J. Heat Mass Transfer,* vol. 26, p. 1357, 1983.

78 Lienhard, J. H.: On the Commonality of Equations for Natural Convection from Immersed Bodies, *Int. J. Heat Mass Transfer,* vol. 16, p. 2121, 1973.

79 Amato, W. S., and C. L. Tien: Free Convection Heat Transfer from Isothermal Spheres in Water, *Int. J. Heat Mass Transfer,* vol. 15, p. 327, 1972.

80 Warrington, R.O., and R. E. Powe: The Transfer of Heat by Natural Convection Between Bodies and Their Enclosures, *Int. J. Heat Mass Transfer,* vol. 28, p. 319, 1985.

81 Sparrow, E. M., and A. J. Stretton: Natural Convection from Bodies of Unity Aspect Ratio, *Int. J. Heat Mass Transfer,* vol. 28, p. 741, 1985.

82 El Sherbing, S. M., G. D. Raithby, and K. G. T. Hollands: Heat Transfer across Vertical and Inclined Air Layers, *J. Heat Transfer,* vol. 104C, p. 96, 1982.

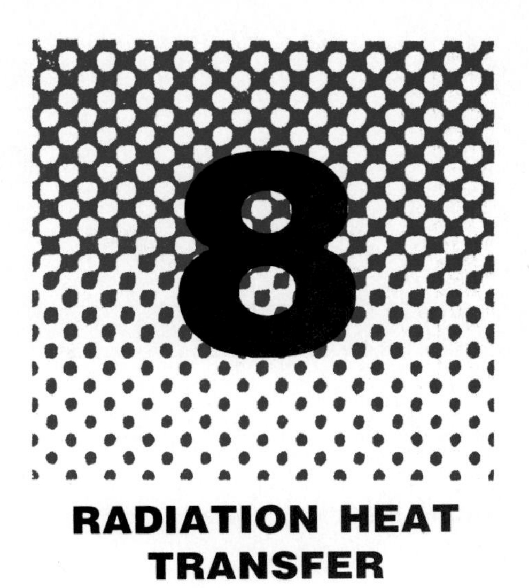

RADIATION HEAT TRANSFER

8-1 INTRODUCTION

Preceding chapters have shown how conduction and convection heat transfer may be calculated with the aid of both mathematical analysis and empirical data. We now wish to consider the third mode of heat transfer—thermal radiation. *Thermal radiation* is that electromagnetic radiation emitted by a body as a result of its temperature. In this chapter, we shall first describe the nature of thermal radiation, its characteristics, and the properties which are used to describe materials insofar as the radiation is concerned. Next, the transfer of radiation through space will be considered. Finally, the overall problem of heat transfer by thermal radiation will be analyzed, including the influence of the material properties and the geometric arrangement of the bodies on the total energy which may be exchanged.

8-2 PHYSICAL MECHANISM

There are many types of electromagnetic radiation; thermal radiation is only one. Regardless of the type of radiation, we say that it is propagated at the speed of light, 3×10^8 m/s. This speed is equal to the product of the wavelength and frequency of the radiation,

$$c = \lambda \nu$$

where c = speed of light
λ = wavelength
ν = frequency

The unit for λ may be centimeters, angstroms ($1\ \text{Å} = 10^{-8}$ cm), or micrometers ($1\ \mu\text{m} = 10^{-6}$ m). A portion of the electromagnetic spectrum is shown in Fig. 8-1. Thermal radiation lies in the range from about 0.1 to 100 μm, while the

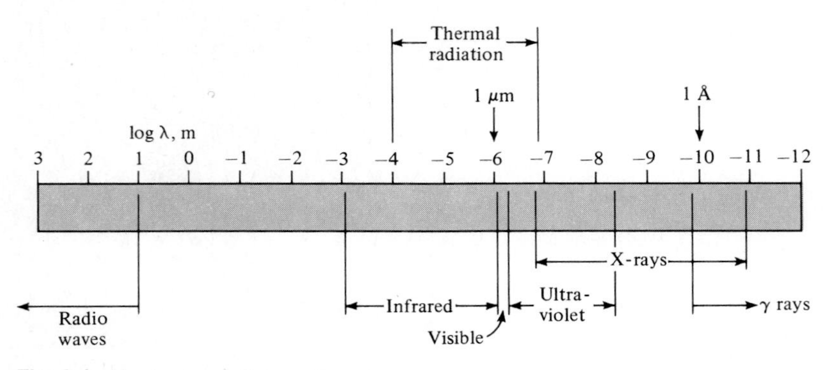

Fig. 8-1 Electromagnetic spectrum.

visible-light portion of the spectrum is very narrow, extending from about 0.35 to 0.75 μm.

The propagation of thermal radiation takes place in the form of discrete quanta, each quantum having an energy of

$$E = h\nu \tag{8-1}$$

where h is Planck's constant and has the value

$$h = 6.625 \times 10^{-34} \text{ J} \cdot \text{s}$$

A very rough physical picture of the radiation propagation may be obtained by considering each quantum as a particle having energy, mass, and momentum, just as we considered the molecules of a gas. So, in a sense, the radiation might be thought of as a "photon gas" which may flow from one place to another. Using the relativistic relation between mass and energy, expressions for the mass and momentum of the "particles" could thus be derived; viz,

$$E = mc^2 = h\nu$$

$$m = \frac{h\nu}{c^2}$$

$$\text{Momentum} = c\,\frac{h\nu}{c^2} = \frac{h\nu}{c}$$

By considering the radiation as such a gas, the principles of quantum-statistical thermodynamics can be applied to derive an expression for the energy density of radiation per unit volume and per unit wavelength as†

$$u_\lambda = \frac{8\pi hc\lambda^{-5}}{e^{hc/\lambda kT} - 1} \tag{8-2}$$

† See, for example, J. P. Holman, "Thermodynamics," 4th ed., McGraw-Hill Book Company, New York, 1988, p. 350.

where k is Boltzmann's constant, 1.38066×10^{-23} J/molecule · K. When the energy density is integrated over all wavelengths, the total energy emitted is proportional to absolute temperature to the fourth power:

$$E_b = \sigma T^4 \tag{8-3}$$

Equation (8-3) is called the Stefan-Boltzmann law, E_b is the energy radiated per unit time and per unit area by the ideal radiator, and σ is the Stefan-Boltzmann constant, which has the value

$$\sigma = 5.669 \times 10^{-8}\ \text{W/m}^2 \cdot \text{K}^4 \quad [0.1714 \times 10^{-8}\ \text{Btu/h} \cdot \text{ft}^2 \cdot {}^\circ\text{R}^4]$$

where E_b is in watts per square meter and T is in degrees Kelvin. In the thermodynamic analysis the energy density is related to the energy radiated from a surface per unit time and per unit area. Thus the heated interior surface of an enclosure produces a certain energy density of thermal radiation in the enclosure. We are interested in radiant exchange with surfaces—hence the reason for the expression of radiation from a surface in terms of its temperature. The subscript b in Eq. (8-3) denotes that this is the radiation from a blackbody. We call this *blackbody radiation* because materials which obey this law appear black to the eye; they appear black because they do not reflect any radiation. Thus a blackbody is also considered as one which absorbs all radiation incident upon it. E_b is called the *emissive power* of a blackbody.

It is important to note at this point that the "blackness" of a surface-to-thermal radiation can be quite deceiving insofar as visual observations are concerned. A surface coated with lampblack appears black to the eye and turns out to be black for the thermal-radiation spectrum. On the other hand, snow and ice appear quite bright to the eye but are essentially "black" for long-wavelength thermal radiation. Many white paints are also essentially black for long-wavelength radiation. This point will be discussed further in later sections.

8-3 RADIATION PROPERTIES

When radiant energy strikes a material surface, part of the radiation is reflected, part is absorbed, and part is transmitted, as shown in Fig. 8-2. We define the reflectivity ρ as the fraction reflected, the absorptivity α as the fraction absorbed, and the transmissivity τ as the fraction transmitted. Thus

$$\rho + \alpha + \tau = 1 \tag{8-4}$$

Most solid bodies do not transmit thermal radiation, so that for many applied problems the transmissivity may be taken as zero. Then

$$\rho + \alpha = 1$$

Two types of reflection phenomena may be observed when radiation strikes a surface. If the angle of incidence is equal to the angle of reflection, the reflection is called *specular*. On the other hand, when an incident beam is distributed uniformly in all directions after reflection, the reflection is called

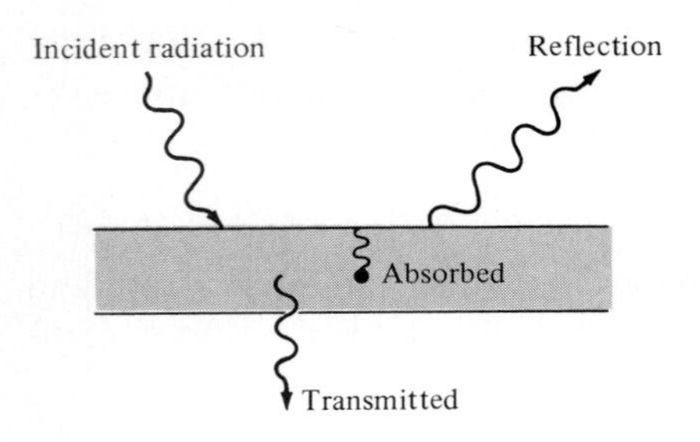

Fig. 8-2 Sketch showing effects of incident radiation.

diffuse. These two types of reflection are depicted in Fig. 8-3. Note that a specular reflection presents a mirror image of the source to the observer. No real surface is either specular or diffuse. An ordinary mirror is quite specular for visible light, but would not necessarily be specular over the entire wavelength range of thermal radiation. Ordinarily, a rough surface exhibits diffuse behavior better than a highly polished surface. Similarly, a polished surface is more specular than a rough surface. The influence of surface roughness on thermal-radiation properties of materials is a matter of serious concern and remains a subject for continuing research.

The emissive power of a body E is defined as the energy emitted by the body per unit area and per unit time. One may perform a thought experiment to establish a relation between the emissive power of a body and the material properties defined above. Assume that a perfectly black enclosure is available, i.e., one which absorbs all the incident radiation falling upon it, as shown schematically in Fig. 8-4. This enclosure will also emit radiation according to the T^4 law. Let the radiant flux arriving at some area in the enclosure be q_i W/m^2. Now suppose that a body is placed inside the enclosure and allowed to come into temperature equilibrium with it. At equilibrium the energy absorbed by the body must be equal to the energy emitted; otherwise there would

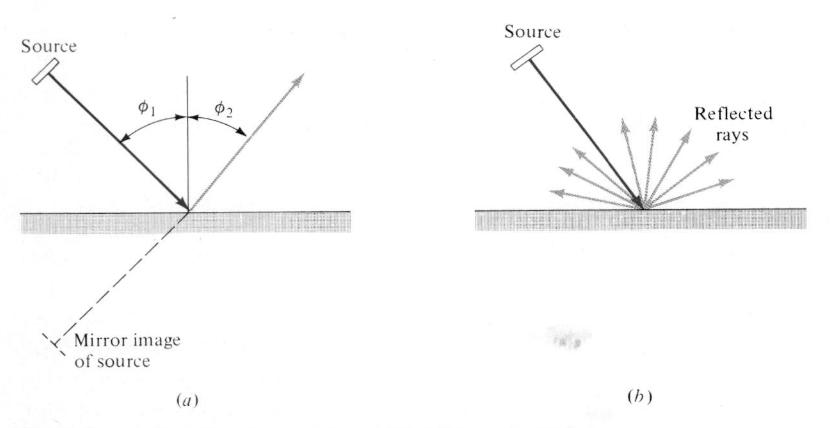

Fig. 8-3 (*a*) Specular ($\phi_1 = \phi_2$) and (*b*) diffuse reflection.

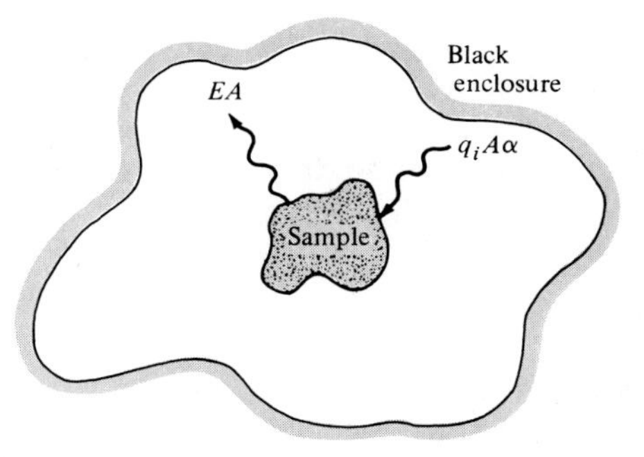

Fig. 8-4 Sketch showing model used for deriving Kirchhoff's law.

be an energy flow into or out of the body which would raise or lower its temperature. At equilibrium we may write

$$EA = q_iA\alpha \tag{8-5}$$

If we now replace the body in the enclosure with a blackbody of the same size and shape and allow it to come to equilibrium with the enclosure *at the same temperature*,

$$E_bA = q_iA(1) \tag{8-6}$$

since the absorptivity of a blackbody is unity. If Eq. (8-5) is divided by Eq. (8-6),

$$\frac{E}{E_b} = \alpha$$

and we find that the ratio of the emissive power of a body to the emissive power of a blackbody *at the same temperature* is equal to the absorptivity of the body. This ratio is defined as the *emissivity* ϵ of the body,

$$\epsilon = \frac{E}{E_b} \tag{8-7}$$

so that

$$\epsilon = \alpha \tag{8-8}$$

Equation (8-8) is called Kirchhoff's identity. At this point we note that the emissivities and absorptivities which have been discussed are the *total* properties of the particular material; i.e., they represent the integrated behavior of the material over all wavelengths. Real substances emit less radiation than ideal black surfaces as measured by the emissivity of the material. In reality, the emissivity of a material varies with temperature and the wavelength of the radiation.

The Gray Body

A *gray body* is defined such that the monochromatic emissivity ϵ_λ of the body is independent of wavelength. The *monochromatic emissivity* is defined as the ratio of the monochromatic-emissive power of the body to the monochromatic-emissive power of a blackbody at the same wavelength and temperature. Thus

$$\epsilon_\lambda = \frac{E_\lambda}{E_{b\lambda}}$$

The total emissivity of the body may be related to the monochromatic emissivity by noting that

$$E = \int_0^\infty \epsilon_\lambda E_{b\lambda}\, d\lambda \qquad \text{and} \qquad E_b = \int_0^\infty E_{b\lambda}\, d\lambda = \sigma T^4$$

so that

$$\epsilon = \frac{E}{E_b} = \frac{\int_0^\infty \epsilon_\lambda E_{b\lambda}\, d\lambda}{\sigma T^4} \tag{8-9}$$

where $E_{b\lambda}$ is the emissive power of a blackbody per unit wavelength. If the gray-body condition is imposed, that is, ϵ_λ = constant, Eq. (8-9) reduces to

$$\epsilon = \epsilon_\lambda \tag{8-10}$$

The emissivities of various substances vary widely with wavelength, temperature, and surface condition. Some typical values of the total emissivity of various surfaces are given in Appendix A. A very complete survey of radiation properties is given in Ref. 14.

The functional relation for $E_{b\lambda}$ was derived by Planck by introducing the quantum concept for electromagnetic energy. The derivation is now usually performed by methods of statistical thermodynamics, and $E_{b\lambda}$ is shown to be related to the energy density of Eq. (8-2) by

$$E_{b\lambda} = \frac{u_\lambda c}{4} \tag{8-11}$$

or

$$E_{b\lambda} = \frac{C_1 \lambda^{-5}}{e^{C_2/\lambda T} - 1} \tag{8-12}$$

where λ = wavelength, μm
T = temperature, K
$C_1 = 3.743 \times 10^8\ \text{W} \cdot \mu\text{m}^4/\text{m}^2$ $[1.187 \times 10^8\ \text{Btu} \cdot \mu\text{m}^4/\text{h} \cdot \text{ft}^2]$
$C_2 = 1.4387 \times 10^4\ \mu\text{m} \cdot \text{K}$ $[2.5896 \times 10^4\ \mu\text{m} \cdot {}^\circ\text{R}]$

A plot of $E_{b\lambda}$ as a function of temperature and wavelength is given in Fig. 8-5*a*. Notice that the peak of the curve is shifted to the shorter wavelengths for the higher temperatures. These maximum points in the radiation curves are related by Wien's displacement law,

$$\lambda_{max} T = 2897.6\ \mu\text{m} \cdot \text{K} \quad [5215.6\ \mu\text{m} \cdot {}^\circ\text{R}] \tag{8-13}$$

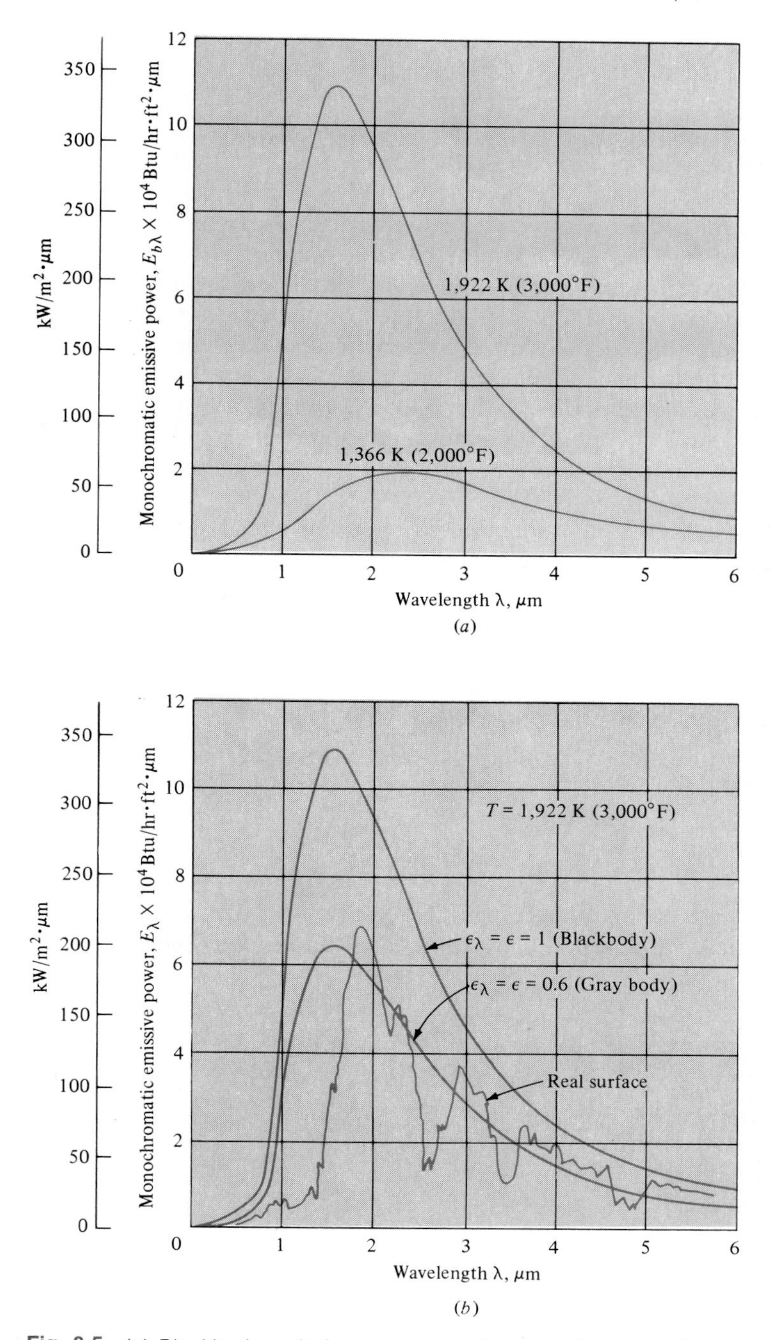

Fig. 8-5 (*a*) Blackbody emissive power as a function of wavelength and temperature; (*b*) comparison of emissive power of ideal blackbodies and gray bodies with that of a real surface.

Figure 8-5*b* indicates the relative radiation spectra from a blackbody at 3000°F and a corresponding ideal gray body with emissivity equal to 0.6. Also shown is a curve indicating an approximate behavior for a real surface, which may differ considerably from that of either an ideal blackbody or an ideal gray body. For analysis purposes surfaces are usually considered as gray bodies, with emissivities taken as the integrated average value.

The shift in the maximum point of the radiation curve explains the change in color of a body as it is heated. Since the band of wavelengths visible to the eye lies between about 0.3 and 0.7 μm, only a very small portion of the radiant-energy spectrum at low temperatures is detected by the eye. As the body is heated, the maximum intensity is shifted to the shorter wavelengths, and the first visible sign of the increase in temperature of the body is a dark-red color. With further increase in temperature, the color appears as a bright red, then bright yellow, and finally white. The material also appears much brighter at higher temperatures because a larger portion of the total radiation falls within the visible range.

We are frequently interested in the amount of energy radiated from a blackbody in a certain specified wavelength range. The fraction of the total energy radiated between 0 and λ is given by

$$\frac{E_{b_{0-\lambda}}}{E_{b_{0-\infty}}} = \frac{\int_0^\lambda E_{b\lambda}\, d\lambda}{\int_0^\infty E_{b\lambda}\, d\lambda} \tag{8-14}$$

Equation (8-12) may be rewritten by dividing both sides by T^5, so that

$$\frac{E_{b\lambda}}{T^5} = \frac{C_1}{(\lambda T)^5(e^{C_2/\lambda T} - 1)} \tag{8-15}$$

Now, for any specified temperature, the integrals in Eq. (8-14) may be expressed in terms of the single variable λT. The results have been tabulated by Dunkle

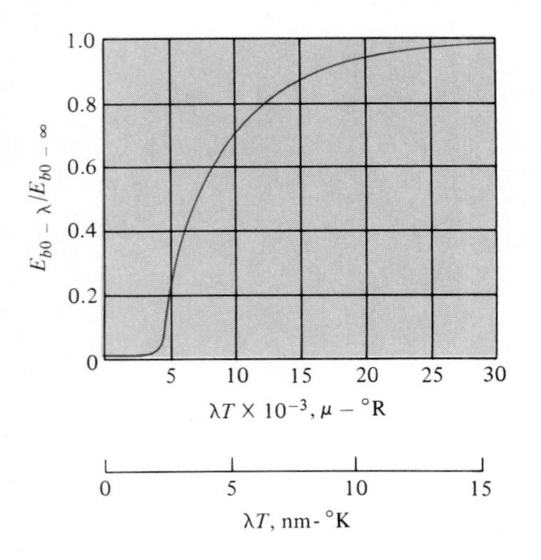

Fig. 8-6 Fraction of blackbody radiation in wavelength interval.

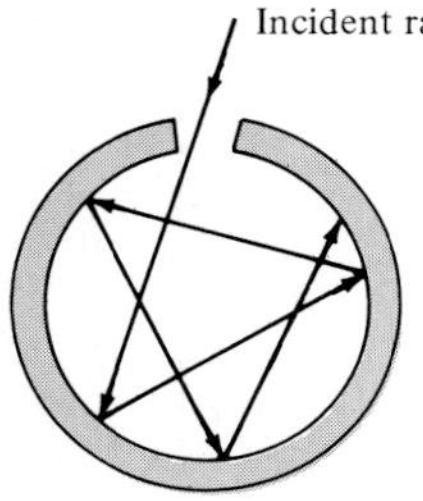

Incident radiation

Fig. 8-7 Method of constructing a blackbody enclosure.

[2]. The ratio in Eq. (8-14) is plotted in Fig. 8-6 and tabulated in Table 8-1. If the radiant energy emitted between wavelengths λ_1 and λ_2 is desired, then

$$E_{b_{\lambda 1-\lambda 2}} = E_{b_{0-\infty}} \left(\frac{E_{b_{0-\lambda 2}}}{E_{b_{0-\infty}}} - \frac{E_{b_{0-\lambda 1}}}{E_{b_{0-\infty}}} \right) \tag{8-16}$$

where $E_{b_{0-\infty}}$ is the total radiation emitted over all wavelengths,

$$E_{b_{0-\infty}} = \sigma T^4 \tag{8-17}$$

and is obtained by integrating the Planck distribution formula of Eq. (8-12) over all wavelengths.

The concept of a blackbody is an idealization; i.e., a perfect blackbody does not exist—all surfaces reflect radiation to some extent, however slight. A blackbody may be approximated very accurately, however, in the following way. A cavity is constructed, as shown in Fig. 8-7, so that it is very large compared with the size of the opening in the side. An incident ray of energy is reflected many times on the inside before finally escaping from the side opening. With each reflection there is a fraction of the energy absorbed corresponding to the absorptivity of the inside of the cavity. After the many absorptions, practically all the incident radiation at the side opening is absorbed. It should be noted that the cavity of Fig. 8-7 behaves approximately as a blackbody emitter as well as an absorber.

■ **EXAMPLE 8-1** Transmission and absorption in glass plate

A glass plate 30 cm square is used to view radiation from a furnace. The transmissivity of the glass is 0.5 from 0.2 to 3.5 μm. The emissivity may be assumed to be 0.3 up to 3.5 μm and 0.9 above that. The transmissivity of the glass is zero, except in the range from 0.2 to 3.5 μm. Assuming that the furnace is a blackbody at 2000°C, calculate the energy absorbed in the glass and the energy transmitted.

Solution

$$T = 2000°\text{C} = 2273 \text{ K}$$

$$\lambda_1 T = (0.2)(2273) = 454.6 \ \mu\text{m} \cdot \text{K}$$

$$\lambda_2 T = (3.5)(2273) = 7955.5 \ \mu\text{m} \cdot \text{K}$$

$$A = (0.3)^2 = 0.09 \text{ m}^2$$

Table 8-1 Radiation Functions.

λT		$E_{b\lambda}/T^5$		
		Btu	W	
$\mu m \cdot °R$	$\mu m \cdot K$	$h \cdot ft^2 \cdot °R^5 \cdot \mu m$ $\times 10^{15}$	$m^2 \cdot K^5 \cdot \mu m$ $\times 10^{11}$	$\frac{E_{b0-\lambda T}}{\sigma T^4}$
1,000	555.6	0.000671	0.400×10^{-5}	0.170×10^{-7}
1,200	666.7	0.0202	0.120×10^{-3}	0.756×10^{-6}
1,400	777.8	0.204	0.00122	0.106×10^{-4}
1,600	888.9	1.057	0.00630	0.738×10^{-4}
1,800	1,000.0	3.544	0.02111	0.321×10^{-3}
2,000	1,111.1	8.822	0.05254	0.00101
2,200	1,222.2	17.776	0.10587	0.00252
2,400	1,333.3	30.686	0.18275	0.00531
2,600	1,444.4	47.167	0.28091	0.00983
2,800	1,555.6	66.334	0.39505	0.01643
3,000	1,666.7	87.047	0.51841	0.02537
3,200	1,777.8	108.14	0.64404	0.03677
3,400	1,888.9	128.58	0.76578	0.05059
3,600	2,000.0	147.56	0.87878	0.06672
3,800	2,111.1	164.49	0.97963	0.08496
4,000	2,222.2	179.04	1.0663	0.10503
4,200	2,333.3	191.05	1.1378	0.12665
4,400	2,444.4	200.51	1.1942	0.14953
4,600	2,555.6	207.55	1.2361	0.17337
4,800	2,666.7	212.32	1.2645	0.19789
5,000	2,777.8	215.06	1.2808	0.22285
5,200	2,888.9	216.00	1.2864	0.24803
5,400	3,000.0	215.39	1.2827	0.27322
5,600	3,111.1	213.46	1.2713	0.29825
5,800	3,222.2	210.43	1.2532	0.32300
6,000	3,333.3	206.51	1.2299	0.34734
6,200	3,444.4	201.88	1.2023	0.37118
6,400	3,555.6	196.69	1.1714	0.39445
6,600	3,666.7	191.09	1.1380	0.41708
6,800	3,777.8	185.18	1.1029	0.43905
7,000	3,888.9	179.08	1.0665	0.46031
7,200	4,000.0	172.86	1.0295	0.48085
7,400	4,111.1	166.60	0.99221	0.50066
7,600	4,222.2	160.35	0.95499	0.51974
7,800	4,333.3	154.16	0.91813	0.53809
8,000	4,444.4	148.07	0.88184	0.55573
8,200	4,555.6	142.10	0.84629	0.57267
8,400	4,666.7	136.28	0.81163	0.58891
8,600	4,777.8	130.63	0.77796	0.60449
8,800	4,888.9	125.15	0.74534	0.61941
9,000	5,000.0	119.86	0.71383	0.63371
9,200	5,111.1	114.76	0.68346	0.64740
9,400	5,222.2	109.85	0.65423	0.66051
9,600	5,333.3	105.14	0.62617	0.67305
9,800	5,444.4	100.62	0.59925	0.68506
10,000	5,555.6	96.289	0.57346	0.69655

Table 8-1 Radiation Functions (*Continued*).

λT		$E_{b\lambda}/T^5$		
		Btu	W	
$\mu m \cdot {}^\circ R$	$\mu m \cdot K$	$\frac{Btu}{h \cdot ft^2 \cdot {}^\circ R^5 \cdot \mu m} \times 10^{15}$	$\frac{W}{m^2 \cdot K^5 \cdot \mu m} \times 10^{11}$	$\frac{E_{b0-\lambda T}}{\sigma T^4}$
10,200	5,666.7	92.145	0.54877	0.70754
10,400	5,777.8	88.181	0.52517	0.71806
10,600	5,888.9	84.394	0.50261	0.72813
10,800	6,000.0	80.777	0.48107	0.73777
11,000	6,111.1	77.325	0.46051	0.74700
11,200	6,222.2	74.031	0.44089	0.75583
11,400	6,333.3	70.889	0.42218	0.76429
11,600	6,444.4	67.892	0.40434	0.77238
11,800	6,555.6	65.036	0.38732	0.78014
12,000	6,666.7	62.313	0.37111	0.78757
12,200	6,777.8	59.717	0.35565	0.79469
12,400	6,888.9	57.242	0.34091	0.80152
12,600	7,000.0	54.884	0.32687	0.80806
12,800	7,111.1	52.636	0.31348	0.81433
13,000	7,222.2	50.493	0.30071	0.82035
13,200	7,333.3	48.450	0.28855	0.82612
13,400	7,444.4	46.502	0.27695	0.83166
13,600	7,555.6	44.645	0.26589	0.83698
13,800	7,666.7	42.874	0.25534	0.84209
14,000	7,777.8	41.184	0.24527	0.84699
14,200	7,888.9	39.572	0.23567	0.85171
14,400	8,000.0	38.033	0.22651	0.85624
14,600	8,111.1	36.565	0.21777	0.86059
14,800	8,222.2	35.163	0.20942	0.86477
15,000	8,333.3	33.825	0.20145	0.86880
16,000	8,888.9	27.977	0.16662	0.88677
17,000	9,444.4	23.301	0.13877	0.90168
18,000	10,000.0	19.536	0.11635	0.91414
19,000	10,555.6	16.484	0.09817	0.92462
20,000	11,111.1	13.994	0.08334	0.93349
21,000	11,666.7	11.949	0.07116	0.94104
22,000	12,222.2	10.258	0.06109	0.94751
23,000	12,777.8	8.852	0.05272	0.95307
24,000	13,333.3	7.676	0.04572	0.95788
25,000	13,888.9	6.687	0.03982	0.96207
26,000	14,444.4	5.850	0.03484	0.96572
27,000	15,000.0	5.139	0.03061	0.96892
28,000	15,555.6	4.532	0.02699	0.97174
29,000	16,111.1	4.012	0.02389	0.97423
30,000	16,666.7	3.563	0.02122	0.97644
40,000	22,222.2	1.273	0.00758	0.98915
50,000	27,777.8	0.560	0.00333	0.99414
60,000	33,333.3	0.283	0.00168	0.99649
70,000	38,888.9	0.158	0.940×10^{-3}	0.99773
80,000	44,444.4	0.0948	0.564×10^{-3}	0.99845
90,000	50,000.0	0.0603	0.359×10^{-3}	0.99889
100,000	55,555.6	0.0402	0.239×10^{-3}	0.99918

From Table 8-1

$$\frac{E_{b_{0-\lambda 1}}}{\sigma T^4} = 0 \qquad \frac{E_{b_{0-\lambda 2}}}{\sigma T^4} = 0.85443$$

$$\sigma T^4 = (5.669 \times 10^{-8})(2273)^4 = 1513.3 \text{ kW/m}^2$$

Total incident radiation is

$$0.2\ \mu\text{m} < \lambda < 3.5\ \mu\text{m} = (1.5133 \times 10^6)(0.85443 - 0)(0.3)^2$$
$$= 116.4 \text{ kW} \quad [3.97 \times 10^5 \text{ Btu/h}]$$

$$\text{Total radiation transmitted} = (0.5)(116.5) = 58.2 \text{ kW}$$

$$\text{Radiation absorbed} = \begin{cases} (0.3)(116.4) = 34.92 \text{ kW} & \text{for } 0 < \lambda < 3.5\ \mu\text{m} \\ (0.9)(1 - 0.85443)(1513.3)(0.09) = 17.84 \text{ kW} & \text{for } 3.5\ \mu\text{m} < \lambda < \infty \end{cases}$$

$$\text{Total radiation absorbed} = 34.92 + 17.84 = 52.76 \text{ kW} \quad [180{,}000 \text{ Btu/h}]$$

8-4 RADIATION SHAPE FACTOR

Consider two black surfaces A_1 and A_2, as shown in Fig. 8-8. We wish to obtain a general expression for the energy exchange between these surfaces when they are maintained at different temperatures. The problem becomes essentially one of determining the amount of energy which leaves one surface and reaches the other. To solve this problem the *radiation shape factors* are defined as

F_{1-2} = fraction of energy leaving surface 1 which reaches surface 2

F_{2-1} = fraction of energy leaving surface 2 which reaches surface 1

F_{m-n} = fraction of energy leaving surface m which reaches surface n

Other names for the radiation shape factor are *view factor, angle factor,* and *configuration factor*. The energy leaving surface 1 and arriving at surface 2 is

$$E_{b1}A_1F_{12}$$

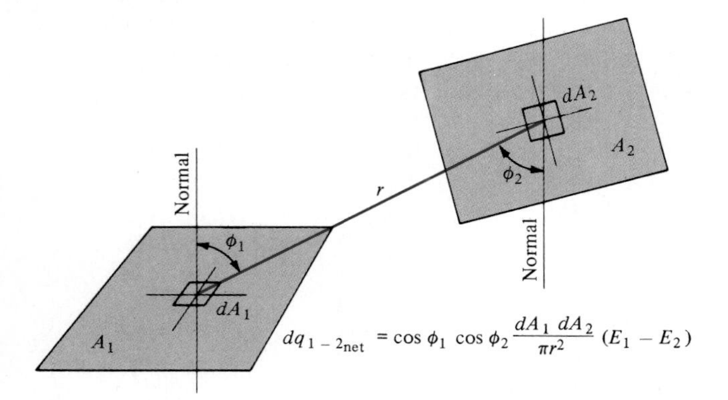

Fig. 8-8 Sketch showing area elements used in deriving radiation shape factor.

and that energy leaving surface 2 and arriving at surface 1 is

$$E_{b2}A_2F_{21}$$

Since the surfaces are black, all the incident radiation will be absorbed, and the net energy exchange is

$$E_{b1}A_1F_{12} - E_{b2}A_2F_{21} = Q_{1-2}$$

If both surfaces are at the same temperature, there can be no heat exchange, that is, $Q_{1-2} = 0$. Also

$$E_{b1} = E_{b2}$$

so that

$$A_1F_{12} = A_2F_{21} \tag{8-18}$$

The net heat exchange is therefore

$$Q_{1-2} = A_1F_{12}(E_{b1} - E_{b2}) = A_2F_{21}(E_{b1} - E_{b2}) \tag{8-19}$$

Equation (8-18) is known as a reciprocity relation, and it applies in a general way for any two surfaces m and n:

$$A_mF_{mn} = A_nF_{nm} \tag{8-18a}$$

Although the relation is derived for black surfaces, it holds for other surfaces also as long as diffuse radiation is involved.

We now wish to determine a general relation for F_{12} (or F_{21}). To do this, we consider the elements of area dA_1 and dA_2 in Fig. 8-8. The angles ϕ_1 and ϕ_2 are measured between a normal to the surface and the line drawn between the area elements r. The projection of dA_1 on the line between centers is

$$dA_1 \cos \phi_1$$

This may be seen more clearly in the elevation drawing shown in Fig. 8-9. We assume that the surfaces are diffuse, i.e., that the intensity of the radiation is the same in all directions. The intensity is the radiation emitted per unit area and per unit of solid angle in a certain specified direction. So, in order to obtain the energy emitted by the element of area dA_1 in a certain direction, we must multiply the intensity by the projection of dA_1 in the specified direction. Thus the energy leaving dA_1 in the direction given by the angle ϕ_1 is

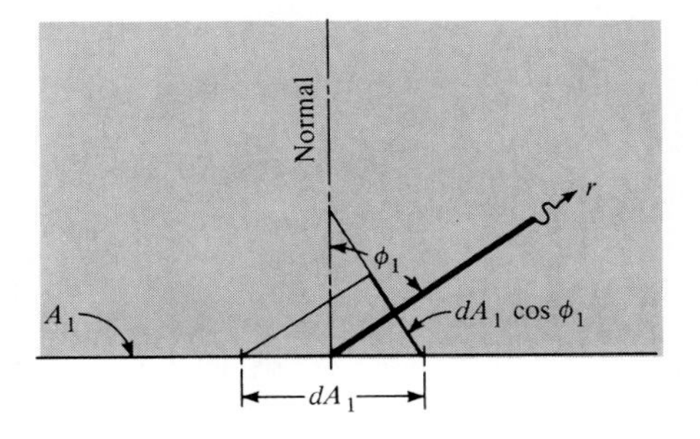

Fig. 8-9 Elevation view of area shown in Fig. 8-8.

$$I_b \, dA_1 \cos \phi_1 \tag{a}$$

where I_b is the blackbody intensity. The radiation arriving at some area element dA_n at a distance r from A_1 would be

$$I_b \, dA_1 \cos \phi_1 \frac{dA_n}{r^2} \tag{b}$$

where dA_n is constructed normal to the radius vector. The quantity dA_n/r^2 represents the solid angle subtended by the area dA_n. The intensity may be obtained in terms of the emissive power by integrating expression (b) over a hemisphere enclosing the element of area dA_1. In a spherical coordinate system like that in Fig. 8-10,

$$dA_n = r^2 \sin \phi \, d\psi \, d\phi$$

Then

$$E_b \, dA_1 = I_b \, dA_1 \int_0^{2\pi} \int_0^{\pi/2} \sin \phi \cos \phi \, d\phi \, d\psi$$

$$= \pi I_b \, dA_1$$

so that

$$E_b = \pi I_b \tag{8-20}$$

We may now return to the energy-exchange problem indicated in Fig. 8-8. The area element dA_n is given by

$$dA_n = \cos \phi_2 \, dA_2$$

so that the energy leaving dA_1 which arrives at dA_2 is

$$dq_{1-2} = E_{b1} \cos \phi_1 \cos \phi_2 \frac{dA_1 \, dA_2}{\pi r^2}$$

That energy leaving dA_2 and arriving at dA_1 is

$$dq_{2-1} = E_{b2} \cos \phi_2 \cos \phi_1 \frac{dA_2 \, dA_1}{\pi r^2}$$

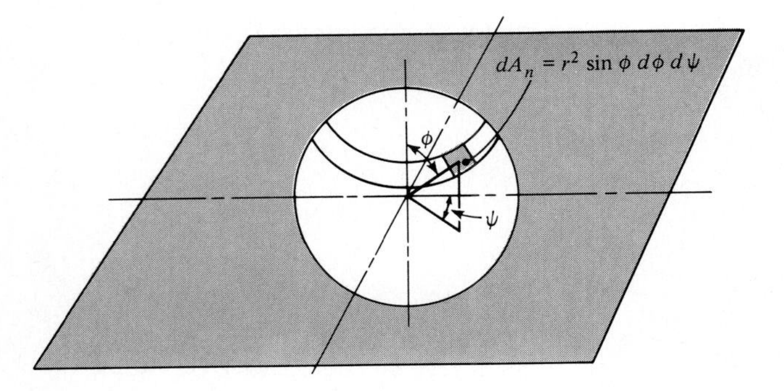

Fig. 8-10 Spherical coordinate system used in derivation of radiation shape factor.

and the net energy exchange is

$$q_{\text{net}_{1-2}} = (E_{b1} - E_{b2}) \int_{A_2} \int_{A_1} \cos \phi_1 \cos \phi_2 \frac{dA_1 \, dA_2}{\pi r^2} \tag{8-21}$$

The integral is either A_1F_{12} or A_2F_{21} according to Eq. (8-19). To evaluate the integral, the specific geometry of the surfaces A_1 and A_2 must be known. We shall work out an elementary problem and then present the results for more complicated geometries in graphical form.

Consider the radiation from the small area dA_1 to the flat disk A_2, as shown in Fig. 8-11. The element of area dA_2 is chosen as the circular ring of radius x. Thus

$$dA_2 = 2\pi x \, dx$$

We note that $\phi_1 = \phi_2$ and apply Eq. (8-21), integrating over the area A_2:

$$dA_1 F_{dA_1-A_2} = dA_1 \int_{A_2} \cos^2 \phi_1 \frac{2\pi x \, dx}{\pi r^2}$$

Making the substitutions

$$r = (R^2 + x^2)^{1/2} \qquad \text{and} \qquad \cos \phi_1 = \frac{R}{(R^2 + x^2)^{1/2}}$$

we have
$$dA_1 F_{dA_1-A_2} = dA_1 \int_0^{D/2} \frac{2R^2 x \, dx}{(R^2 + x^2)^2}$$

Performing the integration gives

$$dA_1 F_{dA_1-A_2} = -dA_1 \left(\frac{R^2}{R^2 + x^2} \right) \Bigg]_0^{D/2} = dA_1 \frac{D^2}{4R^2 + D^2}$$

so that
$$F_{dA_1-A_2} = \frac{D^2}{4R^2 + D^2} \tag{8-22}$$

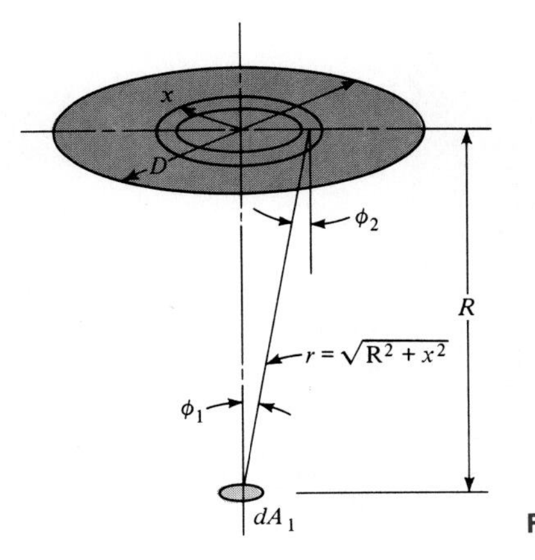

Fig. 8-11 Radiation from a small-area element to a disk.

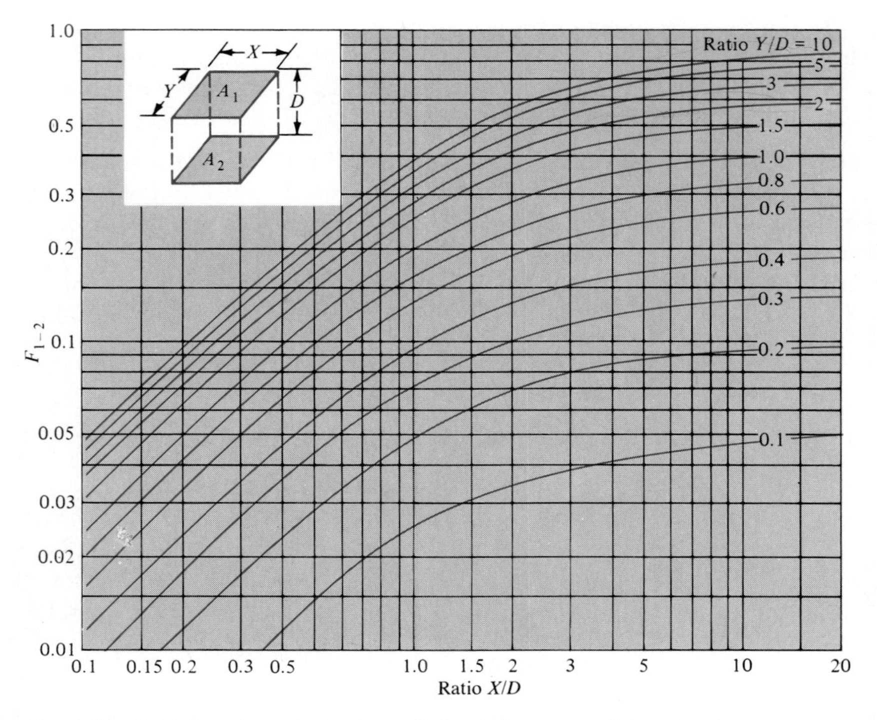

Fig. 8-12 Radiation shape factor for radiation between parallel rectangles.

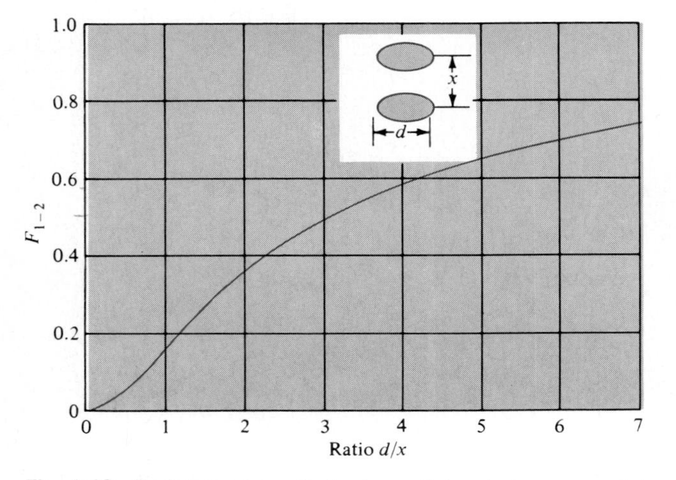

Fig. 8-13 Radiation shape factor for radiation between parallel disks.

The calculation of shape factors may be extended to more complex geometries, as described in Refs. 3, 5, 24, and 38; 38 gives a very complete catalog of analytical relations and graphs for shape factors. For our purposes we give only the results of a few geometries as shown in Figs. 8-12 through 8-16.

Real-Surface Behavior

Real surfaces exhibit interesting deviations from the ideal surfaces described in the preceding paragraphs. Real surfaces, for example, are not perfectly diffuse, and hence the intensity of emitted radiation is not constant over all directions. The directional-emittance characteristics of several types of surfaces are shown in Fig. 8-17. These curves illustrate the characteristically different behavior of electric conductors and nonconductors. Conductors emit more energy in a direction having a large azimuth angle. This behavior may be satisfactorily explained with basic electromagnetic wave theory, and is discussed in Ref. 24. As a result of this basic behavior of conductors and nonconductors, we may anticipate the appearance of a sphere which is heated to incandescent temperatures, as shown in Fig. 8-18. An electric conducting sphere will appear bright around the rim since more energy is emitted at large angles ϕ. A sphere constructed of a nonconducting material will have the opposite behavior and will appear bright in the center and dark around the edge.

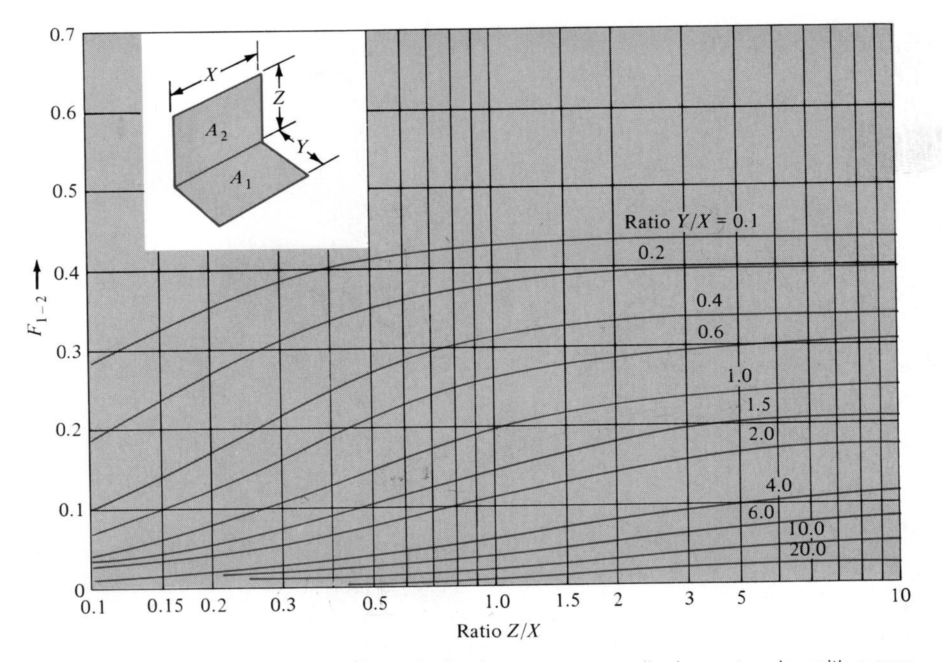

Fig. 8-14 Radiation shape factor for radiation between perpendicular rectangles with a common edge.

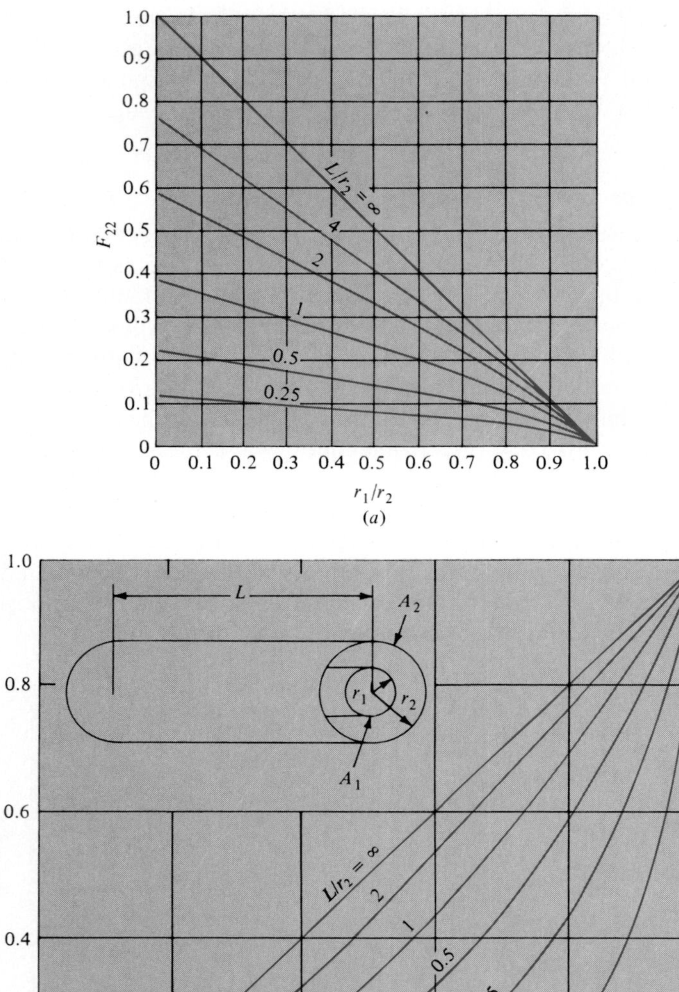

(a)

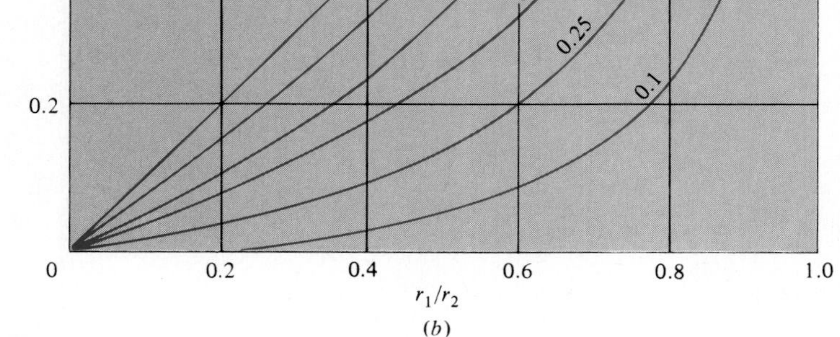

(b)

Fig. 8-15 Radiation shape factors for two concentric cylinders of finite length. (a) Outer cylinder to itself; (b) outer cylinder to inner cylinder.

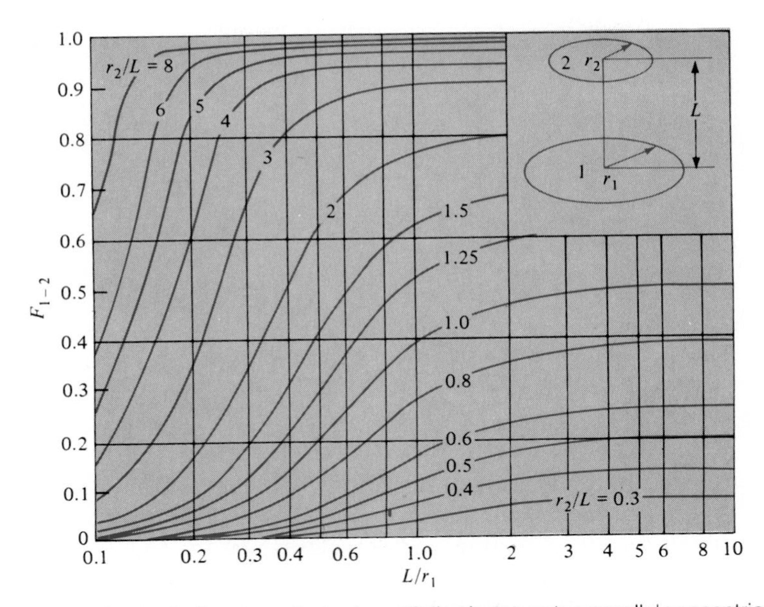

Fig. 8-16 Radiation shape factor for radiation between two parallel concentric disks.

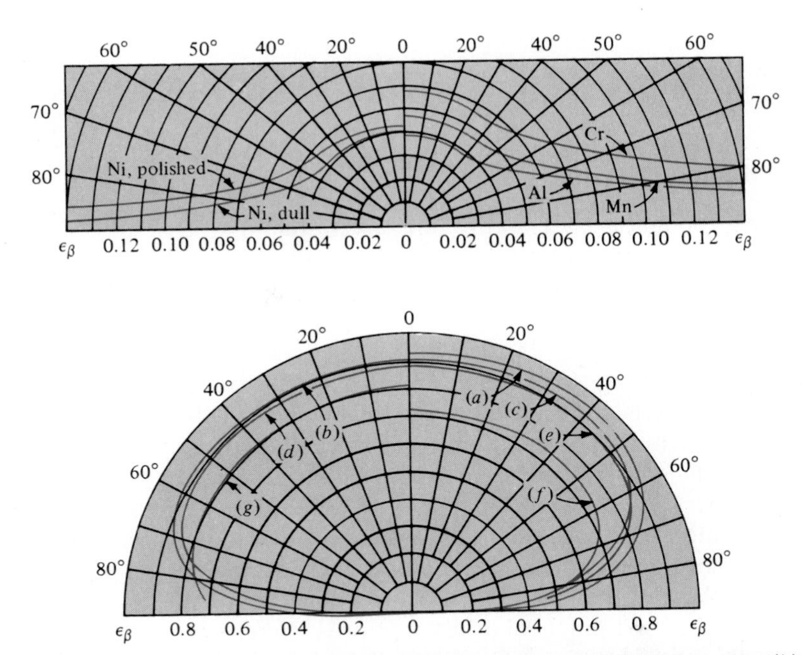

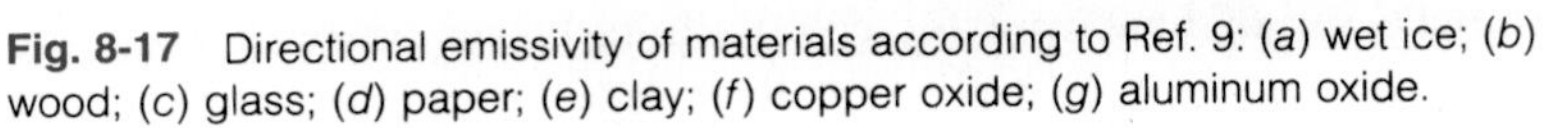

Fig. 8-17 Directional emissivity of materials according to Ref. 9: (*a*) wet ice; (*b*) wood; (*c*) glass; (*d*) paper; (*e*) clay; (*f*) copper oxide; (*g*) aluminum oxide.

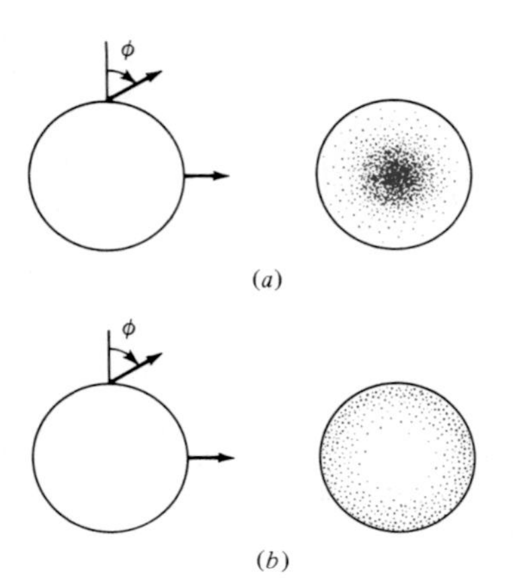

Fig. 8-18 Effect of directional emittance on appearance of an incandescent sphere: (*a*) electrical conductor; (*b*) electrical nonconductor.

Reflectance and absorptance of thermal radiation from real surfaces are a function not only of the surface itself but also of the surroundings. These properties are dependent on the direction and wavelength of the incident radiation. But the distribution of the intensity of incident radiation with wavelength may be a very complicated function of the temperatures and surface characteristics of all the surfaces which incorporate the surroundings. Let us denote the total incident radiation on a surface per unit time, per unit area, and per unit wavelength as G_λ. Then the total absorptivity will be given as the ratio of the total energy absorbed to the total energy incident on the surface,

or

$$\alpha = \frac{\int_0^\infty \alpha_\lambda G_\lambda \, d\lambda}{\int_0^\infty G_\lambda \, d\lambda} \tag{8-23}$$

If we are fortunate enough to have a gray body such that $\epsilon_\lambda = \epsilon =$ constant, this relation simplifies considerably. It may be shown that Kirchoff's law [Eq. (8-8)] may be written for monochromatic radiation as

$$\epsilon_\lambda = \alpha_\lambda \tag{8-24}$$

Therefore, for a gray body, $\alpha_\lambda =$ constant, and Eq. (8-23) expresses the result that the total absorptivity is also constant and independent of the wavelength distribution of incident radiation. Furthermore, since the emissivity and absorptivity are constant over all wavelengths for a gray body, they must be independent of temperature as well. Unhappily, real surfaces are not always "gray" in nature, and significant errors may ensue by assuming gray-body behavior. On the other hand, analysis of radiation exchange using real-surface behavior is so complicated that the ease and simplification of the gray-body assumption is justified by the practical utility it affords. References 10, 11, and 24 present comparisons of heat-transfer calculations based on both gray and nongray analyses.

■ **EXAMPLE 8-2** Heat transfer between black surfaces

Two parallel black plates 0.5 by 1.0 m are spaced 0.5 m apart. One plate is maintained at 1000°C and the other at 500°C. What is the net radiant heat exchange between the two plates?

Solution

The ratios for use with Fig. 8-12 are

$$\frac{Y}{D} = \frac{0.5}{0.5} = 1.0 \qquad \frac{X}{D} = \frac{1.0}{0.5} = 2.0$$

so that $F_{12} = 0.285$. The heat transfer is calculated from

$$\begin{aligned} q &= A_1F_{12}(E_{b_1} - E_{b_2}) = \sigma A_1F_{12}(T_1^4 - T_2^4) \\ &= (5.669 \times 10^{-8})(0.5)(0.285)(1273^4 - 773^4) \\ &= 18.33 \text{ kW} \quad [62{,}540 \text{ Btu/h}] \end{aligned}$$

■ 8-5 RELATIONS BETWEEN SHAPE FACTORS

Some useful relations between shape factors may be obtained by considering the system shown in Fig. 8-19. Suppose that the shape factor for radiation from A_3 to the combined area $A_{1,2}$ is desired. This shape factor must be given very simply as

$$F_{3-1,2} = F_{3-1} + F_{3-2} \tag{8-25}$$

i.e., the total shape factor is the sum of its parts. We could also write Eq. (8-25) as

$$A_3F_{3-1,2} = A_3F_{3-1} + A_3F_{3-2} \tag{8-26}$$

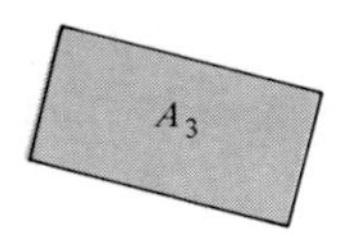

Fig. 8-19 Sketch showing some relations between shape factors.

and making use of the reciprocity relations

$$A_3F_{3-1,2} = A_{1,2}F_{1,2-3}$$

$$A_3F_{3-1} = A_1F_{1-3}$$

$$A_3F_{3-2} = A_2F_{2-3}$$

the expression could be rewritten

$$A_{1,2}F_{1,2-3} = A_1F_{1-3} + A_2F_{2-3} \tag{8-27}$$

which simply states that the total radiation arriving at surface 3 is the sum of the radiations from surfaces 1 and 2. Suppose we wish to determine the shape factor F_{1-3} for the surfaces in Fig. 8-20 in terms of known shape factors for perpendicular rectangles with a common edge. We may write

$$F_{1-2,3} = F_{1-2} + F_{1-3}$$

in accordance with Eq. (8-25). Both $F_{1-2,3}$ and F_{1-2} may be determined from Fig. 8-14, so that F_{1-3} is easily calculated when the dimensions are known. Now consider the somewhat more complicated situation shown in Fig. 8-21. An expression for the shape factor F_{1-4} is desired in terms of known shape factors for perpendicular rectangles with a common edge. We write

$$A_{1,2}F_{1,2-3,4} = A_1F_{1-3,4} + A_2F_{2-3,4} \tag{a}$$

in accordance with Eq. (8-25). Both $F_{1,2-3,4}$ and $F_{2-3,4}$ can be obtained from Fig. 8-14, and $F_{1-3,4}$ may be expressed

$$A_1F_{1-3,4} = A_1F_{1-3} + A_1F_{1-4} \tag{b}$$

Also,

$$A_{1,2}F_{1,2-3} = A_1F_{1-3} + A_2F_{2-3} \tag{c}$$

Solving for A_1F_{1-3} from (c), inserting this in (b), and then inserting the resultant expression for $A_1F_{1-3,4}$ in (a) gives

$$A_{1,2}F_{1,2-3,4} = A_{1,2}F_{1,2-3} - A_2F_{2-3} + A_1F_{1-4} + A_2F_{2-3,4} \tag{d}$$

Notice that all shape factors except F_{1-4} may be determined from Fig. 8-14.

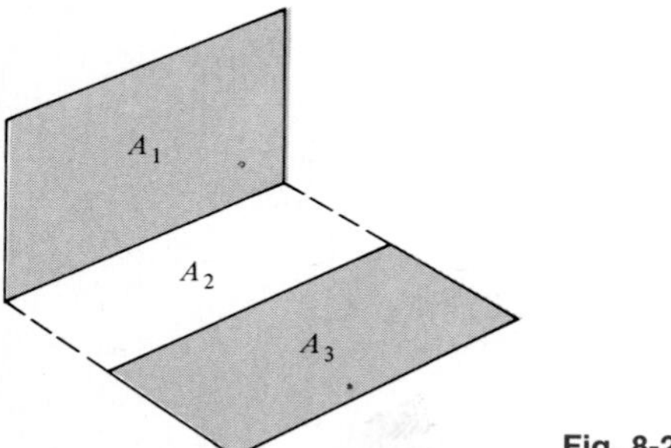

Fig. 8-20

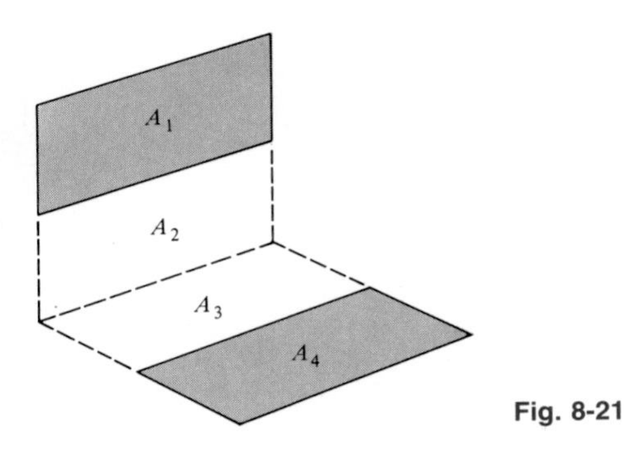

Fig. 8-21

Thus

$$F_{1-4} = \frac{1}{A_1}(A_{1,2}F_{1,2-3,4} + A_2F_{2-3} - A_{1,2}F_{1,2-3} - A_2F_{2-3,4}) \qquad (8\text{-}28)$$

In the foregoing discussion the tacit assumption has been made that the various bodies do not see themselves, i.e.,

$$F_{11} = F_{22} = F_{33} = 0 \cdots$$

To be perfectly general, we must include the possibility of concave curved surfaces, which may then see themselves. The general relation is therefore

$$\sum_{j=1}^{n} F_{ij} = 1.0 \qquad (8\text{-}29)$$

where F_{ij} is the fraction of the total energy leaving surface i which arrives at surface j. Thus for a three-surface enclosure we would write

$$F_{11} + F_{12} + F_{13} = 1.0$$

and F_{11} represents the fraction of energy leaving surface 1 which strikes itself. A certain amount of care is required in analyzing radiation exchange between curved surfaces.

Hamilton and Morgan [5] have presented generalized relations for parallel and perpendicular rectangles in terms of shape factors which may be obtained from Figs. 8-12 and 8-14. The two situations of interest are shown in Figs. 8-22 and 8-23. For the perpendicular rectangles of Fig. 8-22 it can be shown that the following reciprocity relations apply [5]:

$$A_1F_{13'} = A_3F_{31'} = A_{3'}F_{3'1} = A_{1'}F_{1'3} \qquad (8\text{-}30)$$

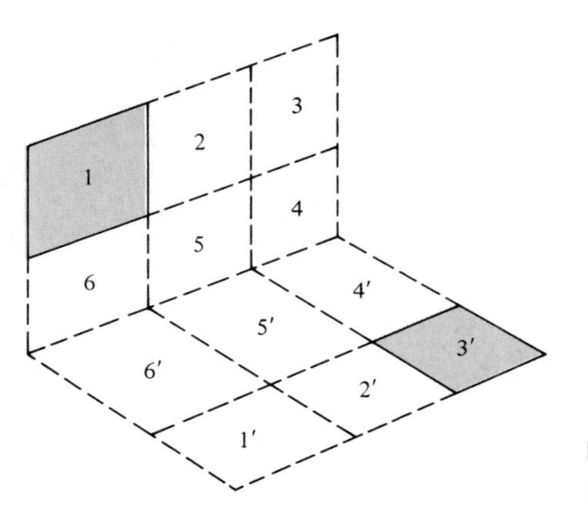

Fig. 8-22 Generalized perpendicular-rectangle arrangement.

By making use of these reciprocity relations, the radiation shape factor $F_{13'}$ may be expressed by

$$\begin{aligned} A_1F_{13'} = \tfrac{1}{2}[&K_{(1,2,3,4,5,6)^2} - K_{(2,3,4,5)^2} - K_{(1,2,5,6)^2} \\ &+ K_{(4,5,6)^2} - K_{(4,5,6)-(1',2',3',4',5',6')} \\ &- K_{(1,2,3,4,5,6)-(4',5',6')} + K_{(1,2,5,6)-(5'6')} \\ &+ K_{(2,3,4,5)-(4',5')} + K_{(5,6)-(1',2',5',6')} \\ &+ K_{(4,5)-(2',3',4',5')} + K_{(2,5)^2} - K_{(2,5)-5'} \\ &- K_{(5,6)^2} - K_{(4,5)^2} - K_{5-(2',5')} + K_{5^2}] \end{aligned} \tag{8-31}$$

where the K terms are defined by

$$K_{m-n} = A_mF_{m-n} \tag{8-32}$$

$$K_{(m)^2} = A_mF_{m-m'} \tag{8-33}$$

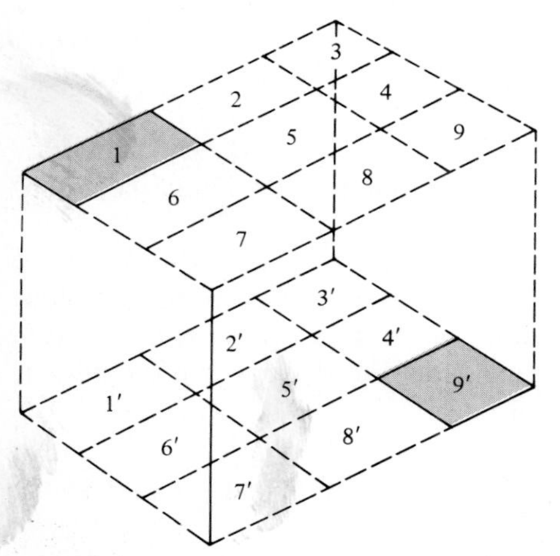

Fig. 8-23 Generalized parallel-rectangle arrangement.

The generalized parallel-rectangle arrangement is depicted in Fig. 8-23. The reciprocity relations which apply to this situation are given in Ref. 5 as

$$A_1F_{19'} = A_3F_{37'} = A_9F_{91'} = A_7F_{73'} \tag{8-34}$$

Making use of these relations, it is possible to derive the shape factor $F_{19'}$ as

$$\begin{aligned} A_1F_{19'} = \tfrac{1}{4}[&K_{(1,2,3,4,5,6,7,8,9)^2} - K_{(1,2,5,6,7,8)^2} \\ &- K_{(2,3,4,5,8,9)^2} - K_{(1,2,3,4,5,6)^2} + K_{(1,3,5,6)^2} \\ &+ K_{(2,3,4,5)^2} + K_{(4,5,8,9)^2} - K_{(4,5)^2} - K_{(5,8)^2} \\ &- K_{(5,6)^2} - K_{(4,5,6,7,8,9)^2} + K_{(5,6,7,8)^2} \\ &+ K_{(4,5,6)^2} + K_{(2,5,8)^2} - K_{(2,5)^2} + K_{(5)^2}] \end{aligned} \tag{8-35}$$

The nomenclature for the K terms is the same as given in Eqs. (8-32) and (8-33).

■ **EXAMPLE 8-3** Shape-factor algebra

Two concentric cylinders having diameters of 10 and 20 cm have a length of 20 cm. Calculate the shape factor between the open ends of the cylinders.

Solution

We use the nomenclature of Fig. 8-15 for this problem and designate the open ends as surfaces 3 and 4. We have $L/r_2 = 20/10 = 2.0$ and $r_1/r_2 = 0.5$; so from Fig. 8-15 we obtain

$$F_{21} = 0.43 \qquad F_{22} = 0.33$$

Using the reciprocity relation [Eq. (8-18)] we have

$$A_1F_{12} = A_2F_{21} \qquad \text{and} \qquad F_{12} = (d_2/d_1)F_{21} = (20/10)(0.43) = 0.86$$

For surface 2 we have

$$F_{21} + F_{22} + F_{23} + F_{24} = 1.0$$

From symmetry $F_{23} = F_{24}$ so that

$$F_{23} = F_{24} = (\tfrac{1}{2})(1 - 0.43 - 0.33) = 0.12$$

Using reciprocity again,

$$A_2F_{23} = A_3F_{32}$$

and

$$F_{32} = \frac{\pi(20)(20)}{\pi(20^2 - 10^2)/4}0.12 = 0.64$$

We observe that $F_{11} = F_{33} = F_{44} = 0$ and for surface 3

$$F_{31} + F_{32} + F_{34} = 1.0 \tag{a}$$

So, if F_{31} can be determined, we can calculate the desired quantity F_{34}. For surface 1

$$F_{12} + F_{13} + F_{14} = 1.0$$

and from symmetry $F_{13} = F_{14}$ so that

$$F_{13} = (\tfrac{1}{2})(1 - 0.86) = 0.07$$

Using reciprocity gives

$$A_1F_{13} = A_3F_{31}$$

$$F_{31} = \frac{\pi\,(10)(20)}{\pi\,(20^2 - 10^2)/4}\,0.07 = 0.187$$

Then, from Eq. (*a*)

$$F_{34} = 1 - 0.187 - 0.64 = 0.173$$

EXAMPLE 8-4 Shape-factor algebra

A truncated cone has top and bottom diameters of 10 and 20 cm and a height of 10 cm. Calculate the shape factor between the top surface and the side and also the shape factor between the side and itself.

Solution

We employ Fig. 8-16 for solution of this problem and take the nomenclature as shown, designating the top as surface 2, the bottom as surface 1, and the side as surface 3. Thus, the desired quantities are F_{23} and F_{33}. We have $L/r_1 = 10/10 = 1.0$ and $r_2/L = 5/10 = 0.5$. Thus, from Fig. 8-16

$$F_{12} = 0.12$$

From reciprocity [Eq. (8-18)]

$$A_1F_{12} = A_2F_{21}$$

$$F_{21} = (20/10)^2(0.12) = 0.48$$

and

$$F_{22} = 0$$

so that

$$F_{21} + F_{23} = 1.0$$

and

$$F_{23} = 1 - 0.48 = 0.52$$

For surface 3,

$$F_{31} + F_{32} + F_{33} = 1.0 \qquad (a)$$

so we must find F_{31} and F_{32} in order to evaluate F_{33}. Since $F_{11} = 0$ we have

$$F_{12} + F_{13} = 1.0 \qquad \text{and} \qquad F_{13} = 1 - 0.12 = 0.88$$

and from reciprocity

$$A_1F_{13} = A_3F_{31} \qquad (b)$$

The surface area of the side is

$$A_3 = \pi\,(r_1 + r_2)[(r_1 - r_2)^2 + L^2]^{1/2}$$

$$= \pi\,(5 + 10)(5^2 + 10^2)^{1/2} = 526.9\ \text{cm}^2$$

So, from Eq. (*b*)

$$F_{31} = \frac{\pi\,(10^2)}{526.9}\,0.88 = 0.525$$

A similar procedure applies with surface 2 so that

$$F_{32} = \frac{\pi\,(5)^2}{526.9}\,0.52 = 0.0775$$

Finally, from Eq. (*a*)

$$F_{33} = 1 - 0.525 - 0.0775 = 0.397$$

■ **EXAMPLE 8-5** Shape-factor algebra

The circular half-cylinder shown has a diameter of 60 cm and a square rod 20 by 20 cm placed along the geometric centerline. Both are surrounded by a large enclosure. Find F_{12}, F_{13}, and F_{11} in accordance with the nomenclature in the figure.

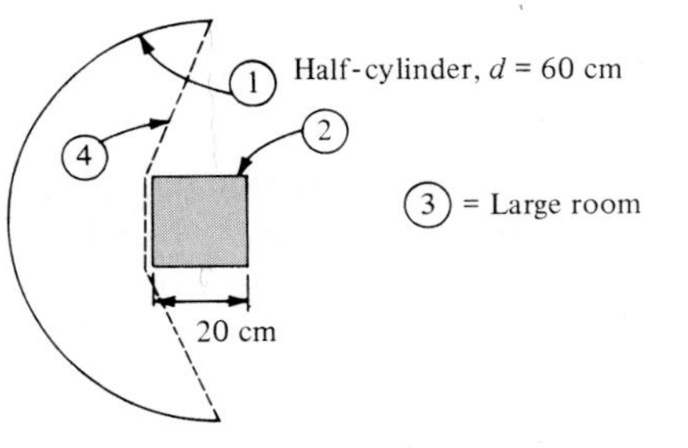

Fig. Ex. 8-5

Solution

From symmetry we have

$$F_{21} = F_{23} = 0.5 \tag{a}$$

In general, $F_{11} + F_{12} + F_{13} = 1.0$. To aid in the analysis we create the ficticious surface 4 shown as the dashed line. For this surface, $F_{41} = 1.0$. Now, all radiation leaving surface 1 will arrive either at 2 or at 3. Likewise, this radiation will arrive at the imaginery surface 4, so that

$$F_{14} = F_{12} + F_{13} \tag{b}$$

From reciprocity,

$$A_1 F_{14} = A_4 F_{41}$$

The areas are, for unit length,

$$A_1 = \pi d/2 = \pi\,(0.6)/2 = 0.942$$

$$A_4 = 0.2 + (2)[(0.1)^2 + (0.2)^2]^{1/2} = 0.647$$

$$A_2 = (4)(0.2) = 0.8$$

so that

$$F_{14} = \frac{A_4}{A_1} F_{41} = \frac{(0.647)(1.0)}{0.942} = 0.686 \qquad (c)$$

We also have, from reciprocity,

$$A_2 F_{21} = A_1 F_{12}$$

so

$$F_{12} = \frac{A_2}{A_1} F_{21} = \frac{(0.8)(0.5)}{0.942} = 0.425 \qquad (d)$$

Combining (*b*), (*c*), and (*d*) gives

$$F_{13} = 0.686 - 0.425 = 0.261$$

Finally,

$$F_{11} = 1 - F_{12} - F_{13} = 1 - 0.425 - 0.261 = 0.314$$

This example illustrates how one may make use of clever geometric considerations to calculate the radiation shape factors.

8-6 HEAT EXCHANGE BETWEEN NONBLACKBODIES

The calculation of the radiation heat transfer between black surfaces is relatively easy because all the radiant energy which strikes a surface is absorbed. The main problem is one of determining the geometric shape factor, but once this is accomplished, the calculation of the heat exchange is very simple. When nonblackbodies are involved, the situation is much more complex, for all the energy striking a surface will not be absorbed; part will be reflected back to another heat-transfer surface, and part may be reflected out of the system entirely. The problem can become complicated because the radiant energy can be reflected back and forth between the heat-transfer surfaces several times. The analysis of the problem must take into consideration these multiple reflections if correct conclusions are to be drawn.

We shall assume that all surfaces considered in our analysis are diffuse and uniform in temperature and that the reflective and emissive properties are constant over all the surface. Two new terms may be defined:

G = irradiation

= total radiation incident upon a surface per unit time and per unit area

J = radiosity

= total radiation which leaves a surface per unit time and per unit area

In addition to the assumptions stated above, we shall also assume that the radiosity and irradiation are uniform over each surface. This assumption is not strictly correct, even for ideal gray diffuse surfaces, but the problems become exceedingly complex when this analytical restriction is not imposed. Sparrow and Cess [10] give a discussion of such problems. The radiosity is the sum of the energy emitted and the energy reflected when no energy is transmitted, or

$$J = \epsilon E_b + \rho G \tag{8-36}$$

where ϵ is the emissivity and E_b is the blackbody emissive power. Since the transmissivity is assumed to be zero, the reflectivity may be expressed as

$$\rho = 1 - \alpha = 1 - \epsilon$$

so that

$$J = \epsilon E_b + (1 - \epsilon)G \tag{8-37}$$

The net energy leaving the surface is the difference between the radiosity and the irradiation:

$$\frac{q}{A} = J - G = \epsilon E_b + (1 - \epsilon)G - G$$

Solving for G in terms of J from Eq. (8-37),

$$q = \frac{\epsilon A}{1 - \epsilon}(E_b - J)$$

or

$$q = \frac{E_b - J}{(1 - \epsilon)/\epsilon A} \tag{8-38}$$

At this point we introduce a very useful interpretation for Eq. (8-38). If the denominator of the right side is considered as the surface resistance to radiation heat transfer, the numerator as a potential difference, and the heat flow as the "current," then a network element could be drawn as in Fig. 8-24 to represent the physical situation. This is the first step in the network method of analysis originated by Oppenheim [20].

Now consider the exchange of radiant energy by two surfaces A_1 and A_2. Of that total radiation which leaves surface 1, the amount that reaches surface 2 is

$$J_1 A_1 F_{12}$$

and of that total energy leaving surface 2, the amount that reaches surface 1 is

$$J_2 A_2 F_{21}$$

The net interchange between the two surfaces is

$$q_{1-2} = J_1 A_1 F_{12} - J_2 A_2 F_{21}$$

But

$$A_1 F_{12} = A_2 F_{21}$$

so that

$$q_{1-2} = (J_1 - J_2) A_1 F_{12} = (J_1 - J_2) A_2 F_{21}$$

or

$$q_{1-2} = \frac{J_1 - J_2}{1/A_1 F_{12}} \tag{8-39}$$

We may thus construct a network element which represents Eq. (8-39), as

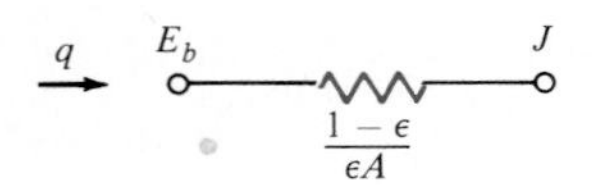

Fig. 8-24 Element representing "surface resistance" in radiation-network method.

q_{1-2} J_1 J_2 $\frac{1}{A_1F_{12}}$

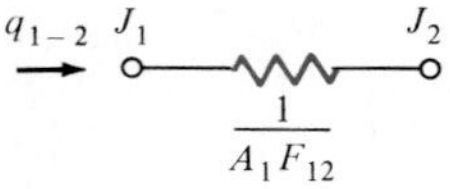

Fig. 8-25 Element representing "space resistance" in radiation-network method.

shown in Fig. 8-25. The two network elements shown in Figs. 8-24 and 8-25 represent the essentials of the radiation-network method. To construct a network for a particular radiation heat-transfer problem we need only connect a "surface resistance" $(1 - \epsilon)/\epsilon A$ to each surface and a "space resistance" $1/A_mF_{m-n}$ between the radiosity potentials. For example, two surfaces which exchange heat with each other *and nothing else* would be represented by the network shown in Fig. 8-26. In this case the net heat transfer would be the overall potential difference divided by the sum of the resistances:

$$q_{\text{net}} = \frac{E_{b_1} - E_{b_2}}{(1 - \epsilon_1)/\epsilon_1A_1 + 1/A_1F_{12} + (1 - \epsilon_2)/\epsilon_2A_2}$$

$$= \frac{\sigma(T_1^4 - T_2^4)}{(1 - \epsilon_1)/\epsilon_1A_1 + 1/A_1F_{12} + (1 - \epsilon_2)/\epsilon_2A_2} \tag{8-40}$$

A three-body problem is shown in Fig. 8-27. In this case each of the bodies exchanges heat with the other two. The heat exchange between body 1 and body 2 would be

$$q_{1-2} = \frac{J_1 - J_2}{1/A_1F_{12}}$$

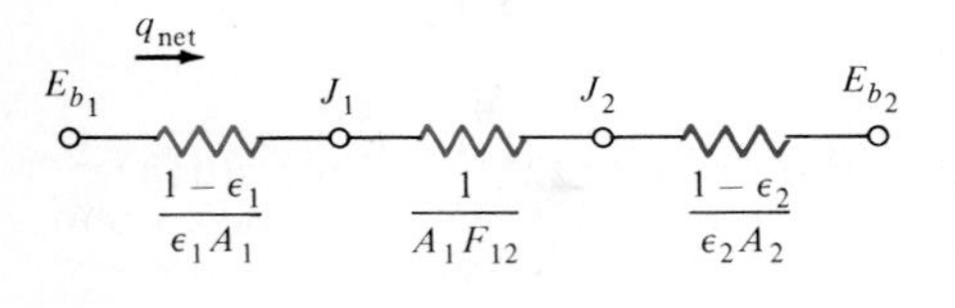

Fig. 8-26 Radiation network for two surfaces which see each other and nothing else.

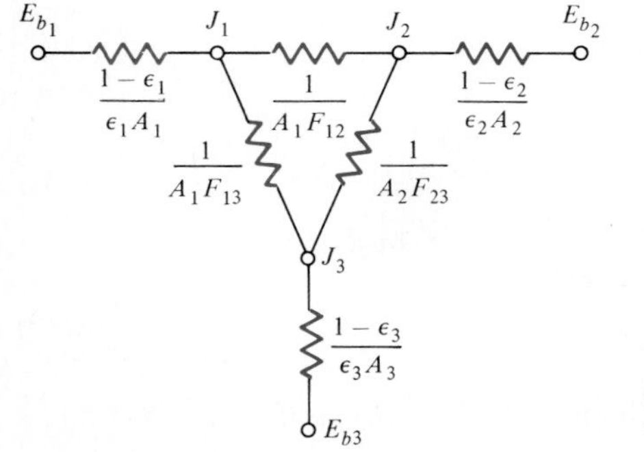

Fig. 8-27 Radiation network for three surfaces which see each other and nothing else.

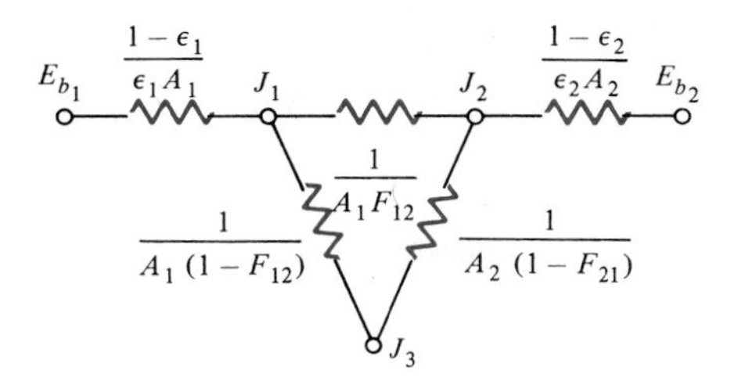

Fig. 8-28 Radiation network for two surfaces enclosed by a third surface which is nonconducting but re-radiating.

and that between body 1 and body 3,

$$q_{1-3} = \frac{J_1 - J_3}{1/A_1F_{13}}$$

To determine the heat flows in a problem of this type, the values of the radiosities must be calculated. This may be accomplished by performing standard methods of analysis used in dc circuit theory. The most convenient method is an application of Kirchhoff's current law to the circuit, which states that the sum of the currents entering a node is zero. Example 8-5 illustrates the use of the method for the three-body problem.

A problem which may be easily solved with the network method is that of two flat surfaces exchanging heat with one other but connected by a third surface which does not exchange heat, i.e., one which is perfectly insulated. This third surface nevertheless influences the heat-transfer process because it absorbs and re-radiates energy to the other two surfaces which exchange heat. The network for this system is shown in Fig. 8-28. Notice that node J_3 is not connected to a radiation surface resistance because surface 3 does not exchange energy. Notice also that the values for the space resistances have been written

$$F_{13} = 1 - F_{12}$$

$$F_{23} = 1 - F_{21}$$

since surface 3 completely surrounds the other two surfaces. For the special case where surfaces 1 and 2 are convex, i.e., they do not see themselves and $F_{11} = F_{22} = 0$, Fig. 8-28 is a simple series-parallel network which may be solved for the heat flow as

$$q_{\text{net}} = \frac{\sigma A_1(T_1^4 - T_2^4)}{\dfrac{A_1 + A_2 - 2A_1F_{12}}{A_2 - A_1(F_{12})^2} + \left(\dfrac{1}{\epsilon_1} - 1\right) + \dfrac{A_1}{A_2}\left(\dfrac{1}{\epsilon_2} - 1\right)} \tag{8-41}$$

where the reciprocity relation

$$A_1F_{12} = A_2F_{21}$$

has been used to simplify the expression. *It is to be noted again that Eq. (8-41) applies only to surfaces which do not see themselves, that is,* $F_{11} = F_{22} = 0$.

This network, and others which follow, assume that the only heat exchange is by radiation. Conduction and convection are neglected for now.

■ **EXAMPLE 8-6** Hot plates in a room

Two parallel plates 0.5 by 1.0 m are spaced 0.5 m apart. One plate is maintained at 1000°C and the other at 500°C. The emissivities of the plates are 0.2 and 0.5, respectively. The plates are located in a very large room, the walls of which are maintained at 27°C. The plates exchange heat with each other and with the room, but only the plate surfaces facing each other are to be considered in the analysis. Find the net transfer to each plate and to the room.

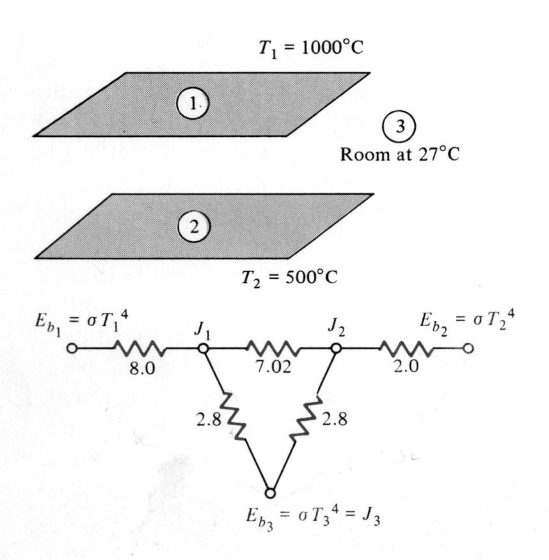

Fig. Ex. 8-6 (*a*) Schematic. (*b*) Network.

Solution

This is a three-body problem, the two plates and the room, so the radiation network is shown in Fig. 8-27. From the data of the problem

$$T_1 = 1000°\text{C} = 1273\ \text{K} \qquad A_1 = A_2 = 0.5\ \text{m}^2$$

$$T_2 = 500°\text{C} = 773\ \text{K} \qquad \epsilon_1 = 0.2$$

$$T_3 = 27°\text{C} = 300\ \text{K} \qquad \epsilon_2 = 0.5$$

Because the area of the room A_3 is very large, the resistance $(1 - \epsilon_3)/\epsilon_3 A_3$ may be taken as zero and we obtain $E_{b_3} = J_3$. The shape factor was given in Example 8-2:

$$F_{12} = 0.285 = F_{21}$$

$$F_{13} = 1 - F_{12} = 0.715$$

$$F_{23} = 1 - F_{21} = 0.715$$

The resistances in the network are calculated as

$$\frac{1-\epsilon_1}{\epsilon_1 A_1} = \frac{1-0.2}{(0.2)(0.5)} = 8.0 \qquad \frac{1-\epsilon_2}{\epsilon_2 A_2} = \frac{1-0.5}{(0.5)(0.5)} = 2.0$$

$$\frac{1}{A_1 F_{12}} = \frac{1}{(0.5)(0.285)} = 7.018 \qquad \frac{1}{A_1 F_{13}} = \frac{1}{(0.5)(0.715)} = 2.797$$

$$\frac{1}{A_2 F_{23}} = \frac{1}{(0.5)(0.715)} = 2.797$$

Taking the resistance $(1 - \epsilon_3)/\epsilon_3 A_3$ as zero, we have the network as shown. To calculate the heat flows at each surface we must determine the radiosities J_1 and J_2. The network is solved by setting the sum of the heat currents entering nodes J_1 and J_2 to zero:

node J_1:
$$\frac{E_{b1} - J_1}{8.0} + \frac{J_2 - J_1}{7.018} + \frac{E_{b3} - J_1}{2.797} = 0 \qquad (a)$$

node J_2:
$$\frac{J_1 - J_2}{7.018} + \frac{E_{b3} - J_2}{2.797} + \frac{E_{b2} - J_2}{2.0} = 0 \qquad (b)$$

Now
$$E_{b1} = \sigma T_1^4 = 148.87 \text{ kW/m}^2 \quad [47{,}190 \text{ Btu/h} \cdot \text{ft}^2]$$
$$E_{b2} = \sigma T_2^4 = 20.241 \text{ kW/m}^2 \quad [6416 \text{ Btu/h} \cdot \text{ft}^2]$$
$$E_{b3} = \sigma T_3^4 = 0.4592 \text{ kW/m}^2 \quad [145.6 \text{ Btu/h} \cdot \text{ft}^2]$$

Inserting the values of E_{b_1}, E_{b_2}, and E_{b_3} into Eqs. (*a*) and (*b*), we have two equations and two unknowns J_1 and J_2 which may be solved simultaneously to give

$$J_1 = 33.469 \text{ kW/m}^2 \qquad J_2 = 15.054 \text{ kW/m}^2$$

The total heat lost by plate 1 is

$$q_1 = \frac{E_{b1} - J_1}{(1-\epsilon_1)/\epsilon_1 A_1} = \frac{148.87 - 33.469}{8.0} = 14.425 \text{ kW}$$

and the total heat lost by plate 2 is

$$q_2 = \frac{E_{b2} - J_2}{(1-\epsilon_2)/\epsilon_2 A_2} = \frac{20.241 - 15.054}{2.0} = 2.594 \text{ kW}$$

The total heat received by the room is

$$q_3 = \frac{J_1 - J_3}{1/A_1 F_{13}} + \frac{J_2 - J_3}{1/A_2 F_{23}}$$
$$= \frac{33.469 - 0.4592}{2.797} + \frac{15.054 - 0.4592}{2.797} = 17.020 \text{ kW} \quad [58{,}070 \text{ Btu/h}]$$

From an overall-balance standpoint we must have

$$q_3 = q_1 + q_2$$

because the net energy lost by both plates must be absorbed by the room.

EXAMPLE 8-7 Surface in radiant balance

Two rectangles 50 by 50 cm are placed perpendicularly with a common edge. One surface has $T_1 = 1000$ K, $\epsilon_1 = 0.6$, while the other surface is insulated and in radiant balance with a large surrounding room at 300 K. Determine the temperature of the insulated surface and the heat lost by the surface at 1000 K.

Solution

The radiation network is shown in the accompanying figure where surface 3 is the room and surface 2 is the insulated surface. Note that $J_3 = E_{b3}$ because the room is large and $(1 - \epsilon_3)/\epsilon_3 A_3$ approaches zero. Because surface 2 is insulated it has zero heat transfer and $J_2 = E_{b2}$. J_2 "floats" in the network and is determined from the overall radiant balance. From Fig. 8-14 the shape factors are

$$F_{12} = 0.2 = F_{21}$$

Because $F_{11} = 0$ and $F_{22} = 0$ we have

$$F_{12} + F_{13} = 1.0 \qquad \text{and} \qquad F_{13} = 1 - 0.2 = 0.8 = F_{23}$$

$$A_1 = A_2 = (0.5)^2 = 0.25 \text{ m}^2$$

The resistances are

$$\frac{1 - \epsilon_1}{\epsilon_1 A_1} = \frac{0.4}{(0.6)(0.25)} = 2.667$$

$$\frac{1}{A_1 F_{13}} = \frac{1}{A_2 F_{23}} = \frac{1}{(0.25)(0.8)} = 5.0$$

$$\frac{1}{A_1 F_{12}} = \frac{1}{(0.25)(0.2)} = 20.0$$

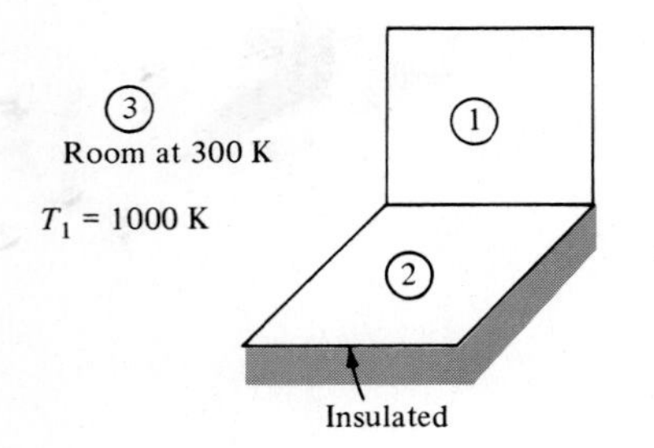

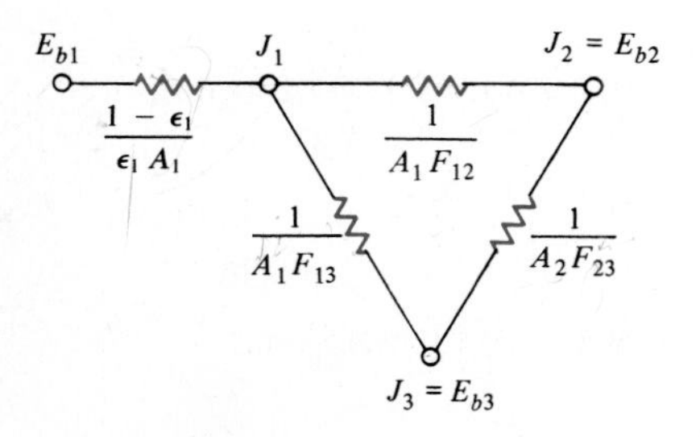

Fig. Ex. 8-7 (*a*) Schematic. (*b*) Network.

We also have

$$E_{b1} = (5.669 \times 10^{-8})(1000)^4 = 5.669 \times 10^4 \text{ W/m}^2$$

$$J_3 = E_{b3} = (5.669 \times 10^{-8})(300)^4 = 459.2 \text{ W/m}^2$$

The overall circuit is a series-parallel arrangement and the heat transfer is

$$q = \frac{E_{b1} - E_{b3}}{R_{\text{equiv}}}$$

We have

$$R_{\text{equiv}} = 2.667 + \frac{1}{\frac{1}{5} + 1/(20 + 5)} = 6.833$$

and

$$q = \frac{56{,}690 - 459.2}{6.833} = 8.229 \text{ kW} \quad [28{,}086 \text{ Btu/h}]$$

This heat transfer can also be written

$$q = \frac{E_{b1} - J_1}{(1 - \epsilon_1)/\epsilon_1 A_1}$$

Inserting the values we obtain

$$J_1 = 34{,}745 \text{ W/m}^2$$

The value of J_2 is determined from proportioning the resistances between J_1 and J_3, so that

$$\frac{J_1 - J_2}{20} = \frac{J_1 - J_3}{20 + 5}$$

and

$$J_2 = 7316 = E_{b2} = \sigma T_2^4$$

Finally, we obtain the temperature as

$$T_2 = \left(\frac{7316}{5.669 \times 10^{-8}}\right)^{1/4} = 599.4 \text{ K} \quad [619°\text{F}]$$

■ 8-7 INFINITE PARALLEL PLANES

When two infinite parallel planes are considered, A_1 and A_2 are equal; and the radiation shape factor is unity since all the radiation leaving one plane reaches the other. The network is the same as in Fig. 8-26, and the heat flow per unit area may be obtained from Eq. (8-40) by letting $A_1 = A_2$ and $F_{12} = 1.0$. Thus

$$\frac{q}{A} = \frac{\sigma(T_1^4 - T_2^4)}{1/\epsilon_1 + 1/\epsilon_2 - 1} \tag{8-42}$$

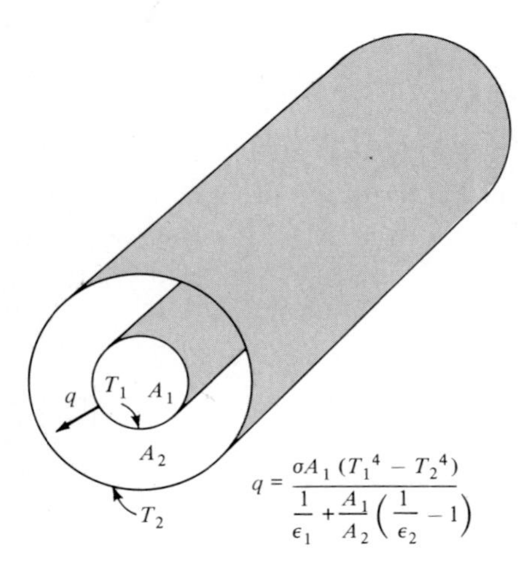

Fig. 8-29 Radiation exchange between two cylindrical surfaces.

When two long concentric cylinders as shown in Fig. 8-29 exchange heat we may again apply Eq. (8-40). Rewriting the equation and noting that $F_{12} = 1.0$,

$$q = \frac{\sigma A_1(T_1^4 - T_2^4)}{1/\epsilon_1 + (A_1/A_2)(1/\epsilon_2 - 1)} \tag{8-43}$$

The area ratio A_1/A_2 may be replaced by the diameter ratio d_1/d_2 when cylindrical bodies are concerned.

Equation (8-43) is particularly important when applied to the limiting case of a convex object completely enclosed by a very large concave surface. In this instance $A_1/A_2 \rightarrow 0$ and the following simple relation results:

$$q = \sigma A_1 \epsilon_1 (T_1^4 - T_2^4) \tag{8-43a}$$

This equation is readily applied to calculate the radiation-energy loss from a hot object in a large room.

EXAMPLE 8-8

The 30-cm-diameter hemisphere in the accompanying figure is maintained at a constant temperature of 500°C and insulated on its back side. The surface emissivity is 0.4. The opening exchanges radiant energy with a large enclosure at 30°C. Calculate the net radiant exchange.

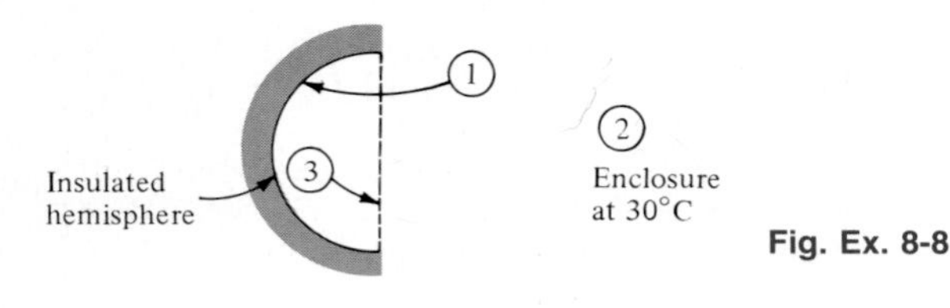

Fig. Ex. 8-8

Solution

This is an object completely surrounded by a large enclosure but the inside surface of the sphere is *not convex;* i.e., it sees itself, and therefore we are *not* permitted to use Eq. (8-43*a*). In the figure we take the inside of the sphere as surface 1 and the enclosure as surface 2. We also create an imaginary surface 3 covering the opening. We actually have a two-surface problem (surfaces 1 and 2) and therefore may use Eq. (8-40) to calculate the heat transfer. Thus,

$$E_{b1} = \sigma T_1^4 = \sigma(773)^4 = 20{,}241 \text{ W/m}^2$$

$$E_{b2} = \sigma T_2^4 = \sigma(303)^4 = 478 \text{ W/m}^2$$

$$A_1 = 2\pi r^2 = (2)\pi(0.15)^2 = 0.1414 \text{ m}^2$$

$$\frac{1 - \epsilon_1}{\epsilon_1 A_1} = \frac{0.6}{(0.4)(0.1414)} = 10.61$$

$$A_2 \rightarrow \infty$$

so that

$$\frac{1 - \epsilon_2}{\epsilon_2 A_2} \rightarrow 0$$

Now, at this point we recognize that all of the radiation leaving surface 1 which will eventually arrive at enclosure 2 will also hit the imaginary surface 3; i.e., $F_{12} = F_{13}$. We also recognize that

$$A_1 F_{13} = A_3 F_{31}$$

But, $F_{31} = 1.0$ so that

$$F_{13} = F_{12} = \frac{A_3}{A_1} = \frac{\pi r^2}{2\pi r^2} = 0.5$$

Then $1/A_1 F_{12} = 1/(0.1414)(0.5) = 14.14$

and we can calculate the heat transfer by inserting the quantities in Eq. (8-40):

$$q = \frac{20{,}241 - 478}{10.61 + 14.14 + 0} = 799 \text{ W}$$

8-8 RADIATION SHIELDS

One way of reducing radiant heat transfer between two particular surfaces is to use materials which are highly reflective. An alternative method is to use radiation shields between the heat-exchange surfaces. These shields do not deliver or remove any heat from the overall system; they only place another resistance in the heat-flow path so that the overall heat transfer is retarded. Consider the two parallel infinite planes shown in Fig. 8-30*a*. We have shown that the heat exchange between these surfaces may be calculated with Eq. (8-42). Now consider the same two planes, but with a radiation shield placed between them, as in Fig. 8-30*b*. The heat transfer will be calculated for this latter case and compared with the heat transfer without the shield.

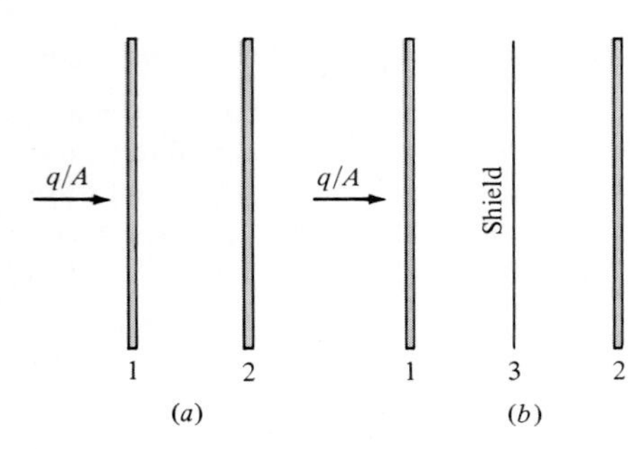

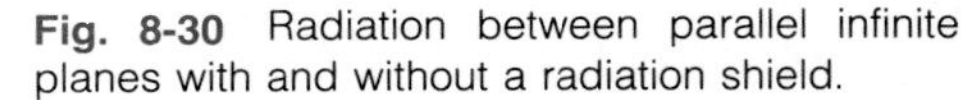
Fig. 8-30 Radiation between parallel infinite planes with and without a radiation shield.

Since the shield does not deliver or remove heat from the system, the heat transfer between plate 1 and the shield must be precisely the same as that between the shield and plate 2, and this is the overall heat transfer. Thus

$$\left(\frac{q}{A}\right)_{1-3} = \left(\frac{q}{A}\right)_{3-2} = \frac{q}{A}$$

$$\frac{q}{A} = \frac{\sigma(T_1^4 - T_3^4)}{1/\epsilon_1 + 1/\epsilon_3 - 1} = \frac{\sigma(T_3^4 - T_2^4)}{1/\epsilon_3 + 1/\epsilon_2 - 1} \tag{8-44}$$

The only unknown in Eq. (8-44) is the temperature of the shield T_3. Once this temperature is obtained, the heat transfer is easily calculated. If the emissivities of all three surfaces are equal, that is, $\epsilon_1 = \epsilon_2 = \epsilon_3$, we obtain the simple relation

$$T_3^4 = \tfrac{1}{2}(T_1^4 + T_2^4) \tag{8-45}$$

and the heat transfer is

$$\frac{q}{A} = \frac{\frac{1}{2}\sigma(T_1^4 - T_2^4)}{1/\epsilon_1 + 1/\epsilon_3 - 1}$$

But since $\epsilon_3 = \epsilon_2$, we observe that this heat flow is just one-half of that which would be experienced if there were no shield present. The radiation network corresponding to the situation in Fig. 8-30*b* is given in Fig. 8-31.

Multiple-radiation-shield problems may be treated in the same manner as that outlined above. When the emissivities of all surfaces are different, the overall heat transfer may be calculated most easily by using a series radiation network with the appropriate number of elements, similar to the one in Fig.

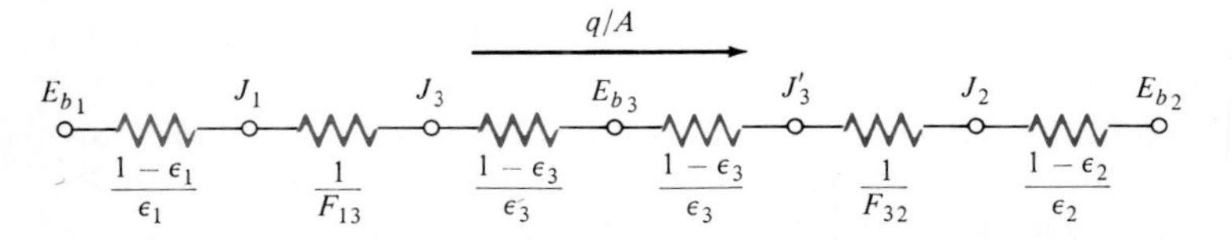

Fig. 8-31 Radiation network for two parallel planes separated by one radiation shield.

8-31. If the emissivities of all surfaces are equal, a rather simple relation may be derived for the heat transfer when the surfaces may be considered as infinite parallel planes. Let the number of shields be n. Considering the radiation network for the system, all the "surface resistances" would be the same since the emissivities are equal. There would be two of these resistances for each shield and one for each heat-transfer surface. There would be $n + 1$ "space resistances," and these would all be unity since the radiation shape factors are unity for the infinite parallel planes. The total resistance in the network would thus be

$$R\ (n\ \text{shields}) = (2n + 2)\frac{1 - \epsilon}{\epsilon} + (n + 1)(1) = (n + 1)\left(\frac{2}{\epsilon} - 1\right)$$

The resistance when no shield is present is

$$R\ (\text{no shield}) = \frac{1}{\epsilon} + \frac{1}{\epsilon} - 1 = \frac{2}{\epsilon} - 1$$

We note that the resistance with the shields in place is $n + 1$ times as large as when the shields are absent. Thus

$$\left(\frac{q}{A}\right)_{\substack{\text{with}\\ \text{shields}}} = \frac{1}{n + 1}\left(\frac{q}{A}\right)_{\substack{\text{without}\\ \text{shields}}} \tag{8-46}$$

if the temperatures of the heat-transfer surfaces are maintained the same in both cases. The radiation-network method may also be applied to shield problems involving cylindrical systems. In these cases the proper area relations must be used.

Notice that the analyses above, dealing with infinite parallel planes, have been carried out on a per-unit-area basis because all areas are the same.

■ **EXAMPLE 8-9** Heat-transfer reduction with shield

Two very large parallel planes with emissivities 0.3 and 0.8 exchange heat. Find the percentage reduction in heat transfer when a polished-aluminum radiation shield ($\epsilon = 0.04$) is placed between them.

Solution

The heat transfer without the shield is given by

$$\frac{q}{A} = \frac{\sigma(T_1^4 - T_2^4)}{1/\epsilon_1 + 1/\epsilon_2 - 1} = 0.279\sigma(T_1^4 - T_2^4)$$

The radiation network for the problem with the shield in place is shown in Fig. 8-31. The resistances are

$$\frac{1 - \epsilon_1}{\epsilon_1} = \frac{1 - 0.3}{0.3} = 2.333$$

$$\frac{1 - \epsilon_3}{\epsilon_3} = \frac{1 - 0.04}{0.04} = 24.0$$

$$\frac{1-\epsilon_2}{\epsilon_2} = \frac{1-0.8}{0.8} = 0.25$$

The total resistance is

$$2.333 + (2)(24.0) + (2)(1) + 0.25 = 52.583$$

and the heat transfer is

$$\frac{q}{A} = \frac{\sigma(T_1^4 - T_2^4)}{52.583} = 0.01902\sigma(T_1^4 - T_2^4)$$

so that the heat transfer is *reduced* by 93.2 percent.

■ **EXAMPLE 8-10** Open cylindrical shield

The two concentric cylinders of Example 8-3 have $T_1 = 1000$ K, $\epsilon_1 = 0.8$, $\epsilon_2 = 0.2$ and are located in a large room at 300 K. The outer cylinder is in radiant balance. Calculate the temperature of the outer cylinder and the total heat lost by the inner cylinder.

Solution

The network for this problem is shown in the accompanying figure. The room is designated as surface 3 and $J_3 = E_{b3}$, because the room is very large. In this problem we must consider the inside and outside of surface 2 and thus have subscripts *i* and *o* to designate the respective quantities. The shape factors can be obtained from Example 8-3 as

$$F_{12} = 0.86 \qquad F_{13} = (2)(0.07) = 0.14$$

$$F_{23i} = (2)(0.12) = 0.24 \qquad F_{23o} = 1.0$$

Also,

$$A_1 = \pi(0.1)(0.2) = 0.06283 \text{ m}^2$$

$$A_2 = \pi(0.2)(0.2) = 0.12566 \text{ m}^2$$

$$E_{b1} = (5.669 \times 10^{-8})(1000)^4 = 5.669 \times 10^4 \text{ W/m}^2$$

$$E_{b3} = (5.669 \times 10^{-8})(300)^4 = 459.2 \text{ W/m}^2$$

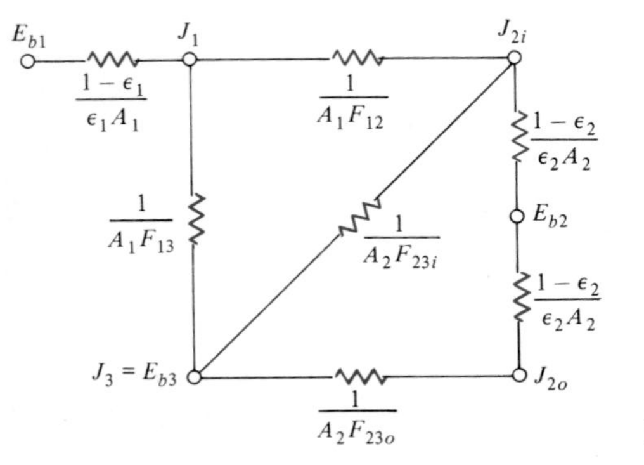

Fig. Ex. 8-10

and the resistances may be calculated as

$$\frac{1-\epsilon_1}{\epsilon_1 A_1} = 3.979 \qquad \frac{1-\epsilon_2}{\epsilon_2 A_2} = 31.83$$

$$\frac{1}{A_1 F_{12}} = 18.51 \qquad \frac{1}{A_2 F_{23i}} = 33.16$$

$$\frac{1}{A_2 F_{23o}} = 7.958 \qquad \frac{1}{A_1 F_{13}} = 113.7$$

The network could be solved as a series-parallel circuit to obtain the heat transfer but we will need the radiosities anyway so we set up three nodal equations to solve for J_1, J_{2i}, and J_{2o}. We sum the currents into each node and set them equal to zero:

node J_1:
$$\frac{E_{b_1} - J_1}{3.979} + \frac{E_{b_3} - J_1}{113.7} + \frac{J_{2i} - J_1}{18.51} = 0$$

node J_{2i}:
$$\frac{J_1 - J_{2i}}{18.51} + \frac{E_{b_3} - J_{2i}}{33.16} + \frac{J_{2o} - J_{2i}}{(2)(31.83)} = 0$$

node J_{2o}:
$$\frac{E_{b_3} - J_{2o}}{7.958} + \frac{J_{2i} - J_{2o}}{(2)(31.83)} = 0$$

These equations have the solution

$$J_1 = 50{,}148 \text{ W/m}^2$$
$$J_{2i} = 27{,}811 \text{ W/m}^2$$
$$J_{2o} = 3498 \text{ W/m}^2$$

The heat transfer is then calculated from

$$q = \frac{E_{b_1} - J_1}{(1-\epsilon_1)/\epsilon_1 A_1} = \frac{56{,}690 - 50.148}{3.979} = 1644 \text{ W} \quad [5611 \text{ Btu/h}]$$

From the network we see that

$$E_{b_2} = \frac{J_{2i} + J_{2o}}{2} = \frac{27{,}811 + 3498}{2} = 15{,}655 \text{ W/m}^2$$

and

$$T_2 = \left(\frac{15{,}655}{5.669 \times 10^{-8}}\right)^{1/4} = 724.9 \text{ K} \quad [845°\text{F}]$$

If the outer cylinder had not been in place acting as a "shield" the heat loss from cylinder 1 could have been calculated from Eq. (8-43*a*) as

$$\begin{aligned} q &= \epsilon_1 A_1 (E_{b_1} - E_{b_3}) \\ &= (0.8)(0.06283)(56{,}690 - 459.2) = 2856 \text{ W} \quad [9645 \text{ Btu/h}] \end{aligned}$$

■ 8-9 GAS RADIATION

Radiation exchange between a gas and a heat-transfer surface is considerably more complex than the situations described in the preceding sections. Unlike most solid bodies, gases are in many cases transparent to radiation. When they absorb and emit radiation, they usually do so only in certain narrow wavelength bands. Some gases, such as N_2, O_2, and others of nonpolar symmetrical molecular structure, are essentially transparent at low temperatures, while CO_2, H_2O, and various hydrocarbon gases radiate to an appreciable extent.

The absorption of radiation in gas layers may be described analytically in the following way, considering the system shown in Fig. 8-32. A monochromatic beam of radiation having an intensity I_λ impinges on the gas layer of thickness dx. The decrease in intensity resulting from absorption in the layers is assumed to be proportional to the thickness of the layer and the intensity of radiation at that point. Thus

$$dI_\lambda = -a_\lambda I_\lambda \, dx \tag{8-47}$$

where the proportionality constant a_λ is called the *monochromatic absorption coefficient*. Integrating this equation gives

$$\int_{I_{\lambda 0}}^{I_{\lambda x}} \frac{dI_\lambda}{I_\lambda} = \int_0^x -a_\lambda \, dx$$

or

$$\frac{I_{\lambda_x}}{I_{\lambda 0}} = e^{-a_\lambda x} \tag{8-48}$$

Equation (8-48) is called Beer's law and represents the familiar exponential-decay formula experienced in many types of radiation analyses dealing with absorption. In accordance with our definitions in Sec. 8-3, the monochromatic transmissivity will be given as

$$\tau_\lambda = e^{-a_\lambda x} \tag{8-49}$$

If the gas is nonreflecting, then

$$\tau_\lambda + \alpha_\lambda = 1$$

and

$$\alpha_\lambda = 1 - e^{-a_\lambda x} \tag{8-50}$$

As mentioned above, gases frequently absorb only in narrow wavelength bands, as indicated for water vapor in Fig. 8-33. These curves also indicate the effect of thickness of the gas layer on monochromatic absorptivity.

The calculation of gas-radiation properties is quite complicated, and Refs.

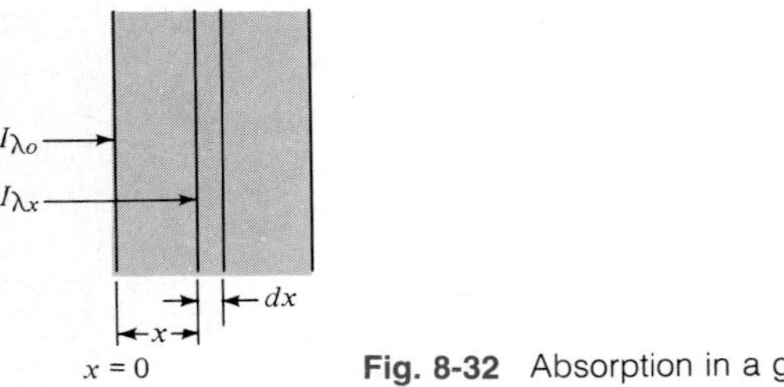

Fig. 8-32 Absorption in a gas layer.

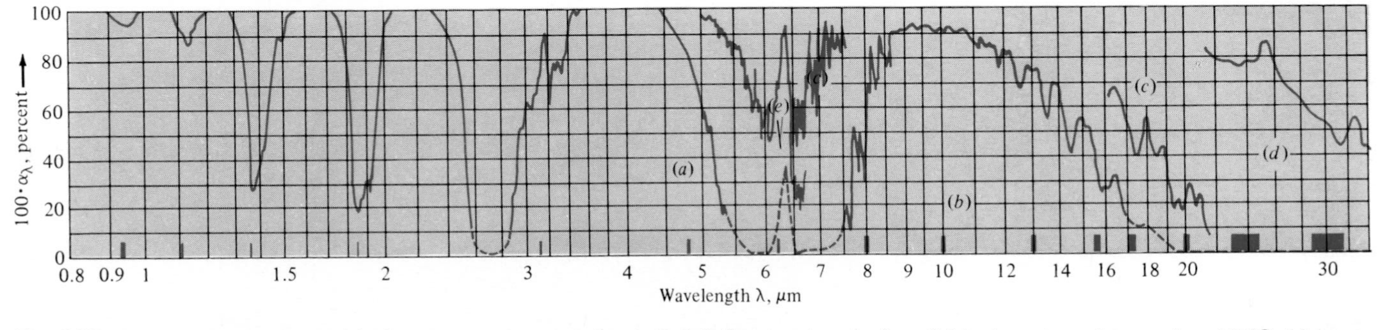

Fig. 8-33 Monochromatic absorptivity for water vapor according to Ref. 8. For wavelengths from 0.8 to 4 m, steam temperature 127°C, thickness of layer 109 cm; wavelengths from 4 to 34 m; (*a*) temperature 127°C, thickness of layer 109 cm; (*b*) temperature 127°C, thickness of layer 104 cm; (*c*) temperature 127°C, thickness of layer 32.4 cm; (*d*) temperature 81°C, thickness of layer 32.4 cm, airstream mixture corresponding to a steam layer approximately 4 cm thick; (*e*) room temperature, layer of moist air 200 cm thick corresponding to a layer of steam at atmospheric pressure approximately 7 cm thick.

23 to 25 should be consulted for detailed information. For engineering calculations Hottel [23] has presented a simplified procedure which may be used to calculate the emittances of water vapor and carbon dioxide gases. Methods for evaluating the radiant exchange between these gases and enclosures are also available.

Mean Beam Length

Equations (8-48) and (8-50) describe the intensity variation and absorptivity for a gas layer of thickness x. These are the values we might expect to measure in a laboratory experiment with radiation passing straight through the layer. If we imagine a practical problem of a gas contained between two large parallel plates which emit radiation diffusely, we see that the radiant energy transmitted through the gas travels many distances; the energy transmitted normal to the surface travels a distance equal to the plane spacing; energy emitted at shallow angles is absorbed in the gas over a much longer distance; and so on. By a

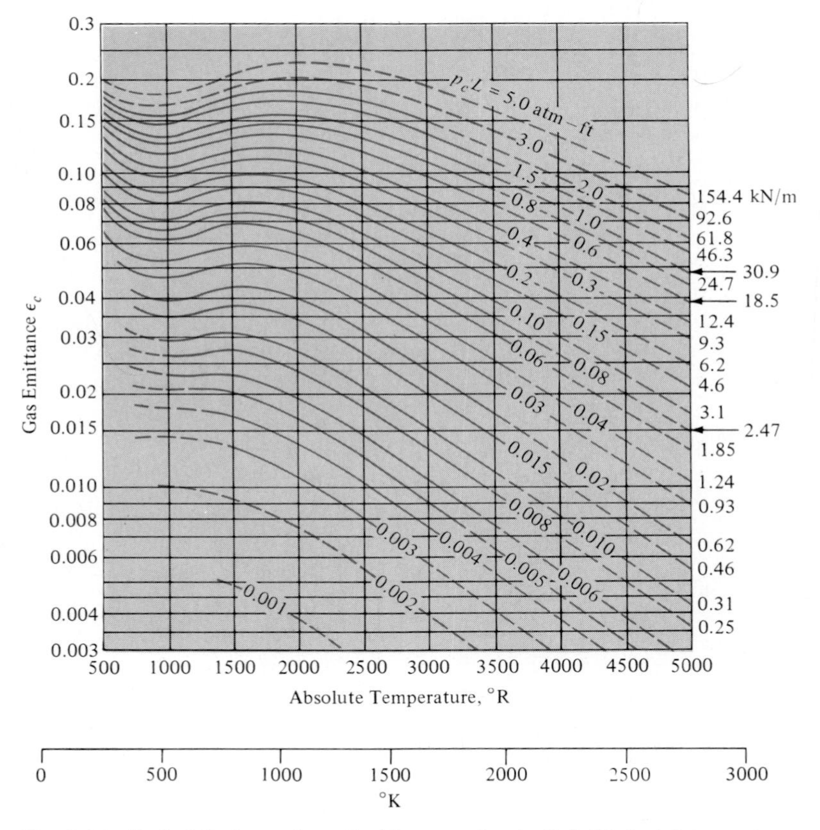

Fig. 8-34 Emissivity for carbon dioxide according to Ref. 22, for total pressure = 1 atm; 1.0 kN/m = 0.03238 atm·ft.

careful correlation of several sources of experimental data Hottel and Egbert [29] were able to present the gas emittances for carbon dioxide and water vapor as shown in Figs. 8-34 and 8-35. In these figures L_e is a characteristic dimension of the system called the *mean beam length*. Tabulations of these lengths are presented in Table 8-2 according to Hottel [22] and Eckert and Drake [25]. In the absence of mean-beam-length information for a specific geometry, a satisfactory approximation can be obtained from

$$L_e = 3.6 \frac{V}{A} \tag{8-51}$$

where V is the total volume of the gas and A is the total surface area. In Figs. 8-34 and 8-35 the total pressure of the mixture is 1 atm, and p_c and p_w represent the partial pressures of carbon dioxide and water vapor, respectively. For total

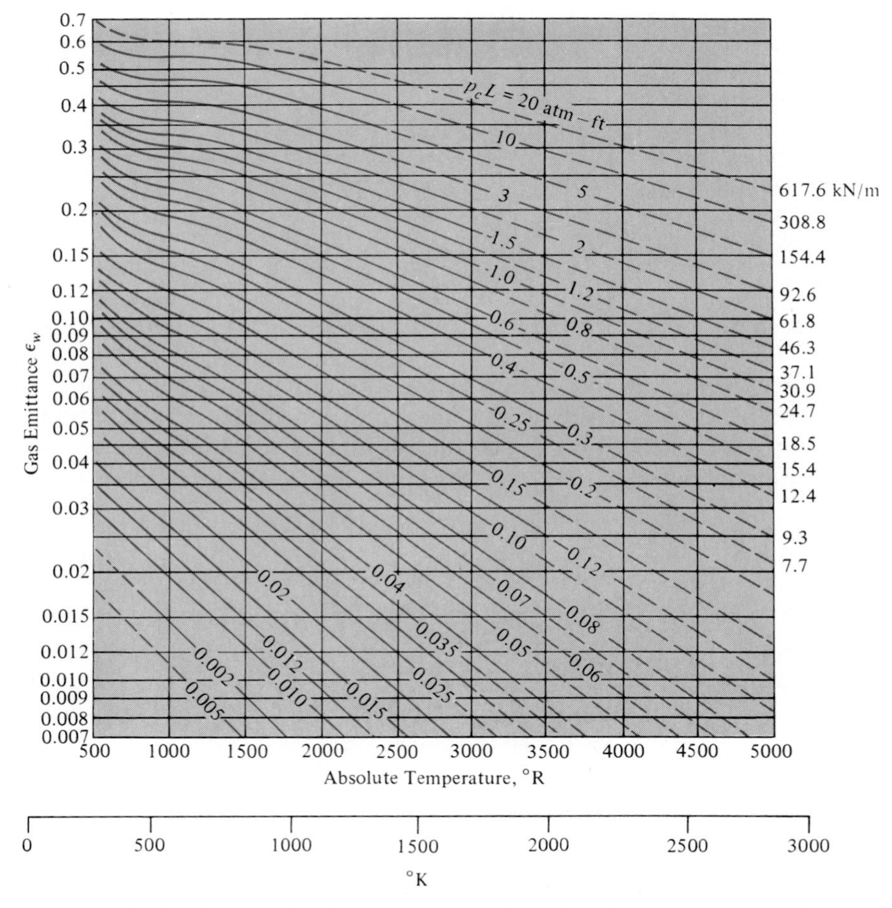

Fig. 8-35 Emissivity for water vapor according to Ref. 22 for total pressure = 1 atm; 1.0 kN/m = 0.03238 atm·ft.

Table 8-2 Mean Equivalent Length L_e for Radiation from Entire Gas Volume, According to Refs. 22 and 25.

Gas volume	*Characteristic dimension*	L_e
Volume between two infinite planes	Separation distance L	$1.8L$
Circular cylinder with the height = diameter, radiation to center of base	Diameter D	$0.71D$
Hemisphere, radiation to element in center of base	Radius R	R
Sphere, radiation to entire surface	Diameter D	$0.65D$
Infinite circular cylinder, radiation to convex bounding surface	Diameter D	$0.95D$
Circular cylinder with height = diameter, radiation to entire surface	Diameter D	$0.60D$
Circular cylinder, semi-infinite height, radiation to entire base	Diameter D	$0.65D$
Cube, radiation to any face	Edge L	$0.60L$
Volume surrounding infinite tube bundle, radiation to a single tube	Tube diameter D, distance between tube centers S	
Equilateral-triangle arrangement:		
$S = 2D$		$3.0(S - D)$
$S = 3D$		$3.8(S - D)$
Square arrangement		$3.5(S - D)$

pressures other than 1 atm, correction factors are provided in Figs. 8-36 and 8-37. When both carbon dioxide and water vapor are present, an additional correction $\Delta\epsilon$ from Fig. 8-38 must be subtracted from the total of the emittances of the two components; thus the total gas emittance ϵ_g of the mixture is expressed as

$$\epsilon_g = C_c\epsilon_c + C_w\epsilon_w - \Delta\epsilon \tag{8-52}$$

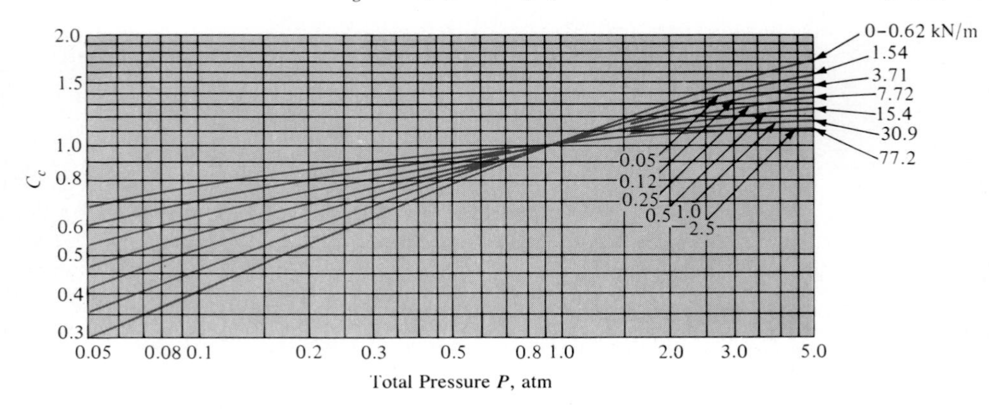

Fig. 8-36 Correction factor for CO_2 emissivity according to Ref. 22.

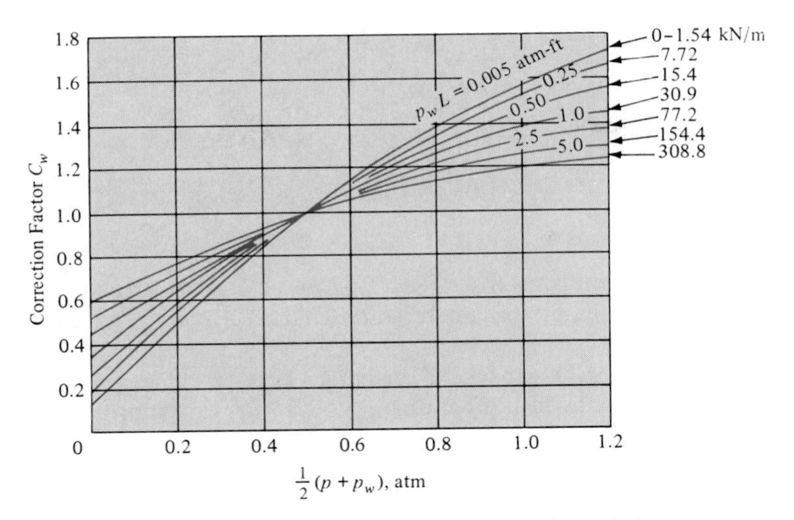

Fig. 8-37 Correction factor for H_2O emissivity according to Ref. 22.

☐ Heat Exchange between Gas Volume and Black Enclosure

Now consider a gas volume at uniform temperature T_g enclosed by a black surface at temperature T_w. Because of the band structure of the gas the absorption of energy emitted by the wall at T_w will differ from the energy emitted by the gas at T_g. The net heat transfer from the gas to the enclosure is therefore

$$\frac{q}{A} = \text{energy emitted by gas} - \text{energy from enclosure absorbed by gas}$$

$$= \epsilon_g(T_g)\sigma T_g^4 - \alpha_g(T_w)\sigma T_w^4 \tag{8-53}$$

where $\epsilon_g(T_g)$ is the gas emittance at T_g which is evaluated as discussed above and $\alpha_g(T_w)$ is the gas absorptance for the radiation from the black enclosure at T_w and is a function of both T_w and T_g. For a mixture of carbon dioxide and water vapor an empirical relation for α_g is

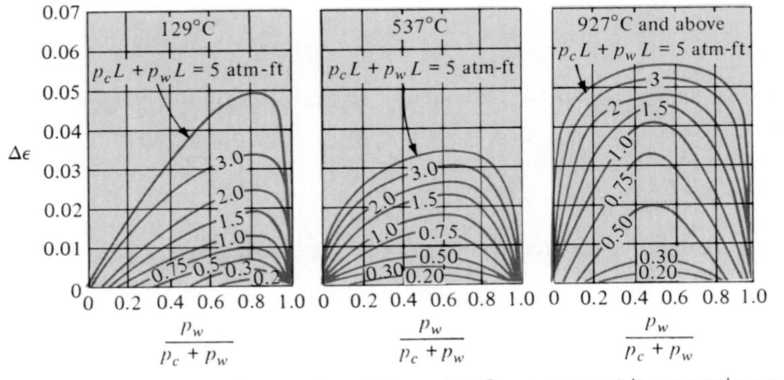

Fig. 8-38 Correction factor when CO_2 and H_2O are present in an enclosure according to Ref. 22: 1.0 atm·ft = 30.881 kN/m.

$$\alpha_g(T_w) = \alpha_c + \alpha_w - \Delta\alpha \tag{8-54}$$

where

$$\alpha_c = C_c \epsilon_c' \left(\frac{T_g}{T_w}\right)^{0.65} \tag{8-55}$$

$$\alpha_w = C_w \epsilon_w' \left(\frac{T_g}{T_w}\right)^{0.45} \tag{8-56}$$

$$\Delta\alpha = \Delta\epsilon \qquad \text{at } T_w \tag{8-57}$$

The values of ϵ_c' and ϵ_w' are evaluated from Figs. 8-34 and 8-35 with an abscissa of T_w but with pressure-beam-length parameters of $p_c L_e(T_w/T_g)$ and $p_w L_e(T_w/T_g)$, respectively.

Equation (8-53) was written for the heat exchange between a gas volume and a black enclosure at a uniform temperature. Now let us consider a more complicated case in which the gas volume is enclosed by two black parallel planes at different temperatures T_1 and T_2. In this case we must make an energy balance on each plate. For plate 1 the net energy *gain* is

$$q_1 = G_1 A_1 - E_{b_1} A_1 \tag{8-58}$$

Similarly, for plate 2

$$q_2 = G_2 A_2 - E_{b_2} A_2 \tag{8-59}$$

The irradiations G_1 and G_2 must be evaluated in terms of the total energy arriving at each surface. For surface 1

Irradiation on surface 1 = irradiation from gas
+ irradiation from surface 2 transmitted through gas

or

$$G_1 A_1 = A_g F_{g1} \epsilon_g(T_g) E_{b_g} + A_2 F_{21} \tau_g(T_2) E_{b_2} \tag{8-60}$$

The transmissivity for the radiation from T_2 is obtained from

$$\tau_g(T_2) = 1 - \alpha_g(T_2)$$

where the α_g is evaluated as before.

In a similar manner, the irradiation on surface 2 would be

$$G_2 A_2 = A_g F_{g2} \epsilon_g(T_g) E_{b_g} + A_1 F_{12} \tau_g(T_1) E_{b_1} \tag{8-61}$$

If we take the gas areas as $A_g = A_1 = A_2$ for the parallel-plate system, then $F_{g_1} = F_{g_2} = 1.0$. Also, $F_{12} = F_{21} = 1.0$. Once G_1 and G_2 are determined, the heat transfer to each surface can be calculated from Eqs. (8-58) and (8-59).

Heat Exchange between Gas Volume and Gray Enclosure

Up to this point, the gas-radiation calculation methods discussed were concerned only with black surfaces exchanging heat with the gas. In many engineering applications the enclosure walls are dirty and sooty, the wall emissivity is very high, and the heat-transfer calculation by Eq. (8-53) may be a reasonable approximation. An analysis of gray enclosures is very complex because the multiple reflection phenomena must take into account the band-absorption characteristics of the gas. For engineering calculations Hottel [22] has shown that the net heat transfer to a gray enclosure having an emissivity ϵ_w may be handled approximately by multiplying Eq. (8-53) by a factor such that

$$\frac{q_{\text{gray}}}{q_{\text{black}}} = \frac{\epsilon_w + 1}{2} \quad \text{for } \epsilon_w > 0.8 \tag{8-62}$$

For values of $\epsilon_w < 0.8$ much more elaborate procedures must be employed to calculate the heat transfer.

EXAMPLE 8-11 Gas radiation in a furnace

A cubical furnace 0.5 m on a side has interior walls that are essentially black. Inside the furnace the gas is 20 percent carbon dioxide by volume and 80 percent nitrogen at a total pressure of 1 atm and a temperature of 1500 K. The walls of the furnace are to be maintained at 300°C. Calculate the amount of cooling required to maintain the walls at the specified temperature.

Solution

For this calculation we make use of Eq. (8-53)

$$\frac{q}{A} = \epsilon_g(T_g)\sigma T_g^{\,4} - \alpha_g(T_w)\sigma T_w^{\,4} \tag{a}$$

where $\sigma T_g^{\,4} = (5.669 \times 10^{-8})(1500)^4 = 287 \text{ kW/m}^2 \quad [90{,}980 \text{ Btu/h} \cdot \text{ft}^2]$
$\sigma T_w^{\,4} = (5.669 \times 10^{-8})(573)^4 = 6.11 \text{ kW/m}^2 \quad [1937 \text{ Btu/h} \cdot \text{ft}^2]$

From Table 8-2 the equivalent beam length is

$$L_e = (0.6)(0.5) = 0.3 \text{ m} \quad [0.984 \text{ ft}]$$

and the partial pressure of CO_2 is 0.2 atm, or

$$p_c = (0.2)(1.0132 \times 10^5) = 20.26 \text{ kPa} \quad [2.94 \text{ lb/in}^2 \text{ abs}]$$

Then $p_cL_e = 6.08$ kN/m, and we enter Fig. 8-34 at 1500 K to obtain

$$\epsilon_c = 0.072 = \epsilon_g(T_g)$$

There is no correction factor because the total pressure is 1 atm. To evaluate $\alpha_g(T_w)$ we use a temperature of 573 K and a pressure-beam-length parameter of

$$p_cL_e\frac{T_w}{T_g} = 6.08\,\frac{573}{1500} = 2.32 \text{ kN/m}$$

From Fig. 8-34, $\epsilon_c' = 0.065$, $C_c = 1.0$, and we use Eq. (8-55) to obtain

$$\alpha_g(T_w) = \alpha_c = (0.065)\left(\frac{1500}{573}\right)^{0.65} = 0.121$$

Now, making use of Eq. (*a*), we have

$$\frac{q}{A} = (0.072)(287) - (0.121)(6.11) = 19.92 \text{ kW/m}^2$$

There are six interior sides, and so the total heat transfer is

$$q = (19.92)(1.5) = 29.88 \text{ kW} \quad [102{,}000 \text{ Btu/h}]$$

EXAMPLE 8-12

Two parallel black plates are separated by a distance of 0.7 m and maintained at temperatures of 200 and 500°C. Between the planes is a gas mixture of 20 percent CO_2, 15 percent water vapor, and 65 percent N_2 by volume at a total pressure of 3 atm. The gas temperature is 1000°C. Calculate the heat exchange with each plate.

Solution

For this problem we must make use of Eqs. (8-58) and (8-59) which require evaluation of a number of properties. Setting

$$T_g = 1000°\text{C} = 1273 \text{ K} \qquad T_1 = 200°\text{C} = 473 \text{ K} \qquad T_2 = 500°\text{C} = 773 \text{ K}$$

we have

$$\sigma T_g^4 = E_{b_g} = 148.9 \text{ kW/m}^2$$

$$\sigma T_1^4 = E_{b_1} = 2.84 \text{ kW/m}^2$$

$$\sigma T_2^4 = E_{g2} = 20.24 \text{ kW/m}^2$$

Using Table 8-2, we calculate the mean beam length as

$$L_e = (1.8)(0.7) = 1.26 \text{ m}$$

For the mixture at 3 atm the partial pressures of CO_2 and H_2O are

$$p_c = (0.20)(3)(1.0132 \times 10^5) = 60.8 \text{ kPa} \quad [8.82 \text{ lb/in}^2 \text{ abs}]$$

$$p_w = (0.15)(3)(1.0132 \times 10^5) = 45.6 \text{ kPa} \quad [6.61 \text{ lb/in}^2 \text{ abs}]$$

Then

$$p_cL_e = (60.8)(1.26) = 76.6 \text{ kN/m} = 2.48 \text{ atm} \cdot \text{ft}$$

$$p_wL_e = (45.6)(1.26) = 57.5 \text{ kN/m} = 1.87 \text{ atm} \cdot \text{ft}$$

$$p_cL_e + p_wL_e = 134.1 \text{ kN/m} = 4.35 \text{ atm} \cdot \text{ft}$$

Also, $(\frac{1}{2})(p + p_w) = (0.5)(3 + 0.45) = 1.725$ atm, and

$$\frac{p_w}{p_c + p_w} = \frac{0.45}{0.6 + 0.45} = 0.429$$

Consulting the various calculation charts, we obtain

$$\epsilon_w = 0.22 \qquad C_w \approx 1.4$$

$$\epsilon_c = 0.17 \qquad C_c = 1.1 \qquad \Delta\epsilon = 0.055$$

so that, from Eq. (8-52)

$$\epsilon_g(T_g) = (1.1)(0.17) + (1.4)(0.22) - 0.055 = 0.44$$

We must now determine the values of α_g at T_1 and T_2. At $T = T_1 = 473$ K

$$p_cL_e\left(\frac{T_1}{T_g}\right) = 76.6\,\frac{473}{1273} = 28.5 \text{ kN/m}$$

$$p_wL_e\left(\frac{T_1}{T_g}\right) = 57.5\,\frac{473}{1273} = 21.4 \text{ kN/m} = 0.69 \text{ atm} \cdot \text{ft}$$

From the calculation charts

$$\epsilon'_w = 0.26 \qquad C_w \approx 1.5$$

$$\epsilon'_c = 0.13 \qquad C_c = 1.15 \qquad \Delta\epsilon = \Delta\alpha = 0.02$$

Now, from Eqs. (8-55) and (8-56)

$$\alpha_c = (1.15)(0.13)\left(\frac{1273}{473}\right)^{0.65} = 0.285$$

$$\alpha_w = (1.5)(0.26)\left(\frac{1273}{473}\right)^{0.45} = 0.608$$

and $$\alpha_g(T_1) = \alpha_c + \alpha_w - \Delta\alpha = 0.285 + 0.608 - 0.02 = 0.874$$

$$\tau_g(T_1) = 1 - \alpha_g(T_1) = 0.126$$

At $T = T_2 = 773$ K,

$$p_cL_e\frac{T_2}{T_g} = 76.6\frac{773}{1273} = 46.5 \text{ kN/m} = 1.51 \text{ atm} \cdot \text{ft}$$

$$p_wL_e\frac{T_2}{T_g} = 57.5\frac{773}{1273} = 34.9 \text{ kN/m} = 1.13 \text{ atm} \cdot \text{ft}$$

From the calculation charts

$$\epsilon'_w = 0.24 \qquad C_w \approx 1.45$$

$$\epsilon'_c = 0.17 \qquad C_c = 1.13 \qquad \Delta\epsilon = \Delta\alpha = 0.028$$

$$\alpha_c = (1.13)(0.17)\left(\frac{1273}{473}\right)^{0.65} = 0.266$$

$$\alpha_w = (1.45)(0.24)\left(\frac{1273}{473}\right)^{0.45} = 0.436$$

Then $$\alpha_g(T_2) = \alpha_c + \alpha_w - \Delta\alpha = 0.266 + 0.436 - 0.028 = 0.674$$

and $$\tau_g(T_2) = 1 - \alpha_g(T_2) = 0.326$$

For the parallel-plate system all the areas are equal and all the shape factors are unity, so Eq. (8-60) becomes

$$\begin{aligned} G_1 &= \epsilon_g(T_g)E_{b_g} + \tau_g(T_2)E_{b_2} \\ &= (0.44)(148.9) + (0.326)(20.24) = 72.1 \text{ kW/m}^2 \quad [22{,}860 \text{ Btu/h} \cdot \text{ft}^2] \end{aligned}$$

Similarly,

$$\begin{aligned} G_2 &= \epsilon_g E_{b_g} + \tau_g(T_1)E_{b_1} \\ &= (0.44)(148.9) + (0.126)(2.84) = 65.9 \text{ kW/m}^2 \end{aligned}$$

Both surfaces are black so the heat *gain* by each surface is

$$\frac{q_1}{A} = G_1 - E_{b_1} = 72.1 - 2.84 = 69.3 \text{ kW/m}^2$$

$$\frac{q_2}{A} = G_2 - E_{b_2} = 65.9 - 20.24 = 45.7 \text{ kW/m}^2$$

The net energy lost by the gas is the sum of these two figures, or 115 kW/m^2 of plate area.

8-10 RADIATION NETWORK FOR AN ABSORBING AND TRANSMITTING MEDIUM

The foregoing discussions have shown the methods that may be used to calculate radiation heat transfer between surfaces separated by a completely transparent medium. The radiation-network method is used to great advantage in these types of problems.

Many practical problems involve radiation heat transfer through a medium which is both absorbing and transmitting. The various glass substances are one example of this type of medium; gases are another. We have already seen some of the complications which arise with gas radiation. We shall now examine a radiation-network method for analyzing absorbing-transmitting systems, keeping in mind the many problems which may be involved with gases.

To begin, let us consider a simple case, that of two nontransmitting surfaces which see each other and nothing else. In addition, we let the space between these surfaces be occupied by a transmitting and absorbing medium. The practical problem might be that of two large planes separated by either an absorbing gas or a transparent sheet of glass or plastic. The situation is shown schematically in Fig. 8-39. The transparent medium is designated by the subscript m. We make the assumption that the medium is nonreflecting and that Kirchoff's identity applies, so that

$$\alpha_m + \tau_m = 1 = \epsilon_m + \tau_m \tag{8-63}$$

The assumption that the medium is nonreflecting is a valid one when gases are considered. For glass or plastic plates this is not necessarily true, and reflectivities of the order of 0.1 are common for many glass substances. In addition, the transmissive properties of glasses are usually limited to a narrow wavelength band between about 0.2 and 4 μm. Thus the analysis which follows is highly idealized and serves mainly to furnish a starting point for the solution of problems in which transmission of radiation must be considered. Other complications with gases are mentioned later in the discussion. When both

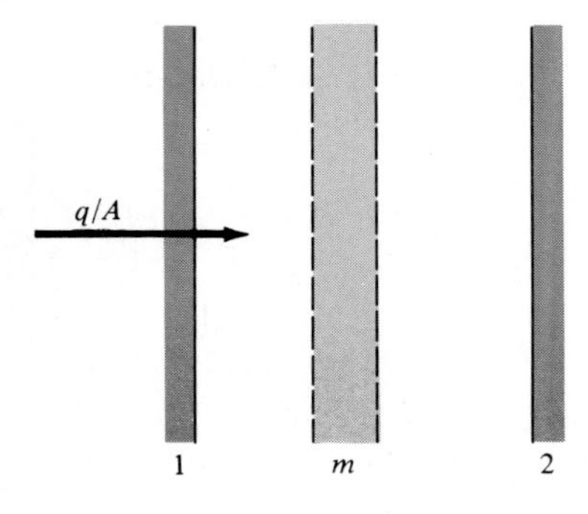

Fig. 8-39 Radiation system consisting of a transmitting medium between two planes.

reflection and transmission must be taken into account, the analysis techniques discussed in Sec. 8-12 must be employed.

Returning to the analysis, we note that the medium can emit and transmit radiation from one surface to the other. Our task is to determine the network elements to use in describing these two types of exchange processes. The transmitted energy may be analyzed as follows. The energy leaving surface 1 which is transmitted through the medium and arrives at surface 2 is

$$J_1A_1F_{12}\tau_m$$

and that which leaves surface 2 and arrives at surface 1 is

$$J_2A_2F_{21}\tau_m$$

The net exchange in the transmission process is therefore

$$q_{1-2\text{transmitted}} = A_1F_{12}\tau_m(J_1 - J_2) = A_1F_{12}(J_1 - J_2)$$

$$q_{1-2\text{transmitted}} = \frac{J_1 - J_2}{1/A_1F_{12}(1 - \epsilon_m)} \tag{8-64}$$

and the network element which may be used to describe this process is shown in Fig. 8-40.

Now consider the exchange process between surface 1 and the transmitting medium. Since we have assumed that this medium is nonreflecting, the energy leaving the medium (other than the transmitted energy, which we have already considered) is precisely that energy which is emitted by the medium

$$J_m = \epsilon_m E_{bm}$$

And of the energy leaving the medium, the amount which reaches surface 1 is

$$A_mF_{m1}J_m = A_mF_{m1}\epsilon_m E_{bm}$$

Of that energy which leaves surface 1, the quantity which reaches the transparent medium is

$$J_1A_1F_{1m}\alpha_m = J_1A_1F_{1m}\epsilon_m$$

At this point we note that absorption in the medium means that the incident radiation has "reached" the medium. Consistent with the above relations, the net energy exchange between the medium and surface 1 is the difference between the amount emitted by the medium toward surface 1 and that absorbed which emanated from surface 1. Thus

$$q_{m-1\text{net}} = A_mF_{m1}\epsilon_m E_{bm} - J_1A_1F_{1m}\epsilon_m$$

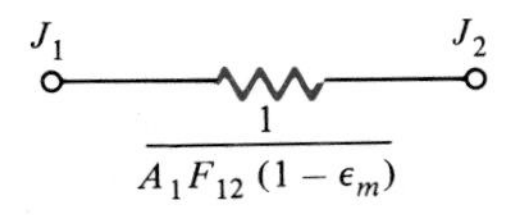

Fig. 8-40 Network element for transmitted radiation through medium.

Using the reciprocity relation

$$A_1F_{1m} = A_mF_{m1}$$

we have

$$q_{m-1_{\text{net}}} = \frac{E_{bm} - J_1}{1/A_1F_{1m}\epsilon_m} \tag{8-65}$$

This heat-exchange process is represented by the network element shown in Fig. 8-41. The total network for the physical situation of Fig. 8-39 is shown in Fig. 8-42.

If the transport medium is maintained at some fixed temperature, then the potential E_{bm} is fixed according to

$$E_{bm} = \sigma T_m^{\ 4}$$

On the other hand, if no net energy is delivered to the medium, then E_{bm} becomes a floating node, and its potential is determined by the other network elements.

In reality, the radiation shape factors F_{1-2}, F_{1-m}, and F_{2-m} are unity for this example, so that the expression for the heat flow could be simplified to some extent; however, these shape factors are included in the network resistances for the sake of generality in the analysis.

When the practical problem of heat exchange between gray surfaces through an absorbing gas is encountered, the major difficulty is that of determining the transmissivity and emissivity of the gas. These properties are functions not only of the temperature of the gas, but also of the thickness of the gas layer; i.e., thin gas layers transmit more radiation than thick layers. The usual practical problem almost always involves more than two heat-transfer surfaces, as in the simple example given above. As a result, the transmissivities between the various heat-transfer surfaces can be quite different, depending on their geometric orientation. Since the temperature of the gas will vary, the transmissive and emissive properties will vary with their location in the gas. One way of handling this situation is to divide the gas body into layers and set up a radiation network accordingly, letting the potentials of the various nodes "float," and thus arriving at the gas-temperature distribution. Even with this procedure, an iterative method must eventually be employed because the radiation properties of the gas are functions of the unknown "floating potentials." Naturally, if the temperature of the gas is uniform, the solution is much easier.

We shall not present the solution of a complex-gas-radiation problem since the tedious effort required for such a solution is beyond the scope of our present discussion; however, it is worthwhile to analyze a two-layer transmitting system in order to indicate the general scheme of reasoning which might be applied to more complex problems.

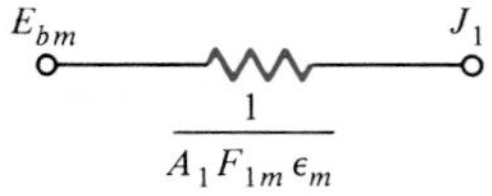

E_{bm} J_1 $\frac{1}{A_1F_{1m}\epsilon_m}$

Fig. 8-41 Network element for radiation exchange between medium and surface.

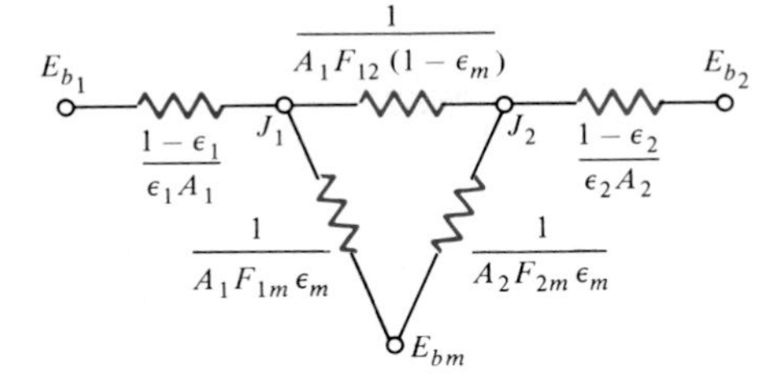

Fig. 8-42 Total radiation network for system of Fig. 8-39.

Consider the physical situation shown in Fig. 8-43. Two radiating and absorbing surfaces are separated by two layers of transmitting and absorbing media. These two layers might represent two sheets of transparent media, such as glass, or they might represent the division of a separating gas into two parts for purposes of analysis. We designate the two transmitting and absorbing layers with the subscripts m and n. The energy exchange between surface 1 and m is given by

$$q_{1-m} = A_1F_{1m}\epsilon_m J_1 - A_mF_{m1}\epsilon_m E_{bm} = \frac{J_1 - E_{bm}}{1/A_1F_{1m}\epsilon_m} \tag{8-66}$$

and that between surface 2 and n is

$$q_{2-n} = A_1F_{2n}\epsilon_n J_2 - A_nF_{n2}\epsilon_n E_{bn} = \frac{J_2 - E_{bn}}{1/A_2F_{2n}\epsilon_n} \tag{8-67}$$

Of that energy leaving surface 1, the amount arriving at surface 2 is

$$q_{1-2} = A_1F_{12}J_1\tau_m\tau_n = A_1F_{12}J_1(1 - \epsilon_m)(1 - \epsilon_n)$$

and of that energy leaving surface 2, the amount arriving at surface 1 is

$$q_{2-1} = A_2F_{21}J_2\tau_n\tau_m = A_2F_{12}J_2(1 - \epsilon_n)(1 - \epsilon_m)$$

so that the net energy exchange by transmission between surfaces 1 and 2 is

$$q_{1-2\text{transmitted}} = A_1F_{12}(1 - \epsilon_m)(1 - \epsilon_n)(J_1 - J_2) = \frac{J_1 - J_2}{1/A_1F_{12}(1 - \epsilon_m)(1 - \epsilon_n)} \tag{8-68}$$

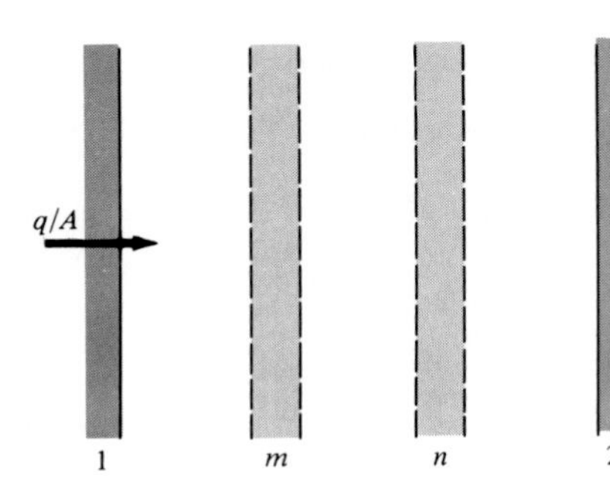

Fig. 8-43 Radiation system consisting of two transmitting layers between two planes.

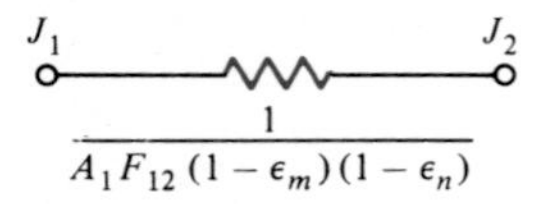

Fig. 8-44 Network element for transmitted radiation between planes.

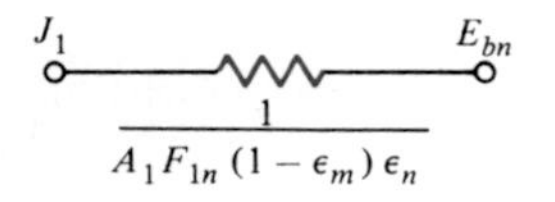

Fig. 8-45 Network element for transmitted radiation for medium n to plane 1.

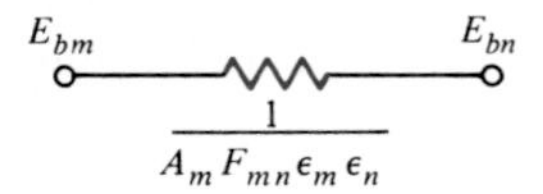

Fig. 8-46 Network element for radiation exchange between two transparent layers.

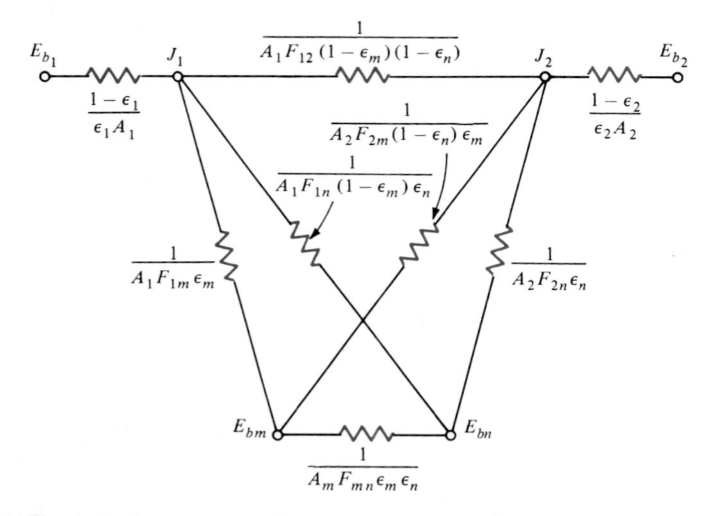

Fig. 8-47 Total radiation network for system of Fig. 8-43.

and the network element representing this transmission is shown in Fig. 8-44. Of that energy leaving surface 1, the amount which is absorbed in n is

$$q_{1-n} = A_1F_{1n}J_1\tau_m\epsilon_n = A_1F_{1n}J_1(1 - \epsilon_m)\epsilon_n$$

Also,

$$q_{n-1} = A_nF_{n1}J_n\tau_m = A_nF_{n1}\epsilon_nE_{bn}(1 - \epsilon_m)$$

since

$$J_n = \epsilon_nE_{bn}$$

The net exchange between surface 1 and n is therefore

$$q_{1-n_{\text{net}}} = A_1F_{1n}(1 - \epsilon_m)\epsilon_n(J_1 - E_{bn}) = \frac{J_1 - E_{bn}}{1/A_1F_{1n}(1 - \epsilon_m)\epsilon_n} \tag{8-69}$$

and the network element representing this situation is shown in Fig. 8-45. In like manner, the net exchange between surface 2 and m is

$$q_{2-m_{\text{net}}} = \frac{J_2 - E_{bm}}{1/A_2F_{2m}(1 - \epsilon_n)\epsilon_m} \tag{8-70}$$

Of that radiation leaving m, the amount absorbed in n is

$$q_{m-n} = J_m A_m F_{mn} \alpha_n = A_m F_{mn} \epsilon_m \epsilon_n E_{bm}$$

and

$$q_{n-m} = A_n F_{nm} \epsilon_n \epsilon_m E_{bn}$$

so that the net energy exchange between m and n is

$$q_{m-n_{\text{net}}} = A_m F_{mn} \epsilon_m \epsilon_n (E_{bm} - E_{bn}) = \frac{E_{bm} - E_{bn}}{1/A_m F_{mn} \epsilon_m \epsilon_n} \tag{8-71}$$

and the network element representing this energy transfer is given in Fig. 8-46.

The final network for the entire heat-transfer process is shown in Fig. 8-47, with the surface resistances added. If the two transmitting layers m and n are maintained at given temperatures, the solution to the network is relatively easy to obtain because only two unknown potentials J_1 and J_2 need be determined to establish the various heat-flow quantities. In this case the two transmitting layers will either absorb or lose a certain quantity of energy, depending on the temperature at which they are maintained.

When no net energy is delivered to the transmitting layers, nodes E_{bm} and E_{bn} must be left "floating" in the analysis; and for this particular system four nodal equations would be required for a solution of the problem.

■ EXAMPLE 8-13 Network for gas radiation

Two large parallel planes are at $T_1 = 800$ K, $\epsilon_1 = 0.3$, $T_2 = 400$ K, $\epsilon_2 = 0.7$ and are separated by a gray gas having $\epsilon_g = 0.2$, $\tau_g = 0.8$. Calculate the heat-transfer rate between the two planes and the temperature of the gas using a radiation network. Compare with the heat transfer without presence of the gas.

Solution

The network shown in Fig. 8-42 applies to this problem. All the shape factors are unity for large planes and the various resistors can be computed on a unit-area basis as

$$\frac{1 - \epsilon_1}{\epsilon_1} = \frac{0.7}{0.3} = 2.333 \qquad \frac{1}{F_{12}(1 - \epsilon_g)} = \frac{1}{1 - 0.2} = 1.25$$

$$\frac{1 - \epsilon_2}{\epsilon_2} = \frac{0.3}{0.7} = 0.4286 \qquad \frac{1}{F_{1g}\epsilon_g} = \frac{1}{F_{2g}\epsilon_g} = \frac{1}{0.2} = 5.0$$

$$E_{b_1} = \sigma T_1^4 = 23{,}220 \text{ W/m}^2 \qquad E_{b_2} = \sigma T_2^4 = 1451 \text{ W/m}^2$$

The equivalent resistance of the center "triangle" is

$$R = \frac{1}{1/1.25 + 1/(5.0 + 5.0)} = 1.1111$$

The total heat transfer is then

$$\frac{q}{A} = \frac{E_{b_1} - E_{b_2}}{\Sigma R} = \frac{23{,}200 - 1451}{2.333 + 1.111 + 0.4286} = 5616 \text{ W/m}^2$$

If there were no gas present the heat transfer would be given by Eq. (8-42):

$$\frac{q}{A} = \frac{23{,}200 - 1451}{1/0.3 + 1/0.7 - 1} = 5781 \text{ W/m}^2$$

The radiosities may be computed from

$$\frac{q}{A} = (E_{b_1} - J_1)\left(\frac{\epsilon_1}{1 - \epsilon_1}\right) = (J_2 - E_{b_2})\left(\frac{\epsilon_2}{1 - \epsilon_2}\right) = 5616 \text{ W/m}^2$$

which gives $J_1 = 10.096$ W/m² and $J_2 = 3858$ W/m². For the network E_{bg} is just the mean of these values

$$E_{bg} = \tfrac{1}{2}(10.096 + 3858) = 6977 = \sigma T_g^4$$

so that the temperature of the gas is

$$T_g = 592.3 \text{ K}$$

8-11 RADIATION EXCHANGE WITH SPECULAR SURFACES

All the preceding discussions have considered radiation exchange between diffuse surfaces. In fact, the radiation shape factors defined by Eq. (8-21) hold only for diffuse radiation because the radiation was assumed to have no preferred direction in the derivation of this relation. In this section we extend the analysis to take into account some simple geometries containing surfaces that may have a specular type of reflection. No real surface is completely diffuse or completely specular. We shall assume, however, that all the surfaces to be considered *emit* radiation diffusely but that they may *reflect* radiation partly in a specular manner and partly in a diffuse manner. We therefore take the reflectivity to be the sum of a specular component and a diffuse component:

$$\rho = \rho_s + \rho_D \tag{8-72}$$

It is still assumed that Kirchhoff's identity applies so that

$$\epsilon = \alpha = 1 - \rho \tag{8-73}$$

The net heat lost by a surface is the difference between the energy emitted and absorbed:

$$q = A\,(\epsilon E_b - \alpha G) \tag{8-74}$$

We define the *diffuse radiosity* J_D as the total *diffuse* energy leaving the surface per unit area and per unit time, or

$$J_D = \epsilon E_b + \rho_D G \tag{8-75}$$

Solving for the irradiation G from Eq. (8-75) and inserting in Eq. (8-74) gives

$$q = \frac{\epsilon A}{\rho_D}[E_b(\epsilon + \rho_D) - J_D]$$

or, written in a different form,

$$q = \frac{E_b - J_D/(1 - \rho_s)}{\rho_D/[\epsilon A(1 - \rho_s)]} \tag{8-76}$$

where $1 - \rho_s$ has been substituted for $\epsilon + \rho_D$. It is easy to see that Eq. (8-76) may be represented with the network element shown in Fig. 8-48. A quick inspection will show that this network element reduces to that in Fig. 8-24 for the case of a surface which reflects in only a diffuse manner, i.e., for $\rho_s = 0$.

Now let us compute the radiation exchange between two specular-diffuse surfaces. For the moment, we assume that the surfaces are oriented as shown in Fig. 8-49. In this arrangement any diffuse radiation leaving surface 1 which is specularly reflected by 2 will not be reflected directly back to 1. This is an important point, for in eliminating such reflections we are considering only the *direct* diffuse exchange between the two surfaces. In subsequent paragraphs we shall show how the specular reflections must be analyzed. For the surfaces in Fig. 8-49 the *diffuse* exchanges are given by

$$q_{1\to2} = J_{1D}A_1F_{12}(1 - \rho_{2s}) \tag{8-77}$$

$$q_{2\to1} = J_{2D}A_2F_{21}(1 - \rho_{1s}) \tag{8-78}$$

Equation (8-77) expresses the diffuse radiation leaving 1 which arrives at 2 *and* which may contribute to a diffuse radiosity of surface 2. The factor $1 - \rho_s$ represents the fraction absorbed plus the fraction reflected diffusely. The inclusion of this factor is most important because we are considering only diffuse direct exchange, and thus must leave out the specular-reflection contribution

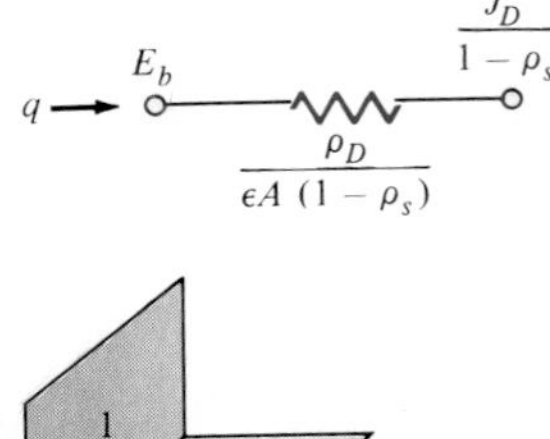

Fig. 8-48 Network element representing Eq. (8-76).

Fig. 8-49

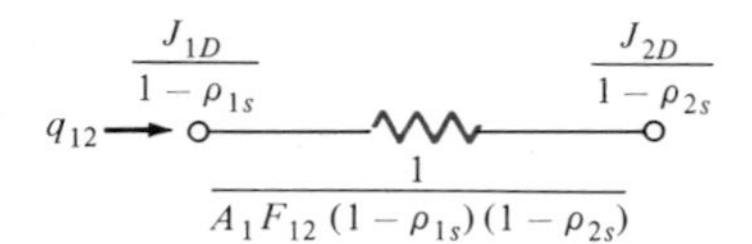

Fig. 8-50 Network element representing Eq. (8-79).

for now. The net exchange is given by the difference between Eqs. (8-77) and (8-78), according to Ref. 21.

$$q_{12} = \frac{J_{1D}/(1 - \rho_{1s}) - J_{2D}/(1 - \rho_{2s})}{1/[A_1F_{12}(1 - \rho_{1s})(1 - \rho_{2s})]} \tag{8-79}$$

The network element representing Eq. (8-79) is shown in Fig. 8-50.

To analyze specular reflections we utilize a technique presented in Refs. 12 and 13. Consider the enclosure with four long surfaces shown in Fig. 8-51. Surfaces 1, 2, and 4 reflect diffusely, while surface 3 has both a specular and a diffuse component of reflection. The dashed lines represent mirror images of the surfaces 1, 2, and 4 in surface 3. (A specular reflection produces a mirror image.) The nomenclature 2(3) designates the mirror image of surface 2 in mirror 3.

Now consider the radiation leaving 2 which arrives at 1. There is a direct diffuse radiation of

$$(q_{2\rightarrow 1})^{\text{direct}}_{\text{diffuse}} = J_2A_2F_{21} \tag{8-80}$$

Part of the diffuse radiation from 2 is specularly reflected in 3 and strikes 1. This specularly reflected radiation acts like *diffuse* energy coming from the image surface 2(3). Thus we may write

$$(q_{2\rightarrow 1})^{\text{specular}}_{\text{reflected}} = J_2A_{2(3)}F_{2(3)1}\rho_{3s} \tag{8-81}$$

The radiation shape factor $F_{2(3)1}$ is the one between surface 2(3) and surface 1. The reflectivity ρ_{3s} is inserted because only this fraction of the radiation gets to 1. Of course, $A_2 = A_{2(3)}$. We now have

$$q_{2\rightarrow 1} = J_2A_2(F_{21} + \rho_{3s}F_{2(3)1}) \tag{8-82}$$

Similar reasoning leads to

$$q_{1\rightarrow 2} = J_1A_1(F_{12} + \rho_{3s}F_{1(3)2}) \tag{8-83}$$

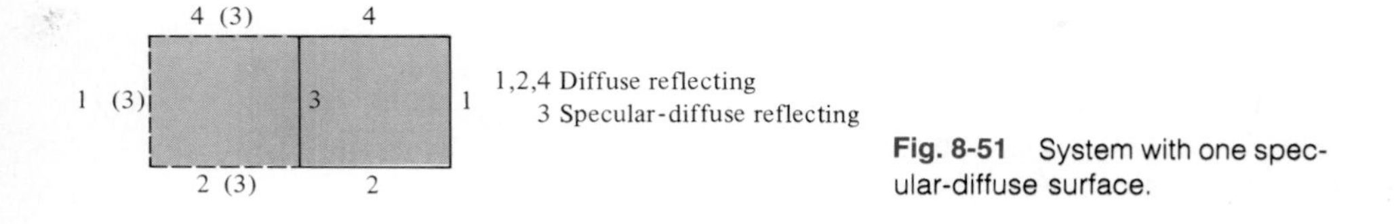

Fig. 8-51 System with one specular-diffuse surface.

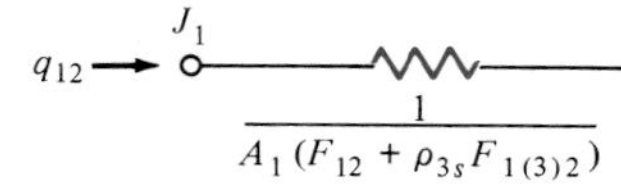

Fig. 8-52 Network element for Eq. (8-84).

Combining Eqs. (8-82) and (8-83) and making use of the reciprocity relation $A_1F_{12} = A_2F_{21}$ gives

$$q_{12} = \frac{J_1 - J_2}{1/[A_1(F_{12} + \rho_{3s}F_{1(3)2})]} \tag{8-84}$$

The network element represented by Eq. (8-84) is shown in Fig. 8-52.

Analogous network elements may be developed for radiation between the other surfaces in Fig. 8-51, so that the final complete network becomes as shown in Fig. 8-53. It is to be noted that the elements connecting to J_{3D} are simple modifications of the one shown in Fig. 8-50 since $\rho_{1s} = \rho_{2s} = \rho_{4s} = 0$. An interesting observation can be made about this network for the case where $\rho_{3D} = 0$. In this instance surface 3 is completely specular and

$$J_{3D} = \epsilon_3 E_{b3}$$

so that we are left with only three unknowns, J_1, J_2, and J_4, when surface 3 is completely specular-reflecting.

Now let us complicate the problem a step further by letting the enclosure have two specular-diffuse surfaces, as shown in Fig. 8-54. In this case multiple images may be formed as shown. Surface 1(3, 2) represents the image of 1 after it is viewed first through 3 and then through 2. In other words, it is the image of surface 1(3) in mirror 2. At the same location is surface 1(2, 3), which is the image of surface 1(2) in mirror 3.

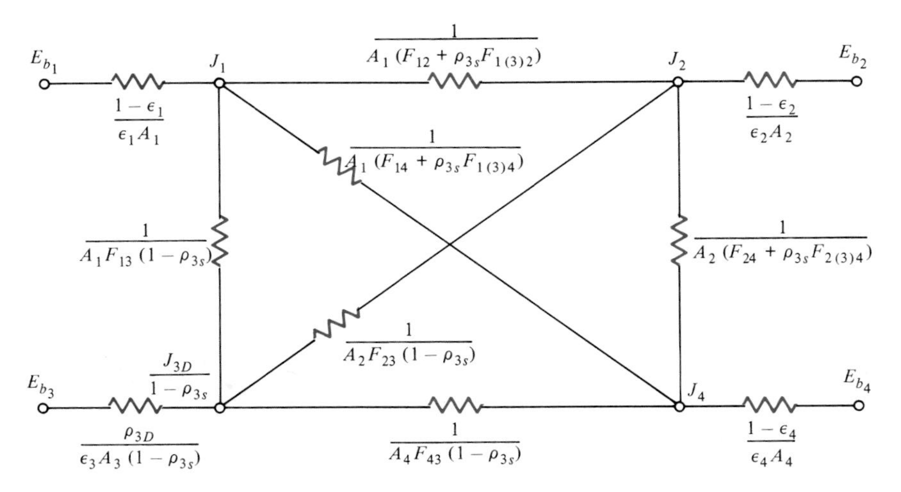

Fig. 8-53 Complete radiation network for system in Fig. 8-51.

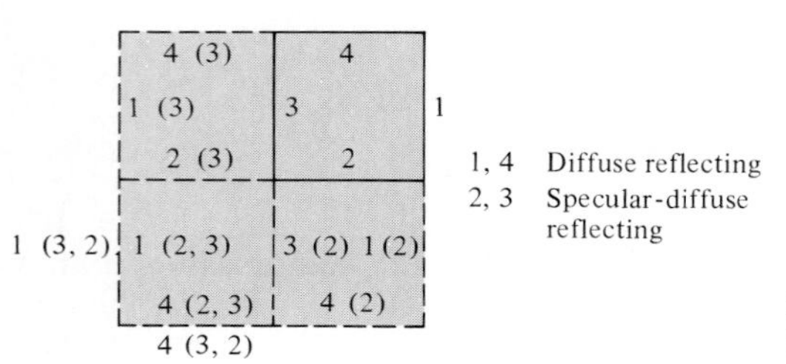

Fig. 8-54 System with two specular-diffuse surfaces.

This problem is complicated because multiple specular reflections must be considered. Consider the exchange between surfaces 1 and 4. Diffuse energy leaving 1 can arrive at 4 in five possible ways:

direct: $J_1A_1F_{14}$

reflection in 2 only: $J_1A_1F_{1(2)4}\rho_{2s}$

reflection in 3 only: $J_1A_1F_{1(3)4}\rho_{3s}$

reflection first in 2 and then in 3: (8-85)

$$J_1A_1\rho_{3s}\rho_{2s}F_{1(2,3)4}$$

reflection first in 3 and then in 2:

$$J_1A_1\rho_{2s}\rho_{3s}F_{1(3,2)4}$$

The last shape factor $F_{1(3,2)4}$, is zero because surface 1(3, 2) cannot see surface 4 when looking *through* mirror 2. On the other hand, $F_{1(2,3)4}$ is not zero because surface 1(2, 3) can see surface 4 when looking through mirror 3. The sum of the above terms is given as

$$q_{1\to4} = J_1A_1(F_{14} + \rho_{2s}F_{1(2)4} + \rho_{3s}F_{1(3)4} + \rho_{3s}\rho_{2s}F_{1(2,3)4}) \tag{8-86}$$

In a similar manner,

$$q_{4\to1} = J_4A_4(F_{41} + \rho_{2s}F_{4(2)1} + \rho_{3s}F_{4(3)1} + \rho_{3s}\rho_{2s}F_{4(3,2)1}) \tag{8-87}$$

Subtracting these two equations and applying the usual reciprocity relations gives the network element shown in Fig. 8-55.

Now consider the diffuse exchange between surfaces 1 and 3. Of the energy leaving 1, the amount which contributes to the diffuse radiosity of surface 3 is

$$q_{1\to3} = J_1A_1F_{13}(1 - \rho_{3s}) + J_1A_1\rho_{2s}F_{1(2)3}(1 - \rho_s) \tag{8-88}$$

The first term represents the direct exchange, and the second term represents the exchange after one specular reflection in mirror 2. As before, the factor

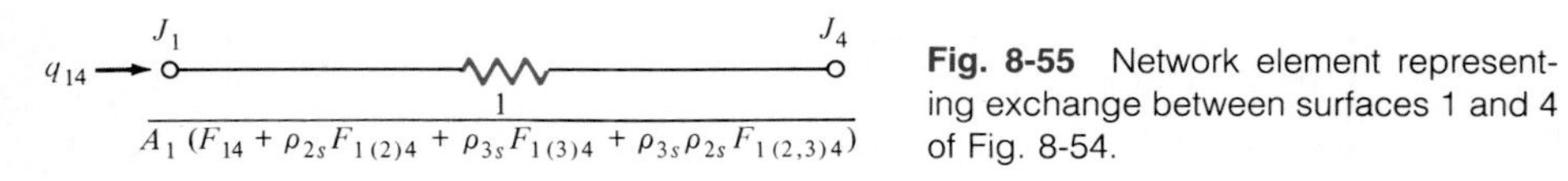

Fig. 8-55 Network element representing exchange between surfaces 1 and 4 of Fig. 8-54.

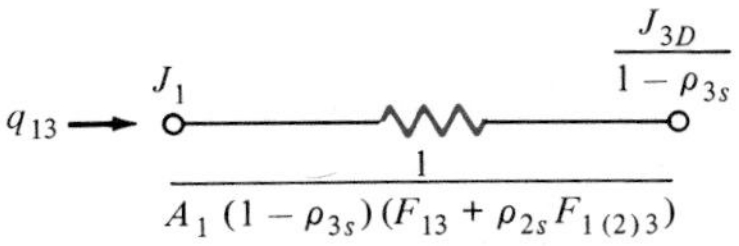

q_{13}
J_1
$\frac{J_{3D}}{1-\rho_{3s}}$
$\frac{1}{A_1(1-\rho_{3s})(F_{13}+\rho_{2s}F_{1(2)3})}$

Fig. 8-56 Network element representing exchange between surfaces 1 and 3 of Fig. 8-54.

$1 - \rho_{3s}$ is included to leave out of consideration the specular reflection from 3. This reflection, of course, is taken into account in other terms. The *diffuse* energy going from 3 to 1 is

$$q_{3\to1} = J_{3D}A_3F_{31} + J_{3D}A_3\rho_{2s}F_{3(2)1} \tag{8-89}$$

The first term is the direct radiation, and the second term is that which is specularly reflected in mirror 2. Combining Eqs. (8-88) and (8-89) gives the network element shown in Fig. 8-56.

The above two elements are typical for the enclosure of Fig. 8-54 and the other elements may be drawn by analogy. Thus the final complete network is given in Fig. 8-57.

If both surfaces 2 and 3 are pure specular reflectors, that is,

$$\rho_{2D} = \rho_{3D} = 0$$

we have $$J_{2D} = \epsilon_2 E_{b2} \qquad J_{3D} = \epsilon_3 E_{b3}$$

and the network involves only two unknowns, J_1 and J_4, under these circumstances.

We could complicate the calculation further by installing the specular surfaces opposite each other. In this case there would be an infinite number of images, and a series solution would have to be obtained; however, the series

E_{b_1}
J_1
$\frac{1}{A_1(F_{14}+\rho_{2s}F_{1(2)4}+\rho_{3s}F_{1(3)4}+\rho_{3s}\rho_{2s}F_{1(2,3)4})}$
J_4
E_{b_4}
$\frac{1-\epsilon_1}{\epsilon_1A_1}$
$\frac{1}{A_4(1-\rho_{2s})(F_{42}+\rho_{3s}F_{4(3)2})}$
$\frac{1-\epsilon_4}{\epsilon_4A_4}$
$\frac{1}{A_1(1-\rho_{3s})(F_{13}+\rho_{2s}F_{1(2)3})}$
$\frac{1}{A_4(1-\rho_{2s})(F_{43}+\rho_{3s}F_{4(2)3})}$
$\frac{1}{A_1(1-\rho_{2s})(F_{12}+\rho_{3s}F_{1(3)2})}$
E_{b_3}
E_{b_2}
$\frac{\rho_{3D}}{\epsilon_3A_3(1-\rho_{3s})}$
$\frac{J_{3D}}{1-\rho_{3s}}$
$\frac{1}{A_3F_{32}(1-\rho_{2s})(1-\rho_{3s})}$
$\frac{J_{2D}}{1-\rho_{2s}}$
$\frac{\rho_{2D}}{\epsilon_2A_2(1-\rho_{2s})}$

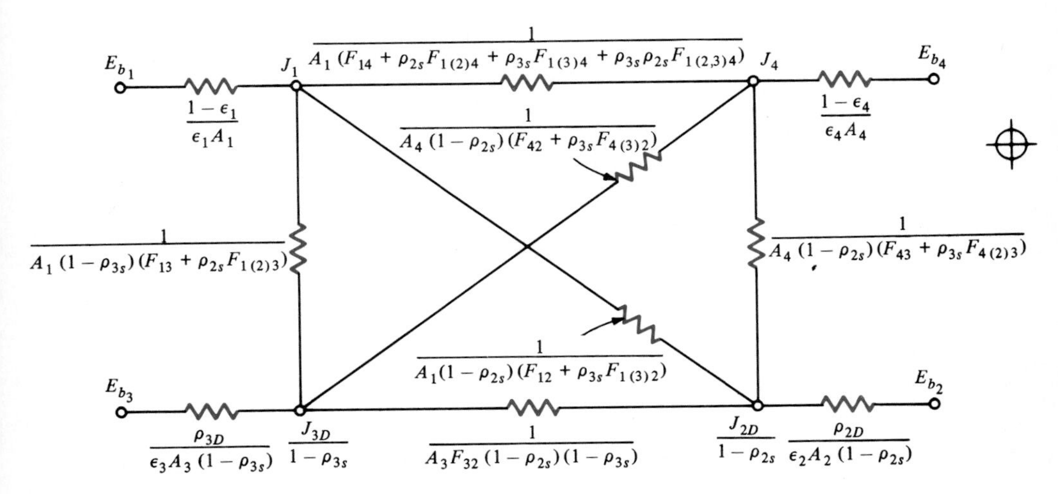

Fig. 8-57 Complete radiation network for system in Fig. 8-54.

for such problems usually converge rather rapidly. The reader should consult Ref. 13 for further information on this aspect of radiation exchange between specular surfaces.

8-12 RADIATION EXCHANGE WITH TRANSMITTING, REFLECTING, AND ABSORBING MEDIA

We now consider a simple extension of the presentations in Secs. 8-10 and 8-11 to analyze a medium where reflection, transmission, and absorption modes are all important. As in Sec. 8-10, we shall analyze a system consisting of two parallel diffuse planes with a medium in between which may absorb, transmit, and reflect radiation. For generality we assume that the surface of the transmitting medium may have both a specular and a diffuse component of reflection. The system is shown in Fig. 8-58.

For the transmitting medium m we have

$$\alpha_m + \rho_{mD} + \rho_{ms} + \tau_m = 1 \tag{8-90}$$

Also,

$$\epsilon_m = \alpha_m$$

The diffuse radiosity of a particular surface of the medium is defined by

$$J_{mD} = \epsilon_m E_{bm} + \rho_{mD} G \tag{8-91}$$

where G is the irradiation on the particular surface. Note that J_{mD} no longer represents the total diffuse energy leaving a surface. Now it represents only emission and diffuse reflection. The transmitted energy will be analyzed with additional terms. As before, the heat exchange is written

$$q = A(\epsilon E_b - \alpha G) \tag{8-92}$$

Solving for G from Eq. (8-91) and making use of Eq. (8-90) gives

$$q = \frac{E_{bm} - J_{mD}/(1 - \tau_m - \rho_{ms})}{\rho_{mD}/[\epsilon_m A_m(1 - \tau_m - \rho_{ms})]} \tag{8-93}$$

The network element representing Eq. (8-93) is shown in Fig. 8-59. This element is quite similar to the one shown in Fig. 8-48, except that here we must take the transmissivity into account.

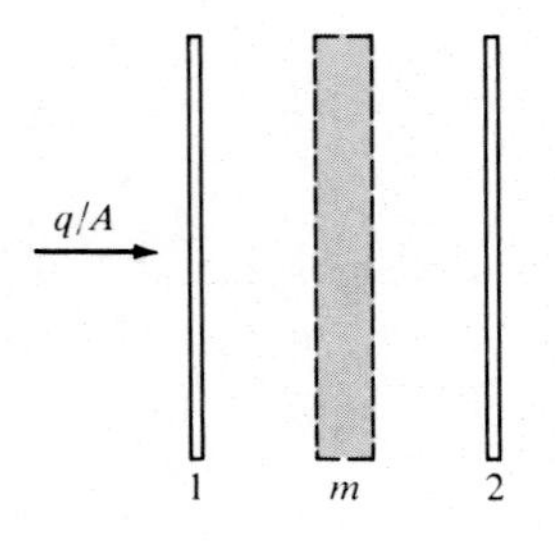

Fig. 8-58 Physical system for analysis of transmitting and reflecting layers.

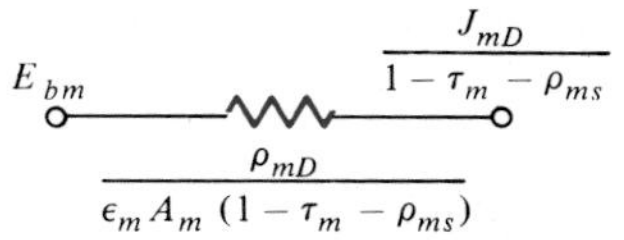

E_{bm} $\frac{J_{mD}}{1-\tau_m-\rho_{ms}}$

$\frac{\rho_{mD}}{\epsilon_m A_m(1-\tau_m-\rho_{ms})}$

Fig. 8-59 Network element representing Eq. (8-93).

The transmitted heat exchange between surfaces 1 and 2 is the same as in Sec. 8-10; that is,

$$q = \frac{J_1 - J_2}{1/A_1F_{12}\tau_m} \tag{8-94}$$

The heat exchange between surface 1 and m is computed in the following way. Of that energy leaving surface 1, the amount which arrives at m and contributes to the diffuse radiosity of m is

$$q_{1\to m} = J_1A_1F_{1m}(1 - \tau_m - \rho_{ms}) \tag{8-95}$$

The diffuse energy leaving m which arrives at 1 is

$$q_{m\to 1} = J_{mD}A_mF_{m1} \tag{8-96}$$

Subtracting (8-96) from (8-95) and using the reciprocity relation

$$A_1F_{1m} = A_mF_{m1}$$

gives

$$q_{1m} = \frac{J_1 - J_{mD}/(1 - \tau_m - \rho_{ms})}{1/[A_1F_{1m}(1 - \tau_m - \rho_{ms})]} \tag{8-97}$$

The network element corresponding to Eq. (8-97) is quite similar to the one shown in Fig. 8-50. An equation similar to Eq. (8-97) can be written for the radiation exchange between surface 2 and m. Finally, the complete network may be drawn as in Fig. 8-60. It is to be noted that J_{mD} represents the diffuse

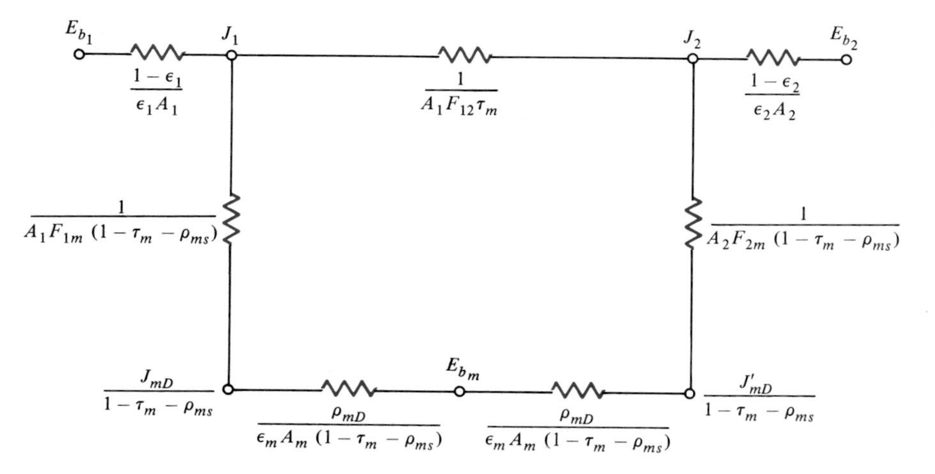

Fig. 8-60 Complete radiation network for system in Fig. 8-58.

radiosity of the left side of m, while J'_{mD} represents the diffuse radiosity of the right side of m.

If m is maintained at a fixed temperature, then J_1 and J_2 must be obtained as a solution to nodal equations for the network. On the other hand, if no net energy is delivered to m, then E_{bm} is a floating node, and the network reduces to a simple series-parallel arrangement. In this latter case the temperature of m must be obtained by solving the network for E_{bm}.

We may extend the analysis a few steps further by distinguishing between specular and diffuse transmission. A specular transmission is one where the incident radiation goes "straight through" the material, while a diffuse transmission is encountered when the incident radiation is scattered in passing through the material, so that it emerges from the other side with a random spatial orientation. As with reflected energy, the assumption is made that the transmissivity may be represented with a specular and a diffuse component:

$$\tau = \tau_s + \tau_D \tag{8-98}$$

The diffuse radiosity is still defined as in Eq. (8-91), and the net energy exchange with a transmitting surface is given by Eq. (8-93). The analysis of transmitted energy exchange with other surfaces must be handled somewhat differently, however.

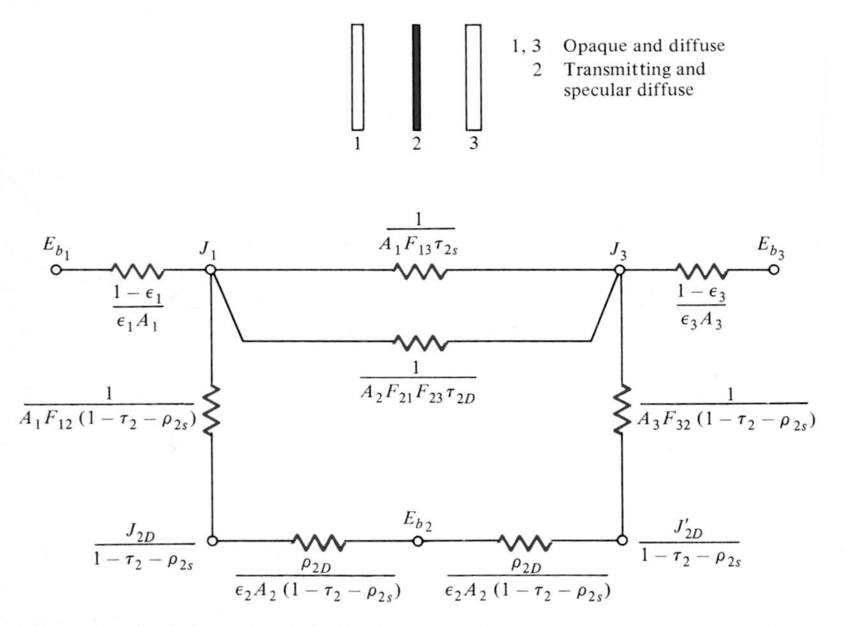

Fig. 8-61 Radiation network for infinite parallel planes separated by a transmitting specular-diffuse plane.

Consider, for example, the arrangement in Fig. 8-61. The two diffuse opaque surfaces are separated by a specular-diffuse transmitting and reflecting plane. For this example all planes are assumed to be infinite in extent. The *specular*-transmitted exchange between surfaces 1 and 3 may be calculated immediately with

$$(q_{13})_{\text{specular-transmitted}} = \frac{J_1 - J_3}{1/A_1F_{13}\tau_{2s}} \tag{8-99}$$

The *diffuse*-transmitted exchange between 1 and 3 is a bit more complicated. The energy leaving 1 which is transmitted diffusely through 2 is

$$J_1A_1F_{12}\tau_{2D}$$

Of this amount transmitted through 2, the amount which arrives at 3 is

$$(q_{13})_{\text{diffuse-transmitted}} = J_1A_1F_{12}\tau_{2D}F_{23} \tag{8-100}$$

Similarly, the amount leaving 3 which is diffusely transmitted to 1 is

$$(q_{31})_{\text{diffuse-transmitted}} = J_3A_3F_{32}\tau_{2D}F_{21} \tag{8-101}$$

Now, by making use of the reciprocity relations, $A_1F_{12} = A_2F_{21}$ and $A_3F_{32} = A_2F_{23}$, subtraction of Eq. (8-101) from Eq. (8-100) gives

$$(q_{13})_{\text{net diffuse-transmitted}} = \frac{J_1 - J_3}{1/A_1F_{21}F_{23}\tau_{2D}} \tag{8-102}$$

Making use of Eqs. (8-99) and (8-102) gives the complete network for the system as shown in Fig. 8-61. Of course, all the radiation shape factors in the above network are unity, but they have been included for the sake of generality. In this network J_{2D} refers to the diffuse radiosity on the left side of 2, while J'_{2D} is the diffuse radiosity on the right side of this surface.

It is easy to extend the network method to the case of two finite parallel planes separated by a transmitting plane inside a large enclosure. Such an arrangement is shown in Fig. 8-62. In this network the notation F_{24R} means the radiation shape factor for radiation leaving the right side of surface 2, while F_{24L} refers to the radiation leaving the left side of surface 2. The notation $F'_{14(2)}$ means the fraction of energy leaving 1 which arrives at the enclosure 4 after specular transmission through 2. The notation $F'_{34(2)}$ has a similar meaning with respect to surface 3. The factors $F_{1(2)4}$ and $F_{3(2)4}$ designate the radiation shape factors between the image surfaces 1(2) and 3(2) and the enclosure 4.

The above networks require information on the specular and diffuse properties of material which are to be analyzed. Unfortunately, such information is quite meager at this time. Some data are available in Refs. 17 to 19.

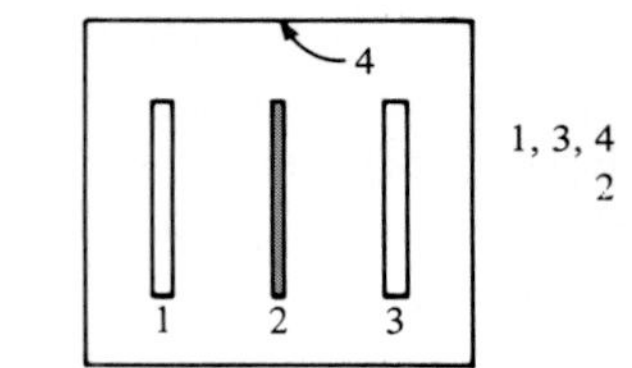

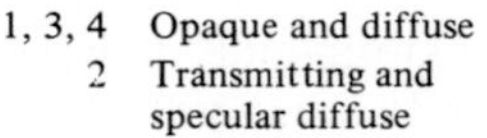

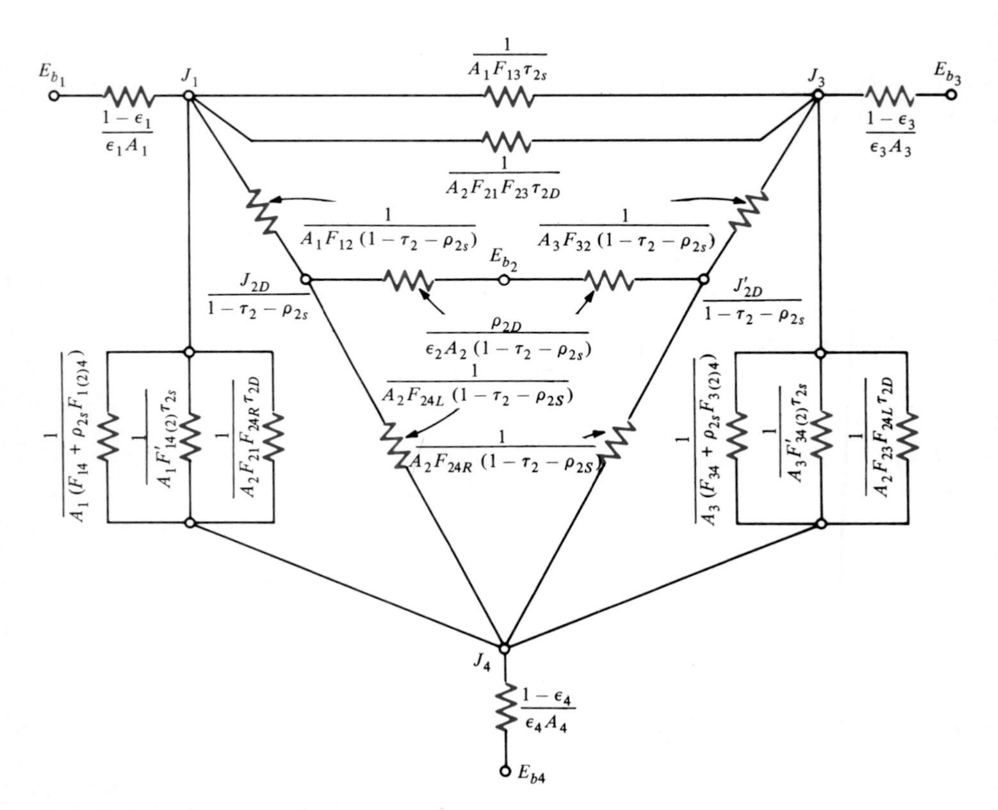

Fig. 8-62 Radiation network for system of finite specular-diffuse planes in an enclosure.

■ **EXAMPLE 8-14 Transmitting and reflecting system**

A furnace of 1000°C has a small opening in the side which is covered with a quartz window having the following properties:

$$0 < \lambda < 4\ \mu\text{m} \qquad \tau = 0.9 \qquad \epsilon = 0.1 \qquad \rho = 0$$

$$4 < \lambda < \infty \qquad \tau = 0 \qquad \epsilon = 0.8 \qquad \rho = 0.2$$

The interior of the furnace may be treated as a blackbody. Calculate the radiation lost through the quartz window to a room at 30°C. Diffuse surface behavior is assumed.

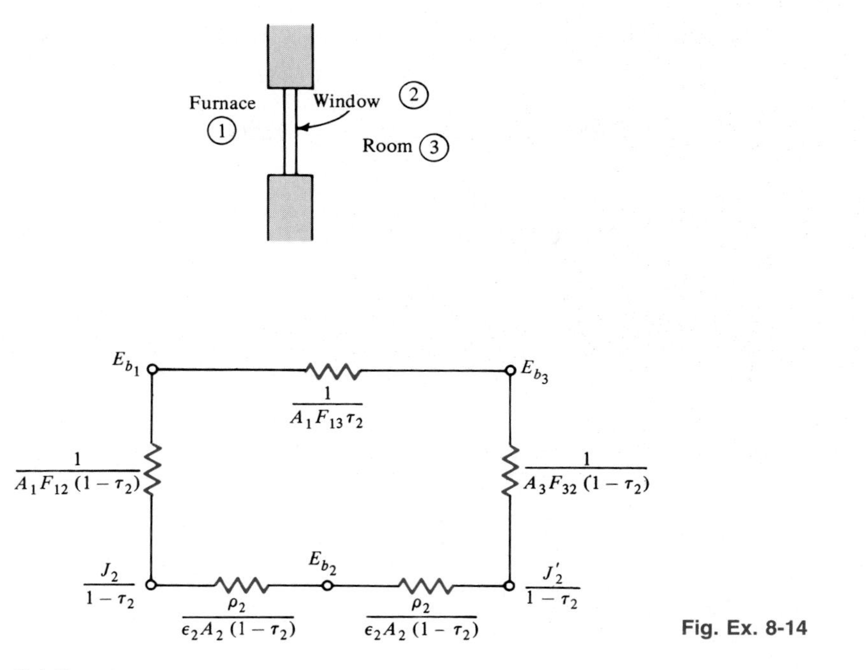

Fig. Ex. 8-14

Solution

The diagram for this problem is shown in the sketch. Because the room is large it may be treated as a blackbody also. We shall analyze the problem by calculating the heat transfer for each wavelength band and then adding them together to obtain the total. The network for each band is a modification of Fig. 8-60, as shown below for black furnace and room. We shall make the calculation for unit area; then

$$A_1 = A_2 = A_3 = 1.0$$

$$F_{12} = 1.0 \qquad F_{13} = 1.0 \qquad F_{32} = 1.0$$

The total emissive powers are

$$E_{b_1} = (5.669 \times 10^{-8})(1273)^4 = 1.4887 \times 10^5 \text{ W/m}^2$$

$$E_{b_3} = (5.669 \times 10^{-8})(303)^4 = 477.8 \text{ W/m}^2$$

To determine the fraction of radiation in each wavelength band, we calculate

$$\lambda T_1 = (4)(1273) = 5092 \ \mu\text{m} \cdot \text{K}$$

$$\lambda T_3 = (4)(303) = 1212 \ \mu\text{m} \cdot \text{K}$$

Consulting Table 8-1, we find

$$E_{b_1}(0 - 4 \ \mu\text{m}) = 0.6450E_{b_1} = 96{,}021 \text{ W/m}^2$$

$$E_{b_3}(0 - 4 \ \mu\text{m}) = 0.00235E_{b_3} = 1.123 \text{ W/m}^2$$

$$E_{b_1}(4 - \infty) = (1 - 0.6450)E_{b_1} = 52{,}849 \text{ W/m}^2$$

$$E_{b_3}(4 - \infty) = (1 - 0.00235)E_{b_3} = 476.7 \text{ W/m}^2$$

We now apply these numbers to the network for the two wavelength bands, with unit areas.

$0 < \lambda < 4$ λm *band:*

$$\frac{1}{F_{13}\tau_2} = \frac{1}{0.9} \qquad \frac{1}{F_{32}(1 - \tau_2)} = \frac{1}{0.1} = \frac{1}{F_{12}(1 - \tau_2)}$$

$$\frac{\rho_2}{\epsilon_2(1 - \tau_2)} = 0$$

The net heat transfer from the network is then

$$q = \frac{E_{b1} - E_{b3}}{R_{\text{equiv}}} = \frac{96{,}021 - 1.123}{1.0526} = 91{,}219 \text{ W/m}^2 \qquad 0 < \lambda < 4\ \mu\text{m}$$

$4\ \mu$m $< \lambda < +\infty$ *band:*

$$\frac{1}{F_{13}\tau_2} = \infty \qquad \frac{1}{F_{32}(1 - \tau_2)} = \frac{1}{F_{12}(1 - \tau_2)} = 1.0$$

$$\frac{\rho_2}{\epsilon_2(1 - \tau_2)} = \frac{0.2}{0.8} = 0.25$$

The net heat transfer from the network is

$$q = \frac{E_{b1} - E_{b3}}{1 + 0.25 + 0.25 + 1} = \frac{52{,}849 - 476.7}{2.5} = 20.949 \text{ W/m}^2 \qquad 4 < \lambda < \infty$$

The total heat loss is then

$$q_{\text{total}} = 91{,}219 + 20.949 = 112{,}167 \text{ W/m}^2 \quad [35{,}560 \text{ Btu/h} \cdot \text{ft}^2]$$

With no windows at all, the heat transfer would have been the difference in blackbody emissive powers,

$$q - E_{b1} - E_{b3} = 1.4887 \times 10^5 - 477.8 = 1.4839 \times 10^5 \text{ W/m}^2 \quad [47{,}040 \text{ Btu/h} \cdot \text{ft}^2]$$

8-13 FORMULATION FOR NUMERICAL SOLUTION

The network method which we have used to analyze radiation problems is an effective artifice for visualizing radiant exchange between surfaces. For simple problems which do not involve too many surfaces the network method affords a solution that can be obtained quite easily. When many heat-transfer surfaces are involved, it is to our advantage to formalize the procedure for writing the nodal equations. For this procedure we consider only opaque, gray, diffuse surfaces. The reader should consult Ref. 10 for information on transmitting and specular surfaces. The radiant-energy balance on a particular opaque surface can be written

$$\text{Net heat lost by surface} = \text{energy emitted} - \text{energy absorbed}$$

or on a unit-area basis with the usual gray-body assumptions,

$$\frac{q}{A} = \epsilon E_b - \alpha G$$

Considering the ith surface, the total irradiation is the sum of all irradiations G_j from the other j surfaces. Thus, for $\epsilon = \alpha$,

$$\frac{q_i}{A_i} = \epsilon_i \left(E_{b_i} - \sum_j G_j \right) \tag{8-103}$$

But, the irradiations can be expressed by

$$A_j J_j F_{ji} = G_j A_i \tag{8-104}$$

From reciprocity, we have

$$A_j F_{ji} = A_i F_{ij}$$

so that we can combine the equations to give

$$\frac{q_i}{A_i} = \epsilon_i \left(E_{bi} - \sum_j F_{ij} J_j \right) \tag{8-105}$$

The heat transfer at each surface is then evaluated in terms of the radiosities J_j. These parameters are obtained by recalling that the heat transfer can also be expressed as

$$\frac{q_i}{A_i} = J_i - G_i = J_i - \sum_j F_{ij} J_j \tag{8-106}$$

Combining Eqs. (8-105) and (8-106) gives

$$J_i - (1 - \epsilon_i) \sum_j F_{ij} J_j = \epsilon_i E_{bi} \tag{8-107}$$

In the equations above it must be noted that the summations must be performed over *all* surfaces in the enclosure. For a three-surface enclosure, with $i = 1$, the summation would then become

$$\sum_j F_{ij} J_j = F_{11} J_1 + F_{12} J_2 + F_{13} J_3$$

Of course, if surface 1 is convex, $F_{11} = 0$ and some simplification could be effected.

The nodal equations for the radiosities may also be derived from the nodes in the network formulation. At each J_i node an energy balance gives

$$\frac{\epsilon_i}{1 - \epsilon_i} (E_{b_i} - J_i) + \sum_j F_{ij} (J_j - J_i) = 0 \tag{8-108}$$

Again, an equation will be obtained for each J_i which is entirely equivalent to Eq. (8-107). Once all the equations are written out they can be expressed in the matrix form

$$[A][J] = [C] \tag{8-109}$$

$$\text{where } [A] = \begin{bmatrix} a_{11} & a_{12} & \cdots & a_{1i} \\ a_{21} & a_{22} & \cdots & a_{2i} \\ \cdots & \cdots & \cdots & \cdots \\ a_{i1} & a_{i2} & \cdots & a_{ii} \end{bmatrix} \qquad [J] = \begin{bmatrix} J_1 \\ J_2 \\ \cdot \\ \cdot \\ \cdot \\ J_i \end{bmatrix} \qquad [C] = \begin{bmatrix} C_1 \\ C_2 \\ \cdot \\ \cdot \\ \cdot \\ C_i \end{bmatrix}$$

The solution for the radiosities is found by obtaining the inverse to $[A]$ such that

$$[J] = [A]^{-1}[C]$$

The inverse $[A]^{-1}$ is written as

$$[A]^{-1} = \begin{bmatrix} b_{11} & b_{12} & \cdots & b_{1i} \\ b_{21} & \cdots & \cdots & \cdots \\ \cdots & \cdots & \cdots & \cdots \\ b_{i1} & b_{i2} & \cdots & b_{ii} \end{bmatrix}$$

so that the unknown radiosities are written as

$$J_1 = b_{11}C_1 + b_{12}C_2 + \cdots + b_{1i}C_i$$
$$\cdots\cdots\cdots\cdots\cdots\cdots$$
$$J_i = b_{i1}C_1 + b_{i2}C_2 + \cdots + b_{ii}C_i$$

Standard computer subroutines are available to obtain the inverse matrix and perform the final calculations of the J_i. The heat-transfer rate at each ith surface having an area A_i is then calculated from

$$\frac{q_i}{A_i} = \frac{\epsilon_i}{1 - \epsilon_i}(E_{b_i} - J_i) \tag{8-110}$$

In formulating the nodal equations one must take note of the consequence of Eq. (8-110) for an *insulated* surface, i.e., one for which there is no net heat transfer. Equation (8-110) thus requires that

$$E_{b_i} = J_i \qquad \text{for insulated surface} \tag{8-111}$$

From a practical point of view, a Gauss-Seidel iteration scheme may be the most efficient numerical procedure to follow in solving the set of equations for the J_i's. For the Gauss-Seidel scheme the above equations must be organized in explicit form for J_i. Solving for J_i in Eq. (8-107) and breaking out the F_{ii} term gives

$$J_i = (1 - \epsilon_i)\sum_{j \neq i} F_{ij}J_j + (1 - \epsilon_i)F_{ii}J_i + \epsilon_i E_{b_i}$$

$$J_i = \frac{1}{1 - F_{ii}(1 - \epsilon_i)}\left[(1 - \epsilon_i)\sum_{j \neq i} F_{ij}J_j + \epsilon_i E_{b_i}\right] \tag{8-112}$$

For a surface in radiant equilibrium, $q_i/A_i = 0$ and $J_i = E_{b_i}$ may be substituted into Eq. (8-112) to give

$$J_i = \frac{1}{1 - F_{ii}} \sum_{j \neq i} F_{ij} J_j \qquad \text{for } \frac{q_i}{A_i} = 0 \tag{8-113}$$

If the problem formulation is to include a specified heat flux q_i/A_i at one of the ith surfaces, we can solve for E_{bi} from Eq. (8-110) to give

$$E_{b_i} = J_i + \frac{1 - \epsilon_i}{\epsilon_i} \frac{q_i}{A_i} \tag{8-114}$$

Substituting this value into Eq. (8-107) and then solving for J_i give

$$J_i = \frac{1}{1 - F_{ii}} \left(\sum_{j \neq i} F_{ij} J_j + \frac{q_i}{A_i} \right) \tag{8-115}$$

In many cases the radiation solution must take conduction and convection into account at the ith surface. The appropriate energy balance is then, for steady state,

Heat conducted into surface + heat convected into surface
= radiant heat *lost* from surface

or

$$q_{\text{cond}, i} + q_{\text{conv}, i} = q_{i, \text{rad}} \tag{8-116}$$

This energy balance may then be used in conjunction with Eq. (8-115) to obtain the proper nodal equation for J_i.

While the above formulations may appear rather cumbersome at first glance they are easily solved by computer, with either matrix inversion or iteration. For many practical radiation problems, the number of equations is small and programmable calculators may be employed for solution. In most cases one will not know the surface properties (ϵ_i) within better than a few percent, so an iterative solution need not be carried out to unreasonable limits of precision.

In summary, we outline the computational procedure to be followed for numerical solution of radiation heat transfer between diffuse, gray surfaces. This basic procedure is the same for a hand computation, calculation with a minicomputer, or a large computer.

1. Evaluate F_{ij} and ϵ_i for all surfaces.

2. Evaluate E_{b_i} for all surfaces with specified temperature.

3. Formulate nodal equations for the J_i using:
   a. Eq. (8-112) for surfaces with specified T_i.
   b. Eq. (8-113) for surfaces in radiant balance ($J_i = E_{b_i}$).
   c. Eq. (8-115) for surfaces with specified q_i.

4. Solve the equations for the J_i's. If a Gauss-Seidel iteration is performed, use the following steps:
   a. Assume initial values of the J_i's. For large-machine calculation, initial values may be taken as zero. For minicomputer or abbreviated computation scheme, best initial guess should be made.

b. Recompute all J_i's in accordance with equations of step 3, always using most recent values for the calculation.
c. Stop calculations when an acceptable precision δ is attained such that

$$J_i^{n+1} - J_i^n \leq \delta$$

where n designates the number of iterations.

5. Compute the q_i's and T_i's, using:
a. q_i from Eq. (8-110) for gray surfaces and Eq. (8-106) for black surfaces with specified T_i.
b. T_i from $J_i = E_{b_i} = \sigma T_i^4$ for surfaces in radiant balance.
c. T_i using E_{b_i} obtained from Eq. (8-114) for surfaces with specified q_i.

Of course, the above equations may be put in the following form if direct matrix inversion is preferred over an iteration scheme:

$$J_i[1 - F_{ii}(1 - \epsilon_i)] - (1 - \epsilon_i) \sum_{j \neq i} F_{ij} J_j = \epsilon_i E_{b_i} \tag{8-112a}$$

$$J_i(1 - F_{ii}) - \sum_{j \neq i} F_{ij} J_j = 0 \tag{8-113a}$$

$$J_i(1 - F_{ii}) - \sum_{j \neq i} F_{ij} J_j = \frac{q_i}{A_i} \tag{8-115a}$$

EXAMPLE 8-15

The geometry of Example 8-5 is used for radiant exchange with a large enclosure. Surface 2 is diffuse with $\epsilon = 0.5$ while surface 1 is perfectly insulated. $T_2 = 1000$ K, and $T_3 = 300$ K. Calculate the heat lost to the large room per unit length of surface 2, using the numerical formulation.

Solution

For unit length we have:

$$E_{b2} = \sigma T_2^4 = 5.669 \times 10^4 \qquad E_{b3} = \sigma T_3^4 = 459$$

$$A_1 = (4)(0.2) = 0.8 \text{ m}^2/\text{m} \qquad A_2 = \pi(0.60)/2 = 0.94 \text{ m}^2/\text{m}$$

We will use the numerical formulation. We find from Example 8-5, using the nomenclature of the figure, $F_{11} = 0.314$, $F_{12} = 0.425$, $F_{13} = 0.261$, $F_{21} = 0.5$, $F_{22} = 0$, $F_{23} = 0.5$, $F_{31} \to 0$, $F_{32} \to 0$, $F_{33} \to 1.0$. We now write the equations. Surface 1 is insulated so we use Eq. (8-113*a*):

$$J_1(1 - 0.314) - 0.425 J_2 - 0.261 J_3 = 0$$

Surface 2 is constant temperature so we use Eq. (8-112*a*):

$$J_2(1 - 0) - (1 - 0.5)[0.5 J_1 + 0.5 J_3] = (0.5)(56{,}690)$$

Because surface 3 is so large,

$$J_3 = E_{b3} = 459 \text{ W/m}^2$$

Rearranging the equations gives

$$0.686J_1 - 0.425J_2 = 119.8$$
$$-0.25J_1 + J_2 = 28{,}460$$

which have the solution

$$J_1 = 21{,}070 \text{ W/m}^2$$
$$J_2 = 33{,}727 \text{ W/m}^2$$

The heat transfer is thus

$$q = \frac{E_{b2} - J_2}{\dfrac{1 - \epsilon_2}{\epsilon_2 A_2}} = \frac{56{,}690 - 33{,}727}{(1 - 0.5)/(0.5)(4)(0.2)} = 18{,}370 \text{ W/m length}$$

Because surface 1 is insulated, $J_1 = E_{b1}$, and we could calculate the temperature as

$$T_1 = \left(\frac{21{,}070}{5.669 \times 10^{-8}}\right)^{1/4} = 781 \text{ K}$$

■ **EXAMPLE 8-16** Numerical solutions

Two 1-m-square surfaces are separated by a distance of 1 m with $T_1 = 1000$ K, $T_2 = 400$ K, $\epsilon_1 = 0.8$, $\epsilon_2 = 0.5$. Obtain the numerical solutions for this system when (*a*) the plates are surrounded by a large room at 300 K and (*b*) the surfaces are connected by a re-radiating wall perfectly insulated on its outer surface.

Solution

Consulting Fig. 8-12, we obtain

$$F_{12} = 0.2 \qquad F_{21} = 0.2 \qquad F_{11} = 0 = F_{22}$$
$$F_{13} = 0.8 \qquad F_{23} = 0.8$$
$$A_1 = A_2 = 1 \text{ m}^2$$

(surface 3 is the surroundings or insulated surface). For part (*a*)

$$E_{b_1} = \sigma T_1^4 = 56.69 \text{ kW/m}^2 \quad [17{,}970 \text{ Btu/h} \cdot \text{ft}^2]$$
$$E_{b_2} = \sigma T_2^4 = 1.451 \text{ kW/m}^2$$
$$E_{b_3} = \sigma T_3^4 = 0.459 \text{ kW/m}^2$$

Because $A_3 \to \infty$, F_{31} and F_{32} must approach zero since $A_1F_{13} = A_3F_{31}$ and $A_2F_{23} = A_3F_{32}$. The nodal equations are written in the form of Eq. (8-107):

surface 1: $\quad J_1 - (1 - \epsilon_1)(F_{11}J_1 + F_{12}J_2 + F_{13}J_3) = \epsilon_1 E_{b_1}$

surface 2: $\quad J_2 - (1 - \epsilon_2)(F_{21}J_1 + F_{22}J_2 + F_{23}J_3) = \epsilon_2 E_{b_2} \qquad (a)$

surface 3: $\quad J_3 - (1 - \epsilon_3)(F_{31}J_1 + F_{32}J_2 + F_{33}J_3) = \epsilon_3 E_{b_3}$

Because F_{31} and F_{32} approach zero, F_{33} must be 1.0.

Inserting the numerical values for the various terms, we have

$$J_1 - (1 - 0.8)[(0)J_1 + (0.2)J_2 + (0.8)J_3] = (0.8)(56.69)$$
$$J_2 - (1 - 0.5)[(0.2)J_1 + (0)J_2 + (0.8)J_3] = (0.5)(1.451) \qquad (b)$$
$$J_3 - (1 - \epsilon_3)[(0)J_1 + (0)J_2 + (1.0)J_3] = \epsilon_3(0.459)$$

The third equation yields $J_3 = 0.459 \text{ kW/m}^2 \cdot \text{K}$. Because the room is so large it acts like a hohlraum, or blackbody. But *it does not have zero heat transfer*.

Finally, the equations are written in compact form as

$$J_1 - 0.04J_2 - 0.16J_3 = 45.352$$
$$-0.1J_1 + J_2 - 0.4J_3 = 0.7255 \qquad (c)$$
$$J_3 = 0.459$$

Of course, there only remain two unknowns, J_1 and J_2, in this set.

For part (*b*), A_3 for the enclosing wall is 4.0 m², and we set $J_3 = E_{b_3}$ because surface 3 is insulated. From reciprocity we have

$$A_1F_{13} = A_3F_{31} \qquad F_{31} = \frac{(1.0)(0.8)}{4.0} = 0.2$$

$$A_2F_{23} = A_3F_{32} \qquad F_{32} = \frac{(1.0)(0.8)}{4.0} = 0.2$$

Then from $F_{31} + F_{32} + F_{33} = 1.0$ we have $F_{33} = 0.6$.

The set of equations in (*a*) still applies, so we insert the numerical values to obtain (with $J_3 = E_{b_3}$)

$$J_1 - (1 - 0.8)[(0)J_1 + (0.2)J_2 + (0.8)J_3] = (0.8)(56.69)$$
$$J_2 - (1 - 0.5)[(0.2)J_1 + (0)J_2 + (0.8)J_3] = (0.5)(1.451) \qquad (d)$$
$$J_3 - (1 - \epsilon_3)[(0.2)J_1 + (0.2)J_2 + (0.6)J_3] = \epsilon_3 J_3$$

Notice that the third equation of set (*d*) can be written as

$$J_3(1 - \epsilon_3) - (1 - \epsilon_3)[(0.2)J_1 + (0.2)J_2 + (0.6)J_3] = 0$$

so that the $1 - \epsilon_3$ term drops out, and we obtain our final set of equations as

$$J_1 - 0.04J_2 - 0.16J_3 = 45.352$$
$$-0.1J_1 + J_2 - 0.4J_3 = 0.7255 \qquad (e)$$
$$-0.2J_1 - 0.2J_2 + 0.4J_3 = 0$$

To obtain the heat transfers the set of equations is first solved for the radiosities. For set (*c*),

$$J_1 = 45.644 \text{ kW/m}^2 \quad [14{,}470 \text{ Btu/h} \cdot \text{ft}^2]$$
$$J_2 = 5.474 \text{ kW/m}^2$$
$$J_3 = 0.459 \text{ kW/m}^2$$

The heat transfers are obtained from Eq. (8-110):

$$q_1 = \frac{A_1\epsilon_1}{1 - \epsilon_1}(E_{b_1} - J_1) = \frac{(1.0)(0.8)}{1 - 0.8}(56.69 - 45.644) = 44.184 \text{ kW} \quad [150{,}760 \text{ Btu/h}]$$

$$q_2 = \frac{A_2\epsilon_2}{1 - \epsilon_2}(E_{b_2} - J_2) = \frac{(1.0)(0.5)}{1 - 0.5}(1.451 - 5.474) = -4.023 \text{ kW} \quad [-13{,}730 \text{ Btu/h}]$$

The net heat *absorbed* by the room is the algebraic sum of q_1 and q_2 or

$$q_{3,\text{absorbed}} = 44.184 - 4.023 = 40.161 \text{ kW} \qquad [137{,}030 \text{ Btu/h}]$$

For part (*b*) the solutions to set (*e*) are

$$J_1 = 51.956 \text{ kW/m}^2 \qquad J_2 = 20.390 \text{ kW/m}^2 \qquad J_3 = 36.173 \text{ kW/m}^2$$

The heat transfers are

$$q_1 = \frac{A_1\epsilon_1}{1 - \epsilon_1}(E_{b_1} - J_1) = \frac{(1.0)(0.8)}{1 - 0.8}(56.69 - 51.965) = 18.936 \text{ kW}$$

$$q_2 = \frac{A_2\epsilon_2}{1 - \epsilon_2}(E_{b_2} - J_2) = \frac{(1.0)(0.5)}{1 - 0.5}(1.451 - 20.390) = -18.936 \text{ kW}$$

Of course, these heat transfers should be equal in magnitude with opposite sign because the insulated wall exchanges no heat. The temperature of the insulated wall is obtained from

$$J_3 = E_{b_3} = \sigma T_3^4 = 36.173 \text{ kW/m}^2$$

and

$$T_3 = 894 \text{ K} \quad [621°\text{C}, 1150°\text{F}]$$

■ **EXAMPLE 8-17** Radiation from a hole

To illustrate the radiation formulation for numerical solution we consider the circular hole 2 cm in diameter and 3 cm deep, as shown in the accompanying figure. The hole is machined in a large block of metal, which is maintained at 1000°C and has a surface emissivity of 0.6. The temperature of the large surrounding room is 20°C. A simple approach to this problem would assume the radiosity uniform over the entire heated internal surface. In reality, the radiosity varies over the surface, and we break it into segments 1 (bottom of the hole), 2, 3, and 4 (sides of the hole) for analysis.

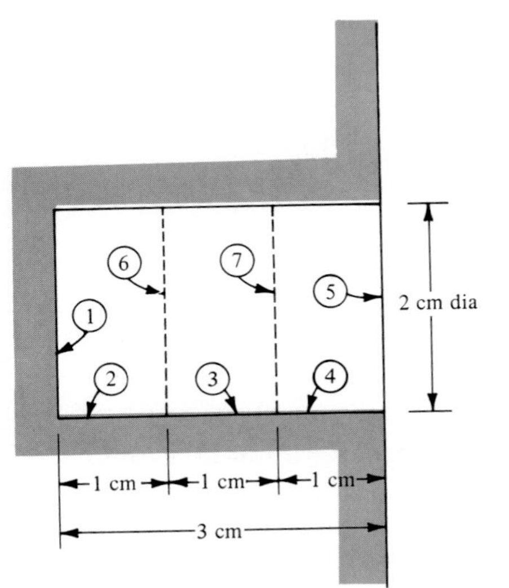

Fig. Ex. 8-17

The large room acts like a blackbody at 20°C, so for analysis purposes we can assume the hole is covered by an imaginary black surface 5 at 20°C. We shall set the problem up for a numerical solution for the radiosities and then calculate the heat-transfer rates. After that, we shall examine an insulated-surface case for this same geometry.

All the shape factors can be obtained with the aid of Fig. 8-13 and the imaginary disk surfaces 6 and 7. We have

$$E_{b1} = \sigma T_1^4 = (5.669 \times 10^{-8})(1273)^4 = 1.48874 \times 10^5 \text{ W/m}^2$$

$$= E_{b2} = E_{b3} = E_{b4}$$

$$E_{b5} = \sigma T_5^4 = (5.669 \times 10^{-8})(293)^4 = 417.8 \text{ W/m}^2$$

$$\epsilon_1 = \epsilon_2 = \epsilon_3 = \epsilon_4 = 0.6 \qquad \epsilon_5 = 1.0$$

$$A_1 = A_5 = \pi(1)^2 = \pi \text{ cm}^2 = A_6 = A_7$$

$$A_2 = A_3 = A_4 = \pi(2)(1) = 2\pi$$

$$F_{11} = F_{55} = 0 \qquad F_{16} = 0.37 \qquad F_{17} = 0.175 \qquad F_{15} = 0.1$$

$$F_{12} = 1 - F_{16} = 0.63 = F_{54}$$

$$F_{13} = F_{16} - F_{17} = 0.195 = F_{53}$$

$$F_{14} = F_{17} - F_{15} = 0.075 = F_{52}$$

$$F_{21} = F_{26} = F_{16}\frac{A_1}{A_2} = 0.315 = F_{45} = F_{36} = F_{37}$$

$$F_{22} = 1 - F_{21} - F_{26} = 0.37 = F_{33} = F_{44}$$

$$F_{31} = F_{13}\frac{A_1}{A_3} = 0.0975$$

$$F_{32} = F_{36} - F_{31} = 0.2175 = F_{34} = F_{43} = F_{23}$$

$$F_{27} = F_{26} - F_{23} = F_{21} - F_{23} = 0.0975 = F_{46}$$

$$F_{41} = F_{14}\frac{A_1}{A_4} = 0.0375 = F_{25}$$

$$F_{42} = F_{46} - F_{41} = 0.06 = F_{24}$$

The equations for the radiosities are now written in the form of Eq. (8-112), noting that $F_{11} = 0$ and $J_5 = E_{b5}$:

$$J_1 = (1 - \epsilon_1)(F_{12}J_2 + F_{13}J_3 + F_{14}J_4 + F_{15}E_{b5}) + \epsilon_1 E_{b1}$$

$$J_2 = \frac{1}{1 - F_{22}(1 - \epsilon_2)}[(1 - \epsilon_2)(F_{21}J_1 + F_{23}J_3 + F_{24}J_4 + F_{25}E_{b5}) + \epsilon_2 E_{b2}]$$

$$J_3 = \frac{1}{1 - F_{33}(1 - \epsilon_3)}[(1 - \epsilon_3)(F_{31}J_1 + F_{32}J_2 + F_{34}J_4 + F_{35}E_{b5}) + \epsilon_3 E_{b3}]$$

$$J_4 = \frac{1}{1 - F_{44}(1 - \epsilon_4)}[(1 - \epsilon_4)(F_{41}J_1 + F_{42}J_2 + F_{43}J_3 + F_{45}E_{b5}) + \epsilon_4 E_{b4}]$$

When all the numerical values are inserted, we obtain

$$J_1 = 0.252J_2 + 0.078J_3 + 0.03J_4 + 89{,}341$$

$$J_2 = 0.1479J_1 + 0.1021J_3 + 0.02817J_4 + 104{,}848$$

$$J_3 = 0.04577J_1 + 0.1021J_2 + 0.1021J_4 + 104{,}859$$

$$J_4 = 0.01761J_1 + 0.02817J_2 + 0.1021J_3 + 104{,}902$$

These equations may be solved to give

$$J_1 = 1.4003 \times 10^5 \text{ W/m}^2$$

$$J_2 = 1.4326 \times 10^5 \text{ W/m}^2$$

$$J_3 = 1.3872 \times 10^5 \text{ W/m}^2$$

$$J_4 = 1.2557 \times 10^5 \text{ W/m}^2$$

The heat transfers can be calculated from Eq. (8-110):

$$q_i = \frac{\epsilon_i A_i}{1 - \epsilon_i}(E_{b_i} - J_i)$$

$$q_1 = \frac{(0.6)(\pi \times 10^{-4})}{1 - 0.6}(1.4887 - 1.4003)(10^5) = 4.1658 \text{ W}$$

$$q_2 = \frac{(0.6)(2\pi \times 10^{-4})}{1 - 0.6}(1.4887 - 1.4326)(10^5) = 5.2873 \text{ W}$$

$$q_3 = \frac{(0.6)(2\pi \times 10^{-4})}{1 - 0.6}(1.4887 - 1.3872)(10^5) = 9.5661 \text{ W}$$

$$q_4 = \frac{(0.6)(2\pi \times 10^{-4})}{1 - 0.6}(1.4887 - 1.2557)(10^5) = 21.959 \text{ W}$$

The total heat transfer is the sum of these four quantities or

$$q_{\text{total}} = 40.979 \text{ W} \quad [139.8 \text{ Btu/h}]$$

It is of interest to compare this heat transfer with the value we would obtain by assuming uniform radiosity on the hot surface. We would then have a two-body problem with

$$A_1 = \pi + 3(2\pi) = 7\pi \text{ cm}^2 \qquad A_5 = \pi \qquad F_{51} = 1.0 \qquad \epsilon_1 = 0.6 \qquad \epsilon_5 = 1.0$$

The heat transfer is then calculated from Eq. (8-43), with appropriate change of nomenclature:

$$q = \frac{(E_{b1} - E_{b5})A_5}{1/\epsilon_5 + (A_5/A_1)(1/\epsilon_1 - 1)} = \frac{(\pi \times 10^{-4})(1.4887 \times 10^5 - 417.8)}{1 + (\frac{1}{7})(1/0.6 - 1)}$$

$$= 42.581 \text{ W} \quad [145.3 \text{ Btu/h}]$$

Thus, the simple assumption of uniform radiosity gives a heat transfer which is 3.9 percent above the value obtained by breaking the hot surface into four parts for the calculation. This indicates that the uniform-radiosity assumption we have been using is a rather good one for engineering calculations.

Let us now consider the case where surface 1 is still radiating at 1000°C with $\epsilon = 0.6$ but the side walls 2, 3, and 4 are insulated. The radiation is still to the large room at 20°C. The nodal equation for J_1 is the same as before but now the equations for J_2, J_3, and J_4 must be written in the form of Eq. (8-113). When that is done and the numerical values are inserted, we obtain

$$J_1 = 0.252J_2 + 0.078J_3 + 0.03J_4 + 89{,}341$$

$$J_2 = 0.5J_1 + 0.3452J_3 + 0.09524J_4 + 24.869$$

$$J_3 = 0.1548J_1 + 0.3452J_2 + 0.3452J_4 + 64.66$$

$$J_4 = 0.05952J_1 + 0.0952J_2 + 0.3452J_3 + 208.9$$

When these equations are solved, we obtain

$$J_1 = 1.1532 \times 10^5 \text{ W/m}^2 \quad [36{,}560 \text{ Btu/h} \cdot \text{ft}^2]$$

$$J_2 = 0.81019 \times 10^5 \text{ W/m}^2$$

$$J_3 = 0.57885 \times 10^5 \text{ W/m}^2$$

$$J_4 = 0.34767 \times 10^5 \text{ W/m}^2$$

The heat transfer at surface 1 is

$$q_1 = \frac{\epsilon_1 A_1}{1 - \epsilon_1}(E_{b_1} - J_1) = \frac{(0.6)(\pi \times 10^{-4})}{1 - 0.6}(1.4887 - 1.1532)(10^5)$$

$$= 15.81 \text{ W} \quad [53.95 \text{ Btu/h}]$$

The temperatures of the insulated surface elements are obtained from

$$J_i = E_{b_i} = \sigma T_i^4$$

$$T_2 = 1093 \text{ K} = 820°\text{C} \quad [1508°\text{F}]$$

$$T_3 = 1005 \text{ K} = 732°\text{C} \quad [1350°\text{F}]$$

$$T_4 = 895 \text{ K} = 612°\text{C} \quad [1134°\text{F}]$$

It is of interest to compare the heat transfer calculated above with that obtained by assuming surfaces 2, 3, and 4 uniform in temperature and radiosity. Equation (8-41) applies for this case:

$$q = \frac{A_1(E_{b_1} - E_{b_5})}{\dfrac{A_1 + A_2 + 2A_1F_{15}}{A_5 - A_1(F_{15})^2} + \left(\dfrac{1}{\epsilon_1} - 1\right) + \dfrac{A_1}{A_5}\left(\dfrac{1}{\epsilon_5} - 1\right)}$$

and

$$q = \frac{(\pi \times 10^{-4})(1.4887 \times 10^5 - 417.8)}{\dfrac{\pi + \pi - 2\pi(0.1)}{\pi - \pi(0.1)^2} + \dfrac{1}{0.6} - 1} = 18.769 \text{ W} \quad [64.04 \text{ Btu/h}]$$

In this case the assumption of uniform radiosity at the insulated surface gives an overall heat transfer with surface 1 (bottom of hole) that is 18.7 percent too high.

■ EXAMPLE 8-18 Heater with constant heat flux

In the figure shown an electric heater is installed in surface 1 such that a constant heat flux of 100 kW/m² is generated at the surface. The four surrounding surfaces are in radiant balance with surface 1 and the large room at 20°C. The surface properties are $\epsilon_1 = 0.8$ and $\epsilon_2 = \epsilon_3 = \epsilon_4 = \epsilon_5 = 0.4$. Determine the temperatures of all surfaces. The back side of surface 1 is insulated. Repeat the calculation assuming surfaces 2, 3, 4, and 5 are just one surface uniform in temperature.

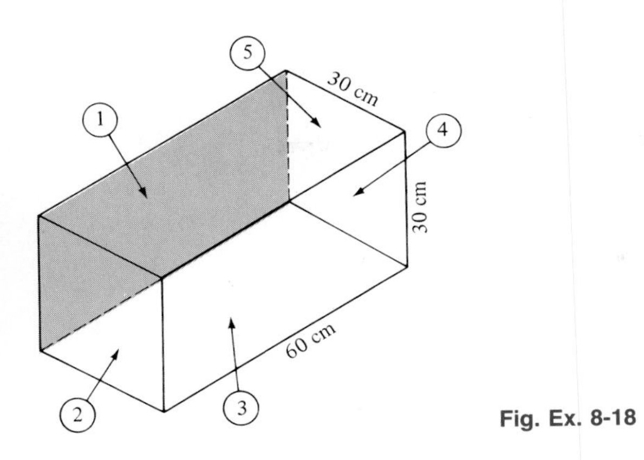

Fig. Ex. 8-18

Solution

In reality, surfaces 2, 3, 4, and 5 have *two* surfaces each; an inside and an outside surface. We thus have *nine* surfaces plus the room, so a 10-body problem is involved. Of course, from symmetry we can see that $T_2 = T_4$ and $T_3 = T_5$, but we set up the problem in the general numerical formulation. We designate the large room as surface 6 and it behaves as if $\epsilon_6 = 1.0$. So, it is as if the opening were covered with a black surface at 20°C. The shape factors of the inside surfaces are obtained from Figs. 8-12 and 8-14:

$$F_{16} = F_{61} = 0.285 \qquad F_{13} = F_{15} = 0.24 = F_{31} = F_{51}$$

$$F_{12} = F_{14} = 0.115 \qquad F_{24} = F_{42} = 0.068$$

$$F_{35} = F_{53} = 0.285 \qquad F_{32} = F_{52} = F_{34} = 0.115$$

$$F_{25} = F_{23} = F_{45} = F_{43} = F_{21} = F_{41} = F_{26} = F_{46} = 0.23$$

$$F_{11} = F_{22} = F_{33} = F_{44} = F_{55} = 0$$

For the *outside* surfaces,

$$F'_{26} = F'_{36} = F'_{46} = F'_{56} = 1.0$$

where the primes indicate the outside surfaces. We shall also use primes to designate the radiosities of the outside surfaces. For the room, $J_6 = E_{b6} = (5.669 \times 10^{-8})(293)^4 = 417.8 \text{ W/m}^2$.

For surface 1 with constant heat flux, we use Eq. (8-115*a*) and write

$$J_1 - (F_{12}J_2 + F_{13}J_3 + F_{14}J_4 + F_{15}J_5 + F_{16}J_6) = 1.0 \times 10^5 \tag{a}$$

Because of the radiant balance condition we have

$$(J_2 - E_{b2})\frac{\epsilon_2 A_2}{1 - \epsilon_2} = (E_{b2} - J'_2)\frac{\epsilon_2 A_2}{1 - \epsilon_2}$$

and

$$E_{b2} = \frac{J_2 + J'_2}{2} \tag{b}$$

where the prime designates the outside radiosity. A similar relation applies for surfaces 3, 4, and 5. Thus, we can use Eq. (8-112*a*) for *inside* surface 2

$$J_2 - (1 - \epsilon_2)(F_{21}J_1 + F_{23}J_3 + F_{24}J_4 + F_{25}J_5 + F_{26}J_6) = \frac{\epsilon_2}{2}(J_2 + J_2') \qquad (c)$$

and for the *outside* surface 2

$$J_2' - (1 - \epsilon_2)(F_{26}'J_6) = \frac{\epsilon_2}{2}(J_2 + J_2') \qquad (d)$$

Equations like (*c*) and (*d*) are written for surfaces 3, 4, and 5 also, and with the shape factors and emissivities inserted the following set of equations is obtained:

$$\begin{aligned}
J_1 - 0.115J_2 - 0.24J_3 - 0.115J_4 - 0.24J_5 &= 1.0012 \times 10^5 \\
-0.138J_1 + 0.8J_2 - 0.2J_2' - 0.138J_3 - 0.0408J_4 - 0.138J_5 &= 57.66 \\
0.2J_2 - 0.8J_2' &= -250.68 \\
-0.144J_1 - 0.069J_2 + 0.8J_3 - 0.2J_3' - 0.069J_4 - 0.05J_5 &= 60.16 \\
0.2J_3 - 0.8J_3' &= -250.68 \\
-0.138J_1 - 0.0408J_2 - 0.138J_3 + 0.8J_4 - 0.2J_4' - 0.138J_5 &= 57.66 \\
0.2J_4 - 0.8J_4' &= -250.68 \\
-0.144J_1 - 0.069J_2 - 0.057J_3 - 0.069J_4 + 0.8J_5 - 0.2J_5' &= 60.16 \\
0.2J_5 - 0.8J_5' &= -250.68
\end{aligned}$$

We thus have nine equations and nine unknowns which may be solved to give

$$\begin{aligned}
J_1 &= 1.24887 \times 10^5 \text{ W/m}^2 \\
J_2 = J_4 &= 37{,}549 \\
J_2' = J_4' &= 9701 \\
J_3 = J_5 &= 33{,}605 \\
J_3' = J_5' &= 8714
\end{aligned}$$

The temperatures are thus computed from Eq. (*b*):

$$E_{b2} = \frac{37{,}549 + 9701}{2} = 23{,}625 \qquad T_2 = T_4 = 803.5 \text{ K}$$

$$E_{b3} = \frac{33{,}605 + 8714}{2} = 21{,}160 \qquad T_3 = T_5 = 781.6 \text{ K}$$

For surface 1 we observed that

$$\frac{q}{A} = \frac{\epsilon}{1 - \epsilon}(E_{b_1} - J_1)$$

so that

$$E_{b_1} = \frac{(1.0 \times 10^5)(1 - 0.8)}{0.8} + 1.24887 \times 10^5 = 1.49887 \times 10^5$$

and $T_1 = 1275$ K

We note again that we could have observed the symmetry of the problem and set $J_2 = J_4$, $J_2' = J_4'$, and so on. By so doing, we could have had only five equations with five unknowns.

Surfaces 2, 3, 4, and 5 as one surface

We now go back and take surfaces 2, 3, 4, and 5 as one surface which we choose to call surface 7. The shape factors are then

$$F_{16} = F_{61} = 0.285 \qquad F_{17} = 1 - 0.285 = 0.715$$

$$A_1 = 2.0 \qquad A_7 = 6.0$$

Thus

$$F_{71} = (0.715)(\tfrac{2}{6}) = 0.2383 = F_{76}$$

$$F_{77} = 1 - (2)(0.2383) = 0.5233 \qquad F_{76}' = 1.0$$

Then for surface 1 we use Eq. (8-115*a*) to obtain

$$J_1 - (F_{17}J_7 + F_{16}J_6) = 1.0 \times 10^5$$

Using $E_{b7} = (J_7 + J_7')/2$, we have for the *inside* of surface 7

$$J_7[1 - F_{77}(1 - \epsilon_7)] - (1 - \epsilon_7)(F_{71}J_1 + F_{76}J_6) = \frac{\epsilon_7}{2}(J_7 + J_7')$$

while for the *outside* we have

$$J_7' - (1 - \epsilon_7)F_{76}'J_6 = \frac{\epsilon_7}{2}(J_7 + J_7')$$

When the numerical values are inserted, we obtain the set of three equations:

$$J_1 - 0.715J_7 = 1.0012 \times 10^5$$

$$-0.143J_1 + 0.486J_7 - 0.2J_7' = 59.74$$

$$0.2J_7 - 0.8J_7' = -250.68$$

which have the solution

$$J_1 = 1.31054 \times 10^5 \text{ W/m}^2$$

$$J_7 = 43{,}264$$

$$J_7' = 11{,}129$$

The temperatures are then calculated as before:

$$E_{b7} = \frac{43{,}264 + 11{,}129}{2} = 27{,}197 \qquad T_7 = 832.2 \text{ K}$$

$$E_{b1} = \frac{(1.0 \times 10^5)(1 - 0.8)}{0.8} + 1.31054 \times 10^5 = 1.65054 \times 10^5$$

$$T_1 = 1306 \text{ K}$$

So, there is about a 30 K temperature difference between the two methods.

Comment

With such a small difference between the solutions we may conclude that the extra complexity of choosing each surface at a different radiosity is probably not worth the effort, particularly when one recognizes the uncertainties which are present in the surface emissivities. *This points out that our assumptions of uniform irradiation and radiosity, though strictly not correct, give answers which are quite satisfactory.*

EXAMPLE 8-19 Combined convection and radiation (nonlinear system)

A 0.5 by 0.5 m plate is maintained at 1300 K and exposed to a convection and radiation surrounding at 300 K. Attached to the top are two radiation shields 0.5 by 0.5 m as shown in the accompanying figure. The convection heat-transfer coefficient for all surfaces is 50 W/m² · K, and $\epsilon_1 = 0.8$, $\epsilon_2 = 0.3 = \epsilon_3$. Determine the total heat lost from the 1300 K surface and the temperature of the shields.

Solution

This example illustrates how it is possible to handle convection-radiation problems with the numerical formulation and an iterative computational procedure. Nomenclature is shown in the figure. Using Figs. 8-12 and 8-14, we can evaluate the shape factors as

$$F_{12} = F_{13} = 0.2 \qquad F_{14} = 1 - 0.2 - 0.2 = 0.6$$

$$F_{23} = F_{32} = 0.2 \qquad F_{24L} = F_{34R} = 1.0$$

$$F_{21} = F_{12} = F_{31} = 0.2 \qquad F_{24R} = F_{34L} = 0.6$$

$$F_{11} = F_{22} = F_{33} = 0$$

$$J_{2R} = J_{3L} \qquad J_{2L} = J_{3R} \qquad \text{from symmetry}$$

$$J_4 = E_{b4}$$

We now use Eq. (8-112) to obtain a relation for J_1:

$$J_1 = (1 - \epsilon_1)[F_{12}J_{2R} + F_{13}J_{3L} + F_{14}J_4] + \epsilon_1 E_{b1}$$

But $J_{2R} = J_{3L}$ and $F_{12} = F_{13}$ so that

$$J_1 = (1 - \epsilon_1)(2F_{13}J_{2R} + F_{14}J_4) + \epsilon_1 E_{b1} \tag{a}$$

We use Eq. (8-115) for the overall energy balance on surface 2:

$$\begin{aligned} 2h(T_\infty - T_2) &= \frac{\epsilon_2}{1 - \epsilon_2}(E_{b2} - J_{2R}) + \frac{\epsilon_2}{1 - \epsilon_2}(E_{b2} - J_{2L}) \\ &= \frac{\epsilon_2}{1 - \epsilon_2}(2E_{b2} - J_{2R} - J_{2L}) \end{aligned} \tag{b}$$

Equation (8-112) is used for surface J_{2R}.

$$J_{2R} = (1 - \epsilon_2)(F_{21}J_1 + F_{23}J_{3L} + F_{24R}J_4) + \epsilon_2 E_{b2}$$

But $J_{2R} = J_{3L}$ so that

$$J_{2R} = \frac{1}{1 - (1 - \epsilon_2)F_{23}}[(1 - \epsilon_2)(F_{21}J_1 + F_{24R}J_4) + \epsilon_2 E_{b2}] \tag{c}$$

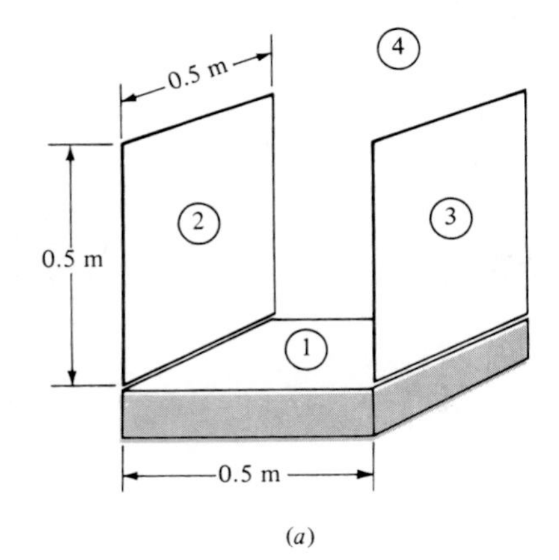

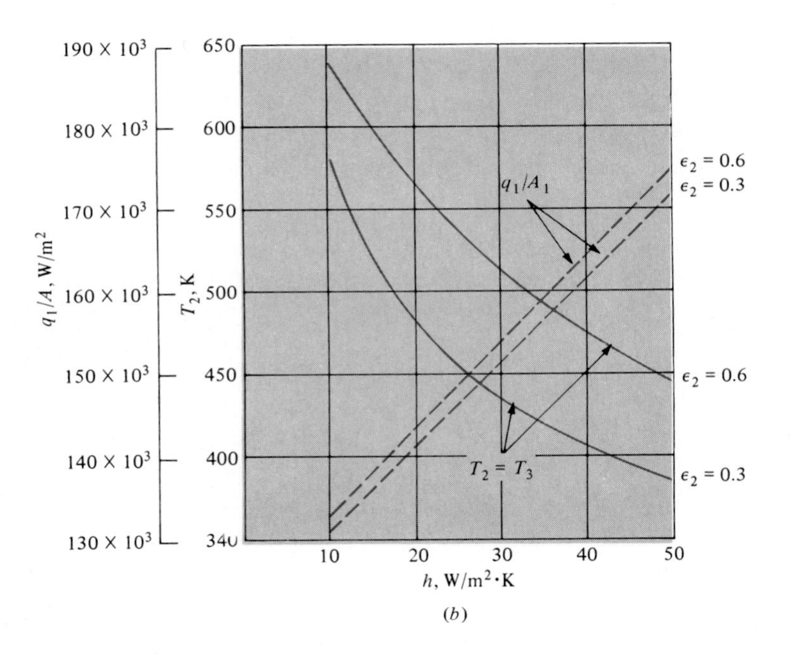

Fig. Ex. 8-19

For surface J_{2L} the equation is

$$J_{2L} = (1 - \epsilon_2)(F_{24L}J_4) + \epsilon_2 E_{b2} \tag{d}$$

We now have four equations with four unknowns, J_1, J_{2R}, J_{2L}, E_{b2}, with $T_2 = (E_{b2}/\sigma)^{1/4}$. However, Eq. (*b*) is nonlinear in E_b so we must use a special procedure to solve the set. Before embarking on the procedure, we can insert the various numerical quantities into Eqs. (*a*) to (*d*) to obtain the matrix coefficients:

J_1	J_{2R}	J_{2L}	E_{b_2}	C
1	−0.08	0	0	129,584
0	−0.4286	−0.4286	0.8571	30,000 − 100T_2
0.162791	−1	0	0.348837	−224.2605
0	0	1	−0.3	321.4

The computational algorithm which we employ iterates on values of T_2 until the value of E_{b_2} obtained from solving the set of equations agrees with $E_{b_2} = \sigma T_2^4$. The sequence of steps is as follows.

1. Assume a reasonable value for T_2.
2. Using this value in Eq. (*b*) solve the matrix for an estimated value of $E_{b_2}(e)$.
3. Compute $E_{b_2}(T_2) = \sigma T_2^4$ from the assumed value of T_2.
4. Compute $\Delta E_{b_2} = E_{b_2}(e) - E_{b_2}(T_2)$.
5. If $\Delta E_{b_2} > 0$, assume a higher value for T_2 and go to step 2. If $\Delta E_{b_2} < 0$, assume a lower value for T_2 and go to step 2.
6. Continue until ΔE_{b_2} is sufficiently small, at which point the solution to the matrix is the one desired.

After two iterations have been completed, interpolation or extrapolation is used to bring $\Delta E_{b_2} = 0$ based on the last two values of ΔE_{b_2} to speed the calculation. The results of the iteration are:

Iteration	T_2	$E_{b_2}(e)$	$E_{b_2}(T_2)$	ΔE_{b_2}
1	300	16,283	459	15,824
2	500	−18,377	3,543	−21,920
3	384	1,725.7	1,232.6	493.1
4	386.6	1,275	1,266	9

Clearly the method converges very rapidly, and the final solution is

$$J_1 = 1.3135 \times 10^5 \qquad J_{2R} = 22{,}051$$

$$J_{2L} = 710 \qquad E_{b_2} = 1275 \qquad T_2 = 386.6 \text{ K}$$

The total heat flux lost by surface 1 is

$$\frac{q_1}{A_1} = h(T_1 - T_\infty) + (E_{b_1} - J_1)\frac{\epsilon_1}{1 - \epsilon_1}$$

$$= 1.7226 \times 10^5 \text{ W/m}^2 \qquad (e)$$

For a 0.5 by 0.5 m surface the heat lost is thus

$$q_1 = (1.7226 \times 10^5)(0.5)^2 = 43{,}065 \text{ W}$$

Other cases may be computed, and the influence which h and ϵ_2 have on the results is shown in the accompanying figure.

Comment

This example illustrates how nonlinear equations with combined convection and radiation can be solved with a straightforward iterative procedure. Recognizing the uncertainties in knowledge of surface emissivity and convection coefficient, only a few iterations are necessary to achieve an acceptable solution.

8-14 SOLAR RADIATION

Solar radiation is a form of thermal radiation having a particular wavelength distribution. Its intensity is strongly dependent on atmospheric conditions, time of year, and the angle of incidence for the sun's rays on the surface of the earth. At the outer limit of the atmosphere the total solar irradiation when the earth is at its mean distance from the sun is 1395 W/m^2. This number is called the *solar constant* and is subject to modification upon collection of more precise experimental data.

Not all the energy expressed by the solar constant reaches the surface of the earth, because of strong absorption by carbon dioxide and water vapor in the atmosphere. The solar radiation incident on the earth's surface is also dependent on the atmospheric content of dust and other pollutants. The maximum solar energy reaches the surface of the earth when the rays are directly incident on the surface since (1) a larger view area is presented to the incoming solar flux and (2) the solar rays travel a smaller distance through the atmosphere so that there is less absorption than there would be for an incident angle tilted from the normal. Figure 8-63 indicates the atmospheric absorption effects for a sea-level location on clear days in a moderately dusty atmosphere with moderate water-vapor content.

It is quite apparent from Fig. 8-63 that solar radiation which arrives at the surface of the earth does not behave like the radiation from an ideal gray body, while outside the atmosphere the distribution of energy follows more of an ideal pattern. To determine an equivalent blackbody temperature for the solar radiation, we might employ the wavelength at which the maximum in the spectrum occurs (about 0.5 μm, according to Fig. 8-63) and Wien's displacement law [Eq. (8-13)]. This estimate gives

$$T \approx \frac{2987.6}{0.5} = 5795 \text{ K} \qquad (10{,}431°\text{R})$$

The equivalent solar temperature for thermal radiation is therefore about 5800 K (10,000°R).

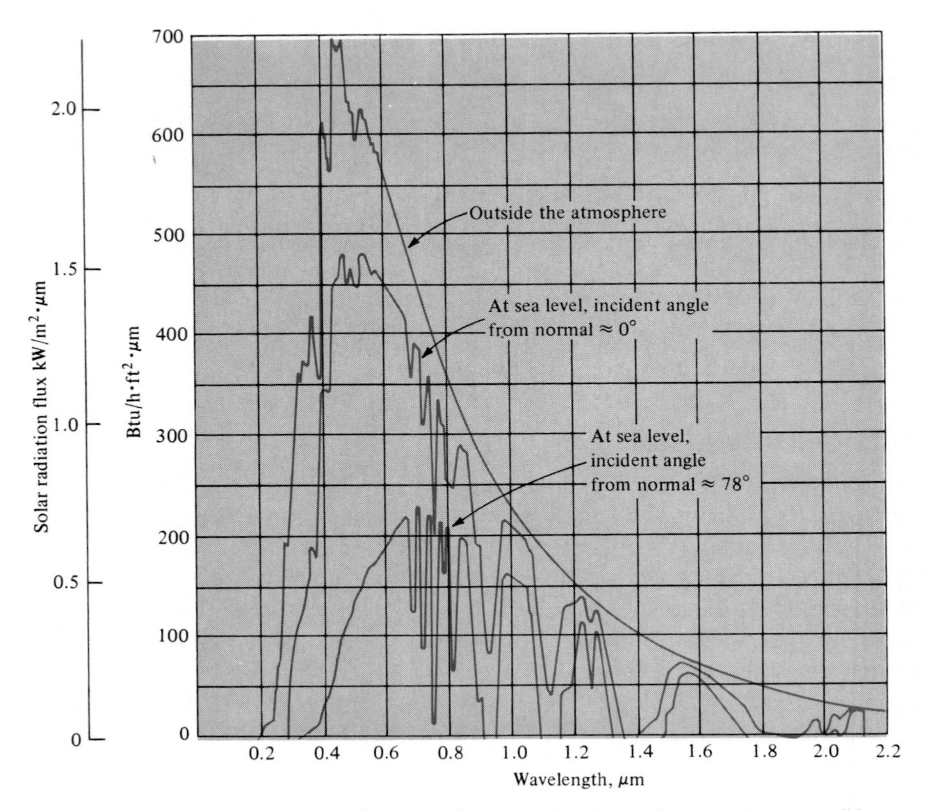

Fig. 8-63 Spectral distribution of solar radiation as functions of atmospheric conditions and angle of incidence according to Ref. 15.

If all materials exhibited gray-body behavior, solar-radiation analysis would not present a particularly unusual problem; however, since solar radiation is concentrated at short wavelengths, as opposed to much longer wavelengths for most "earth-bound" thermal radiation, a particular material may exhibit entirely different absorptance and transmittance properties for the two types of radiation. The classic example of this behavior is a greenhouse. Ordinary glass transmits radiation very readily at wavelengths below 2 μm; thus it transmits the large part of solar radiation incident upon it. This glass, however, is essentially opaque to long-wavelength radiation above 3 or 4 μm. Practically all the low-temperature radiation emitted by objects in the greenhouse is of such a long-wavelength character that it remains trapped in the greenhouse. Thus the glass allows much more radiation to come in than can escape, thereby producing the familiar heating effect. The solar radiation absorbed by objects in the greenhouse must eventually be dissipated to the surroundings by convection from the outside walls of the greenhouse.

Similar behavior is observed for the absorptance and reflectance of solar as

Table 8-3 Comparisons of Absorptivities of Various Surfaces to Solar and Low-Temperature Thermal Radiation as Compiled from Ref. 14.

	Absorptivity	
Surface	*For solar radiation*	*For low-temperature radiation ~25°C*
Aluminum, highly polished	0.15	0.04
Copper, highly polished	0.18	0.03
Tarnished	0.65	0.75
Cast iron	0.94	0.21
Stainless steel, no. 301, polished	0.37	0.60
White marble	0.46	0.95
Asphalt	0.90	0.90
Brick, red	0.75	0.93
Gravel	0.29	0.85
Flat black lacquer	0.96	0.95
White paints, various types of pigments	0.12–0.16	0.90–0.95

opposed to low-temperature radiation from opaque metal or painted surfaces. In many instances, the total absorptivity for solar radiation can be quite different from the absorptivity for blackbody radiation at some moderate temperature like 25°C. Table 8-3 gives a brief comparison of the absorptivities for some typical surfaces for both solar and low-temperature radiation. As will be noted, rather striking differences can occur.

This brief discussion of solar radiation is not intended to be comprehensive. Rather, it has the purpose of alerting the reader to some of the property information (like that of Ref. 14) when making calculations for solar radiation.

■ **EXAMPLE 8-20**

Calculate the radiation equilibrium temperature for a plate exposed to a solar flux of 700 W/m² and a surrounding temperature of 25°C if the surface is coated with (*a*) white paint or (*b*) flat black lacquer. Neglect convection.

Solution

At radiation equilibrium the net energy absorbed from the sun must equal the long-wavelength radiation exchange with the surroundings, or

$$\left(\frac{q}{A}\right)_{\text{sun}} \alpha_{\text{sun}} = \alpha_{\text{low temp}}\, \sigma(T^4 - T_{\text{surr}}^4) \qquad (a)$$

For white paint we obtain from Table 8-3

$$\alpha_{\text{sun}} = 0.12 \qquad \alpha_{\text{low temp}} = 0.9$$

so that Eq. (*a*) becomes

$$(700)(0.12) = (0.9)(5.669 \times 10^{-8})(T^4 - 298^4)$$

and

$$T = 312.5 \text{ K} = 39.5°\text{C} \quad [103°\text{F}]$$

For flat black lacquer we obtain

$$\alpha_{\text{sun}} = 0.96 \qquad \alpha_{\text{low temp}} = 0.95$$

so that Eq. (*a*) becomes

$$(700)(0.96) = (0.95)(5.669 \times 10^{-8})(T^4 - 298^4)$$

and

$$T = 377.8 \text{ K} = 104.8°\text{C} \quad [220.6°\text{F}]$$

We conclude from this example what we may have known from the start, that white surfaces are cooler than black surfaces in the sunlight.

■ **EXAMPLE 8-21** A flat-plate solar collector

A flat-plate solar collector is constructed as shown in the figure. A glass plate covers the blackened surface, which is insulated. Solar energy at the rate of 750 W/m² is transmitted through the glass and absorbed in the blackened surface. The surface heats up and radiates to the glass and also loses heat by convection across the air gap, which has $k_e/k = 1.4$. The outside surface of the glass loses heat by radiation and convection to the environment at 30°C with $h = 20$ W/m² · °C. It is assumed that the glass does not transmit any of the thermal radiation and has $\epsilon = 0.9$. The blackened surface is assumed to have $\epsilon = 1.0$ for all radiation. Determine the temperatures of the glass and inside surface.

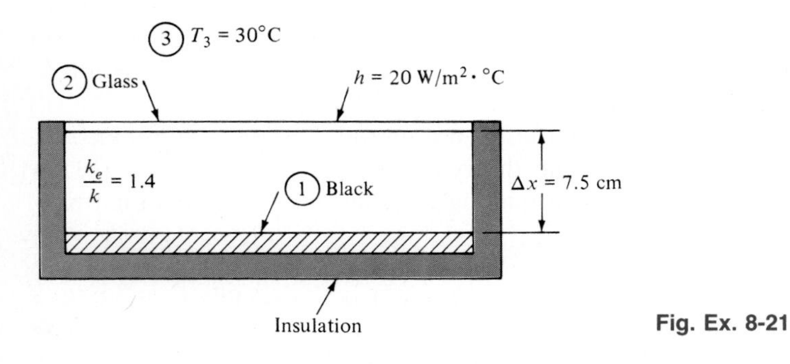

Fig. Ex. 8-21

Solution

This is an interesting example of combined radiation and convection heat-transfer analysis. We designate the black plate as surface 1, the glass as surface 2, and the surroundings as surface 3. We assume no absorption of the solar energy in the glass. For the black plate

$$J_1 = E_{b1}$$

The convection and solar energy delivered to surface 1 is

$$\left.\frac{q}{A}\right]_1 = \frac{k_e}{\Delta x}(T_2 - T_1) + \left.\frac{q}{A}\right]_s \tag{a}$$

If Eq. (8-115) is now applied, we have

$$E_{b1} - F_{12}J_{2i} = \frac{k_e}{\Delta x}(T_2 - T_1) + \left.\frac{q}{A}\right]_s \tag{b}$$

where J_{2i} is the inside radiosity for the glass. The overall energy balance for surface 2 is

$$\frac{\epsilon_2}{1 - \epsilon_2}(2E_{b2} - J_{2i} - J_{2o}) = \frac{k_e}{\Delta x}(T_1 - T_2) + h(T_3 - T_2) \tag{c}$$

where, now, J_{2o} is the outside radiosity of the glass.

For the overall system, the solar energy absorbed must eventually be lost by convection and radiation from the outside surface of the glass. Thus,

$$\left.\frac{q}{A}\right]_s = h(T_2 - T_3) + \epsilon_2(E_{b2} - E_{b3}) \tag{d}$$

Finally, the radiation lost from the outside of the glass can be written two ways:

$$\left.\frac{q}{A}\right]_{\text{rad}} = \epsilon_2(E_{b2} - E_{b3}) = (E_{b2} - J_{2o})\frac{\epsilon_2}{1 - \epsilon_2} \tag{e}$$

The area of the collector is very large compared to the spacing so $F_{12} \approx 1.0$. We now have four equations and four unknowns: E_{b1}, E_{b2}, J_{2i}, and J_{2o}. Of course, T_1 and T_2 are expressed in terms of E_{b1} and E_{b2}. The solution procedure is as follows:

1. Solve Eq. (*d*) for E_{b2} and T_2 by iteration.
2. Solve Eq. (*e*) for J_{2o}.
3. Solve Eq. (*b*) for J_{2i} and substitute in Eq. (*c*) giving (with $F_{12} = 1.0$)

$$\frac{\epsilon_2}{1 - \epsilon_2}\left[2E_{b2} - E_{b1} - \frac{k_e}{\Delta x}(T_1 - T_2) + \left.\frac{q}{A}\right]_s - J_{2o}\right] = \frac{k_e}{\Delta x}(T_1 - T_2) + h(T_3 - T_2) \tag{f}$$

4. Solve Eq. (*f*) by iteration for T_1 and E_{b1}.

When this procedure is followed, with $k = 0.03$ W/m · °C, the results are

$$E_{b2} = 682.96 \text{ W/m}^2 \qquad T_2 = 331.3 \text{ K}$$

$$J_{2o} = 662.45 \text{ W/m}^2 \qquad J_{2i} = 760.94 \text{ W/m}^2$$

$$E_{b1} = 1471.68 \text{ W/m}^2 \qquad T_1 = 401.4 \text{ K} \quad [262.5°\text{F}]$$

and we see that a rather substantial temperature can be achieved inside the collector. This result is for the insulated case. In normal practice energy will be extracted from the hot surface and lower temperatures will be experienced, depending on the rate of energy removal. Extensive information on solar collectors is given in Ref. 37.

8-15 RADIATION PROPERTIES OF THE ENVIRONMENT

We have already described the radiation spectrum of the sun and noted that the major portion of solar energy is concentrated in the short-wavelength region. It was also noted that as a consequence of this spectrum, real surfaces may exhibit substantially different absorption properties for solar radiation than for long-wavelength "earthbound" radiation.

Meteorologists and hydrologists use the term *insolation* to describe the intensity of direct solar radiation incident on a horizontal surface per unit area and per unit time, designated with the symbol I. Although we shall emphasize other units, it will be helpful to mention a unit that appears in the meterological literature:

$$1 \text{ langley (Ly)} = 1 \text{ cal/cm}^2 \quad [3.687 \text{ Btu/ft}^2]$$

Insolation and radiation intensity are frequently expressed in langleys per unit time; e.g., the Stefan-Boltzmann constant would be

$$\sigma = 0.826 \times 10^{-10} \text{ Ly/min} \cdot \text{K}^4$$

Radiation heat transfer in the environment is governed by the absorption, scattering, and reflection properties of the atmosphere and natural surfaces. Two types of scattering phenomena occur in the atmosphere. *Molecular scattering* is observed because of the interaction of radiation with individual molecules. The blue color of the sky results from the scattering of the violet (short) wavelengths by the air molecules. *Particulate scattering* in the atmosphere results from the interaction of radiation with the many types of particles that may be suspended in the air. Dust, smog, and water droplets are all important types of particulate scattering centers. The scattering process is governed mainly by the size of the particle in comparison with the wavelength of radiation. Maximum scattering occurs when wavelength and particle size are equal and decreases progressively for longer wavelengths. For wavelengths smaller than the particle size, the radiation tends to be reflected.

Reflection phenomena in the atmosphere occur for wavelengths less than the particle size and are fairly independent of wavelength in this region. The term *albedo* is used to describe the reflective properties of surfaces and is defined by

$$A = \text{albedo} = \frac{\text{reflected energy}}{\text{incident energy}} \tag{8-117}$$

The albedos of some natural surfaces are given in Table 8-4. The effect of solar incident angle on the albedo of water is shown in Fig. 8-64, where α is the angle which the incoming rays make with the horizontal.

The atmosphere absorbs radiation quite selectively in narrow-wavelength bands. The absorption for solar radiation occurs in entirely different bands from the absorption of the radiation from the earth because of the different spectrums for the two types of radiation. In Fig. 8-65 we see the approximate spectrums for solar and earth radiation with some important absorption bands

Table 8-4 Albedo of Natural Surfaces, According to Ref. 32.

Surface	*Albedo A*	*Surface*	*Albedo A*
Water	0.03–0.40	Spring wheat	0.10–0.25
Black, dry soil	0.14	Winter wheat	0.16–0.23
Black, moist soil	0.08	Winter rye	0.18–0.23
Gray, dry soil	0.25–0.30	High, dense grass	0.18–0.20
Gray, moist soil	0.10–0.12	Green grass	0.26
Blue, dry loam	0.23	Grass dried in sun	0.19
Blue, moist loam	0.16	Tops of oak	0.18
Desert loam	0.29–0.31	Tops of pine	0.14
Yellow sand	0.35	Tops of fir	0.10
White sand	0.34–0.40	Cotton	0.20–0.22
River sand	0.43	Rice field	0.12
Bright, fine sand	0.37	Lettuce	0.22
Rock	0.12–0.15	Beets	0.18
Densely urbanized areas	0.15–0.25	Potatoes	0.19
Snow	0.40–0.85	Heather	0.10
Sea ice	0.36–0.50		

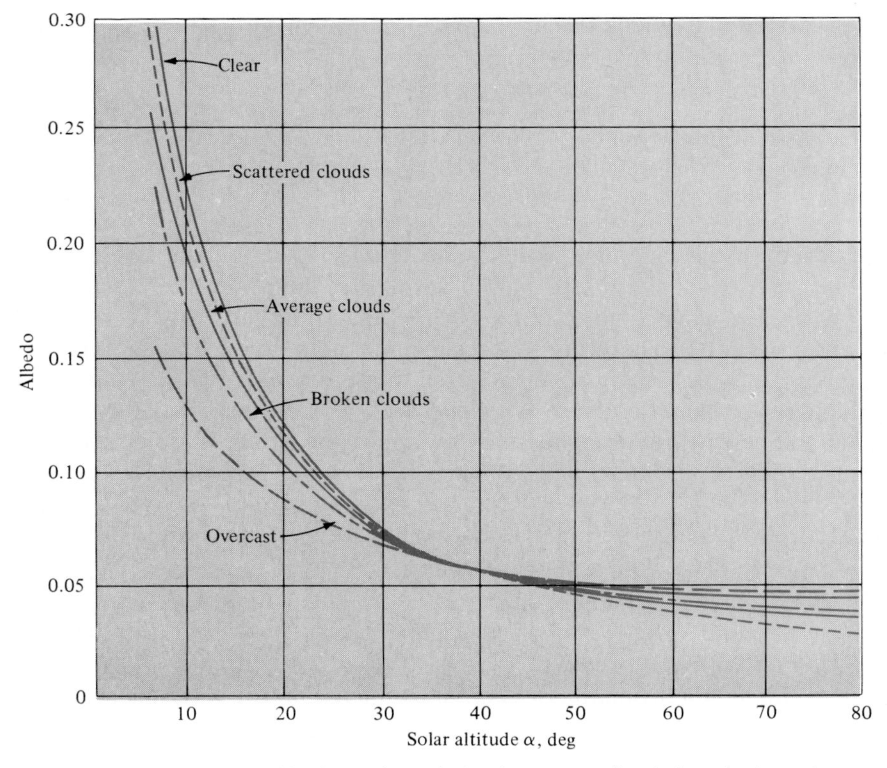

Fig. 8-64 Effect of solar altitude angle and cloud cover on albedo for a horizontal water surface, according to Ref. 33.

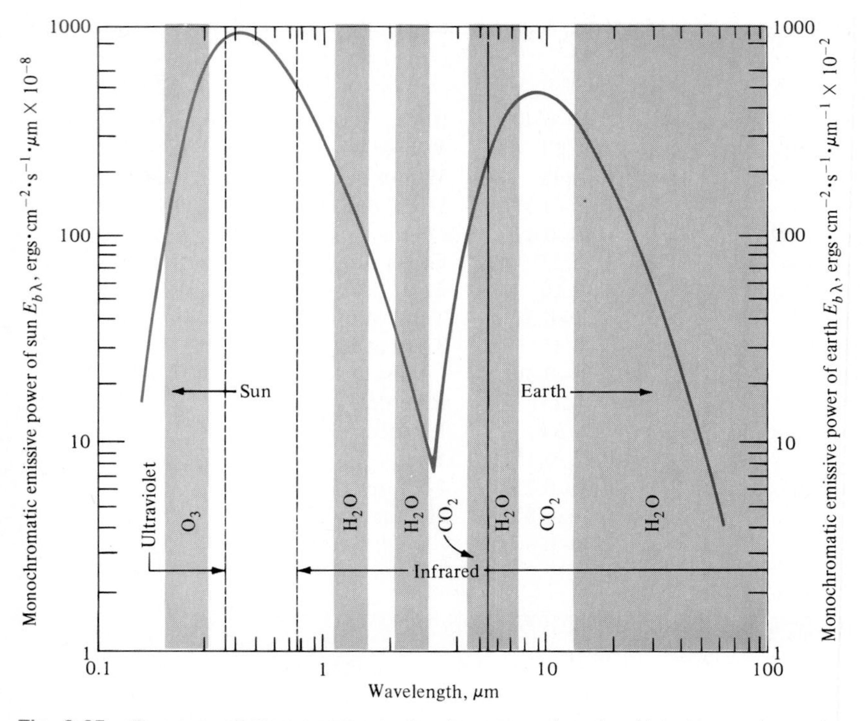

Fig. 8-65 Therma-radiation spectrums for the sun and earth with primary absorption bands indicated by shaded areas. Note different scales; 1 erg/cm^2·s = 1 mW/m^2.

superimposed on the diagram. Note the scale differential on the two curves. A quick inspection of these curves will show that the atmosphere transmits most of the short-wavelength radiation while absorbing most of the back radiation from the earth. Therefore, the atmosphere acts very much like a greenhouse, trapping the incoming solar radiation to provide energy and warmth for humans on earth. Some concern is voiced that people may upset the energy budget of the earth through excessive contamination of the atmosphere with pollutants. Such a possibility does exist but is beyond the scope of our discussion.

The absorption and scattering of radiation may be described with Beer's law [Eq. (8-48)], which we repeat here for convenience

$$\frac{I_{\lambda_x}}{I_{\lambda_o}} = e^{-a_\lambda x} \tag{8-118}$$

where a_λ is the monochromatic absorption coefficient and x is the thickness of the layer absorbing the radiation. For a scattering process, we would replace a_λ by a *scattering coefficient* k_λ.

The conventional approach in atmospheric problems is to assume that the absorption and scattering processes are superimposed on each other and may be expressed in the form of Eq. (8-118) over all wavelengths. The appropriate coefficients are defined as

a_{ms} = average molecular scattering coefficient over all wavelengths

a_{ps} = average particulate scattering coefficient over all wavelengths

a = average absorption coefficient over all wavelengths

$a_t = a_{ms} + a_{ps} + a$ = total attenuation coefficient over all wavelengths

With these coefficients, the radiation insolation at the earth's surface is expressed as

$$\frac{I_c}{I_o} = e^{-a_t m} = e^{-n a_{ms} m} \tag{8-119}$$

where m is the relative thickness of the air mass and n is defined as the *turbidity factor* of the air:

$$n = \frac{a_t}{a_{ms}} \tag{8-120}$$

The insolations are defined as

I_c = direct, cloudless sky insolation at earth's surface

I_o = insolation at outer limits of earth's atmosphere

The molecular scattering coefficient for air at atmospheric pressure is given as [32]

$$a_{ms} = 0.128 - 0.054 \log m \tag{8-121}$$

The relative thickness of the air mass is calculated as the cosecant of the solar altitude α. The turbidity factor is thus a convenient means of specifying atmospheric purity and clarity; its value ranges from about 2.0 for very clear air to 4 or 5 for very smoggy, industrial environments.

The insolation at the outer edge of the atmosphere is expressed in terms of the solar constant E_{b_o} by

$$I_o = E_{b_o} \sin \alpha \tag{8-122}$$

where α is again the angle the rays make with the horizontal. We gave the solar constant as

$$\begin{aligned} E_{b_o} &= 442.4 \text{ Btu/h} \cdot \text{ft}^2 \\ &= 2.00 \text{ cal/cm}^2 \cdot \text{min} \\ &= 1395 \text{ W/m}^2 \end{aligned}$$

An average variation of incident solar radiation for cloudy and cloudless situations as a function of solar altitude angle is given in Table 8-5.

Table 8-5 Solar Irradiation (insolation) on a Horizontal Surface under Average Atmospheric Conditions, According to Ref. 33.

Solar altitude α, deg	*Average total insolation* Ly/h	W/m²
5	3.6	41.9
10	9.7	112.8
15	17.2	200.0
20	25.0	290.7
25	32.8	381.4
30	40.6	472.1
35	47.7	554.6
40	54.7	636.0
45	61.1	710.4
50	67.2	781.4
60	77.5	901.1
70	85.3	991.8
80	89.7	1043
90	91.4	1063

EXAMPLE 8-22

A certain smoggy atmosphere has a turbidity of 4.0. Calculate the direct, cloudless-sky insolation for a solar altitude angle of 75°. How much is this reduced from that of a clear sky?

Solution

We shall be using Eq. (8-119) for this calculation, so that we need to make the following intermediate calculations.

$$I_o = E_{b_o} \sin \alpha = 1395 \sin 75° = 1347 \text{ W/m}^2$$

$$m = \csc 75° = 1.035$$

$$a_{ms} = 0.128 - 0.054 \log m = 0.1272$$

We have $n = 4.0$, so from Eq. (8-119),

$$I_c = 1347 \exp [-(0.1272)(4.0)(1.035)]$$

$$= 795.5 \text{ W/m}^2$$

For a very clear day, $n = 2.0$ and we would have

$$I_c = 1347 \exp [-(0.1272)(2.0)(1.035)]$$

$$= 1035 \text{ W/m}^2$$

Thus the insolation has been reduced by 23 percent as a result of the smoggy environment.

The absorption of radiation in natural bodies of water is an important process because of its influence on evaporation rates and the eventual dispersion of water vapor in the atmosphere. Experimental measurements [34] show that solar radiation is absorbed very rapidly in the top layers of the water followed by an approximately exponential decay with depth in the water. The incident radiation follows a variation of

$$\frac{I_z}{I_s} = (1 - \beta)e^{-az} \tag{8-123}$$

where I_s is the intensity at the surface and I_z is the intensity at a depth z; β represents the fraction of energy absorbed at the surface, and a is an absorption of *extinction coefficient* for the material. Beta may be interpreted as a measure of the long-wavelength content of solar radiation, since the shorter wavelengths penetrate the water more readily. This coefficient appears to have the value 0.4 for all lakes for which data are available, and it is assumed to be independent of time. Values of the extinction coefficient can vary considerably, e.g., it has the value of 0.16 m^{-1} for the extremely clear water of Lake Tahoe [35] and a value of 0.89 m^{-1} for the more turbid water of Lake Castle, California [36].

Additional information on radiation in the atmosphere is given by Kondratyev [30] and Geiger [31].

■ EXAMPLE 8-23

Calculate the heat-generation rate resulting from solar-radiation absorption in a lake with an extinction coefficient of 0.328 m^{-1} and a solar altitude of 90° on a clear day. Perform the calculation for a depth of 1 ft [0.3048 m].

Solution

The heat-generation rate is obtained by differentiating Eq. (8-123):

$$\dot{q} = \frac{q_{\text{absorbed}}}{A\,dz} = \frac{dI_z}{dz} = I_s a(1 - \beta)e^{-az}$$

The surface insolation I_s is calculated with Eq. (8-119):

$$I_o = E_{b_o} \sin \alpha = 1395 \text{ W/m}^2$$

$$m = \csc 90° = 1.0$$

$$a_{ms} = 0.128 - 0.054 \log m = 0.128$$

$$n = 2.0$$

$$I_c = I_s = 1395 \exp[-(0.128)(2.0)(1.0)]$$

$$= 1080 \text{ W/m}^2$$

We also have $a = 0.328$ and $\beta = 0.4$, so that

$$\dot{q} = (1080)(0.328)(1 - 0.4) \exp[-(0.328)(0.3048)]$$

$$= 192.3 \text{ W/m}^3$$

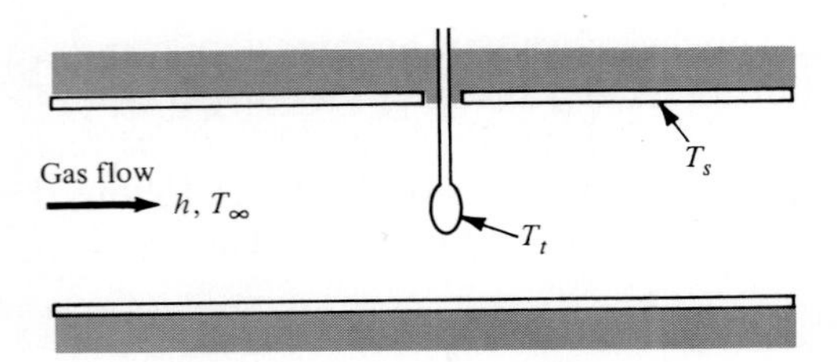

Fig. 8-66 Thermometer element in flow stream.

8-16 EFFECT OF RADIATION ON TEMPERATURE MEASUREMENT

When a thermometer is placed in a gas-flow stream to measure temperature, the temperature indicated by the sensing element is determined by the overall energy balance on the element. Consider the element shown in Fig. 8-66. The temperature of the gas is T_∞, the effective radiation surrounding temperature is T_s, and the temperature indicated by the thermometer is T_t. Assuming that T_∞ is greater than T_s, energy will be transferred by convection to the thermometer and then dissipated by radiation to the surroundings. Thus the energy balance becomes

$$hA(T_\infty - T_t) = \sigma A\epsilon(T_t^4 - T_s^4) \tag{8-124}$$

where A is the surface area of the element and ϵ is its emissivity. Equation (8-124) assumes that the surroundings are either very large or black so that Eq. (8-43a) can be applied for the radiation heat transfer.

From this energy balance we see that the temperature indicated by the thermometer is *not* the true gas temperature but some radiation-convection equilibrium temperature. Very large errors can result in temperature measurements if this energy balance is not properly taken into account. Radiation shields are frequently employed to alleviate this difficulty.

EXAMPLE 8-24

A mercury-in-glass thermometer having $\epsilon = 0.9$ hangs in a metal building and indicates a temperature of 20°C. The walls of the building are poorly insulated and have a temperature of 5°C. The value of h for the thermometer may be taken as 8.3 W/m² · °C. Calculate the true air temperature.

Solution

We employ Eq. (8-124) for the solution:

$$h(T_\infty - T_t) = \sigma\epsilon(T_t^4 - T_s^4)$$

Inserting the numerical values, with $T_t = 20°\text{C} = 293$ K, $T_s = 5°\text{C} = 278$ K, gives

$$(8.3)(T_\infty - 293) = (5.669 \times 10^{-8})(0.9)(293^4 - 278^4)$$

and

$$T_\infty = 301.6 \text{ K} = 28.6°\text{C} \quad [83.5°\text{F}]$$

In this simple example the thermometer is in error by 8.6°C [15.5°F]!

8-17 THE RADIATION HEAT-TRANSFER COEFFICIENT

In the development of convection heat transfer in the preceding chapters, we found it convenient to define a heat-transfer coefficient by

$$q_{\text{conv}} = h_{\text{conv}} A(T_w - T_\infty)$$

Since radiation heat-transfer problems are often very closely associated with convection problems, and the total heat transfer by both convection and radiation is often the objective of an analysis, it is worthwhile to put both processes on a common basis by defining a radiation heat-transfer coefficient h_r as

$$q_{\text{rad}} = h_r A_1(T_1 - T_2)$$

where T_1 and T_2 are the temperatures of the two bodies exchanging heat by radiation. The total heat transfer is then the sum of the convection and radiation,

$$q = (h_c + h_r)A_1(T_w - T_\infty) \tag{8-125}$$

if we assume that the second-radiation-exchange surface is an enclosure and is at the same temperature as the fluid. For example, the heat loss by free convection and radiation from a hot steam pipe passing through a room could be calculated from Eq. (8-125).

In many instances the convection heat-transfer coefficient is not strongly dependent on temperature. However, this is not so with the radiation heat-transfer coefficient. The value of h_r, corresponding to Eq. (8-43), could be calculated from

$$\frac{q}{A_1} = \frac{\sigma(T_1^4 - T_2^4)}{1/\epsilon_1 + (A_1/A_2)(1/\epsilon_2 - 1)} = h_r(T_1 - T_2)$$

$$h_r = \frac{\sigma(T_1^2 + T_2^2)(T_1 + T_2)}{1/\epsilon_1 + (A_1/A_2)(1/\epsilon_2 - 1)} \tag{8-126}$$

Obviously, the radiation coefficient is a very strong function of temperature.

The reader may recall that we used a concept like Eq. (8-126) to obtain a "radiation resistance" for numerical examples in Chaps. 3 and 4.

8-18 SUMMARY

In this chapter we have examined several means for analyzing radiation heat transfer. The gray-body assumption, although not strictly correct, is a viable method for performing heat-transfer calculations. Assumptions of uniform radiosity and irradiation over surfaces are also not strictly correct but provide an approximation which is usually well within the accuracy of knowledge of surface properties. In Table 8-6, we present a tabular summary of a few formulas which are often used.

Table 8-6 Summary of Radiation Formulas

Type	*Equation*	*Equation number*
Energy emitted by a black body	$E_b = \sigma T^4 \quad \text{W/m}^2$	(8-3)
Basic radiation shape factor reciprocity relation	$A_m F_{mn} = A_n F_{nm}$	(8-18*a*)
Net energy lost by a gray surface	$q = \dfrac{E_b - J}{(1 - \epsilon)/\epsilon A} \quad \text{W}$	(8-38)
Net radiant exchange between convex surface 1 and large enclosure 2	$q = \sigma A_1 \epsilon_1 (T_1^4 - T_2^4) \quad \text{W}$	(8-43*a*)
Radiation-balance equation for constant surface temperature of *i*th surface	$J_i[1 - F_{ii}(1 - \epsilon_i)] - (1 - \epsilon_i) \sum_{j \neq i} F_{ij} J_j = \epsilon_i E_{bi}$	(8-112*a*)
Radiation-balance equation for surface in radiant balance; i.e., $q/A = 0$	$J_i(1 - F_{ii}) - \sum_{j \neq i} F_{ij} J_j = 0$	(8-113*a*)
Radiation-balance equation for surface with specified heat flux	$J_i(1 - F_{ii}) - \sum_{j \neq i} F_{ij} J_j = \dfrac{q_i}{A_i}$	(8-115*a*)

REVIEW QUESTIONS

1. How does thermal radiation differ from other types of electromagnetic radiation?
2. What is the Stefan-Boltzmann law?
3. Distinguish between specular and diffuse surfaces.
4. Define radiation intensity.
5. What is Kirchhoff's identity? When does it apply?
6. What is a gray body?
7. What is meant by the radiation shape factor?
8. Define irradiation and radiosity.
9. What is Beer's law?
10. Why do surfaces absorb differently for solar or earthbound radiation?
11. Explain the greenhouse effect.
12. Why is the sky blue?
13. Define albedo.
14. What is meant by the atmospheric greenhouse effect?
15. What is meant by the turbidity factor?

PROBLEMS

8-1 Fused quartz transmits 90 percent of the incident thermal radiation between 0.2 and 4 μm. Suppose a certain heat source is viewed through a quartz window. What heat flux in watts will be transmitted through the material from blackbody radiation sources at (*a*) 800°C, (*b*) 550°C, (*c*) 250°C, and (*d*) 70°C?

8-2 Repeat Prob. 8-1 for synthetic sapphire, which has a transmissivity of 0.85 between 0.2 and 5.5 μm.

8-3 Repeat Prob. 8-1 for cesium iodide, which has a transmissivity of approximately 0.92 between 0.3 and 52 μm.

8-4 Calculate the energy emitted between 4 and 15 μm by a gray body at 100°F with $\epsilon = 0.6$.

8-5 A furnace with black interior walls maintained at 1100°C has an opening in the side covered with a glass window having the following properties:

$$0 < \lambda < 3\ \mu\text{m}: \qquad \tau = 0.8 \qquad \epsilon = 0.2 \qquad \rho = 0$$

$$3\ \mu\text{m} < \lambda < \infty: \qquad \tau = 0 \qquad \epsilon = 0.8 \qquad \rho = 0.2$$

Assume diffuse behavior and calculate the radiation lost through the window to a large room at 25°C.

8-6 A certain surface has the following absorption properties:

$$\alpha_\lambda = 0.05 \qquad 0 < \lambda < 1.2\ \mu\text{m}$$

$$\alpha_\lambda = 0.5 \qquad 1.2 < \lambda < 3\ \mu\text{m}$$

$$\alpha_\lambda = 0.4 \qquad 3 < \lambda < 6\ \mu\text{m}$$

$$\alpha_\lambda = 0.2 \qquad 6 < \lambda < 20\ \mu\text{m}$$

$$\alpha_\lambda = 0 \qquad 20 < \lambda < \infty\ \mu\text{m}$$

Calculate the total absorptivity of the surface if it is irradiated with blackbody radiation at (*a*) 300 K, (*b*) 500 K, (*c*) 1000 K, (*d*) 2000 K, (*e*) 5000 K.

8-7 Assuming solar radiation is like a blackbody at 5795 K, calculate the fraction of energy in the following wavelength bands: (*a*) 0 to 0.2 μm, (*b*) 0.2 to 0.4 μm, (*c*) 0.4 to 1.0 μm, (*d*) 1.0 to 2.0 μm, (*e*) over 2.0 μm.

8-8 Color photographic films are designed for particular wavelength sensitivities such as "daylight" and "indoor incandescent lighting." An incandescent light bulb radiates approximately as a blackbody at 3200 K while daylight is approximately like a blackbody at 5800 K. Make some calculations and comment on the results which might be obtained if a film is used for other than its intended light source, viz., daylight film with indoor lighting.

8-9 A black surface is at 800°C. Calculate the fraction of the total energy emitted between (*a*) 1 and 2 μm, (*b*) 2 and 3 μm, (*c*) 3 and 4 μm, (*d*) 5 and 6 μm.

8-10 A black radiation source is at 1100°C. Calculate the upper wavelength in micrometers for emissions of (*a*) 25, (*b*) 50, (*c*) 75, and (*d*) 98 percent of the total radiation.

8-11 Find the radiation shape factors F_{1-2} for the situations shown.

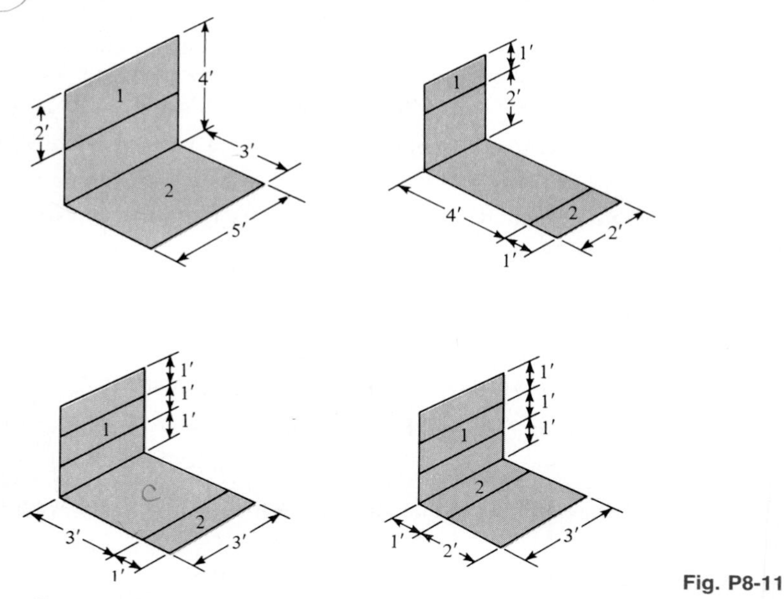

Fig. P8-11

8-12 Two parallel concentric disks have $d_1 = 10$ cm, $d_2 = 5$ cm and are spaced 10 cm apart. Determine F_{12} and F_{21}.

8-13 A 10-cm-diameter disk is placed at the center and parallel to a 25-cm-diameter hemisphere. Designating the sphere as surface 1, the disk as surface 2, and a large surrounding room as surface 3, calculate all the shape factors.

8-14 Two concentric cylinders have diameters of 15 and 25 cm and a length of 7 cm. Calculate the shape factor between the open ends.

8-15 Two parallel square plates 30 cm on a side are separated by a distance of 10 cm but the edge of the lower plate lies halfway between the edges of the top plate. Calculate the shape factor for this geometry.

8-16 Two parallel disks having diameters of 50 cm are separated by a distance of 10 cm. One disk also has a 20-cm-diameter hole cut in the center. Find the shape factor from this disk to the one without the hole.

8-17 A room 3 by 3 by 3 m has one side wall maintained at 260°C; the floor is maintained at 90°C. The other four surfaces are perfectly insulated. Assume that all surfaces are black. Calculate the net heat transfer between the hot wall and the cool floor.

8-18 Two perfectly black parallel planes 1.2 by 1.2 m are separated by a distance of 1.2 m. One plane is maintained at 800 K and the other at 500 K. The planes are located in a large room whose walls are at 300 K. What is the net heat transfer between the planes?

8-19 Two parallel disks, 60 cm in diameter, are separated by a distance of 15 cm and completely enclosed by a large room at 30°C. The properties of the surfaces are $T_1 = 540$°C, $\epsilon_1 = 0.7$, $T_2 = 300$°C, $\epsilon_2 = 0.5$. What is the net radiant heat transfer with each surface? (Do not include back side exchange, only that from the surfaces facing each other.)

8-20 Rework Prob. 8-19 for the case where disk 2 (300°C, $\epsilon_2 = 0.5$) is perfectly specular-reflecting.

8-21 A square room 3 by 3 m has a floor heated to 300 K, a ceiling at 290 K, and walls that are assumed perfectly insulated. The height of the room is 2.5 m. The emissivity of all surfaces is 0.8. Using the network method, find the net interchange between floor and ceiling and the wall temperature.

8-22 It is desired to transmit energy from one spaceship to another. A 1.5-m-square plate is available on each ship to accomplish this. The ships are guided so that the plates are parallel and 30 cm apart. One plate is maintained at 800°C and the other at 280°C. The emissivities are 0.5 and 0.8, respectively. Find (*a*) the net heat transferred between the spaceships in watts and (*b*) the total heat lost by the hot plate in watts. Assume that outer space is a blackbody at 0 K.

8-23 Two parallel planes 90 by 60 cm are separated by a distance of 60 cm. One plane is maintained at a temperature of 800 K and has an emissivity of 0.6. The other plane is insulated. The planes are placed in a large room which is maintained at 290 K. Calculate the temperature of the insulated plane and the energy lost by the heated plane.

8-24 A long pipe 5 cm in diameter passes through a room and is exposed to air at atmospheric pressure and temperature of 20°C. The pipe surface temperature is 93°C. Assuming that the emissivity of the pipe is 0.6, calculate the radiation heat loss per foot of length of pipe.

8-25 A vertical plate 60 cm high and 30 cm wide is maintained at a temperature of 95°C in a room where the air is 20°C and 1 atm. The walls of the rooms are also at 20°C. Assume that $\epsilon = 0.8$ for the plate. How much radiant heat is lost by the plate?

8-26 A horizontal pipe 6 m long and 12.5 cm in diameter is maintained at a temperature of 150°C in a large room where the air is 20°C and 1 atm. The walls of the room are at 38°C. Assume that $\epsilon = 0.7$ for the pipe. How much heat is lost by the pipe through both convection and radiation?

8-27 An oven with a radiant heater for drying painted metal parts on a moving conveyor belt is designed as shown. The length of the heating section is 3 m, and

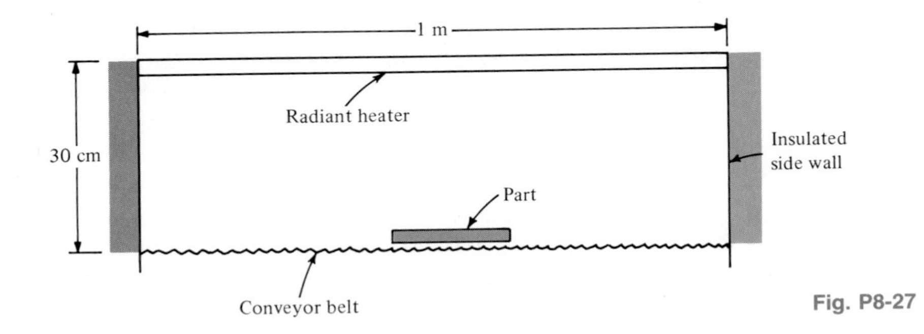

Fig. P8-27

the heater temperature is 425°C. The side walls are insulating, and it is determined experimentally that the conveyor belt and parts attain a temperature of 120°C. The belt-part combination has an effective emissivity of 0.8, and the radiant-heater surface has $\epsilon = 0.7$. Calculate the energy supplied to the heater. Be sure to consider the radiation that is lost from the ends of the channel. Take the surroundings as a blackbody at 25°C.

8-28 The top surface of the truncated cone in Example 8-4 has $T_1 = 1200$ K and $\epsilon_1 = 0.75$. The side surface is insulated and has $\epsilon = 0.3$. The bottom is open and exposed to a large room at 20°C. Calculate the temperature of the side surface and the heat lost to the room.

8-29 Two concentric cylinders have lengths of 30 cm. The inner cylinder has a diameter of 8.0 cm. What must the outer-cylinder diameter be such that $F_{12} = 0.8$ where the inner cylinder is considered as surface 1?

8-30 The bottom surface of a cylindrical furnace has an electric heating element which produces a constant heat flux of 7 kW/m^2 over a diameter of 75 cm. The side walls are 50 cm high and may be considered as nonconducting and re-radiating. The bottom has $\epsilon = 0.8$. Over the top is placed a surface with $\epsilon = 0.6$ which is maintained at 400 K. Determine the temperature of the bottom and side surfaces.

8-31 A long rod heater with $\epsilon = 0.8$ is maintained at 980°C and is placed near a half-cylinder reflector as shown. The diameter of the rod is 7.5 cm, and the diameter of the reflector is 50 cm. The reflector is insulated, and the combined heater reflector is placed in a large room whose walls are maintained at 15°C. Calculate the radiant heat loss per unit length of the heater rod. How does this compare with the energy which would be radiated by the rod if it were used without the reflector?

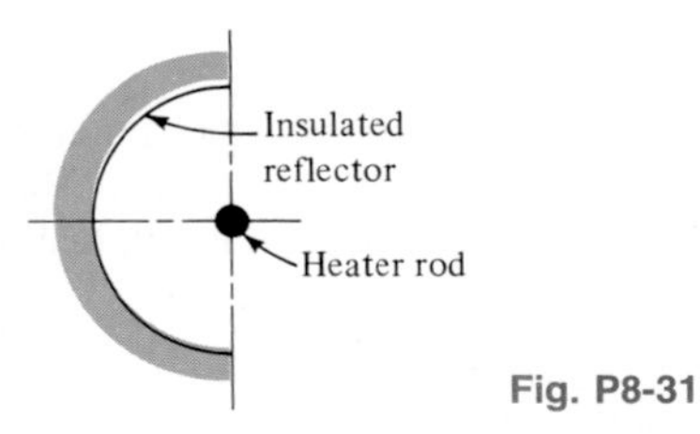

Fig. P8-31

8-32 A conical hole is machined in a block of metal whose emissivity is 0.5. The hole is 2.5 cm in diameter at the surface and 5 cm deep. If the metal block is heated to 800 K, calculate the radiation emitted by the hole. Calculate the value of an *apparent* emissivity of the hole, defined as the ratio of the actual energy emitted by the hole to that energy which would be emitted by a black surface having an area equal to that of the opening and a temperature equal to that of the inside surfaces.

8-33 A hole 2.5 cm in diameter is drilled in a 7.5-cm metal plate which is maintained at 550 K. The hole is lined with a thin foil having an emissivity of 0.07. A heated surface at 700 K having an emissivity of 0.5 is placed over the hole on one side of the plate, and the hole is left open on the other side of the plate. The 700° K surface is insulated from the plate insofar as conduction is concerned. Calculate the energy emitted from the open hole.

8-34 A cylindrical hole of depth x and diameter d is drilled in a block of metal having an emissivity of ϵ. Using the definition given in Prob. 8-32, plot the apparent emissivity of the hole as a function of x/d and ϵ.

8-35 A cylindrical hole of diameter d is drilled in a metal plate of thickness x. Assuming that the radiation emitted through the hole on each side of the plate is due only to the temperature of the plate, plot the apparent emissivity of the hole as a function of x/d and the emissivity of the plate material ϵ.

8-36 A 1-m-diameter cylinder, 1 m long, is maintained at 800 K and has an emissivity of 0.65. Another cylinder, 2 m in diameter and 1 m long, encloses the first cylinder and is perfectly insulated. Both cylinders are placed in a large room maintained at 300 K. Calculate the heat lost by the inner cylinder.

8-37 A heated plate with $T = 700°C$ and emissivity of 0.8 is placed as shown. The plate is 2 by 3 m and 2-m-high walls are placed on each side. Each of these walls is insulated. The whole assembly is placed in a large room at 30°C. Draw the network for this problem assuming that the four walls act as one surface (insulated). Then calculate the heat transfer to the large room.

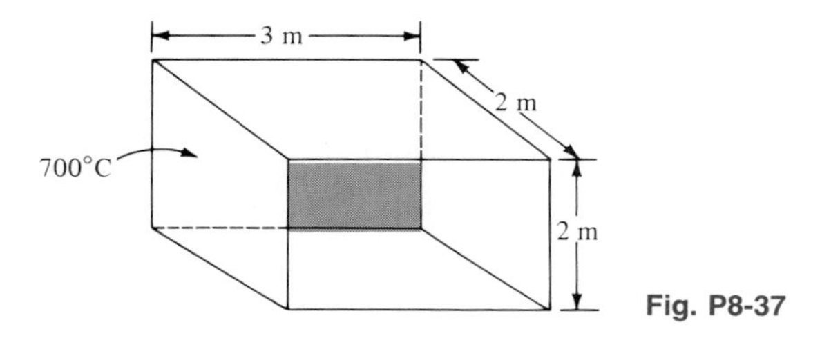

Fig. P8-37

8-38 A 3.0-cm-diameter sphere is maintained at 900°C and has an emissivity of 0.6. It is enclosed by another sphere having a diameter of 9.0 cm with an emissivity of 0.3 (inside and outside). Both spheres are enclosed in a large room at 30°C. Calculate the heat lost from the small sphere.

8-39 Two 60 by 60 cm plates are perpendicular with a common edge and are placed in a large room at 30°C. One plate has $T = 600°C$ and an emissivity of 0.65. The other plate is insulated with an emissivity of 0.45. Calculate the temperature of the insulated plate and the heat lost by the 600°C plate.

8-40 Two concentric cylinders having diameters of 10 cm and 20 cm are placed in a large room maintained at 30°C. The length of the cylinders is 10 cm and the inner cylinder is maintained at 700°C with an emissivity of 0.6. The outer cylinder is perfectly insulated and has an emissivity of 0.7. Calculate the heat lost by the inner cylinder.

8-41 A 5-m-square room has a ceiling maintained at 28°C and a floor maintained at 20°C. The connecting walls are 4 m high and perfectly insulated. Emissivity of the ceiling is 0.62 and that of the floor is 0.75. Calculate the heat transfer from ceiling to floor, and the temperature of the connecting walls.

8-42 The inside temperature of a half cylinder is maintained at 1000 K and is enclosed by a large room. The surface emissivity is 0.7. The diameter is 50 cm and the length may be assumed to be very long. The back side of the cylinder is insulated. Calculate the heat loss per unit length if the room temperature is 300 K.

8-43 Two parallel disks having diameters of 50 cm are separated by a distance of 12.5 cm and placed in a large room at 300 K. One disk is at 1000 K and the other is maintained at 500 K. Both have emissivities of 0.8. Calculate the heat-transfer rate for each disk.

8-44 A 50-cm-disk is maintained at a temperature of 1000°C with an emissivity of 0.55. Extending from the disk is a radiation shield having an emissivity of 0.1 as shown in the figure. The arrangement is placed in a large room maintained at 30°C. Calculate the heat lost by the disk and the temperature of the shield.

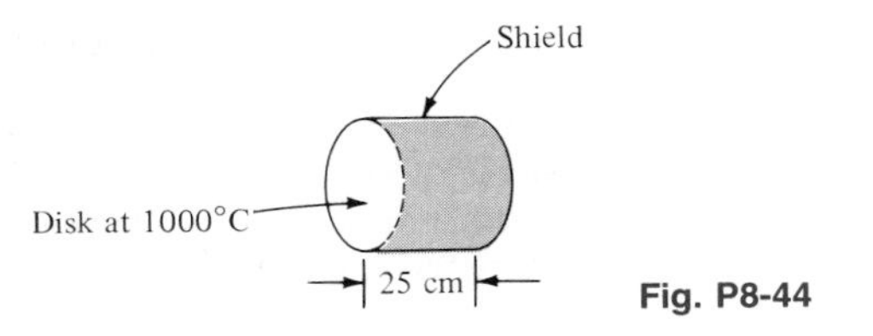

Fig. P8-44

8-45 Two concentric cylinders have properties of $d_1 = 20$ cm, $\epsilon_1 = 0.6$, $T_1 = 500°C$ and $d_2 = 40$ cm, $\epsilon_2 = 0.7$, $T_2 = 100°C$. Both are enclosed in a large room at 20°C. Calculate the length of the cylinder that would enable one to use Eq. (8-43) for calculating the heat transfer with no more than a 10 percent error.

8-46 A 30 by 30 cm plate whose emissivity is 0.5 is attached to the side of a spaceship so that it is perfectly insulated from the inside of the ship. Assuming that outer space is a blackbody at 0 K, determine the equilibrium temperature for the plate at a point in space where the radiant heat flux from the sun is 1500 W/m^2. Assume gray-body behavior.

8-47 An artificial satellite 1 m in diameter circles the earth at an altitude of 400 km. Assuming that the diameter of the earth is 12.9 Mm and the outer surface of the satellite is polished aluminum, calculate the radiation equilibrium temperature of the satellite when it is on the "dark" side of the earth. Take the earth as a blackbody at 15°C and outer space as a blackbody at 0 K. The geometric shape factor from the satellite to the earth may be taken as the ratio of the solid angle subtended by the earth to the total solid angle for radiation from the satellite. When the satellite is on the "bright" side of the earth, it is irradiated with a heat flux of approximately 1400 W/m^2 from the sun. Recalculate the equilibrium temperature of the satellite under these conditions, assuming gray-body behavior. Assume that the satellite receives radiation from the sun as a disk and radiates to space as a sphere.

8-48 Two 30 by 30 cm vertical plates are separated by a distance of 10 cm and placed in room air at 20°C. One plate is maintained at 150°C while the other plate attains a temperature in accordance with its radiant and convection energy exchange with the 150°C plate and the surroundings. Both plates have $\epsilon = 0.8$. Using the approximate free-convection relations of Chap. 7, calculate the temperature of the other plate.

8-49 A heated rod protrudes from a spaceship. The rod loses heat to outer space by radiation. Assuming that the emissivity of the rod is ϵ and that none of the radiation leaving the rod is reflected, set up the differential equation for the temperature distribution in the rod. Also set up the boundary conditions which the differential equation must satisfy. The length of the rod is L, its cross-sectional area is A, its perimeter is P, and its base temperature is T_0. Assume that outer space is a blackbody at 0 K.

8-50 Three infinite parallel plates are arranged as shown. Plate 1 is maintained at 1200 K, and plate 3 is maintained at 300 K; $\epsilon_1 = 0.2$, $\epsilon_2 = 0.5$, and $\epsilon_3 = 0.8$. Plate 2 receives no heat from external sources. What is the temperature of plate 2?

1 2 3 **Fig. P8-50**

8-51 Two large parallel planes having emissivities of 0.3 and 0.5 are maintained at temperatures of 800 K and 400 K, respectively. A radiation shield having an emissivity of 0.05 on both sides is placed between the two planes. Calculate (*a*) the heat-transfer rate per unit area if the shield were not present, (*b*) the heat-transfer rate per unit area with the shield present, and (*c*) the temperature of the shield.

8-52 Two parallel planes 1.2 by 1.2 m are separated by a distance of 1.2 m. The emissivities of the planes are 0.4 and 0.6, and the temperatures are 760 and 300°C, respectively. A 1.2 by 1.2 m radiation shield having an emissivity of 0.05 on both sides is located equidistant between the two planes. The combined arrangement is placed in a large room which is maintained at 40°C. Calculate (*a*) the heat-transfer rate from each of the two planes if the shield were not present, (*b*) the heat-transfer rate from each of the two planes with the shield present, (*c*) the temperature of the shield.

8-53 A long cylindrical heater 2.5 cm in diameter is maintained at 650°C and has a surface emissivity of 0.8. The heater is located in a large room whose walls are at 25°C. How much will the radiant heat transfer from the heater be reduced if it is surrounded by a 30-cm-diameter radiation shield of aluminum having an emissivity of 0.2? What is the temperature of the shield?

8-54 Two long concentric cylinders have diameters of 4 and 8 cm, respectively. The inside cylinder is at 800°C and the outer cylinder is at 100°C. The inside and outside emissivities are 0.8 and 0.4, respectively. Calculate the percent reduction in heat transfer if a cylindrical radiation shield having a diameter of 6 cm and emissivity of 0.3 is placed between the two cylinders.

8-55 Two finite-length concentric cylinders are placed in a large room maintained at 20°C. The inner cylinder has a diameter of 5.0 cm and the outer cylinder has a diameter of 10 cm. The length of the cylinders is 10 cm. The inner cylinder is newly turned cast iron and maintained at a temperature of 400°C. The outer cylinder is Monel metal oxidized at 1110°F. Calculate the heat lost by the inner cylinder.

8-56 An annular space is filled with a gas whose emissivity and transmissivity are 0.3 and 0.7, respectively. The inside and outside diameters of the annular space are 30 and 60 cm, and the emissivities of the surface are 0.5 and 0.3, respectively. The inside surface is maintained at 760°C, while the outside surface is maintained at 370°C. Calculate the net heat transfer per unit length from the hot surface to the cooler surface. What is the temperature of the gas? Neglect convection heat transfer.

8-57 For the conditions in Prob. 8-56, plot the net heat transfer per unit length of the annulus as a function of the gas emissivity, assuming that $\epsilon_m + \tau_m = 1$.

8-58 Repeat Prob. 8-56 for two infinite parallel planes with the same temperatures and emissivities. Calculate the heat-transfer rates per unit area of the parallel planes.

8-59 The gas of Prob. 8-56 is forced through the annular space at a velocity of 6.0 m/s and is maintained at a temperature of 1100°C. The properties of the gas are

$$\rho = 1.6 \text{ kg/m}^3 \qquad c_p = 1.67 \text{ kJ/kg} \cdot {}^\circ\text{C}$$

$$\mu = 5.4 \times 10^{-5} \text{ kg/m} \cdot \text{s} \qquad k = 0.11 \text{ W/m} \cdot {}^\circ\text{C}$$

Assuming the same temperatures and emissivities of the surfaces as in Prob. 8-56, estimate the heating or cooling required for the inner and outer surfaces to maintain them at these temperatures. Assume that the convection heat-transfer coefficient may be estimated with the Dittus-Boelter equation (6-4).

8-60 Two long concentric cylinders have $T_1 = 900$ K, $\epsilon_1 = 0.4$, $d_1 = 5$ cm, and $T_2 = 400$ K, $\epsilon_2 = 0.6$, and $d_2 = 10$ cm. They are separated by a gray gas having $\epsilon_g = 0.15$, $\tau_g = 0.85$. Calculate the heat-transfer rate between the two cylinders and the gas temperature using a radiation-network approach.

8-61 A cubical furnace has interior walls that are black and measure 70 cm on a side. The gas inside the furnace is 15 percent CO_2 by volume and 85 percent N_2 at a total pressure of 1 atm. The gas temperature is 1600 K, and the walls of the furnace are to be maintained at 250°C by a suitable cooling process. How much cooling is required?

8-62 Two parallel black plates are separated by a distance of 50 cm and maintained at temperatures of 250 and 600°C. Between the planes is a gas mixture of 15 percent CO_2, 20 percent water vapor, and 65 percent N_2 by volume at a total pressure of 2.5 atm. The gas temperature is 1400 K. Calculate the heat exchange with each plate per unit surface area. What would the heat transfer be if the gas were not present?

8-63 A cylindrical furnace has a height and diameter of 80 cm and a surface emissivity of 0.85. The gas inside the furnace is 10 percent CO_2, 20 percent water vapor, and 70 percent N_2 by volume and is at a temperature of 1500 K. The pressure is 1 atm. Calculate the cooling required to maintain the furnace walls at 350°C.

8-64 A gas mixture at 3 atm and 1600 K contains 17 percent CO_2, 22 percent water vapor, and 61 percent N_2 by volume and is enclosed between two black planes at 100 and 500°C. The planes are separated by a distance of 90 cm. Calculate the radiant heat exchange with each plate. What would the heat transfer be if the gas were not present?

8-65 Two gray plates having $\epsilon = 0.8$ are separated by a distance of 6 cm and CO_2 at 1 atm. The plates are maintained at 900 and 600 K and the CO_2 may be assumed gray. Calculate the heat transfer between the plates with and without the CO_2 present.

8-66 A mixture of 20 percent CO_2 and 80 percent N_2 by volume is contained in a spherical enclosure at 1 atm and 1200 K. Calculate the gas emissivity for a diameter of 1.5 m.

8-67 A mixture of air and water vapor is contained in a spherical enclosure at 2 atm and 900 K. The partial pressure of the water vapor is 0.15 atm. Calculate the gas emissivity for a diameter of 1.0 m.

8-68 A mixture of 40 percent CO_2 and 60 percent H_2O by volume is contained in a cubical enclosure 1 m on a side at 1 atm and 800 K. The walls of the enclosure are maintained at 400 K and have $\epsilon = 0.8$. Calculate the cooling required to maintain the walls at 400 K.

8-69 Two parallel disks 10 cm in diameter are separated by a distance of 2.5 cm. One disk is maintained at 540°C, is completely diffuse-reflecting, and has an emissivity of 0.3. The other disk is maintained at 260°C but is a specular-diffuse reflector such that $\rho_D = 0.2$, $\rho_s = 0.4$. The surroundings are maintained at 20°C. Calculate the heat lost by the inside surface of each disk.

8-70 Rework Prob. 8-23, assuming that the 550°C plane reflects in only a specular manner. The insulated plane is diffuse.

8-71 Draw the radiation network for a specular-diffuse surface losing heat to a large enclosure. Obtain an expression for the heat transfer under these circumstances. How does this heat transfer compare with that which would be lost by a completely diffuse surface with the same emissivity?

8-72 A 30 by 60 cm plate with $\epsilon = 0.6$ is placed in a large room and heated to 370°C. Only one side of the plate exchanges heat with the room. A highly reflecting plate ($\rho_s = 0.7$, $\rho_D = 0.1$) of the same size is placed perpendicular to the heated plate with the room. The room temperature is 90°C. Calculate the energy lost by the hot plate both with and without the reflector. What is the temperature of the reflector? Neglect convection.

8-73 A 5 by 5 by 2.5 cm cavity is constructed of stainless steel ($\epsilon = 0.6$) and heated to 260°C. Over the top is placed a special ground-glass window ($\rho_s = 0.1$, $\rho_D = 0.1$, $\tau_D = 0.3$, $\tau_s = 0.3$, $\epsilon = 0.2$) 5 by 5 cm. Calculate the heat lost to a very large room at 20°C, and compare with the energy which would be lost to the room if the glass window were not in place.

8-74 Repeat Prob. 8-73 for the case of a window which is all diffuse-reflecting and all specular-transmitting, that is, $\rho_D = 0.2$, $\tau_s = 0.6$, $\epsilon = 0.2$.

8-75 The cavity of Prob. 8-73 has a fused-quartz window placed over it, and the cavity is assumed to be perfectly insulated with respect to conduction and convection loss to the surroundings. The cavity is exposed to a solar irradiation flux of 900 W/m^2. Assuming that the quartz is nonreflecting and $\tau = 0.9$, calculate the equilibrium temperature of the inside surface of the cavity. Recall that the transmission range for quartz is 0.2 to 4 μm. Neglect convection loss from the window. The surroundings may be assumed to be at 20°C.

8-76 A circular cavity, 5 cm in diameter and 1.4 cm deep, is formed in a material with an emissivity of 0.8. The cavity is maintained at 200°C, and the opening is covered with a transparent material having $\tau = 0.7$, $\epsilon = 0.3$, $\rho = 0$. The outside surface of the transparent material experiences a convection heat-transfer coefficient of 17 W/m² · °C. The surrounding air and room are at 20°C. Calculate the net heat lost by the cavity and the temperature of the transparent covering.

8-77 Two parallel infinite plates are maintained at 800 and 35°C with emissivities of 0.5 and 0.8, respectively. To reduce the heat-transfer rate a radiation shield is placed between the two plates. Both sides of the shield are specular-diffuse-reflecting and have $\rho_D = 0.4$, $\rho_s = 0.4$. Calculate the heat-transfer rate with and without the shield. Compare this result with that obtained when the shield is completely diffuse-reflecting with $\rho = 0.8$.

8-78 Apply Eqs. (8-105) and (8-107) to the problem described by Eq. (8-43). Apply the equations directly to obtain Eq. (8-43*a*) for a convex object completely enclosed by a very large concave surface.

8-79 Rework Example 8-3 using the formulation of Eqs. (8-112) and (8-115).

8-80 Rework Prob. 8-21 using the formulation of Eqs. (8-112) and (8-115). Recall that $J = E_b$ for the insulated surface.

8-81 One way of constructing a blackbody cavity is to drill a hole in a metal plate. As a result of multiple reflections on the inside of the hole, the interior walls appear to have a higher emissivity than a flat surface in free space would have. A strict analysis of the cavity must take into account the fact that the irradiation is nonuniform over the interior surface. Thus the specific location at which a radiant-flux-measuring device sights on this surface must be known in order to say how "black" the surface may be. Consider a 1.25-cm-diameter hole 2.5 cm deep. Divide the interior surface into three sections, as shown. Assume, as an approximation, that the irradiation is uniform over each of these three surfaces and that the temperature and emissivity are uniform inside the entire cavity. A radiometer will detect the total energy leaving a surface (radiosity). Calculate the ratio J/E_b for each of the three surfaces, assuming $\epsilon = 0.6$ and no appreciable radiation from the exterior surroundings.

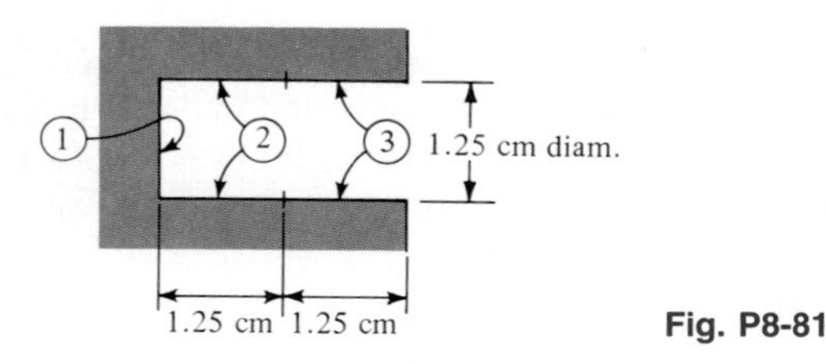

Fig. P8-81

8-82 Repeat Prob. 8-81, assuming a cavity temperature of 120°C and a surrounding temperature of 93°C.

8-83 Two infinite strips, 30 cm wide, are separated by a distance of 10 cm. The strips have surface emissivities of 0.25 and 0.5, and the respective temperatures are constant over the surfaces at 200 and 1000°C. The strips are completely enclosed by a large surrounding at 0°C. Using the numerical method, divide each strip into three equal segments and calculate the net heat transfer from each strip. What would be the heat transfer of each strip if uniform radiosity were assumed for each? Assume gray-body behavior.

8-84 Repeat Prob. 8-83, using one strip at 1000°C with $\epsilon = 0.6$ and the other strip perfectly insulated (radiant balance) with $\epsilon = 0.25$.

8-85 Two 10 by 30 cm rectangular plates are spaced 10 cm apart and connected by four insulated and re-radiating walls. The plate temperatures are uniform at 1000 and 300°C, and their emissivities are 0.6 and 0.4, respectively. Using the numerical method, determine the net heat transfer under the assumptions that (*a*) the four re-radiating surfaces act as one surface and have uniform radiosity and (*b*) the four re-radiating surfaces have radiosities determined from the radiant balance with all other surfaces. Assume that the 1000 and 300°C surfaces have uniform radiosity. Also calculate the temperatures for the re-radiating surfaces for each case above.

8-86 Two parallel disks 30 cm in diameter are separated by a distance of 5 cm in a large room at 20°C. One disk contains an electric heater that produces a constant heat flux of 100 kW/m² and $\epsilon = 0.9$ on the surface facing the other disk. Its back surface is insulated. The other disk has $\epsilon = 0.5$ on both sides and is in radiant balance with the other disk and room. Calculate the temperatures of both disks.

8-87 Verify the results of Example 8-19 for $\epsilon_2 = \epsilon_3 = 0.6$ and $h = 25\ \text{W/m}^2 \cdot {}^\circ\text{C}$.

8-88 A long duct has an equilateral triangle shape as shown. The surface conditions are $T_1 = 1100$ K, $\epsilon_1 = 0.6$, $T_2 = 2100$ K, $\epsilon_2 = 0.8$, $(q/A)_3 = 1000\ \text{W/m}^2$, $\epsilon_3 = 0.7$. Calculate the heat fluxes for surfaces 1 and 2 and the temperature of surface 3.

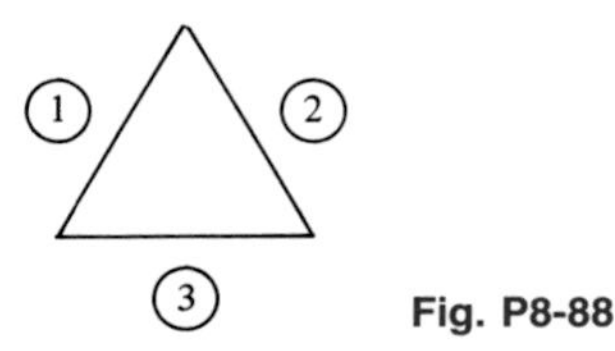

Fig. P8-88

8-89 Two 1-m-square plates are placed in a perpendicular position and joined together. One is maintained at 300°C with $\epsilon = 0.5$ while the other is perfectly insulated with $\epsilon = 0.7$. Both plates are placed in a large room maintained at 30°C. Calculate the heat lost by the 300°C plate and the temperature of the insulated plate using the numerical method.

8-90 Solve Prob. 8-19 using the numerical formulation.

8-91 Two parallel disks, each 1 m in diameter, are spaced 25 cm apart. One disk is maintained at 300°C while the other disk is insulated on the back side. Both disks have an emissivity of 0.5 and are placed in a large room which is maintained at 30°C. Calculate the radiation energy lost by the 300°C disk.

8-92 Two parallel disks 50 cm in diameter are separated by a distance of 10 cm. An electric heater supplies a constant heat flux of 10 kW/m² to one disk while the other disk is maintained at a constant temperature of 350 K. The two disks are connected by an insulated surface. Both disks have surface emissivity of 0.6. Using the numerical formulation set up the equations for the problem and solve for the temperatures of the hot disk and surrounding insulated surface.

8-93 In the figure shown calculate the heat loss by radiation from disk surface 1 and the irradiation on spherical surface 2. Surface 3 is insulated.

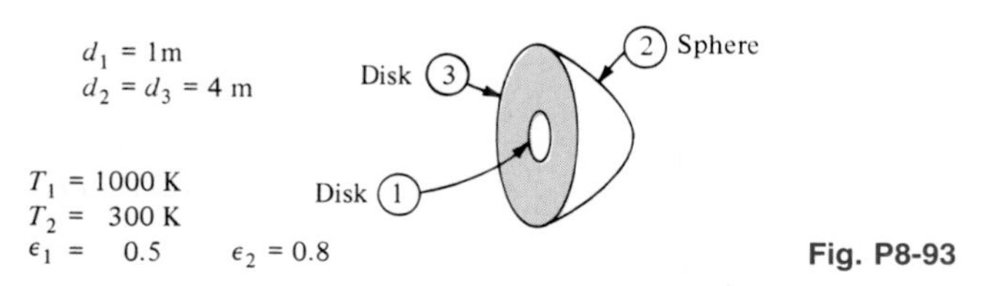

Fig. P8-93

8-94 A 10-cm black disk maintained at 500°C is installed at the end of a 10-cm-diameter cylinder having a length of 10 cm. The cylinder has an emissivity of 0.3 on both the inside and outside surfaces. The open end of the cylinder is exposed to a large room at 40°C. Calculate the energy lost by the 500°C disk.

8-95 A "focusing" radiation heater is constructed as shown with the hot surface at 1000°C having an emissivity of 0.55, and a diameter of 0.5 m. The walls surrounding the heater surface are insulated and have $\epsilon = 0.4$. The arrangement is open to a large room at 20°C. Calculate the heat transfer to the room.

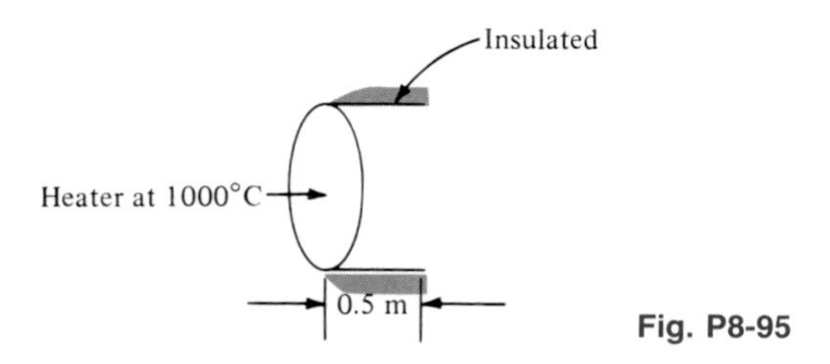

Fig. P8-95

8-96 A tapered cylindrical hole is machined in a block of insulation such that the depth is 3 cm. The diameter at the bottom is 1 cm and the diameter at the opening is 2 cm. An electric heater element is installed at the bottom of the hole such that the temperature is maintained at 500°C with a metal surface emissivity of 0.6. Outside the hole the large surroundings are at 30°C. Assuming the insulation is "perfect" calculate the power which must be supplied to the heater. The emissivity for the insulation is 0.3.

8-97 A cubical box is 1 m on a side. The top surface has $T = 100°C$, $\epsilon = 0.7$ while the bottom surface has $T = 40°C$, $\epsilon = 0.5$. The four sides have $\epsilon = 0.6$. Two sides opposite each other have $T = 70°C$ while the other two are at 60°C. Calculate the net radiant heat transfer for each side using the numerical method.

8-98 Two parallel concentric disks have diameters of 10 and 20 cm and are separated by a distance of 10 cm. The smaller disk has $T = 700$ K and $\epsilon = 0.8$ while the larger disk has $\epsilon = 0.4$ on both sides and is in radiant balance with the small disk and a large surrounding room at 25°C. Calculate the heat lost by the small disk and the temperature of the large disk.

8-99 Two parallel disks have diameters of 50 cm and are separated by a distance of 12.5 cm. One disk is maintained at a constant temperature of 800°C and has $\epsilon = 0.63$ while the other generates a constant heat flux of 80 kW/m² and has $\epsilon = 0.75$. The disks are surrounded by a large room at 30°C. Calculate the heat absorbed by the room and the temperature of the constant-heat-flux disk.

8-100 Rework Prob. 8-99 such that the 800°C disk is replaced with a disk having constant heat flux of 100 kW/m² and $\epsilon = 0.63$. What are the temperatures of the two disks under these conditions?

8-101 Two concentric cylinders have $d_1 = 10$ cm, $\epsilon_1 = 0.4$ and $d_2 = 20$ cm, $\epsilon_2 = 0.6$ respectively. The length is 10 cm. At one end of the cylinders a heater is placed which has $\epsilon = 0.8$ and covers the open space between the cylinders. The heater has a constant heat flux of 90 kW/m². The entire assembly is placed in a large room at 30°C. Using both the network and numerical methods, calculate the temperatures of the two cylinders.

8-102 Two long concentric cylinders have the properties of $d_1 = 2$ cm, $\epsilon_1 = 0.75$, $T_1 = 600$ K and $d_2 = 5$ cm, $\epsilon_2 = 0.8$ on both inner and outer surfaces. An airstream is blown across the outer cylinder with $T_\infty = 35$°C and $h = 180$ W/m² · °C. The assembly is located in a large room at 25°C. Calculate the heat lost by the inner cylinder per unit length and the temperature of the outer cylinder.

8-103 Surface 1 in the accompanying figure has $T_1 = 500$ K and $\epsilon_1 = 0.8$. Surface 2 is very thin, has $\epsilon = 0.4$, and is in radiant balance with surface 1 and a large enclosure at 300 K. Determine the temperature of surface 2 assuming it is uniform in temperature.

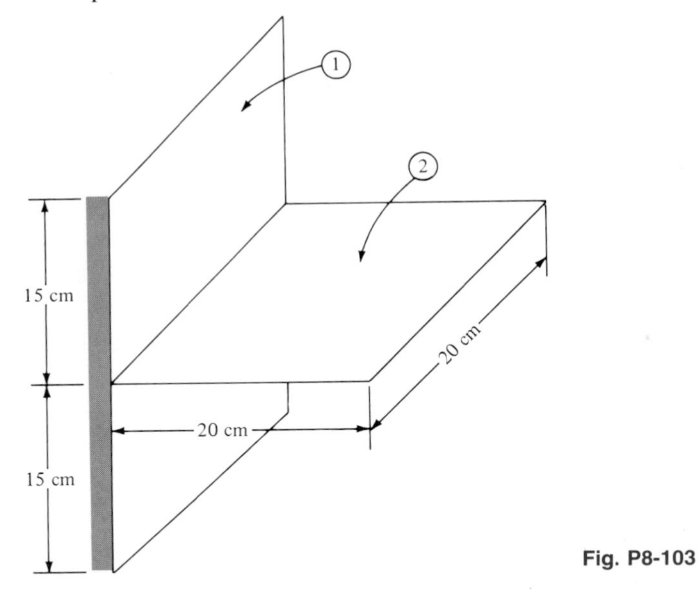

Fig. P8-103

8-104 Suppose a convection coefficient of $h = 10$ W/m² · °C at $T = 300$ K was experienced on surface 2 (top and bottom) of Prob. 8-103. Obtain the temperature distribution in this circumstance. Repeat for $h = 50$ W/m$^2 \cdot$ °C. Assume the surface is so thin that conduction is negligible.

8-105 Suppose surface 2 in Prob. 8-103 is oxidized copper with $\epsilon = 0.78$ and has a thickness of 0.5 mm. Surface 1 has $T_1 = 1000$ K and $\epsilon_1 = 0.5$. The convection and radiation environment is at 300 K and $h = 75$ W/m$^2 \cdot$ °C. Calculate the temperatures along the fin, taking into account conduction. Consult Chap. 3 for setting up conduction heat-transfer terms.

8-106 Rework Prob. 8-46 for a polished-aluminum plate having the radiation characteristics given in Table 8-3.

8-107 A slab of white marble is exposed to a solar radiation flux of 1070 W/m^2. Assuming the effective radiation temperature of the sky is -70°C, calculate the radiation equilibrium temperature of the slab, using the properties given in Table 8-3. For this calculation neglect all conduction and convection losses.

8-108 The plate of Example 5-4 is 30 cm by 30 cm square and is sprayed with a white paint having a solar absorptivity of 0.16 and a low-temperature absorptivity of 0.09. The plate is exposed to a solar radiation flux of 1100 W/m^2 and allowed to reach equilibrium with the convection surroundings. Assuming that the underside of the plate is insulated, calculate the equilibrium temperature of the plate.

8-109 A solar collector consists of an insulated black surface covered by a glass plate having the following properties:

$$0 < \lambda < 2.5\ \mu\text{m}: \qquad \tau = 0.9 \qquad \epsilon = 0.1 \qquad \rho = 0$$

$$2.5\ \mu\text{m} < \lambda < \infty: \qquad \tau = 0 \qquad \epsilon = 0.95 \qquad \rho = 0.05$$

Assume that the radiant flux from the sun is 700 W/m^2 and lies entirely below 2.5 μm. Calculate the radiation equilibrium temperature of the black surface neglecting convection and with a surrounding environment at 35°C.

8-110 A large surface has $\alpha = 0.6$ for long-wavelength radiation and $\alpha = 0.95$ for solar radiation. Calculate the radiation equilibrium temperature of the surface if its back side is insulated and top side is exposed to a solar radiation flux of 950 W/m^2 and an environment at 300 K with $h = 12$ W/m$^2 \cdot$ °C.

8-111 A cast iron plate is placed in an environment at 25°C and exposed to a solar flux of 800 W/m^2. Calculate the radiation equilibrium temperature of the plate neglecting convection.

8-112 Calculate the absorption rate for solar radiation on bright fine sand for a solar altitude angle of 50° and a turbidity factor of 3.5.

8-113 Calculate the ground insolation for a solar altitude of 30° in a smoggy environment with a turbidity factor of 4.0. How does this compare with the ground insolation for clear air at a solar altitude of 90°?

8-114 Plot the ground solar insolation as a function of turbidity factor for a solar altitude of 80°.

8-115 Using the spectrum of Fig. 8-65, estimate the fraction of earthbound radiation that may be transmitted by the atmosphere. What is the magnitude of this transmitted energy?

8-116 Using the spectrum of Fig. 8-65, estimate the magnitude of the solar radiation flux at the earth's surface. How does this calculation compare with one based on Eq. (8-119)?

8-117 A pollution-control enthusiast claims that a certain metropolitan area has such a high concentration of atmospheric contaminants that the solar insolation is attenuated by 50 percent. How do you evaluate this claim? Assuming a turbidity factor of 4.5, what solar altitude would be necessary to produce the 50 percent attenuation factor?

8-118 Solar radiation arrives at the surface of a lake when the solar altitude is 90°. The lake is quite turbid, with an extinction coefficient of 0.82 m^{-1}, and the atmosphere has a turbidity factor of 2.8. What fraction of the incident radiation has been absorbed at depths of (*a*) 0 ft, (*b*) 0.25 ft, (*c*) 0.5 ft, (*d*) 1.0 ft?

8-119 Calculate the heat-generation rates for the depths given in Prob. 8-118.

8-120 Using the data of Table 8-5, estimate the values of the turbidity factor at 5, 20, 30, 45, 60, and 90° solar altitudes upon which these "average" computations were based.

8-121 Plot the solar insolation versus solar altitude angle for a smoggy atmosphere having a turbidity factor of (*a*) 3.5 and (*b*) 4.5.

8-122 At what solar altitude angle will the solar insolation equal 50 percent of the insolation at the outer limits of the atmosphere when the turbidity factor is 4.0?

8-123 What turbidity factor would be necessary to reduce the solar insolation by 50 percent from that of a clear sky at $\alpha = 90°$?

8-124 A thermocouple is placed in a large heated duct to measure the temperature of the gas flowing through the duct. The duct walls are at 425°C, and the thermocouple indicates a temperature of 170°C. The heat-transfer coefficient from the gas to the thermocouple is 150 W/m² · °C. The emissivity of the thermocouple material is 0.43. What is the temperature of the gas?

8-125 Suppose a flat plate is exposed to a high-speed environment. We define the radiation equilibrium temperature of the plate as that temperature it would attain if insulated so that the energy received by aerodynamic heating is just equal to the heat lost by radiation to the surroundings, i.e.,

$$hA(T_w - T_{aw}) = -\sigma A\epsilon(T_w^4 - T_s^4)$$

where the surroundings are supposed to be infinite and at the temperature T_s and ϵ is the emissivity of the plate surface. Assuming an emissivity of 0.8 for the surface, calculate the radiation equilibrium temperature for the flow conditions of Example 5-9. Assume that the effective radiation temperature for the suroundings is -40°C.

8-126 On a clear night the effective radiation temperature of the sky may be taken as -70°C. Assuming that there is no wind and the convection heat-transfer coefficient from the air to the dew which has collected on the grass is 28 W/m² · °C, estimate the minimum temperature which the air must have to prevent formation of frost. Neglect evaporation of the dew, and assume that the grass is insulated from the ground insofar as conduction is concerned. Take the emissivity as unity for the water.

8-127 A thermocouple enclosed in a 3.2-mm stainless-steel sheath ($\epsilon = 0.6$) is inserted horizontally into a furnace to measure the air temperature inside. The walls of the furnace are at 650°C, and the true air temperature is 560°C. What temperature will be indicated by the thermocouple? Assume free convection from the thermocouple.

8-128 The thermocouple of Prob. 8-127 is placed horizontally in an air-conditioned room. The walls of the room are at 32°C, and the air temperature in the room is 20°C. What temperature is indicated by the thermocouple? What would be the effect on the reading if the thermocouple were enclosed by a polished-aluminum radiation shield?

8-129 A horizontal metal thermocouple with a diameter of 3 mm and an emissivity of 0.6 is inserted in a duct to measure the temperature of an airstream flowing at 7 m/s. The walls of the duct are at 400°C, and the thermocouple indicates a temperature of 100°C. Using the convection relations of Chap. 6, calculate the true gas temperature.

8-130 Suppose the outer cylinder of Example 8-8 loses heat by convection to the room with $h = 50\ \text{W/m}^2 \cdot °\text{C}$ on its outside surface. Neglecting convection on other surfaces determine the temperature of the outer cylinder and the total heat lost by the inner cylinder.

8-131 A thermocouple is used to measure the air temperature inside an electrically heated metallurgical furnace. The surface of the thermocouple has an emissivity of 0.7 and a convection heat-transfer coefficient of 20 W/m² · °C. The thermocouple indicates a temperature of 750°C. The true temperature of the air is only 650°C. Estimate the temperature of the walls of the furnace.

8-132 A mercury-in-glass thermometer is inserted in a duct to measure the temperature of an air flow stream. The thermometer indicates a temperature of 55°C and the temperature of the walls of the duct is measured with a thermocouple as 100°C. By the methods of Chap. 6 the convection heat-transfer coefficient from the thermometer to the air is calculated as 30 W/m² · °C. Calculate the temperature of the air.

8-133 Air at 20°C flows across a 50-cm-diameter cylinder at a velocity of 25 m/s. The cylinder is maintained at a temperature of 150°C and has a surface emissivity of 0.7. Calculate the total heat loss from the cylinder per unit length if the effective radiation temperature of the surroundings is 20°C.

8-134 Determine the shape factors for the geometries shown.

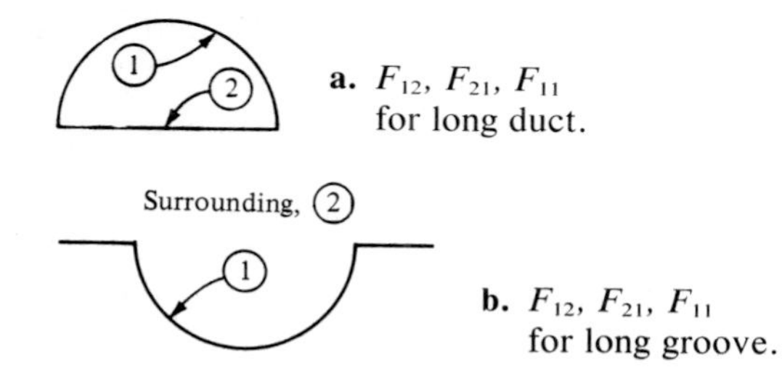

a. F_{12}, F_{21}, F_{11} for long duct.

b. F_{12}, F_{21}, F_{11} for long groove.

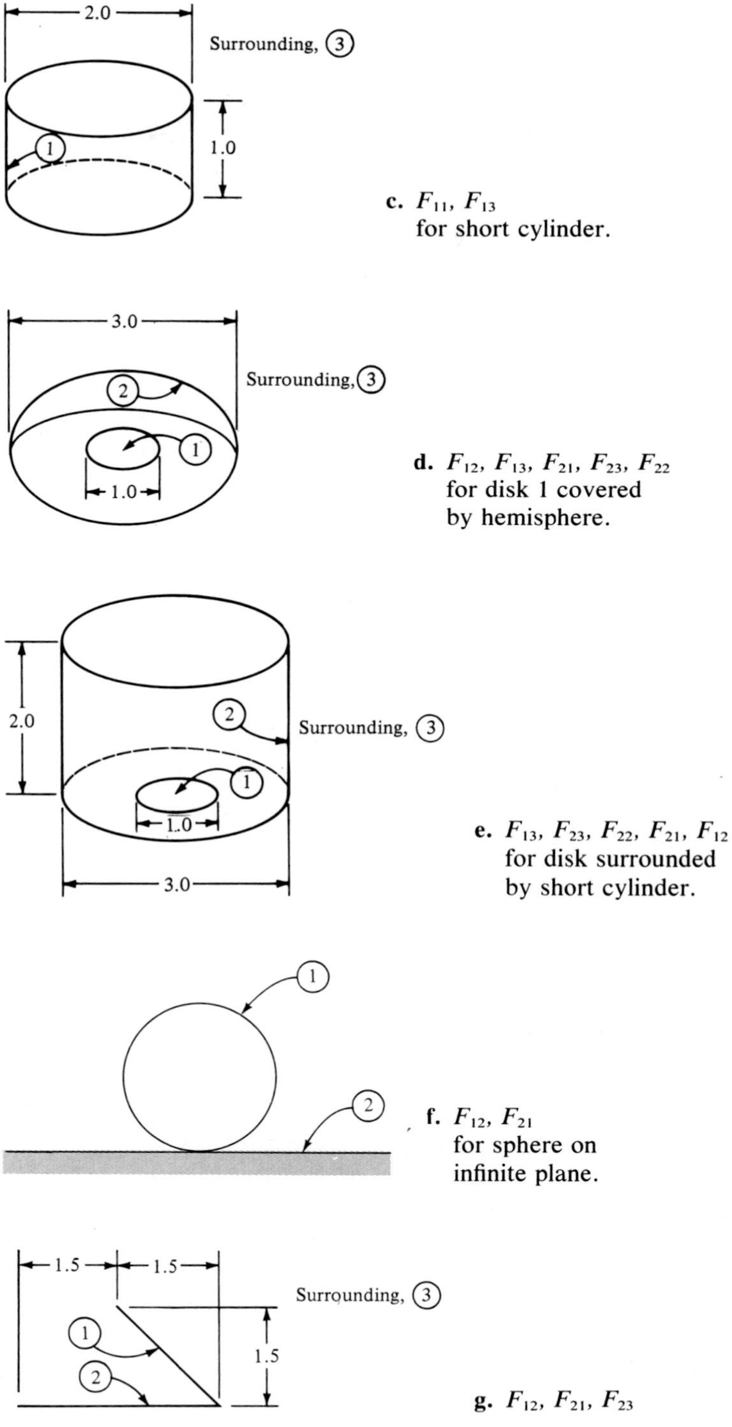

c. F_{11}, F_{13}
for short cylinder.

d. F_{12}, F_{13}, F_{21}, F_{23}, F_{22}
for disk 1 covered
by hemisphere.

e. F_{13}, F_{23}, F_{22}, F_{21}, F_{12}
for disk surrounded
by short cylinder.

f. F_{12}, F_{21}
for sphere on
infinite plane.

g. F_{12}, F_{21}, F_{23}

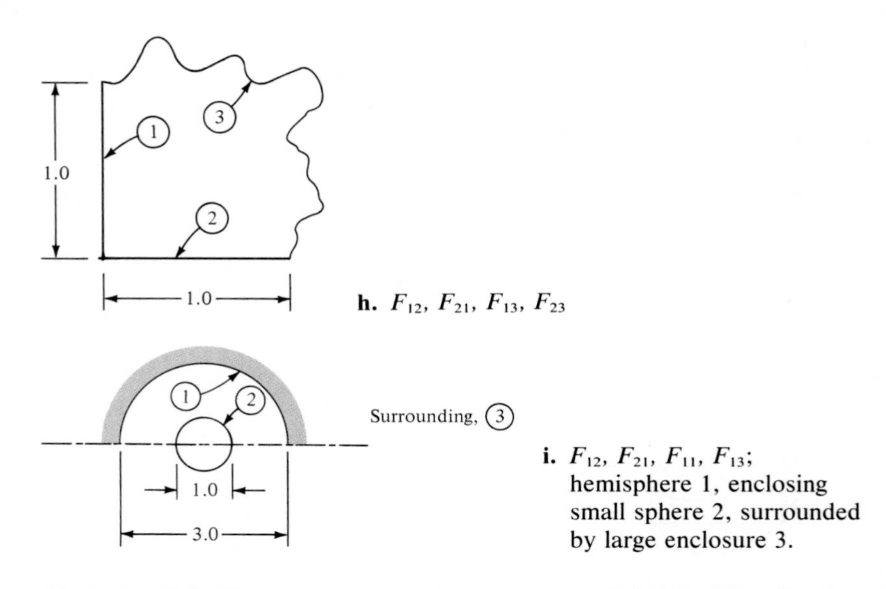

h. F_{12}, F_{21}, F_{13}, F_{23}

i. F_{12}, F_{21}, F_{11}, F_{13}; hemisphere 1, enclosing small sphere 2, surrounded by large enclosure 3.

8-135 A tungsten light filament operates at a temperature of 3400 K. What fraction of the radiation lies in the visible range, 0.4 to 0.7 μm?

8-136 Calculate all the shape factors for a cubical enclosure.

8-137 A regular tetrahedron has sides in the shape of equilateral triangles. Calculate all the shape factors.

8-138 Suppose the tungsten light filament in Prob. 8-135 consumes 400 W of electric power and all the visible light is transmitted through the enclosing glass bulb. Calculate the area of the filament to dissipate the total of 400 W of radiant energy. What is the efficiency of the light bulb?

8-139 A certain material has an emissivity of 0.6 between 0 and 2 μm and 0.2 between 2 and 8 μm. Calculate the emissive power of such a material maintained at 3000 K.

8-140 Show that the Stefan-Boltzmann constant can be expressed in terms of the radiation constants C_1 and C_2 by the relation

$$\sigma = \left(\frac{\pi}{C_2}\right)^4 \left(\frac{C_1}{15}\right)$$

8-141 A plastic tinted coating is frequently applied to glass windows to reduce the transmission of solar energy. The transmissivity for the uncoated glass is 0.9 from about 0.25 to 2.5 μm, while the tinted glass has the same transmissivity from 0.5 to 1.5 μm. Assuming solar radiation to be that of a blackbody at 5800 K, calculate the fraction of the incident solar energy transmitted through each glass. Also calculate the visible energy (0.4 to 0.7 μm) transmitted through each glass.

8-142 Air flows across a 0.2-m-square plate producing a convection heat-transfer coefficient of 12 W/m^2 · °C. The air temperature is 400 K. The plate is insulated on the back side and exposed to an irradiation of 2200 W/m², of which 450 W/m² is reflected. When the surface temperature is 510 K, its emissive power is 1100 W/m². Calculate the emissivity and radiosity of the plate surface.

8-143 A certain surface maintained at 1600 K has the following spectral emissive characteristics:

$$\epsilon_\lambda = 0.08 \qquad 0 < \lambda < 0.6\ \mu\text{m}$$

$$\epsilon_\lambda = 0.4 \qquad 0.6 < \lambda < 5\ \mu\text{m}$$

$$\epsilon_\lambda = 0.7 \qquad 5 < \lambda < \infty$$

Calculate the emissive power of the surface.

8-144 A surface has the emissive characteristics shown in the figure. Calculate the emissive power for the surface maintained at 2000 K.

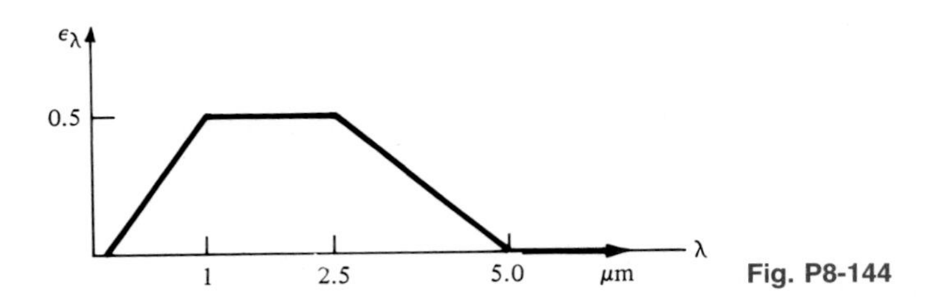

Fig. P8-144

8-145 A cubical container, 3 m on a side, has the bottom surface at 500 K with $\epsilon = 0.5$ and the top surface at 300 K with $\epsilon = 0.7$. The other four sides are perfectly insulated. Calculate the heat transfer from the bottom surface.

8-146 A 10-cm-diameter disk at 800 K with $\epsilon = 0.6$ is placed at the bottom of an insulated cylinder 15 cm high. On top of the cylinder is a hemisphere 10 cm in diameter with $T = 300$ K and $\epsilon = 0.3$. Calculate the heat lost by the disk.

8-147 A 5.0-cm-diameter long cylinder with $T = 800$ K and $\epsilon = 0.7$ is enclosed by a large room at 300 K. How much will the radiant heat loss from the cylinder be decreased if it is enclosed by a cylindrical radiation shield having $d = 10$ cm and $\epsilon = 0.2$ on the inner and outer surfaces?

8-148 A cylindrical furnace has diameter and lengths of 50 cm. The bottom disk surface is at 450 K with $\epsilon = 0.5$, the top disk is at 600 K with $\epsilon = 0.6$, and the vertical cylinder surface is at 1000 K with $\epsilon = 0.7$. Calculate the emissive power, radiosity, irradiation, and heat loss for the three surfaces.

8-149 Repeat Prob. 8-148 with the top disk surface removed and the open cylindrical cavity exposed to a large room at 300 K.

8-150 An evacuated insulating material consists of outer layers of aluminum with emissivity of 0.11 and three inner polished aluminum shields having emissivity of 0.04 and spaced 2 mm apart. The total thickness, neglecting the thickness of the outer layers, is 8 mm. Calculate the radiation heat transfer across the assembly for temperatures of 300 and 85 K. Also calculate an effective thermal conductivity and R value for these temperatures.

8-151 A cylinder 40 cm in diameter and 30 cm high has the bottom disk surface maintained at 750 K with $\epsilon = 0.75$. The vertical cylindrical surface is perfectly insulated. The top disk has a 20-cm-diameter hole in the center with a surface temperature of 600 K and $\epsilon = 0.4$. The open hole is exposed to a large room at 300 K. Calculate the heat loss, emissive power, radiosity, and irradiation for all three surfaces and the temperature of the vertical cylindrical surface.

8-152 Consider a cavity having an inside surface area of A_1 and emissivity of ϵ_1. In the side of the cavity is a hole having an area A_2. Show that the apparent emissivity of the hole as defined in Prob. 8-32 can be calculated from

$$\epsilon_{app} = \frac{\epsilon_1 A_1}{A_2 + \epsilon_1 (A_1 - A_2)}$$

REFERENCES

1. Sears, F. W.: "Introduction to Thermodynamics, Kinetic Theory, and Statistical Mechanics," pp. 123–124, Addison-Wesley Publishing Company, Inc., Reading, Mass., 1953.
2. Dunkle, R. V.: Thermal Radiation Tables and Applications, *Trans. ASME,* vol. 76, p. 549, 1954.
3. Mackey, C. O., L. T. Wright, Jr., R. E. Clark, and N. R. Gay: Radiant Heating and Cooling, part I, *Cornell Univ. Eng. Exp. Stn. Bull.,* vol. 32, 1943.
4. Chapman, A. J.: "Heat Transfer," pp. 319–323, The Macmillan Company, New York, 1960.
5. Hamilton, D. C., and W. R. Morgan: Radiant Interchange Configuration Factors, *NACA Tech. Note* 2836, 1952.
6. Eckert, E. R. G., and R. M. Drake: "Heat and Mass Transfer," 2d ed., pp. 381–393, McGraw-Hill Book Company, New York, 1959.
7. McAdams, W. H.: "Heat Transmission," 3d ed., chap. 2, McGraw-Hill Book Company, New York, 1954.
8. Schmidt, E.: Messung der Gesamtstrahlung das Wasserdampfes bei Temperaturen bis 1000C, *Forsch. Geb. Ingenieurwes.,* vol. 3, p. 57, 1932.
9. Schmidt, E., and E. R. G. Eckert: Über die Richtungsverteilung der Wärmestrahlung von Oberflachen, *Forsch. Geb. Ingenieurwes.,* vol. 6, p. 175, 1935.
10. Sparrow, E. M., and R. D. Cess: "Radiation Heat Transfer," Wadsworth Publishing Co., Inc., Englewood Cliffs, N.J., 1966.
11. Wiebelt, J. A.: "Engineering Radiation Heat Transfer," Holt, Rinehart and Winston, Inc., New York, 1966.
12. Eckert, E. R. G., and E. M. Sparrow: Radiative Heat Exchange between Surfaces with Specular Reflection, *Int. J. Heat Mass Transfer,* vol. 3, pp. 43–54, 1961.
13. Sparrow, E. M., E. R. G. Eckert, and V. K. Jonsson: An Enclosure Theory for Radiative Exchange between Specular and Diffusely Reflecting Surfaces, *J. Heat Transfer,* ser. C., vol. 84, pp. 294–299, 1962.

14 Gubareff, G. G., J. E. Janssen, and R. H. Torborg: "Thermal Radiation Properties Survey," 2d ed., Minneapolis Honeywell Regulator Co., Minneapolis, Minn., 1960.

15 Threlkeld, J. L., and R. C. Jordan: Direct Solar Radiation Available on Clear Days, *ASHAE Trans.*, vol. 64, pp. 45–56, 1958.

16 Jakob, M.: "Heat Transfer," vol. 2, John Wiley & Sons, Inc., New York, 1957.

17 Birkebak, R. D., and E. R. G. Eckert: Effects of Roughness of Metal Surfaces on Angular Distribution of Monochromatic Radiation, *J. Heat Transfer,* vol. 87, p. 85, 1965.

18 Torrance, K. E., and E. M. Sparrow: Off Specular Peaks in the Directional Distribution of Reflected Thermal Radiation, *J. Heat Transfer,* vol. 88, p. 223, 1966.

19 Hering, R. G., and T. F. Smith: Surface Roughness Effects on Radiant Transfer between Surfaces, *Int. J. Heat Mass Transfer,* vol. 13, p. 725, 1970.

20 Oppenheim, A. K.: Radiation Analysis by the Network Method, *Trans. ASME,* vol. 78, pp. 725–735, 1956.

21 Holman, J. P.: Radiation Networks for Specular-Diffuse Transmitting and Reflecting Surfaces, *ASME Pap.* 66 WA/HT-9, December 1966.

22 Hottel, H. C.: Radiant Heat Transmission, chap. 4, in W. H. McAdams (ed.), "Heat Transmission," 3d ed., McGraw-Hill Book Company, New York, 1954.

23 Hottel, H. C., and A. F. Sarofim: "Radiative Transfer," McGraw-Hill Book Company, New York, 1967.

24 Siegel, R., and J. R. Howell: "Thermal Radiation Heat Transfer," 2d ed., McGraw-Hill Book Company, New York, 1980.

25 Eckert, E. R. G., and R. M. Drake: "Analysis of Heat and Mass Transfer," McGraw-Hill Book Company, New York, 1972.

26 Dunkle, R. V.: "Geometric Mean Beam Lengths for Radiant Heat Transfer Calculations," *J. Heat Transfer,* vol. 86, p. 75, February 1964.

27 Edwards, D. K., and K. E. Nelson, "Rapid Calculation of Radiant Energy Transfer between Nongray Walls and Isothermal H_2O and CO_2 Gas, *J. Heat Transfer,* vol. 84, p. 273, 1962.

28 Edwards, D. K.: Radiation Interchange in a Nongray Enclosure Containing an Isothermal Carbon Dioxide–Nitrogen Gas Mixture, *J. Heat Transfer,* vol. 84, p. 1, 1962.

29 Hottel, H. C., and R. B. Egbert: Radiant Heat Transmission from Water Vapor, *Trans. AIChE,* vol. 38, p. 531, 1942.

30 Kondratyev, K. Y.: "Radiative Heat Exchange in the Atmosphere," Pergamon Press, New York, 1965.

31 Geiger, R.: "The Climate Near the Ground," rev. ed., Harvard University Press, Cambridge, Mass., 1965.

32 Eagleson, P. S.: "Dynamic Hydrology," McGraw-Hill Book Company, New York, 1970.

33 Raphael, J. M.: Prediction of Temperature in Rivers and Reservoirs, *Proc. ASCE Power Div.,* no. PO2, pap. 3200, July 1962.

34 Dake, J. M. K., and D. R. F. Harleman: Thermal Stratification in Lakes: Analytical and Laboratory Studies, *Water Resour. Res.*, vol. 5, no. 2, p. 484, April 1969.

35 Goldman, C. R., and C. R. Carter: An Investigation by Rapid Carbon-14 Bioassay of Factors Affecting the Cultural Entrophication of Lake Tahoe, California, *J. Water Pollut. Control Fed.*, p. 1044, July 1965.

36 Bachmann, R. W., and C. R. Goldman: Hypolimnetic Heating in Castle Lake, California, *Limnol. Oceanog.*, vol. 10, p. 2, April 1965.

37 Duffie, J. A., and W. A. Beckman: "Solar Energy Thermal Process," John Wiley & Son, Inc., New York, 1974.

38 Howell, J. R.: "A Catalog of Radiation Configuration Factors," McGraw-Hill Book Company, New York, 1982.

CONDENSATION AND BOILING HEAT TRANSFER

9-1 INTRODUCTION

Our preceding discussions of convection heat transfer have considered homogeneous single-phase systems. Of equal importance are the convection processes associated with a change of phase of a fluid. The two most important examples are condensation and boiling phenomena, although heat transfer with solid-gas changes has become important because of a number of applications.

In many types of power or refrigeration cycles one is interested in changing a vapor to a liquid, or a liquid to a vapor, depending on the particular part of the cycle under study. These changes are accomplished by boiling or condensation, and the engineer must understand the processes involved in order to design the appropriate heat-transfer equipment. High heat-transfer rates are usually involved in boiling and condensation, and this fact has also led designers of compact heat exchangers to utilize the phenomena for heating or cooling purposes not necessarily associated with power cycles.

9-2 CONDENSATION HEAT-TRANSFER PHENOMENA

Consider a vertical flat plate exposed to a condensable vapor. If the temperature of the plate is below the saturation temperature of the vapor, condensate will form on the surface and under the action of gravity will flow down the plate. If the liquid wets the surface, a smooth film is formed, and the process is called *film condensation*. If the liquid does not wet the surface, droplets are formed which fall down the surface in some random fashion. This process is called *dropwise condensation*. In the film-condensation process the surface is blanketed by the film, which grows in thickness as it moves down the plate. A temperature gradient exists in the film, and the film represents a thermal resistance to heat transfer. In dropwise condensation a large portion of the area of the plate is directly exposed to the vapor; there is no film barrier to heat flow, and higher heat-transfer rates are experienced. In fact, heat-transfer rates in dropwise condensation may be as much as 10 times higher than in film condensation.

Because of the higher heat-transfer rates, dropwise condensation would be preferred to film condensation, but it is extremely difficult to maintain since most surfaces become wetted after exposure to a condensing vapor over an extended period of time. Various surface coatings and vapor additives have been used in attempts to maintain dropwise condensation, but these methods have not met with general success to date. Some of the pioneer work on drop condensation was conducted by Schmidt [26] and a good summary of the overall problem is presented in Ref. 27. Measurements of Ref. 35 indicate that the drop conduction is the main resistance to heat flow for atmospheric pressure and above. Nucleation site density on smooth surfaces can be of the order of 10^8 sites per square centimeter, and heat-transfer coefficients in the range of 170 to 290 $kW/m^2 \cdot °C$ [30,000 to 50,000 $Btu/h \cdot ft^2 \cdot °F$] have been reported by a number of investigators.

Figure 9-1 is a photograph illustrating the different appearances of dropwise and filmwise condensation. The figure caption explains the phenomena.

Fig. 9-1 Dropwise and filmwise condensation of steam on a copper plate from Westwater [47]. The right side of the plate is clean copper and the steam condenses as a continuous film. The left side has a thin coating of cupric oleate which causes dropwise condensation. The heat transfer coefficient for the dropwise condensation is about seven times the value for filmwise condensation. The diameter of the horizontal thermocouple probe is 1.7 mm. (Photograph courtesy of Professor J.W. Westwater.)

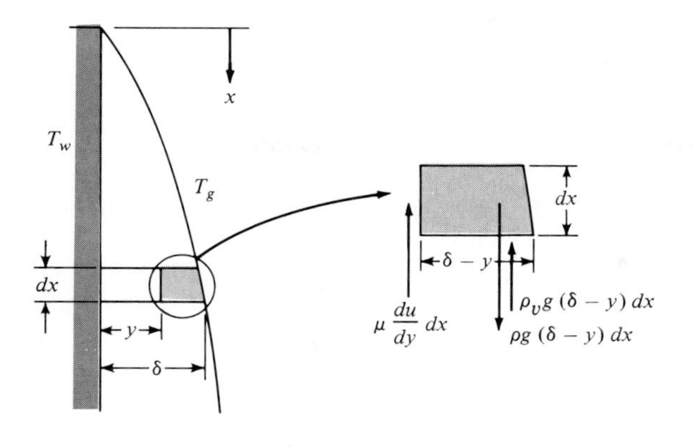

Fig. 9-2 Film condensation on a vertical flat plate.

Film condensation on a vertical plate may be analyzed in a manner first proposed by Nusselt [1]. Consider the coordinate system shown in Fig. 9-2. The plate temperature is maintained at T_w, and the vapor temperature at the edge of the film is the saturation temperature T_g. The film thickness is represented by δ, and we choose the coordinate system with the positive direction of x measured downward, as shown. It is assumed that the viscous shear of the vapor on the film is negligible at $y = \delta$. It is further assumed that a linear temperature distribution exists between wall and vapor conditions. The weight of the fluid element of thickness dx between y and δ is balanced by the viscous-shear force at y and the buoyancy force due to the displaced vapor. Thus

$$\rho g(\delta - y)\, dx = \mu \frac{du}{dy}\, dx + \rho_v g(\delta - y)\, dx \tag{9-1}$$

Integrating and using the boundary condition that $u = 0$ at $y = 0$ gives

$$u = \frac{(\rho - \rho_v)g}{\mu}\left(\delta y - \tfrac{1}{2}y^2\right) \tag{9-2}$$

The mass flow of condensate through any x position of the film is thus given by

$$\text{Mass flow} = \dot{m} = \int_0^\delta \rho \left[\frac{(\rho - \rho_v)g}{\mu}\left(\delta y - \tfrac{1}{2}y^2\right)\right] dy$$

$$= \frac{\rho(\rho - \rho_v)g\,\delta^3}{3\mu} \tag{9-3}$$

when unit depth is assumed. The heat transfer at the wall in the area dx is

$$q_x = -k\, dx \left.\frac{\partial T}{\partial y}\right]_{y=0} = k\, dx \frac{T_g - T_w}{\delta} \tag{9-4}$$

since a linear temperature profile was assumed. As the flow proceeds from x to $x + dx$, the film grows from δ to $\delta + d\delta$ as a result of the influx of additional condensate. The amount of condensate added between x and $x + dx$ is

$$\frac{d}{dx}\left[\frac{\rho(\rho - \rho_v)g\,\delta^3}{3\mu}\right]dx = \frac{d}{d\delta}\left[\frac{\rho(\rho - \rho_v)g\,\delta^3}{3\mu}\right]\frac{d\delta}{dx}\,dx$$

$$= \frac{\rho(\rho - \rho_v)g\,\delta^2\,d\delta}{\mu}$$

The heat removed by the wall must equal this incremental mass flow times the latent heat of condensation of the vapor. Thus

$$\frac{\rho(\rho - \rho_v)g\,\delta^2\,d\delta}{\mu}h_{fg} = k\,dx\,\frac{T_g - T_w}{\delta} \tag{9-5}$$

Equation (9-5) may be integrated witl. the boundary condition $\delta = 0$ at $x = 0$ to give

$$\delta = \left[\frac{4\mu kx(T_g - T_w)}{gh_{fg}\rho(\rho - \rho_v)}\right]^{1/4} \tag{9-6}$$

The heat-transfer coefficient is now written

$$h\,dx\,(T_w - T_g) = -k\,dx\,\frac{T_g - T_w}{\delta}$$

or

$$h = \frac{k}{\delta}$$

so that

$$h_x = \left[\frac{\rho(\rho - \rho_v)gh_{fg}k^3}{4\mu x(T_g - T_w)}\right]^{1/4} \tag{9-7}$$

Expressed in dimensionless form in terms of the Nusselt number, this is

$$\mathrm{Nu}_x = \frac{hx}{k} = \left[\frac{\rho(\rho - \rho_v)gh_{fg}x^3}{4\mu k(T_g - T_w)}\right]^{1/4} \tag{9-8}$$

The average value of the heat-transfer coefficient is obtained by integrating over the length of the plate:

$$\bar{h} = \frac{1}{L}\int_0^L h_x\,dx = \tfrac{4}{3}h_{x=L} \tag{9-9}$$

or

$$\bar{h} = 0.943\left[\frac{\rho(\rho - \rho_v)gh_{fg}k_f^3}{L\mu_f(T_g - T_w)}\right]^{1/4} \tag{9-10}$$

More refined analyses of film condensation are presented in detail by Rohsenow [37]. The most significant refinements take into account a nonlinear temperature profile in the film and modifications to the energy balance to include additional energy to cool the film below the saturation temperature. Both effects can be handled by replacing h_{fg} with h'_{fg}, defined by

$$h'_{fg} = h_{fg} + 0.68c(T_g - T_w) \tag{9-11}$$

where c is the specific heat of the liquid. Otherwise, properties in Eqs. (9-7) and (9-10) should be evaluated at the film temperature

$$T_f = \frac{T_g + T_w}{2}$$

With these substitutions Eq. (9-10) may be used for vertical plates and cylinders and fluids with $\Pr > 0.5$ and $cT/h_{fg} \leq 1.0$.

For laminar film condensation on horizontal tubes Nusselt obtained the relation

$$\bar{h} = 0.725 \left[\frac{\rho(\rho - \rho_v) g h_{fg} k_f^3}{\mu_f d (T_g - T_w)} \right]^{1/4} \tag{9-12}$$

where d is the diameter of the tube. When condensation occurs on a horizontal tube bank with n tubes placed directly over one another in the vertical direction, the heat-transfer coefficient may be calculated by replacing the diameter in Eq. (9-12) with nd. The analysis of Ref. 48 has shown that Eq. (9-12) can be used for an isothermal sphere if the constant is changed to 0.815.

When a plate on which condensation occurs is sufficiently large or there is a sufficient amount of condensate flow, turbulence may appear in the condensate film. This turbulence results in higher heat-transfer rates. As in forced-convection flow problems, the criterion for determining whether the flow is laminar or turbulent is the Reynolds number, and for the condensation system it is defined as

$$\mathrm{Re}_f = \frac{D_H \rho V}{\mu_f} = \frac{4A\rho V}{P\mu_f}$$

where D_H = hydraulic diameter
A = flow area
P = shear, or "wetted," perimeter
V = average velocity in flow

But

$$\dot{m} = \rho A V$$

so that

$$\mathrm{Re}_f = \frac{4\dot{m}}{P\mu_f} \tag{9-13}$$

where $\dot{m}$ is the mass flow through the particular section of the condensate film. For a vertical plate of unit depth, $P = 1$; for a vertical tube, $P = \pi d$. The critical Reynolds number is approximately 1800, and turbulent correlations for heat transfer must be used at Reynolds numbers greater than this value. The Reynolds number is sometimes expressed in terms of the mass flow per unit depth of plate Γ, so that

$$\mathrm{Re}_f = \frac{4\Gamma}{\mu_f} \tag{9-14}$$

In calculating the Reynolds numbers the mass flow may be related to the total heat transfer and the heat-transfer coefficient by

$$q = \bar{h} A (T_{\mathrm{sat}} - T_w) = \dot{m} h_{fg} \tag{9-15}$$

where A is the total surface area for heat transfer. Thus

$$\dot{m} = \frac{q}{h_{fg}} = \frac{\bar{h}A(T_{sat} - T_w)}{h_{fg}}$$

$$\text{Re}_f = \frac{4\bar{h}A(T_{sat} - T_w)}{h_{fg}P\mu_f} \tag{9-16}$$

But $\quad A = LW \quad$ and $\quad P = W$

where L and W are the height and width of the plate, respectively, so that

$$\text{Re}_f = \frac{4\bar{h}L(T_{sat} - T_w)}{h_{fg}\mu_f} \tag{9-17}$$

The laminar condensation equations presented above match experimental data very well as long as the film remains smooth and well behaved. In practice, it has been found that ripples will develop in the film for Reynolds numbers as slow as 30 or 40. When this occurs, the experimental values of $\bar{h}$ can be 20 percent higher than predicted by Eq. (9-12). Because this is such a common occurrence, McAdams [3] was prompted to suggest that the 20 percent increase be adopted for design purposes. For our discussions here we shall use Eq. (9-10) without the increase, recognizing that this is a conservative approach which provides a safety factor in design problems. If one wishes to employ the 20 percent higher coefficient, the resulting equation for vertical plates is

$$\bar{h} = 1.13\left[\frac{\rho(\rho - \rho_v)gh_{fg}k^3}{L\mu(T_g - T_w)}\right]^{1/4} \tag{9-18}$$

If the vapor to be condensed is superheated, the preceding equations may be used to calculate the heat-transfer coefficient, provided the heat flow is calculated on the basis of the temperature difference between the surface and the saturation temperature corresponding to the system pressure. When a noncondensable gas is present along with the vapor, there may be an impediment of the heat transfer since the vapor must diffuse through the gas before it can condense on the surface. The reader should consult Refs. 3 and 4 for more information on this subject.

Inclined Surfaces

If a plate or cylinder is inclined at an angle ϕ with the horizontal, the net effect on the above analysis is to replace the gravitational force with its component parallel to the heat-transfer surface, or

$$g' = g \sin \phi \tag{9-19}$$

Therefore, for laminar flow, we suggest that inclined surfaces be treated with the simple substitution indicated in Eq. (9-19).

9-3 THE CONDENSATION NUMBER

Because the film Reynolds number is so important in determining condensation behavior, it is convenient to express the heat-transfer coefficient directly in terms of Re. We include the effect of inclination and write the heat-transfer equations in the form

$$\bar{h} = C\left[\frac{\rho(\rho - \rho_v)k^3 g \sin\phi\, h_{fg}}{\mu L(T_g - T_w)}\right]^{1/4} \tag{9-20}$$

where the constant is evaluated for a plate or cylindrical geometry. From Eq. (9-15) we can solve for $T_g - T_w$ as

$$T_g - T_w = \frac{\dot{m} h_{fg}}{\bar{h} A} \tag{9-21}$$

where A, once again, is the surface area for heat transfer. Substituting Eq. (9-21) in (9-20) and solving for $\bar{h}$ gives

$$\bar{h}^{3/4} = C\left[\frac{\rho(\rho - \rho_v)g \sin\phi\, k^3\, A/L}{\mu \dot{m}}\right]^{1/4} \tag{9-22}$$

This may be restructured as

$$\bar{h}^{3/4} = C\left[\frac{\rho(\rho - \rho_v)gk^3}{\mu^2}\,\frac{\mu P}{4\dot{m}}\,\frac{4 \sin\phi\, A/P}{L}\right]^{1/4}$$

and we may solve for $\bar{h}$ as

$$\bar{h} = C^{4/3}\left[\frac{\rho(\rho - \rho_v)gk^3}{\mu^2}\,\frac{\mu P}{4\dot{m}}\,\frac{4 \sin\phi\, A/P}{L}\right]^{1/3} \tag{9-23}$$

We now define a new dimensionless group, the *condensation number* Co, as

$$\text{Co} = \bar{h}\left[\frac{\mu^2}{k^3\rho(\rho - \rho_v)g}\right]^{1/3} \tag{9-24}$$

so that Eq. (9-23) can be expressed in the form

$$\text{Co} = C^{4/3}\left(\frac{4 \sin\phi\, A/P}{L}\right)^{1/3} \text{Re}_f^{-1/3} \tag{9-25}$$

For a vertical plate $A/PL = 1.0$, and we obtain, using the constant from Eq. (9-10),

$$\text{Co} = 1.47\, \text{Re}_f^{-1/3} \qquad \text{for } \text{Re}_f < 1800 \tag{9-26}$$

For a horizontal cylinder $A/PL = \pi$ and

$$\text{Co} = 1.514\, \text{Re}_f^{-1/3} \qquad \text{for } \text{Re}_f < 1800 \tag{9-27}$$

When turbulence is encountered in the film, an empirical correlation by Kirkbride [2] may be used:

$$\text{Co} = 0.0077\, \text{Re}_f^{0.4} \qquad \text{for } \text{Re}_f > 1800 \tag{9-28}$$

9-4 FILM CONDENSATION INSIDE HORIZONTAL TUBES

Our discussion of film condensation so far has been limited to *exterior surfaces*, where the vapor and liquid condensate flows are not restricted by some overall flow-channel dimensions. Condensation inside tubes is of considerable practical interest because of applications to condensers in refrigeration and air-conditioning systems, but unfortunately these phenomena are quite complicated and not amenable to a simple analytical treatment. The overall flow rate of vapor strongly influences the heat-transfer rate in the forced convection-condensation system, and this in turn is influenced by the rate of liquid accumulation on the walls. Because of the complicated flow phenomena involved we shall present only two empirical relations for heat transfer and refer the reader to Rohsenow [37] for more complete information.

Chato [38] obtained the following expression for condensation of refrigerants at low vapor velocities inside horizontal tubes:

$$\bar{h} = 0.555 \left[\frac{\rho(\rho - \rho_v) g k^3 h'_{fg}}{\mu d(T_g - T_w)} \right]^{1/4} \tag{9-29}$$

This equation is restricted to low vapor Reynolds numbers such that

$$\text{Re}_v = \frac{dG_v}{\mu_v} < 35{,}000 \tag{9-30}$$

where Re_v is evaluated at *inlet* conditions to the tube. For higher flow rates an approximate empirical expression is given by Akers, Deans, and Crosser [39] as

$$\frac{\bar{h}d}{k_f} = 0.026\, \text{Pr}_f^{1/3}\, \text{Re}_m^{0.8} \tag{9-31}$$

where now Re_m is a mixture Reynolds number, defined as

$$\text{Re}_m = \frac{d}{\mu_f} \left[G_f + G_v \left(\frac{\rho_f}{\rho_v} \right)^{1/2} \right] \tag{9-32}$$

The mass velocities for the liquid G_f and vapor G_v are calculated as if each occupied the entire flow area. Equation (9-31) correlates experimental data within about 50 percent when

$$\text{Re}_v = \frac{dG_v}{\mu_v} > 20{,}000 \qquad \text{Re}_f = \frac{dG_f}{\mu_f} > 5000$$

EXAMPLE 9-1 Condensation on vertical plate

A vertical square plate, 30 by 30 cm, is exposed to steam at atmospheric pressure. The plate temperature is 98°C. Calculate the heat transfer and the mass of steam condensed per hour.

Solution

The Reynolds number must be checked to determine if the condensate film is laminar or turbulent. Properties are evaluated at the film temperature:

$$T_f = \frac{100 + 98}{2} = 99°\text{C} \qquad \rho_f = \quad 960 \text{ kg/m}^3$$

$$\mu_f = 2.82 \times 10^{-4} \text{ kg/m} \cdot \text{s} \qquad k_f = 0.68 \text{ W/m} \cdot °\text{C}$$

For this problem the density of the vapor is very small in comparison with that of the liquid, and we are justified in making the substitution

$$\rho_f(\rho_f - \rho_v) \approx \rho_f^2$$

In trying to calculate the Reynolds number we find that it is dependent on the mass flow of condensate. But this is dependent on the heat-transfer coefficient, which is dependent on the Reynolds number. To solve the problem we assume either laminar or turbulent flow, calculate the heat-transfer coefficient, and then check the Reynolds number to see if our assumption was correct. Let us assume laminar film condensation. At atmospheric pressure we have

$$T_{\text{sat}} = 100°\text{C} \qquad h_{fg} = 2255 \text{ kJ/kg}$$

$$\begin{aligned}\bar{h} &= 0.943\left[\frac{\rho_f^2 g h_{fg} k_f^3}{L\mu_f(T_g - T_w)}\right]^{1/4} \\ &= 0.943\left[\frac{(960)^2(9.8)(2.255 \times 10^6)(0.68)^3}{(0.3)(2.82 \times 10^{-4})(100 - 98)}\right]^{1/4} \\ &= 13{,}150 \text{ W/m}^2 \cdot °\text{C} \quad [2316 \text{ Btu/h} \cdot \text{ft}^2 \cdot °\text{F}]\end{aligned}$$

Checking the Reynolds number with Eq. (9-17), we have

$$\begin{aligned}\text{Re}_f &= \frac{4hL(T_{\text{sat}} - T_w)}{h_{fg}\mu_f} \\ &= \frac{(4)(13{,}150)(0.3)(100 - 98)}{(2.255 \times 10^6)(2.82 \times 10^{-4})} = 49.6\end{aligned}$$

so that the laminar assumption was correct. The heat transfer is now calculated from

$$q = \bar{h}A(T_{\text{sat}} - T_w) = (13{,}150)(0.3)^2(100 - 98) = 2367 \text{ W} \quad [8079 \text{ Btu/h}]$$

The total mass flow of condensate is

$$\dot{m} = \frac{q}{h_{fg}} = \frac{2367}{2.255 \times 10^6} = 1.05 \times 10^{-3} \text{ kg/s} = 3.78 \text{ kg/h} \quad [8.33 \text{ lb}_m\text{/h}]$$

■ **EXAMPLE 9-2** Condensation on tube bank

One hundred tubes of 0.50-in (1.27-cm) diameter are arranged in a square array and exposed to atmospheric steam. Calculate the mass of steam condensed per unit length of tubes for a tube wall temperature of 98°C.

Solution

The condensate properties are obtained from Example 9-1. We employ Eq. (9-12) for the solution, replacing d by nd, where $n = 10$. Thus,

$$\bar{h} = 0.725 \left[\frac{\rho_f^2 g h_{fg} k_f^3}{\mu_f n d (T_g - T_w)} \right]^{1/4}$$

$$= 0.725 \left[\frac{(960)^2(9.8)(2.255 \times 10^6)(0.68)^3}{(2.82 \times 10^{-4})(10)(0.0127)(100 - 98)} \right]^{1/4}$$

$$= 12{,}540 \text{ W/m}^2 \cdot {}^\circ\text{C} \quad [2209 \text{ Btu/h} \cdot \text{ft}^2 \cdot {}^\circ\text{F}]$$

The total surface area is

$$\frac{A}{L} = n \pi d = (100)\pi(0.0127) = 3.99 \text{ m}^2\text{/m}$$

so the heat transfer is

$$\frac{q}{L} = h \frac{A}{L} (T_g - T_w)$$

$$= (12{,}540)(3.99)(100 - 98) = 100.07 \text{ kW/m}$$

The total mass flow of condensate is then

$$\frac{\dot{m}}{L} = \frac{q/L}{h_{fg}} = \frac{1.0007 \times 10^5}{2.255 \times 10^6} = 0.0444 \text{ kg/s} = 159.7 \text{ kg/h} \quad [352 \text{ lb}_m\text{/h}]$$

9-5 BOILING HEAT TRANSFER

When a surface is exposed to a liquid and is maintained at a temperature above the saturation temperature of the liquid, boiling may occur, and the heat flux will depend on the difference in temperature between the surface and the saturation temperature. When the heated surface is submerged below a free surface of liquid, the process is referred to as *pool boiling*. If the temperature of the liquid is below the saturation temperature, the process is called *subcooled*, or *local, boiling*. If the liquid is maintained at saturation temperature, the process is known as *saturated*, or *bulk, boiling*.

The different regimes of boiling are indicated in Fig. 9-3, where heat-flux data from an electrically heated platinum wire submerged in water are plotted against temperature excess $T_w - T_{sat}$. In region I free-convection currents are responsible for motion of the fluid near the surface. In this region the liquid near the heated surface is superheated slightly, and it subsequently evaporates when it rises to the surface. The heat transfer in this region can be calculated with the free-convection relations presented in Chap. 7. In region II bubbles begin to form on the surface of the wire and are dissipated in the liquid after breaking away from the surface. This region indicates the beginning of *nucleate boiling*. As the temperature excess is increased further, bubbles form more rapidly and rise to the surface of the liquid, where they are dissipated. This is indicated in region III. Eventually, bubbles are formed so rapidly that they blanket the heating surface and prevent the inflow of fresh liquid from taking their place. At this point the bubbles coalesce and form a vapor film which covers the surface. The heat must be conducted through this film before it can

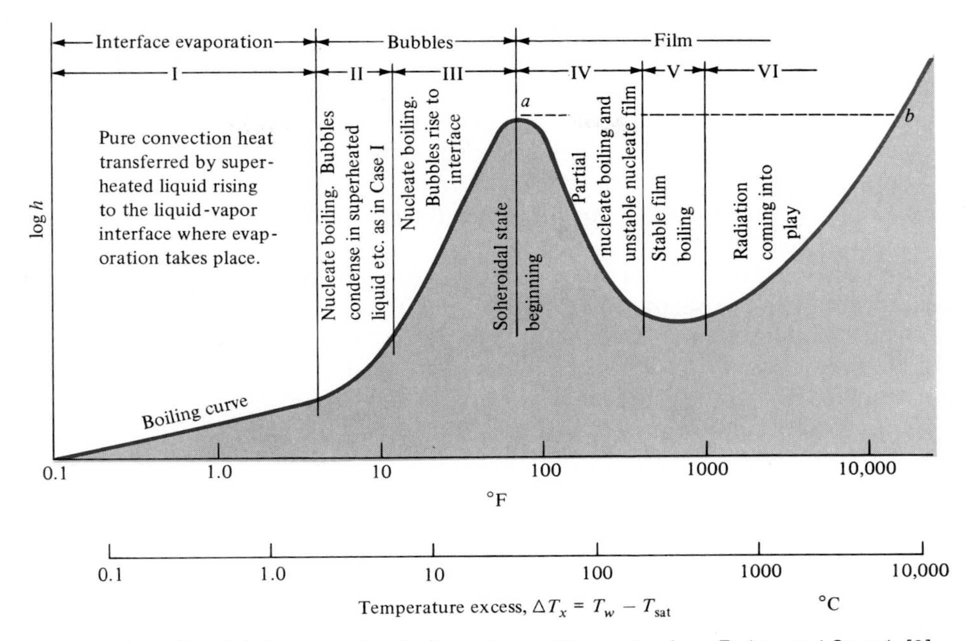

Fig. 9-3 Heat-flux data from an electrically heated platinum wire, from Farber and Scorah [9].

reach the liquid and effect the boiling process. The thermal resistance of this film causes a reduction in heat flux, and this phenomenon is illustrated in region IV, the *film-boiling* region. This region represents a transition from nucleate boiling to film boiling and is unstable. Stable film boiling is eventually encountered in region V. The surface temperatures required to maintain stable film boiling are high, and once this condition is attained, a significant portion of the heat lost by the surface may be the result of thermal radiation, as indicated in region VI.

An electrically heated wire is unstable at point a since a small increase in ΔT_x at this point results in a decrease in the boiling heat flux. But the wire still must dissipate the same heat flux, or its temperature will rise, resulting in operation farther down to the boiling curve. Eventually, equilibrium may be reestablished only at point b in the film-boiling region. This temperature usually exceeds the melting temperature of the wire, so that burnout results. If the electric-energy input is quickly reduced when the system attains point a, it may be possible to observe the partial nucleate boiling and unstable film region.

In nucleate boiling, bubbles are created by the expansion of entrapped gas or vapor at small cavities in the surface. The bubbles grow to a certain size, depending on the surface tension at the liquid-vapor interface and the temperature and pressure. Depending on the temperature excess, the bubbles may collapse on the surface, may expand and detach from the surface to be dissipated in the body of the liquid, or at sufficiently high temperatures may rise to the surface of the liquid before being dissipated. When local boiling conditions are observed, the primary mechanism of heat transfer is thought to be the intense agitation at the heat-transfer surface which creates the high heat-transfer rates observed in boiling. In saturated, or bulk, boiling the bubbles may break away from the surface because of the buoyancy action and move into the body of the liquid. In this case the heat-transfer rate is influenced by both the agitation caused by the bubbles and the vapor transport of energy into the body of the liquid.

Experiments have shown that the bubbles are not always in thermodynamic equilibrium with the surrounding liquid; i.e., the vapor inside the bubble is not necessarily at the same temperature as the liquid. Considering a spherical bubble as shown in Fig. 9-4, the pressure forces of the liquid and vapor must be balanced by the surface-tension force at the vapor-liquid interface. The pressure force acts on an area of πr^2, and the surface tension acts on the interface length of $2\pi r$. The force balance is

$$\pi r^2(p_v - p_l) = 2\pi r\sigma$$

or

$$p_v - p_l = \frac{2\sigma}{r} \tag{9-32a}$$

where p_v = vapor pressure inside bubble
p_l = liquid pressure
σ = surface tension of vapor-liquid interface

Now, suppose we consider a bubble in pressure equilibrium, i.e., one which is not growing or collapsing. Let us assume that the temperature of the vapor inside the bubble is the saturation temperature corresponding to the pressure p_v. If the liquid is at the saturation temperature corresponding to the pressure p_l, it is below the temperature inside the bubble. Consequently, heat must be conducted out of the bubble, the vapor inside must condense, and the bubble must collapse. This is the phenomenon which occurs when the bubbles collapse on the heating surface or in the body of the liquid. In order for the bubbles to grow and escape to the surface, they must receive heat from the liquid. This requires that the liquid be in a superheated condition so that the temperature

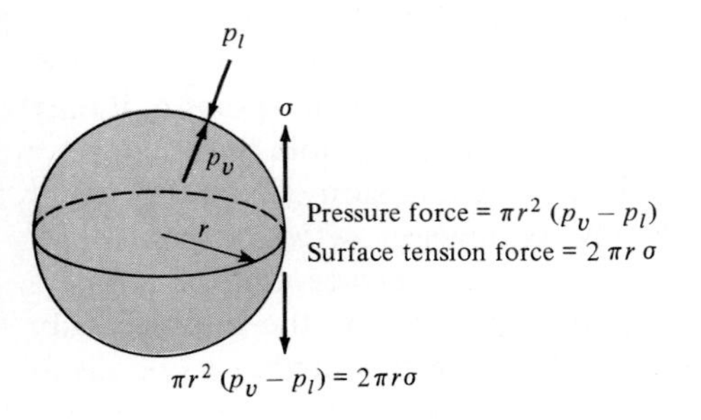

Fig. 9-4 Force balance on a vapor bubble.

of the liquid is greater than the vapor temperature inside the bubble. This is a metastable condition, but it is observed experimentally and accounts for the growth of bubbles after leaving the surface in some regions of nucleate boiling.

Figure 9-5 is a photograph illustrating several boiling regimes. The horizontal 0.25-in-diameter copper rod is heated from the right side and immersed in isopropanol. As a result of the temperature gradient along the rod, it was possible to observe the different regimes simultaneously. At the left end of the rod, the surface temperature is only slightly greater than the bulk fluid temperature, so that free-convection boiling is observed. Farther to the right, higher surface temperatures are experienced, and nucleate boiling is observed. Still farther to the right, transition boiling takes place; finally, film boiling is observed at the wall. Note the blanketing action of the vapor film on the right-hand portion of the rod.

More detailed photographs of the different boiling regimes using methanol are given in Fig. 9-6. The vigorous action of nucleate boiling is illustrated in Fig. 9-6*a*. At higher surface temperatures the bubbles start to coalesce, and transition boiling is observed, as in Fig. 9-6*b*. Finally, at still higher temperatures the heat-transfer surface is completely covered by a vapor film, and large vapor bubbles break away from the surface. A more vigorous film-boiling phenomenon is illustrated in Fig. 9-7 for methanol on a *vertical* tube. The vapor film rises up to the surface and develops into very active turbulent behavior at the top.

The process of bubble growth is a complex one, but a simple qualitative explanation of the physical mechanism may be given. Bubble growth takes place when heat is conducted to the liquid-vapor interface from the liquid. Evaporation then takes place at the interface, thereby increasing the total vapor volume. Assuming that the liquid pressure remains constant, Eq. (9-32*a*) requires that the pressure inside the bubble be reduced. Corresponding to a reduction in pressure inside the bubble will be a reduction in the vapor temperature and a larger temperature difference betwen the liquid and vapor if the bubble stays at its same spatial position in the liquid. However, the bubble will likely rise from the heated surface, and the farther away it moves, the lower the liquid temperature will be. Once the bubble moves into a region where the liquid temperature is below that of the vapor, heat will be conducted out, and the bubble will collapse. Hence the bubble growth process may reach a balance at some location in the liquid, or if the liquid is superheated enough, the bubbles may rise to the surface before being dissipated.

There is considerable controversy as to exactly how bubbles are initially formed on the heat-transfer surface. Surface conditions—both roughness and type of material—can play a central role in the bubble formation-and-growth drama. The mystery has not been completely solved, and remains a subject of intense research. Excellent summaries of the status of knowledge of boiling heat transfer are presented in Refs. 18, 23, 49, and 50. The interested reader is referred to these discussions for more extensive information than is presented

Fig. 9-5 Photograph of 0.25-in-diameter copper rod heated on the right side and immersed in isopropanol, from Haley and Westwater [17]. (*Photograph courtesy of Professor J. W. Westwater.*)

in this chapter. Heat-transfer problems in two-phase flow are discussed by Wallis [28].

Before specific relations for calculating boiling heat transfer are presented, it is suggested that the reader review the discussion of the last few pages and correlate it with some simple experimental observations of boiling. For this purpose a careful study of the boiling process in a pan of water on the kitchen stove can be quite enlightening.

Rohsenow [5] correlated experimental data for nucleate pool boiling with the following relation.

$$\frac{C_l \Delta T_x}{h_{fg}\,\mathrm{Pr}_l^s} = C_{sf}\left[\frac{q/A}{\mu_l h_{fg}}\sqrt{\frac{g_c \sigma}{g(\rho_l - \rho_v)}}\right]^{0.33} \tag{9-33}$$

Fig. 9-6 Boiling of methanol on a horizontal 9.53-mm-diameter steam-heated copper tube, from Ref. 40: (*a*) nucleate boiling, q/A = 242.5 kW/m², ΔT_x = 37°C; (*b*) transition boiling, q/A = 217.6 kW/m², ΔT_x = 62°C; (*c*) film boiling, q/A = 40.9 kW/m², ΔT_x = 82°C. (*Photographs courtesy of Professor J. W. Westwater.*)

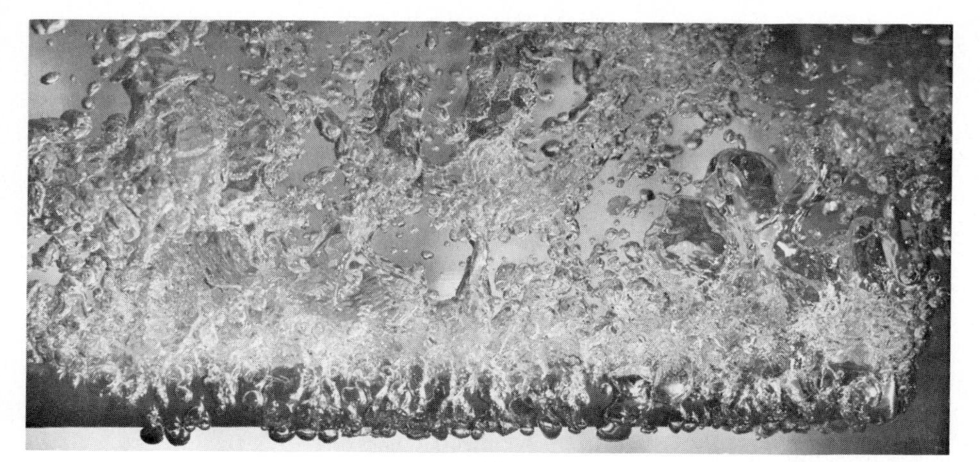

(*a*)

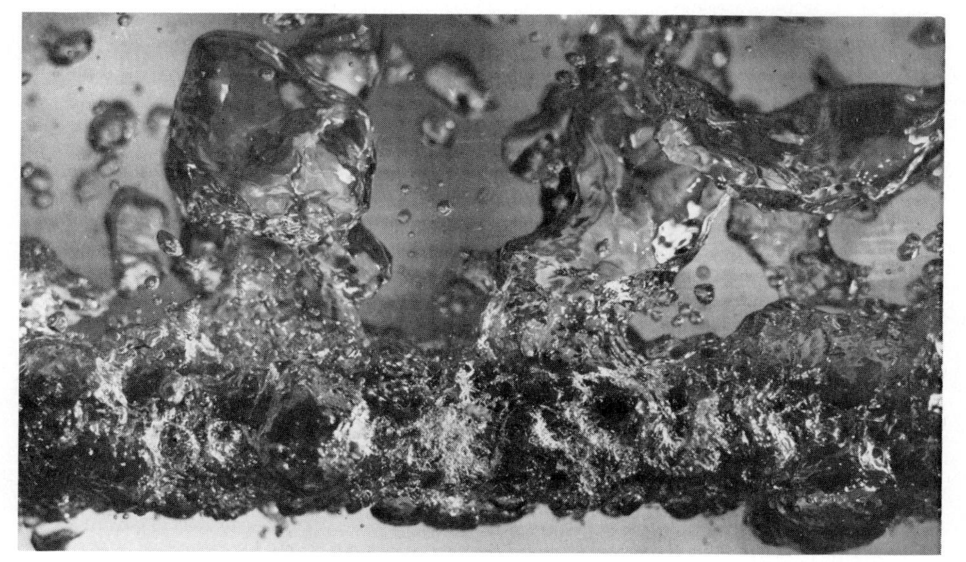

(*b*)

(*c*)

Fig. 9-7 Turbulent film boiling of methanol on a vertical 19.05-mm-diameter steam-heated copper tube, from Ref. 41; q/A = 38.8 kW/m², ΔT_x = 138°C. (*Photograph courtesy Professor J. W. Westwater.*)

where C_l = specific heat of saturated liquid, Btu/lb$_m$ · °F or J/kg · °C
ΔT_x = temperature excess = $T_w - T_{sat}$, °F or °C
h_{fg} = enthalpy of vaporization, Btu/lb$_m$ or J/kg
$\Pr_l$ = Prandtl number of saturated liquid
q/A = heat flux per unit area, Btu/h · ft² or W/m² · °C
μ_l = liquid viscosity, lb$_m$/h · ft, or kg/m·s
σ = surface tension of liquid-vapor interface, lb$_f$/ft or N/m
g = gravitational acceleration, ft/s² or m/s²
ρ_l = density of saturated liquid, lb$_m$/ft³ or kg/m³
ρ_v = density of saturated vapor, lb$_m$/ft³ or kg/m³
C_{sf} = constant, determined from experimental data
s = 1.0 for water and 1.7 for other liquids

Values of the surface tension are given in Ref. 10, and a brief tabulation of the vapor-liquid surface tension for water is given in Table 9-1.

The functional form of Eq. (9-33) was determined by analyzing the significant parameters in bubble growth and dissipation. Experimental data for nucleate

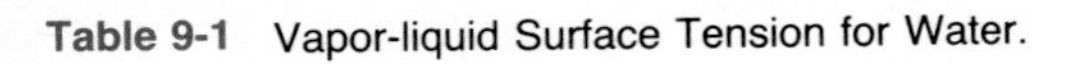
Table 9-1 Vapor-liquid Surface Tension for Water.

Saturation temperature		*Surface tension*	
°F	°C	$\sigma \times 10^4$, lb_f/ft	σ, mN/m
32	0	51.8	75.6
60	15.56	50.2	73.3
100	37.78	47.8	69.8
140	60	45.2	66.0
200	93.33	41.2	60.1
212	100	40.3	58.8
320	160	31.6	46.1
440	226.67	21.9	32.0
560	293.33	11.1	16.2
680	360	1.0	1.46
705.4	374.1	0	0

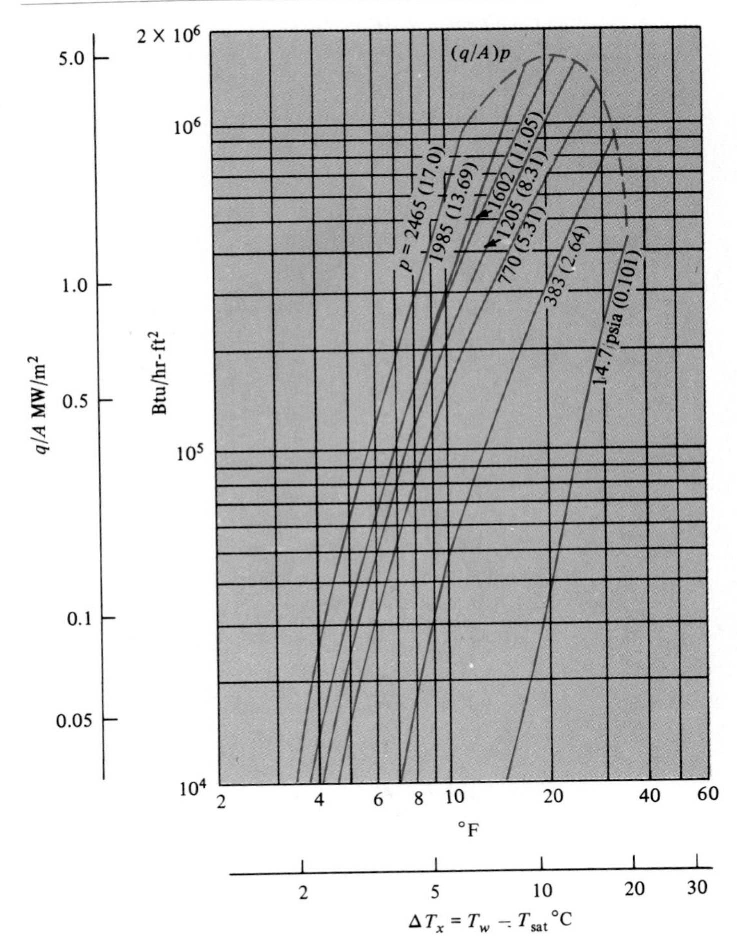

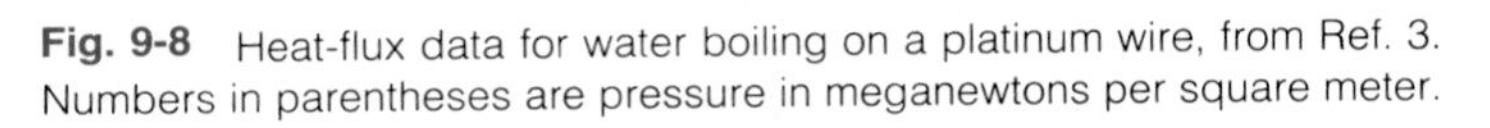
Fig. 9-8 Heat-flux data for water boiling on a platinum wire, from Ref. 3. Numbers in parentheses are pressure in meganewtons per square meter.

boiling of water on a platinum wire are shown in Fig. 9-8, and a correlation of these data by the Rohsenow equation is shown in Fig. 9-9, indicating good agreement. The value of the constant C_{sf} for the water-platinum combination is 0.013. Values for other fluid-surface combinations are given in Table 9-2. Equation (9-33) may be used for geometries other than horizontal wires, and in general it is found that geometry is not a strong factor in determining heat transfer for pool boiling. This would be expected because the heat transfer is primarily dependent on bubble formation and agitation, which is dependent on surface area, and not surface shape. Vachon, Nix, and Tanger [29] have determined values of the constants in the Rohsenow equation for a large number of surface-fluid combinations. There are several extenuating circumstances that influence the determination of the constants.

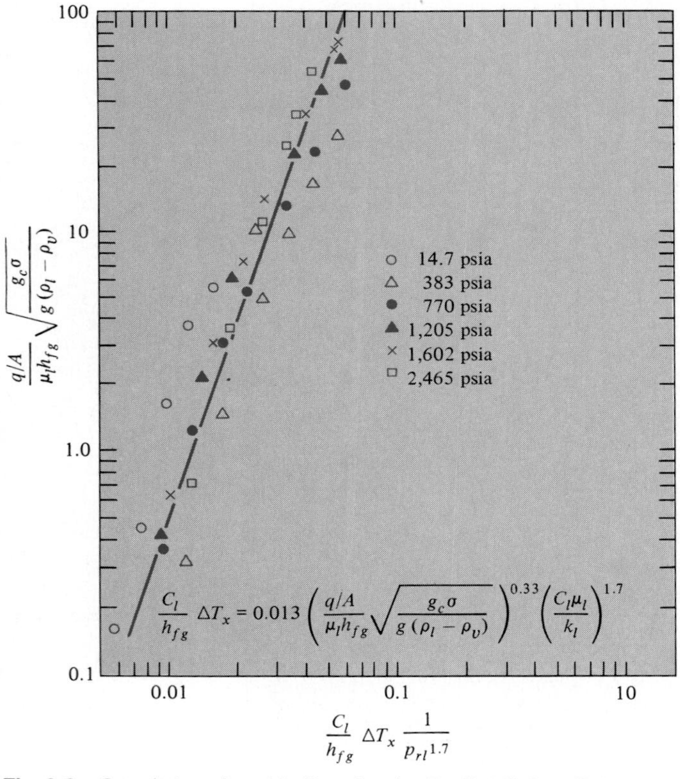

Fig. 9-9 Correlation of pool-boiling data by Eq. (9-33), from Rohsenow [5].

Table 9-2 Values of the Coefficient C_{sf} for Various Liquid-surface Combinations.

Fluid-heating-surface combination	C_{sf}
Water-copper [11]†	0.013
Water-platinum [12]	0.013
Water-brass [13]	0.0060
Water–emery-polished copper [29]	0.0128
Water–ground and polished stainless steel [29]	0.0080
Water–chemically etched stainless steel [29]	0.0133
Water–mechanically polished stainless steel [29]	0.0132
Water–emery-polished and paraffin-treated copper [29]	0.0147
Water–scored copper [29]	0.0068
Water–Teflon pitted stainless steel [29]	0.0058
Carbon tetrachloride-copper [11]	0.013
Carbon tetrachloride–emery-polished copper [29]	0.0070
Benzene-chromium [14]	0.010
n-Butyl alcohol–copper [11]	0.00305
Ethyl alcohol–chromium [14]	0.027
Isopropyl alcohol–copper [11]	0.00225
n-Pentane–chromium [14]	0.015
n-Pentane–emery-polished copper [29]	0.0154
n-Pentane–emery-polished nickel [29]	0.0127
n-Pentane–lapped copper [29]	0.0049
n-Pentane–emery-rubbed copper [29]	0.0074
35% K_2CO_3–copper [11]	0.0054
50% K_2CO_3–copper [11]	0.0027

†Numbers in brackets refer to source of data.

■ **EXAMPLE 9-3** Boiling on brass plate

A heated brass plate is submerged in a container of water at atmospheric pressure. The plate temperature is 242°F. Calculate the heat transfer per unit area of plate.

Solution

We could solve this problem by determining all the properties for use in Eq. (9-33) and subsequently determining the heat flux. An alternative method is to use the data of Fig. 9-8 in conjunction with Table 9-2. Upon writing Eq. (9-33), we find that if the heat flux for *one* particular water-surface combination is known, the heat flux for some other surface may easily be determined in terms of the constants C_{sf} for the two surfaces since the fluid properties at any given temperature and pressure are the same regardless of the surface material. From Fig. 9-8 the heat flux for the water-platinum combination is

$$\frac{q}{A} = 3 \times 10^5 \text{ Btu/h} \cdot \text{ft}^2 \quad [946.1 \text{ kW/m}^2]$$

since

$$T_w - T_{\text{sat}} = 242 - 212 = 30°\text{F} \quad [16.7°\text{C}]$$

From Table 9-2

$$C_{sf} = \begin{cases} 0.013 & \text{for water-platinum} \\ 0.006 & \text{for water-brass} \end{cases}$$

Accordingly, $$\frac{(q/A)_{\text{water-brass}}}{(q/A)_{\text{water-platinum}}} = \left(\frac{C_{sf\ \text{water-platinum}}}{C_{sf\ \text{water-brass}}}\right)^3$$

and $$\left(\frac{q}{A}\right)_{\text{water-brass}} = (3 \times 10^5)\left(\frac{0.013}{0.006}\right)^3$$

$$= 3.4 \times 10^6\ \text{Btu/h} \cdot \text{ft}^2 \quad [1.072 \times 10^7\ \text{W/m}^2]$$

When a liquid is forced through a channel or over a surface maintained at a temperature greater than the saturation temperature of the liquid, forced-convection boiling may result. For forced-convection boiling in smooth tubes Rohsenow and Griffith [6] recommended that the forced-convection effect be computed with the Dittus-Boelter relation of Chap. 6 [Eq. (6-4)] and that this effect be added to the boiling heat flux computed from Eq. (9-33). Thus

$$\left(\frac{q}{A}\right)_{\text{total}} = \left(\frac{q}{A}\right)_{\text{boiling}} + \left(\frac{q}{A}\right)_{\text{forced convection}} \tag{9-34}$$

For computing the forced-convection effect, it is recommended that the coefficient 0.023 be replaced by 0.019 in the Dittus-Boelter equation. The temperature difference between wall and liquid bulk temperature is used to compute the forced-convection effect.

The concept of adding the forced convection and boiling heat fluxes has been developed further in Ref. 46 with good results; however, the terms in the equations are much more complicated and too elaborate to present here. An individual working in this field should consult this reference.

Forced-convection boiling is not necessarily as simple as might be indicated by Eq. (9-34). This equation is generally applicable to forced-convection situations where the bulk liquid temperature is subcooled, in other words, for *local* forced-convection boiling. Once saturated or bulk boiling conditions are reached, the situation changes rapidly. A fully developed nucleate boiling phenomenon is eventually encountered which is independent of the flow velocity or forced-convection effects. Various relations have been presented for calculating the heat flux in the fully developed boiling state. McAdams et al. [21] suggested the following relation for low-pressure boiling water:

$$\frac{q}{A} = 2.253\,(\Delta T_x)^{3.96} \quad \text{W/m}^2 \quad \text{for } 0.2 < p < 0.7 \text{ MPa} \tag{9-35}$$

For higher pressures Levy [22] recommends the relation

$$\frac{q}{A} = 283.2p^{4/3}\,(\Delta T_x)^3 \quad \text{W/m}^2 \quad \text{for } 0.7 < p < 14 \text{ MPa} \tag{9-36}$$

In these equations ΔT_x is in degrees Celsius and p is in mega pascals.

If boiling is maintained for a sufficiently long length of tube, the majority of the flow area will be occupied by vapor. In this instance the vapor may flow rapidly in the central portion of the tube while a liquid film is vaporized along the outer surface. This situation is called *forced-convection vaporization* and is normally treated as a subject in two-phase flow and heat transfer. Several complications arise in this interesting subject, many of which are summarized by Tong [23] and Wallis [28].

The peak heat flux for nucleate pool boiling is indicated as point a in Fig. 9-3 and by a dashed line in Fig. 9-8. Zuber [7] has developed an analytical expression for the peak heat flux in nucleate boiling by considering the stability requirements of the interface between the vapor film and liquid. This relation is

$$\left(\frac{q}{A}\right)_{\text{max}} = \frac{\pi}{24} h_{fg}\rho_v \left[\frac{\sigma g(\rho_l - \rho_v)}{\rho_v^2}\right]^{1/4} \left(1 + \frac{\rho_v}{\rho_l}\right)^{1/2} \tag{9-37}$$

where σ is the vapor-liquid surface tension. This relation is in good agreement with experimental data. In general, the type of surface material does not affect the peak heat flux, although surface cleanliness can be an influence, dirty surfaces causing increases of approximately 15 percent in the peak value.

The peak heat flux in flow boiling is a more complicated situation because the rapid generation of vapor produces a complex two-phase flow system which strongly influences the maximum heat flux that may be attained at the heat-transfer surface. Near the heated surface a thin layer of superheated liquid is formed, followed by a layer containing both bubbles and liquid. The core of the flow is occupied, for the most part, by vapor. The heat transfer at the wall is influenced by the boundary-layer development in that region and also by the rate at which diffusion of vapor and bubbles can proceed radially. Still further complications may arise from flow oscillations which are generated under certain conditions. Gambill [24] has suggested that the critical heat flux in flow boiling may be calculated by a superposition of the critical heat flux for pool boiling [Eq. (9-37)] and a forced-convection effect similar to the technique employed in Eq. (9-34). Levy [25] has considered the effects of vapor diffusion on the peak heat flux in flow boiling, and Tong [23] presents a summary of available data on the subject.

An interesting peak heat-flux phenomenon is observed when liquid droplets impinge on hot surfaces. Experiments with water, acetone, alcohol, and some of the Freons indicate that the maximum heat transfer is observed for temperature excesses of about 165°C, for all the fluids. The peak flux is a function of the fluid properties and the normal component of the impact velocity. A correlation of experimental data is given in Ref. 30 as

$$\frac{Q_{\text{max}}}{\rho_L\, d^3\lambda} = 1.83 \times 10^{-3} \left(\frac{\rho_L{}^2 V^2 d}{\rho_{vf}\sigma g_c}\right)^{0.341} \tag{9-38}$$

where Q_{max} = maximum heat transfer per drop
ρ_L = density of liquid droplet
V = normal component of impact velocity
ρ_{vf} = vapor density evaluated at film temperature $(T_w + T_{\text{sat}})/2$
σ = surface tension

d = drop diameter
λ = modified heat of vaporization, defined by

$$\lambda = h_{fg} + c_{pv}\left(\frac{T_w - T_{sat}}{2}\right)$$

While not immediately apparent from this equation, the heat-transfer rates in droplet impingement are quite high, and as much as 50 percent of the droplet is evaporated during the short time interval of impact and bouncing. The case of zero impact velocity is of historical note and is called the *Leidenfrost phenomenon* [31]. This latter case can be observed by watching water droplets sizzle and dance about on a hot plate. Very high heat-transfer rates are also experienced when a liquid jet impinges on a hot surface maintained at temperatures significantly greater than the saturation temperature. Both nucleate- and film-boiling phenomena can be observed, and relations for calculating the heat-transfer rates are presented in Ref. 36.

Sun and Lienhard [34] have presented a relation for the peak boiling heat flux on horizontal cylinders which is in good agreement with experimental data. The relation is

$$\frac{q''_{max}}{q''_{maxF}} = 0.89 + 2.27 \exp\left(-3.44\sqrt{R'}\right) \qquad \text{for } 0.15 < R' \tag{9-39}$$

where R' is a dimensionless radius defined by

$$R' = R\left[\frac{g(\rho_l - \rho_v)}{\sigma}\right]^{1/2}$$

and q''_{maxF} is the peak heat flux on an infinite horizontal plate derived in Ref. 33 as

$$q''_{maxF} = 0.131\sqrt{\rho_v}\,h_{fg}[\sigma g(\rho_l - \rho_v)]^{1/4} \tag{9-40}$$

Here, σ is the surface tension.

Bromley [8] suggests the following relation for calculation of heat-transfer coefficients in the stable film-boiling region on a horizontal tube:

$$h_b = 0.62\left[\frac{k_g^3\rho_v(\rho_l - \rho_v)g(h_{fg} + 0.4c_{pv}\,\Delta T_x)}{d\mu_v\,\Delta T_x}\right]^{1/4} \tag{9-41}$$

where d is the tube diameter. This heat-transfer coefficient considers only the conduction through the film and does not include the effects of radiation. The total heat-transfer coefficient may be calculated from the empirical relation

$$h = h_b\left(\frac{h_b}{h}\right)^{1/3} + h_r \tag{9-42}$$

where h_r is the radiation heat-transfer coefficient and is calculated by assuming an emissivity of unity for the liquid. Thus

$$h_r = \frac{\sigma\epsilon(T_w{}^4 - T_{sat}^4)}{T_w - T_{sat}} \tag{9-43}$$

where σ is the Stefan-Boltzmann constant and ϵ is the emissivity of the surface. Note that Eq. (9-42) will require an iterative solution for the total heat-transfer coefficient.

The properties of the vapor in Eq. (9-41) are to be evaluated at the film temperature defined by

$$T_f = \tfrac{1}{2}(T_w + T_{sat})$$

while the enthalpy of vaporization h_{fg} is to be evaluated at the saturation temperature.

9-6 SIMPLIFIED RELATIONS FOR BOILING HEAT TRANSFER WITH WATER

Many empirical relations have been developed to estimate the boiling heat-transfer coefficients for water. Some of the simplest relations are those presented by Jakob and Hawkins [15] for water boiling on the outside of submerged surfaces at atmospheric pressure (Table 9-3). These heat-transfer coefficients may be modified to take into account the influence of pressure by using the empirical relation

$$h_p = h_1 \left(\frac{p}{p_1}\right)^{0.4} \tag{9-44}$$

where h_p = heat-transfer coefficient at some pressure p
h_1 = heat-transfer coefficient at atmospheric pressure as determined from Table 9-3
p = system pressure
p_1 = standard atmospheric pressure

Table 9-3 Simplified Relations for Boiling Heat-Transfer Coefficients to Water at Atmospheric Pressure, Adapted from Ref. 15. $\Delta T_x = T_w - T_{sat}$, °C.

Surface	$\frac{q}{A}$, kW/m²	*h*, W/m² · °C
Horizontal	$\frac{q}{A} < 16$	$1042\ (\Delta T_x)^{1/3}$
	$16 < \frac{q}{A} < 240$	$5.56\ (\Delta T_x)^3$
Vertical	$\frac{q}{A} < 3$	$537\ (\Delta T_{ax})^{1/7}$
	$3 < \frac{q}{A} < 63$	$7.96\ (\Delta T_x)^3$

For forced-convection local boiling inside vertical tubes the following relation is recommended [16]:

$$h = 2.54(\Delta T_x)^3 e^{p/1.551} \quad (9\text{-}45)$$

where ΔT_x is the temperature difference between the surface and saturated liquid in degrees Celsius and p is the pressure in mega pascals. The heat-transfer coefficient has the units of watts per square meter per degree Celsius. Equation (9-45) is valid over a pressure range of 5 to 170 atm.

■ **EXAMPLE 9-4** Flow boiling

Water at 5 atm flows inside a tube of 1-in [2.54-cm] diameter under local boiling conditions where the tube wall temperature is 10°C above the saturation temperature. Estimate the heat transfer in a 1.0-m length of tube.

Solution

For this calculation we use Eq. (9-45), noting that

$$\Delta T_x = 10°\text{C}$$

$$p = (5)(1.0132 \times 10^5\ \text{N/m}^2) = 0.5066\ \text{MPa}$$

The heat-transfer coefficient is then calculated as

$$h = (2.54)(10)^3 e^{0.5066/1.551}$$

$$= 3521\ \text{W/m}^2 \cdot °\text{C} \quad [620\ \text{Btu/h} \cdot \text{ft}^2 \cdot °\text{F}]$$

The surface area for a 1-m length of tube is

$$A = \pi\, dL = \pi(0.0254)(1.0) = 0.0798\ \text{m}^2$$

so the heat transfer is

$$q = hA(T_w - T_{\text{sat}})$$

$$= (3521)(0.0798)(10) = 2810\ \text{W/m}$$

■ 9-7 SUMMARY AND DESIGN INFORMATION

Boiling and condensation phenomena are very complicated, as we have shown in the preceding sections. The equations presented in these sections may be used to calculate heat-transfer coefficients for various geometries and fluid-surface combinations. For many preliminary design applications only approximate values of heat flux or heat-transfer coefficient are required, and Tables 9-4 to 9-6 give summaries of such information. Of course, more accurate values should be obtained for the final design of heat-transfer equipment.

Table 9-4 Approximate Values of Condensation Heat-transfer Coefficients for Vapors at 1 atm, According to Refs. 3 and 45.

Fluid	*Geometry*	*h* Btu/h · ft² · °F	*h* W/m² · °C	$T_g - T_w$, °C
Steam	Vertical surface	700–2000	4000–11,300	22–3
	Horizontal tubes, 0.6 to 3.0 in diameter	1700–4300	9600–24,400	20–2
Diphenyl	Vertical surface, turbulent, 12 ft	120–430	680–2400	72–13
	Horizontal tube, 1.7-in diameter	225–400	1280–2270	15–5
Dowtherm A	Vertical surface, 12 ft, turbulent	120–540	680–3060	40–20
Ethanol	Vertical surface, 6 in	200–350	1130–2000	55–11
	Horizontal tube, 2.0-in diameter	320–450	1800–2550	22–6
Propanol	Horizontal tube, 2.0-in diameter	250–300	1400–1700	26–13
Butanol	Horizontal tube, 2.0-in diameter	250–300	1400–1700	26–13
Benzene	Horizontal tubes, 0.6- to 1.3-in diameter	230–380	1300–2150	45–13

Table 9-5 Relative Magnitudes of Nucleate-Boiling Heat-transfer Coefficients at 1 atm Referenced to Value for Water, According to Ref. 42.

Fluid	$\frac{h}{h_w}$
Water	1.0
20% sugar	0.87
10% Na_2SO_4	0.94
26% glycerin	0.83
55% glycerin	0.75
24% NaCl	0.61
Isopropanol	0.70
Methanol	0.53
Toluene	0.36
Carbon tetrachloride	0.35
n-Butanol	0.32

Table 9-6 Approximate Burnout Heat Flux at 1 atm, According to Refs. 3, 43, and 44.

Fluid-surface combination	$(q/A)_{max}$ Btu/h · ft^2 × 10^{-3}	kW/m^2	ΔT_x, °C
Water, copper	200–270	620–850	
Copper-chrome plated	300–400	940–1260	23–28
Steel	410	1290	30
Benzene, copper	43.5	130	
Aluminum	50.5	160	
Propanol, nickel-plated copper	67–110	210–340	42–50
Butanol, nickel-plated copper	79–105	250–330	33–39
Ethanol, aluminum	55	170	
Copper	80.5	250	
Methanol, copper	125	390	
Chrome-plated copper	111	350	
Steel	125	390	
Liquid H_2	9.53	30	2
Liquid N_2	31.7	100	11
Liquid O_2	47.5	150	11

REVIEW QUESTIONS

1. Why are higher heat-transfer rates experienced in dropwise condensation than in film condensation?
2. How is the Reynolds number defined for film condensation?
3. What is meant by subcooled and saturated boiling?
4. Distinguish between nucleate and film boiling.
5. How is forced-convection boiling calculated?
6. Why does radiation play a significant role in film-boiling heat transfer?

PROBLEMS

9-1 Using Eq. (9-28) as a starting point, develop an expression for the average heat-transfer coefficient in turbulent condensation as a function of only the fluid properties, length of the plate, and temperature difference; i.e., eliminate the Reynolds number from Eq. (9-28) to obtain a relation similar to Eq. (9-10) for laminar condensation.

9-2 Show that the condensation Reynolds number for laminar condensation on a vertical plate may be expressed as

$$\mathrm{Re}_f = 3.77\left[\frac{L^3(T_g - T_w)^3\rho_f(\rho_f - \rho_v)gk_f^3}{\mu_f^5 h_{fg}^3}\right]^{1/4}$$

9-3 Develop an expression for the total condensate flow in a turbulent film in terms of the fluid properties, the temperature difference, and the dimensions of the plate.

9-4 Plot Eqs. (9-26) and (9-28) as

$$\log\left\{\bar{h}\left[\frac{\mu_f^2}{k_f^3\rho_f(\rho_f - \rho_v)g}\right]^{1/3}\right\}$$

versus log Re_f. Discuss this plot.

9-5 A vertical plate 30 cm wide and 1.2 m high is maintained at 70°C and exposed to saturated steam at 1 atm. Calculate the heat transfer and the total mass of steam condensed per hour.

9-6 A 40 by 40 cm plate is inclined at an angle of 30° with the vertical and exposed to water vapor at 1 atm. The plate is maintained at 98°C. Calculate the heat-transfer and mass-flow rate of condensate.

9-7 A 50 by 50 cm square vertical plate is maintained at 95°C and exposed to saturated steam at 1 atm pressure. Calculate the amount of steam condensed per hour.

9-8 Calculate the rate of condensation on a 1.5 by 1.5 m vertical plate maintained at 40°F and exposed to saturated water vapor at 55°F; h_{fg} = 2376 kJ/kg at 55°F.

9-9 A vertical plate 40 by 40 cm is exposed to saturated vapor ammonia at 100°F and the plate surface is maintained constant at 30°C. Calculate the condensation rate if h_{fg} = 477.8 Btu/lb_m at 100°F.

9-10 Saturated steam at 100 lb/in² abs condenses on the outside of a horizontal 1-in-diameter tube. The tube wall temperature is maintained at 280°F. Calculate the heat-transfer coefficient and the condensate flow per unit length of tube. Take T_{sat} = 328°F and h_{fg} = 889 Btu/lb.

9-11 An uninsulated, chilled water pipe carrying water at 2°C passes through a hot, humid factory area where the temperature is 35°C and the relative humidity is 80 percent because of steam-operated equipment in the factory. If the pipe is 5 cm in diameter and the exposed length is 7.5 m, estimate the condensate which will drip off the pipe. For this estimate assume that the pipe is exposed to saturated vapor at the partial pressure of the water vapor in the air.

9-12 A certain pressure cooker is designed to operate at 20 lb/in^2 gauge. It is well known that an item of food will cook faster in such a device because of the higher steam temperature at the higher pressure. Consider a certain item of food as a horizontal 4-in-diameter cylinder at a temperature of 95°F when placed in the cooker. Calculate the percentage increase in heat transfer to this cylinder for the 20-lb/in^2-gauge condition compared with condensation on the cylinder at standard atmospheric pressure.

9-13 Saturated steam at 100 lb/in^2 abs condenses on the outside of a horizontal 1-in-diameter tube. The tube wall temperature is maintained at 280°F. Calculate the heat-transfer coefficient and the condensate flow per unit length of tube.

9-14 Saturated steam at 1 atm pressure condenses on the outside of a 30-cm-diameter tube whose surface is maintained at 95°C. The tube is 15 m long. Calculate the amount of steam condensed per hour.

9-15 Condensing carbon dioxide at 20°C is in contact with a horizontal 10-cm-diameter tube maintained at 15°C. Calculate the condensation rate per meter of length if h_{fg} = 153.2 kJ/kg at 20°C.

9-16 A square array of four hundred $\frac{1}{4}$-in tubes is used to condense steam at atmospheric pressure. The tube walls are maintained at 88°C by a coolant flowing inside the tubes. Calculate the amount of steam condensed per hour per unit length of the tubes.

9-17 A condenser is to be designed to condense 1.3 kg/s of steam at atmospheric pressure. A square array of 1.25-cm-OD tubes is to be used with the outside tube walls maintained at 93°C. The spacing of the tubes is to be 1.9 cm between centers, and their length is 3 times the square dimension. How many tubes are required for the condenser, and what are the outside dimensions?

9-18 A heat exchanger is to be designed to condense 600 kg/h of steam at atmospheric pressure. A square array of four hundred 1.0-cm-diameter tubes is available for the task, and the tube wall temperature is to be maintained at 97°C. Estimate the length of tubes required.

9-19 Saturated water vapor at 1 atm enters a horizontal 5-cm-diameter tube 1.5 m long. Estimate the condensation for a tube wall temperature of 98°C.

9-20 Steam at 1 atm is to be condensed on the outside of a bank of 10 × 10 horizontal tubes 1.0 in in diameter. The tube surface temperature is maintained at 95°C. Calculate the quantity of steam condensed for a tube length of 2.0 ft.

9-21 A square array of 1.0-in-diameter tubes contains 100 tubes each having a length of 3 ft. The distance between centers is 1.8 in and the tube wall temperature is 97°C. The tubes are exposed to steam at 1 atm. Calculate the condensation rate in kilograms per hour.

9-22 In a large cold-storage plant ammonia is used as the refrigerant and in one application 1.2×10^7 kJ/h must be removed by condensing the ammonia at 85°F with an array of tubes with walls maintained at 78°F. Select several tube sizes, lengths, and array dimensions which might be used to accomplish the task. h_{fg} = 1148 kJ/kg at 85°F.

9-23 An ammonia condenser uses a 20 by 20 array of 0.25-in-diameter tubes 1.0 ft long. The ammonia condenses at 90°F and the tube walls are maintained at 82°F by a water flow inside. Calculate the rate of condensation of ammonia. h_{fg} at 90°F is 488.5 Btu/lb_m.

9-24 A condenser is to be designed to condense 10,000 kg/h of refrigerant 12 (CCl_2F_2) at 100°F. A square 25 by 25 array of 12-mm-diameter tubes is to be used, with water flow inside the tubes maintaining the wall temperature at 90°F. Calculate the length of the tubes. h_{fg} = 55.93 Btu/lb_m at 100°F.

9-25 Refrigerant 12 (CCl_2F_2) is condensed inside a horizontal 12-mm-diameter tube at a low vapor velocity. The condensing temperature is 90°F and the tube wall is at 80°F. Calculate the mass condensed per meter of tube length. h_{fg} = 57.46 Btu/lb_m at 90°F.

9-26 A heated vertical plate at a temperature of 107°C is immersed in a tank of water exposed to atmospheric pressure. The temperature of the water is 100°C, and boiling occurs at the surface of the plate. The area of the plate is 0.3 m^2. What is the heat lost from the plate in watts?

9-27 The surface tension of water at 212°F is 58.8 dyn/cm for the vapor in contact with the liquid. Assuming that the saturated vapor inside a bubble is at 213°F while the surrounding liquid is saturated at 212°F, calculate the size of the bubble.

9-28 Assuming that the bubble in Prob. 9-27 moves through the liquid at a velocity of 4.5 m/s, estimate the time required to cool the bubble 0.3°C by calculating the heat-transfer coefficient for flow over a sphere and using this in a lumped-capacity analysis as described in Chap. 4.

9-29 A heated 30 by 30 cm square copper plate serves as the bottom for a pan of water at 1 atm pressure. The temperature of the plate is maintained at 117°C. Estimate the heat transferred per hour by the plate.

9-30 Compare the heat flux calculated from the simple relations of Table 9-3 with the curve for atmospheric pressure in Fig. 9-8. Make the comparisons for two or three values of the temperature excess.

9-31 Water at 4 atm pressure flows inside a 2-cm-diameter tube under local boiling conditions where the tube wall temperature is 12°C above the saturation temperature. Estimate the heat transfer in a 60-cm length of tube.

9-32 Compare the heat-transfer coefficients for nucleate boiling of water, as shown in Fig. 9-8, with the simplified relations given in Table 9-3.

9-33 Using Eqs. (9-14) and (9-7), develop Eq. (9-26).

9-34 A platinum wire is submerged in saturated water at 5.3 MPa. What is the heat flux for a temperature excess of 11°C?

9-35 Water at 1 atm flows in a 1.25-cm-diameter brass tube at a velocity of 1.2 m/s. The tube wall is maintained at 110°C, and the average bulk temperature of the water is 96°C. Calculate the heat-transfer rate per unit length of tube.

9-36 A kettle with a flat bottom 30 cm in diameter is available. It is desired to boil 2.3

kg/h of water at atmospheric pressure in this kettle. At what temperature must the bottom surface of the kettle be maintained to accomplish this?

9-37 A 5-mm-diameter copper heater rod is submerged in water at 1 atm. The temperature excess is 11°C. Estimate the heat loss per unit length of the rod.

9-38 Compare the heat-transfer coefficients for boiling water and condensing steam on a horizontal tube for normal atmospheric pressure.

9-39 A certain boiler employs one hundred 2-cm-diameter tubes 1 m long. The boiler is designed to produce local forced-convection boiling of water at 3 MPa pressure with ΔT_x = 10°C. Estimate the total heat-transfer rate and the amount of saturated vapor which can be produced at 3 MPa.

9-40 A horizontal tube 3 mm in diameter and 7.5 cm long is submerged in water at 1.6 atm. Calculate the surface temperature necessary to generate a heat flux of 0.2 MW/m^2.

9-41 Copper electric heating rods 1.0 in diameter are used to produce steam at 5-lb/in^2-gauge pressure in a nucleate-pool-boiling arrangement where ΔT_x = 4°C. Estimate the length of rod necessary to produce 2000 lb_m/h of saturated vapor steam.

9-42 Estimate the nucleate-pool-boiling heat-transfer coefficient for a water–26% glycerin mixture at 1 atm in contact with a copper surface and ΔT_x = 15°C.

9-43 Compare Eqs. (9-35) and (9-36) with Eq. (9-45).

9-44 Estimate the nucleate-pool-boiling heat flux for water at 1 atm in contact with ground and polished stainless steel and ΔT_x = 30°F.

9-45 Calculate the peak heat flux for boiling water at atmospheric pressure on a horizontal cylinder of 1.25-cm OD. Use the Lienhard relation.

9-46 How much heat would be lost from a horizontal platinum wire, 1.0 mm in diameter and 12 cm long, submerged in water at atmospheric pressure if the surface temperature of the wire is 232°F?

9-47 A horizontal pipe at 94°C is exposed to steam at atmospheric pressure and 100°C. The pipe is 4.0 cm in diameter. Calculate the condensation rate per meter of length.

9-48 Liquid water at 1 atm and 98°C flows in a horizontal 2.5-cm-diameter brass tube maintained at 110°C. Calculate the heat-transfer coefficient if the Reynolds number based on liquid bulk conditions is 40,000.

9-49 A steel bar 1.25 cm in diameter and 5 cm long is removed from a 1200°C furnace and placed in a container of water at atmospheric pressure. Estimate the heat-transfer rate from the bar when it is first placed in the water.

9-50 Estimate the peak heat flux for boiling water at normal atmospheric pressure.

9-51 Heat-transfer coefficients for boiling are usually large compared with those for ordinary convection. Estimate the flow velocity which would be necessary to produce a value of h for forced convection through a smooth 6.5-mm-diameter brass tube comparable with that which could be obtained by pool boiling with ΔT_x = 16.7°C, p = 100 lb/in^2 abs, and water as the fluid. See Prob. 9-10 for data on properties.

9-52 A horizontal tube having a 1.25-cm OD is submerged in water at 1 atm and 100°C. Calculate the heat flux for surface temperatures of (*a*) 540°C, (*b*) 650°C, and (*c*) 800°C. Assume $\epsilon = 0.8$, and use Eq. (9-41).

9-53 Water at a rate of 1.0 liter/h in the form of 0.4-mm droplets at 25°C is sprayed on a hot surface at 280°C with an impact velocity of 3 m/s. Estimate the maximum heat transfer which can be achieved with this arrangement.

9-54 A square array of 196 tubes 1.25 cm in diameter is used to condense steam at 1 atm pressure. Water flowing inside the tubes maintains the outside surface temperature at 92°C. Calculate the condensation rate for tube lengths of 2.0 m.

9-55 A vertical plate maintained at 91°C is exposed to saturated steam at 1 atm pressure. Determine the height of plate to just produce a Reynolds number of 1800. What would the condensation rate be under these conditions?

9-56 Determine the boiling heat flux for a water-ground and polished stainless-steel combination for a temperature excess of 15°C at 1 atm.

9-57 Calculate the heat-transfer coefficient for forced-convection local boiling of water at 5 atm in a vertical tube. The temperature excess is 12°C.

9-58 Calculate the condensation rate for saturated steam at 1 atm exposed to a 30-cm-diameter cylinder maintained at 94°C.

9-59 A 20-cm-square vertical plate is maintained at 93°C and exposed to saturated water vapor at 1 atm pressure. Calculate the condensation rate and film thickness at the bottom of the plate.

REFERENCES

1. Nusselt, W.: Die Oberflachenkondensation des Wasserdampfes, *VDI Z.*, vol. 60, p. 541, 1916.
2. Kirkbride, C. G.: Heat Transfer by Condensing Vapors on Vertical Tubes, *Trans. AIChE,* vol. 30, p. 170, 1934.
3. McAdams, W. H.: "Heat Transmission," 3d ed., McGraw-Hill Book Company, New York, 1954.
4. Kern, D. Q.: "Process Heat Transfer," McGraw-Hill Book Company, New York, 1950.
5. Rohsenow, W. M.: A Method of Correlating Heat Transfer Data for Surface Boiling Liquids, *Trans. ASME,* vol. 74, p. 969, 1952.
6. Rohsenow, W. M., and P. Griffith: Correlation of Maximum Heat Flux Data for Boiling of Saturated Liquids, *AIChE-ASME Heat Transfer Symp.*, Louisville, Ky., 1955.
7. Zuber, N.: On the Stability of Boiling Heat Transfer, *Trans. ASME,* vol. 80, p. 711, 1958.
8. Bromley, L. A.: Heat Transfer in Stable Film Boiling, *Chem. Eng. Prog.* vol. 46, p. 221, 1950.
9. Farber, E. A., and E. L. Scorah: Heat Transfer to Water Boiling under Pressure, *Trans. ASME,* vol. 70, p. 369, 1948.

10 "Handbook of Chemistry and Physics," Chemical Rubber Publishing Company, Cleveland, Ohio, 1960.

11 Piret, E. L., and H. S. Isbin: Natural Circulation Evaporation Two-Phase Heat Transfer, *Chem. Eng. Prog.*, vol. 50, p. 305, 1954.

12 Addoms, J. N.: Heat Transfer at High Rates to Water Boiling outside Cylinders, Sc.D. thesis, Massachusetts Institute of Technology, Cambridge, Mass., 1948.

13 Cryder, D. S., and A. C. Finalbargo: Heat Transmission from Metal Surfaces to Boiling Liquids: Effect of Temperature of the Liquid on Film Coefficient, *Trans. AIChE,* vol. 33, p. 346, 1937.

14 Cichelli, M. T., and C. F. Bonilla: Heat Transfer to Liquids Boiling under Pressure, *Trans. AIChE,* vol. 41, p. 755, 1945.

15 Jakob, M., and G. Hawkins: "Elements of Heat Transfer," 3rd ed., John Wiley & Sons, Inc., New York, 1957.

16 Jakob, M.: "Heat Transfer," vol. 2, p. 584, John Wiley & Sons, Inc., New York, 1957.

17 Haley, K. W., and J. W. Westwater: Heat Transfer from a Fin to a Boiling Liquid, *Chem. Eng. Sci.,* vol. 20, p. 711, 1965.

18 Rohsenow, W. M. (ed.): "Developments in Heat Transfer," The M.I.T. Press, Cambridge, Mass., 1964.

19 Leppert, G., and C. C. Pitts: Boiling, *Adv. Heat Transfer,* vol. 1, 1964.

20 Gebhart, C.: "Heat Transfer," McGraw-Hill Book Company, New York, 1961.

21 McAdams, W. H., et al.: Heat Transfer at High Rates to Water with Surface Boiling, *Ind. Eng. Chem.,* vol. 41, pp. 1945–1955, 1949.

22 Levy, S.: Generalized Correlation of Boiling Heat Transfer, *J. Heat Transfer,* vol. 81C, pp. 37–42, 1959.

23 Tong, L. S.: "Boiling Heat Transfer and Two-Phase Flow," John Wiley & Sons, Inc., New York, 1965.

24 Gambill, W. R.: Generalized Prediction of Burnout Heat Flux for Flowing, Subcooled, Wetting Liquids, *AIChE Rep.* 17, *5th Nat. Heat Transfer Conf., Houston, 1962.*

25 Levy, S.: Prediction of the Critical Heat Flux in Forced Convection Flow, *USAEC Rep.* 3961, 1962.

26 Schmidt, E., W. Schurig, and W. Sellschop: Versuche über die Kondensation von Wasserdampf in Film- und Tropfenform, *Tech. Mech. Thermodyn. Bull.,* vol. 1, p. 53, 1930.

27 Citakoglu, E., and J. W. Rose: Dropwise Condensation: Some Factors Influencing the Validity of Heat Transfer Measurements, *Int. J. Heat Mass Transfer,* vol. 11, p. 523, 1968.

28 Wallis, G. B.: "One-dimensional Two-Phase Flow," McGraw-Hill Book Company, New York, 1969.

29 Vachon, R. I., G. H. Nix, and G. E. Tanger: Evaluation of Constants for the Rohsenow Pool-Boiling Correlation, *J. Heat Transfer,* vol. 90, p. 239, 1968.

30 McGinnis, F. K., and J. P. Holman: Individual Droplet Heat Transfer Rates for Splattering on Hot Surfaces, *Int. J. Heat Mass Transfer,* vol. 12, p. 95, 1969.

31 Bell, K. J.: The Leidenfrost Phenomenon: A Survey, *Chem. Eng. Prog. Symp. Ser.,* no. 79, p. 73, 1967.

32 Rohsenow, W. M.: Nucleation with Boiling Heat Transfer, *ASME Pap.* 70-HT-18.

33 Zuber, N., M. Tribus, and J. W. Westwater: The Hydrodynamic Crises in Pool Boiling of Saturated and Subcooled Liquids, *Int. Dev. Heat Transfer,* pp. 230–235, 1963.

34 Sun, K. H., and J. H. Lienhard: The Peak Boiling Heat Flux on Horizontal Cylinders, *Int. J. Heat Mass Trans.,* vol. 13, p. 1425, 1970.

35 Graham, C., and P. Griffith, Drop Size Distributions and Heat Transfer in Dropwise Condensation, *Int. J. Heat Mass Transfer,* vol. 16, p. 337, 1973.

36 Ruch, M. A., and J. P. Holman: Boiling Heat Transfer to a Freon-113 Jet Impinging upward onto a Flat Heated Surface, *Int. J. Heat Mass Transfer,* vol. 18, p. 51, 1974.

37 Rohsenow, W. M.: Film Condensation, chap. 12 in "Handbook of Heat Transfer," McGraw-Hill Book Company, New York, 1973.

38 Chato, J. C.: *J. Am. Soc. Refrig. Air Cond. Eng.,* February 1962, p. 52.

39 Akers, W. W., H. A. Deans, and O. K. Crosser: Condensing Heat Transfer within Horizontal Tubes, *Chem. Eng. Prog. Symp. Ser.,* vol. 55, no. 29, p. 171, 1958.

40 Westwater, J. W., and J. G. Santangelo: *Ind. Eng. Chem.,* vol. 47, p. 1605, 1955.

41 Westwater, J. W.: *Am. Sci.,* vol. 47, p. 427, 1959, photo by Y. Y. Hsu.

42 Fritz, W.: Verdampfen und Kondensieren, *Z. VDI, Beih, Verfahrenstech.,* no. 1, 1943.

43 Sauer, E. T., Cooper, H. B., and W. H. McAdams: Heat Transfer to Boiling Liquids, *Mech. Eng.,* vol. 60, p. 669, 1938.

44 Weil, L.: Échanges thermiques dans les liquides bouillants, *Fourth Int. Congr. Ind. Heating,* group I, sec. 13, Rep. 210, Paris, 1952.

45 Chilton, T. H., A. P. Colburn, R. P. Genereaux, and H. C. Vernon: *Trans. ASME, Pet. Mech. Eng.,* vol. 55, p. 7, 1933.

46 Bjorge, R. W., G. R. Hall, and W. M. Rohsenow: Correlations of Forced Convection Boiling Heat Transfer Data, *Int. J. Heat Mass Transfer,* vol. 25, p. 753, 1982.

47 Westwater, J. W.: "Gold Surfaces for Condensation Heat Transfer," *Gold Bulletin,* vol. 14, pp. 95–101, 1981.

48 Dhir, V. K., and J. H. Lienhard: Laminar Film Condensation on Plane and Axisymmetric Bodies in Non-Uniform Gravity, *J. Heat Trans.,* vol. 93, p. 97, 1971.

49 Rohsenow, W. M., and J. P. Hartnett (eds), "Handbook of Heat Transfer," Chap. 13, McGraw-Hill Book Company, New York, 1973.

50 Lienhard, J. H.: "A Heat Transfer Textbook," chap. 10, Prentice-Hall, Inc., Englewood Cliffs, N.J., 1981.

HEAT EXCHANGERS

10-1 INTRODUCTION

The application of the principles of heat transfer to the design of equipment to accomplish a certain engineering objective is of extreme importance, for in applying the principles to design, the individual is working toward the important goal of product development for economic gain. Eventually, economics plays a key role in the design and selection of heat-exchange equipment, and the engineer should bear this in mind when embarking on any new heat-transfer design problem. The weight and size of heat exchangers used in space or aeronautical applications are very important parameters, and in these cases cost considerations are frequently subordinated insofar as material and heat-exchanger construction costs are concerned; however, the weight and size are important cost factors in the overall application in these fields and thus may still be considered as economic variables.

A particular application will dictate the rules which one must follow to obtain the best design commensurate with economic considerations, size, weight, etc. An analysis of all these factors is beyond the scope of our present discussion, but it is well to remember that they all must be considered in practice. Our discussion of heat exchangers will take the form of technical analysis; i.e., the methods of predicting heat-exchanger performance will be outlined, along with a discussion of the methods which may be used to estimate the heat-exchanger size and type necessary to accomplish a particular task. In this respect, we limit our discussion to heat exchangers where the primary modes of heat transfer are conduction and convection. This is not to imply that radiation is not important in heat-exchanger design, for in many space applications it is the predominant means available for effecting an energy transfer. The reader is referred to the discussions by Siegal and Howell [1] and Sparrow and Cess [7] for detailed consideration of radiation heat-exchanger design.

10-2 THE OVERALL HEAT-TRANSFER COEFFICIENT

We have already discussed the overall heat-transfer coefficient in Sec. 2-4 with the heat transfer through the plane wall of Fig. 10-1 expressed as

$$q = \frac{T_A - T_B}{1/h_1A + \Delta x/kA + 1/h_2A} \tag{10-1}$$

where T_A and T_B are the fluid temperatures on each side of the wall. The overall heat-transfer coefficient U is defined by the relation

$$q = UA\,\Delta T_{\text{overall}} \tag{10-2}$$

From the standpoint of heat-exchanger design the plane wall is of infrequent application; a more important case for consideration would be that of a double-pipe heat exchanger, as shown in Fig. 10-2. In this application one fluid flows on the inside of the smaller tube while the other fluid flows in the annular space between the two tubes. The convection coefficients are calculated by the methods described in previous chapters, and the overall heat transfer is obtained from the thermal network of Fig. 10-2*b* as

$$q = \frac{T_A - T_B}{\dfrac{1}{h_iA_i} + \dfrac{\ln(r_o/r_i)}{2\pi kL} + \dfrac{1}{h_oA_o}} \tag{10-3}$$

where the subscripts i and o pertain to the inside and outside of the smaller inner tube. The overall heat-transfer coefficient may be based on either the inside or outside area of the tube at the discretion of the designer. Accordingly,

$$U_i = \frac{1}{\dfrac{1}{h_i} + \dfrac{A_i\ln(r_o/r_i)}{2\pi kL} + \dfrac{A_i}{A_o}\dfrac{1}{h_o}} \tag{10-4a}$$

$$U_o = \frac{1}{\dfrac{A_o}{A_i}\dfrac{1}{h_i} + \dfrac{A_o\ln(r_o/r_i)}{2\pi kL} + \dfrac{1}{h_o}} \tag{10-4b}$$

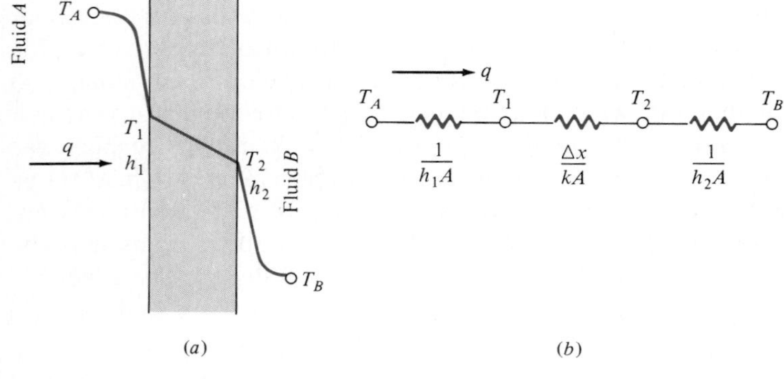

Fig. 10-1 Overall heat transfer through a plane wall.

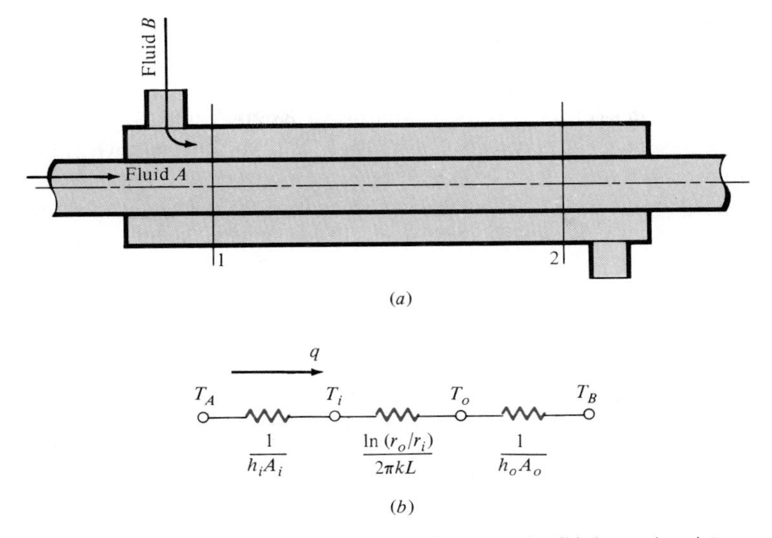

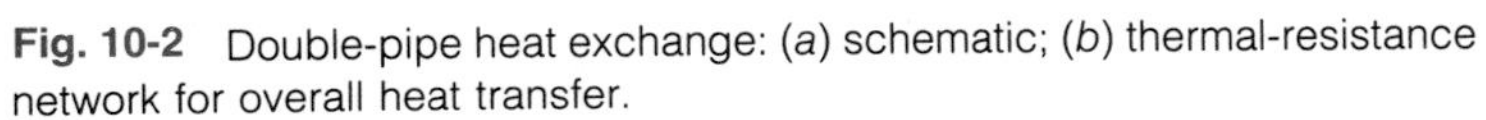

Fig. 10-2 Double-pipe heat exchange: (*a*) schematic; (*b*) thermal-resistance network for overall heat transfer.

Table 10-1 Approximate Values of Overall Heat-transfer Coefficients.

Physical situation	*U* Btu/h · ft² · °F	W/m² · °C
Brick exterior wall, plaster interior, uninsulated	0.45	2.55
Frame exterior wall, plaster interior:		
Uninsulated	0.25	1.42
With rock-wool insulation	0.07	0.4
Plate-glass window	1.10	6.2
Double plate-glass window	0.40	2.3
Steam condenser	200–1000	1100–5600
Feedwater heater	200–1500	1100–8500
Freon-12 condenser with water coolant	50–150	280–850
Water-to-water heat exchanger	150–300	850–1700
Finned-tube heat exchanger, water in tubes, air across tubes	5–10	25–55
Water-to-oil heat exchanger	20–60	110–350
Steam to light fuel oil	30–60	170–340
Steam to heavy fuel oil	10–30	56–170
Steam to kerosone or gasoline	50–200	280–1140
Finned-tube heat exchanger, steam in tubes, air over tubes	5–50	28–280
Ammonia condenser, water in tubes	150–250	850–1400
Alcohol condenser, water in tubes	45–120	255–680
Gas-to-gas heat exchanger	2–8	10–40

Although final heat-exchanger designs will be made on the basis of careful calculations of U, it is helpful to have a tabulation of values of the overall heat-transfer coefficient for various situations which may be encountered in practice. Comprehensive information of this sort is available in Refs. 5 and 6, and an abbreviated list of values of U is given in Table 10-1. We should remark that the value of U is governed in many cases by only one of the convection heat-transfer coefficients. In most practical problems the conduction resistance is small compared with the convection resistances. Then, if one value of h is markedly lower than the other value, it will tend to dominate the equation for U. Examples 10-1 and 10-2 illustrate this concept.

■ EXAMPLE 10-1

Hot water at 98°C flows through a 2-in schedule 40 horizontal steel pipe [k = 54 W/m · °C] and is exposed to atmospheric air at 20°C. The water velocity is 25 cm/s. Calculate the overall heat-transfer coefficient for this situation, based on the outer area of pipe.

Solution

From Appendix A the dimensions of 2-in schedule 40 pipe are

$$\text{ID} = 2.067 \text{ in} = 0.0525 \text{ m}$$

$$\text{OD} = 2.375 \text{ in} = 0.06033 \text{ m}$$

The heat-transfer coefficient for the water flow on the inside of the pipe is determined from the flow conditions with properties evaluated at the bulk temperature. The free-convection heat-transfer coefficient on the outside of the pipe depends on the temperature difference between the surface and ambient air. This temperature difference depends on the overall energy balance. First, we evaluate h_i and then formulate an iterative procedure to determine h_o.

The properties of water at 98°C are

$$\rho = 960 \text{ kg/m}^3 \qquad \mu = 2.82 \times 10^{-4} \text{ kg/m} \cdot \text{s}$$

$$k = 0.68 \text{ W/m} \cdot {}^\circ\text{C} \qquad \text{Pr} = 1.76$$

The Reynolds number is

$$\text{Re} = \frac{\rho u d}{\mu} = \frac{(960)(0.25)(0.0525)}{2.82 \times 10^{-4}} = 44{,}680 \qquad (a)$$

and since turbulent flow is encountered, we may use Eq. (6-4):

$$\begin{aligned} \text{Nu} &= 0.023 \text{ Re}^{0.8} \text{ Pr}^{0.4} \\ &= (0.023)(44{,}680)^{0.8}(1.76)^{0.4} = 151.4 \end{aligned}$$

$$h_i = \text{Nu}\,\frac{k}{d} = \frac{(151.4)(0.68)}{0.0525} = 1961 \text{ W/m}^2 \cdot \text{C} \quad [345 \text{ Btu/h} \cdot \text{ft}^2 \cdot {}^\circ\text{F}] \qquad (b)$$

For unit length of the pipe the thermal resistance of the steel is

$$R_s = \frac{\ln (r_o/r_i)}{2\pi k} = \frac{\ln (0.06033/0.0525)}{2\pi(54)} = 4.097 \times 10^{-4} \qquad (c)$$

Again, on a unit-length basis the thermal resistance on the inside is

$$R_i = \frac{1}{h_i A_i} = \frac{1}{h_i 2\pi r_i} = \frac{1}{(1961)\pi(0.0525)} = 3.092 \times 10^{-3} \tag{d}$$

The thermal resistance for the outer surface is as yet unknown but is written, for unit lengths,

$$R_o = \frac{1}{h_o A_o} = \frac{1}{h_o 2\pi r_o} \tag{e}$$

From Table 7-2, for laminar flow, the simplified relation for h_o is

$$h_o = 1.32\left(\frac{\Delta T}{d}\right)^{1/4} = 1.32\left(\frac{T_o - T_\infty}{d}\right)^{1/4} \tag{f}$$

where T_o is the unknown outside pipe surface temperature. We designate the inner pipe surface as T_i and the water temperature as T_w; then the energy balance requires

$$\frac{T_w - T_i}{R_i} = \frac{T_i - T_o}{R_s} = \frac{T_o - T_\infty}{R_o} \tag{g}$$

Combining Eqs. (*e*) and (*f*) gives

$$\frac{T_o - T_\infty}{R_o} = 2\pi r_o \frac{1.32}{d^{1/4}}(T_o - T_\infty)^{5/4} \tag{h}$$

This relation may be introduced into Eq. (*g*) to yield two equations with the two unknowns T_i and T_o:

$$\frac{98 - T_i}{3.092 \times 10^{-3}} = \frac{T_i - T_o}{4.097 \times 10^{-4}}$$

$$\frac{T_i - T_o}{4.097 \times 10^{-4}} = \frac{(\pi)(0.06033)(1.32)(T_o - 20)^{5/4}}{(0.06033)^{1/4}}$$

This is a nonlinear set which may be solved by iteration to give

$$T_o = 97.6°\text{C} \qquad T_i = 97.65°\text{C}$$

As a result, the outside heat-transfer coefficient and thermal resistance are

$$h_o = \frac{(1.32)(97.6 - 20)^{1/4}}{(0.06033)^{1/4}} = 7.91\ \text{W/m}^2 \cdot °\text{C} \quad [1.39\ \text{Btu/h} \cdot \text{ft}^2 \cdot °\text{F}]$$

$$R_o = \frac{1}{(0.06033)(7.91)\pi} = 0.667$$

The calculation clearly illustrates the fact that the free convection controls the overall heat-transfer because R_o is much larger than R_i or R_s. The overall heat-transfer coefficient based on the outer area is written in terms of these resistances as

$$U_o = \frac{1}{A_o[R_i + R_s + R_o]} \tag{i}$$

With numerical values inserted,

$$U_o = \frac{1}{\pi(0.06033)[3.092 \times 10^{-3} + 4.097 \times 10^{-4} + 0.667]}$$
$$= 7.87 \text{ W/Area} \cdot °\text{C}$$

In this calculation we used the outside area for a 1.0-m length as

$$A_o = \pi(0.06033) = 0.1895 \text{ m}^2/\text{m}$$
$$U_o = 7.87 \text{ W/m}^2 \cdot °\text{C}$$

Thus, we find that the overall heat-transfer coefficient is almost completely controlled by the value of h_o. We might have expected this result strictly on the basis of our experience with the relative magnitude of convection coefficients; free-convection values for air are very low compared with forced convection with liquids.

■ **EXAMPLE 10-2**

The pipe and hot-water system of Example 10-1 is exposed to steam at 1 atm and 100°C. Calculate the overall heat-transfer coefficient for this situation based on the outer area of pipe.

Solution

We have already determined the inside convection heat-transfer coefficient in Example 10-1 as

$$h_i = 1961 \text{ W/m}^2 \cdot °\text{C}$$

The convection coefficient for condensation on the outside of the pipe is obtained by using Eq. (9-12),

$$h_o = 0.725 \left[\frac{\rho(\rho - \rho_v)gh_{fg}k_f^3}{\mu_f d(T_g - T_o)}\right]^{1/4} \tag{a}$$

where T_o is the outside pipe-surface temperature. The water film properties are

$\rho = 960 \text{ kg/m}^3$ $\quad \mu_f = 2.82 \times 10^{-4} \text{ kg/m} \cdot \text{s}$ $\quad k_f = 0.68 \text{ W/m} \cdot °\text{C}$ $\quad h_{fg} = 2255 \text{ kJ/kg}$

so Eq. (*a*) becomes

$$h_o = 0.725 \left[\frac{(960)^2(9.8)(2.255 \times 10^6)(0.68)^3}{(2.82 \times 10^{-4})(0.06033)(100 - T_o)}\right]^{1/4}$$
$$= 17{,}960(100 - T_o)^{-1/4} \tag{b}$$

and the outside thermal resistance per unit length is

$$R_o = \frac{1}{h_o A_o} = \frac{(100 - T_o)^{1/4}}{(17{,}960)\pi(0.06033)} = \frac{(100 - T_o)^{1/4}}{3403} \tag{c}$$

The energy balance requires

$$\frac{100 - T_o}{R_o} = \frac{T_o - T_i}{R_s} = \frac{T_i - T_w}{R_i} \tag{d}$$

From Example 10-1

$$R_i = 3.092 \times 10^{-3} \qquad R_s = 4.097 \times 10^{-4} \qquad T_w = 98°\text{C}$$

and Eqs. (*c*) and (*d*) may be combined to give

$$3403(100 - T_o)^{3/4} = \frac{T_o - T_i}{4.097 \times 10^{-4}}$$

$$\frac{T_o - T_i}{4.097 \times 10^{-4}} = \frac{T_i - 98}{3.092 \times 10^{-3}}$$

This is a nonlinear set of equations which may be solved to give

$$T_o = 99.91°\text{C} \qquad T_i = 99.69°\text{C}$$

The exterior heat-transfer coefficient and thermal resistance then become

$$h_o = 17{,}960(100 - 99.91)^{-1/4} = 32{,}790 \text{ W/m}^2 \cdot °\text{C} \tag{e}$$

$$R_o = \frac{(100 - 99.91)^{1/4}}{3403} = 1.610 \times 10^{-4} \tag{f}$$

Based on unit length of pipe, the overall heat-transfer coefficient is

$$U_o = \frac{1}{A_o(R_i + R_s + R_o)}$$

$$= \frac{1}{\pi(0.06033)[3.092 \times 10^{-3}\ 4.097 \times 10^{-4} + 1.610 \times 10^{-4}]} \tag{g}$$

$$= 1441 \text{ W/°C} \cdot \text{Area}$$

Since A_o and the R's were both per unit length,

$$U_o = 1441 \text{ W/m}^2 \cdot °\text{C} \quad [254 \text{ Btu/h} \cdot \text{ft}^2 \cdot °\text{F}]$$

In this problem the water-side convection coefficient is the main controlling factor because h_o is so large for a condensation process. In fact, the outside thermal resistance is smaller than the conduction resistance of the steel. The approximate *relative* magnitudes of the resistances are

$$R_o \sim 1 \qquad R_s \sim 2.5 \qquad R_i \sim 19$$

10-3 FOULING FACTORS

After a period of operation the heat-transfer surfaces for a heat exchanger may become coated with various deposits present in the flow systems, or the surfaces may become corroded as a result of the interaction between the fluids and the material used for construction of the heat exchanger. In either event, this coating represents an additional resistance to the heat flow, and thus results in decreased performance. The overall effect is usually represented by a *fouling*

Table 10-2 Table of Normal Fouling Factors, According to Ref. 2.

Type of fluid	*Fouling factor,* h · ft² · °F/Btu	m² · °C/W
Seawater, below 125°F	0.0005	0.00009
Above 125°F	0.001	0.002
Treated boiler feedwater above 125°F	0.001	0.0002
Fuel oil	0.005	0.0009
Quenching oil	0.004	0.0007
Alcohol vapors	0.0005	0.00009
Steam, non-oil-bearing	0.0005	0.00009
Industrial air	0.002	0.0004
Refrigerating liquid	0.001	0.0002

factor, or fouling resistance, R_f, which must be included along with the other thermal resistances making up the overall heat-transfer coefficient.

Fouling factors must be obtained experimentally by determining the values of U for both clean and dirty conditions in the heat exchanger. The fouling factor is thus defined as

$$R_f = \frac{1}{U_{\text{dirty}}} - \frac{1}{U_{\text{clean}}}$$

An abbreviated list of recommended values of the fouling factor for various fluids is given in Table 10-2, and a very complete treatment of the subject is available in Ref.[9].

EXAMPLE 10-3

Suppose the water in Example 10-2 is seawater above 125°F and a fouling factor of 0.0002 m² · °C/W is experienced. What is the percent reduction in the convection heat-transfer coefficient?

Solution

The fouling factor influences the heat-transfer coefficient on the inside of the pipe. We have

$$R_f = 0.0002 = \frac{1}{h_{\text{dirty}}} - \frac{1}{h_{\text{clean}}}$$

Using $h_{\text{clean}} = 1961$ W/m² · °C we obtain

$$h_i = 1409 \text{ W/m}^2 \cdot {}^\circ\text{C}$$

This is a 28 percent reduction because of the fouling factor.

10-4 TYPES OF HEAT EXCHANGERS

One type of heat exchanger has already been mentioned, that of a double-pipe arrangement as shown in Fig. 10-2. Either counterflow or parallel flow may be

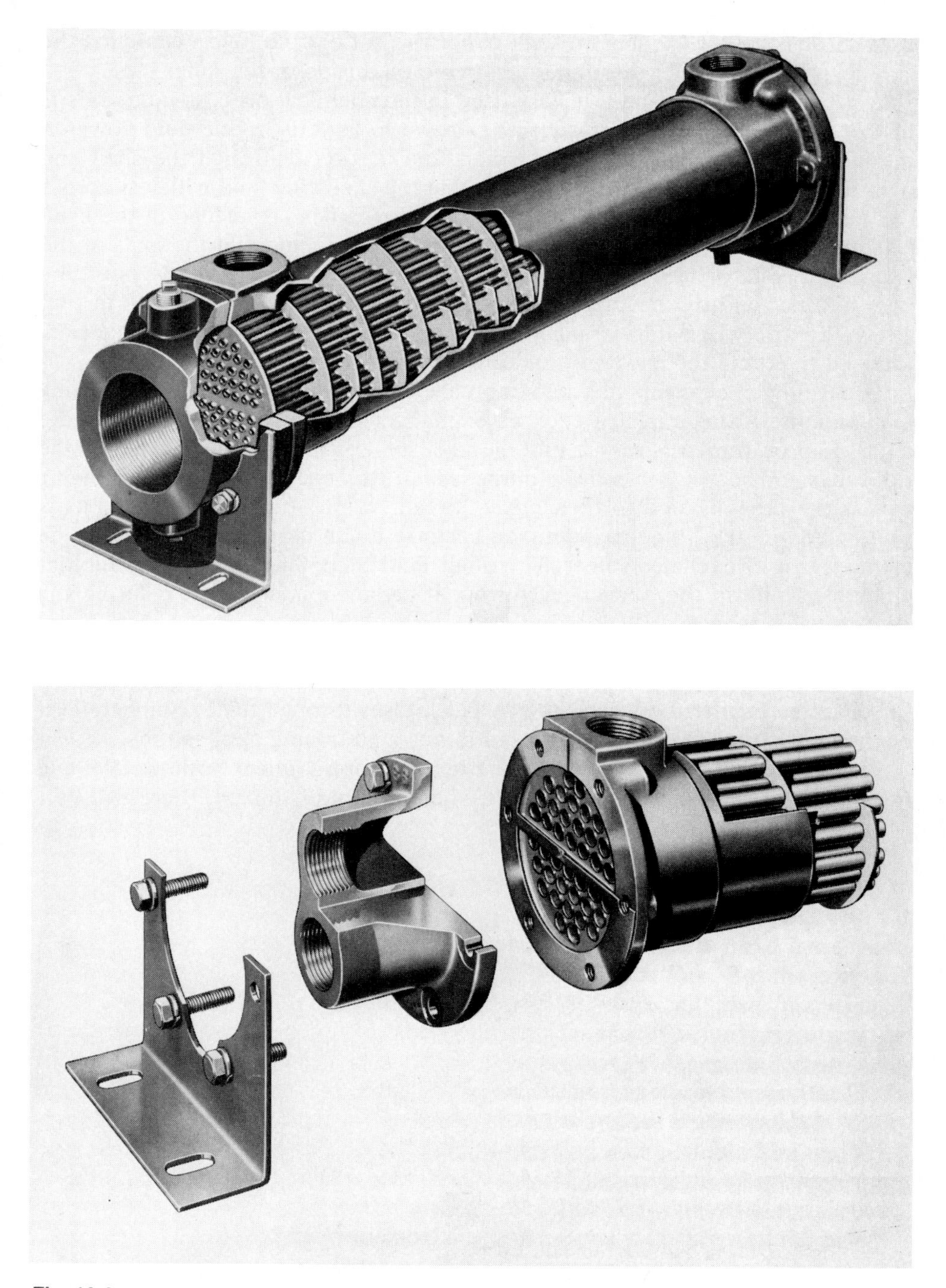

Fig. 10-3 (*a*) Shell-and-tube heat exchanger with one tube pass. (*b*) Heat arrangement for shell-and-tube heat exchanger with two tube passes. (*Young Radiator Company.*)

used in this type of exchanger, with either the hot or cold fluid occupying the annular space and the other fluid occupying the inside of the inner pipe.

A type of heat exchanger widely used in the chemical-process industries is that of the shell-and-tube arrangement shown in Fig. 10-3. One fluid flows on the inside of the tubes, while the other fluid is forced through the shell and over the outside of the tubes. To ensure that the shell-side fluid will flow across the tubes and thus induce higher heat transfer, baffles are placed in the shell as shown in the figure. Depending on the head arrangement at the ends of the exchanger, one or more tube passes may be utilized. In Fig. 10-3*a* one tube pass is used, and the head arrangement for two tube passes is shown in Fig. 10-3*b*. A variety of baffle arrangements are used in practice, and the reader is referred to Ref. 2 for more information on this matter.

Cross-flow exchangers are commonly used in air or gas heating and cooling applications. An example of such an exchanger is shown in Fig. 10-4, where a gas may be forced across a tube bundle, while another fluid is used inside the tubes for heating or cooling purposes. In this exchanger the gas flowing across the tubes is said to be a *mixed* stream, while the fluid in the tubes is said to be *unmixed*. The gas is mixed because it can move about freely in the exchanger as it exchanges heat. The other fluid is confined in separate tubular channels while in the exchanger so that it cannot mix with itself during the heat-transfer process.

A different type of cross-flow exchanger is shown in Fig. 10-5. In this case the gas flows across finned-tube bundles and thus is *unmixed* since it is confined in separate channels between the fins as it passes through the exchanger. This exchanger is typical of the types used in air-conditioning applications.

If a fluid is unmixed, there can be a temperature gradient both parallel and normal to the flow direction, whereas when the fluid is mixed, there will be a

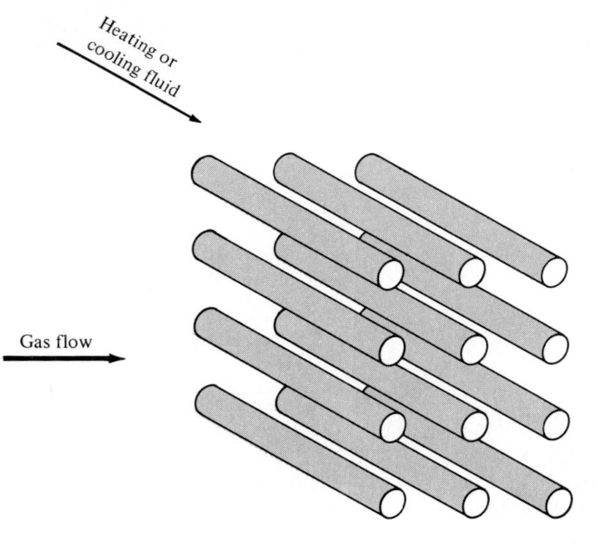

Fig. 10-4 Cross-flow heat exchanger, one fluid mixed and one unmixed.

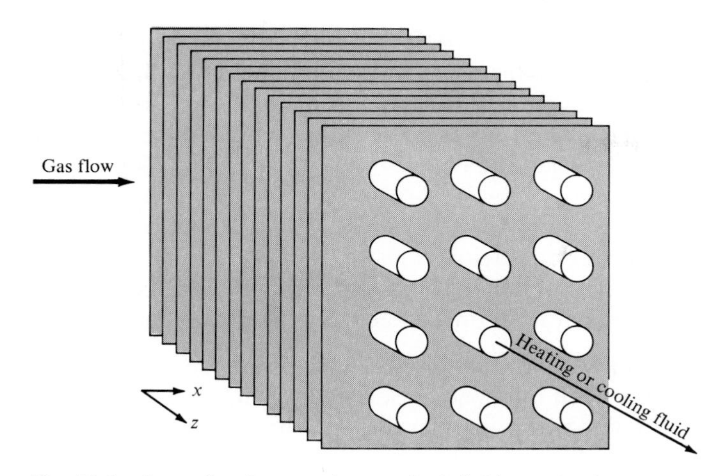

Fig. 10-5 Cross-flow heat exchanger, both fluids unmixed.

tendency for the fluid temperature to equalize in the direction normal to the flow as a result of the mixing. An approximate temperature profile for the gas flowing in the exchanger of Fig. 10-5 is indicated in Fig. 10-6, assuming that the gas is being heated as it passes through the exchanger. The fact that a fluid is mixed or unmixed influences the overall heat transfer in the exchanger because this heat transfer is dependent on the temperature difference between the hot and cold fluids.

There are a number of other configurations called *compact heat exchangers* which are primarily used in gas-flow systems where the overall heat-transfer coefficients are low and it is desirable to achieve a large surface area in a small volume. These exchangers generally have surface areas of greater than 650 m^2 per cubic meter of volume and will be given a fuller discussion in Sec. 10-7.

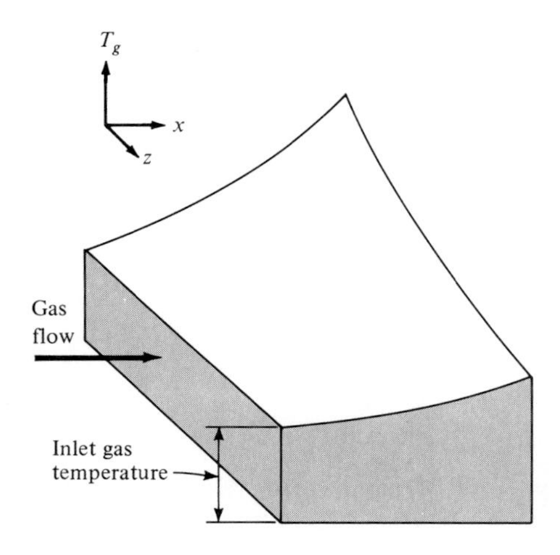

Fig. 10-6 Typical temperature profile for cross-flow heat exchanger of Fig. 10-5.

10-5 THE LOG MEAN TEMPERATURE DIFFERENCE

Consider the double-pipe heat exchanger shown in Fig. 10-2. The fluids may flow in either parallel flow or counterflow, and the temperature profiles for these two cases are indicated in Fig. 10-7. We propose to calculate the heat transfer in this double-pipe arrangement with

$$q = UA\,\Delta T_m \tag{10-5}$$

where U = overall heat-transfer coefficient
A = surface area for heat transfer consistent with definition of U
ΔT_m = suitable mean temperature difference across heat exchanger

An inspection of Fig. 10-7 shows that the temperature difference between the hot and cold fluids varies between inlet and outlet, and we must determine the average value for use in Eq. (10-5). For the parallel-flow heat exchanger shown in Fig. 10-7, the heat transferred through an element of area dA may be written

$$dq = -\dot{m}_h c_h\,dT_h = \dot{m}_c c_c\,dT_c \tag{10-6}$$

where the subscripts h and c designate the hot and cold fluids, respectively.

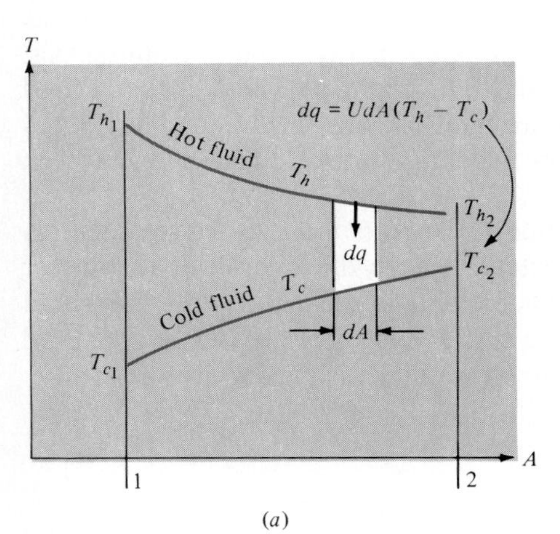

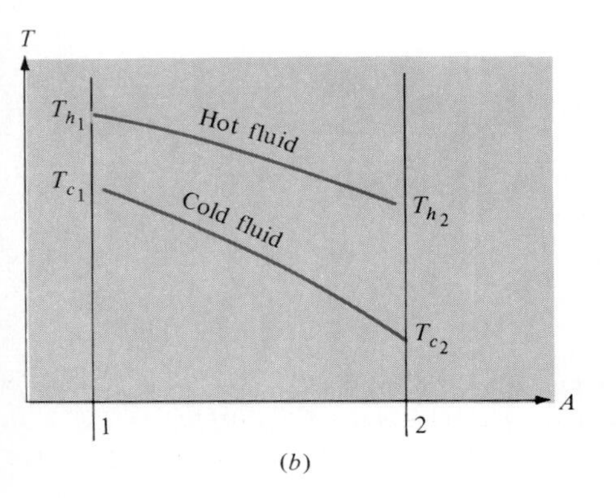

Fig. 10-7 Temperature profiles for parallel flow and counterflow in double-pipe heat exchanger.

The heat transfer could also be expressed.

$$dq = U(T_h - T_c)\, dA \tag{10-7}$$

From Eq. (10-6)

$$dT_h = \frac{-dq}{\dot{m}_h c_h}$$

$$dT_c = \frac{dq}{\dot{m}_c c_c}$$

where $\dot{m}$ represents the mass-flow rate and c is the specific heat of the fluid. Thus

$$dT_h - dT_c = d(T_h - T_c) = -dq\left(\frac{1}{\dot{m}_h c_h} + \frac{1}{\dot{m}_c c_c}\right) \tag{10-8}$$

Solving for dq from Eq. (10-7) and substituting into Eq. (10-8) gives

$$\frac{d(T_h - T_c)}{T_h - T_c} = -U\left(\frac{1}{\dot{m}_h c_h} + \frac{1}{\dot{m}_c c_c}\right) dA \tag{10-9}$$

This differential equation may now be integrated between conditions 1 and 2 as indicated in Fig. 10-7. The result is

$$\ln \frac{T_{h2} - T_{c2}}{T_{h1} - T_{c1}} = -UA\left(\frac{1}{\dot{m}_h c_h} + \frac{1}{\dot{m}_c c_c}\right) \tag{10-10}$$

Returning to Eq. (10-6), the products $\dot{m}_c c_c$ and $\dot{m}_h c_h$ may be expressed in terms of the total heat transfer q and the overall temperature differences of the hot and cold fluids. Thus

$$\dot{m}_h c_h = \frac{q}{T_{h1} - T_{h2}}$$

$$\dot{m}_c c_c = \frac{q}{T_{c2} - T_{c1}}$$

Substituting these relations into Eq. (10-10) gives

$$q = UA\,\frac{(T_{h2} - T_{c2}) - (T_{h1} - T_{c1})}{\ln\,[(T_{h2} - T_{c2})/(T_{h1} - T_{c1})]} \tag{10-11}$$

Comparing Eq. (10-11) with Eq. (10-5), we find that the mean temperature difference is the grouping of terms in the brackets. Thus

$$\Delta T_m = \frac{(T_{h2} - T_{c2}) - (T_{h1} - T_{c1})}{\ln\,[(T_{h2} - T_{c2})/(T_{h1} - T_{c1})]} \tag{10-12}$$

This temperature difference is called the *log mean temperature difference* (LMTD). Stated verbally, it is the temperature difference at one end of the heat exchanger less the temperature difference at the other end of the exchanger divided by the natural logarithm of the ratio of these two temperature differences. It is left as an exercise for the reader to show that this relation may also be used to calculate the LMTDs for counterflow conditions.

The above derivation for LMTD involves two important assumptions: (1) the fluid specific heats do not vary with temperature, and (2) the convection heat-transfer coefficients are constant throughout the heat exchanger. The second assumption is usually the more serious one because of entrance effects, fluid viscosity, and thermal-conductivity changes, etc. Numerical methods must normally be employed to correct for these effects. Section 10-8 describes one way of performing a variable-properties analysis.

If a heat exchanger other than the double-pipe type is used, the heat transfer is calculated by using a correction factor applied to the LMTD ***for a counterflow double-pipe arrangement with the same hot and cold fluid temperatures***. The heat-transfer equation then takes the form

$$q = UAF\,\Delta T_m \tag{10-13}$$

Values of the correction factor F according to Ref. 4 are plotted in Figs. 10-8 to 10-11 for several different types of heat exchangers. When a phase change is involved, as in condensation or boiling (evaporation), the fluid normally

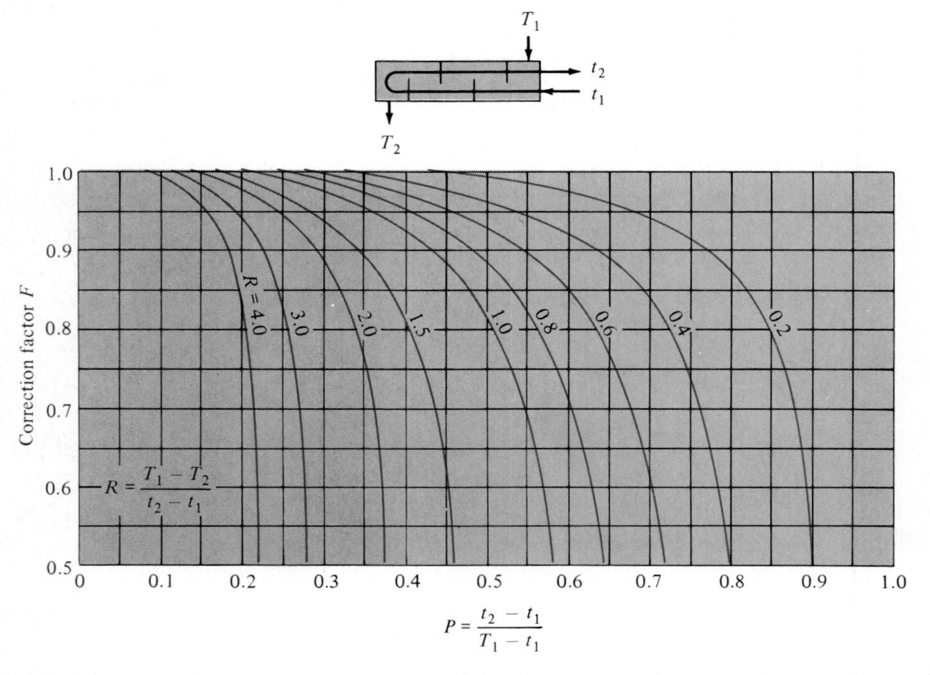

Fig. 10-8 Correction-factor plot for exchanger with one shell pass and two, four, or any multiple of tube passes.

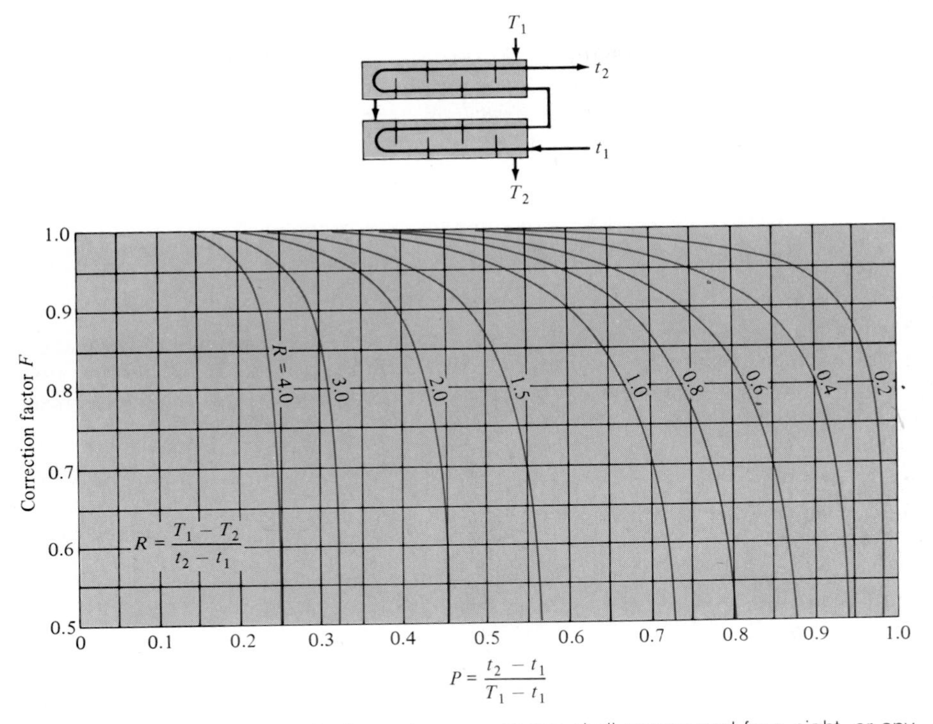

Fig. 10-9 Correction-factor plot for exchanger with two shell passes and four, eight, or any multiple of tube passes.

remains at essentially constant temperature and the relations are simplified. For this condition, P or R becomes zero and we obtain

$$F = 1.0 \qquad \text{for boiling or condensation}$$

Examples 10-4 to 10-8 illustrate the use of the LMTD for calculation of heat-exchanger performance.

■ EXAMPLE 10-4

Water at the rate of 68 kg/min is heated from 35 to 75°C by an oil having a specific heat of 1.9 kJ/kg · °C. The fluids are used in a counterflow double-pipe heat exchanger, and the oil enters the exchanger at 110°C and leaves at 75°C. The overall heat-transfer coefficient is 320 W/m² · °C. Calculate the heat-exchanger area.

Solution

The total heat transfer is determined from the energy absorbed by the water:

$$q = \dot{m}_w c_w \, \Delta T_w = (68)(4180)(75 - 35) = 11.37 \text{ MJ/min} \qquad (a)$$
$$= 189.5 \text{ kW} \quad [6.47 \times 10^5 \text{ Btu/h}]$$

Since all the fluid temperatures are known, the LMTD can be calculated by using the temperature scheme in Fig. 10-7*b*:

$$\Delta T_m = \frac{(110 - 75) - (75 - 35)}{\ln [(110 - 75)/(75 - 35)]} = 37.44°\text{C} \tag{b}$$

Then, since $q = UA\ \Delta T_m$,

$$A = \frac{1.895 \times 10^5}{(320)(37.44)} = 15.82 \text{ m}^2 \quad [170 \text{ ft}^2]$$

■ **EXAMPLE 10-5**

Instead of the double-pipe heat exchanger of Example 10-4, it is desired to use a shell-and-tube exchanger with the water making one shell pass and the oil making two tube passes. Calculate the area required for this exchanger, assuming that the overall heat-transfer coefficient remains at 320 W/m² · °C.

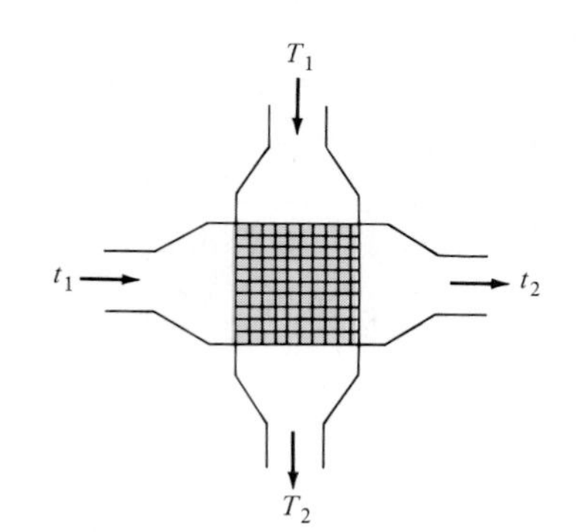

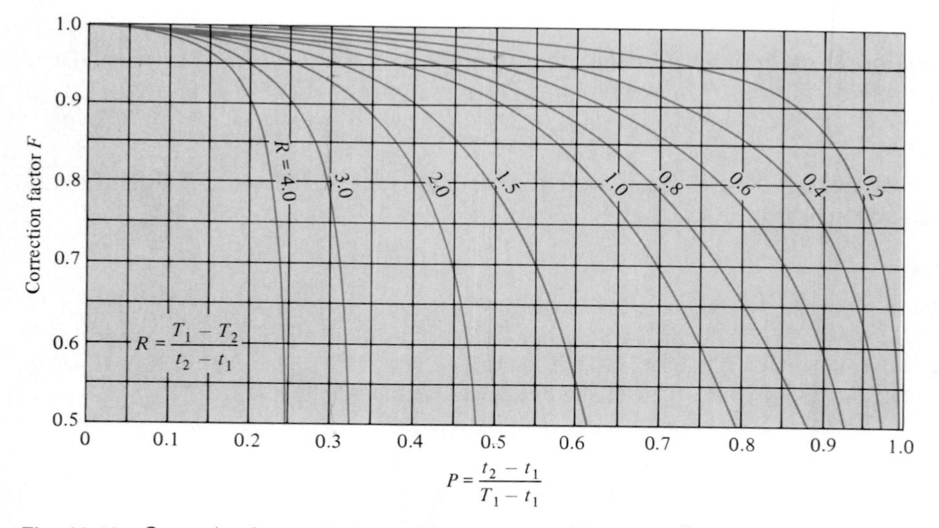

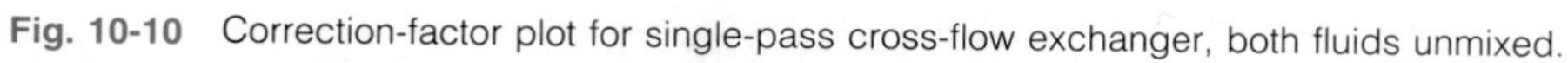

Fig. 10-10 Correction-factor plot for single-pass cross-flow exchanger, both fluids unmixed.

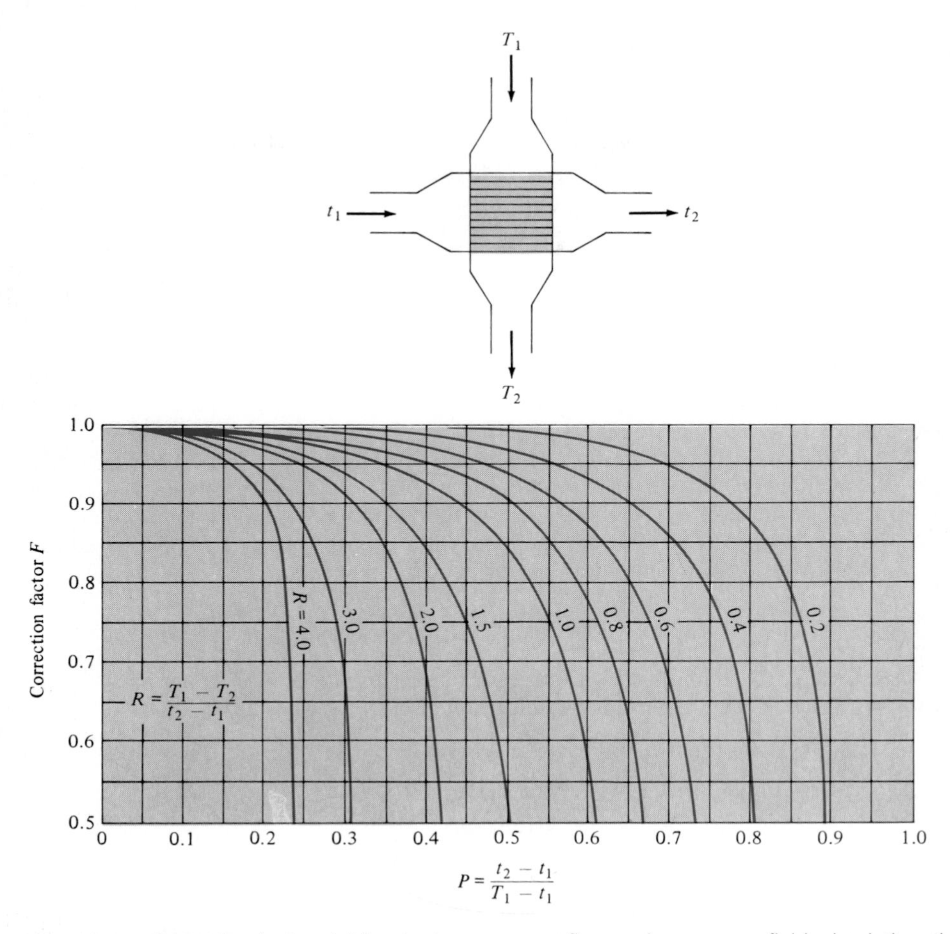

Fig. 10-11 Correction-factor plot for single-pass cross-flow exchanger, one fluid mixed, the other unmixed.

Solution

To solve this problem, we determine a correction factor from Fig. 10-8 to be used with the LMTD calculated on the basis of a counterflow exchanger. The parameters according to the nomenclature of Fig. 10-8 are

$$T_1 = 35°\text{C} \qquad T_2 = 75°\text{C} \qquad t_1 = 110°\text{C} \qquad t_2 = 75°\text{C}$$

$$P = \frac{t_2 - t_1}{T_1 - t_1} = \frac{75 - 110}{35 - 110} = 0.467$$

$$R = \frac{T_1 - T_2}{t_2 - t_1} = \frac{35 - 75}{75 - 110} = 1.143$$

so the correction factor is

$$F = 0.81$$

and the heat transfer is

$$q = UAF\,\Delta T_m$$

so that

$$A = \frac{1.895 \times 10^5}{(320)(0.81)(37.44)} = 19.53\ \text{m}^2 \quad [210\ \text{ft}^2]$$

■ EXAMPLE 10-6

Water at the rate of 30,000 lb_m/h [3.783 kg/s] is heated from 100 to 130°F [37.78 to 54.44°C] in a shell-and-tube heat exchanger. On the shell side one pass is used with water as the heating fluid, 15,000 lb_m/h [1.892 kg/s], entering the exchanger at 200°F [93.33°C]. The overall heat-transfer coefficient is 250 Btu/h · ft² · °F [1419 W/m² · °C], and the average water velocity in the $\frac{3}{4}$-in [1.905-cm] diameter tubes is 1.2 ft/s [0.366 m/s]. Because of space limitations the tube length must not be longer than 8 ft [2.438 m]. Calculate the number of tube passes, the number of tubes per pass, and the length of the tubes, consistent with this restriction.

Solution

We first assume one tube pass and check to see if it satisfies the conditions of this problem. The exit temperature of the hot water is calculated from

$$q = \dot{m}_c c_c\,\Delta T_c = \dot{m}_h c_h\,\Delta T_h$$

$$\Delta T_h = \frac{(30{,}000)(1)(130 - 100)}{(15{,}000)(1)} = 60°\text{F} = 33.33°\text{C} \qquad (a)$$

so

$$T_{h,\text{exit}} = 93.33 - 33.33 = 60°\text{C}$$

The total required heat transfer is obtained from Eq. (*a*) for the cold fluid:

$$q = (3.783)(4182)(54.44 - 37.78) = 263.6\ \text{kW} \quad [8.08 \times 10^5\ \text{Btu/h}]$$

For a counterflow exchanger

$$\text{LMTD} = \Delta T_m = \frac{(93.33 - 54.44) - (60 - 37.78)}{\ln\left[(93.33 - 54.44)/(60 - 37.78)\right]} = 29.78°\text{C}$$

$$q = UA\,\Delta T_m$$

$$A = \frac{2.636 \times 10^5}{(1419)(29.78)} = 6.238\ \text{m}^2 \quad [67.1\ \text{ft}^2] \qquad (b)$$

Using the average water velocity in the tubes and the flow rate, we calculate the total flow area with

$$\dot{m}_c = \rho A u$$

$$A = \frac{3.783}{(1000)(0.366)} = 0.01034\ \text{m}^2 \qquad (c)$$

This area is the product of the number of tubes and the flow area per tube:

$$0.01034 = n\frac{\pi d^2}{4}$$

$$n = \frac{(0.01034)(4)}{\pi(0.01905)^2} = 36.3$$

or $n = 36$ tubes. The surface area per tube per meter of length is

$$\pi d = \pi(0.01905) = 0.0598 \text{ m}^2/\text{tube}\cdot\text{m}$$

We recall that the total surface area required for a one-tube-pass exchanger was calculated in Eq. (*b*) as 6.238 m². We may thus compute the length of tube for this type of exchanger from

$$n\pi dL = 6.238$$

$$L = \frac{6.238}{(36)(0.0598)} = 2.898 \text{ m}$$

This length is greater than the allowable 2.438 m, so we must use more than one tube pass. When we increase the number of passes, we correspondingly increase the total surface area required because of the reduction in LMTD caused by the correction factor F. We next try two tube passes. From Fig. 10-8, $F = 0.88$, and thus

$$A_{\text{total}} = \frac{q}{UF\,\Delta T_m} = \frac{2.636\times10^5}{(1419)(0.88)(29.78)} = 7.089 \text{ m}^2$$

The number of tubes per pass is still 36 because of the velocity requirement. For the two-tube-pass exchanger the total surface area is now related to the length by

$$A_{\text{total}} = 2n\pi dL$$

so that

$$L = \frac{7.089}{(2)(36)(0.0598)} = 1.646 \text{ m} \quad [5.4 \text{ ft}]$$

This length is within the 2.438-m requirement, so the final design choice is

Number of tubes per pass = 36

Number of passes = 2

Length of tube per pass = 1.646 m [5.4 ft]

■ **EXAMPLE 10-7**

A heat exchanger like that shown in Fig. 10-4 is used to heat an oil in the tubes (c = 1.9 kJ/kg · °C) from 15°C to 85°C. Blowing across the outside of the tubes is steam which enters at 130°C and leaves at 110°C with a mass-flow of 5.2 kg/sec. The overall heat-transfer coefficient is 275 W/m² · °C and c for steam is 1.86 kJ/kg · °C. Calculate the surface area of the heat exchanger.

Solution

The total heat transfer may be obtained from an energy balance on the steam

$$q = \dot{m}_s c_s\,\Delta T_s = (5.2)(1.86)(130 - 110) = 193 \text{ kW}$$

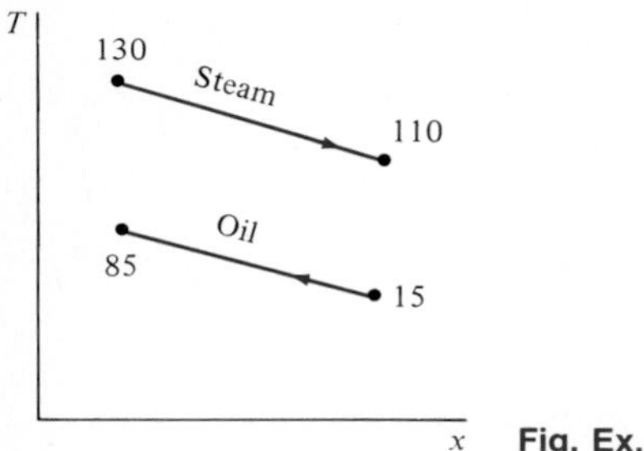

Fig. Ex. 10-7

We can solve for the area from Eq. (10-13). The value of ΔT_m is calculated *as if the exchanger were counterflow double pipe,* i.e., as shown in the accompanying figure. Thus,

$$\Delta T_m = \frac{(130 - 85) - (110 - 15)}{\ln \left(\dfrac{130 - 85}{110 - 15} \right)} = 66.9°\text{C}$$

Now, from Fig. 10-11, t_1 and t_2 will represent the unmixed fluid (the oil) and T_1 and T_2 will represent the mixed fluid (the steam) so that

$$T_1 = 130 \qquad T_2 = 110 \qquad t_1 = 15 \qquad t_2 = 85°\text{C}$$

and we calculate

$$R = \frac{130 - 110}{85 - 15} = 0.286 \qquad P = \frac{85 - 15}{130 - 15} = 0.609$$

Consulting Fig. 10-11 we find

$$F = 0.97$$

so the area is calculated from

$$A = \frac{q}{UF\,\Delta T_m} = \frac{193{,}000}{(275)(0.97)(66.9)} = 10.82 \text{ m}^2$$

■ **EXAMPLE 10-8**

Investigate the heat-transfer performance of the exchanger in Example 10-7 if the oil flow rate is reduced in half while the steam flow remains the same. Assume U remains constant at 275 W/m² · °C.

Solution

We did not calculate the oil flow in Example 10-7 but can do so now from

$$q = \dot{m}_o c_o \, \Delta T_o$$

$$\dot{m}_o = \frac{193}{(1.9)(85 - 15)} = 1.45 \text{ kg/s}$$

The new flow rate will be half this value or 0.725 kg/s. We are assuming the inlet temperatures remain the same at 130°C for the steam and 15°C for the oil. The new relation for the heat transfer is

$$q = \dot{m}_o c_o (T_{e,o} - 15) = \dot{m}_s c_{p_s} (130 - T_{e,s}) \tag{a}$$

but the exit temperatures, $T_{e,o}$ and $T_{e,s}$ are unknown. Furthermore, ΔT_m is unknown without these temperatures, as are the values of R and P from Fig. 10-11. This means we must use an iterative procedure to solve for the exit temperatures using Eq. (*a*) and

$$q = UAF\, \Delta T_m \tag{b}$$

The general procedure is to assume values of the exit temperatures until the q's agree between Eqs. (*a*) and (*b*).

The objective of this example is to show that an iterative procedure is required when the inlet and outlet temperatures are not known or easily calculated. There is no need to go through this iteration because it can be avoided by using the techniques described in Sec. 10-6.

10-6 EFFECTIVENESS-NTU METHOD

The LMTD approach to heat-exchanger analysis is useful when the inlet and outlet temperatures are known or are easily determined. The LMTD is then easily calculated, and the heat flow, surface area, or overall heat-transfer coefficient may be determined. When the inlet or exit temperatures are to be evaluated for a given heat exchanger, the analysis frequently involves an iterative procedure because of the logarithmic function in the LMTD. In these cases the analysis is performed more easily by utilizing a method based on the effectiveness of the heat exchanger in transferring a given amount of heat. The effectiveness method also offers many advantages for analysis of problems in which a comparison between various types of heat exchangers must be made for purposes of selecting the type best suited to accomplish a particular heat-transfer objective.

We define the heat-exchanger effectiveness as

$$\text{Effectiveness} = \epsilon = \frac{\text{actual heat transfer}}{\text{maximum possible heat transfer}}$$

The actual heat transfer may be computed by calculating either the energy lost by the hot fluid or the energy gained by the cold fluid. Consider the parallel-flow and counterlow heat exchangers shown in Fig. 10-7. For the parallel-flow exchanger

$$q = \dot{m}_h c_h (T_{h_1} - T_{h_2}) = \dot{m}_c c_c (T_{c_2} - T_{c_1}) \tag{10-14}$$

and for the counterflow exchanger

$$q = \dot{m}_h c_h (T_{h_1} - T_{h_2}) = \dot{m}_c c_c (T_{c_1} - T_{c_2}) \tag{10-15}$$

To determine the maximum possible heat transfer for the exchanger, we first recognize that this maximum value could be attained if one of the fluids were to undergo a temperature change equal to the maximum temperature difference present in the exchanger, which is the difference in the entering temperatures for the hot and cold fluids. The fluid which might undergo this maximum temperature difference is the one having the *minimum* value of $\dot{m}c$ because the energy balance requires that the energy received by one fluid be equal to that given up by the other fluid; if we let the fluid with the larger value of $\dot{m}c$ go through the maximum temperature difference, this would require that the other fluid undergo a temperature difference greater than the maximum, and this is impossible. So, maximum possible heat transfer is expressed as

$$q_{\max} = (\dot{m}c)_{\min}(T_{h\,\text{inlet}} - T_{c\,\text{inlet}}) \tag{10-16}$$

The minimum fluid may be either the hot or cold fluid, depending on the mass-flow rates and specific heats. For the parallel-flow exchanger

$$\epsilon_h = \frac{\dot{m}_h c_h(T_{h_1} - T_{h_2})}{\dot{m}_h c_h(T_{h_1} - T_{c_1})} = \frac{T_{h_1} - T_{h_2}}{T_{h_1} - T_{c_1}} \tag{10-17}$$

$$\epsilon_c = \frac{\dot{m}_c c_c(T_{c_2} - T_{c_1})}{\dot{m}_c c_c(T_{h_1} - T_{c_1})} = \frac{T_{c_2} - T_{c_1}}{T_{h_1} - T_{c_1}} \tag{10-18}$$

The subscripts on the effectiveness symbols designate the fluid which has the minimum value of $\dot{m}c$. For the counterflow exchanger:

$$\epsilon_h = \frac{\dot{m}_h c_h(T_{h_1} - T_{h_2})}{\dot{m}_h c_h(T_{h_1} - T_{c_2})} = \frac{T_{h_1} - T_{h_2}}{T_{h_1} - T_{c_2}} \tag{10-19}$$

$$\epsilon_c = \frac{\dot{m}_c c_c(T_{c_1} - T_{c_2})}{\dot{m}_c c_c(T_{h_1} - T_{c_2})} = \frac{T_{c_1} - T_{c_2}}{T_{h_1} - T_{c_2}} \tag{10-20}$$

In a general way the effectiveness is expressed as

$$\epsilon = \frac{\Delta T\ (\text{minimum fluid})}{\text{Maximum temperature difference in heat exchanger}} \tag{10-21}$$

We may derive an expression for the effectiveness in parallel flow as follows. Rewriting Eq. (10-10), we have

$$\ln \frac{T_{h_2} - T_{c_2}}{T_{h_1} - T_{c_1}} = -UA\left(\frac{1}{\dot{m}_h c_h} + \frac{1}{\dot{m}_c c_c}\right) = \frac{-UA}{\dot{m}_c c_c}\left(1 + \frac{\dot{m}_c c_c}{\dot{m}_h c_h}\right) \tag{10-22}$$

or

$$\frac{T_{h_2} - T_{c_2}}{T_{h_1} - T_{c_1}} = \exp\left[\frac{-UA}{\dot{m}_c c_c}\left(1 + \frac{\dot{m}_c c_c}{\dot{m}_h c_h}\right)\right] \tag{10-23}$$

If the cold fluid is the minimum fluid,

$$\epsilon = \frac{T_{c_2} - T_{c_1}}{T_{h_1} - T_{c_1}}$$

Rewriting the temperature ratio in Eq. (10-23) gives

$$\frac{T_{h_2} - T_{c_2}}{T_{h_1} - T_{c_1}} = \frac{T_{h_1} + (\dot{m}_c c_c/\dot{m}_h c_h)(T_{c_1} - T_{c_2}) - T_{c_2}}{T_{h_1} - T_{c_1}} \tag{10-24}$$

when the substitution

$$T_{h_2} = T_{h_1} + \frac{\dot{m}_c c_c}{\dot{m}_h c_h}(T_{c_1} - T_{c_2})$$

is made from Eq. (10-6). Equation (10-24) may now be rewritten

$$\frac{(T_{h_1} - T_{c_1}) + (\dot{m}_c c_c/\dot{m}_h c_h)(T_{c_1} - T_{c_2}) + (T_{c_1} - T_{c_2})}{T_{h_1} - T_{c_1}} = 1 - \left(1 + \frac{\dot{m}_c c_c}{\dot{m}_h c_h}\right)\epsilon$$

Inserting this relation back in Eq. (10-23) gives for the effectiveness

$$\epsilon = \frac{1 - \exp\left[(-UA/\dot{m}_c c_c)(1 + \dot{m}_c c_c/\dot{m}_h c_h)\right]}{1 + \dot{m}_c c_c/\dot{m}_h c_h} \tag{10-25}$$

It may be shown that the same expression results for the effectiveness when the hot fluid is the minimum fluid, except that $\dot{m}_c c_c$ and $\dot{m}_h c_h$ are interchanged. As a consequence, the effectiveness is usually written

$$\epsilon = \frac{1 - \exp\left[(-UA/C_{\min})(1 + C_{\min}/C_{\max})\right]}{1 + C_{\min}/C_{\max}} \tag{10-26}$$

where $C = \dot{m}c$ is defined as the capacity rate.

A similar analysis may be applied to the counterflow case, and the following relation for effectiveness results:

$$\epsilon = \frac{1 - \exp\left[(-UA/C_{\min})(1 - C_{\min}/C_{\max})\right]}{1 - (C_{\min}/C_{\max})\exp\left[(-UA/C_{\min})(1 - C_{\min}/C_{\max})\right]} \tag{10-27}$$

The grouping of terms $UA/C_{\min}$ is called the *number of transfer units* (NTU) since it is indicative of the size of the heater exchanger.

Kays and London [3] have presented effectiveness ratios for various heat-exchanger arrangements, and some of the results of their analyses are available in chart form in Figs. 10-12 to 10-17. Examples 10-9 to 10-14 illustrate the use of the effectiveness-NTU method in heat-exchanger analysis.

While the effectiveness-NTU charts can be of great practical utility in design problems, there are applications where more precision is desired than can be obtained by reading the graphs. In addition, more elaborate design procedures may be computer-based, requiring analytical expressions for these curves. Table 10-3 summarizes the effectiveness relations. In some cases the objective of the analysis is a determination of NTU, and it is possible to give an explicit relation for NTU in terms of effectiveness and capacity ratio. Some of these relations are listed in Table 10-4.

The formulas for one shell pass and 2, 4, 6 tube passes are strictly correct for 2 tube passes but may produce a small error for higher multiples. The error is usually less than 1 percent for C less than 0.5 and N less than 3.0. The formulas may overpredict by about 6.5 percent at N = 6.0 and C = 1.0. Further information is given by Kraus and Kern (10).

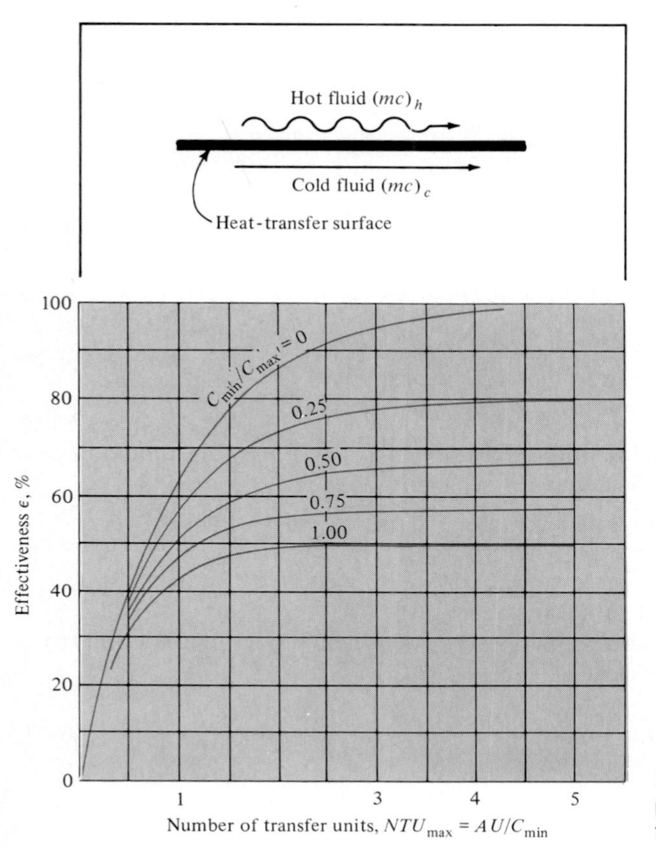

Fig. 10-12 Effectiveness for parallel-flow exchanger performance.

Boilers and Condensers

We noted earlier that in a boiling or condensation process the fluid temperature stays essentially constant, or the fluid acts as if it had infinite specific heat. In these cases $C_{min}/C_{max} \to 0$ and all the heat-exchanger effectiveness relations approach a single simple equation,

$$\epsilon = 1 - e^{-\text{NTU}}$$

The equation is shown as the last entry in Table 10-3.

EXAMPLE 10-9

Complete Example 10-8 using the effectiveness method.

Solution

For the steam

$$C_s = \dot{m}_s c_s = (5.2)(1.86) = 9.67 \text{ kW/°C}$$

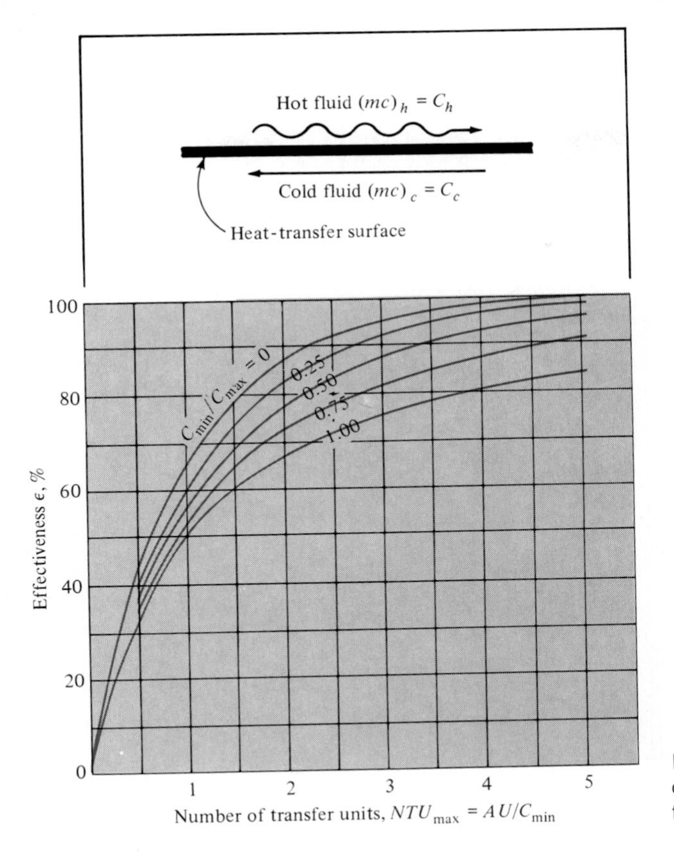

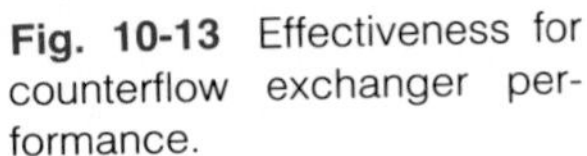
Fig. 10-13 Effectiveness for counterflow exchanger performance.

and for the oil

$$C_o = \dot{m}_o c_o = (0.725)(1.9) = 1.38 \text{ kW/°C}$$

so the oil is the minimum fluid. We thus have

$$C_{min}/C_{max} = 1.38/9.67 = 0.143$$

and

$$\text{NTU} = UA/C_{min} = (275)(10.82)/1380 = 2.156$$

We choose to use the Table and note that C_{min} (oil) is unmixed and C_{max} (steam) is mixed so that the first relation in the table applies. We therefore calculate ϵ as

$$\epsilon = (1/0.143)\{1 - \exp[-(0.143)(1 - e^{-2.156})]\} = 0.831$$

If we were using Fig. 10-14 we would have to evaluate

$$C_{mixed}/C_{unmixed} = 9.67/1.38 = 7.01$$

and would still determine $\epsilon \approx 0.8$. Now, using the effectiveness we can determine the temperature difference of the minimum fluid (oil) as

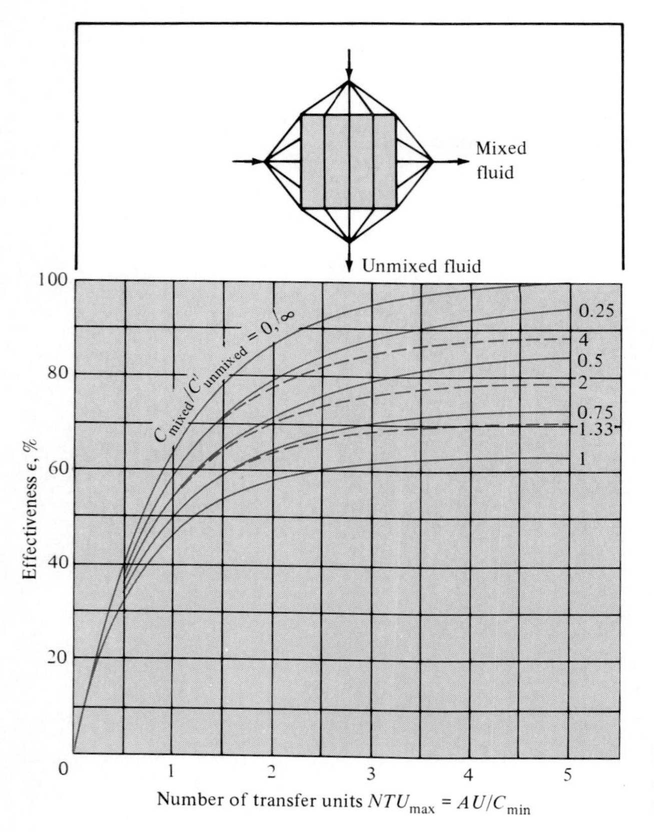

Fig. 10-14 Effectiveness for cross-flow exchanger with one fluid mixed.

$$\Delta T_o = \epsilon(\Delta T_{max}) = (0.831)(130 - 15) = 95.5°C$$

so that the heat transfer is

$$q = \dot{m}_o c_o \, \Delta T_o = (1.38)(95.5) = 132 \text{ kW}$$

and we find a reduction in the oil flow rate of 50 percent causes a reduction in heat transfer of only 32 percent.

■ **EXAMPLE 10-10**

The heat exchanger of Example 10-4 is used for heating water as described in the example. Using the same entering-fluid temperatures, calculate the exit water temperature when only 40 kg/min of water is heated but the same quantity of oil is used. Also calculate the total heat transfer under these new conditions.

Solution

The flow rate of oil is calculated from the energy balance for the original problem:

$$\dot{m}_h c_h \, \Delta T_h = \dot{m}_c c_c \, \Delta T_c \qquad (a)$$

$$\dot{m}_h = \frac{(68)(4180)(75 - 35)}{(1900)(110 - 75)} = 170.97 \text{ kg/min}$$

The capacity rates for the new conditions are now calculated as

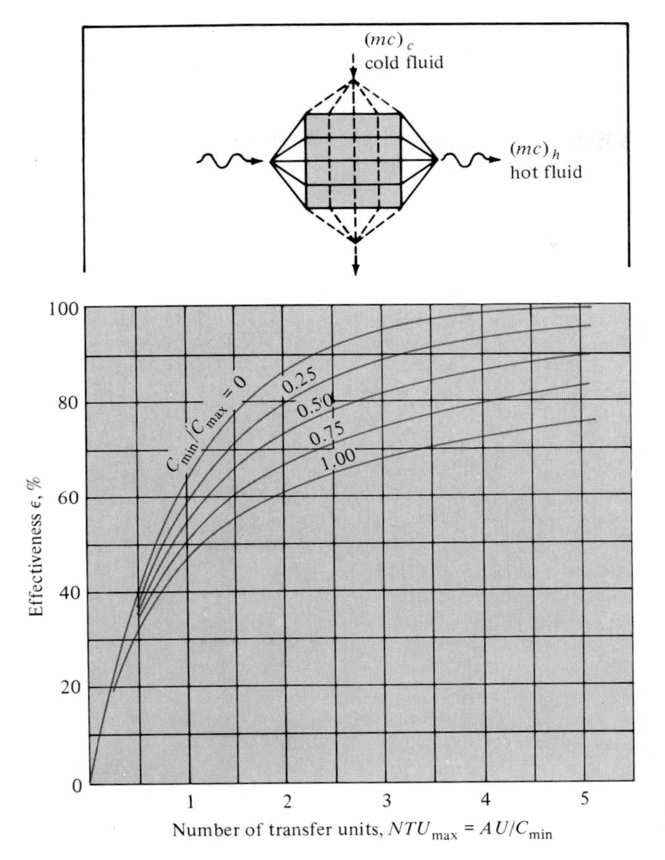

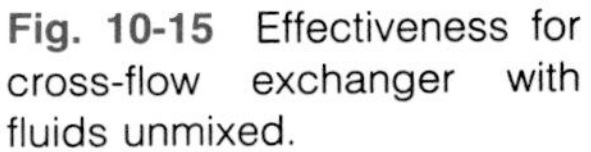

Fig. 10-15 Effectiveness for cross-flow exchanger with fluids unmixed.

$$\dot{m}_h c_h = \frac{170.97}{60}(1900) = 5414 \text{ W/°C}$$

$$\dot{m}_c c_c = \frac{40}{60}(4180) = 2787 \text{ W/°C}$$

so that the water (cold fluid) is the minimum fluid, and

$$\frac{C_{\text{min}}}{C_{\text{max}}} = \frac{2787}{5414} = 0.515$$

$$\text{NTU}_{\text{max}} = \frac{UA}{C_{\text{min}}} = \frac{(320)(15.82)}{2787} = 1.816 \qquad (b)$$

where the area of 15.82 m² is taken from Example 10-4. From Fig. 10-13 or Table 10-3 the effectiveness is

$$\epsilon = 0.744$$

and because the cold fluid is the minimum, we can write

$$\epsilon = \frac{\Delta T_{\text{cold}}}{\Delta T_{\text{max}}} = \frac{\Delta T_{\text{cold}}}{110 - 35} = 0.744 \qquad (c)$$

$$\Delta T_{\text{cold}} = 55.8°\text{C}$$

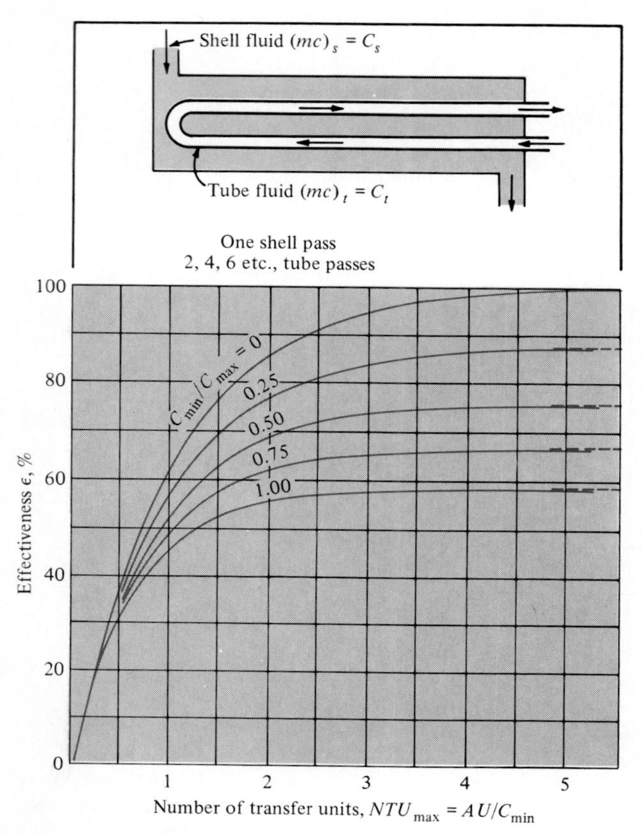

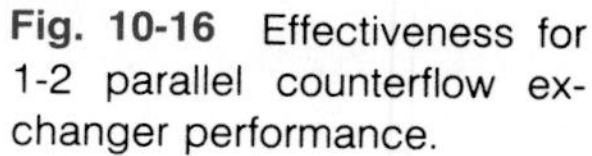
Fig. 10-16 Effectiveness for 1-2 parallel counterflow exchanger performance.

and the exit water temperature is

$$T_{w,\text{exit}} = 35 + 55.8 = 90.8°\text{C}$$

The total heat transfer under the new flow conditions is calculated as

$$q = \dot{m}_c c_c \, \Delta T_c = \frac{40}{60}(4180)(55.8) = 155.5 \text{ kW} \quad [5.29 \times 10^5 \text{ Btu/h}] \qquad (d)$$

Notice that although the flow rate has been reduced by 41 percent (68 to 40 kg/min), the heat transfer is reduced by only 18 percent (189.5 to 155.5 kW) because the exchanger is more effective at the lower flow rate.

■ **EXAMPLE 10-11**

A finned-tube heat exchanger like that shown in Fig. 10-5 is used to heat 5000 ft³/min [2.36 m³/s] of air at 1 atm from 60 to 85°F (15.55 to 29.44°C). Hot water enters the tubes at 180°F [82.22°C], and the air flows across the tubes, producing an average overall heat-transfer coefficient of 40 Btu/h · ft² · °F [227 W/m² · °C]. The total surface area of the exchanger is 100 ft² [9.29 m²]. Calculate the exit water temperature and the heat-transfer rate.

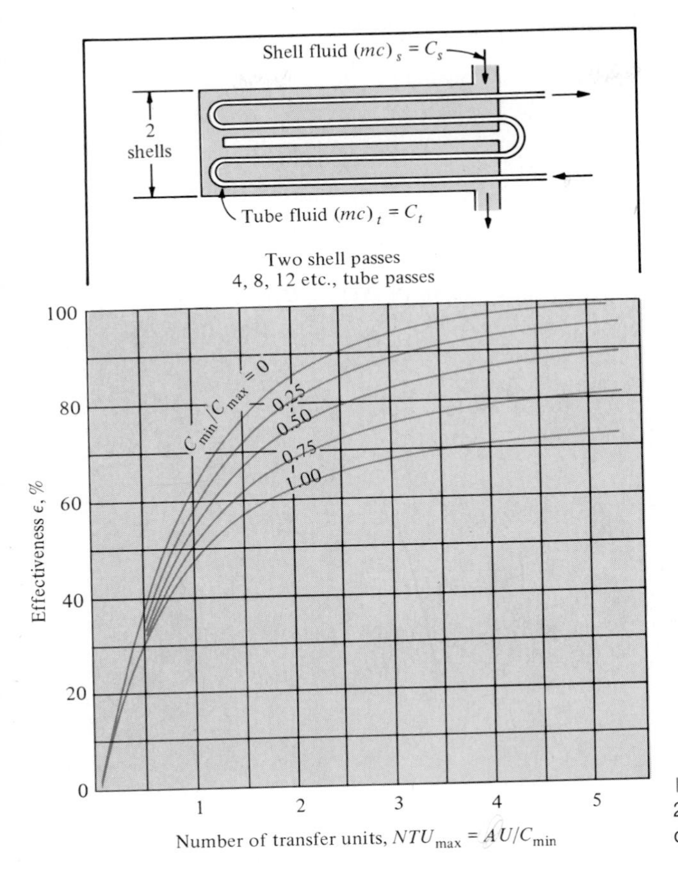

Fig. 10-17 Effectiveness for 2-4 multipass counterflow exchanger performance.

Solution

The heat transfer is calculated from the energy balance on the air. First, the inlet air density is

$$\rho = \frac{p}{RT} = \frac{1.0132 \times 10^5}{(287)(288.7)} = 1.223 \text{ kg/m}^3$$

so the mass flow of air (the cold fluid) is

$$\dot{m}_c = (2.36)(1.223) = 2.887 \text{ kg/s}$$

The heat transfer is then

$$q = \dot{m}_c c_c \, \Delta T_c = (2.887)(1006)(29.44 - 15.55)$$
$$= 40.34 \text{ kW} \quad [1.38 \times 10^5 \text{ Btu/h}] \tag{a}$$

From the statement of the problem we do not know whether the air or water is the minimum fluid. If the air is the minimum fluid, we may immediately calculate NTU and use Fig. 10-15 to determine the water-flow rate and hence the exit water temperature. If the water is the minimum fluid, a trial-and-error procedure must be used with Fig. 10-15 or Table 10-3. We assume that the air is the minimum fluid and then check our assumption. Then

Table 10-3 Heat-exchanger Effectiveness Relations.
$N = \text{NTU} = \frac{UA}{C_{\min}} \qquad C = \frac{C_{\min}}{C_{\max}}$

Flow geometry	*Relation*
Double pipe:	
Parallel flow	$\epsilon = \frac{1 - \exp[-N(1 + C)]}{1 + C}$
Counterflow	$\epsilon = \frac{1 - \exp[-N(1 - C)]}{1 - C\exp[-N(1 - C)]}$
Counterflow, $C = 1$	$\epsilon = \frac{N}{N + 1}$
Cross flow:	
Both fluids unmixed	$\epsilon = 1 - \exp\left[\frac{\exp(-NCn) - 1}{Cn}\right]$ where $n = N^{-0.22}$
Both fluids mixed	$\epsilon = \left[\frac{1}{1 - \exp(-N)} + \frac{C}{1 - \exp(-NC)} - \frac{1}{N}\right]^{-1}$
$C_{\max}$ mixed, $C_{\min}$ unmixed	$\epsilon = (1/C)\{1 - \exp[-C(1 - e^{-N})]\}$
$C_{\max}$ unmixed, $C_{\min}$ mixed	$\epsilon = 1 - \exp\{-(1/C)[1 - \exp(-NC)]\}$
Shell and tube:	
One shell pass, 2, 4, 6, tube passes	$\epsilon = 2\left\{1 + C + (1 + C^2)^{1/2} \times \frac{1 + \exp[-N(1 + C^2)^{1/2}]}{1 - \exp[-N(1 + C^2)^{1/2}]}\right\}^{-1}$
Multiple shell passes, $2n$, $4n$, $6n$ tube passes (ϵ_p = effectiveness of each shell pass, n = number of shell passes)	$\epsilon = \frac{[(1 - \epsilon_p C)/(1 - \epsilon_p)]^n - 1}{[(1 - \epsilon_p C)/(1 - \epsilon_p)]^n - C}$
Special case for $C = 1$	$\epsilon = \frac{n\epsilon_p}{1 + (n - 1)\epsilon_p}$
All exchangers with $C = 0$	$\epsilon = 1 - e^{-N}$

$$\dot{m}_c c_c = (2.887)(1006) = 2904 \text{ W/°C}$$

and

$$\text{NTU}_{\max} = \frac{UA}{C_{\min}} = \frac{(227)(9.29)}{2904} = 0.726$$

and the effectiveness based on the air as the minimum fluid is

$$\epsilon \frac{\Delta T_{\text{air}}}{\Delta T_{\max}} = \frac{29.44 - 15.55}{82.22 - 15.55} = 0.208 \tag{b}$$

Table 10-4 NTU Relations for Heat Exchangers.
$C = C_{min}/C_{max}$ ϵ = effectiveness $N = \text{NTU} = UA/C_{min}$

Flow geometry	*Relation*
Double pipe:	
Parallel flow	$N = \dfrac{-\ln[1-(1+C)\epsilon]}{1+C}$
Counterflow	$N = \dfrac{1}{C-1}\ln\left(\dfrac{\epsilon-1}{C\epsilon-1}\right)$
Counterflow, $C = 1$	$N = \dfrac{\epsilon}{1-\epsilon}$
Cross flow:	
C_{max} mixed, C_{min} unmixed	$N = -\ln\left[1 + \dfrac{1}{C}\ln(1 - C\epsilon)\right]$
C_{max} unmixed, C_{min} mixed	$N = \dfrac{-1}{C}\ln[1 + C\ln(1-\epsilon)]$
Shell and tube:	
One shell pass, 2, 4, 6, tube passes	$N = -(1+C^2)^{-1/2} \times \ln\left[\dfrac{2/\epsilon - 1 - C - (1+C^2)^{1/2}}{2/\epsilon - 1 - C + (1+C^2)^{1/2}}\right]$
All exchangers, $C = 0$	$N = -\ln(1-\epsilon)$

Entering Fig. 10-15, we are unable to match these quantities with the curves. This requires that the hot fluid be the minimum. We must therefore assume values for the water-flow rate until we are able to match the performance as given by Fig. 10-5 or Table 10-3. We first note that

$$C_{max} = \dot{m}_c c_c = 2904 \text{ W/°C} \tag{c}$$

$$\text{NTU}_{max} = \frac{UA}{C_{min}} \tag{d}$$

$$\epsilon = \frac{\Delta T_h}{\Delta T_{max}} = \frac{\Delta T_h}{82.22 - 15.55} \tag{e}$$

$$\Delta T_h = \frac{4.034 \times 10^4}{C_{min}} = \frac{4.034 \times 10^4}{C_h} \tag{f}$$

The iterations are:

				ϵ	
$\dfrac{C_{min}}{C_{max}}$	$C_{min} = \dot{m}_h c_h$	NTU_{max}	ΔT_h	*From Fig. 10-15 or Table 10-3*	*Calculated from Eq. (e)*
0.5	1452	1.452	27.78	0.65	0.417
0.25	726	2.905	55.56	0.89	0.833
0.22	639	3.301	63.13	0.92	0.947

We thus estimate the water-flow rate as about

$$\dot{m}_h c_h = 645 \text{ W/°C}$$

and

$$\dot{m}_h = \frac{645}{4180} = 0.154 \text{ kg/s} \quad [1221 \text{ lb}_m/\text{h}]$$

The exit water temperature is accordingly

$$T_{w,\text{exit}} = 82.22 - \frac{4.034 \times 10^4}{645} = 19.68°\text{C}$$

■ EXAMPLE 10-12

A counterflow double-pipe heat exchanger is used to heat 1.25 kg/s of water from 35 to 80°C by cooling an oil [c_p = 2.0 kJ/kg · °C] from 150 to 85°C. The overall heat-transfer coefficient is 150 Btu/h · ft² · °F. A similar arrangement is to be built at another plant location, but it is desired to compare the performance of the single counterflow heat exchanger with two smaller counterflow heat exchangers connected in series on the water side and in parallel on the oil side, as shown in the sketch. The oil flow is split equally between the two exchangers, and it may be assumed that the overall heat-transfer coefficient for the smaller exchangers is the same as for the large exchanger. If the smaller exchangers cost 20 percent more per unit surface area, which would be the most economical arrangement—the one large exchanger or two equal-sized small exchangers?

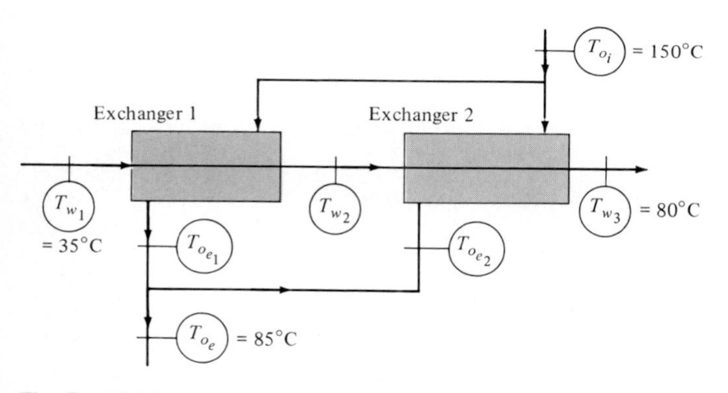

Fig. Ex. 10-12

Solution

We calculate the surface area required for both alternatives and then compare costs. For the one large exchanger

$$\begin{aligned} q &= \dot{m}_c c_c \, \Delta T_c = \dot{m}_h c_h \, \Delta T_h \\ &= (1.25)(4180)(80 - 35) = \dot{m}_h c_h (150 - 85) \\ &= 2.351 \times 10^5 \text{ W} \quad [8.02 \times 10^5 \text{ Btu/h}] \\ \dot{m}_c c_c &= 5225 \text{ W/°C} \quad \dot{m}_h c_h = 3617 \text{ W/°C} \end{aligned}$$

so that the oil is the minimum fluid:

$$\epsilon_h = \frac{\Delta T_h}{150 - 35} = \frac{150 - 85}{150 - 35} = 0.565$$

$$\frac{C_{\min}}{C_{\max}} = \frac{3617}{5225} = 0.692$$

From Fig. 10-13 or Table 10-4, $\text{NTU}_{\max} = 1.09$, so that

$$A = \text{NTU}_{\max}\frac{C_{\min}}{U} = \frac{(1.09)(3617)}{850} = 4.649 \text{ m}^2 \quad [50.04 \text{ ft}^2]$$

We now wish to calculate the surface-area requirement for the two small exchangers shown in the sketch. We have

$$\dot{m}_h c_h = \frac{3617}{2} = 1809 \text{ W/°C}$$

$$\dot{m}_c c_c = 5225 \text{ W/°C}$$

$$\frac{C_{\min}}{C_{\max}} = \frac{1809}{5225} = 0.347$$

The number of transfer units is the same for each heat exchanger because UA and $C_{\min}$ are the same for each exchanger. This requires that the effectiveness be the same for each exchanger. Thus,

$$\epsilon_1 = \frac{T_{oi} - T_{oe,1}}{T_{oi} - T_{w,1}} = \epsilon_2 = \frac{T_{oi} - T_{oe,2}}{T_{oi} - T_{w,2}}$$

$$\epsilon_1 = \frac{150 - T_{oe,1}}{150 - 35} = \epsilon_2 = \frac{150 - T_{oe,2}}{150 - T_{w,2}} \tag{a}$$

where the nomenclature for the temperatures is indicated in the sketch. Because the oil flow is the same in each exchanger and the average exit oil temperature must be 85°C, we may write

$$\frac{T_{oe,1} + T_{oe,2}}{2} = 85 \tag{b}$$

An energy balance on the second heat exchanger gives

$$(5225)(T_{w3} - T_{w2}) = (1809)(T_{oi} - T_{oe,2})$$

$$(5225)(80 - T_{w2}) = (1809)(150 - T_{oe,2}) \tag{c}$$

We now have the three equations (a), (b), and (c) which may be solved for the three unknowns $T_{oe,1}$, $T_{oe,2}$, and T_{w2}. The solutions are

$$T_{oe,1} = 76.98\text{°C}$$

$$T_{oe,2} = 93.02\text{°C}$$

$$T_{w2} = 60.26\text{°C}$$

The effectiveness can then be calculated as

$$\epsilon_1 = \epsilon_2 = \frac{150 - 76.98}{150 - 35} = 0.635$$

From Fig. 10-13 or Table 10-4, we obtain $\text{NTU}_{\max} = 1.16$, so that

$$A = \text{NTU}_{\max} \frac{C_{\min}}{U} = \frac{(1.16)(1809)}{850} = 2.47 \text{ m}^2$$

We thus find that 2.47 m^2 of area is required for each of the small exchangers, of a total of 4.94 m^2. This is greater than the 4.649 m^2 required in the one larger exchanger; in addition, the cost per unit area is greater so that the most economical choice would be the single larger exchanger. It may be noted, however, that the pumping costs for the oil would probably be less with the two smaller exchangers, so that this could precipitate a decision in favor of the smaller exchangers if pumping costs represented a sizable economic factor.

EXAMPLE 10-13

Hot oil at 100°C is used to heat air in a shell-and-tube heat exchanger. The oil makes six tube passes and the air makes one shell pass; 2.0 kg/s of air are to be heated from 20 to 80°C. The specific heat of the oil is 2100 J/kg · °C, and its flow rate is 3.0 kg/s. Calculate the area required for the heat exchanger for $U = 200$ W/m^2 · °C.

Solution

The basic energy balance is

$$\dot{m}_o c_o \, \Delta T_o = \dot{m}_a c_{pa} \, \Delta T_a$$

or

$$(3.0)(2100)(100 - T_{oe}) = (2.0)(1009)(80 - 20)$$

$$T_{oe} = 80.78°\text{C}$$

We have

$$\dot{m}_h c_h = (3.0)(2100) = 6300 \text{ W/°C}$$

$$\dot{m}_c c_c = (2.0)(1009) = 2018 \text{ W/°C}$$

so the air is the minimum fluid and

$$C = \frac{C_{\min}}{C_{\max}} = \frac{2018}{6300} = 0.3203$$

The effectiveness is

$$\epsilon = \frac{\Delta T_c}{\Delta T_{\max}} = \frac{80 - 20}{100 - 20} = 0.75$$

Now, we may use either Fig. 10-16 or the analytical relation from Table 10-4 to obtain NTU. For this problem we choose to use the table.

$$N = -(1 + 0.3203^2)^{-1/2} \ln \left[\frac{2/0.75 - 1 - 0.3203 - (1 + 0.3203^2)^{1/2}}{2/0.75 - 1 - 0.3203 + (1 + 0.3203^2)^{1/2}}\right]$$

$$= 1.99$$

Now, with $U = 200$ we calculate the area as

$$A = \text{NTU} \frac{C_{\min}}{U} = \frac{(1.99)(2018)}{200} = 20.09 \text{ m}^2$$

■ **EXAMPLE 10-14**

A shell-and-tube heat exchanger is used as an ammonia condenser with ammonia vapor entering the shell at 50°C as a saturated vapor. Water enters the single-pass tube arrangement at 20°C and the total heat transfer required is 200 kW. The overall heat-transfer coefficient is estimated from Table 10-1 as 1000 W/m² · °C. Determine the area to achieve a heat exchanger effectiveness of 60 percent with an exit water temperature of 40°C. What percent reduction in heat transfer would result if the water flow is reduced in half while keeping the heat exchanger area and U the same?

Solution

The mass flow can be calculated from the heat transfer with

$$q = 200 \text{ kW} = \dot{m}_w c_w \, \Delta T_w$$

so

$$\dot{m}_w = \frac{200}{(4.18)(40 - 20)} = 2.39 \text{ kg/s}$$

Because this is a condenser the water is the minimum fluid and

$$C_{\min} = \dot{m}_w c_w = (2.39)(4.18) = 10 \text{ kW/°C}$$

The value of NTU is obtained from the last entry of Table 10-4, with $\epsilon = 0.6$:

$$N = -\ln(1 - \epsilon) = -\ln(1 - 0.6) = 0.916$$

so that the area is calculated as

$$A = \frac{C_{\min} N}{U} = \frac{(10{,}000)(0.916)}{1000} = 9.16 \text{ m}^2$$

When the flow rate is reduced in half the new value of NTU is

$$N = \frac{UA}{C_{\min}} = \frac{(1000)(9.16)}{(10{,}000/2)} = 1.832$$

And the effectiveness is computed from the last entry of Table 10-3:

$$\epsilon = 1 - e^{-N} = 0.84$$

The new water temperature difference is computed as

$$\Delta T_w = \epsilon(\Delta T_{\max}) = (0.84)(50 - 20) = 25.2°\text{C}$$

so the new heat transfer is

$$q = C_{\min} \, \Delta T_w = \left(\frac{10{,}000}{2}\right)(25.2) = 126 \text{ kW}$$

So, by reducing the flow rate in half we have lowered the heat transfer from 200 to 126 kW, or by 37 percent.

■ 10-7 COMPACT HEAT EXCHANGERS

A number of heat-exchanger surfaces do not fall into the categories discussed in the foregoing sections. Most notable are the compact exchangers which achieve a very high surface area per unit volume. These exchangers are most adaptable to applications where gas flows and low values of h are to be encountered. Kays and London [3] have studied these types of exchangers very extensively, and four typical configurations are shown in Fig. 10-18. In Fig. 10-18a a finned-tube exchanger is shown with flat tubes, Fig. 10-18b shows a circular finned-tube array in a different configuration, and Figs. 10-18c and d offer ways to achieve very high surface areas on both sides of the exchanger. These last two configurations are applicable to processes where gas-to-gas heat transfer is involved.

The heat transfer and friction factor for two typical compact exchangers are shown in Figs. 10-19 and 10-20. The Stanton and Reynolds numbers are based on the mass velocities in the minimum flow cross-sectional area and a hydraulic diameter stated in the figure.

$$G = \frac{\dot{m}}{A_c} \tag{10-28}$$

The ratio of the free-flow area to frontal area

$$\sigma = \frac{A_c}{A} \tag{10-29}$$

is also given in the figure. Thus,

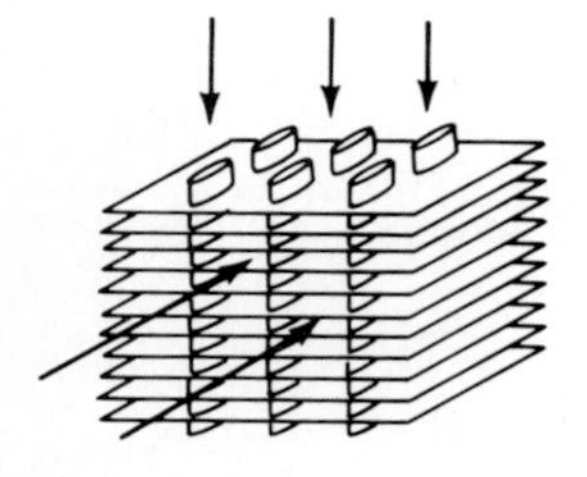

(*a*)

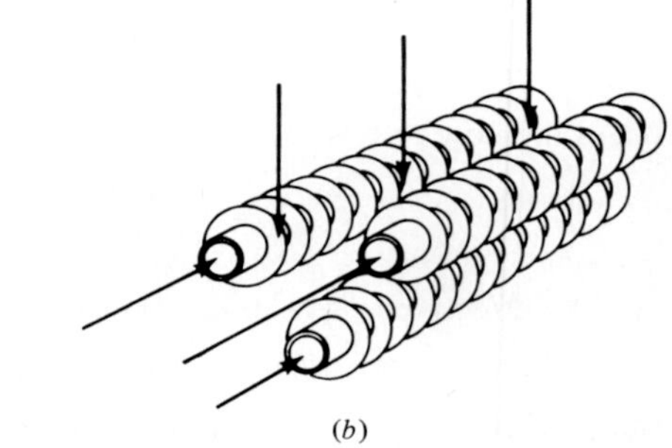

(*b*)

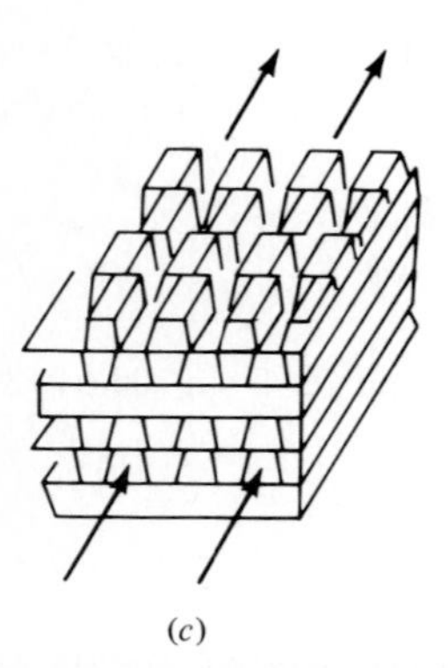

(*c*)

Fig. 10-18 Examples of compact heat-exchanger configurations according to Ref. 3.

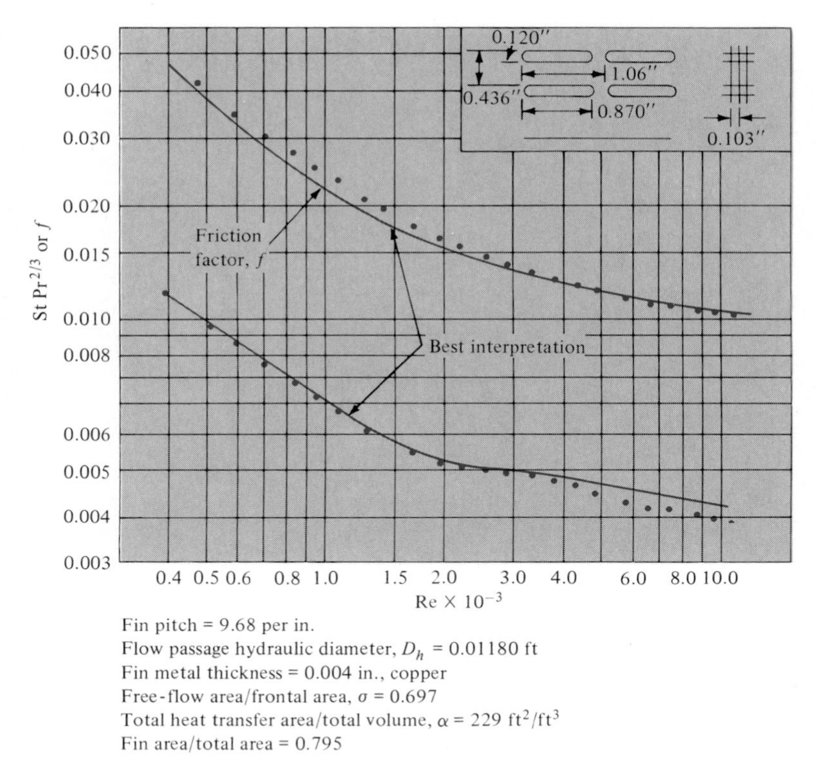

Fig. 10-19 Heat transfer and friction factor for finned flat-tube heat exchanger according to Ref. 3.

$$\text{St} = \frac{h}{Gc_p} \qquad \text{Re} = \frac{D_h G}{\mu}$$

Fluid properties are evaluated at the average bulk temperature. Heat transfer and fluid friction *inside* the tubes are evaluated with the hydraulic diameter method discussed in Chap. 6. Pressure drop is calculated with the chart friction factor f and the following relation:

$$\Delta p = \frac{v_1 G^2}{2g_c}\left[(1 + \sigma^2)\left(\frac{v_2}{v_1} - 1\right) + f\frac{A}{A_c}\frac{v_m}{v_1}\right] \tag{10-30}$$

where v_1 and v_2 are the entrance and exit specific volumes, respectively, and v_m is the mean specific volume in the exchanger, normally taken as $v_m = (v_1 + v_2)/2$.

Rather meticulous design procedures are involved with compact heat exchangers, and these are given a full discussion in Ref. 3.

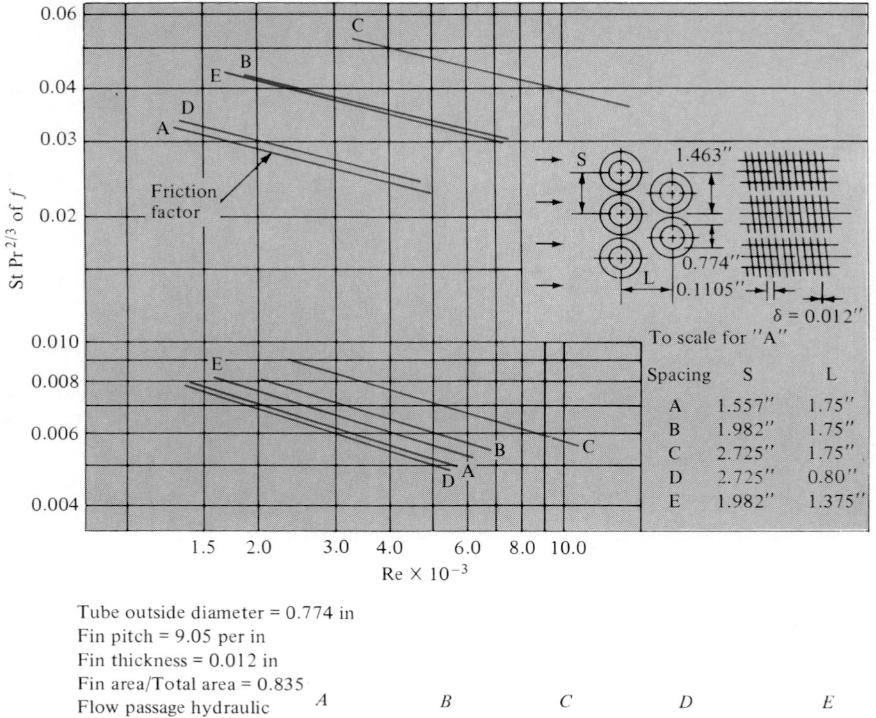

Tube outside diameter = 0.774 in
Fin pitch = 9.05 per in
Fin thickness = 0.012 in
Fin area/Total area = 0.835

	A	*B*	*C*	*D*	*E*
Flow passage hydraulic diameter, D_h =	0.01681	0.02685	0.0445	0.01587	0.02108 ft
Free-flow area/frontal area, σ =	0.455	0.572	0.688	0.537	0.572
Heat transfer area/total volume, α =	108	85.1	61.9	135	108 ft^2/ft^3

Note: Minimum free-flow area in all cases occurs in the spaces transverse to the flow, except for *D*, in which the minimum area is in the diagonals.

Fig. 10-20 Heat transfer and friction factor for finned circulator-tube heat exchanger according to Ref. 3.

■ **EXAMPLE 10-15**

Air at 1 atm and 300 K enters an exchanger like that shown in Fig. 10-19 with a velocity of 15 m/s. Calculate the heat-transfer coefficient.

Solution

We obtain the air properties from Table A-5 as

$$\rho = 1.1774 \text{ kg/m}^3 \qquad c_p = 1.0057 \text{ kJ/kg} \cdot {}^\circ\text{C}$$

$$\mu = 1.983 \times 10^{-5} \text{ kg/m} \cdot \text{s} \qquad \text{Pr} = 0.708$$

From Fig. 10-19 we have

$$\sigma = \frac{A_c}{A} = 0.697 \qquad D_h = 0.0118 \text{ ft} = 3.597 \text{ mm}$$

The mass velocity is thus

$$G = \frac{\dot{m}}{A_c} = \frac{\rho u_\infty A}{A_c} = \frac{(1.1774)(15)}{0.697} = 38.18 \text{ kg/m}^2 \cdot \text{s}$$

and the Reynolds number is

$$\text{Re} = \frac{D_h G}{\mu} = \frac{(3.597 \times 10^{-3})(38.18)}{1.983 \times 10^{-5}} = 6.926 \times 10^3$$

From Fig. 10-19 we can read

$$\text{St Pr}^{2/3} = 0.0036 = \frac{h}{Gc_p} \text{Pr}^{2/3}$$

and the heat-transfer coefficient is

$$\begin{aligned} h &= (0.0036)(38.18)(1005.7)(0.708)^{-2/3} \\ &= 174 \text{ W/m}^2 \cdot {}^\circ\text{C} \quad [30.64 \text{ Btu/h} \cdot \text{ft}^2 \cdot {}^\circ\text{F}] \end{aligned}$$

10-8 ANALYSIS FOR VARIABLE PROPERTIES

The convection heat-transfer coefficient is dependent on the fluid being considered. Correspondingly, the overall heat-transfer coefficient for a heat exchanger may vary substantially through the exchanger if the fluids are such that their properties are strongly temperature-dependent. In this circumstance the analysis is best performed on a numerical or finite-difference basis. To illustrate the technique, let us consider the simple parallel-flow double-pipe heat exchanger of Sec. 10-5. The heat exchanger is divided into increments of surface area ΔA_j. For this incremental surface area the hot and cold temperatures are T_{h_j} and T_{c_j}, respectively, and we shall assume that the overall heat-transfer coefficient can be expressed as a function of these temperatures. Thus

$$U_j = U_j(T_{h_j}, T_{c_j})$$

The incremental heat transfer in ΔA_j is, according to Eq. (10-6),

$$\Delta q_j = -(\dot{m}_h c_c)_j (T_{h_{j+1}} - T_{h_j}) = (\dot{m}_c c_c)_j (T_{c_{j+1}} - T_{c_j}) \tag{10-31}$$

Also,

$$\Delta q_j = U_j \, \Delta A_j (T_h - T_c)_j \tag{10-32}$$

The finite-difference equation analogous to Eq. (10-9) is

$$\begin{aligned} \frac{(T_h - T_c)_{j+1} - (T_h - T_c)_j}{(T_h - T_c)_j} &= -U_j \left[\frac{1}{(\dot{m}_h c_h)_j} + \frac{1}{(\dot{m}_c c_c)_j} \right] \Delta A_j \\ &= -K_j(T_h, T_c) \, \Delta A_j \end{aligned} \tag{10-33}$$

where we have introduced the indicated definition for K_j. Reducing Eq. (10-33), we obtain

$$\frac{(T_h - T_c)_{j+1}}{(T_h - T_c)_j} = 1 - K_j \Delta A_j \tag{10-34}$$

The numerical-analysis procedure is now clear when the inlet temperatures and flows are given:

1. Choose a convenient value of ΔA_j for the analysis.
2. Calculate the value of U for the inlet conditions and through the initial ΔA increment.
3. Calculate the value of q for this increment from Eq. (10-32).
4. Calculate the values of T_h, T_c, and $T_h - T_c$ for the *next* increment, using Eqs. (10-31) and (10-34).
5. Repeat the foregoing steps until all the increments in ΔA are employed.

The total heat-transfer rate is then calculated from

$$q_{\text{total}} = \sum_{j=1}^{n} \Delta q_j$$

where n is the number of increments in ΔA.

A numerical analysis such as the one discussed above is best performed with a computer. Heat-transfer rates calculated from a variable-properties analysis can frequently differ by substantial amounts from a constant-properties analysis. The most difficult part of the analysis is, of course, a determination of the values of h. The interested reader is referred to the heat-transfer literature for additional information on this complicated but important subject.

EXAMPLE 10-16 Transient response of thermal-energy storage system

A rock-bed thermal-energy storage unit is employed to remove energy from a hot airstream and store for later use. The schematic for the device is shown in the sketch. The surface is covered with a material having an overall R value of 2 h · °F · ft²/Btu. The inlet flow area is 5 × 5 = 25 ft², and the rock-bed length is 10 ft. Properties of the rock are

$$\rho_r = 80 \text{ lb}_m/\text{ft}^3 \quad [1281.4 \text{ kg/m}^3]$$

$$c_r = 0.21 \text{ Btu/lb}_m \cdot {}^\circ\text{F} \quad [0.88 \text{ kJ/kg} \cdot {}^\circ\text{C}]$$

$$k_r = 0.5 \text{ Btu/h} \cdot \text{ft} \cdot {}^\circ\text{F} \quad [0.87 \text{ W/m} \cdot {}^\circ\text{C}]$$

As the air flows through the rock, it is in such intimate contact with the rock that the air and rock temperatures may be assumed equal at any x position.

The rock bed is initially at 40°F and the air enters at 1 atm and 100°F. The surroundings remain at 40°F. Calculate the energy storage relative to 40°F as a function of time for inlet velocities of 1.0 and 3.0 ft/s.

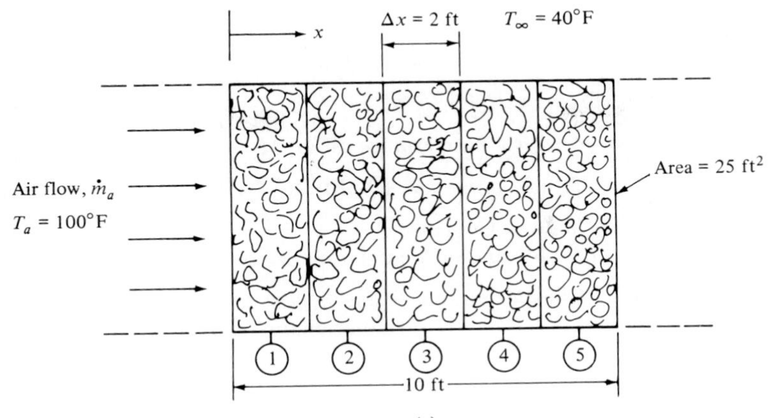

(a)

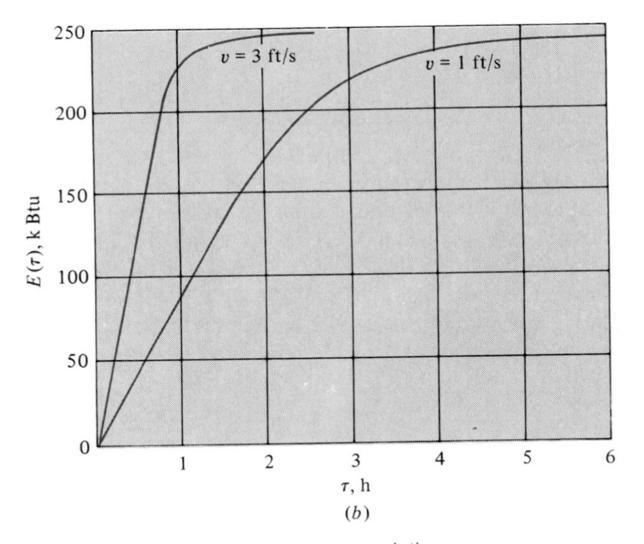

(b)

Fig. Ex. 10-16 (*a*) Schematic, (*b*) energy accumulation.

Solution

It can be shown that the axial energy conduction is small compared to the mass-energy transport. For a 60°F temperature difference over a 2-ft length

$$q_{\text{cond}} = kA\,\frac{\Delta T}{\Delta x} = (0.5)(25)\,\frac{60}{2} = 375 \text{ Btu/h} \quad [109.9 \text{ W}] \tag{a}$$

The density of the air at 100°F is

$$\rho_a = \frac{(14.696)(144)}{(53.35)(560)} = 0.07083 \text{ lb}_m/\text{ft}^3 \quad [1.1346 \text{ kg/m}^3] \tag{b}$$

and the mass flow at 1.0 ft/s is

$$\dot{m}_a = \rho A v = (0.07083)(25)(1.0) = 1.7708 \text{ lb}_m/\text{s}$$

$$= 6375 \text{ lb}_m/\text{h} \quad [2891.7 \text{ kg/h}] \tag{c}$$

The corresponding energy transport for a temperature difference of 60°F is

$$q = \dot{m}c_{p_a}\,\Delta T = (6375)(0.24)(60) = 91{,}800 \text{ Btu/h} \quad [26{,}904 \text{ W}] \tag{d}$$

and this is much larger than the value in Eq. (a).

We now write an energy balance for one of the axial nodes as

Energy transported in − energy transported out − energy lost to surroundings = rate of energy accumulation of node

or

$$\dot{m}_a c_{p_a}\,(T^p_{m-1} - T^p_m) - \frac{(T^p_m - T_\infty)P\,\Delta x}{R_\infty} = \rho_r c_r\,\Delta V_r\,\frac{(T^{p+1}_m - T^p_m)}{\Delta\tau} \tag{e}$$

where the air exit temperature from node m is assumed to be the rock temperature of that node (T^p_m). Equation (e) may be solved to give

$$T^{p+1}_m = F\dot{m}_a c_{p_a} T^p_{m-1} + \left[1 - F\left(\dot{m}_a c_{p_a} + \frac{P\,\Delta x}{R_\infty}\right)\right] T^p_m + \frac{FP\,\Delta x}{R_\infty}\,T_\infty \tag{f}$$

where

$$F = \frac{\Delta r}{\rho_r c_r\,\Delta V_r}$$

Here P is the perimeter and Δx is the x increment ($P = 4 \times 5 = 20$ ft for this problem). We are thus in a position to calculate the temperatures in the rock bed as time progresses.

The stability requirement is such that the coefficient on the T^p_m terms cannot be negative. Using $\Delta x = 2$ ft, we find that the maximum value of F is 6.4495×10^{-4}, which yields a maximum time increment of 0.54176 h. With a velocity of 3 ft/s the maximum time increment for stability is 0.1922 h. For the calculations we select the following values of $\Delta\tau$ with the resultant calculated values of F:

v	$\Delta\tau$, h	F
1.0	0.2	2.38095×10^{-4}
3.0	0.1	1.190476×10^{-4}

With the appropriate properties and these values inserted into Eq. (f) there results

$$T^{p+1}_m = 0.3642943T^p_{m-1} + 0.630943T^p_m + 0.1904762 \qquad \text{for } v = 1.0 \text{ ft/s} \tag{g}$$

$$T^{p+1}_m = 0.546430633T^p_{m-1} + 0.451188T^p_m + 0.0952381 \qquad \text{for } v = 3.0 \text{ ft/s} \tag{h}$$

The energy storage relative to 40°F can then be calculated from

$$E(\tau) = \sum_{m=1}^{5} \rho_r c_r\,\Delta V_r[T_m(\tau) - 40] \tag{i}$$

as a function of time. The computation procedure is as follows.

1. Initialize all T_m at 40°F with T_{m-1} for node 1 at 100°F for all time increments.
2. Compute new values of T_m from either Eq. (g) or (h), progressing forward in time until a desired stopping point is reached or the temperature attains steady-state conditions.
3. Using computed values of $T_m(\tau)$, evaluate $E(\tau)$ from Eq. (i).

Results of the calculations are shown in the accompanying figure. For v = 3.0 ft/s, steady state is reached at about τ = 1.5 h while for v = 1.0 ft/s it is reached at about τ = 5.5 h. Note that the steady-state value of E for v = 1.0 ft/s is lower than for v = 3.0 ft/s because a longer time is involved and more of the energy ''leaks out'' through the insulation.

This example shows how a rather complex problem can be solved in a straightforward way by using a numerical formulation.

■ EXAMPLE 10-17 Variable-properties analysis of a duct heater

A 600-ft-long duct having a diameter of 1 ft serves as a space heater in a warehouse area. Hot air enters the duct at 800°F, and the emissivity of the outside duct surface is 0.6. Determine duct air temperature, wall temperature, and outside heat flux along the duct for flow rates of 0.3, 1.0, and 1.5 lb_m/s. Take into account variations in air properties. The room temperature for both convection and radiation is 70°F.

Solution

This is a problem where a numerical solution must be employed. We choose a typical section of the duct with length Δx and perimeter P as shown and make the energy balances. We assume that the conduction resistance of the duct wall is negligible. Inside the duct the energy balance is

$$\dot{m}_a c_p T_{m,a} = h_i P\,\Delta x(T_{m,a} - T_{m,w}) + \dot{m}_a c_p T_{m+1,a} \tag{a}$$

where h_i is the convection heat-transfer coefficient on the inside which may be calculated from (the flow is turbulent)

$$\text{Nu} = \frac{h_i d}{k} = 0.023\,\text{Re}_d^{0.8}\,\text{Pr}^{0.3} \tag{b}$$

with properties evaluated at the bulk temperature of air ($T_{m,a}$). The energy balance for the heat flow through the wall is

$$q_{\text{conv},i} = q_{\text{conv},o} + q_{\text{rad},o}$$

or, by using convection coefficients and radiation terms per unit area,

$$h_i(T_{m,a} - T_{m,w}) = h_c(T_{m,w} - T_\infty) + \sigma\epsilon(T_{m,w}^4 - T_\infty{}^4) \tag{c}$$

where the outside convection coefficient can be calculated from the free-convection relation

$$h_c = 0.27\left(\frac{T_{m,w} - T_\infty}{d}\right)^{1/4} \qquad \text{Btu/h}\cdot\text{ft}^2\cdot{}^\circ\text{F} \tag{d}$$

Inserting this relation in Eq. (c) gives

$$h_i(T_{m,a} - T_{m,w}) = \frac{0.27}{d^{1/4}}(T_{m,w} - T_\infty)^{5/4} + \sigma\epsilon(T_{m,w}^4 - T_\infty^4) \tag{e}$$

Equation (a) may be solved for $T_{m+1,a}$ to give

$$T_{m+1,a} = \left(1 - \frac{h_i P\,\Delta x}{\dot m_a c_p}\right)_m T_{m,a} + \left(\frac{h_i P\,\Delta x}{\dot m_a c_p}\right)_m T_{m,w} \tag{f}$$

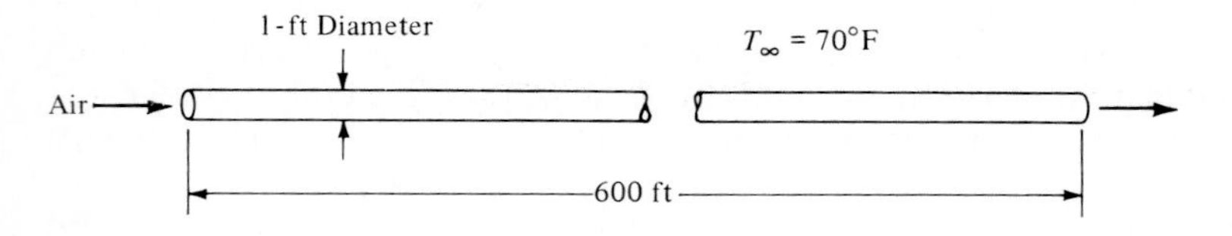

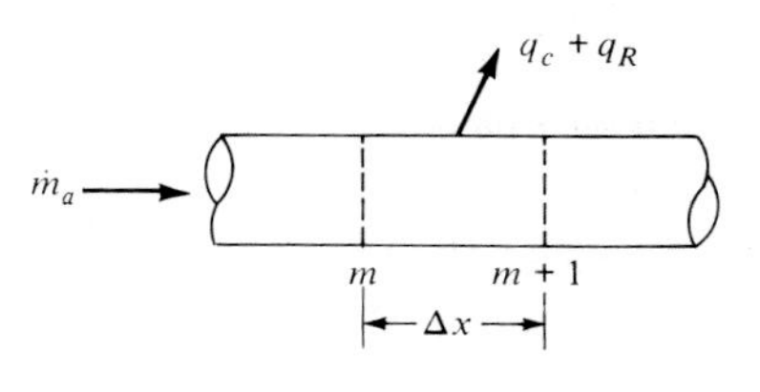

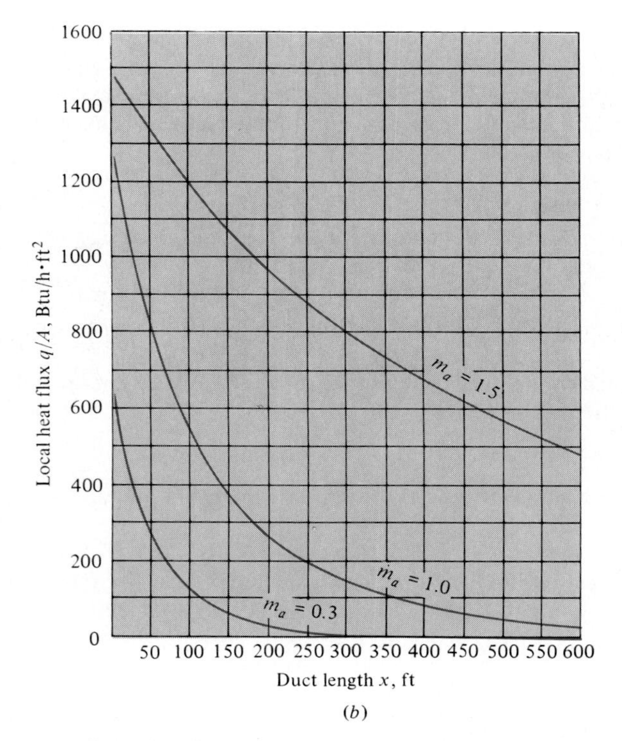

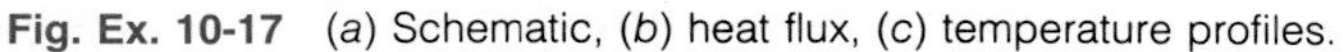
Fig. Ex. 10-17 (a) Schematic, (b) heat flux, (c) temperature profiles.

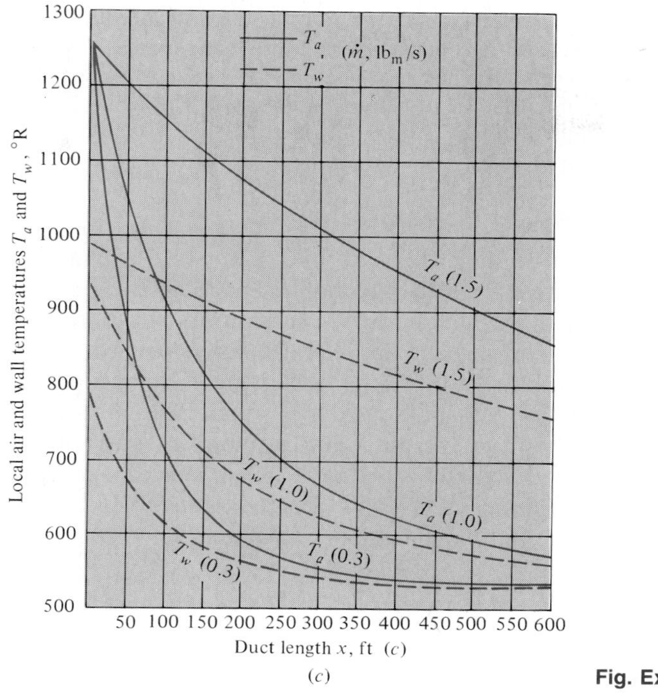

Fig. Ex. 10-17 *(Continued)*

With these equations at hand, we may now formulate the computational algorithm as follows. Note that all temperatures must be in degrees Rankine because of the radiation term.

1. Select Δx.
2. Starting at $x = 0$, entrance conditions, evaluate h_i from Eq. (b) with properties evaluated at $T_{m,a}$. (At entrance $T_{m,a} = 800°\text{F} = 1260°\text{R}$.)
3. Solve (by iteration) Eq. (e) for $T_{m,w}$.
4. Solve for $T_{m+1,a}$ from Eq. (f).
5. Repeat for successive increments until the end of the duct ($x = 600$ ft) is reached.
6. The heat lost at each increment is

$$q = P\,\Delta x\, h_i(T_{m,a} - T_{m,w})$$

or the heat flux is

$$\frac{q}{A} = h_i(T_{m,a} - T_{m,w}) \tag{g}$$

7. The results for $T_{m,a}$, $T_{m,w}$, and $(q/A)_m$ may be plotted as in the accompanying figures.

For these calculations we have selected $\Delta x = 50$ ft. For the low flow rate (0.3 lb_m/s) we note that the air essentially attains the room temperature halfway along the length of the duct, so that little heating is provided past that point. With the 1.0-lb_m/s flow rate there is still some heating at the end of the duct, although it is small. The 1.5-lb_m/s flow rate contributes substantial heating all along the length of the duct.

10-9 HEAT-EXCHANGER DESIGN CONSIDERATIONS

In the process and power industries, or related activities, many heat exchangers are purchased as off-the-shelf items, and a selection is made on the basis of cost and specifications furnished by the various manufacturers. In more specialized applications, such as the aerospace and electronics industries, a particular design is frequently called for. Where a heat exchanger forms a part of an overall machine or device to be manufactured, a standard item may be purchased; or if cost considerations and manufacturing quantities warrant, the heat exchanger may be specially designed for the application. Whether the heat exchanger is selected as an off-the-shelf item or designed especially for the application, the following factors are almost always considered:

1. Heat-transfer requirements
2. Cost
3. Physical size
4. Pressure-drop characteristics

The heat-transfer requirements must be met in the selection or design of any heat exchanger. The way that the requirements are met depends on the relative weights placed on items 2 to 4. By forcing the fluids through the heat exchanger at higher velocities the overall heat-transfer coefficient may be increased, but this higher velocity results in a larger pressure drop through the exchanger and correspondingly larger pumping costs. If the surface area of the exchanger is increased, the overall heat-transfer coefficient, and hence the pressure drop, need not be so large; however, there may be limitations on the physical size which can be accommodated, and a larger physical size results in a higher cost for the heat exchanger. Prudent judgment and a consideration of all these factors will result in the proper design. A practitioner in the field will find the extensive information of Ref. 8 to be very useful.

REVIEW QUESTIONS

1 Define the overall heat-transfer coefficient.

2 What is a fouling factor?

3 Why does a "mixed" or "unmixed" fluid arrangement influence heat-exchanger performance?

4 When is the LMTD method most applicable to heat-exchanger calculations?

5 Define effectiveness.

6 What advantage does the effectiveness-NTU method have over the LMTD method?

7 What is meant by the "minimum" fluid?

8 Why is a counterflow exchanger more effective than a parallel-flow exchanger?

PROBLEMS

10-1 A long steel pipe with a 5-cm ID and 3.2-mm wall thickness passes through a large room maintained at 30°C and atmospheric pressure; 0.6 kg/s of hot water enters one end of the pipe at 82°C. If the pipe is 15 m long, calculate the exit water temperature, considering both free convection and radiation heat loss from the outside of the pipe.

10-2 Some of the brine from a large refrigeration system is to be used to furnish chilled water for air-conditioning part of an office building. The brine is available at −15°C, and 105 kW of cooling is required. The chilled water from the conditioned air coolers enters a shell-and-tube heat exchanger at 10°C, and the exchanger is to be designed so that the exit chilled-water temperature is not below 5°C. The overall heat-transfer coefficient for the heat-exchanger is 850 $W/m^2 \cdot °C$. If the chilled water is used on the tube side and two tube passes are employed, plot the heat-exchanger area required as a function of the brine exit temperature.

10-3 Air at 207 kPa and 200°C enters a 2.5-cm-ID tube at 6 m/s. The tube is constructed of copper with a thickness of 0.8 mm and a length of 3 m. Atmospheric air at 1 atm and 20°C flows normal to the outside of the tube with a free-stream velocity of 12 m/s. Calculate the air temperature at exit from the tube. What would be the effect of reducing the hot-air flow in half?

10-4 Repeat Prob. 10-3 for water entering the tube at 1 m/s and 95°C. What would be the effect of reducing the water flow in half?

10-5 Hot water at 90°C flows on the inside of a 2.5-cm-ID steel tube with 0.8-mm wall thickness at a velocity of 4 m/s. Engine oil at 20°C is forced across the tube at a velocity of 7 m/s. Calculate the overall heat-transfer coefficient for this arrangement.

10-6 Hot water at 90°C flows on the inside of a 2.5-cm-ID steel tube with 0.8-mm wall thickness at a velocity of 4 m/s. This tube forms the inside of a double-pipe heat exchanger. The outer pipe has a 3.75-cm ID, and engine oil at 20°C flows in the annular space at a velocity of 7 m/s. Calculate the overall heat-transfer coefficient for this arrangement. The tube length is 6.0 m.

10-7 Air at 2 atm and 200°C flows inside a 1-in schedule 80 steel pipe with $h = 65$ $W/m^2 \cdot °C$. A hot gas with $h = 180$ $W/m^2 \cdot °C$ flows across the outside of the pipe at 400°C. Calculate the overall heat-transfer coefficient.

10-8 Hot engine oil enters a 1-in schedule 40 steel pipe at 80°C with a velocity of 5 m/s. The pipe is submerged horizontally in water at 20°C so that it loses heat by free convection. Calculate the length of pipe necessary to lower the oil temperature to 60°C. Perform some kind of calculation to indicate the effect variable properties have on the results.

10-9 Hot exhaust gases are used in a finned-tube cross-flow heat exchanger to heat 2.5 kg/s of water from 35 to 85°C. The gases [c_p = 1.09 kJ/kg · °C] enter at 200 and leave at 93°C. The overall heat-transfer coefficient is 180 W/m² · °C. Calculate the area of the heat exchanger using (*a*) the LMTD approach and (*b*) the effectiveness-NTU method.

10-10 Derive Eq. (10-12), assuming that the heat exchanger is a counterflow double-pipe arrangement.

10-11 Derive Eq. (10-27).

10-12 Water at the rate of 230 kg/h at 35°C is available for use as a coolant in a double-pipe heat exchanger whose total surface area is 1.4 m². The water is to be used to cool oil [c_p = 2.1 kJ/kg · °C] from an initial temperature of 120°C. Because of other circumstances, an exit water temperature greater than 99°C cannot be allowed. The exit temperature of the oil must not be below 60°C. The overall heat-transfer coefficient is 280 W/m² · °C. Estimate the maximum flow rate of oil which may be cooled, assuming the flow rate of water is fixed at 230 kg/h.

10-13 The condenser on a certain automobile air conditioner is designed to remove 60,000 Btu/h from Freon 12 when the automobile is moving at 40 mi/h and the ambient temperature is 95°F. The Freon 12 temperature is 150°F under these conditions, and it may be assumed that the air-temperature rise across the exchanger is 10°F. The overall heat-transfer coefficient for the finned-tube heat exchanger under these conditions is 35 Btu/h · ft² · °F. If the overall heat-transfer coefficient varies as the seven-tenths power of velocity and air-mass flow varies directly as the velocity, plot the percentage reduction in performance of the condenser as a function of velocity between 10 and 40 mi/h. Assume that the Freon temperature remains constant at 150°F.

10-14 A small shell-and-tube exchanger with one tube pass [A = 4.64 m² and U = 280 W/m² · °C] is to be used to heat high-pressure water initially at 20°C with hot air initially at 260°C. If the exit water temperature is not to exceed 93°C and the air flow rate is 0.45 kg/s, calculate the water flow rate.

10-15 A counterflow double-pipe heat exchanger is to be used to heat 0.6 kg/s of water from 35 to 90°C with an oil flow of 0.9 kg/s. The oil has a specific heat of 2.1 kJ/kg · °C and enters the heat exchanger at a temperature of 175°C. The overall heat-transfer coefficient is 425 W/m² · °C. Calculate the area of the heat exchanger and the effectiveness.

10-16 A shell-and-tube heat exchanger is to be designed to heat 7.5 kg/s of water from 85 to 99°C. The heating process is accomplished by condensing steam at 345 kPa. One shell pass is used along with two tube passes, each consisting of thirty 2.5-cm-OD tubes. Assuming a value of U of 2800 W/m², calculate the length of tubes required in the heat exchanger.

10-17 Suppose the heat exchanger in Prob. 10-16 has been in operation an extended period of time so that the fouling factors in Table 10-2 apply. Calculate the exit water temperature for fouled conditions, assuming the same total flow rate.

10-18 Rework Example 6-4, using the LMTD concept. Repeat for an inlet air temperature of 37°C.

10-19 A shell-and-tube heat exchanger operates with two shell passes and four tube passes. The shell fluid is ethylene glycol, which enters at 140°C and leaves at 80°C with a flow rate of 4500 kg/h. Water flows in the tubes, entering at 35°C and leaving at 85°C. The overall heat-transfer coefficient for this arrangement is 850 W/m^2 · °C. Calculate the flow rate of water required and the area of the heat exchanger.

10-20 The flow rate of glycol to the exchanger in Prob. 10-19 is reduced in half with the entrance temperatures of both fluids remaining the same. What is the water exit temperature under these new conditions, and by how much is the heat-transfer rate reduced?

10-21 For the exchanger in Prob. 10-9 the water flow rate is reduced by half, while the gas flow rate is maintained constant along with the fluid inlet temperatures. Calculate the percentage reduction in heat transfer as a result of this reduced flow rate. Assume that the overall heat-transfer coefficient remains the same.

10-22 Repeat Prob. 10-9 for a shell-and-tube exchanger with two tube passes. The gas is the shell fluid.

10-23 Repeat Prob. 10-21, using the shell-and-tube exchanger of Prob. 10-22.

10-24 It is desired to heat 230 kg/h of water from 35 to 93°C with oil [c_p = 2.1 kJ/kg · °C] having an initial temperature of 175°C. The mass flow of oil is also 230 kg/h. Two double-pipe heat exchangers are available:

exchanger 1:	U = 570 W/m^2 · °C	A = 0.47 m^2
exchanger 2:	U = 370 W/m^2 · °C	A = 0.94 m^2

Which exchanger should be used?

10-25 A small steam condenser is designed to condense 0.76 kg/min of steam at 83 kPa with cooling water at 10°C. The exit water temperature is not to exceed 57°C. The overall heat-transfer coefficient is 3400 W/m^2 · °C. Calculate the area required for a double-pipe heat exchanger.

10-26 Suppose the inlet water temperature in the exchanger of Prob. 10-25 is raised to 30°C. What percentage increase in flow rate would be necessary to maintain the same rate of condensation?

10-27 A counterflow double-pipe heat exchanger is used to heat water from 20 to 40°C by cooling an oil from 90 to 55°C. The exchanger is designed for a total heat transfer of 29 kW with an overall heat-transfer coefficient of 340 W/m^2 · °C. Calculate the surface area of the exchanger.

10-28 A feedwater heater uses a shell-and-tube exchanger with condensing steam in one shell pass at 120°C. Water enters the tubes at 30°C and makes four passes to produce an overall U value of 2000 W/m^2 · °C. Calculate the area of the exchanger for 2.5-kg/s mass flow of the water, with a water exit temperature of 100°C.

10-29 Suppose the exchanger in Prob. 10-28 has been in service a long time such that a fouling factor of 0.0002 m² · °C/W is experienced. What would be the exit water temperature under these conditions?

10-30 An air-to-air heat recovery unit uses a cross-flow exchanger with both fluids unmixed and an air flow rate of 0.5 kg/s on both sides. The hot air enters at 400°C while the cool air enters at 20°C. Calculate the exit temperatures for $U = 40$ W/m² · °C and a total exchanger area of 20 m².

10-31 In a large air-conditioning application 1500 m³/min of air at 1 atm and 10°C are to be heated in a finned tube heat exchanger with hot water entering the exchanger at 80°C. The overall heat-transfer coefficient is 50 W/m² · °C. Calculate the required area for the heat exchanger for an exit air temperature of 35°C and exit water temperature of 50°C.

10-32 A cross-flow finned-tube heat exchanger uses hot water to heat air from 20 to 45°C. The entering water temperature is 75°C and its exit temperature is 45°C. The total heat transfer rate is to be 100,000 Btu/h. If the overall heat transfer coefficient is 50 W/m² · °C, calculate the area of the heat exchanger.

10-33 Hot oil at 120°C with a flow rate of 95 kg/min is used in a shell-and-tube heat exchanger with one shell pass and two tube passes to heat 55 kg/min of water which enters at 30°C. The area of the exchanger is 14 m². Calculate the heat transfer and exit temperature of both fluids if the overall heat-transfer coefficient is 250 W/m² · °C.

10-34 A double-pipe heat exchanger is constructed of copper and operated in a counterflow mode. It is designed to heat 100 lb_m/min of water from 50°F to 175°F. The water flows through the inner pipe. The heating is accomplished by condensing steam in the outer pipe at a temperature of 250°F. The water-side heat-transfer coefficient is 250 Btu/h · ft² · °F. Assume a reasonable value for the steam-side coefficient and then calculate the area of the heat exchanger. Estimate what the exit water temperature would be if the water flow rate were reduced by 60 percent for this exchanger.

10-35 A counterflow double-pipe heat exchanger is used to heat liquid ammonia from 10 to 30°C with hot water that enters the exchanger at 60°C. The flow rate of the water is 5.0 kg/s and the overall heat-transfer coefficient is 800 W/m² · °C. The area of the heat exchanger is 30 m². Calculate the flow rate of ammonia.

10-36 A shell-and-tube heat exchanger has condensing steam at 100°C in the shell side with one shell pass. Two tube passes are used with air in the tubes entering at 10°C. The total surface area of the exchanger is 30 m² and the overall heat-transfer coefficient may be taken as 150 W/m² · °C. If the effectiveness of the exchanger is 85 percent, what is the total heat-transfer rate?

10-37 Suppose a fouling factor of 0.0002 m² · °C/W is used for the water in Prob. 10-36 and 0.0004 m² · °C/W for the air. What percent increase in area should be included in the design to take these factors into account for future operation?

10-38 Suppose both flow rates in Prob. 10-30 were cut in half. What would be the exit temperatures in this case, assuming no change in U? What if the flow rates were doubled?

10-39 Hot water at 90°C is used in the tubes of a finned-tube heat exchanger. Air flows across the fins and enters at 1 atm, 30°C, with a flow rate of 65 kg/min. The overall heat-transfer coefficient is 52 W/m² · °C, and the exit air temperature is to be 45°C. Calculate the exit water temperature if the total area is 8.0 m².

10-40 A microprocessor is to be programmed to control the exchanger in Prob. 10-39 by varying the water flow rate to maintain the same exit air temperature for changes in inlet water temperature. Calculate the percentage changes necessary for the water flow rate for inlet water temperatures of 60, 70, 80, and 100°C. Assume U remains constant.

10-41 Hot water enters a counterflow heat exchanger at 99°C. It is used to heat a cool stream of water from 4 to 32°C. The flow rate of the cool stream is 1.3 kg/s, and the flow rate of the hot stream is 2.6 kg/s. The overall heat-transfer coefficient is 830 W/m² · °C. What is the area of the heat exchanger? Calculate the effectiveness of the heat exchanger.

10-42 Starting with a basic energy balance, derive an expression for the effectiveness of a heat exchanger in which a condensing vapor is used to heat a cooler fluid. Assume that the hot fluid (condensing vapor) remains at a constant temperature throughout the process.

10-43 Water at 75°C enters a counterflow heat exchanger. It leaves at 30°C. The water is used to heat an oil from 25 to 48°C. What is the effectiveness of the heat exchanger?

10-44 Saturated steam at 100 lb/in² abs is to be used to heat carbon dioxide in a cross-flow heat exchanger consisting of four hundred $\frac{1}{4}$-in-OD brass tubes in a square in-line array. The distance between tube centers is $\frac{3}{8}$ in, in both the normal- and parallel-flow directions. The carbon dioxide flows across the tube bank, while the steam is condensed on the inside of the tubes. A flow rate of 1 lb_m/s of CO_2 at 15 lb/in² abs and 70°F is to be heated to 200°F. Estimate the length of the tubes to accomplish this heating. Assume that the steam-side heat-transfer coefficient is 1000 Btu/h · ft² · °F, and neglect the thermal resistance of the tube wall.

10-45 Repeat Prob. 10-44 with the CO_2 flowing on the inside of the tubes and the steam condensing on the outside of the tubes. Compare these two designs on the basis of CO_2 pressure drop through the exchanger.

10-46 Replot Figs. 10-12 and 10-13 as ϵ versus log NTU_{max} over the range $0.1 < NTU_{max} < 100$. If possible use a computer.

10-47 A counterflow double-pipe heat exchanger is currently used to heat 2.5 kg/s of water from 25 to 65°C by cooling an oil [c_p = 2.1 kJ/kg · °C] from 138 to 93°C. It is desired to "bleed off" 0.62 kg/s of water at 50°C so that the single exchanger will be replaced by a two-exchanger arrangement which will permit this. The overall heat-transfer coefficient is 450 W/m² · °C for the single exchanger and may be taken as this same value for each of the smaller exchangers. The same total oil flow is used for the two-exchanger arrangement, except that the flow is split between the two exchangers. Determine the areas of each of the smaller exchangers and the oil flow rates through each. Assume that the water flows in series through the two exchangers, with the bleed-off taking place between them. Assume the smaller exchangers have the same areas.

10-48 Repeat Prob. 10-47, assuming that condensing steam at 138°C is used instead of the hot oil and that the exchangers are of the shell-and-tube type with the water making two passes on the tube side. The overall heat-transfer coefficient may be taken as 1700 W/m^2 · °C for this application.

10-49 A shell-and-tube heat exchanger with four tube passes is used to heat 2.5 kg/s of water from 25 to 70°C. Hot water at 93°C is available for the heating process, and a flow rate of 5 kg/s may be used. The cooler fluid is used on the tube side of the exchanger. The overall heat-transfer coefficient is 800 W/m^2 · °C. Assuming that the flow rate of the hot fluid and the overall heat-transfer coefficient remain constant, plot the percentage reduction in heat transfer as a function of the mass-flow rate of the cooler fluid.

10-50 Two identical double-pipe heat exchangers are constructed of a 2-in standard schedule 40 pipe placed inside a 3-in standard pipe. The length of the exchangers is 10 ft; 40 gal/min of water initially at 80°F is to be heated by passing through the inner pipes of the exchangers in a series arrangement, and 30 gal/min of water at 120°F and 30 gal/min of water at 200°F are available to accomplish the heating. The two heating streams may be mixed in any way desired before and after they enter the heat exchangers. Determine the flow arrangement for optimum performance (maximum heat transfer) and the total heat transfer under these conditions.

10-51 Suppose that the oil in Prob. 10-27 is sufficiently dirty for a fouling factor of 0.004 to be necessary in the analysis. What is the surface area under these conditions? How much would the heat transfer be reduced if the exchanger in Prob. 10-27 were used with this fouling factor and the same inlet fluid temperatures?

10-52 A shell-and-tube exchanger with one shell pass and two tube passes is used as a water-to-water heat-transfer system with the hot fluid in the shell side. The hot water is cooled from 80 to 60°C, and the cool fluid is heated from 5 to 60°C. Calculate the surface area for a heat transfer of 60 kW and a heat-transfer coefficient of 1100 W/m^2 · °C.

10-53 What is the heat transfer for the exchanger in Prob. 10-52 if the flow rate of the hot fluid is reduced in half while the inlet conditions and heat-transfer coefficient remain the same?

10-54 High-temperature flue gases at 450°C [c_p = 1.2 kJ/kg · °C] are employed in a cross-flow heat exchanger to heat an engine oil from 30 to 80°C. Using the information given in this chapter, obtain an approximate design for the heat exchanger for an oil flow rate of 0.6 kg/s.

10-55 A cross-flow finned-tube heat exchanger uses hot water to heat an appropriate quantity of air from 15 to 25°C. The water enters the heat exchanger at 70°C and leaves at 40°C, and the total heat-transfer rate is to be 29 kW. The overall heat-transfer coefficient is 45 W/m^2 · °C. Calculate the area of the heat exchanger.

10-56 Calculate the heat-transfer rate for the exchanger in Prob. 10-55 when the water flow rate is reduced to one-third that of the design value.

10-57 A gas-turbine regenerator is a heat exchanger which uses the hot exhaust gases from the turbine to preheat the air delivered to the combustion chamber. In an air-standard analysis of gas-turbine cycles, it is assumed that the mass of fuel is small in comparison with the mass of air, and consequently the hot-gas flow through the turbine is essentially the same as the airflow into the combustion chamber. Using this assumption, and also assuming that the specific heat of the hot exhaust gases is the same as that of the incoming air, derive an expression for the effectiveness of a regenerator under both counterflow and parallel-flow conditions.

10-58 Water at 90°C enters a double-pipe heat exchanger and leaves at 55°C. It is used to heat a certain oil from 25 to 50°C. Calculate the effectiveness of the heat exchanger.

10-59 Because of priority requirements the hot fluid flow rate for the exchanger in Probs. 10-19 and 10-20 must be reduced by 40 percent. The same water flow must be heated from 35 to 85°C. To accomplish this, a shell-and-tube steam preheater is added, with steam condensing at 150°C and an overall heat-transfer coefficient of 2000 W/m^2 · °C. What surface area and steam flow are required for the preheater?

10-60 An engine-oil heater employs ethylene glycol at 100°C entering a tube bank consisting of 50 copper tubes, five rows high with an OD of 2.5 cm and a wall thickness of 0.8 mm. The tubes are 70 cm long with $S_p = S_n = 3.75$ cm in an in-line arrangement. The oil enters the tube bank at 20°C and a velocity of 1 m/s. The glycol enters the tubes with a velocity of 1.5 m/s. Calculate the total heat transfer and the exit oil temperature. Repeat for an inlet glycol velocity of 1.0 m/s.

10-61 An air preheater for a power plant consists of a cross-flow heat exchanger with hot exhaust gases used to heat incoming air at 1 atm and 300 K. The gases enter at 375°C with a flow rate of 5 kg/s. The air flow rate is 5.0 kg/s, and the heat exchanger has $A = 110$ m^2 and $U = 50$ W/m^2 · °C. Calculate the heat-transfer rate and exit temperatures for two cases, both fluids unmixed and one fluid mixed. Assume the hot gases have the properties of air.

10-62 Condensing steam at 100°C with $h = 5000$ W/m^2 · °C is used inside the tubes of the exchanger in Example 10-16. If the heat exchanger has a frontal area of 0.5 m^2 and a depth of 40 cm in the air-flow direction, calculate the heat-transfer rate and exit air temperature. State assumptions.

10-63 A counterflow double-pipe heat exchanger is employed to heat 30 kg/s of water from 20 to 40°C with a hot oil at 200°C. The overall heat-transfer coefficient is 275 W/m^2 · °C. Determine effectiveness and NTU for exit oil temperatures of 190, 180, 140, and 80°C.

10-64 A shell-and-tube heat exchanger is designed for condensing steam at 200°C in the shell with one shell pass; 50 kg/s of water are heated from 60 to 90°C. The overall heat-transfer coefficient is 4500 W/m^2 · °C. A controller is installed on the steam inlet to vary the temperature by controlling the pressure, and the effect on the outlet water temperature is desired. Calculate the effectiveness and outlet water temperature for steam inlet temperatures of 180, 160, 140, and 120°C. Use the analytical expressions to derive a relation for the outlet water temperature as a function of steam inlet temperature.

10-65 A shell-and-tube heat exchanger with one shell pass and two tube passes is used to heat 5.0 kg/s of water from 30°C to 80°C. The water flows in the tubes. Condensing steam at 1 atm is used in the shell side. Calculate the area of the heat exchanger, if the overall heat-transfer coefficient is 900 W/m² · °C. Suppose this same exchanger is used with entering water at 30°C, $U = 900$, but with a water flow rate of 1.3 kg/s. What would be the exit water temperature under these conditions?

10-66 An ammonia condenser is constructed of a five by five array of horizontal tubes having an outside diameter of 2.5 cm and a wall thickness of 1.0 mm. Water enters the tubes at 20°C and 5 m/s and leaves at 40°C. The ammonia condensing temperature is 60°C. Calculate the length of tubes required. How much ammonia is condensed? Consult thermodynamics tables for the needed properties of ammonia.

10-67 Rework Prob. 10-66 for a 10 by 10 array of tubes. If the length of tubes is reduced by half, what reduction in condensate results for the same inlet water temperature? (The exit water temperature is not the same.)

10-68 A counterflow double-pipe heat exchanger is used to heat water from 20°C to 40°C with a hot oil which enters the exchanger at 180°C and leaves at 140°C. The flow rate of water is 3.0 kg/s and the overall heat-transfer coefficient is 130 W/m² · °C. Assume the specific heat for oil is 2100 J/kg · °C. Suppose the water flow rate is cut in half. What new oil flow rate would be necessary to maintain a 40°C outlet water temperature? (The oil flow is *not* cut in half.)

10-69 A home air-conditioning system uses a cross-flow finned-tube heat exchanger to cool 0.8 kg/s of air from 85°F to 45°F. The cooling is accomplished with 0.75 kg/s of water entering at 3°C. Calculate the area of the heat exchanger assuming an overall heat-transfer coefficient of 55 W/m² · °C. If the water flow rate is cut in half while the same air flow rate is maintained, what percent reduction in heat transfer will occur?

10-70 The same airflow as in Prob. 10-69 is to be cooled in a finned-tube exchanger with evaporating Freon in the tubes. It may be assumed that the Freon temperature remains constant at 35°F and that the overall heat-transfer coefficient is 125 W/m² · °C. Calculate the exchanger area required in this case. Also calculate the reduction in heat transfer which would result from cutting the air flow rate by one-third.

10-71 A shell-and-tube heat exchanger with one shell pass and four tube passes is designed to heat 4000 kg/h of engine oil from 40°C to 80°C with the oil in the tube side. On the shell side is condensing steam at 1-atm pressure, and the overall heat-transfer coefficient is 1200 W/m² · °C. Calculate the mass flow of condensed steam if the flow of oil is reduced in half while the inlet temperature and U value are kept the same.

10-72 A heat exchanger with an effectiveness of 80 percent is used to heat 5 kg/s of water from 50°C with condensing steam at 1 atm. Calculate the area for $U =$ 1200 W/m² · °C.

10-73 If the flow rate of water for the exchanger in Prob. 10-72 is reduced in half, what is the water exit temperature and the overall heat transfer?

10-74 Hot water at 180°F is used to heat air from 45°F to 115°F in a finned tube cross-flow heat exchanger. The water exit temperature is 125°F. Calculate the effectiveness of this heat exchanger.

10-75 If the mass flow of water in the exchanger in Prob. 10-74 is 150 kg/min and the overall heat-transfer coefficient is 50 $W/m^2 \cdot °C$, calculate the area of the heat exchanger.

10-76 A shell-and-tube heat exchanger having one shell pass and four tube passes is used to heat 10 kg/s of ethylene glycol from 20 to 40°C on the shell side; 15 kg/s of water entering at 70°C is used in the tubes. The overall heat-transfer coefficient is 40 $W/m^2 \cdot °C$. Calculate the area of the heat exchanger.

10-77 The same exchanger as in Prob. 10-76 is used with the same inlet temperature conditions but the water flow reduced to 10 kg/s. Because of the reduced water flow rate the overall heat-transfer coefficient drops to 35 $W/m^2 \cdot °C$. Calculate the exit glycol temperature.

10-78 A large condenser is designed to remove 800 MW of energy from condensing steam at 1-atm pressure. To accomplish this task, cooling water enters the condenser at 25°C and leaves at 30°C. The overall heat-transfer coefficient is 2000 $W/m^2 \cdot °C$. Calculate the area required for the heat exchanger.

10-79 Suppose the water flow rate for the exchanger in Prob. 10-78 is reduced in half from the design value. What will be the steam condensation rate in kilograms per hour under these conditions if U remains the same?

10-80 A shell-and-tube heat exchanger with one shell pass and two tube passes is used to heat ethylene glycol in the tubes from 20°C to 60°C. The flow rate of glycol is 1.2 kg/s. Water is used in the shell side to supply the heat and enters the exchanger at 90°C and leaves at 50°C. The overall heat-transfer coefficient is 600 $W/m^2 \cdot °C$. Calculate the area of the heat exchanger.

10-81 A compact heat exchanger like that shown in Fig. 10-19 is to be designed to cool water from 200°F to 160°F with an airflow which enters the exchanger at 1 atm and 70°F. The inlet air flow velocity is 50 ft/s. The total heat transfer is to be 240,000 Btu/h. Select several alternative designs and investigate each in terms of exchanger size (area and/or volume) and the pressure drop.

10-82 Air at 300 K enters a compact heat exchanger like that shown in Fig. 10-19. Inside the tubes steam is condensing at a constant temperature of 100°C with $h = 9000$ $W/m^2 \cdot °C$. If the entering air velocity is 10 m/s, calculate the amount of steam condensed with an array 30 by 30 cm square and 60 cm long.

10-83 Repeat Prob. 10-82 for the same air flow stream in configuration D of Fig. 10-20.

10-84 If one wishes to condense half as much steam as in Prob. 10-82, how much smaller an array can be used while keeping the length at 60 cm?

10-85 A double-pipe heat exchanger is to be designed to cool water from 80 to 60°C with ethylene glycol entering the exchanger at 20°C. The flow rate of glycol is 0.7 kg/s, and the water flow rate is 0.5 kg/s. Calculate the effectiveness of the heat exchanger. If the overall heat-transfer coefficient is 1000 $W/m^2 \cdot °C$, calculate the area required for the heat exchanger.

10-86 A cross-flow heat exchanger uses oil ($c_p = 2.1$ kJ/kg · °C) in the tube bank with an entering temperature of 100°C. The flow rate of oil is 1.2 kg/s. Water flows across the unfinned tubes and is heated from 20 to 50°C with a flow rate of 0.6 kg/s. If the overall heat-transfer coefficient is 250 W/m^2 · °C, calculate the area required for the heat exchanger.

10-87 Rework Prob. 10-86 with the water flowing inside the tubes and the oil flowing across the tubes.

10-88 A shell-and-tube heat exchanger with three shell passes and six tube passes is used to heat 2 kg/s of water from 10 to 70°C in the shell side; 3 kg/s of hot oil ($c_p = 2.1$ kJ/kg · °C) at 120°C is used inside the tubes. If the overall heat-transfer coefficient is 300 W/m^2 · °C, calculate the area required for the heat exchanger.

10-89 The water flow rate for the exchanger in Prob. 10-88 is reduced to 1.0 kg/s while the temperature of the entering fluid remains the same, as does the value of U. Calculate the exit fluid temperatures under this new condition.

10-90 A shell-and-tube heat exchanger with two shell passes and four tube passes is used to condense Freon 12 in the shell at 100°F. Water enters the tubes at 70°F and leaves at 80°F. Freon is to be condensed at a rate of 1800 lb$_m$/h with an enthalpy of vaporization $h_{fg} = 56$ Btu/lb$_m$. If the overall heat-transfer coefficient is 700 W/m^2 · °C, calculate the area of the heat exchanger.

10-91 Calculate the percent reduction in condensation of Freon in the exchanger of Prob. 10-90 if the water flow is reduced in half but the inlet temperature and value of U remain the same.

10-92 A shell-and-tube heat exchanger with four shell passes and eight tube passes is used to heat 3 kg/s of water from 10 to 30°C in the shell side by cooling 3 kg/s of water from 80 to 60°C in the tube side. If $U = 1000$ W/m^2 · °C, calculate the area of the heat exchanger.

10-93 For the area of heat exchanger found in Prob. 10-92 calculate the percent reduction in heat transfer if the cold-fluid flow rate is reduced to half while keeping the inlet temperature and value of U the same.

10-94 Show that for $C = 0.5$ and 1.0 the effectiveness given in Fig. 10-17 can be calculated from an effectiveness read from Fig. 10-16 and the equation for n shell passes given in Table 10-3. Note that

$$\text{NTU}(n \text{ shell passes}) = n \times \text{NTU(one shell pass)}$$

10-95 Show that for an n-shell-pass exchanger the effectiveness for each shell pass is given by

$$\epsilon_p = \frac{[(1 - \epsilon C)/(1 - \epsilon)]^{1/n} - 1}{[(1 - \epsilon C)/(1 - \epsilon)]^{1/n} - C}$$

where ϵ is the effectiveness for the multishell-pass exchanger.

10-96 Hot oil ($c_p = 2.1$ kJ/kg · °C) at a rate of 7.0 kg/s and 100°C is used to heat 3.5 kg/s of water at 20°C in a cross-flow heat exchanger with the oil inside the tubes and the water flowing across the tubes. The effectiveness of the heat exchanger is 60 percent. Calculate the exit temperatures for both fluids and the product UA for the heat exchanger.

10-97 A single-shell-pass heat exchanger with two tube passes is used to condense steam at 100°C (1 atm) on the shell side. Water is used on the tube side and enters the exchanger at 20°C with a flow rate of 1.0 kg/s. The overall heat-transfer coefficient is 2500 W/m^2 · °C, and the surface area of the exchanger is 0.8 m^2. Calculate the exit temperature of the water.

10-98 A shell-and-tube heat exchanger with four shell passes and eight tube passes uses 3.0 kg/s of ethylene glycol in the shell to heat 1.5 kg/s of water from 20 to 50°C. The glycol enters at 80°C, and the overall heat-transfer coefficient is 900 W/m^2 · °C. Determine the area of the heat exchanger.

10-99 After the heat exchanger in Prob. 10-98 is sized, i.e., its area determined, it is operated with a glycol flow of only 1.5 kg/s and all other parameters the same. What would the exit water temperature be under these conditions?

10-100 A shell-and-tube heat exchanger with three shell passes and six tube passes is used to heat an oil (c_p = 2.1 kJ/kg · °C) in the shell side from 30 to 60°C. In the tube side high-pressure water is cooled from 110 to 90°C. Calculate the effectiveness for each shell pass.

10-101 If the water flow rate in Prob. 10-100 is 3 kg/s and U = 230 W/m^2 · °C, calculate the area of the heat exchanger. Using the area, calculate the exit fluid temperatures when the water flow is reduced to 2 kg/s and all other factors remain the same.

■ REFERENCES

1. Siegal, R., and J. R. Howell: "Thermal Radiation Heat Transfer," 2d ed., Hemisphere Publishing Corp., New York, 1980.
2. "Standards of Tubular Exchanger Manufacturers Association," latest edition.
3. Kays, W. M., and A. L. London: "Compact Heat Exchangers," 2d ed., McGraw-Hill Book Company, New York, 1964.
4. Bowman, R. A., A. E. Mueller, and W. M. Nagle: Mean Temperature Difference in Design, *Trans. ASME,* vol. 62, p. 283, 1940.
5. Perry, J. H. (ed.): "Chemical Engineers' Handbook," 4th ed., McGraw-Hill Book Company, New York, 1963.
6. American Society of Heating and Air Conditioning Engineers Guide, annually.
7. Sparrow, E. M., and R. D. Cess: "Radiation Heat Transfer," Wadsworth Publishing Co., Inc., New York, 1966.
8. Schlunder, E. W.: "Heat Exchanger Design Handbook," Hemisphere Publishing Corp., New York, 1982.
9. Somerscales, E. F. C. and J. G. Knudsen (eds): "Fouling of Heat Transfer Equipment," Hemisphere Publishing Corp., New York, 1981.
10. Kraus, A. D., and D. Q. Kern: The Effectiveness of Heat Exchangers with One Shell Pass and Even Number of Tube Passes, *ASME Paper* 65 - HT - 18, presented at National Heat Transfer Conference, August 1965.

MASS TRANSFER

11-1 INTRODUCTION

Mass transfer can result from several different phenomena. There is a mass transfer associated with convection in that mass is transported from one place to another in the flow system. This type of mass transfer occurs on a macroscopic level and is usually treated in the subject of fluid mechanics. When a mixture of gases or liquids is contained such that there exists a concentration gradient of one or more of the constituents across the system, there will be a mass transfer on a microscopic level as the result of diffusion from regions of high concentration to regions of low concentration. In this chapter we are primarily concerned with some of the simple relations which may be used to calculate mass diffusion and their relation to heat transfer. Nevertheless, one must remember that the general subject of mass transfer encompasses both mass diffusion on a molecular scale and the bulk mass transport, which may result from a convection process.

Not only may mass diffusion occur on a molecular basis, but also in turbulent-flow systems accelerated diffusion rates will occur as a result of the rapid-eddy mixing processes, just as these mixing processes created increased heat transfer and viscous action in turbulent flow.

Although beyond the scope of our discussion, it is well to mention that mass diffusion may also result from a temperature gradient in a system; this is called *thermal diffusion*. Similarly, a concentration gradient can give rise to a temperature gradient and a consequent heat transfer. These two effects are termed *coupled phenomena* and may be treated by the methods of irreversible thermodynamics. The reader is referred to the monographs by Prigogine [1] and de Groot [2] for a discussion of irreversible thermodynamics and coupled phenomena and their application to diffusion processes.

11-2 FICK'S LAW OF DIFFUSION

Consider the system shown in Fig. 11-1. A thin partition separates the two gases A and B. When the partition is removed, the two gases diffuse through each other until equilibrium is established and the concentration of the gases is uniform throughout the box. The diffusion rate is given by Fick's law of diffusion, which states that the mass flux of a constituent per unit area is proportional to the concentration gradient. Thus

$$\frac{\dot{m}_A}{A} = -D\frac{\partial C_A}{\partial x} \tag{11-1}$$

where D = proportionality constant-diffusion coefficient, m^2/s
$\dot{m}_A$ = mass flux per unit time, kg/s
C_A = mass concentration of component A per unit volume, kg/m^3

An expression similar to Eq. (11-1) could also be written for the diffusion of constituent A in either the y or z direction.

Notice the similarity between Eq. (11-1) and the Fourier law of heat conduction,

$$\left(\frac{q}{A}\right)_x = -k\frac{\partial T}{\partial x}$$

and the equation for shear stress between fluid layers,

$$\tau = \mu\frac{\partial u}{\partial y}$$

The heat-conduction equation describes the transport of energy, the viscous-shear equation describes the transport of momentum across fluid layers, and the diffusion law describes the transport of mass.

To understand the physical mechanism of diffusion, consider the imaginary plane shown by the dashed line in Fig. 11-2. The concentration of component A is greater on the left side of this plane than on the right side. A higher concentration means that there are more molecules per unit volume. If the system is a gas or a liquid, the molecules move about in a random fashion, and the higher the concentration, the more molecules will cross a given plane per unit time. Thus, on the average, more molecules are moving from left to right across the plane than in the opposite direction. This results in a net mass transfer from the region of high concentration to the region of low concentration. The fact that the molecules collide with each other influences the diffusion process

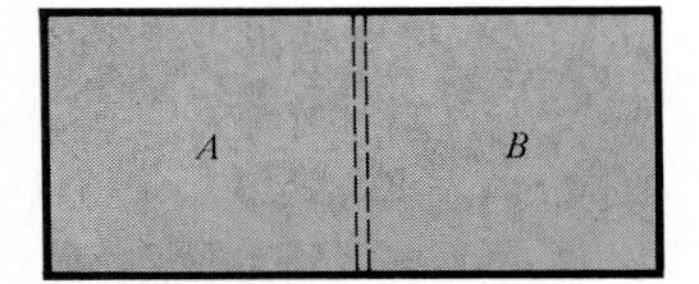

Fig. 11-1 Diffusion of component A into component B.

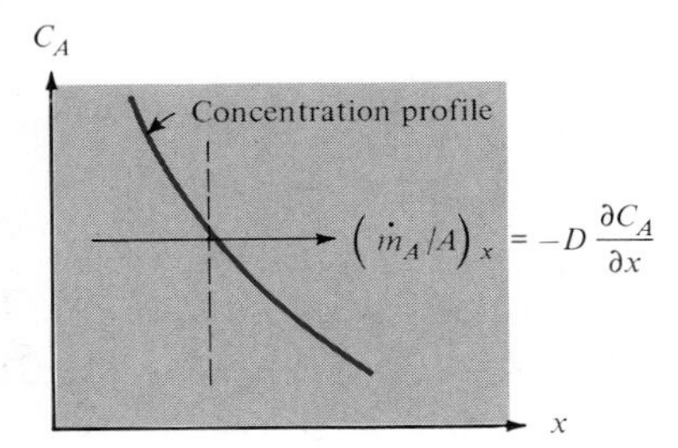

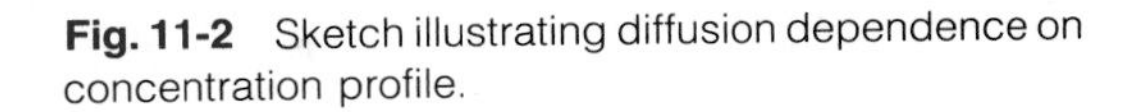
Fig. 11-2 Sketch illustrating diffusion dependence on concentration profile.

strongly. In a mixture of gases there is a decided difference between a collision of like molecules and a collision of unlike molecules. The collision between like molecules does not appreciably alter the basic molecular movement, because the two molecules are identical and it does not make any difference whether one or the other of the two molecules crosses a certain plane. The collision of two unlike molecules, say, molecules A and B, might result in molecule B crossing some particular plane instead of molecule A. The molecules would, in general, have different masses; thus the mass transfer would be influenced by the collision. By using the kinetic theory of gases it is possible to predict analytically the diffusion rates for some systems by taking into account the collision mechanism and molecular weights of the constituent gases.

In gases the diffusion rates are clearly dependent on the molecular speed, and consequently we should expect a dependence of the diffusion coefficient on temperature since the temperature indicates the average molecular speed.

11-3 DIFFUSION IN GASES

Gilliland [4] has proposed a semiempirical equation for the diffusion coefficient in gases:

$$D = 435.7 \frac{T^{3/2}}{p(V_A^{1/3} + V_B^{1/3})^2} \sqrt{\frac{1}{M_A} + \frac{1}{M_B}} \tag{11-2}$$

where D is in square centimeters per second, T is in degrees Kelvin, p is the total system pressure in pascals, and V_A and V_B are the molecular volumes of constituents A and B as calculated from the atomic volumes in Table 11-1; M_A and M_B are the molecular weights of constituents A and B. Example 11-1 illustrates the use of Eq. (11-2) for calculation of diffusion coefficients.

Equation (11-2) offers a convenient expression for calculating the diffusion coefficient for various compounds and mixtures, but it should not be used as a substitute for experimental values of the diffusion coefficient when they are available for a particular system. References 3 and 5 to 9 present more information on calculation of diffusion coefficients. An abbreviated table of diffusion coefficients is given in Appendix A.

Table 11-1 Atomic Volumes.†

Air	29.9	In secondary amines	1.20
Bromine	27.0	Oxygen, molecule (O_2)	7.4
Carbon	14.8	Coupled to two other elements:	
Carbon dioxide	34.0	In aldehydes and ketones	7.4
Chlorine		In methyl esters	9.1
Terminal as in R—Cl	21.6	In ethyl esters	9.9
Medial as in R—CHCl—R	24.6	In higher esters and ethers	11.0
Fluorine	8.7	In acids	12.0
Hydrogen, molecule (H_2)	14.3	In union with S, P, N	8.3
In compounds	3.7	Phosphorus	27.0
Iodine	37.0	Sulfur	25.6
Nitrogen, molecule (N_2)	15.6	Water	18.8
In primary amines	10.5		

†For three-membered ring, deduct 6.0. For four-membered ring, deduct 8.5. For five-membered ring, deduct 11.5. For six-membered ring, deduct 15.0. For naphthalene ring, deduct 30.0.

■ **EXAMPLE 11-1**

Calculate the diffusion coefficient for CO_2 in air at atmospheric pressure and 25°C using Eq. (11-2), and compare this value with that in Table A-8.

Solution

From Table 11-1

$$V_{CO_2} = 34.0 \qquad M_{CO_2} = 44$$

$$V_{air} = 29.9 \qquad M_{air} = 28.9$$

$$D = \frac{(435.7)(298)^{3/2}}{(1.0132 \times 10^5)[(34.0)^{1/3} + (29.9)^{1/3}]^2} \sqrt{\frac{1}{44} + \frac{1}{28.9}}$$

$$= 0.132 \text{ cm}^2/\text{s}$$

From Table A-8

$$D = 0.164 \text{ cm}^2/\text{s} = 0.62 \text{ ft}^2/\text{h}$$

so that the two values are in fair agreement.

We realize from the discussion pertaining to Fig. 11-1 that the diffusion process is occurring in two ways at the same time; i.e., gas A is diffusing into gas B at the same time that gas B is diffusing into gas A. We thus could refer to the diffusion coefficient for either of these processes.

In working with Fick's law, one may use mass flux per unit area and mass concentration as in Eq. (11-1), or the equation may be expressed in terms of molal concentrations and fluxes. There is no general rule to say which type of expression will be most convenient, and the specific problem under consideration will determine the one to be used. For gases, Fick's law may be expressed in terms of partial pressures by making use of the perfect-gas equation of state. (This transformation holds only for gases at low pressures or at a state where the perfect-gas equation of state applies.)

$$p = \rho RT \tag{11-3}$$

The density ρ represents the mass concentration to be used in Fick's law. The gas constant R for a particular gas may be expressed in terms of the universal gas constant R_0 and the molecular weight of the gas:

$$R_A = \frac{R_0}{M_A} \tag{11-4}$$

where $$R_0 = 8315 \text{ J/kg} \cdot \text{mol} \cdot \text{K}$$

Then $$C_A = \rho_A = \frac{p_A M_a}{R_0 T}$$

Consequently, Fick's law of diffusion for component A into component B could be written

$$\frac{\dot{m}_A}{A} = -D_{AB} \frac{M_A}{R_0 T} \frac{dp_A}{dx} \tag{11-5}$$

if isothermal diffusion is considered. For the system in Fig. 11-1 we could also write for the diffusion of component B into component A

$$\frac{\dot{m}_B}{A} = -D_{BA} \frac{M_B}{R_0 T} \frac{dp_B}{dx} \tag{11-6}$$

still considering isothermal conditions. Notice the different subscripts on the diffusion coefficient. Now consider a physical situation called *equimolal counterdiffusion*, as indicated in Fig. 11-3; N_A and N_B represent the steady-state molal diffusion rates of components A and B, respectively. In this steady-state situation each molecule of A is replaced by a molecule of B, and vice versa. The molal diffusion rates are given by

$$N_A = \frac{\dot{m}_A}{M_A} = -D_{AB} \frac{A}{R_0 T} \frac{dp_A}{dx}$$

$$N_B = \frac{\dot{m}_B}{M_B} = -D_{BA} \frac{A}{R_0 T} \frac{dp_B}{dx}$$

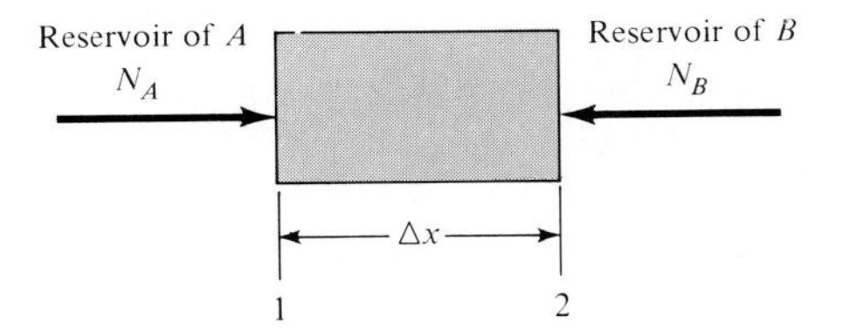

Fig. 11-3 Sketch illustrating equimolal diffusion.

The total pressure of the system remains constant at steady state, so that

$$p = p_A + p_B$$

and

$$\frac{dp_A}{dx} + \frac{dp_B}{dx} = 0$$

or

$$\frac{dp_A}{dx} = \frac{-dp_B}{dx} \tag{11-7}$$

Since each molecule of A is replacing a molecule of B, we may set the molal diffusion rates equal:

$$-N_A = N_B$$

or

$$-D_{AB}\frac{A}{R_0T}\frac{dp_A}{dx} = -D_{BA}\frac{A}{R_0T}\frac{dp_A}{dx}$$

where Eq. (11-7) has been used to express the pressure gradient of component B. We thus find

$$D_{AB} = D_{BA} = D \tag{11-8}$$

The calculation of D may be made with Eq. (11-2).

We may integrate Eq. (11-5) to obtain the mass flux of component A as

$$\frac{\dot{m}_A}{A} = \frac{-DM_A}{R_0T}\frac{p_{A_2} - p_{A_1}}{\Delta x} \tag{11-9}$$

corresponding to the nomenclature of Fig. 11-3.

Now consider the isothermal evaporation of water from a surface and the subsequent diffusion through a stagnant air layer, as shown in Fig. 11-4. The free surface of the water is exposed to air in the tank, as shown. We assume that the system is isothermal and that the total pressure remains constant. We further assume that the system is in steady state. This requires that there be a slight air movement over the top of the tank to remove the water vapor which

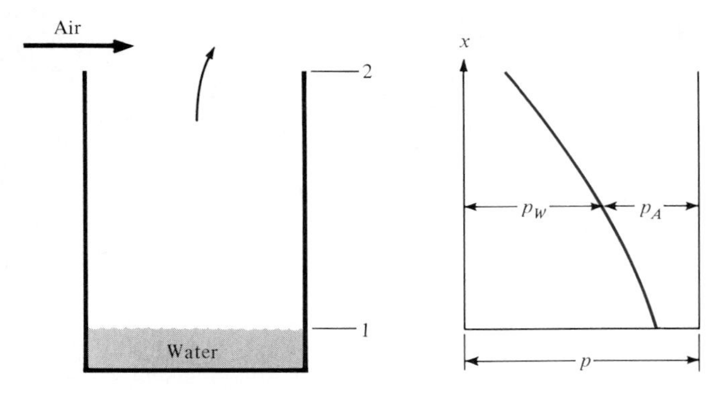

Fig. 11-4 Diffusion of water vapor into air.

diffuses to that point. Whatever air movement may be necessary to accomplish this, it is assumed that it does not create turbulence or otherwise alter the concentration profiles in the air in the tank. We further assume that both the air and water vapor behave as ideal gases.

As the water evaporates, it will diffuse upward through the air, and at steady state this upward movement must be balanced by a downward diffusion of air so that the concentration at any x position will remain constant. But at the surface of the water there can be no net mass movement of air downward. Consequently there must be a bulk mass movement upward with a velocity just large enough to balance the diffusion of air downward. This bulk mass movement then produces an *additional* mass flux of water vapor upward.

The diffusion of air downward is given by

$$\dot{m}_A = \frac{-DAM_A}{R_0T}\frac{dp_A}{dx} \tag{11-10}$$

where A denotes the cross-sectional area of the tank. This must be balanced by the bulk-mass transfer upward so that

$$-\rho_A A v = -\frac{p_A M_a}{R_0T} Av \tag{11-11}$$

where v is the bulk mass velocity, upward.

Combining Eqs. (11-10) and (11-11), we find

$$v = \frac{D}{p_A}\frac{dp_A}{dx} \tag{11-12}$$

The mass diffusion of water vapor upward is

$$\dot{m}_w = -DA\frac{M_w}{R_0T}\frac{dp_w}{dx} \tag{11-13}$$

and the bulk transport of water vapor is

$$\rho_w A v = \frac{p_w M_w}{R_0T} Av \tag{11-14}$$

The total mass transport is the sum of those given in Eqs. (11-13) and (11-14). Adding these quantities and making use of Eq. (11-12) gives

$$\dot{m}_{w\text{total}} = -\frac{DAM_w}{R_0T}\frac{dp_w}{dx} + \frac{p_w M_w}{R_0T}A\frac{D}{p_A}\frac{dp_A}{dx}$$

The partial pressure of the water vapor may be related to the partial pressure of the air by making use of Dalton's law,

$$p_A + p_w = p$$

or

$$\frac{dp_A}{dx} = -\frac{dp_w}{dx}$$

since the total pressure is constant. The total mass flow of water vapor then becomes

$$\dot{m}_{\text{total}} = -\frac{DM_wA}{R_0T}\frac{p}{p - p_w}\frac{dp_w}{dx} \tag{11-15}$$

This relation is called Stefan's law. It may be integrated to give

$$\dot{m}_{w\text{total}} = \frac{DpM_wA}{R_0T(x_2 - x_1)}\ln\frac{p - p_{w2}}{p - p_{w1}} = \frac{DpM_wA}{R_0T(x_2 - x_1)}\ln\frac{p_{A2}}{p_{A1}} \tag{11-16}$$

EXAMPLE 11-2

Estimate the diffusion rate of water from the bottom of a test tube 10 mm in diameter and 15 cm long into dry atmospheric air at 25°C.

Solution

We use Eq. (11-16) to calculate the mass flux. The partial pressure at the bottom of the test tube is the saturation pressure corresponding to 25°C [77°F], and the vapor pressure may be taken as zero at the top of the test tube since it is diffusing into dry air. Accordingly,

$$p_{A1} = p - p_{w1} = 14.696 - 0.4593 = 14.237 \text{ lb/in}^2 \text{ abs} = 9.8155 \times 10^4 \text{ Pa}$$

$$p_{A2} = p - p_{w2} = 14.696 - 0 = 14.696 \text{ lb/in}^2 \text{ abs} = 1.0132 \times 10^5 \text{ Pa}$$

From Table A-8

$$D = 0.256 \text{ cm}^2/\text{s} \qquad \dot{m}_w = \frac{DpM_wA}{R_0T(x_2 - x_1)}\ln\frac{p_{A1}}{p_{A1}}$$

$$m_w = \frac{(0.256 \times 10^{-4})(1.0132 \times 10^5)(18)(\pi)(5 \times 10^{-3})^2}{(8315)(298)(0.15)}\ln\frac{1.0132}{0.98155}$$

$$= 3.131 \times 10^{-10} \text{ kg/s} \quad [0.00113 \text{ g/h}]$$

11-4 DIFFUSION IN LIQUIDS AND SOLIDS

Fick's law of diffusion is also used for problems involving liquid and solid diffusion, and the main difficulty is one of determining the value of the diffusion coefficient for the particular liquid or solid. Unfortunately, only approximate theories are available for predicting diffusion coefficients in these systems. Bird, Stewart, and Lightfoot [9] discuss the calculation of diffusion in liquids, and Jost [6] gives a discussion of the various theories which have been employed to predict values of the diffusion coefficient. The reader is referred to these books for more information on diffusion in liquids and solids.

Diffusion in solids is complex because of the strong influence of the molecular force fields on the process. For these systems Fick's law [Eq. (11-1)] is often used, along with an experimentally determined diffusion coefficient, although there is some indication that this relation may not adequately describe the physical processes. The numerical value of the diffusion coefficient for liquids and solids is much smaller than for gases, primarily because of the larger molecular force fields, the increased number of collisions, and the consequent reduction in the freedom of movement of the molecules.

11-5 THE MASS-TRANSFER COEFFICIENT

We may define a mass-transfer coefficient in a manner similar to that used for defining the heat-transfer coefficient. Thus

$$\dot{m}_A = h_{D_A} A (C_{A_1} - C_{A_2}) \tag{11-17}$$

where $\dot{m}_A$ = diffusive mass flux of component A

h_{D_A} = mass-transfer coefficient

C_{A_1}, C_{A_2} = concentrations through which diffusion occurs

If one considers a steady-state diffusion across a layer of thickness Δx,

$$\dot{m}_A = -\frac{DA(C_{A_2} - C_{A_1})}{\Delta x} = h_{D_A} A (C_{A_2} - C_{A_1})$$

and

$$h_{D_A} = \frac{D}{\Delta x} \tag{11-18}$$

For the water-vaporization example discussed above,

$$C_{w_1} - C_{w_2} = \frac{M_w}{R_0 T}(p_{w_1} - p_{w_2})$$

so that the mass-transfer coefficient for this situation could be written

$$h_{D_w} = \frac{Dp}{(x_2 - x_1)(p_{w_1} - p_{w_2})} \ln \frac{p - p_{w_2}}{p - p_{w_1}} \tag{11-19}$$

Note that the units of the mass-transfer coefficients are in meters per second in SI units.

We have already seen that the phenomenological laws governing heat, mass, and momentum transfer are similar. In Chap. 5 it was shown that the energy and momentum equations of a laminar boundary layer are similar, viz.,

$$u \frac{\partial u}{\partial x} + v \frac{\partial u}{\partial y} = \nu \frac{\partial^2 u}{\partial y^2} \tag{11-20}$$

$$u \frac{\partial T}{\partial x} + v \frac{\partial T}{\partial y} = \alpha \frac{\partial^2 T}{\partial y^2} \tag{11-21}$$

We further observed that the ratio ν/α, the Prandtl number, was the connecting link between the velocity and temperature field and was thus an important parameter in all convection heat-transfer problems. If we considered a laminar boundary layer on a flat plate in which diffusion was occurring as a result of some mass-transfer condition at the surface, we could derive an equation for the concentration of a particular component in the boundary layer. This equation would be

$$u \frac{\partial C_A}{\partial x} + v \frac{\partial C_A}{\partial y} = D \frac{\partial^2 C_A}{\partial^2 y} \tag{11-22}$$

where C_A is the concentration of the component which is diffusing through the boundary layer. Note the similarity between Eq. (11-22) and Eqs. (11-20) and (11-21). The concentration and velocity profiles will have the same shape when $\nu = D$ or $\nu/D = 1$. The dimensionless ratio ν/D is called the Schmidt number,

$$\text{Sc} = \frac{\nu}{D} = \frac{\mu}{\rho D} \tag{11-23}$$

and is important in problems where both convection and mass transfer are important. Thus the Schmidt number plays a role similar to that of the Prandtl number in convection heat-transfer problems. Whereas in convection heat-transfer problems we write for the functional dependence of the heat-transfer coefficient

$$\frac{hx}{k} = f(\text{Re, Pr})$$

in convection mass-transfer problems we should write the functional relation

$$\frac{h_D x}{D} = f(\text{Re, Sc})$$

The temperature and concentration profiles will be similar when $\alpha = D$ or $\alpha/D = 1$, and the ratio α/D is called the Lewis number

$$\text{Le} = \frac{\alpha}{D} \tag{11-24}$$

The similarities between the governing equations for heat, mass, and momentum transfer suggest that empirical correlations for the mass-transfer coefficient would be similar to those for the heat-transfer coefficient. This turns out to be the case, and some of the empirical relations for mass-transfer coefficients are presented below. Gilliland [4] presented the equation

$$\frac{h_D d}{D} = 0.023 \left(\frac{\rho u_m d}{\mu}\right)^{0.83} \left(\frac{\nu}{D}\right)^{0.44} \tag{11-25}$$

for the vaporization of liquids into air inside circular columns where the liquid wets the surface and the air is forced through the column. The grouping of terms $h_D x/D$ or $h_D d/D$ is called the Sherwood number

$$\text{Sh} = \frac{h_D x}{D} \tag{11-26}$$

Note the similarity between Eq. (11-25) and the Dittus-Boelter equation (6-4). Equation (11-25) is valid for

$$2000 < \mathrm{Re}_d < 35{,}000 \qquad \text{and} \qquad 0.6 < \mathrm{Sc} < 2.5$$

and is applicable to flow in smooth tubes.

The Reynolds analogy for pipe flow may be extended to mass-transfer problems to express the mass-transfer coefficient in terms of the friction factor. The analogy is written

$$\frac{h_D}{u_m}\,\mathrm{Sc}^{2/3} = \frac{f}{8} \tag{11-27}$$

which may be compared with the analogy for heat transfer [Eq. (6-10)]:

$$\frac{h}{u_m c_p \rho}\,\mathrm{Pr}^{2/3} = \frac{f}{8} \tag{11-28}$$

For flow over smooth flat plates, the Reynolds analogy for mass transfer becomes

Laminar:

$$\frac{h_D}{u_\infty}\,\mathrm{Sc}^{2/3} = \frac{C_f}{2} = 0.332\,\mathrm{Re}_x^{-1/2} \tag{11-29}$$

Turbulent:

$$\frac{h_D}{u_\infty}\,\mathrm{Sc}^{2/3} = \frac{C_f}{2} = 0.0296\,\mathrm{Re}_x^{-1/5} \tag{11-30}$$

Equations (11-29) and (11-30) are analogous to Eqs. (5-55) and (5-81).

When both heat and mass transfer are occurring simultaneously, the mass- and heat-transfer coefficients may be related by dividing Eq. (11-28) by Eq. (11-27):

$$\frac{h}{h_D} = \rho c_p \left(\frac{\mathrm{Sc}}{\mathrm{Pr}}\right)^{2/3} = \rho c_p \left(\frac{\alpha}{D}\right)^{2/3} = \rho c_p\,\mathrm{Le}^{2/3} \tag{11-31}$$

■ **EXAMPLE 11-3**

Dry air at atmospheric pressure blows across a thermometer which is enclosed in a dampened cover. This is the classical wet-bulb thermometer. The thermometer reads a temperature of 18.3°C. What is the temperature of the dry air?

Solution

We solve this problem by first noting that the thermometer exchanges no net energy transfer at steady-state conditions and that the heat which must be used to evaporate the water from the cover must come from the air. We therefore make the energy balance

$$hA(T_\infty - T_w) = \dot{m}_w h_{fg}$$

where h is the heat-transfer coefficient and $\dot{m}_w$ is the mass of water evaporated. Now

$$\dot{m}_w = h_D A(C_w - C_\infty)$$

so that

$$hA(T_\infty - T_w) = h_D A(C_w - C_\infty)h_{fg}$$

Using Eq. (11-31), we have

$$\rho c_p \left(\frac{\alpha}{D}\right)^{2/3} (T_\infty - T_w) = (C_w - C_\infty)h_{fg} \qquad (a)$$

The concentration at the surface C_w is that corresponding to saturation conditions at the temperature measured by the thermometer. From the steam tables at 65°F [18.3°C]

$$p_g = 0.3056 \text{ lb/in}^2 \text{ abs} = 2107 \text{ Pa}$$

and

$$C_w = \frac{p_w}{R_w T_w} = \frac{(2107)(18)}{(8315)(291.3)} = 0.01566 \text{ kg/m}^3$$

The other properties are

$$C_\infty = 0 \text{ (since the free steam is dry air)}$$

$$\rho = \frac{p}{RT} = \frac{1.0132 \times 10^5}{(287)(291.3)} = 1.212 \text{ kg/m}^3$$

$$c_p = 1.004 \text{ kJ/kg} \cdot {}^\circ\text{C} \qquad \frac{\alpha}{D} = \frac{\text{Sc}}{\text{Pr}} = 0.845$$

$$h_{fg} = 1057 \text{ Btu/lb}_m = 2.456 \text{ MJ/kg}$$

Then, from (*a*)

$$T_\infty - T_W = \frac{(0.01566 - 0)(2.456 \times 10^6)}{(1.212)(1004)(0.845)^{2/3}} = 35.36^\circ\text{C}$$

$$T_\infty = 53.69^\circ\text{C}$$

The calculation should now be corrected by recalculating the density at the arithmetic-average temperature between wall and free-stream conditions. With this adjustment there results

$$\rho = 1.143 \text{ kg/m}^3 \qquad T_\infty = 55.8^\circ\text{C}$$

It is not necessary to correct the ratio Sc/Pr because this parameter does not change appreciably over this temperature range.

■ EXAMPLE 11-4

If the airstream in Example 11-3 is at 32.2°C [90°F] while the wet-bulb temperature remains at 18.3°C, calculate the relative humidity of the airstream.

Solution

From thermodynamics we recall that the relative humidity is defined as the ratio of concentration of vapor to the concentration at saturation conditions for the airstream. We therefore calculate the actual water-vapor concentration in the airstream from

$$\rho c_p \left(\frac{\alpha}{D}\right)^{2/3} (T_\infty - T_w) = (C_w - C_\infty)h_{fg} \qquad (a)$$

and then compare this with the saturation concentration to determine the relative humidity. The properties are taken from Example 11-3:

$$\rho = 1.212 \text{ kg/m}^3 \qquad c_p = 1.004 \text{ kJ/kg}\cdot{}^\circ\text{C} \qquad \frac{\alpha}{D} = \frac{\text{Sc}}{\text{Pr}} = 0.845$$

$$T_w = 18.3^\circ\text{C} \quad [65^\circ\text{F}] \qquad T_\infty = 32.2^\circ\text{C} \quad [90^\circ\text{F}]$$

$$C_w = 0.01566 \text{ kg/m}^3 \qquad h_{fg} = 2.456 \text{ MJ/kg}$$

We insert the numerical values into Eq. (a) and obtain

$$(1.212)(1004)(0.845)^{2/3}(32.2 - 18.3) = (0.01566 - C_\infty)(2.456 \times 10^6)$$

so that

$$C_\infty = 0.0095 \text{ kg/m}^3$$

At 32.2°C [90°F] the saturation concentration for the free stream is obtained from the steam tables:

$$C_g(90^\circ\text{F}) = 2.136 \times 10^{-3} \text{ lb}_m/\text{ft}^3 = 0.0342 \text{ kg/m}^3$$

The relative humidity is therefore

$$\text{RH} = \frac{0.0095}{0.0342} = 27.8\%$$

11-6 EVAPORATION PROCESSES IN THE ATMOSPHERE

We have already described some evaporation processes and indicated relations between heat and mass transfer. In the atmosphere, the continuous evaporation and condensation of water from the soil, oceans, and lakes influences every form of life and provides many of the day-to-day varieties of climate that govern the environment on earth. These processes are very complicated because in practice they are governed by substantial atmospheric convection currents that are difficult to describe analytically.

Let us first consider the diffusion of water vapor from a horizontal surface into quiescent air as indicated in Fig. 11-5. At the surface, the partial pressure of the vapor is p_s. The vapor pressure steadily drops with a rise in elevation z to the "free atmosphere" value of p_∞. The molecular diffusion of the water vapor may be written in the form of Eq. (11-13) as

$$\frac{\dot{m}_w}{A} = -D_w \frac{M_w}{R_0 T} \frac{dp_w}{dz} \tag{11-32}$$

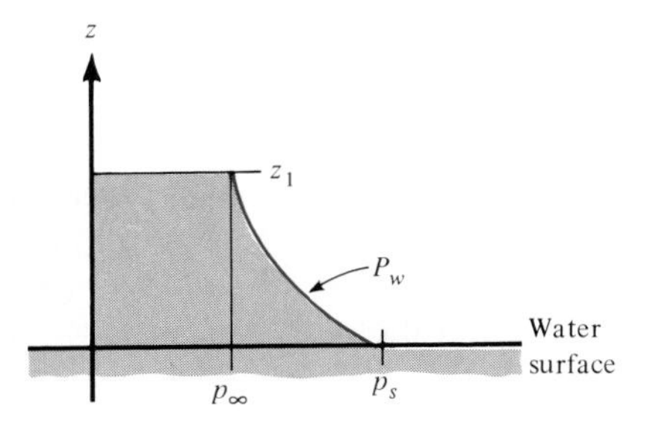

Fig. 11-5 Diffusion of water vapor from a horizontal surface.

where A is the surface area under consideration. In hydrologic applications, it is convenient to express this relation in terms of the local atmospheric density and pressure. The total pressure may be expressed as

$$p = \rho \frac{R_0}{M} T \tag{11-33}$$

where ρ and M are the density and molecular weight of the moist air, respectively. Because the molar concentration of water vapor is so small in atmospheric applications, the molecular weight of the moist air is essentially that of dry air, and Eqs. (11-32) and (11-33) can be combined to give

$$\frac{\dot{m}_w}{A} = -D_w \frac{M_w}{M_a} \frac{\rho}{p} \frac{dp_w}{dz}$$

But, $M_w/M_a = 0.622$, so that

$$\frac{\dot{m}_w}{A} = -0.622 D_w \frac{\rho}{p} \frac{dp_w}{dz} \tag{11-34}$$

By using the boundary conditions

$$p_w = p_s \qquad \text{at } z = 0$$

$$p_w = p_\infty \qquad \text{at } z = z_1$$

Eq. (11-34) may be integrated to give

$$\frac{\dot{m}_w}{A} = 0.622 D_w \frac{\rho}{p} \frac{p_s - p_\infty}{z_1} \tag{11-35}$$

Evaporation processes in the atmosphere are much more complicated than indicated by the simple form of Eq. (11-35) for two reasons:

1. The diffusion process involves substantial turbulent eddy motion so that the diffusion coefficient D_w may vary significantly with the height z.
2. The air is seldom quiescent and wind currents contribute substantially to the evaporation rate.

As in the many convection problems we have encountered previously, the solution to a complicated problem of this sort is frequently obtained by appealing to carefully controlled measurements in search of an empirical relationship to predict evaporation rates.

In this problem, a "standard pan" is used as shown in Fig. 11-6. The mean wind movement is measured 6 in above the pan rim, and water-evaporation rates are measured with the pan placed on the ground (*land pan*) or in a body of water (*floating pan*). For the land pan and with a *convectively stable* atmosphere, the evaporation rate has been correlated experimentally [13] as

$$E_{lp} = (0.37 + 0.0041\bar{u})(p_s - p_w)^{0.88} \tag{11-36}$$

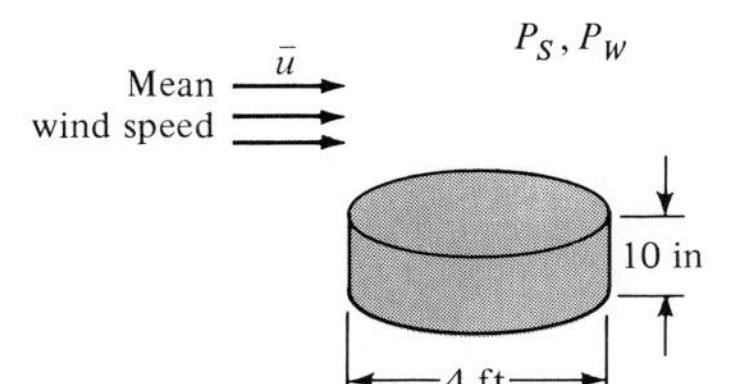

Fig. 11-6 Class A standard pan for measurement of evaporation.

where E_{lp} = land-pan evaporation, in/day

$\bar{u}$ = daily wind movement measured 6 in above the pan rim, mi/day

p_s = saturation vapor pressure at dry-bulb air temperature 5 ft above ground surface, in Hg

p_w = actual vapor pressure of air under temperature and humidity conditions 5 ft above ground surface, in Hg

Heat transfer to the pan influences the evaporation rate differently for the ground or water experiments. To convert the pan measurements to those for a natural surface, Eq. (11-36) is multiplied by a *pan coefficient* that is 0.7 for the land pan and 0.8 for the floating pan. If the atmosphere is not convectively stable, vertical density gradients can cause substantial deviations from Eq. (11-36). These problems are discussed in Refs. 10 to 13.

■ EXAMPLE 11-5

A standard land pan is used to measure the evaporation rate in atmospheric air at 100°F and 30 percent relative humidity. The mean wind speed is 10 mi/h. What is the evaporation rate in $lb_m/h \cdot ft^2$ on the land under these conditions?

Solution

For this calculation we can make use of Eq. (11-36). From thermodynamic steam tables

$$p_s = p_g \quad \text{at } 100°\text{F} = 0.9492 \text{ lb/in}^2 \text{ abs} = 1.933 \text{ in Hg} \quad [6.545 \text{ kPa}]$$

$$\text{Relative humidity} = \frac{p_w}{p_s}$$

$$p_w = (0.30)(1.933) = 0.580 \text{ in Hg} \quad [1.964 \text{ kPa}]$$

Also, $\bar{u}$ = 10 mi/h = 240 mi/day [386.2 km/day].

Equation (11-36) now yields, with the application of the 0.7 factor,

$$E_{lp} = 0.7[0.37 + (0.0041)(240)](1.933 - 0.580)^{0.88}$$
$$= 1.237 \text{ in/day} \quad [31.41 \text{ mm/day}]$$

Noting that the standard pan has a diameter of 4 ft, we can use the figure to calculate the mass evaporation rate per unit area as:

$$\frac{\dot{m}_w}{A} = \frac{E_{lp}}{12}\rho_w$$

$$= \frac{(1.237)(62.4)}{12} = 6.432 \text{ lb}_m/\text{day} \cdot \text{ft}^2 \quad [31.43 \text{ kg/m}^2 \cdot \text{day}]$$

$$= 0.268 \text{ lb}_m/\text{h} \cdot \text{ft}^2 \quad [1.31 \text{ kg/h} \cdot \text{m}^2]$$

As a matter of interest, we might calculate the molecular diffusion rate of water vapor from Eq. (11-35), taking z_1 as the 5-ft dimension above the standard pan. Since

$$\rho = \frac{p}{RT}$$

Eq. (11-35) can be written as

$$\frac{\dot{m}_w}{A} = 0.622\frac{D_w}{RT}\frac{p_s - p_w}{z_1}$$

From Table A-8,

$$D_w = 0.256 \text{ cm}^2/\text{s} = 0.99 \text{ ft}^2/\text{h}$$

so that

$$\frac{\dot{m}_w}{A} = \frac{(0.622)(0.99)(0.9492)(1 - 0.3)}{(53.35)(560)(5.0)}$$

$$= 3.94 \times 10^{-4} \text{ lb}_m/\text{h} \cdot \text{ft}^2 \quad [19.2 \text{ g/h} \cdot \text{m}^2]$$

This number is negligibly small in comparison with the previous calculation. This means that in the actual evaporation process, turbulent diffusion and convection transport play dominant roles in comparison with molecular diffusion.

REVIEW QUESTIONS

1. How is the diffusion coefficient defined?
2. Define the mass-transfer coefficient.
3. Define the Schmidt and Lewis numbers. What is the physical significance of each?
4. Why can not atmospheric evaporation rates be calculated with ordinary molecular-diffusion equations?

PROBLEMS

11-1 Using physical reasoning, justify the $T^{3/2}$ dependence of the diffusion coefficient as shown by Eq. (11-2). *Hint:* Recall that mean molecular velocity is proportional to $T^{1/2}$ and that the density of an ideal gas is inversely proportional to temperature.

11-2 Using Eq. (11-2), calculate the diffusion coefficient for benzene in atmospheric air at 25°C.

11-3 Dry air at atmospheric pressure and 25°C blows across a flat plate at a velocity of 1.5 m/s. The plate is 30 cm square and is covered with a film of water which may evaporate into the air. Plot the heat flow from the plate as a function of the plate temperature between $T_w = 15°C$ and $T_w = 65°C$.

11-4 The cover on a wet-bulb thermometer is soaked in benzene, and the thermometer is exposed to a stream of dry air. The thermometer indicates a temperature of 26°C. Calculate the free-stream air temperature. The vapor pressure of benzene is 13.3 kPa, and the enthalpy of vaporization is 377 kJ/kg at 26°C.

11-5 Dry air at 25°C and atmospheric pressure flows inside a 5-cm-diameter pipe at a velocity of 3 m/s. The wall is coated with a thin film of water, and the wall temperature is 25°C. Calculate the water-vapor concentration in the air at exit of a 3-m length of the pipe.

11-6 An open pan 15 cm in diameter and 7.5 cm deep contains water at 25°C and is exposed to atmospheric air at 25°C and 50 percent relative humidity. Calculate the evaporation rate of water in grams per hour.

11-7 A test tube 1.25 cm in diameter and 15 cm deep contains benzene at 26°C and is exposed to dry atmospheric air at 26°C. Using the properties given in Prob. 11-4, calculate the evaporation rate of benzene in grams per hour.

11-8 Dry air at 25°C and atmospheric pressure blows over a 30-cm-square surface of ice at a velocity of 1.5 m/s. Estimate the amount of moisture evaporated per hour, assuming that the block of ice is perfectly insulated except for the surface exposed to the airstream.

11-9 The temperature of an airstream is to be measured, but the thermometer available does not have a sufficiently high range. Accordingly, a dampened cover is placed around the thermometer before it is placed in the airstream. The thermometer reads 32°C. Estimate the true air temperature, assuming that it is dry at atmospheric pressure.

11-10 Assume that a human forearm may be approximated by a cylinder 4 in in diameter and 1 ft long. The arm is exposed to a dry-air environmental temperature of 115°F in a 10-mi/h breeze on the desert and receives a radiant heat flux from the sun of 350 Btu/h · ft^2 of view area (view area for the cylinder $= Ld$). If the arm is perspiring so that it is covered with a thin layer of water, estimate the arm surface temperature. Neglect internal heat generation of the arm. Assume an emissivity of unity for the water film.

11-11 A 30-cm-square plate is placed on a low-speed wind tunnel; the surface is covered with a thin layer of water. The dry air is at atmospheric pressure and 43°C and blows over the plate at a velocity of 12 m/s. The enclosure walls of the wind tunnel are at 10°C. Calculate the equilibrium temperature of the plate, assuming an emissivity of unity for the water film.

11-12 Calculate the rate of evaporation for the system in Prob. 11-11.

11-13 Refine the analysis of Prob. 11-10 by assuming a body-heat generation of 1860 W/m^3.

11-14 A small tube 6.4 mm in diameter and 13 cm deep contains water with the top open to atmospheric air at 20°C, 1 atm, and 50 percent relative humidity. Heat is added to the bottom of the tube. Plot the diffusion rate of water as a function of water temperature over the range of 20 to 82°C.

11-15 Dry air at 20°C enters a 1.25-cm-ID tube where the interior surface is coated with liquid water. The mean flow velocity is 3 m/s, and the tube wall is maintained at 20°C. Calculate the diffusion rate of water vapor at the entrance conditions. How much moisture is picked up by the air for a tube 1 m long?

11-16 Dry air at 65°C blows over a 30-cm-square plate at a velocity of 6 m/s. The plate is covered with a smooth porous material, and water is supplied to the material at 25°C. Assuming that the underside of the plate is insulated, estimate the amount of water that must be supplied to maintain the plate temperature at 38°C. Assume that the radiation temperature of the surroundings is 65°C and that the porous surface radiates as a blackbody.

11-17 Dry air at atmospheric pressure blows over an insulated flat plate covered with a thin wicking material which has been soaked in ethyl alcohol. The temperature of the plate is 25°C. Calculate the temperature of the airstream assuming that the concentration of alcohol is negligible in the free stream. Also calculate the mass-transfer rate of alcohol for a 30-cm-square plate if the free-stream velocity is 7 m/s.

11-18 Carrier's equation expresses actual water-vapor partial pressure in terms of wet-bulb and dry-bulb temperatures:

$$p_v = p_{gw} - \frac{(p - p_{gw})(T_{DB} - T_{WB})}{2800 - T_{WB}}$$

where p_v = actual partial pressure, lb/in² abs
p_{gw} = saturation pressure corresponding to wet-bulb temperature, lb/in² abs
p = total mixture pressure, lb/in² abs
T_{DB} = dry-bulb temperature, °F
T_{WB} = wet-bulb temperature, °F

Air at 1 atm and 100°F flows across a wet-bulb thermometer, producing a temperature of 70°F. Calculate the relative humidity of the airstream using Carrier's equation and compare with results obtained by methods of this chapter.

11-19 Suppose the wet bulb of Prob. 11-18 is exposed to a black-radiation surrounding at 100°F. What radiation equilibrium temperature would it indicate, assuming an emissivity of unity for the wick?

11-20 A light breeze at 2.2 m/s blows across a standard evaporation pan. The atmospheric conditions are 20°C and 40 percent relative humidity. What is the evaporation rate for a land pan in grams per hour per square meter? What would be the evaporation rate for zero velocity?

11-21 An evaporation rate of 0.3 g/s · m² is experienced for a 4.5-m/s breeze blowing across a standard land pan. What is the relative humidity if the dry-bulb (ambient) air temperature is 40°C?

REFERENCES

1 Prigogine, I.: "Introduction to Thermodynamics of Irreversible Processes," Charles C Thomas, Publisher, Springfield, Ill., 1955.

2 Groot, S. R. de: "Thermodynamics of Irreversible Processes," North-Holland Publishing Company, Amsterdam, 1952.

3 Present, R. D.: "Kinetic Theory of Gases," McGraw-Hill Book Company, New York, 1958.

4 Gilliland, E. R.: Diffusion Coefficients in Gaseous Systems, *Ind. Eng. Chem.,* vol. 26, p. 681, 1934.

5 Perry. J. H. (ed.): "Chemical Engineers' Handbook," 4th ed., McGraw-Hill Book Company, New York, 1963.

6 Jost, W.: "Diffusion in Solids, Liquids and Gases," Academic Press, Inc., New York, 1952.

7 Reid, R. C., and T. K. Sherwood: "The Properties of Gases and Liquids," McGraw-Hill Book Company, New York, 1958.

8 "Handbook of Chemistry and Physics," Chemical Rubber Publishing Company, Cleveland, Ohio, 1960.

9 Bird, R., W. E. Stewart, and E. N. Lightfoot: "Transport Phenomena," John Wiley & Sons, Inc., New York, 1960.

10 Instructions for Climatological Observers, *U.S. Dept. Commerce Weather Bur. Circ. B.,* 10th ed. rev., October 1955.

11 Water-Loss Investigations, vol. 1, *Lake Hefner Studies Tech. Rep., U.S. Geol. Surv. Circ.* 229, 1952.

12 Nordenson, T. J., and D. R. Baker: Comparative Evaluation of Evaporation Instruments, *J. Geophys. Rec.,* vol. 67, no. 2, p. 671, February 1962.

13 Kohler, M. A., T. J. Nordenson, and W. E. Fox: Evaporation from Pans and Lakes, *U.S. Dept. Commerce Weather Bur. Res. Pap.* 38, May 1955.

SPECIAL TOPICS IN HEAT TRANSFER

12-1 INTRODUCTION

A number of specialized subjects in heat transfer are important in modern technology. In this chapter we shall present a brief introduction to five such topics. The treatment here is not intended to be comprehensive, but rather to indicate the basic physical mechanism of the processes involved. It would be possible, of course, to include these topics in other chapters of the book, but we choose to group them together in order to focus more specialized attention on the subject matter. Since the discussions which follow represent only cursory treatments of the subjects, the serious reader will wish to consult the appropriate references for additional information.

12-2 HEAT TRANSFER IN MAGNETOFLUIDYNAMIC (MFD) SYSTEMS

It is known that an electrical conductor moving in a magnetic field generates an emf which is proportional to the speed of motion and the magnetic field strength. Correspondingly, the magnetic field exerts a restraining force on the conductor which tends to impede its motion. The current flow in the conductor generates joulean heat in accordance with the familiar I^2R relation. Similar effects are experienced when a conductive fluid moves through a magnetic field. An ionized gas, whether it occurs as the result of an elevation of temperature or by a suitable seeding process, is electrically conductive and may be influenced by a magnetic field.

Our primary concern is with the heat transfer to a conducting fluid under the influence of a magnetic field. This problem area is important in systems involving high-temperature plasmas applicable to nuclear fusion energy conversion, liquid metal systems, and magnetofluidynamic power-generation systems. Our discussion will be limited to a very simple case which serves to illustrate the influence a magnetic field may have on the flow and heat transfer in a conductive fluid.

First, let us examine some basic electromagnetic concepts. For a neutrally charged system the *current density* $\mathbf{J}$ is given by

$$\mathbf{J} = \sigma \mathbf{E} \tag{12-1}$$

where σ is the *electrical conductivity* and $\mathbf{E}$ is the *electric field vector*. The *magnetic field strength* $\mathbf{B}$ is expressed by

$$\mathbf{B} = \mu_0 \mathbf{H} \tag{12-2}$$

where μ_0 is called the *magnetic permeability* and $\mathbf{H}$ is the *magnetic field intensity*. The force exerted on a system of charged particles by an electric field is given by

$$\mathbf{F}_e = \rho_e \mathbf{E} \tag{12-3}$$

where ρ_e is the *charge density* (charge per unit volume). The magnetic force exerted on a current-carrying conductor is

$$\mathbf{F}_m = \mathbf{J} \times \mathbf{B} \tag{12-4}$$

The total electromagnetic force is given by the sum of Eqs. (12-3) and (12-4):

$$\mathbf{F}_{em} = \rho_e \mathbf{E} + \mathbf{J} \times \mathbf{B} \tag{12-5}$$

The work done on the system per unit time by the electromagnetic force is

$$W_{em} = \mathbf{F}_{em} \cdot \mathbf{V} \tag{12-6}$$

where $\mathbf{V}$ is the velocity of the conductor. The magnetic field induces a voltage in the conductor of magnitude $\mathbf{V} \times \mathbf{B}$ and subsequently induces a current of

$$\mathbf{J}_{\text{ind}} = \sigma(\mathbf{V} \times \mathbf{B}) \tag{12-7}$$

There is a further transport of charge resulting from the macroscopic velocity V, given by

$$\mathbf{J}_{\text{trans}} = \rho_e \mathbf{V} \tag{12-8}$$

The *conduction current* is now defined as

$$\mathbf{J}_c = \sigma(\mathbf{E} + \mathbf{V} \times \mathbf{B}) \tag{12-9}$$

and the total current flow is

$$\mathbf{J} = \mathbf{J}_c + \rho_e \mathbf{V} \tag{12-10}$$

A careful manipulation of the foregoing equations yields the following relation for the electromagnetic work:

$$W_{em} = \mathbf{E} \cdot \mathbf{J} - \frac{\mathbf{J}_c \cdot \mathbf{J}_c}{\sigma} \tag{12-11}$$

The second term in Eq. (12-11) is called the *ohmic heating* and is given as the product of the resistivity and conduction current squared.

We now wish to apply these relations to a very simple boundary-layer flow system. Consider the flow over the flat plate shown in Fig. 12-1. Impressed on the flow is a constant magnetic field B directed in the y direction. We assume that the magnetic field is uniform throughout the boundary layer and that the impressed electric field is zero. In writing a momentum equation for this flow system, we need consider only the magnetic force as given by Eq. (12-4). If all properties are assumed constant, including electrical conductivity, there results

$$\rho\left(u\frac{\partial u}{\partial x}+v\frac{\partial u}{\partial y}\right)=-\frac{\partial p}{\partial y}+\mu\frac{\partial^2 u}{\partial y^2}+(\mathbf{J}_c\times\mathbf{B})_x \tag{12-12}$$

This is the momentum equation, which is equivalent to Eq. (5-11) for conventional flow. Note that only the x component of the magnetic force is considered because the equation represents a force summation in the x direction. The fluid velocity is written

$$\mathbf{V}=u\mathbf{i}+v\mathbf{j} \tag{12-13}$$

so that

$$\begin{aligned}\mathbf{J}_c &= \sigma(\mathbf{E}+\mathbf{V}\times\mathbf{B})=\sigma(\mathbf{V}\times\mathbf{B})\\ &= \sigma(u\mathbf{i}+v\mathbf{j})\times B_y\mathbf{j}=\sigma uB_y\mathbf{k}\end{aligned} \tag{12-14}$$

Thus

$$\mathbf{J}_c\times\mathbf{B}=-\sigma uB_y^2\mathbf{i} \tag{12-15}$$

and

$$(\mathbf{J}_c\times\mathbf{B})_x=-\sigma uB_y^2 \tag{12-16}$$

The momentum equation becomes

$$u\frac{\partial u}{\partial x}+v\frac{\partial u}{\partial y}=-\frac{1}{\rho}\frac{\partial p}{\partial x}+\nu\frac{\partial^2 u}{\partial y^2}-\frac{\sigma B_y^2}{\rho}u \tag{12-17}$$

The equivalent integral momentum equation for zero pressure gradient is

$$\mu\left.\frac{\partial u}{\partial y}\right]_{y=0}+\int_0^\delta \sigma B_y^2 u\,dy=\frac{d}{dx}\left[\int_0^\delta \rho u(u_\infty-u)\,dy\right] \tag{12-18}$$

We are able to write the same conditions on the velocity profile as in Sec. 5-4, so that the cubic-parabola profile is employed to effect the integration,

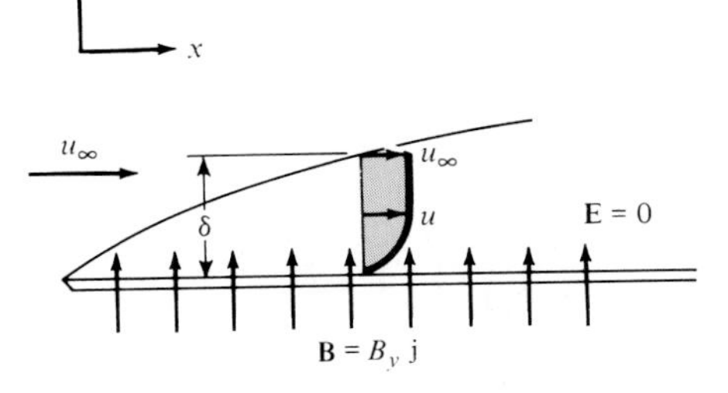

Fig. 12-1 Hydromagnetic boundary layer.

$$\frac{u}{u_\infty} = \frac{3}{2}\frac{y}{\delta} - \frac{1}{2}\left(\frac{y}{\delta}\right)^3 \tag{12-19}$$

Inserting (12-19) in (12-18) and performing the integration gives

$$\frac{3\nu u_\infty}{2\delta} + \frac{5\sigma u_\infty B_y^2}{8\rho} = \frac{39}{280} u_\infty^2 \frac{d\delta}{dx} \tag{12-20}$$

This differential equation is linear in δ^2 and may be solved to give

$$\left(\frac{\delta}{x}\right)^2 = \frac{2.40}{\text{Re}_x N}(e^{8.97N} - 1) \tag{12-21}$$

where N is called the *magnetic influence number,*

$$N = \frac{\sigma B_y^2 x}{\rho u_\infty} \tag{12-22}$$

The Reynolds number is defined in the conventional way; i.e.,

$$\text{Re}_x = \frac{\rho u_\infty x}{\mu}$$

The relation for laminar-boundary-layer thickness from Sec. 5-4 was

$$\frac{\delta_0}{x} = \frac{4.64}{\text{Re}_x^{1/2}} \tag{12-23}$$

We may thus form the comparative ratio

$$\frac{\delta_m}{\delta_0} = \frac{0.334}{N^{1/2}}(e^{8.97N} - 1)^{1/2} \tag{12-24}$$

where now δ_m is the boundary-layer thickness in the presence of the magnetic field and δ_0 is the thickness for zero magnetic field. The functional relationship given in Eq. (12-24) is plotted in Fig. 12-2. The effect of an increased magnetic field is to increase the boundary-layer thickness as a result of the increased retarding force. The effect is analogous to flow against a positive pressure gradient.

We next wish to examine the effect of the magnetic field on the heat transfer from the plate. Again, constant properties are assumed, as well as a zero impressed electric field. The electromagnetic work term then becomes

$$W_{em} = -\frac{\mathbf{J}_c \cdot \mathbf{J}_c}{\sigma} \tag{12-25}$$

and the boundary-layer energy equation is written

$$\rho c_p \left(u\frac{\partial T}{dx} + v\frac{\partial T}{\partial y}\right) = k\frac{\partial^2 T}{\partial y^2} + \mu\left(\frac{\partial u}{\partial y}\right)^2 + \frac{\mathbf{J}_c \cdot \mathbf{J}_c}{\sigma} \tag{12-26}$$

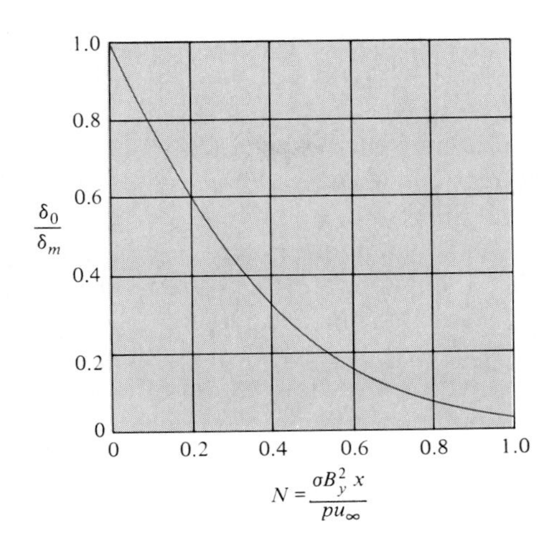

Fig. 12-2 Influence of magnetic field on boundary-layer thickness.

Now consider the flat plate shown in Fig. 12-3. The plate surface is maintained at the constant temperature T_w, the free-stream temperature is T_∞, and the thermal-boundary-layer thickness is designated by the conventional symbol δ_t. To simplify the analysis, we consider low-speed incompressible flow so that the viscous-heating effects are negligible. The integral energy equation then becomes

$$\frac{d}{dx}\left[\int_0^{\delta_t} (T_\infty - T)u\, dy\right] = \alpha \left.\frac{dT}{dy}\right]_{y=0} - \frac{\sigma}{\rho c_p}\int_0^{\delta_t} B_y{}^2 u^2\, dy \qquad (12\text{-}27)$$

since

$$\frac{\mathbf{J}_c \cdot \mathbf{J}_c}{\sigma} = \frac{[\sigma(\mathbf{V} \times \mathbf{B})] \cdot [\sigma(\mathbf{V} \times \mathbf{B})]}{\sigma} = \sigma B_y{}^2 u^2 \qquad (12\text{-}28)$$

In Eq. (12-27) it is assumed that the magnetic heating effects are confined to the boundary-layer region. Again, as in Sec. 5-6, we are able to write a cubic-parabola type of function for the temperature distribution so that

$$\frac{\theta}{\theta_\infty} = \frac{T - T_w}{T_\infty - T_w} = \frac{3}{2}\frac{y}{\delta_t} - \frac{1}{2}\left(\frac{y}{\delta_t}\right)^3 \qquad (12\text{-}29)$$

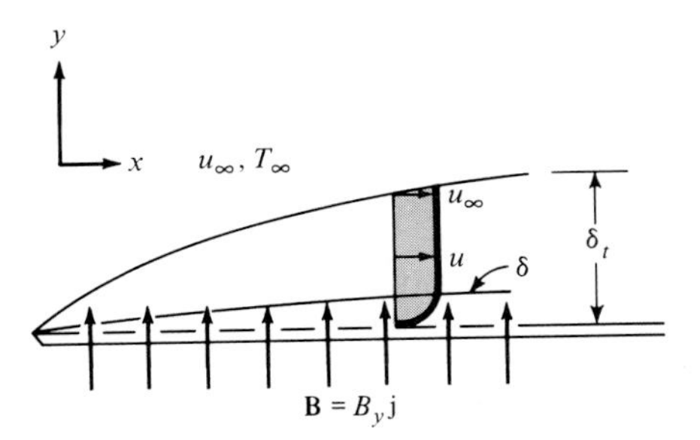

Fig. 12-3 Thermal boundary layer in magnetofluidynamic flow.

If the same procedure were followed at this point as in Sec. 5-6, the temperature and velocity functions given by Eqs. (12-19) and (12-29) would be inserted in (12-27) in order to arrive at a differential equation to be solved for δ_t, the thermal-boundary-layer thickness in the presence of the magnetic field. The problem with this approach is that a nonlinear equation results which must be solved by numerical methods.

Suppose the fluid is highly conducting, such as a liquid metal. In this case, the thermal-boundary-layer thickness will be much greater than the hydrodynamic thickness. This is evidenced by the fact that the Prandtl numbers for liquid metals are very low, of the order of 0.01. For such a fluid, then, we might approximate the actual fluid behavior with a slug-flow model for energy transport in the thermal boundary layer, as outlined in Sec. 6-5. We assume a constant velocity profile

$$u = u_\infty \tag{12-30}$$

Now, inserting Eqs. (12-29) and (12-30) in the integral energy equation gives

$$\frac{3u_\infty\theta_\infty}{8}\frac{d\delta_t}{dx} = \frac{3\alpha\theta_\infty}{2\delta_t} - \frac{\sigma B_y{}^2 u_\infty{}^2}{\rho c_p}\delta_t \tag{12-31}$$

This is a linear differential equation in $\delta_t{}^2$, and has the solution

$$\delta_t{}^2 = \frac{4\alpha}{Ku_\infty}(1 - e^{-2Kx}) \tag{12-32}$$

where

$$K = \frac{8\sigma B_y{}^2 u_\infty}{3\theta_\infty \rho c_p} \tag{12-33}$$

The heat-transfer coefficient in the presence of the magnetic field may now be calculated from

$$h_m = \frac{-k(\partial T/\partial y)_{y=0}}{T_w - T_\infty} = \frac{3k}{2\delta_t} \tag{12-34}$$

In dimensionless form,

$$\mathrm{Nu}_m = \sqrt{\tfrac{3}{2}}\sqrt{N\,\mathrm{Ec}\,\mathrm{Re}_x\,\mathrm{Pr}}\,(1 - e^{-5.33N\,\mathrm{Ec}})^{-1/2} \tag{12-35}$$

where

$$\mathrm{Nu}_m = \text{Nusselt number} = \frac{h_m x}{k} \tag{12-36}$$

$$\mathrm{Ec} = \text{Eckert number} = \frac{u_\infty{}^2}{c_p\theta_\infty} \tag{12-37}$$

and Pr is the Prandtl number of the fluid. If Eq. (12-27) is solved for the case of zero magnetic field but with the same slug-flow model as in Sec. 6-5, there results

$$\mathrm{Nu}_0 = \frac{h_0 x}{k} = \frac{3\sqrt{2}}{8}\sqrt{\mathrm{Re}_x\,\mathrm{Pr}} \tag{12-38}$$

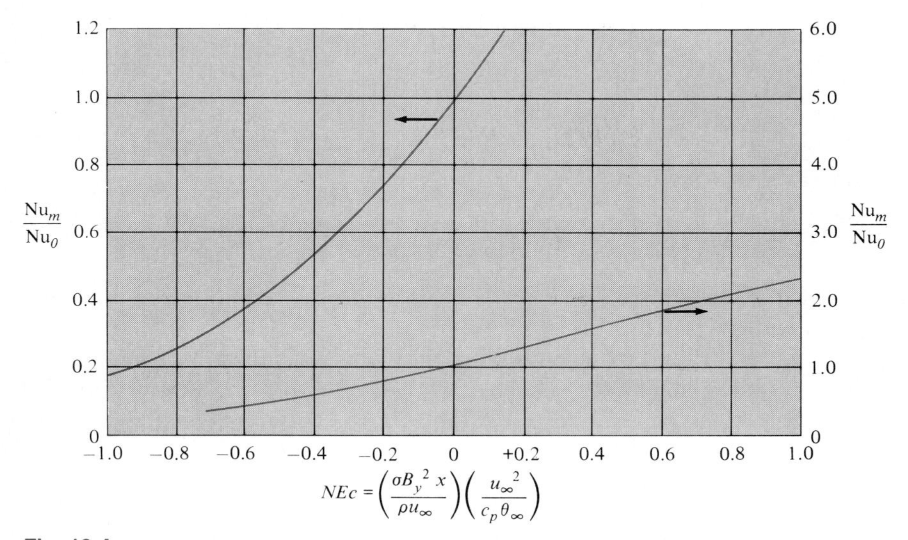

Fig. 12-4 Influence of magnetic field on heat transfer from flat plate.

The comparative ratio of interest in the heat-transfer case is thus

$$\frac{\text{Nu}_m}{\text{Nu}_0} = \left(\frac{5.33N\ \text{Ec}}{1 - e^{-5.33N\ \text{Ec}}}\right)^{1/2} \tag{12-39}$$

This equation is plotted in Fig. 12-4. It may be noted that the magnetic field will increase the heat-transfer rate for positive values of the Eckert number ($T_\infty > T_w$) and decrease the heat transfer for negative Eckert numbers ($T_w > T_\infty$). This behavior results from the fact that the magnetic field tends to heat the fluid, thereby reducing or increasing the temperature gradient between it and the plate.

It is necessary to caution that the foregoing analysis is a highly idealized one, which has been used primarily to illustrate the effects of magnetic fields on heat transfer. A more realistic analysis would consider the variation of electrical conductivity of the fluid and take into account the exact velocity profile rather than the slug-flow model. A survey of more exact relations for heat transfer in MFD systems is given in Refs. 1 and 2.

12-3 TRANSPIRATION COOLING

When high-velocity heat-transfer situations are encountered, as described in Sec. 5-12, the adiabatic wall temperature of the surface exposed to the flow stream can become quite large, and significant amounts of cooling may be required in order to reduce the surface temperature to a reasonable value. One technique for cooling the surface is called *transpiration*, or sweat, cooling. It operates on the principle shown in Fig. 12-5. A porous flat plate is exposed to the high-velocity flow stream while a fluid is forced through the plate into the

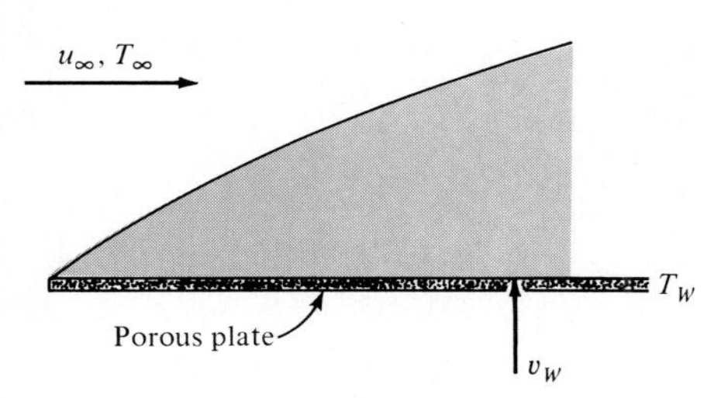

Fig. 12-5 Porous flat plate with fluid injection.

boundary layer. This fluid could be the same as the free stream or some different fluid. The injection process carries additional energy away from the region close to the plate surface, over and above that which would normally be conducted into the boundary layer. There is also an effect on the velocity profile of the boundary layer and, correspondingly, on the frictional-drag characteristics.

For incompressible flow without viscous heating and for zero pressure gradient, the boundary-layer equations to be solved are the familiar ones presented in Chap. 5 when the injected fluid is the same as the free-stream fluid:

$$\begin{aligned} u\frac{\partial u}{\partial x} + v\frac{\partial u}{\partial y} &= \nu\frac{\partial^2 u}{\partial y^2} \\ u\frac{\partial T}{\partial x} + v\frac{\partial T}{\partial y} &= \alpha\frac{\partial^2 T}{\partial y^2} \end{aligned} \tag{12-40}$$

The boundary conditions, however, are different from those used in Chap. 5. Now we must take

$$v = v_w \qquad \text{at } y = 0 \tag{12-41}$$

instead of $v = 0$ at $y = 0$, since a finite injection velocity is involved. We still have

$$\begin{aligned} u &= 0 && \text{at } y = 0 \\ u &= u_\infty && \text{at } y \to \infty \\ \frac{\partial u}{\partial y} &= 0 && \text{at } y \to \infty \end{aligned}$$

The boundary-layer equations may be solved by the technique outlined in Appendix B or by the integral method of Chap. 5. Eckert and Hartnett [3] have developed a comprehensive set of solutions for the transpiration-cooling problem, and we present the results of their analysis without exploring the techniques employed for solution of the equations.

Figure 12-6 shows the boundary-layer velocity profiles which result from various injection rates in a laminar boundary layer. The injection parameter

$$\frac{v_w}{u_\infty}\sqrt{\mathrm{Re}_x}$$

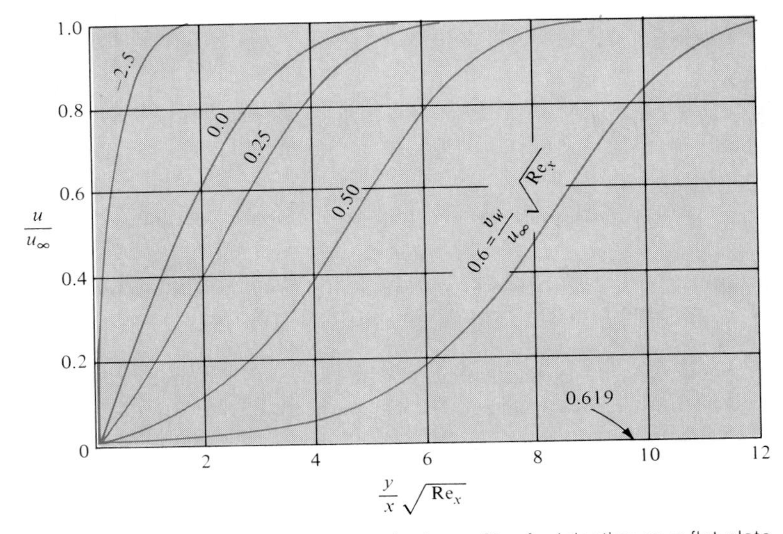

Fig. 12-6 Laminar-boundary-layer velocity profiles for injection on a flat plate, according to Ref. 3.

uses the conventional definition of Reynolds number as $Re_x = \rho u_\infty x/\mu$. Negative values of the injection parameter indicate *suction* on the plate and produce much blunter velocity profiles. For values of the injection parameter greater than 0.619 the boundary layer is completely blown away from the plate. Temperature profiles have a similar shape for the various injection rates. The overall

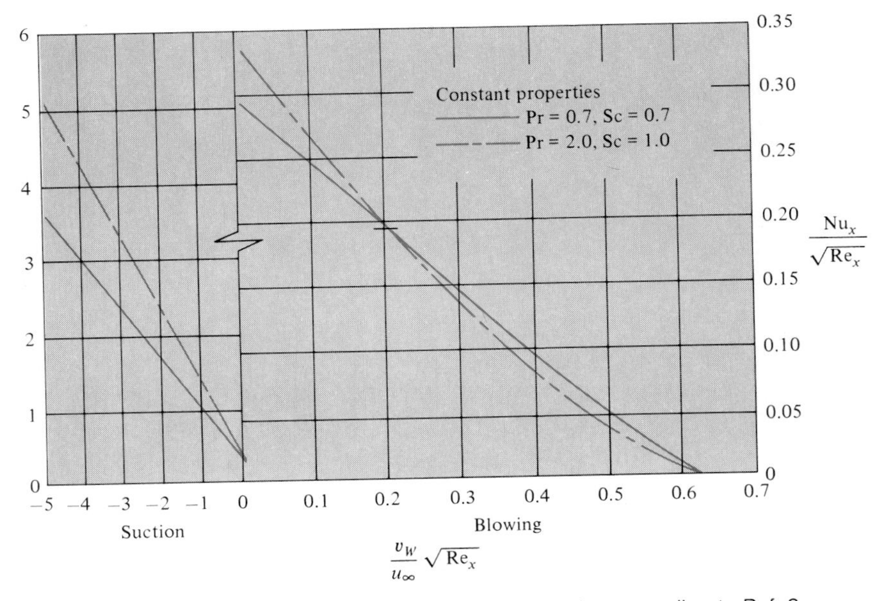

Fig. 12-7 Effect of fluid injection on flat-plate heat transfer, according to Ref. 3.

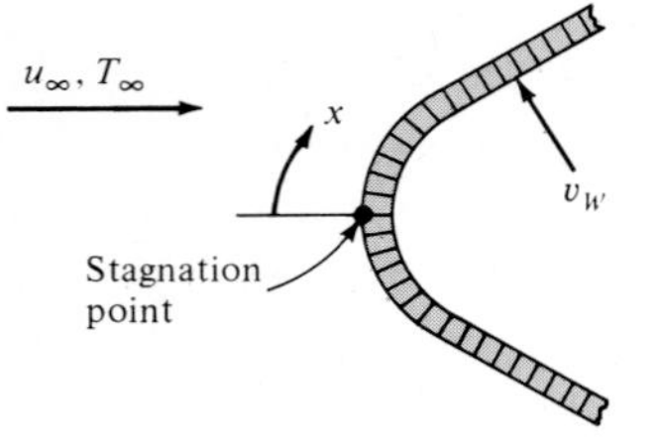

Fig. 12-8 Plane stagnation flow with fluid injection.

effect of the injection (or suction) process on heat transfer is indicated in Fig. 12-7. As expected, blowing causes a reduction in heat transfer, while suction causes an increase.

An important application of transpiration cooling is that of plane stagnation flow, as illustrated in Fig. 12-8. Solutions for the influence of transpiration on heat transfer in the neighborhood of such a stagnation line have also been worked out in Ref. 3, and the results are shown in Fig. 12-9. As would be expected, gas injection or suction can exert a significant effect on the temperature recovery factor for flow over a flat plate. These effects are indicated in Fig. 12-10, where the recovery factor r is defined in the conventional way as

$$r = \frac{T_{aw} - T_\infty}{T_0 - T_\infty} = \frac{T_{aw} - T_\infty}{u_\infty^2/2c_p} \tag{12-42}$$

It is well to remind the reader at this point that the curves presented above all involve the injection of a gas which is identical with the free-stream gas

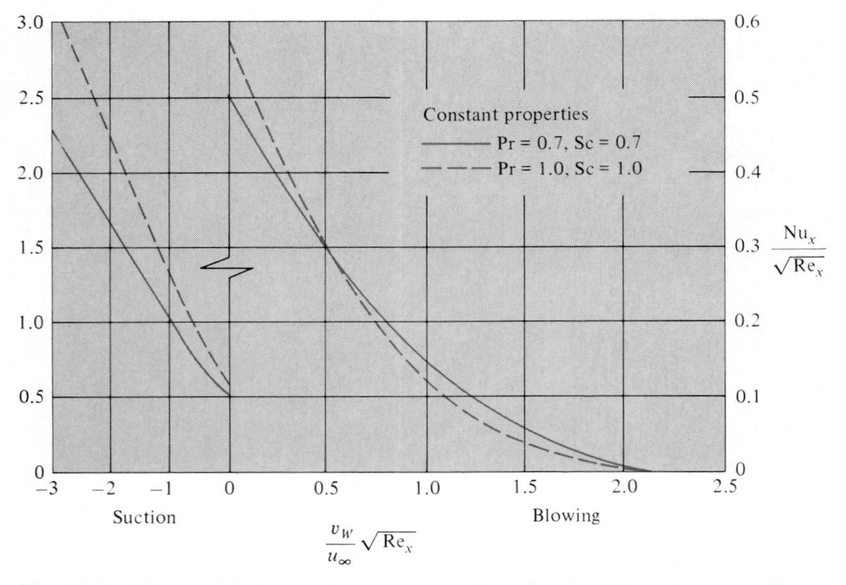

Fig. 12-9 Effects of fluid injection on plane stagnation heat transfer, according to Ref. 3.

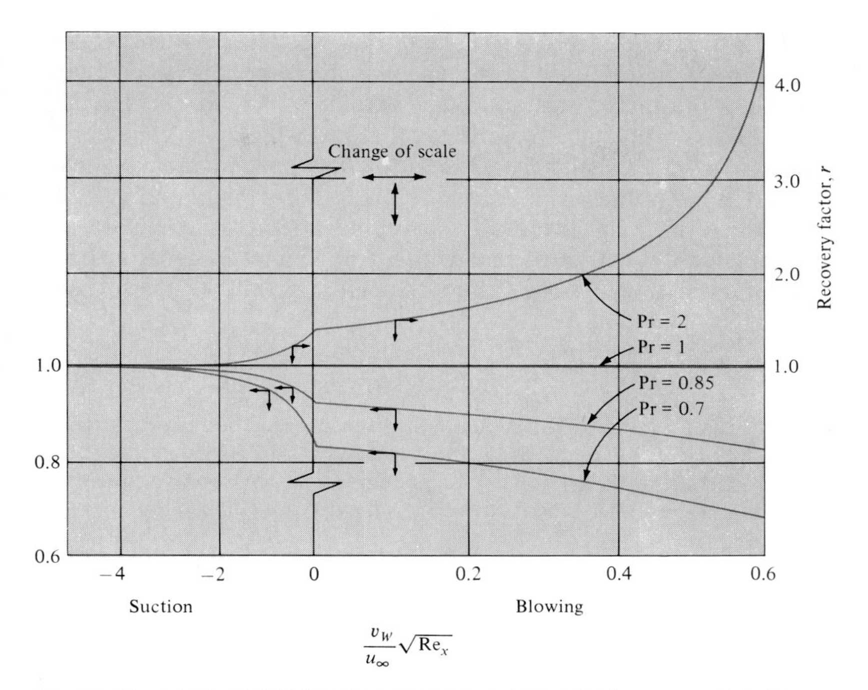

Fig. 12-10 Effects of fluid injection on recovery factor for flow over a flat plate, according to Ref. 3.

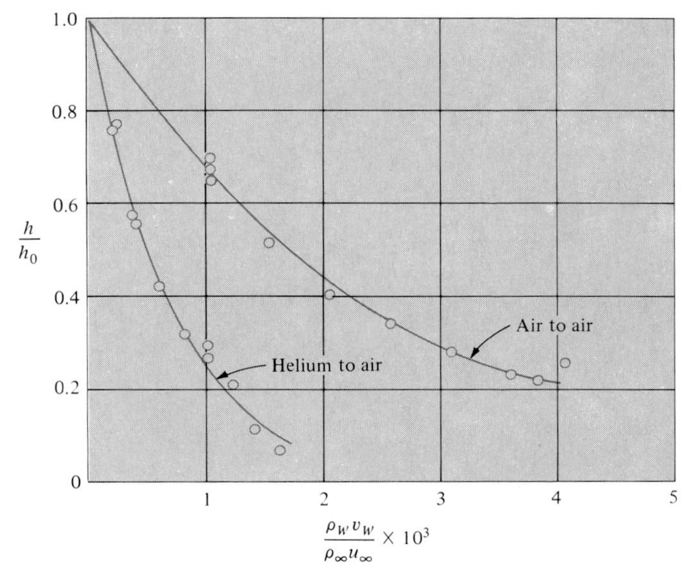

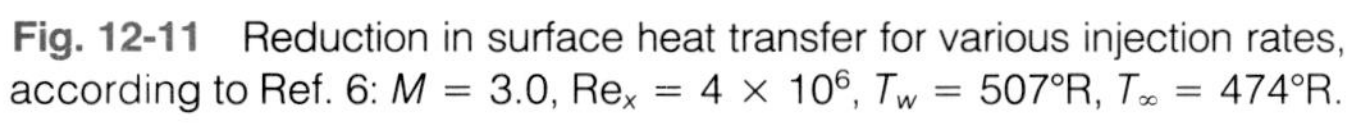

Fig. 12-11 Reduction in surface heat transfer for various injection rates, according to Ref. 6: $M = 3.0$, $Re_x = 4 \times 10^6$, $T_w = 507°R$, $T_\infty = 474°R$.

(usually air). When different injection gases are used, the mass diffusion in the boundary layer must be taken into proper account. Several papers have analyzed this situation, and the interested reader is referred to Refs. 4, 5, and 7 for a discussion of the problems. Experimental results of Ref. 6 indicate the approximate behavior which results from the use of different coolants, as shown in Fig. 12-11. In this figure the product $\rho_w v_w$ refers to the density and velocity of the injection gas at the wall. The ratio h/h_0 represents the ratio of the heat-transfer coefficient with injection to that without injection. In general, lightweight gases exhibit a stronger cooling effect than air injection because of their larger specific heats.

EXAMPLE 12-1

For the flow conditions of Example 5-10 calculate the percent reduction in heat transfer at a position where $\mathrm{Re}_x = 5 \times 10^4$ and the injection parameter is 0.2. For this calculation assume constant properties evaluated at the average temperature between the recovery and free-stream temperatures. Also calculate the mass flow coolant air required per unit area at this location.

Solution

We first calculate the recovery (adiabatic-wall) temperature. From Fig. 12-10

$$r = 0.79 \qquad \text{for Pr} = 0.7$$

From Example 5-10

$$T_0 = 652 \text{ K} = 379°\text{C} \quad [714°\text{F}]$$

$$T_\infty = -40°\text{C} = 233 \text{ K} \quad [-40°\text{F}]$$

$$u_\infty = 918 \text{ m/s} \quad [3011 \text{ ft/s}]$$

Thus

$$T_{aw} - T_\infty = 0.79(652 - 233) = 331$$

$$T_{aw} = 233 + 331 = 564 \text{ K}$$

We evaluate properties at

$$T_f = \frac{564 + 233}{2} = 398.5 \text{ K}$$

According to Fig. 12-7, with

$$\frac{v_w}{u_\infty}\sqrt{\mathrm{Re}_x} = 0.2 \qquad \text{and} \qquad \mathrm{Pr} = 0.7$$

we obtain

$$\frac{\mathrm{Nu}_x}{\sqrt{\mathrm{Re}_x}} = 0.19$$

For zero injection we have

$$\frac{\mathrm{Nu}_x}{\sqrt{\mathrm{Re}_x}} = 0.29$$

The percent reduction in heat transfer is obtained by comparing these last two numbers:

$$\text{Reduction in heat transfer} = \frac{0.29 - 0.19}{0.29} \times 100 = 34.5\%$$

The mass flow of coolant air at the wall is calculated from

$$\frac{\dot{m}_w}{A} = \rho_w v_w$$

We are assuming a constant-properties analysis evaluated at the reference temperature of 398.5 K, so

$$\rho_w = \frac{(1.0132 \times 10^5)(\frac{1}{20})}{(287)(398.5)} = 0.0443 \text{ kg/m}^3$$

$$v_w = \left(\frac{v_w}{u_\infty}\sqrt{\text{Re}_x}\right)\frac{u_\infty}{\sqrt{\text{Re}_x}}$$

$$= \frac{(0.2)(918)}{(5 \times 10^4)^{1/2}} = 0.821 \text{ m/s} \quad [2.69 \text{ ft/s}]$$

Thus

$$\frac{\dot{m}_w}{A} = (0.0443)(0.821) = 0.0364 \text{ kg/m}^2 \cdot \text{s} \quad [0.0074 \text{ lb}_m/\text{ft}^2 \cdot \text{s}]$$

12-4 LOW-DENSITY HEAT TRANSFER

A number of practical situations involve heat transfer between a solid surface and a low-density gas. In employing the term *low density,* we shall mean those circumstances where the mean free path of the gas molecules is no longer small in comparison with a characteristic dimension of the heat-transfer surface. The mean free path is the distance a molecule travels, on the average, between collisions. The larger this distance becomes, the greater the distance required to communicate the temperature of a hot surface to a gas in contact with it. This means that we shall not necessarily be able to assume that a gas in the immediate neighborhood of the surface will have the same temperature as the heated surface, as was done in the boundary-layer analyses of Chap. 5. Because the mean free path is also related to momentum transport between molecules, we shall also be forced to abandon our assumption of zero fluid velocity near a stationary surface for those cases where the mean free path is not negligible in comparison with the surface dimensions.

Three general flow regimes may be anticipated for the flow over a flat plate shown in Fig. 12-12. First, the continuum flow region is encountered when the mean free path λ is very small in comparison with a characteristic body dimension. This is the convection heat-transfer situation analyzed in preceding chapters. At lower gas pressures, when $\lambda \sim L$, the flow seems to "slip" along the surface and $u \neq 0$ at $y = 0$. This situation is appropriately called *slip flow*. At still lower densities, all momentum and energy exchange is the result of

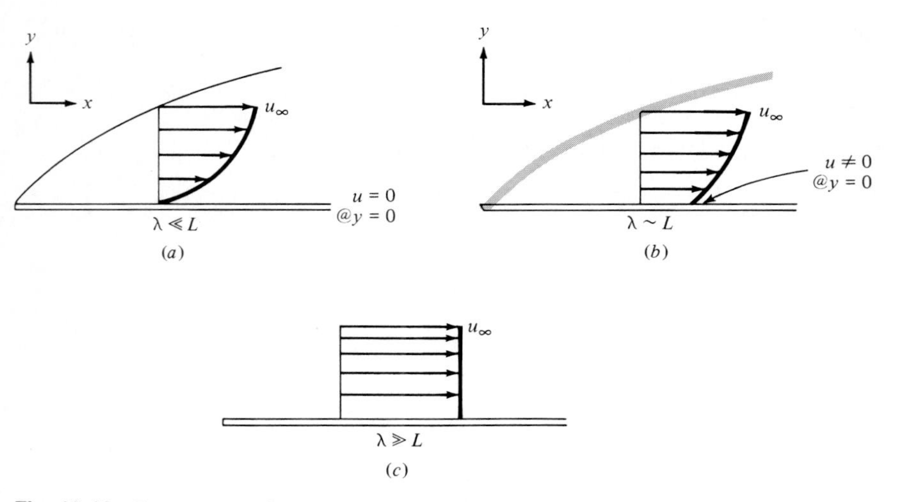

Fig. 12-12 Three types of flow regimes for a flat plate: (*a*) continuum flow; (*b*) slip flow; (*c*) free-molecule flow.

strictly molecular bombardment of the surface. This regime is called *free-molecule flow*, and must be analyzed apart from conventional continuum fluid mechanics.

Evidently, the parameter which is of principal interest is a ratio of the mean free path to a characteristic body dimension. This grouping is called the Knudsen number,

$$\mathrm{Kn} = \frac{\lambda}{L} \tag{12-43}$$

According to the kinetic theory of gases, the mean free path may be calculated from

$$\lambda = \frac{0.707}{4\pi r^2 n} \tag{12-44}$$

where r is the effective molecular radius for collisions and n is the molecular density. An approximate relation for the mean free path of air molecules is given by

$$\lambda = 2.27 \times 10^{-5}\,\frac{T}{p} \qquad \text{meters} \tag{12-45}$$

where T is in degrees Kelvin and p is in pascals.

The speed of sound in a gas is related to mean molecular speed $\bar{v}$ through the relation

$$a = \bar{v}\sqrt{\frac{\pi\gamma}{8}} \tag{12-46}$$

The mean molecular speed may be expressed by

$$\bar{v} = \sqrt{\frac{8RT}{\pi}} \tag{12-47}$$

where R is the gas constant for the particular gas. It can be shown that the transport properties (viscosity, thermal conductivity, diffusion coefficient) of a gas are directly related to the mean molecular speed. Based on this relation, the kinetic-theory representation of the Reynolds number may be derived as

$$\text{Re} = \frac{2u_\infty L}{\bar{v}\lambda} \tag{12-48}$$

Using Eq. (12-46), the Mach number is expressed as

$$M = \frac{u}{a} = \frac{u}{\bar{v}}\sqrt{\frac{8}{\pi\gamma}} \tag{12-49}$$

Now, combining Eqs. (12-43), (12-48), and (12-49), the Knudsen number may be expressed as

$$\text{Kn} = \frac{\lambda}{L} = \sqrt{\frac{\pi\gamma}{2}}\frac{M}{\text{Re}} \tag{12-50}$$

This relation enables us to interpret a low-density flow regime in terms of the conventional flow parameters, Mach and Reynolds numbers. Figure 12-13 gives an approximate representation of these effects.

As a first example of low-density heat transfer let us consider the two parallel infinite plates shown in Fig. 12-14. The plates are maintained at different temperatures and separated by a gaseous medium. Let us neglect natural-convection effects. If the gas density is sufficiently high so that $\lambda \to 0$, a linear temperature profile through the gas will be experienced as shown for the case of λ_1. As the gas density is lowered, the larger mean free paths require a greater distance from the heat-transfer surfaces in order for the gas to accommodate to the surface temperatures. The anticipated temperature profiles are shown in

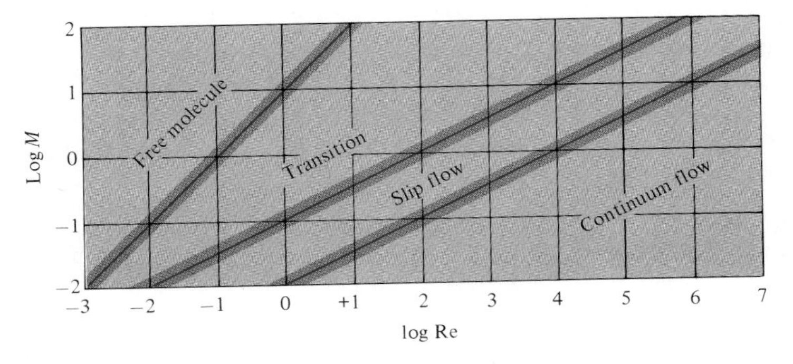

Fig. 12-13 Relation of flow regimes to Mach and Reynolds numbers.

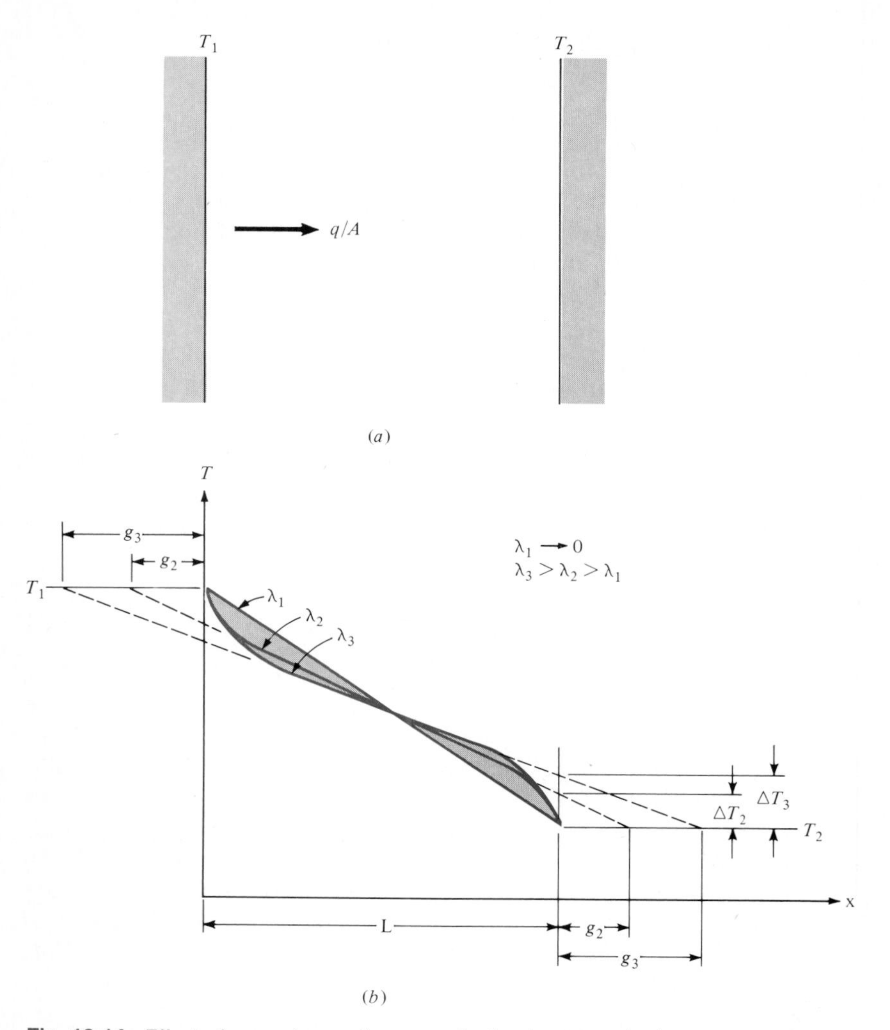

Fig. 12-14 Effect of mean free path on conduction heat transfer between parallel plates: (*a*) physical model; (*b*) anticipated temperature profiles.

Fig. 12-14*b*. Extrapolating the straight portion of the low-density curves to the wall produces a temperature "jump" ΔT, which may be calculated by making the following energy balance:

$$\frac{q}{A} = k\frac{T_1 - T_2}{g + L + g} = k\frac{\Delta T}{g} \tag{12-51}$$

In this equation we are assuming that the extrapolation distance g is the same for both plate surfaces. In general, the temperature jump will depend on the type of surface, and these extrapolation distances will not be equal unless the materials are identical. For different types of materials we should have

$$\frac{q}{A} = k\frac{T_1 - T_2}{g_1 + L + g_2} = k\frac{\Delta T_1}{g_1} = k\frac{\Delta T_2}{g_2} \tag{12-52}$$

where now ΔT_1 and ΔT_2 are the temperature jumps at the two heat-transfer surfaces and g_1 and g_2 are the corresponding extrapolation distances. For identical surfaces the temperature jump would then be expressed as

$$\Delta T = \frac{g}{2g + L}(T_1 - T_2) \tag{12-53}$$

Similar expressions may be developed for low-density conduction between concentric cylinders. In order to predict the heat-transfer rate it is necessary to establish relations for the temperature jump for various gas-to-solid interfaces.

We have already mentioned that the temperature-jump effect arises as a result of the failure of the molecules to "accommodate" to the surface temperature when the mean free path becomes of the order of a characteristic body dimension. The parameter which describes this behavior is called the *accommodation coefficient* α, defined by

$$\alpha = \frac{E_i - E_r}{E_i - E_w} \tag{12-54}$$

where E_i = energy of incident molecules on a surface
E_r = energy of molecules reflected from the surface
E_w = energy molecules would have if they acquired energy of wall at temperature T_w

Values of the accommodation coefficient must be determined from experiment, and a brief summary of such measurements is given in Table 12-1.

It is possible to employ the kinetic theory of gases along with values of α

Table 12-1 Thermal Accommodation Coefficients for Air, According to Ref. 10.†

Surface	*Accommodation coefficient*
Flat black lacquer on bronze	0.88–0.89
Bronze, polished	0.91–0.94
Machined	0.89–0.93
Etched	0.93–0.95
Cast iron, polished	0.87–0.93
Machined	0.87–0.88
Etched	0.89–0.96
Aluminum, polished	0.87–0.95
Machined	0.95–0.97
Etched	0.89–0.97

†See also Refs. 9 and 11.

to determine the temperature jump at a surface. The result of such an analysis is

$$T_{y=0} - T_w = \frac{2 - \alpha}{\alpha} \frac{2\gamma}{\gamma + 1} \frac{\lambda}{\Pr} \frac{\partial T}{\partial y}\bigg]_{y=0} \tag{12-55}$$

The nomenclature for Eq. (12-55) is noted in Fig. 12-15. This temperature jump is denoted by ΔT in Fig. 12-14, and the temperature gradient for use with Fig. 12-14 would be

$$\frac{T_1 - T_2 - 2\Delta T}{L}$$

When free-molecule flow is encountered, Oppenheim [8] has given convenient charts for calculating recovery factors and heat-transfer coefficients for flow over standard geometric shapes. Figures 12-16 and 12-17 give samples of these charts. The molecular speed ratio S used in these charts is defined by

$$S = \frac{u_\infty}{v_m} = \frac{u_\infty}{\sqrt{2RT}} = M\sqrt{\frac{\gamma}{2}} \tag{12-56}$$

where v_m is called the *most probable molecular speed*. The recovery factor r is defined in the usual way as

$$r = \frac{T_{aw} - T_\infty}{T_0 - T_\infty} \tag{12-57}$$

where T_{aw} is the adiabatic wall temperature and T_0 is the free-stream stagnation temperature. The Stanton number is also defined in the conventional way,

$$\mathrm{St} = \frac{h}{\rho c_p u_\infty} \tag{12-58}$$

with the convection heat-transfer coefficient defined in terms of the adiabatic wall temperature,

$$q = hA(T_w - T_{aw}) \tag{12-59}$$

It is well to mention that the equilibrium temperature of a surface in free-molecule flow is usually influenced strongly by radiation heat transfer because

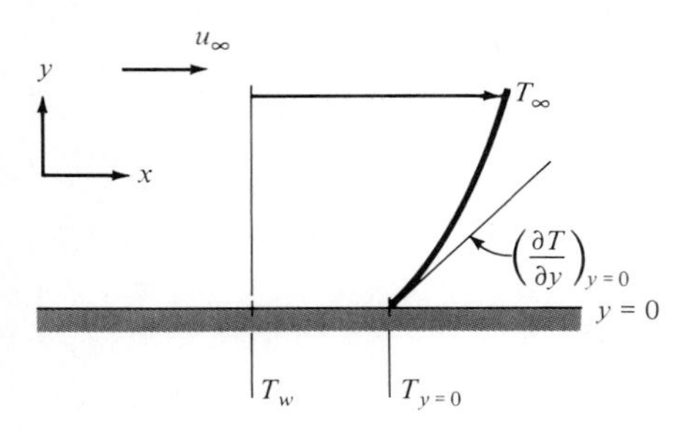

Fig. 12-15 Nomenclature for use with Eq. (12-55).

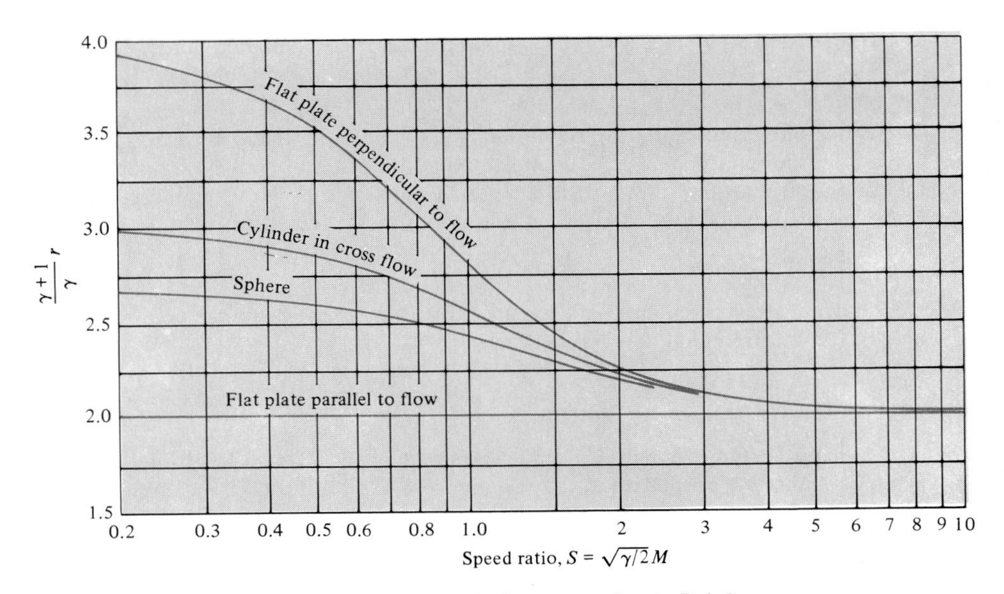

Fig. 12-16 Recovery factors for free-molecule flow, according to Ref. 8.

of the high values of T_{aw} encountered for high flow velocities. The radiation equilibrium temperature for the surface T_{rw} is obtained by equating the convection gain to the radiation loss. Thus

$$hA(T_{aw} - T_{rw}) = \sigma\epsilon A(T_{rw}^4 - T_s^4) \tag{12-60}$$

where ϵ is the surface emissivity and T_s is an effective radiation temperature of the surroundings, usually taken as T_∞.

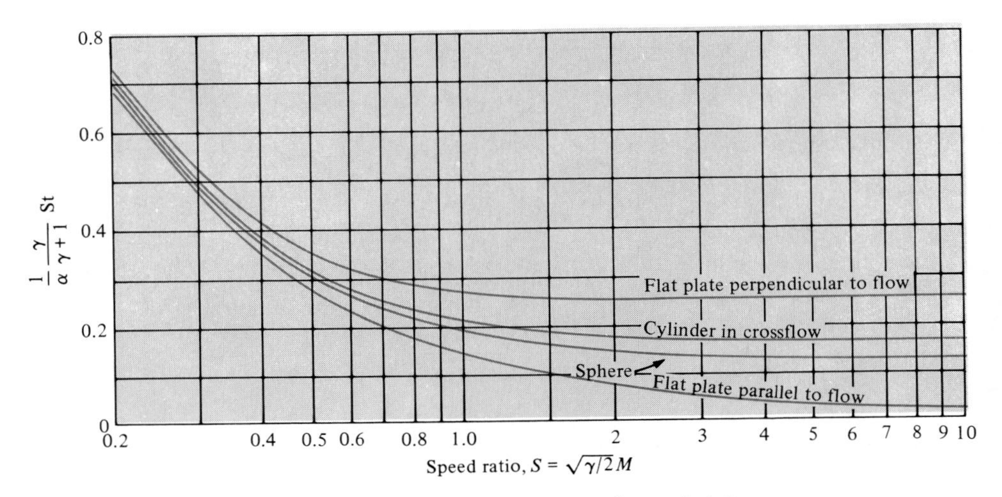

Fig. 12-17 Stanton numbers for free-molecule flow, according to Ref. 8.

A summary of low-density heat transfer is given by Springer [13].

■ **EXAMPLE 12-2**

Two polished-aluminum plates are separated by a distance of 2.5 cm in air at a pressure of 10^{-6} atm. The plates are maintained at 100 and 30°C, respectively. Calculate the conduction heat transfer through the air gap. Compare this with the radiation heat transfer and the conduction for air at normal atmospheric pressure.

Solution

We first calculate the mean free path to determine if low-density effects are important. From Eq. (12-45), at an average temperature of 65°C = 338 K,

$$\lambda = \frac{(2.27 \times 10^{-5})(338)}{(1.0132 \times 10^{-5})(10^{-6})} = 0.0757 \text{ m} = 7.57 \text{ cm} \quad [0.248 \text{ ft}]$$

Since the plate spacing is only 2.5 cm, we should expect low-density effects to be important. Evaluating properties at the mean air temperature of 65°C, we have

$$k = 0.0291 \text{ W/m} \cdot {}^\circ\text{C} \quad [0.0168 \text{ Btu/h} \cdot \text{ft} \cdot {}^\circ\text{F}]$$

$$\gamma = 1.40 \qquad \text{Pr} = 0.7 \qquad \alpha \approx 0.9 \qquad \text{from Table 12-1}$$

Combining Eq. (12-25) with the central-temperature-gradient relation gives

$$\Delta T = \frac{2 - \alpha}{\alpha} \frac{2\gamma}{\gamma + 1} \frac{\lambda}{\text{Pr}} \frac{T_1 - T_2 - 2\,\Delta T}{L}$$

Inserting the appropriate properties gives

$$\Delta T = \frac{2 - 0.9}{0.9} \frac{2.8}{2.4} \frac{0.0757}{0.7} \frac{100 - 30 - 2\,\Delta T}{0.0025}$$

$$= 32.38^\circ\text{C} \quad [58.3^\circ\text{F}]$$

The conduction heat transfer is thus

$$\frac{q}{A} = k\frac{T_1 - T_2 - 2\,\Delta T}{L} = \frac{(0.0291)(70 - 64.76)}{0.025}$$

$$= 6.099 \text{ W/m}^2 \quad [1.93 \text{ Btu/h} \cdot \text{ft}^2]$$

At normal atmospheric pressure the conduction would be

$$\frac{q}{A} = k\frac{T_1 - T_2}{L} = 81.48 \text{ W/m}^2 \quad [25.8 \text{ Btu/h} \cdot \text{ft}^2]$$

The radiation heat transfer is calculated with Eq. (8-42), taking $\epsilon_1 = \epsilon_2 = 0.06$ for polished aluminum:

$$\left(\frac{q}{A}\right)_{\text{rad}} = \frac{\sigma(T_1^4 - T_2^4)}{2/\epsilon - 1} = \frac{(5.669 \times 10^{-8})(393^4 - 303^4)}{2/0.06 - 1}$$

$$= 27.05 \text{ W/m}^2 \quad [8.57 \text{ Btu/h} \cdot \text{ft}^2]$$

Thus, at the low-density condition the radiation heat transfer is almost 5 times as large as the conduction, even with highly polished surfaces.

■ EXAMPLE 12-3

A thermocouple is constructed by welding two wires together so that a spherical bead is formed with a diameter of 1.0 mm. The bead is exposed to a high-velocity airstream at $M = 6$, $p = 10^{-6}$ atm, and $T_\infty = -70°C$. Estimate the temperature of the thermocouple bead assuming a surface emissivity of 0.7. Assume that the bead has an accommodation coefficient equal to that of cast iron.

Solution

We first establish the flow regime. Evaluating properties at free-stream conditions, we have

$$T_\infty = -70°C = 203 \text{ K} \qquad \mu = 1.329 \times 10^{-5} \text{ kg/m} \cdot \text{s}$$

$$k = 0.0181 \text{ W/m} \cdot °C \qquad c_p = 1006 \text{ J/kg} \cdot °C$$

The gas density is

$$\rho = \frac{p}{RT} = \frac{(1.0132 \times 10^5)(10^{-6})}{(287)(203)} = 1.739 \times 10^{-6} \text{ kg/m}^3$$

The acoustic velocity is

$$a = \sqrt{\gamma g_c RT} = [(1.4)(287)(203)]^{1/2} = 285.6 \text{ m/s} \quad [937 \text{ ft/s}]$$

so that

$$u_\infty = (6)(285.6) = 1714 \text{ m/s} \quad [5623 \text{ ft/s}]$$

The Reynolds number is now calculated as

$$\text{Re} = \frac{\rho u_\infty d}{\mu} = \frac{(1.739 \times 10^{-6})(1714)(10^{-3})}{1.329 \times 10^{-5}} = 0.224$$

An inspection of Fig. 12-13 shows that the free-molecule range is encountered. Accordingly, we can make use of Figs. 12-16 and 12-17 to compute the heat-transfer parameters. The molecular speed ratio is calculated as

$$S = \sqrt{\frac{\gamma}{2}}M = (0.7)^{1/2}(6) = 5.02$$

From the figures

$$\frac{\gamma + 1}{\gamma} r = 2.04 \qquad \frac{1}{\alpha}\frac{\gamma}{\gamma + 1} \text{St} = 0.127$$

Then, using $\alpha = 0.9$,

$$r = \frac{(2.04)(1.4)}{2.4} = 1.19 \qquad \text{St} = \frac{(0.127)(2.4)(0.9)}{1.4} = 0.196$$

The stagnation temperature is calculated from Eq. (5-117),

$$T_0 = (203)[1 + (0.2)(6)^2] = 1665 \text{ K} \quad [3000°R]$$

and the adiabatic wall temperature is

$$T_{aw} = T_\infty + r(T_0 - T_\infty) = 203 + (1.19)(1665 - 203) = 1942 \text{ K}$$

The Stanton number is now used to calculate the heat-transfer coefficient from

$$h = \rho c_p u_\infty \, \text{St} = (1.739 \times 10^{-6})(1006)(1714)(0.127) = 0.381 \ \text{W/m}^2 \cdot {}^\circ\text{C}$$

We now employ the energy balance indicated in Eq. (12-60) to calculate the radiation equilibrium temperature T_{rw}, taking the radiation surroundings at −70°C:

$$(0.381)(1942 - T_{rw}) = (5.669 \times 10^{-8})(0.7)[T_{rw}^4 - (203)^4]$$

A solution of this equation gives

$$T_{rw} = 359 \ \text{K} = 86^\circ\text{C} \quad [187^\circ\text{F}]$$

It is easy to see from this example that radiation heat transfer is very important for low-density work, even when high velocities are encountered.

12-5 ABLATION

The very high speeds encountered in missile-reentry situations present many interesting and unusual heat-transfer problems. The major concern is usually with a total energy which must be absorbed in the reentry body rather than with a heat-transfer *rate,* since the reentry times are very short. In these instances a common cooling technique is that of *ablation,* whereby part of the solid body exposed to the hot, high-speed flow is allowed to melt and blow away. Thus, part of the heat is expended in melting the material rather than being conducted into the interior of the vehicle. A number of glasses and plastic materials normally find application as ablators.

We shall employ a simplified analysis of the ablation problem utilizing the coordinate system and nomenclature shown in Fig. 12-18. The solid wall is exposed to a constant heat flux of $(q/A)_0$ at the surface. This heat flux may result from combined convection- and radiation-energy transfer from the high-speed boundary layer. As a result of the high-heat flux the solid body melts, and a portion of the surface is removed at the ablation velocity V_a. We assume that a steady-state situation is attained so that the surface ablates at a constant

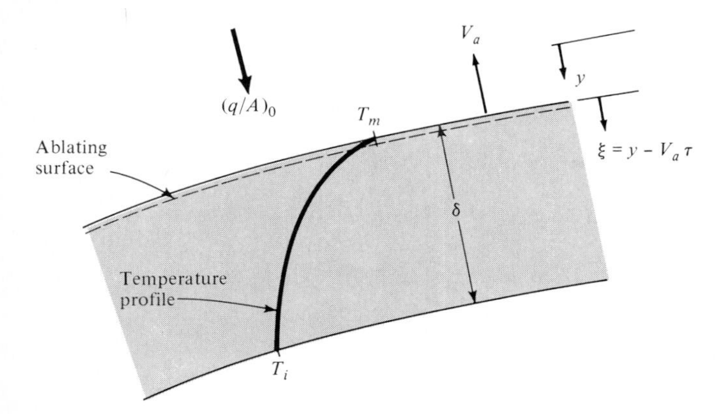

Fig. 12-18 Schematic of ablating solid surface.

rate. It is further assumed that the thickness of the solid is very large compared with the depth of material removed by the melting process. If the process is essentially one-dimensional, the appropriate differential equation for constant properties is

$$\frac{\partial^2 T}{\partial y^2} = \frac{1}{\alpha}\frac{\partial T}{\partial \tau} \tag{12-61}$$

This problem is most easily solved with a simple change of variable into a moving coordinate system. If time is measured from the start of ablation then a convenient variable change is

$$\xi = y - V_a \tau \tag{12-62}$$

The effect of this transformation is to fix the origin at the surface in the moving coordinate system. The derivatives in Eq. (12-61) may now be expressed as

$$\frac{\partial^2 T}{\partial y^2} = \frac{\partial^2 T}{\partial \xi^2} \qquad \frac{\partial T}{\partial \tau} = -V_a \frac{\partial T}{\partial \xi}$$

so that an ordinary differential equation results:

$$\frac{d^2 T}{d\xi^2} + \frac{V_a}{\alpha}\frac{dT}{d\xi} = 0$$

This equation has the solution

$$T = C_1 + C_2 \exp \frac{-V_a \xi}{\alpha} \tag{12-63}$$

At the surface the temperature is the melting temperature T_m, while the temperature in the interior of the body is T_i. The appropriate boundary conditions are thus

$$T = T_m \qquad \text{at } \xi = 0$$

$$T = T_i \qquad \text{as } \xi \to \infty$$

Evaluating the constants in Eq. (12-63) gives the final solution

$$\frac{T - T_i}{T_m - T_i} = \exp \frac{-V_a \xi}{\alpha} \tag{12-64}$$

The total energy incident on the body is either conducted into the slab or used to melt the material. Thus,

$$\begin{aligned}\left(\frac{q}{A}\right)_0 &= \left(\frac{q}{A}\right)_{\text{cond}} + \left(\frac{q}{A}\right)_{\text{abl}} \\ &= -k \left.\frac{\partial T}{\partial \xi}\right]_{\xi=0} + \rho V_a H_{ab}\end{aligned} \tag{12-65}$$

where H_{ab} is the heat of ablation. Evaluating the surface temperature gradient from Eq. (12-64) gives

$$\left(\frac{Q}{A}\right)_0 = \rho V_a c(T_m - T_i) + \rho V_a H_{ab} \tag{12-66}$$

which may be solved for the ablation velocity V_a to give

$$V_a = \frac{(q/A)_0}{\rho H_{ab}[1 + c(T_m - T_i)/H_{ab}]} \tag{12-67}$$

As mentioned previously, the main purpose of ablation is to reduce the quantity of energy conducted into the solid. It is therefore of interest to compare the total energy conducted in time τ with the total incident energy in this same time. The total conduction energy is

$$\begin{aligned}\left(\frac{Q}{A}\right)_{\text{cond}} &= \int_0^\tau - k \left(\frac{\partial T}{\partial \xi}\right)_{\xi=0} d\tau \\ &= \rho V_a c(T_m - T_i)\end{aligned} \tag{12-68}$$

In this time the total energy is $(q/A)_0\tau$. We therefore form the following ratio of interest:

$$\frac{(Q/A)_{\text{cond}}}{(Q/A)_{\text{total}}} = \frac{\rho V_a c(T_m - T_i)}{(q/A)_0} \tag{12-69}$$

Making use of Eq. (12-67) gives

$$\frac{(Q/A)_{\text{cond}}}{(Q/A)_{\text{total}}} = \frac{c(T_m - T_i)}{H_{ab}[1 + c(T_m - T_i)/H_{ab}]} \tag{12-70}$$

A few quick observations are in order at this point. The conducted energy is clearly reduced for large values of the heat of ablation. Similarly, the rate of material removal ρV_a is dependent on the heat of ablation and decreases for increased values of H_{ab}. In order for this very simplified solution to apply, the overall slab thickness δ must be large compared with the depth of penetration of melting. In terms of the above parameters this means that

$$\delta \gg V_a \tau$$

The reader should realize that the above analysis has several serious faults. It does not take into account the transient nature of the incident heat flux or the many complications involved in the melting mass-transfer process which occurs at the surface. Some of these problems are discussed by Dorrance [12].

12-6 THE HEAT PIPE

We have seen that one of the objectives of heat-transfer analysis can be the design of heat exchangers to transfer energy from one location to another. The smaller and more compact the heat exchanger, the better the design. The heat pipe is a novel device that allows the transfer of very substantial quantities of heat through small surface areas. The basic configuration of the device is shown in Fig. 12-19. A circular pipe has a layer of wicking material covering the inside surface, as shown, with a hollow core in the center. A condensible fluid is also contained in the pipe, and the liquid permeates the wicking material by capillary action. When heat is added to one end of the pipe (the evaporator), liquid is vaporized in the wick and the vapor moves to the central core. At the other end of the pipe, heat is removed (the condenser) and the vapor condenses back into the wick. Liquid is replenished in the evaporator section by capillary action.

A variety of fluid and pipe materials have been used for heat-pipe construction, and some typical operating characteristics are summarized in Table 12-2. Very high heat fluxes are obtained; for this reason research efforts are being devoted to optimum wick designs, novel configurations for specialized applications, etc. Two early theoretical analyses of heat pipes are presented by Cotter, et al. [16,17]. The device is in such a state of development that the current research literature should be consulted for latest information. For our purposes we are concerned primarily with applications for the device.

Cooling problems in microelectric circuits are particularly critical because the heat generation must be dissipated from such small surface areas and the performance of the electronic devices is strongly temperature-dependent. The heat-pipe concept offers a convenient way to transfer the heat from the small area to a larger area where it can be dissipated more easily. One way of doing this is shown in Fig. 12-20. Of course, the finned heat-dissipation surface could be water-cooled if required. The advantage of the heat pipe in electronic cooling applications is its nearly isothermal operation regardless of heat flux, within the operating range of the unit.

The basic design of a heat pipe may be modified to operate as a temperature-control device, as shown in Fig. 12-21. A reservoir containing a noncondensible gas is connected to the heat-removal end of the heat pipe. This gas may then form an interface with the vapor and "choke off" part of the condensation to the wick. With increased heat addition, more vapor is generated with an increase in vapor pressure, and the noncondensible gas is forced back into the reservoir, thereby opening up additional condenser area to remove the additional heat. For a reduction in heat addition, just the reverse operation is

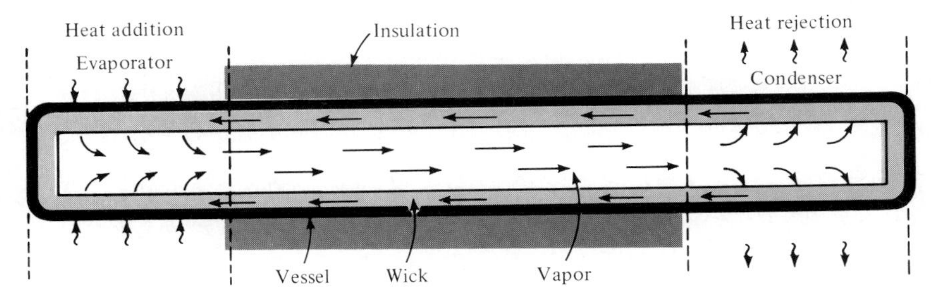

Fig. 12-19 Basic heat-pipe configuration.

Table 12-2 Some Typical Operating Characteristics of Heat Pipes, According to Ref. 15.

Temperature range, °C	*Working fluid*	*Vessel material*	*Measured axial† heat flux,* kW/cm²	*Measured surface† heat flux,* W/cm²
−200 to −80	Liquid nitrogen	Stainless steel	0.067 @ −163°C	1.01 @ −163°C
−70 to +60	Liquid ammonia	Nickel, aluminum, stainless steel	0.295	2.95
−45 to +120	Methanol	Copper, nickel, stainless steel	0.45 @ 100°C‡	75.5 @ 100°C
+5 to +230	Water	Copper, nickel	0.67 @ 200°C	146 @ 170°C
+190 to +550	Mercury§ + 0.02% magnesium	Stainless steel	25.1 @ 360°C¶	181 @ 360°C
+400 to +800	Potassium§	Nickel, stainless steel	5.6 @ 750°C	181 @ 750°C
+500 to +900	Sodium§	Nickel, stainless steel	9.3 @ 850°C	224 @ 760°C
+900 to +1500	Lithium§	Niobium + 1% zirconium	2.0 @ 1250°C	207 @ 1250°C
1500 to +2000	Silver§	Tantalum + 5% tungsten	4.1	413

†Varies with temperature.
‡Using threaded artery wick.
§Tested at Los Alamos Scientific Laboratory
¶Measured value based on reaching the sonic limit of mercury in the heat pipe.

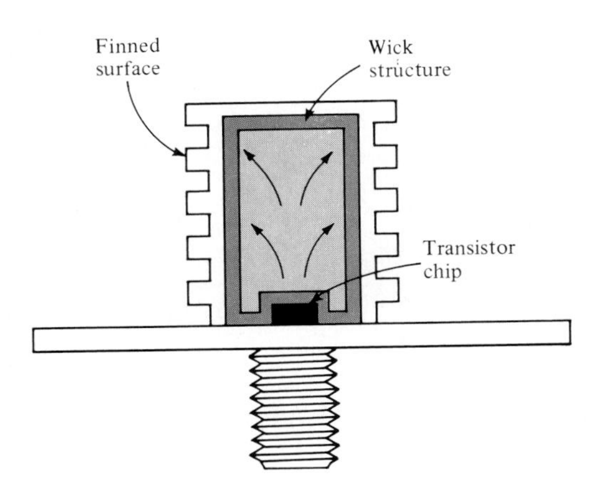

Fig. 12-20 Application of heat-pipe principle to cooling of power transistor, according to Ref. 15.

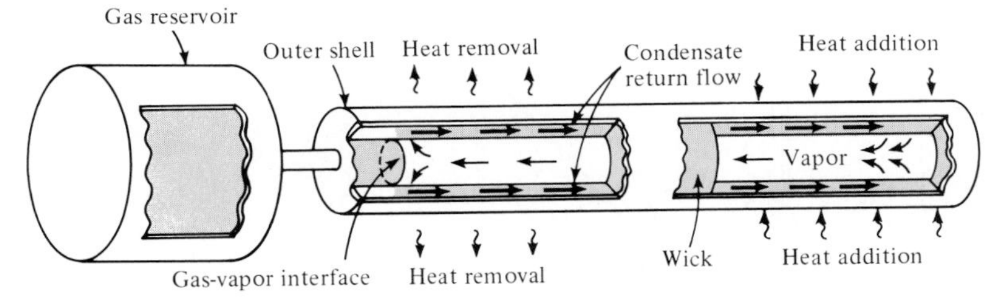

Fig. 12-21 Heat-pipe combined with noncondensible-gas reservoir to provide a temperature-control device, from Ref. 14.

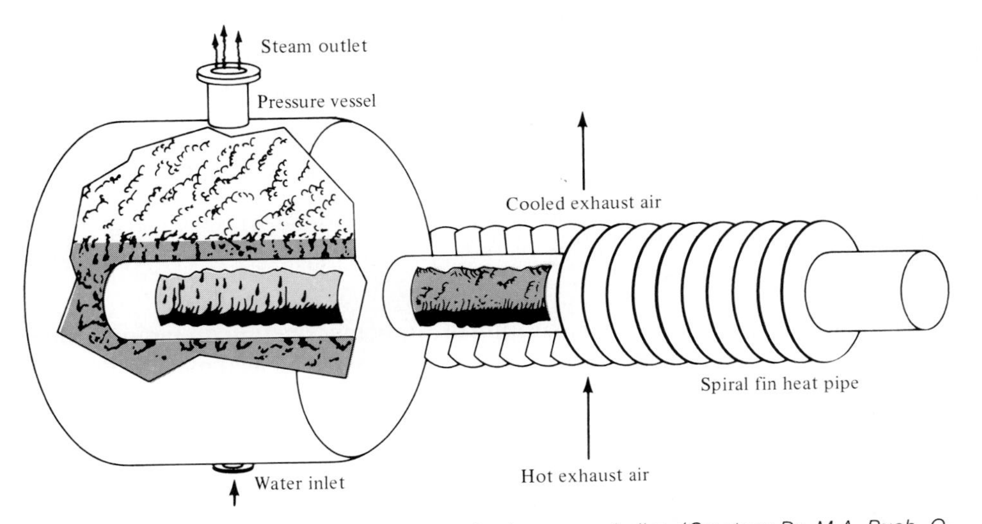

Fig. 12-22 Use of heat pipes to drive waste heat recovery boiler. (*Courtesy Dr. M.A. Ruch, Q-Dot Corporation, Dallas, Texas.*)

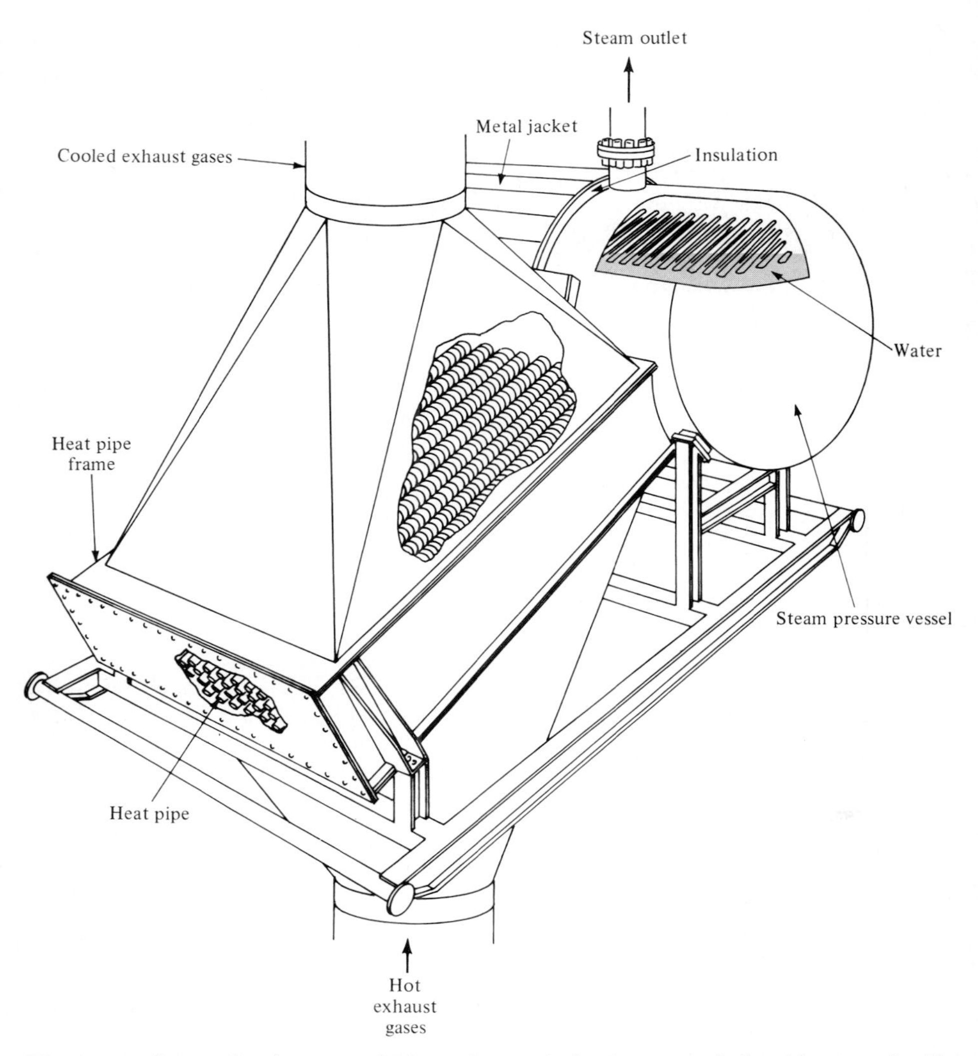

Fig. 12-23 Schematic of commercial heat-pipe waste heat recovery boiler. (*Courtesy Dr. M.A. Ruch, Q-Dot Corporation, Dallas, Texas.*)

observed. If the heat-source temperature drops below a certain minimum value, depending on the specific fluid and gas combinations in the heat pipe, a complete shutoff can occur. So the control feature can be particularly useful for fast warm-up applications in addition to its value as a temperature leveler for variable-load conditions.

Heat pipes are particularly useful in energy-conservation equipment. One example is shown in Fig. 12-22 where hot exhaust gases are used to drive a waste heat recovery boiler. The hot gases from a combustion process, which would otherwise have been discharged to the atmosphere, pass across one end of a finned heat pipe. The other end is submerged in the water of the boiler. A schematic of a commercial unit is shown in Fig. 12-23 where one notes that the heat pipes are inclined to maintain a proper flow of liquid and vapor.

The design of commercial heat pipes depends on many factors and is discussed in Refs. 19 and 20.

■ **EXAMPLE 12-4**

Compare the axial heat flux in a heat pipe using water as the working fluid at 200°C with that in a solid copper rod 8 cm long with a temperature differential of 100°C.

Solution

$$q'' = \frac{q}{A} = -k\frac{\Delta T}{\Delta x}$$

The thermal conductivity of copper is 374 W/m · °C, so that

$$q'' = \frac{(374)(-100)}{0.08} = 467.5 \text{ kW/m}^2 = 0.04675 \text{ kW/cm}^2$$

From Table 12-2 the typical axial heat flux for a water heat pipe is

$$q''_{\text{axial}} = 0.67 \text{ kW/cm}^2 \quad [2.12 \times 10^6 \text{ Btu/h} \cdot \text{ft}^2]$$

Thus this heat pipe transfers more than 10 times the heat of a pure copper rod with a substantial temperature gradient. This example illustrates why the heat pipe enjoys wide application possibilities.

■ REVIEW QUESTIONS

1. Under what conditions will a magnetic field cause an increase in heat transfer when impressed on a conducting fluid? A decrease in heat transfer?
2. How do the boundary conditions for transpiration cooling differ from those for ordinary flow over a flat plate?
3. How is a distinction made between continuum, slip, and free-molecule flows in a physical sense?
4. What is the Knudsen number?
5. What is the accommodation coefficient?
6. What is meant by a radiation equilibrium temperature?
7. How is ablation used for high-speed cooling?
8. What is a heat pipe? How does it work?
9. Why is the heat pipe so useful?
10. Describe how a heat pipe may be used as a temperature-control device.

PROBLEMS

12-1 Air at $-40°C$ and 0.7 kPa flows over a flat plate at $M = 4.0$. Calculate the adiabatic wall temperature for an injection parameter of 0.1 and at a position where $Re_x = 10^5$.

12-2 Calculate the heat-transfer rate q/A for the conditions of Prob. 12-1. Compare this with what would be experienced with no injection.

12-3 An airflow at 540°C and 1 atm impinges on the front side of a porous horizontal cylinder 7.5 cm in diameter at a velocity of 600 m/s. Air is injected through the porous material to maintain the surface temperature at 150°C. Calculate the heat-transfer rate at a distance of 0.75 cm from the stagnation point and with an injection parameter of 0.5.

12-4 Derive an expression for the ratio of the heat conducted through an air layer at low density to that conducted for $\lambda = 0$. Plot this ratio versus λ/L for $\alpha = 0.9$ and air properties evaluated at 35°C.

12-5 Develop an expression similar to Eq. (12-53) for low-density conduction between concentric cylindrical surfaces.

12-6 A fine iron wire 0.025 mm in diameter is exposed to a high-velocity airstream at 10^{-6} atm, $-50°C$, and $M = 5$. Estimate the wire surface temperature, assuming $\epsilon = 0.4$ and $\alpha = 0.9$. Assume the radiation-surroundings temperature is the free-stream temperature.

12-7 Suppose the wire of Prob. 12-6 is employed as a resistance thermometer to sense the free-stream temperature for the flow conditions stated. Plot the indicated temperature of the wire versus the actual free-stream temperature for the range -85 to $-15°C$.

12-8 A superinsulating material is to be constructed of polished aluminum sheets separated by a distance of 0.8 mm. The space between the sheets is sealed and evacuated to a pressure of 10^{-5} atm. Four sheets are used. The two outer sheets are maintained at 35 and 90°C and have a thickness of 0.75 mm, whereas the inner sheets have a thickness of 0.18 mm. Calculate the conduction and radiation transfer across the layered section per unit area. For this calculation, allow the inner sheets to "float" in the determination of the radiation heat transfer. Evaluate properties at 65°C.

12-9 Air flows over a flat plate at $p = 10^{-10}$ atm and $T_\infty = -60°C$. The velocity corresponds to $M = 5.0$, and the plate is 30 cm square. Calculate the radiation equilibrium temperature for an accommodation coefficient of 0.88 and surface emissivities of 0.3 and 0.8. Assume the radiation temperature of the surroundings is $-70°C$.

12-10 What cooling rate would be necessary to maintain the plate of Prob. 12-9 at a surface temperature of 65°C?

12-11 Two large polished plates are separated by a distance of 1.3 mm, and the space between them is evacuated to a pressure of 10^{-5} atm. The surface properties of the plates are $\alpha_1 = 0.87$, $\epsilon_1 = 0.08$, $\alpha_2 = 0.95$, $\epsilon_2 = 0.23$, where α is the accommodation coefficient. The plate temperatures are $T_1 = 70°C$ and $T_2 = 4°C$.

Calculate the total heat transfer between the plates by low-density conduction and radiation.

12-12 What total cross-sectional area of a liquid ammonia-aluminum heat pipe would be required to handle a heat load of 200,000 Btu/h?

12-13 A heated plate is placed in an electrically conducting fluid moving at high velocity such that the Eckert number is unity. A magnetic field is imposed on the flow of sufficient strength to double the hydrodynamic boundary layer in thickness. Estimate the percent reduction or increase in the heat-transfer rate as a result of adding the magnetic field.

12-14 Calculate the injection velocity for Prob. 12-1 at $\mathrm{Re}_x = 10^5$.

REFERENCES

1. Romig, M.: The Influence of Electric and Magnetic Fields on Heat Transfer to Electrically Conducting Fluids, *Adv. Heat Transfer,* vol. 1, pp. 268–352, 1964.
2. Sutton, G. W., and A. Sherman: "Engineering Magnetohydrodynamics," McGraw-Hill Book Company, New York, 1965.
3. Eckert, E. R. G., and J. P. Hartnett: Mass Transfer Cooling in a Laminar Boundary Layer with Constant Fluid Properties, *Trans. ASME,* vol. 79, pp. 247–254, 1957.
4. Sziklas, E. A.: An Analysis of the Compressible Laminar Boundary Layer with Foreign Gas Injection, *United Aircraft Corp. Res. Dept. Rep.* SR-0539-8, 1956.
5. Leontev, A. I.: Heat and Mass Transfer in Turbulent Boundary Layers, *Adv. Heat Transfer,* vol. 3, 1966.
6. Leadon, B. M., and C. J. Scott: Transpiration Cooling Experiments in a Turbulent Boundary Layer at $M = 3.0$, *J. Aeron. Sci.,* vol. 23, pp. 798–799, 1956.
7. Eckert, E. R. G., et al.: Mass Transfer Cooling of a Laminar Boundary Layer by Injection of a Light Weight Foreign Gas, *Jet Propulsion,* vol. 28, pp. 34–39, 1958.
8. Oppenheim, A. K.: Generalized Theory of Convective Heat Transfer in a Free-Molecule Flow, *J. Aeron. Sci.,* vol. 20, p. 49, 1953.
9. Wachman, H. Y.: The Thermal Accommodation Coefficient: A Critical Survey, *ARS J.,* vol. 32, p. 2, 1962.
10. Weidmann, M. L., and P. R. Trumpler: Thermal Accommodation Coefficients, *Trans. ASME,* vol. 68, p. 57, 1946.
11. Devienne, F. M.: Low Density Heat Transfer, *Adv. Heat Transfer,* vol. 2, p. 271, 1965.
12. Dorrance, W. H.: "Viscous Hypersonic Flow," McGraw-Hill Book Company, New York, 1962.
13. Springer, G. S.: Heat Transfer in Rarefied Gases: A Survey, Univ. Mich., *Fluid Dyn. Lab. Publ.* 69-1, May, 1969.
14. Feldman, K. T., and G. H. Whiting: Applications of the Heat Pipe, *Mech. Eng.* vol. 90, p. 48, November 1968.
15. Dutcher, C. H., and M. R. Burke: Heat Pipes: A Cool Way of Cooling Circuits, *Electronics,* pp. 93–100, February 16, 1970.

16 Grover, G. M., T. P. Cotter, and G. F. Erikson: Structures of Very High Thermal Conductance, *J. Appl. Phys.,* vol. 35, p. 1990, 1964.

17 Cotter, T. P.: Theory of Heat Pipes, *Los Alamos Sci. Lab. Rept.* La-3246-MS, February 1965.

18 Basiulis, A., and T. A. Hummel: The Application of Heat Pipe Techniques to Electronic Component Cooling, *ASME Pap.* 72-WA/HT-42.

19 Tien, C. L.: Fluid Mechanics of Heat Pipes, *Ann. Rev. Fluid Mechanics,* vol. 7, p. 167, 1975.

20 Chi, S. W.: ''Heat Pipe Theory and Practice,'' Hemisphere Publishing Co., New York, 1976.

TABLES

Table A-1 The Error Function

$\frac{x}{2\sqrt{\alpha\tau}}$	$\text{erf}\ \frac{x}{2\sqrt{\alpha\tau}}$	$\frac{x}{2\sqrt{\alpha\tau}}$	$\text{erf}\ \frac{x}{2\sqrt{\alpha\tau}}$	$\frac{x}{2\sqrt{\alpha\tau}}$	$\text{erf}\ \frac{x}{2\sqrt{\alpha\tau}}$
0.00	0.00000	0.76	0.71754	1.52	0.96841
0.02	0.02256	0.78	0.73001	1.54	0.97059
0.04	0.04511	0.80	0.74210	1.56	0.97263
0.06	0.06762	0.82	0.75381	1.58	0.97455
0.08	0.09008	0.84	0.76514	1.60	0.97636
0.10	0.11246	0.86	0.77610	1.62	0.97804
0.12	0.13476	0.88	0.78669	1.64	0.97962
0.14	0.15695	0.90	0.79691	1.66	0.98110
0.16	0.17901	0.92	0.80677	1.68	0.98249
0.18	0.20094	0.94	0.81627	1.70	0.98379
0.20	0.22270	0.96	0.82542	1.72	0.98500
0.22	0.24430	0.98	0.83423	1.74	0.98613
0.24	0.26570	1.00	0.84270	1.76	0.98719
0.26	0.28690	1.02	0.85084	1.78	0.98817
0.28	0.30788	1.04	0.85865	1.80	0.98909
0.30	0.32863	1.06	0.86614	1.82	0.98994
0.32	0.34913	1.08	0.87333	1.84	0.99074
0.34	0.36936	1.10	0.88020	1.86	0.99147
0.36	0.38933	1.12	0.88079	1.88	0.99216
0.38	0.40901	1.14	0.89308	1.90	0.99279
0.40	0.42839	1.16	0.89910	1.92	0.99338
0.42	0.44749	1.18	0.90484	1.94	0.99392
0.44	0.46622	1.20	0.91031	1.96	0.99443
0.46	0.48466	1.22	0.91553	1.98	0.99489
0.48	0.50275	1.24	0.92050	2.00	0.995322

Table A-1 The Error Function (*continued*)

$\frac{x}{2\sqrt{\alpha\tau}}$	erf $\frac{x}{2\sqrt{\alpha\tau}}$	$\frac{x}{2\sqrt{\alpha\tau}}$	erf $\frac{x}{2\sqrt{\alpha\tau}}$	$\frac{x}{2\sqrt{\alpha\tau}}$	erf $\frac{x}{2\sqrt{\alpha\tau}}$
0.50	0.52050	1.26	0.92524	2.10	0.997020
0.52	0.53790	1.28	0.92973	2.20	0.998137
0.54	0.55494	1.30	0.93401	2.30	0.998857
0.56	0.57162	1.32	0.93806	2.40	0.999311
0.58	0.58792	1.34	0.94191	2.50	0.999593
0.60	0.60386	1.36	0.94556	2.60	0.999764
0.62	0.61941	1.38	0.94902	2.70	0.999866
0.64	0.63459	1.40	0.95228	2.80	0.999925
0.66	0.64938	1.42	0.95538	2.90	0.999959
0.68	0.66278	1.44	0.95830	3.00	0.999978
0.70	0.67780	1.46	0.96105	3.20	0.999994
0.72	0.69143	1.48	0.96365	3.40	0.999998
0.74	0.70468	1.50	0.96610	3.60	1.000000

Table A-2 Property Values for Metals†

Metal	Properties at 20°C				Thermal conductivity k, W/m · °C									
	ρ, kg/m³	c_p, kJ/kg · °C	k, W/m · °C	α, m²/s × 10⁵	−100°C −148°F	0°C 32°F	100°C 212°F	200°C 392°F	300°C 572°F	400°C 752°F	600°C 1112°F	800°C 1472°F	1000°C 1832°F	1200°C 2192°F
Aluminum:														
Pure	2,707	0.896	204	8.418	215	202	206	215	228	249				
Al-Cu (Duralumin), 94–96% Al, 3–5% Cu, trace Mg	2,787	0.883	164	6.676	126	159	182	194						
Al-Si (Silumin, copper-bearing), 86.5% Al, 1% Cu	2,659	0.867	137	5.933	119	137	144	152	161					
Al-Si (Alusil), 78–80% Al, 20–22% Si	2,627	0.854	161	7.172	144	157	168	175	178					
Al-Mg-Si, 97% Al, 1% Mg, 1% Si, 1% Mn	2,707	0.892	177	7.311		175	189	204						
Lead	11,373	0.130	35	2.343	36.9	35.1	33.4	31.5	29.8					
Iron:														
Pure	7,897	0.452	73	2.034	87	73	67	62	55	48	40	36	35	36
Wrought iron, 0.5% C	7,849	0.46	59	1.626		59	57	52	48	45	36	33	33	33
Steel (C max ≈ 1.5%):														
Carbon steel														
C ≈ 0.5%	7,833	0.465	54	1.474		55	52	48	45	42	35	31	29	31
1.0%	7,801	0.473	43	1.172		43	43	42	40	36	33	29	28	29
1.5%	7,753	0.486	36	0.970		36	36	36	35	33	31	28	28	29
Nickel steel														
Ni ≈ 0%	7,897	0.452	73	2.026										
20%	7,933	0.46	19	0.526										

Table A-2 Property Values for Metals† (*continued*)

Metal	Properties at 20°C				Thermal conductivity k, W/m · °C									
	ρ, kg/m^3	c_p, kJ/kg · °C	k, W/m · °C	α, m^2/s × 10^5	−100°C −148°F	0°C 32°F	100°C 212°F	200°C 392°F	300°C 572°F	400°C 752°F	600°C 1112°F	800°C 1472°F	1000°C 1832°F	1200°C 2192°F
40%	8,169	0.46	10	0.279										
80%	8,618	0.46	35	0.872										
Invar 36% Ni	8,137	0.46	10.7	0.286										
Chrome steel														
Cr = 0%	7,897	0.452	73	2.026	87	73	67	62	55	48	40	36	35	36
1%	7,865	0.46	61	1.665		62	55	52	47	42	36	33	33	
5%	7,833	0.46	40	1.110		40	38	36	36	33	29	29	29	
20%	7,689	0.46	22	0.635		22	22	22	22	24	24	26	29	
Cr-Ni(chrome-nickel): 15% Cr, 10% Ni	7,865	0.46	19	0.527										
18% Cr, 8% Ni (V2A)	7,817	0.46	16.3	0.444		16.3	17	17	19	19	22	27	31	
20% Cr, 15% Ni	7,833	0.46	15.1	0.415										
25% Cr, 20% Ni	7,865	0.46	12.8	0.361										
Tungsten steel														
W = 0%	7,897	0.452	73	2.026										
1%	7,913	0.448	66	1.858										
5%	8,073	0.435	54	1.525										
10%	8,314	0.419	48	1.391										
Copper:														
Pure	8,954	0.3831	386	11.234	407	386	379	374	369	363	353			

Aluminum bronze 95% Cu, 5% Al	8,666	0.410	83	2.330										
Bronze 75% Cu, 25% Sn	8,666	0.343	26	0.859										
Red brass 85% Cu, 9% Sn, 6% Zn	8,714	0.385	61	1.804		59	71							
Brass 70% Cu, 30% Zn	8,522	0.385	111	3.412	88		128	144	147	147				
German silver 62% Cu, 15% Ni, 22% Zn	8,618	0.394	24.9	0.733	19.2		31	40	45	48				
Constantan 60% Cu, 40% Ni	8,922	0.410	22.7	0.612	21		22.2	26						
Magnesium:														
Pure	1,746	1.013	171	9.708	178	171	168	163	157					
Mg-Al (electrolytic) 6–8% Al, 1–2% Zn	1,810	1.00	66	3.605		52	62	74	83					
Molybdenum	10,220	0.251	123	4.790	138	125	118	114	111	109	106	102	99	92
Nickel:														
Pure (99.9%)	8,906	0.4459	90	2.266	104	93	83	73	64	59				
Ni-Cr 90% Ni, 10% Cr	8.666	0.444	17	0.444		17.1	18.9	20.9	22.8	24.6				
80% Ni, 20% Cr	8,314	0.444	12.6	0.343		12.3	13.8	15.6	17.1	18.0	22.5			
Silver:														
Purest	10,524	0.2340	419	17.004	419	417	415	412						
Pure (99.9%)	10,525	0.2340	407	16.563	419	410	415	374	362	360				
Tin, pure	7,304	0.2265	64	3.884	74	65.9	59	57						
Tungsten	19,350	0.1344	163	6.271		166	151	142	133	126	112	76		
Zinc, pure	7,144	0.3843	112.2	4.106	114	112	109	106	100	93				

†Adapted to SI units from E. R. G. Eckert and R. M. Drake, "Heat and Mass Transfer," 2d ed., McGraw-Hill Book Company, New York, 1959.

Table A-3 Properties of Nonmetals†

Substance	*Temperature,* °C	*k,* W/m · °C	*ρ,* kg/m³	*c,* kJ/kg · °C	*α,* m²/s × 10⁷
		Structural and heat-resistant materials			
Asphalt	20–55	0.74–0.76			
Brick:					
Building brick, common	20	0.69	1600	0.84	5.2
Face		1.32	2000		
Carborundum brick	600	18.5			
	1400	11.1			
Chrome brick	200	2.32	3000	0.84	9.2
	550	2.47			9.8
	900	1.99			7.9
Diatomaceous earth, molded and fired	200	0.24			
	870	0.31			
Fireclay brick, burnt 2426°F	500	1.04	2000	0.96	5.4
	800	1.07			
	1100	1.09			
Burnt 2642°F	500	1.28	2300	0.96	5.8
	800	1.37			
	1100	1.40			
Missouri	200	1.00	2600	0.96	4.0
	600	1.47			
	1400	1.77			
Magnesite	200	3.81		1.13	
	650	2.77			
	1200	1.90			
Cement, portland		0.29	1500		
Mortar	23	1.16			
Concrete, cinder	23	0.76			
Stone, 1-2-4 mix	20	1.37	1900–2300	0.88	8.2–6.8
Glass, window	20	0.78 (avg)	2700	0.84	3.4
Corosilicate	30–75	1.09	2200		
Plaster, gypsum	20	0.48	1440	0.84	4.0
Metal lath	20	0.47			
Wood lath	20	0.28			
Stone:					
Granite		1.73–3.98	2640	0.82	8–18
Limestone	100–300	1.26–1.33	2500	0.90	5.6–5.9
Marble		2.07–2.94	2500–2700	0.80	10–13.6
Sandstone	40	1.83	2160–2300	0.71	11.2–11.9
Wood (across the grain):					
Balsa, 8.8 lb/ft³	30	0.055	140		
Cypress	30	0.097	460		
Fir	23	0.11	420	2.72	0.96
Maple or oak	30	0.166	540	2.4	1.28
Yellow pine	23	0.147	640	2.8	0.82
White pine	30	0.112	430		

Table A-3 Properties of Nonmetals† (*continued*)

Substance	*Temperature,* °C	*k,* W/m · °C	*ρ,* kg/m³	*c,* kJ/kg · °C	*α,* m²/s × 10⁷
		Insulating material			
Asbestos:					
Loosely packed	−45	0.149			
	0	0.154	470–570	0.816	3.3–4
	100	0.161			
Asbestos-cement boards	20	0.74			
Sheets	51	0.166			
Felt, 40 laminations/in	38	0.057			
	150	0.069			
	260	0.083			
20 laminations/in	38	0.078			
	150	0.095			
	260	0.112			
Corrugated, 4 plies/in	38	0.087			
	93	0.100			
	150	0.119			
Asbestos cement	—	2.08			
Balsam wool, 2.2 lb/ft³	32	0.04	35		
Cardboard, corrugated	—	0.064			
Celotex	32	0.048			
Corkboard, 10 lb/ft³	30	0.043	160		
Cork, regranulated	32	0.045	45–120	1.88	2–5.3
Ground	32	0.043	150		
Diatomaceous earth (Sil-o-cel)	0	0.061	320		
Felt, hair	30	0.036	130–200		
Wool	30	0.052	330		
Fiber, insulating board	20	0.048	240		
Glass wool, 1.5 lb/ft³	23	0.038	24	0.7	22.6
Insulex, dry	32	0.064			
		0.144			
Kapok	30	0.035			
Magnesia, 85%	38	0.067	270		
	93	0.071			
	150	0.074			
	204	0.080			
Rock wool, 10 lb/ft³	32	0.040	160		
Loosely packed	150	0.067	64		
	260	0.087			
Sawdust	23	0.059			
Silica aerogel	32	0.024	140		
Wood shavings	23	0.059			

†Adapted to SI units from A. I. Brown and S. M. Marco, "Introduction to Heat Transfer," 3d ed., McGraw-Hill Book Company, New York, 1958.

Table A-4 Properties of Saturated Liquids†

T, °C	ρ, kg/m³	c_p, kJ/kg · °C	ν, m²/s	k, W/m · °C	α, m²/s	Pr	β, K⁻¹
			Ammonia, NH_3				
−50	703.69	4.463	0.435×10^{-6}	0.547	1.742×10^{-7}	2.60	
−40	691.68	4.467	0.406	0.547	1.775	2.28	
−30	679.34	4.476	0.387	0.549	1.801	2.15	
−20	666.69	4.509	0.381	0.547	1.819	2.09	
−10	653.55	4.564	0.378	0.543	1.825	2.07	
0	640.10	4.635	0.373	0.540	1.819	2.05	
10	626.16	4.714	0.368	0.531	1.801	2.04	
20	611.75	4.798	0.359	0.521	1.775	2.02	2.45×10^{-3}
30	596.37	4.890	0.349	0.507	1.742	2.01	
40	580.99	4.999	0.340	0.493	1.701	2.00	
50	564.33	5.116	0.330	0.476	1.654	1.99	
			Carbon dioxide, CO_2				
−50	1,156.34	1.84	0.119×10^{-6}	0.0855	0.4021×10^{-7}	2.96	
−40	1,117.77	1.88	0.118	0.1011	0.4810	2.46	
−30	1,076.76	1.97	0.117	0.1116	0.5272	2.22	
−20	1,032.39	2.05	0.115	0.1151	0.5445	2.12	
−10	983.38	2.18	0.113	0.1099	0.5133	2.20	
0	926.99	2.47	0.108	0.1045	0.4578	2.38	
10	860.03	3.14	0.101	0.0971	0.3608	2.80	
20	772.57	5.0	0.091	0.0872	0.2219	4.10	14.00×10^{-3}
30	597.81	36.4	0.080	0.0703	0.0279	28.7	
			Sulfur dioxide, SO_2				
−50	1,560.84	1.3595	0.484×10^{-6}	0.242	1.141×10^{-7}	4.24	
−40	1,536.81	1.3607	0.424	0.235	1.130	3.74	
−30	1,520.64	1.3616	0.371	0.230	1.117	3.31	
−20	1,488.60	1.3624	0.324	0.225	1.107	2.93	
−10	1,463.61	1.3628	0.288	0.218	1.097	2.62	
0	1,438.46	1.3636	0.257	0.211	1.081	2.38	
10	1,412.51	1.3645	0.232	0.204	1.066	2.18	
20	1,386.40	1.3653	0.210	0.199	1.050	2.00	1.94×10^{-3}
30	1,359.33	1.3662	0.190	0.192	1.035	1.83	
40	1,329.22	1.3674	0.173	0.185	1.019	1.70	
50	1,299.10	1.3683	0.162	0.177	0.999	1.61	

Table A-4 Properties of Saturated Liquids† (*continued*)

T, °C	ρ, kg/m³	c_p, kJ/kg · °C	ν, m²/s	k, W/m · °C	α, m²/s	Pr	β, K⁻¹
			Dichlorodifluoromethane (Freon), CCl_2F_2				
−50	1,546.75	0.8750	0.310×10^{-6}	0.067	0.501×10^{-7}	6.2	2.63×10^{-3}
−40	1,518.71	0.8847	0.279	0.069	0.514	5.4	
−30	1,489.56	0.8956	0.253	0.069	0.526	4.8	
−20	1,460.57	0.9073	0.235	0.071	0.539	4.4	
−10	1,429.49	0.9203	0.221	0.073	0.550	4.0	
0	1,397.45	0.9345	0.214×10^{-6}	0.073	0.557×10^{-7}	3.8	
10	1,364.30	0.9496	0.203	0.073	0.560	3.6	
20	1,330.18	0.9659	0.198	0.073	0.560	3.5	
30	1,295.10	0.9835	0.194	0.071	0.560	3.5	
40	1,257.13	1.0019	0.191	0.069	0.555	3.5	
50	1,215.96	1.0216	0.190	0.067	0.545	3.5	
			Glycerin, $C_3H_5(OH)_3$				
0	1,276.03	2.261	0.00831	0.282	0.983×10^{-7}	84.7×10^3	
10	1,270.11	2.319	0.00300	0.284	0.965	31.0	
20	1,264.02	2.386	0.00118	0.286	0.947	12.5	0.50×10^{-3}
30	1,258.09	2.445	0.00050	0.286	0.929	5.38	
40	1,252.01	2.512	0.00022	0.286	0.914	2.45	
50	1,244.96	2.583	0.00015	0.287	0.893	1.63	
			Ethylene glycol, $C_2H_4(OH)_2$				
0	1,130.75	2.294	57.53×10^{-6}	0.242	0.934×10^{-7}	615	
20	1,116.65	2.382	19.18	0.249	0.939	204	0.65×10^{-3}
40	1,101.43	2.474	8.69	0.256	0.939	93	
60	1,087.66	2.562	4.75	0.260	0.932	51	
80	1,077.56	2.650	2.98	0.261	0.921	32.4	
100	1,058.50	2.742	2.03	0.263	0.908	22.4	
			Engine oil (unused)				
0	899.12	1.796	0.00428	0.147	0.911×10^{-7}	47,100	
20	888.23	1.880	0.00090	0.145	0.872	10,400	0.70×10^{-3}
40	876.05	1.964	0.00024	0.144	0.834	2,870	
60	864.04	2.047	0.839×10^{-4}	0.140	0.800	1,050	
80	852.02	2.131	0.375	0.138	0.769	490	
100	840.01	2.219	0.203	0.137	0.738	276	
120	828.96	2.307	0.124	0.135	0.710	175	
140	816.94	2.395	0.080	0.133	0.686	116	
160	805.89	2.483	0.056	0.132	0.663	84	

Table A-4 Properties of Saturated Liquids† (*continued*)

T, °C	ρ, kg/m³	c_p, kJ/kg · °C	ν, m²/s	k, W/m · °C	α, m²/s	Pr	β, K⁻¹
				Mercury, Hg			
0	13,628.22	0.1403	0.124×10^{-6}	8.20	42.99×10^{7}	0.0288	
20	13,579.04	0.1394	0.114	8.69	46.06	0.0249	1.82×10^{-4}
50	13,505.84	0.1386	0.104	9.40	50.22	0.0207	
100	13,384.58	0.1373	0.0928	10.51	57.16	0.0162	
150	13,264.28	0.1365	0.0853	11.49	63.54	0.0134	
200	13,144.94	0.1570	0.0802	12.34	69.08	0.0116	
250	13,025.60	0.1357	0.0765	13.07	74.06	0.0103	
315.5	12,847	0.134	0.0673	14.02	81.5	0.0083	

†Adapted to SI units from E. R. G. Eckert and R. M. Drake, "Heat and Mass Transfer," 2d ed., McGraw-Hill Book Company, New York, 1959.

Table A-5 Properties of Air at Atmospheric Pressure†
The values of μ, k, c_p, and Pr are not strongly pressure-dependent and may be used over a fairly wide range of pressures.

T, K	ρ kg/m^3	c_p, kJ/kg · °C	μ, kg/m · s × 10^5	ν, m^2/s × 10^6	k, W/m · °C	α, m^2/s × 10^4	Pr
100	3.6010	1.0266	0.6924	1.923	0.009246	0.02501	0.770
150	2.3675	1.0099	1.0283	4.343	0.013735	0.05745	0.753
200	1.7684	1.0061	1.3289	7.490	0.01809	0.10165	0.739
250	1.4128	1.0053	1.5990	11.31	0.02227	0.15675	0.722
300	1.1774	1.0057	1.8462	15.69	0.02624	0.22160	0.708
350	0.9980	1.0090	2.075	20.76	0.03003	0.2983	0.697
400	0.8826	1.0140	2.286	25.90	0.03365	0.3760	0.689
450	0.7833	1.0207	2.484	31.71	0.03707	0.4222	0.683
500	0.7048	1.0295	2.671	37.90	0.04038	0.5564	0.680
550	0.6423	1.0392	2.848	44.34	0.04360	0.6532	0.680
600	0.5879	1.0551	3.018	51.34	0.04659	0.7512	0.680
650	0.5430	1.0635	3.177	58.51	0.04953	0.8578	0.682
700	0.5030	1.0752	3.332	66.25	0.05230	0.9672	0.684
750	0.4709	1.0856	3.481	73.91	0.05509	1.0774	0.686
800	0.4405	1.0978	3.625	82.29	0.05779	1.1951	0.689
850	0.4149	1.1095	3.765	90.75	0.06028	1.3097	0.692
900	0.3925	1.1212	3.899	99.3	0.06279	1.4271	0.696
950	0.3716	1.1321	4.023	108.2	0.06525	1.5510	0.699
1000	0.3524	1.1417	4.152	117.8	0.06752	1.6779	0.702
1100	0.3204	1.160	4.44	138.6	0.0732	1.969	0.704
1200	0.2947	1.179	4.69	159.1	0.0782	2.251	0.707
1300	0.2707	1.197	4.93	182.1	0.0837	2.583	0.705
1400	0.2515	1.214	5.17	205.5	0.0891	2.920	0.705
1500	0.2355	1.230	5.40	229.1	0.0946	3.262	0.705
1600	0.2211	1.248	5.63	254.5	0.100	3.609	0.705
1700	0.2082	1.267	5.85	280.5	0.105	3.977	0.705
1800	0.1970	1.287	6.07	308.1	0.111	4.379	0.704
1900	0.1858	1.309	6.29	338.5	0.117	4.811	0.704
2000	0.1762	1.338	6.50	369.0	0.124	5.260	0.702
2100	0.1682	1.372	6.72	399.6	0.131	5.715	0.700
2200	0.1602	1.419	6.93	432.6	0.139	6.120	0.707
2300	0.1538	1.482	7.14	464.0	0.149	6.540	0.710
2400	0.1458	1.574	7.35	504.0	0.161	7.020	0.718
2500	0.1394	1.688	7.57	543.5	0.175	7.441	0.730

†From *Natl. Bur. Stand. (U.S.) Circ.* 564, 1955.

Table A-6 Properties of Gases at Atmospheric Pressure†

Values of μ, k, c_p, and Pr are not strongly pressure-dependent for He, H_2, O_2, and N_2 and may be used over a fairly wide range of pressures.

T, K	ρ, kg/m³	c_p, kJ/kg · °C	μ, kg/m · s	ν, m²/s	k, W/m · °C	α, m²/s	Pr
				Helium			
144	0.3379	5.200	125.5×10^{-7}	37.11×10^{-6}	0.0928	0.5275×10^{-4}	0.70
200	0.2435	5.200	156.6	64.38	0.1177	0.9288	0.694
255	0.1906	5.200	181.7	95.50	0.1357	1.3675	0.70
366	0.13280	5.200	230.5	173.6	0.1691	2.449	0.71
477	0.10204	5.200	275.0	269.3	0.197	3.716	0.72
589	0.08282	5.200	311.3	375.8	0.225	5.215	0.72
700	0.07032	5.200	347.5	494.2	0.251	6.661	0.72
800	0.06023	5.200	381.7	634.1	0.275	8.774	0.72
				Hydrogen			
150	0.16371	12.602	5.595×10^{-6}	34.18×10^{-6}	0.0981	0.475×10^{-4}	0.718
200	0.12270	13.540	6.813	55.53	0.1282	0.772	0.719
250	0.09819	14.059	7.919	80.64	0.1561	1.130	0.713
300	0.08185	14.314	8.963	109.5	0.182	1.554	0.706
350	0.07016	14.436	9.954	141.9	0.206	2.031	0.697
400	0.06135	14.491	10.864	177.1	0.228	2.568	0.690
450	0.05462	14.499	11.779	215.6	0.251	3.164	0.682
500	0.04918	14.507	12.636	257.0	0.272	3.817	0.675
550	0.04469	14.532	13.475	301.6	0.292	4.516	0.668
600	0.04085	14.537	14.285	349.7	0.315	5.306	0.664
700	0.03492	14.574	15.89	455.1	0.351	6.903	0.659
800	0.03060	14.675	17.40	569	0.384	8.563	0.664
900	0.02723	14.821	18.78	690	0.412	10.217	0.676
				Oxygen			
150	2.6190	0.9178	11.490×10^{-6}	4.387×10^{-6}	0.01367	0.05688×10^{-4}	0.773
200	1.9559	0.9131	14.850	7.593	0.01824	0.10214	0.745
250	1.5618	0.9157	17.87	11.45	0.02259	0.15794	0.725
300	1.3007	0.9203	20.63	15.86	0.02676	0.22353	0.709
350	1.1133	0.9291	23.16	20.80	0.03070	0.2968	0.702
400	0.9755	0.9420	25.54	26.18	0.03461	0.3768	0.695
450	0.8682	0.9567	27.77	31.99	0.03828	0.4609	0.694
500	0.7801	0.9722	29.91	38.34	0.04173	0.5502	0.697
550	0.7096	0.9881	31.97	45.05	0.04517	0.641	0.700

Table A-6 Properties of Gases at Atmospheric Pressure† *(continued)*
Values of μ, k, c_p, and Pr are not strongly pressure-dependent for He, H_2, O_2, and N_2 and may be used over a fairly wide range of pressures.

T, K	ρ, kg/m³	c_p, kJ/kg · °C	μ, kg/m · s	ν, m²/s	k, W/m · °C	α, m²/s	Pr
			Nitrogen				
200	1.7108	1.0429	12.947×10^{-6}	7.568×10^{-6}	0.01824	0.10224×10^{-4}	0.747
300	1.1421	1.0408	17.84	15.63	0.02620	0.22044	0.713
400	0.8538	1.0459	21.98	25.74	0.03335	0.3734	0.691
500	0.6824	1.0555	25.70	37.66	0.03984	0.5530	0.684
600	0.5687	1.0756	29.11	51.19	0.04580	0.7486	0.686
700	0.4934	1.0969	32.13	65.13	0.05123	0.9466	0.691
800	0.4277	1.1225	34.84	81.46	0.05609	1.1685	0.700
900	0.3796	1.1464	37.49	91.06	0.06070	1.3946	0.711
1000	0.3412	1.1677	40.00	117.2	0.06475	1.6250	0.724
1100	0.3108	1.1857	42.28	136.0	0.06850	1.8591	0.736
1200	0.2851	1.2037	44.50	156.1	0.07184	2.0932	0.748
			Carbon dioxide				
220	2.4733	0.783	11.105×10^{-6}	4.490×10^{-6}	0.010805	0.05920×10^{-4}	0.818
250	2.1657	0.804	12.590	5.813	0.012884	0.07401	0.793
300	1.7973	0.871	14.958	8.321	0.016572	0.10588	0.770
350	1.5362	0.900	17.205	11.19	0.02047	0.14808	0.755
400	1.3424	0.942	19.32	14.39	0.02461	0.19463	0.738
450	1.1918	0.980	21.34	17.90	0.02897	0.24813	0.721
500	1.0732	1.013	23.26	21.67	0.03352	0.3084	0.702
550	0.9739	1.047	25.08	25.74	0.03821	0.3750	0.685
600	0.8938	1.076	26.83	30.02	0.04311	0.4483	0.668
			Ammonia, NH_3				
273	0.7929	2.177	9.353×10^{-6}	1.18×10^{-5}	0.0220	0.1308×10^{-4}	0.90
323	0.6487	2.177	11.035	1.70	0.0270	0.1920	0.88
373	0.5590	2.236	12.886	2.30	0.0327	0.2619	0.87
423	0.4934	2.315	14.672	2.97	0.0391	0.3432	0.87
473	0.4405	2.395	16.49	3.74	0.0467	0.4421	0.84
			Water vapor				
380	0.5863	2.060	12.71×10^{-6}	2.16×10^{-5}	0.0246	0.2036×10^{-4}	1.060
400	0.5542	2.014	13.44	2.42	0.0261	0.2338	1.040
450	0.4902	1.980	15.25	3.11	0.0299	0.307	1.010
500	0.4405	1.985	17.04	3.86	0.0339	0.387	0.996
550	0.4005	1.997	18.84	4.70	0.0379	0.475	0.991
600	0.3652	2.026	20.67	5.66	0.0422	0.573	0.986
650	0.3380	2.056	22.47	6.64	0.0464	0.666	0.995
700	0.3140	2.085	24.26	7.72	0.0505	0.772	1.000
750	0.2931	2.119	26.04	8.88	0.0549	0.883	1.005
800	0.2739	2.152	27.86	10.20	0.0592	1.001	1.010
850	0.2579	2.186	29.69	11.52	0.0637	1.130	1.019

†Adapted to SI units from E. R. G. Eckert and R. M. Drake, "Heat and Mass Transfer," 2nd ed., McGraw-Hill Book Company, New York, 1959.

Table A-7 Physical Properties of Some Common Low-Melting-Point Metals†

Metal	*Melting point,* °C	*Normal boiling point,* °C	*Temperature,* °C	*Density,* kg/m³ × 10⁻³	*Viscosity* kg/m · s × 10³	*Heat capacity,* kJ/kg · °C	*Thermal conductivity,* W/m · °C	*Prandtl number*
Bismuth	271	1477	316	10.01	1.62	0.144	16.4	0.014
			760	9.47	0.79	0.165	15.6	0.0084
Lead	327	1737	371	10.5	2.40	0.159	16.1	0.024
			704	10.1	1.37	0.155	14.9	0.016
Lithium	179	1317	204	0.51	0.60	4.19	38.1	0.065
			982	0.44	0.42	4.19		
Mercury	−39	357	10	13.6	1.59	0.138	8.1	0.027
			316	12.8	0.86	0.134	14.0	0.0084
Potassium	63.8	760	149	0.81	0.37	0.796	45.0	0.0066
			704	0.67	0.14	0.754	33.1	0.0031
Sodium	97.8	883	204	0.90	0.43	1.34	80.3	0.0072
			704	0.78	0.18	1.26	59.7	0.0038
Sodium potassium: 22% Na	19	826	93.3	0.848	0.49	0.946	24.4	0.019
			760	0.69	0.146	0.883		
56% Na	−11	784	93.3	0.89	0.58	1.13	25.6	0.026
			760	0.74	0.16	1.04	28.9	0.058
Lead bismuth, 44.5% Pb	125	1670	288	10.3	1.76	0.147	10.7	0.024
			649	9.84	1.15			

†Adapted to SI units from J. G. Knudsen and D. L. Katz, "Fluid Dynamics and Heat Transfer," McGraw-Hill Book Company, New York, 1958.

Table A-8 Diffusion coefficients of gases and vapors in air at 25°C and 1 atm†

Substance	*D*, cm²/s	$Sc = \frac{\nu}{D}$	*Substance*	*D*, cm²/s	$Sc = \frac{\nu}{D}$
Ammonia	0.28	0.78	Formic acid	0.159	0.97
Carbon dioxide	0.164	0.94	Acetic acid	0.133	1.16
Hydrogen	0.410	0.22	Aniline	0.073	2.14
Oxygen	0.206	0.75	Benzene	0.088	1.76
Water	0.256	0.60	Toluene	0.084	1.84
Ethyl ether	0.093	1.66	Ethyl benzene	0.077	2.01
Methanol	0.159	0.97	Propyl benzene	0.059	2.62
Ethyl alcohol	0.119	1.30			

†From J. H. Perry (ed.), "Chemical Engineers' Handbook," 4th ed., McGraw-Hill Book Company, New York, 1963.

Table A-9 Properties of Water (Saturated Liquid)†

Note: $Gr_x Pr = \left(\frac{g\beta\rho^2 c_p}{\mu k}\right) x^3 \Delta T$

°F	°C	c_p, kJ/kg · °C	ρ, kg/m³	μ, kg/m · s	k, W/m · °C	Pr	$\frac{g\beta\rho^2 c_p}{\mu k}$, 1/m³ · °C
32	0	4.225	999.8	1.79×10^{-3}	0.566	13.25	
40	4.44	4.208	999.8	1.55	0.575	11.35	1.91×10^{9}
50	10	4.195	999.2	1.31	0.585	9.40	6.34×10^{9}
60	15.56	4.186	998.6	1.12	0.595	7.88	1.08×10^{10}
70	21.11	4.179	997.4	9.8×10^{-4}	0.604	6.78	1.46×10^{10}
80	26.67	4.179	995.8	8.6	0.614	5.85	1.91×10^{10}
90	32.22	4.174	994.9	7.65	0.623	5.12	2.48×10^{10}
100	37.78	4.174	993.0	6.82	0.630	4.53	3.3×10^{10}
110	43.33	4.174	990.6	6.16	0.637	4.04	4.19×10^{10}
120	48.89	4.174	988.8	5.62	0.644	3.64	4.89×10^{10}
130	54.44	4.179	985.7	5.13	0.649	3.30	5.66×10^{10}
140	60	4.179	983.3	4.71	0.654	3.01	6.48×10^{10}
150	65.55	4.183	980.3	4.3	0.659	2.73	7.62×10^{10}
160	71.11	4.186	977.3	4.01	0.665	2.53	8.84×10^{10}
170	76.67	4.191	973.7	3.72	0.668	2.33	9.85×10^{10}
180	82.22	4.195	970.2	3.47	0.673	2.16	1.09×10^{11}
190	87.78	4.199	966.7	3.27	0.675	2.03	
200	93.33	4.204	963.2	3.06	0.678	1.90	
220	104.4	4.216	955.1	2.67	0.684	1.66	
240	115.6	4.229	946.7	2.44	0.685	1.51	
260	126.7	4.250	937.2	2.19	0.685	1.36	
280	137.8	4.271	928.1	1.98	0.685	1.24	
300	148.9	4.296	918.0	1.86	0.684	1.17	
350	176.7	4.371	890.4	1.57	0.677	1.02	
400	204.4	4.467	859.4	1.36	0.665	1.00	
450	232.2	4.585	825.7	1.20	0.646	0.85	
500	260	4.731	785.2	1.07	0.616	0.83	
550	287.7	5.024	735.5	9.51×10^{-5}			
600	315.6	5.703	678.7	8.68			

†Adapted from A. I. Brown and S. M. Marco, "Introduction to Heat Transfer," 3d ed., McGraw-Hill Book Company, New York, 1958.

Table A-10 Normal Total Emissivity of Various Surfaces†

Surface	*T*, °F	*Emissivity* ϵ
Metals and their oxides		
Aluminum:		
Highly polished plate, 98.3% pure	440–1070	0.039–0.057
Commercial sheet	212	0.09
Heavily oxidized	299–940	0.20–0.31
Al-surfaced roofing	100	0.216
Brass:		
Highly polished:		
73.2% Cu, 26.7% Zn	476–674	0.028–0.031
62.4% Cu, 36.8% Zn, 0.4% Pb, 0.3% Al	494–710	0.033–0.037
82.9% Cu, 17.0% Zn	530	0.030
Hard-rolled, polished, but direction of polishing visible	70	0.038
Dull plate	120–660	0.22
Chromium (see nickel alloys for Ni-Cr steels), polished	100–2000	0.08–0.36
Copper:		
Polished	242	0.023
	212	0.052
Plate, heated long time, covered with thick oxide layer	77	0.78
Gold, pure, highly polished	440–1160	0.018–0.035
Iron and steel (not including stainless):		
Steel, polished	212	0.066
Iron, polished	800–1880	0.14–0.38
Cast iron, newly turned	72	0.44
turned and heated	1620–1810	0.60–0.70
Mild steel	450–1950	0.20–0.32
Oxidized surfaces:		
Iron plate, pickled, then rusted red	68	0.61
Iron, dark-gray surface	212	0.31
Rough ingot iron	1700–2040	0.87–0.95
Sheet steel with strong, rough oxide layer	75	0.80
Lead:		
Unoxidized, 99.96% pure	260–440	0.057–0.075
Gray oxidized	75	0.28
Oxidized at 300°F	390	0.63
Magnesium, magnesium oxide	530–1520	0.55–0.20
Molybdenum:		
Filament	1340–4700	0.096–0.202
Massive, polished	212	0.071
Monel metal, oxidized at 1110°F	390–1110	0.41–0.46
Nickel:		
Polished	212	0.072
Nickel oxide	1200–2290	0.59–0.86
Nickel alloys:		
Copper nickel, polished	212	0.059
Nichrome wire, bright	120–1830	0.65–0.79
Nichrome wire, oxidized	120–930	0.95–0.98
Platinum, polished plate, pure	440–1160	0.054–0.104
Silver:		
Polished, pure	440–1160	0.020–0.032
Polished	100–700	0.022–0.031

Table A-10 Normal Total Emissivity of Various Surfaces† *(continued)*

Surface	*T*, °F	*Emissivity* ϵ
Metals and their oxides		
Stainless steels:		
Polished	212	0.074
Type 301; B	450–1725	0.54–0.63
Tin, bright tinned iron	76	0.043 and 0.064
Tungsten, filament	6000	0.39
Zinc, galvanized sheet iron, fairly bright	82	0.23
Refractories, building materials, paints, and miscellaneous		
Alumina (85–99.5%, Al_2O_3, 0–12% SiO_2, 0–1% Ge_2O_3); effect of mean grain size, microns (μm):		
10 μm		0.30–0.18
50 μm		0.39–0.28
100 μm		0.50–0.40
Asbestos, board	74	0.96
Brick:		
Red, rough, but no gross irregularities	70	0.93
Fireclay	1832	0.75
Carbon:		
T-carbon (Gebrüder Siemens) 0.9% ash, started with emissivity of 0.72 at 260°F but on heating changed to values given	260–1160	0.81–0.79
Filament	1900–2560	0.526
Rough plate	212–608	0.77
Lampblack, rough deposit	212–932	0.84–0.78
Concrete tiles	1832	0.63
Enamel, white fused, on iron	66	0.90
Glass:		
Smooth	72	0.94
Pyrex, lead, and soda	500–1000	0.95–0.85
Paints, lacquers, varnishes:		
Snow-white enamel varnish on rough iron plate	73	0.906
Black shiny lacquer, sprayed on iron	76	0.875
Black shiny shellac on tinned iron sheet	70	0.821
Black matte shellac	170–295	0.91
Black or white lacquer	100–200	0.80–0.95
Flat black lacquer	100–200	0.96–0.98
Aluminum paints and lacquers:		
10% Al, 22% lacquer body, on rough or smooth surface	212	0.52
Other Al paints, varying age and Al content	212	0.27–0.67
Porcelain, glazed	72	0.92
Quartz, rough, fused	70	0.93
Roofing paper	69	0.91
Rubber, hard, glossy plate	74	0.94
Water	32–212	9.95–0.963

†Courtesy of H. C. Hottel, from W. H. McAdams, "Heat Transmissions," 3d ed., McGraw-Hill Book Company, New York, 1954.

Table A-11 Steel-pipe Dimensions

Nominal pipe size, in	*OD,* in	*Schedule no.*	*Wall thickness,* in	*ID,* in	*Metal sectional area,* in^2	*Inside cross-sectional area,* ft^2
$\frac{1}{8}$	0.405	40	0.068	0.269	0.072	0.00040
		80	0.095	0.215	0.093	0.00025
$\frac{1}{4}$	0.540	40	0.088	0.364	0.125	0.00072
		80	0.119	0.302	0.157	0.00050
$\frac{3}{8}$	0.675	40	0.091	0.493	0.167	0.00133
		80	0.126	0.423	0.217	0.00098
$\frac{1}{2}$	0.840	40	0.109	0.622	0.250	0.00211
		80	0.147	0.546	0.320	0.00163
$\frac{3}{4}$	1.050	40	0.113	0.824	0.333	0.00371
		80	0.154	0.742	0.433	0.00300
1	1.315	40	0.133	1.049	0.494	0.00600
		80	0.179	0.957	0.639	0.00499
$1\frac{1}{2}$	1.900	40	0.145	1.610	0.799	0.01414
		80	0.200	1.500	1.068	0.01225
		160	0.281	1.338	1.429	0.00976
2	2.375	40	0.154	2.067	1.075	0.02330
		80	0.218	1.939	1.477	0.02050
3	3.500	40	0.216	3.068	2.228	0.05130
		80	0.300	2.900	3.016	0.04587
4	4.500	40	0.237	4.026	3.173	0.08840
		80	0.337	3.826	4.407	0.7986
5	5.563	40	0.258	5.047	4.304	0.1390
		80	0.375	4.813	6.112	0.1263
		120	0.500	4.563	7.953	0.1136
		160	0.625	4.313	9.696	0.1015
6	6.625	40	0.280	6.065	5.584	0.2006
		80	0.432	5.761	8.405	0.1810
10	10.75	40	0.365	10.020	11.90	0.5475
		80	0.500	9.750	16.10	0.5185

Table A-12 Conversion Factors (*See also back inside cover*)

Length:
- 12 in = 1 ft
- 2.54 cm = 1 in
- 1 μm = 10^{-6} m = 10^{-4} cm

Mass:
- 1 kg = 2.205 lb_m
- 1 slug = 32.16 lb_m
- 454 g = 1 lb_{fm}

Force:
- 1 dyn = 2.248 × 10^{-6} lb_f
- 1 lb_f = 4.448 N
- 10^5 dyn = 1 N

Energy:
- 1 ft · lb_f = 1.356 J
- 1 kWh = 3413 Btu
- 1 hp · h = 2545 Btu
- 1 Btu = 252 cal
- 1 Btu = 778 ft · lb_f

Pressure:
- 1 atm = 14.696 lb_f/in^2 = 2116 lb_f/ft^2
- 1 atm = 1.01325 × 10^5 Pa
- 1 in Hg = 70.73 lb_f/ft^2

Viscosity:
- 1 centipoise = 2.42 lb_m/h · ft
- 1 lb_f · s/ft^2 = 32.16 lb_m/s · ft

Thermal conductivity:
- 1 cal/s · cm · °C = 242 Btu/h · ft · °F
- 1 W/cm · °C = 57.79 Btu/h · ft·°F

Useful conversions to SI units

Length:
- 1 in = 0.0254 m
- 1 ft = 0.3048 m
- 1 mi = 1.60934 km

Area:
- 1 in^2 = 645.16 mm^2
- 1 ft^2 = 0.092903 m^2
- 1 mi^2 = 2.58999 km^2

Pressure:
- 1 N/m^2 = 1 Pa
- 1 atm = 1.01325 × 10^5 Pa
- 1 lb_f/in^2 = 6894.76 Pa

Energy:
- 1 erg = 10^{-7} J
- 1 Btu = 1055.04 J
- 1 ft · lb_f = 1.35582 J
- 1 cal (15°C) = 4.1855 J

Power:
- 1 hp = 745.7 W
- 1 Btu/h = 0.293 W

Heat flux:
- 1 Btu/h · ft^2 = 3.15372 W/m^2
- 1 Btu/h · ft = 0.96128 W/m

Thermal conductivity:
- 1 Btu/h · ft · °F = 1.7307 W/m · °C

Heat-transfer coefficient:
- 1 Btu/h · ft^2 · °F = 5.6782 W/m^2 · °C

Volume:
- 1 in^3 = 1.63871 × 10^{-5} m^3
- 1 ft^3 = 0.0283168 m^3
- 1 gal = 231 in^3 = 0.0037854 m^3

Mass:
- 1 lb_m = 0.45359237 kg

Density:
- 1 lb_m/in^3 = 2.76799 × 10^4 kg/m^3
- 1 lb_m/ft^3 = 16.0185 kg/m^3

Force:
- 1 dyn = 10^{-5} N
- 1 lb_f = 4.44822 N

EXACT SOLUTIONS OF LAMINAR-BOUNDARY-LAYER EQUATIONS

We wish to obtain a solution to the laminar-boundary-layer momentum and energy equations, assuming constant fluid properties and zero pressure gradient. We have:

Continuity:
$$\frac{\partial u}{\partial x} + \frac{\partial v}{\partial y} = 0 \tag{B-1}$$

Momentum:
$$u\frac{\partial u}{\partial x} + v\frac{\partial u}{\partial y} = \nu\frac{\partial^2 u}{\partial y^2} \tag{B-2}$$

Energy:
$$u\frac{\partial T}{\partial x} + v\frac{\partial T}{\partial y} = \alpha\frac{\partial^2 T}{\partial y^2} \tag{B-3}$$

It will be noted that the viscous-dissipation term is omitted from the energy equation for the present. In accordance with the order-of-magnitude analysis of Sec. 6-1,

$$\delta \sim \sqrt{\frac{\nu x}{u_\infty}} \tag{B-4}$$

The assumption is now made that the velocity profiles have similar shapes at various distances from the leading edge of the flat plate. The significant variable is then y/δ, and we assume that the velocity may be expressed as a function of this variable. We then have

$$\frac{u}{u_\infty} = g\left(\frac{y}{\delta}\right)$$

Introducing the order-of-magnitude estimate for δ from Eq. (B-4):

$$\frac{u}{u_\infty} = g(\eta) \tag{B-5}$$

where
$$\eta = \frac{y}{\sqrt{\nu x/u_\infty}} = y\sqrt{\frac{u_\infty}{\nu x}} \tag{B-6}$$

Here, η is called the *similarity variable*, and $g(\eta)$ is the function we seek as a solution. In accordance with the continuity equation, a stream function ψ may be defined so that

$$u = \frac{\partial \psi}{\partial y} \tag{B-7}$$

$$v = -\frac{\partial \psi}{\partial x} \tag{B-8}$$

Inserting (B-7) in (B-5) gives

$$\psi = \int u_\infty g(\eta)\,dy = \int u_\infty \sqrt{\frac{\nu x}{u_\infty}} g(\eta)\,d\eta \tag{B-9}$$

or
$$\psi = u_\infty \sqrt{\frac{\nu x}{u_\infty}} f(\eta)$$

where $f(\eta) = \int g(\eta)\,d\eta$.

From (B-8) and (B-9) we obtain

$$v = \frac{1}{2}\sqrt{\frac{\nu u_\infty}{x}}\left(\eta \frac{df}{d\eta} + f\right) \tag{B-10}$$

Making similar transformations on the other terms in Eq. (B-2), we obtain

$$f\frac{d^2 f}{d\eta^2} + 2\frac{d^3 f}{d\eta^3} = 0 \tag{B-11}$$

This is an ordinary differential equation, which may be solved numerically for the function $f(\eta)$. The boundary conditions are

Physical coordinates		*Similarity coordinates*	
$u = 0$	at $y = 0$	$\dfrac{df}{d\eta} = 0$	at $\eta = 0$
$v = 0$	at $y = 0$	$f = 0$	at $\eta = 0$
$\dfrac{\partial u}{\partial y} = 0$	at $y \to \infty$	$\dfrac{df}{d\eta} = 1.0$	at $\eta \to \infty$

The first solution to Eq. (B-11) was obtained by Blasius.† The values of u and v as obtained from this solution are presented in Fig. B-1.

†H. Blasius, *Z. Math. Phys.*, vol. 56, p. 1, 1908.

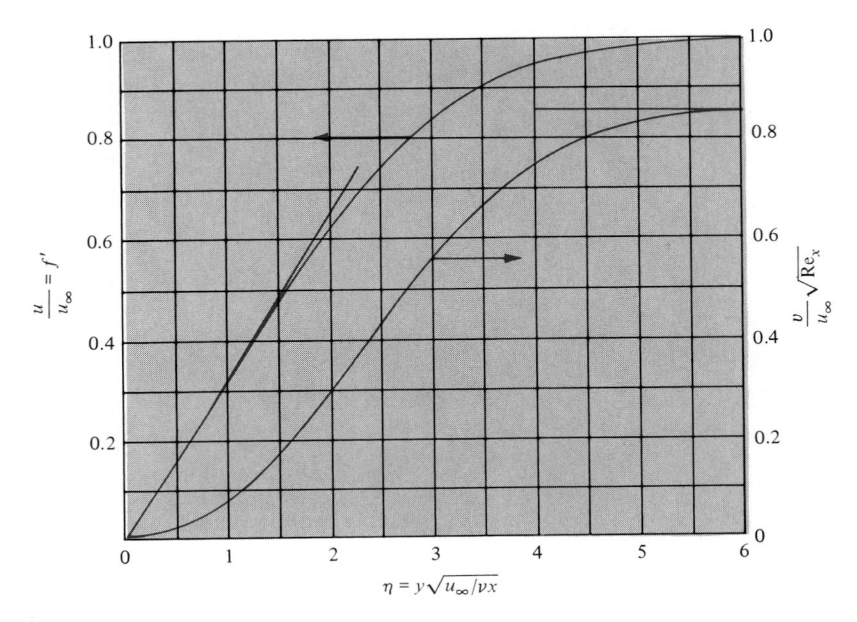

Fig. B-1 Velocity profiles in laminar boundary layer.

The energy equation is solved in a similar manner by first defining a dimensionless temperature variable as

$$\theta(\eta) = \frac{T(\eta) - T_w}{T_\infty - T_w} \tag{B-12}$$

where it is also assumed that θ and T may be expressed as functions of the similarity variable η. Equation (B-3) then becomes

$$\frac{d^2\theta}{d\eta^2} + \frac{1}{2}\,\mathrm{Pr}\,f\frac{d\theta}{d\eta} = 0 \tag{B-13}$$

with the boundary conditions

$$\theta = \begin{cases} 0 & \text{at } y = 0,\ \eta = 0 \\ 1.0 & \text{at } y = \infty,\ \eta = \infty \end{cases}$$

Given the function $f(\eta)$, the solution to Eq. (B-13) may be obtained as

$$\theta(\eta) = \frac{\int_0^\eta \exp\left(-\frac{\mathrm{Pr}}{2}\int_0^\eta f\,d\eta\right)d\eta}{\int_0^\infty \exp\left(-\frac{\mathrm{Pr}}{2}\int_0^\eta f\,d\eta\right)d\eta} \tag{B-14}$$

This solution is given by Pohlhausen† and is shown in Fig. B-2. For Prandtl

†E. Pohlhausen, *Z. Angew. Math. Mech.*, vol. 1, p. 115, 1921.

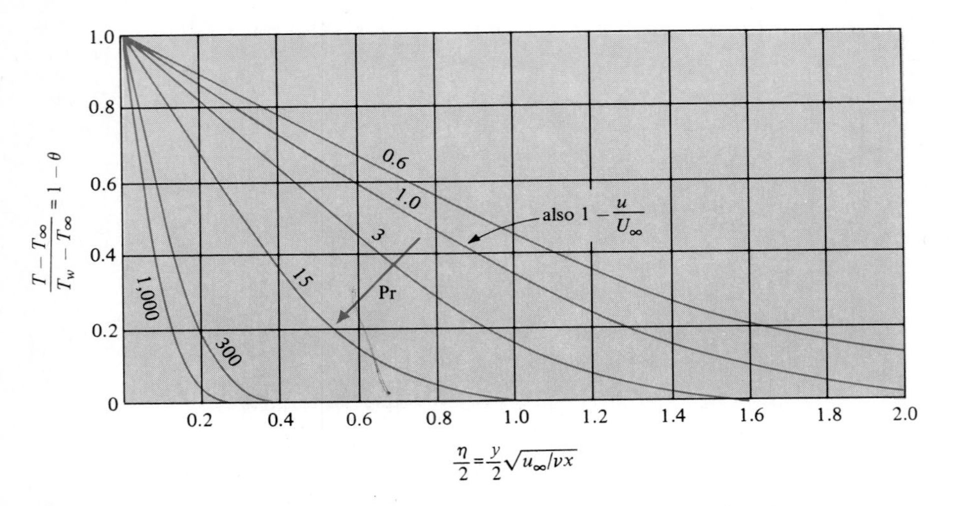

Fig. B-2 Temperature profiles in laminar boundary layer with isothermal wall.

numbers between 0.6 and 15 it was found that the dimensionless temperature gradient at the surface could be represented satisfactorily by

$$\left(\frac{d\theta(\eta)}{d\eta}\right)_{\eta=0} = 0.332\ \mathrm{Pr}^{1/3} \tag{B-15}$$

The heat-transfer coefficient may subsequently be expressed by

$$\mathrm{Nu}_x = 0.332\ \mathrm{Re}_x^{1/2}\ \mathrm{Pr}^{1/3} \tag{B-16}$$

in agreement with the results of Chap. 5.

Now let us consider a solution of the complete energy equation, including the viscous-dissipation term. We have

$$u\frac{\partial T}{ax} + v\frac{\partial T}{ax} = \alpha\frac{\partial^2 T}{\partial y^2} + \frac{\mu}{\rho c_p}\left(\frac{\partial u}{\partial y}\right)^2 \tag{B-17}$$

The solution to this equation is first obtained for the case of an adiabatic plate. By introducing a new dimensionless temperature profile in terms of the stagnation temperature T_0,

$$\theta(\eta) = \frac{T(\eta) - T_\infty}{T_0 - T_\infty} = \frac{T(\eta) - T_\infty}{u_\infty{}^2/2c_p}$$

Eq. (B-17) becomes

$$\frac{d^2\theta}{d\eta^2} + \frac{1}{2}\mathrm{Pr}\,f\frac{d\theta}{d\eta} + 2\ \mathrm{Pr}\left(\frac{d^2 f}{d\eta^2}\right)^2 = 0 \tag{B-18}$$

For the adiabatic-wall case, the boundary conditions are

$$\frac{d\theta}{d\eta} = 0 \qquad \text{at } y = 0,\ \eta = 0$$

$$\theta = 0 \qquad \text{at } y = \infty,\ \eta = \infty$$

The solution to Eq. (B-18) is given by Pohlhausen as

$$\theta_a(\eta, \text{Pr}) = 2\,\text{Pr} \int_\eta^\infty \left(\frac{d^2 f}{d\eta^2}\right)^{\text{Pr}} \left[\int_0^\eta \left(\frac{d^2 f}{d\eta^2}\right)^{2-\text{Pr}} d\eta\right] d\eta \tag{B-19}$$

where the symbol θ_a has been used to indicate the adiabatic-wall solution. A graphical plot of the solution is given in Fig. B-3. The recovery factor is given as

$$r = \theta_a(0, \text{Pr})$$

For Prandtl numbers near unity this reduces to the relation given in Eq. (5-122)

$$r = \text{Pr}^{1/2} \tag{5-122}$$

Now consider the case where the wall is maintained at some temperature other than T_{aw}; that is, there is heat transfer either to or from the fluid. The boundary conditions are now expressed as

$$T = \begin{cases} T_w & \text{at } y = 0,\ \eta = 0 \\ T_\infty & \text{at } y = \infty,\ \eta = \infty \end{cases}$$

We observe that the viscous-heating term in Eq. (B-18) contributes a particular solution to the equation. If there were no viscous heating, the adiabatic-wall solution would yield a uniform temperature profile throughout the boundary layer. We now assume that the temperature profile for the combined case of a heated wall and viscous dissipation can be represented by a linear combination of the solutions given in Eqs. (B-14) and (B-19). This assumption is justified in

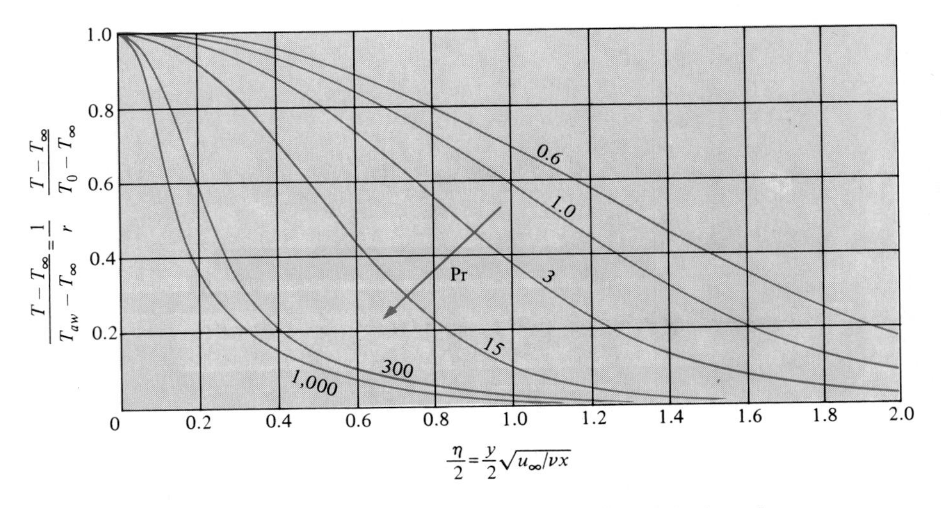

Fig. B-3 Temperature profiles in laminar boundary layer with adiabatic wall.

view of the fact that Eq. (B-18) is linear in the dependent variable θ. We then write

$$T - T_\infty = (T_a - T_\infty) + (T_c - T_{aw}) \tag{B-20}$$

where $T_a - T_\infty$ is the temperature distribution from Eq. (B-19) and $T_c - T_{aw}$ is the solution from Eq. (B-14), with T_{aw} taking the same role as T_∞ in that solution. Equation (B-20) may be written

$$T - T_\infty = \theta_a(T_0 - T_\infty) + \theta(T_{aw} - T_w) + T_w - T_{aw} \tag{B-21}$$

This solution may be tested by inserting it in Eq. (B-18). There results

$$(T_0 - T_\infty)\left[\frac{d^2\theta_a}{d\eta^2} + \frac{1}{2}\,\text{Pr}\,f\frac{d\theta_a}{d\eta} + 2\,\text{Pr}\left(\frac{d^2f}{d\eta^2}\right)^2\right] + (T_{aw} - T_w)\left(\frac{d^2\theta}{d\eta^2} + \frac{1}{2}\,\text{Pr}\,f\frac{d\theta}{d\eta}\right) = 0$$

An inspection of this relation indicates that Eq. (B-21) is a valid solution for the actual boundary conditions of the heated wall. The temperature gradient at the wall may thus be expressed as

$$\left.\frac{\partial T}{\partial \eta}\right]_{\eta=0} = (T_0 - T_\infty)\left.\frac{\partial \theta_a}{\partial \eta}\right]_{\eta=0} + (T_{aw} - T_w)\left.\frac{\partial \theta}{\partial \eta}\right]_{\eta=0}$$

The first term is zero, and Eq. (B-15) may be used to evaluate the second term. There results

$$\left.\frac{\partial T}{\partial \eta}\right]_{\eta=0} = 0.332(T_{aw} - T_w)\,\text{Pr}^{1/3} \tag{B-22}$$

This relation immediately suggests the definition of the heat-transfer coefficient for the case where viscous heating is important by the relation

$$\frac{q}{A} = h(T_w - T_{aw}) \tag{B-23}$$

The analysis then proceeds as discussed in Sec. 5-12.

ANALYTICAL RELATIONS FOR THE HEISLER CHARTS

Analytical solutions for the three transient cases covered in the Heisler charts of Chap. 4 are given in Ref. 1 of Chap. 4. Heisler (Ref. 7 of Chap. 4) was able to show that for $\alpha\tau/L^2$ or $\alpha\tau/r_0^2 > 0.2$, the infinite series solutions for the center temperature ($x = 0$ or $r = 0$) could be approximated within 1 percent with a single term:

$$\frac{\theta_0}{\theta_i} \approx C_B \exp\left(-A_B{}^2 \frac{\alpha\tau}{s^2}\right) \tag{C-1}$$

where $s = L$ for the infinite plate and $s = r_0$ for the cylinder or sphere. C_B and A_B are constants which are functions of the Biot number, hL/k or hr_0/k, and are determined from the following equations.

□ Infinite Plate

A_B is the solution to

$$A_B\,(\tan A_B) = \text{Bi} = \frac{hL}{k} \tag{C-2}$$

C_B is obtained from

$$C_B = \frac{4 \sin A_B}{2A_B + \sin 2A_B}$$

The arguments of the trigonometric functions are in radians.

□ Infinite Cylinder

A_B is the solution to

Table C-1 Abbreviated Tabulation of Bessel Functions of the First Kind†

u	$J_0(u)$	$J_1(u)$	u	$J_0(u)$	$J_1(u)$
0.0	1.0	0.0	6.0	0.15065	−0.27668
0.2	0.99003	0.09950	6.2	0.20175	−0.23292
0.4	0.96040	0.19603	6.4	0.24331	−0.18164
0.6	0.91200	0.28670	6.6	0.27404	−0.12498
0.8	0.84629	0.36884	6.8	0.29310	−0.06522
1.0	0.76520	0.44005	7.0	0.30008	−0.00468
1.2	0.67113	0.49829	7.2	0.29507	0.05432
1.4	0.56686	0.54195	7.4	0.27860	0.10963
1.6	0.45540	0.56990	7.6	0.25160	0.15921
1.8	0.33999	0.58152	7.8	0.21541	0.20136
2.0	0.22389	0.57672	8.0	0.17165	0.23464
2.2	0.11036	0.55596	8.2	0.12222	0.25800
2.4	0.00251	0.52019	8.4	0.06916	0.27079
2.6	−0.09681	0.47082	8.6	0.01462	0.27275
2.8	−0.18504	0.40970	8.8	−0.03923	0.26407
3.0	−0.26005	0.33906	9.0	−0.09033	0.24531
3.2	−0.32019	0.26134	9.2	−0.13675	0.21741
3.4	−0.36430	0.17923	9.4	−0.17677	0.18163
3.6	−0.39177	0.09547	9.6	−0.20898	0.13952
3.8	−0.40256	0.01282	9.8	−0.23228	0.09284
4.0	−0.39715	−0.06604	10.0	−0.24594	0.04347
4.2	−0.37656	−0.13865	10.2	−0.24962	−0.00662
4.4	−0.34226	−0.20278	10.4	−0.24337	−0.05547
4.6	−0.29614	−0.25655	10.6	−0.22764	−0.10123
4.8	−0.24043	−0.29850	10.8	−0.20320	−0.14217
5.0	−0.17760	−0.32760	11.0	−0.17119	−0.17679
5.2	−0.11029	−0.34322	11.2	−0.13299	−0.20385
5.4	−0.04121	−0.34534	11.4	−0.09021	−0.22245
5.6	0.02697	−0.33433	11.6	−0.04462	−0.23200
5.8	0.09170	−0.31103	11.8	0.00197	−0.23229
			12.0	0.04769	−0.22345

† According to Ref. 3 of Chap. 4

$$A_B J_1(A_B)/J_0(A_B) = \text{Bi} = \frac{hr_0}{k} \tag{C-3}$$

where J_0 and J_1 are Bessel functions of the first kind and are tabulated in a number of references such as Refs. 1 and 3 of Chap. 4. An abbreviated tabulation is given in Table C-1.

C_B for the cylinder is then evaluated from

$$C_B = \frac{2}{A_B} \frac{J_1(A_B)}{J_0^2(A_B) + J_1^2(A_B)} \tag{C-4}$$

Sphere

A_B is the solution to

Table C-2 Coefficients for Heisler Solutions

$\frac{hL}{k}$ or $\frac{hr_0}{k}$	Infinite plate		Long cylinder		Sphere	
	A_B	C_B	A_B	C_B	A_B	C_B
0.01	0.0998	1.0017	0.1412	1.0025	0.1730	1.0030
0.02	0.1410	1.0033	0.1995	1.0050	0.2445	1.0060
0.04	0.1987	1.0066	0.2814	1.0099	0.3450	1.0120
0.06	0.2425	1.0098	0.3438	1.0148	0.4217	1.0179
0.08	0.2791	1.0130	0.3960	1.0197	0.4860	1.0239
0.1	0.3111	1.0161	0.4417	1.0246	0.5423	1.0298
0.2	0.4328	1.0311	0.6170	1.0483	0.7593	1.0592
0.3	0.5218	1.0451	0.7465	1.0712	0.9208	1.0880
0.4	0.5932	1.0580	0.8516	1.0931	1.0528	1.1164
0.5	0.6533	1.0701	0.9408	1.1143	1.1656	1.1441
0.6	0.7051	1.0814	1.0185	1.1345	1.2644	1.1713
0.7	0.7506	1.0919	1.0873	1.1539	1.3525	1.1978
0.8	0.7910	1.1016	1.1490	1.1724	1.4320	1.2236
0.9	0.8274	1.1107	1.2048	1.1902	1.5044	1.2488
1.0	0.8603	1.1191	1.2558	1.2071	1.5708	1.2732
2.0	1.0769	1.1785	1.5995	1.3384	2.0288	1.4793
3.0	1.1925	1.2102	1.7887	1.4191	2.2889	1.6227
4.0	1.2646	1.2287	1.9081	1.4698	2.4556	1.7202
5.0	1.3138	1.2403	1.9898	1.5029	2.5704	1.7870
6.0	1.3496	1.2479	2.0490	1.5253	2.6537	1.8338
7.0	1.3766	1.2532	2.0937	1.5411	2.7165	1.8674
8.0	1.3978	1.2570	2.1286	1.5526	2.7654	1.8920
9.0	1.4149	1.2598	2.1566	1.5611	2.8044	1.9106
10.0	1.4289	1.2620	2.1795	1.5677	2.8363	1.9249
20.0	1.4961	1.2699	2.2881	1.5919	2.9857	1.9781
30.0	1.5202	1.2717	2.3261	1.5973	3.0372	1.9898
40.0	1.5325	1.2723	2.3455	1.5993	3.0632	1.9942
50.0	1.5400	1.2727	2.3572	1.6002	3.0788	1.9962
100.0	1.5552	1.2731	2.3809	1.6015	3.1102	1.9990

$$1 - A_B \cot A_B = \text{Bi} = \frac{hr_0}{k} \tag{C-5}$$

and C_B is obtained from

$$C_B = \frac{4(\sin A_B - A_B \cos A_B)}{2A_B - \sin 2A_B} \tag{C-6}$$

For convenience, Table C-2 lists the parameters A_B and C_B as a function of Biot number for the three geometries under consideration.

Off-Center Temperatures

Because of the single-term approximation of the series, solutions for the off-center temperatures may also be expressed in the following simple forms.

Infinite plate:

$$\frac{\theta}{\theta_0} = \cos\left(\frac{A_B x}{L}\right) \tag{C-7}$$

Infinite cylinder:

$$\frac{\theta}{\theta_0} = J_0\left(\frac{A_B r}{r_0}\right) \tag{C-8}$$

Sphere:

$$\frac{\theta}{\theta_0} = \frac{r_0}{rA_B} \sin\left(\frac{A_B r}{r_0}\right) \tag{C-9}$$

Again, arguments of the trigonometric functions are in radians.

□ Total Heat Loss

The total heat loss in time τ corresponding to Figs. 4-16 to 4-18 can also be obtained as follows.

Infinite plate:

$$\frac{Q}{Q_0} = 1 - \left(\frac{\theta_0}{A_B\theta_i}\right) \sin A_B \tag{C-10}$$

Infinite cylinder:

$$\frac{Q}{Q_0} = 1 - \left(\frac{2\theta_0}{A_B\theta_i}\right) J_1(A_B) \tag{C-11}$$

Sphere:

$$\frac{Q}{Q_0} = 1 - \left(\frac{3\theta_0}{A_B^3\theta_i}\right)(\sin A_B - A_B \cos A_B) \tag{C-12}$$

In all of the above equations A_B and C_B are the coefficients *for the particular geometry,* tabulated as a function of Biot number in Table C-2.

The analytical solutions obviously afford the user greater accuracy than the charts, but they may be more cumbersome to handle. The method selected for a problem solution may depend on uncertainties in the convection boundary conditions. In some cases a combination may be used; perhaps the analytical method to find θ_0 and the graphs to evaluate θ/θ_0. This might avoid the need for evaluating Bessel functions.

■ EXAMPLE C-1

A long, steel cylinder having a diameter of 5 cm has an initial uniform temperature of 250°C and is suddenly exposed to a convection environment at 80°C with h = 500 W/m² · °C. Calculate the temperature at a radius of 1.2 cm after a time of 1 min and the heat lost per unit length during this period. Take the properties of steel as ρ = 7800 kg/m³, c = 0.48 kJ/kg · °C, and k = 35 W/m · °C.

Solution

We first calculate the thermal diffusivity as

$$\alpha = \frac{k}{\rho c} = \frac{35}{(7800)(480)} = 9.35 \times 10^{-6} \text{ m}^2/\text{s}$$

and the Biot number as

$$\text{Bi} = \frac{hr_0}{k} = \frac{(500)(0.025)}{35} = 0.357$$

From Table C-2

$$A_B = 0.8064; \qquad C_B = 1.0837$$

From Eq. (C-1)

$$\frac{\theta_0}{\theta_i} = (1.0837)\exp\left[\frac{-(0.8064)^2(9.35 \times 10^{-6})(60)}{(0.025)^2}\right]$$

$$= 0.6045$$

For the off-centerline temperature we must use Eq. (C-8) to obtain

$$\frac{\theta}{\theta_0} = J_0\left[\frac{(0.8064)(0.012)}{0.025}\right] = J_0(0.387) = 0.9623$$

Then

$$\frac{\theta}{\theta_i} = (0.6045)(0.9623) = 0.5817$$

and

$$T = (0.5817)(250 - 80) + 80 = 178.9°\text{C}$$

To obtain the heat loss we employ Eq. (C-11)

$$\frac{Q}{Q_0} = 1 - \left[\frac{(2)(0.6045)}{0.8064}\right] J_1(0.8064)$$

From Table C-1, $J_1(0.8064) = 0.37112$ so that

$$\frac{Q}{Q_0} = 0.4436$$

We have

$$Q_0 = \rho\, cV\theta_i = (7800)(480)\pi(0.025)^2(250 - 80) = 1.25 \times 10^6 \text{ J}$$

so that

$$Q = (0.4436)(1.25 \times 10^6) = 5.54 \times 10^5 \text{ J/m length}$$

HEAT-TRANSFER SOFTWARE

D-1 INTRODUCTION

The heat-transfer software package included with this book was developed by Professor Allan D. Kraus, of the Naval Postgraduate School, and has been offered previously as a separate product by the McGraw-Hill Publishing Company. The programs can be used in two basic modes:

1. As a computational tool for various heat-transfer phenomena. In this mode one need only boot up the system, as directed, and follow the menu.
2. As an instructive adjunct in thermal design problems. The illustrative problems in this appendix give several varied examples demonstrating the power of the computer in heat-transfer design.

Despite the enthusiasm this software may generate, we caution that

> ***The computer programs are no substitute for a basic understanding of the subject of heat transfer.***

This statement means that both students and practitioners must first achieve an understanding of the theory and art behind the equations presented in this text to gain the maximum benefit from the software program.

In this appendix after first discussing how to set up the program for operation, we give a brief documentation for each program, and then we give problems which tie in with the chapters of the book or are of an open-ended design type.

Note that the chapters in the text stand on their own without depending in any way on the software. In contrast, proper use of the software depends on a thorough understanding of the theory of heat transfer as presented in the formal chapters of the text.

D-2 SETTING UP THE PROGRAMS

The program disk is contained in the packet inside the back cover of the book. It requires an IBM PC or IBM-compatible personal computer with 256K of memory, an 80-column monochrome monitor, MS-DOS operating system (version 2.0 or higher), and an optional printer if hard-copy output is desired.

The system consists of two program groups, PROG1, with the programs

A. Evaluation of fin performance

B. Temperatures in a corner

C. Transient temperatures in a slab

D. Transient temperatures in a semi-infinite solid

E. Properties of air and airflow over a flat plate

F. Forced-convection coefficients

and PROG2, with the programs

G. Free or natural convection coefficients

H. Radiation functions

I. Radiation shape-factor determination

J. Radiation network

K. Heat-exchanger effectiveness

L. Heat-exchanger flow arrangements for increased heat recovery

To actuate the programs, first boot the computer with the MS-DOS operating system until the prompt A > appears. Then call either

A > PROG1 or A > PROG2

Some copyright statements will then appear, followed by a menu which is largely self-explanatory. The following sections provide further documentation which may be required or helpful for each of the problems.

D-3 DOCUMENTATION OF THE TWELVE PROGRAMS

Problem A: Evaluation of Fin Performance

This program uses the fin relations of Chap. 2 to compute various performance factors. The menu is self-explanatory, and the fin nomenclature is given in Fig. D-1.

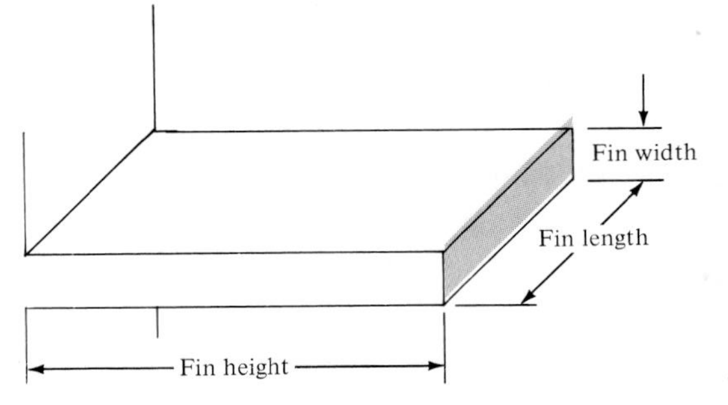

Fig. D-1 Longitudinal fin of rectangular profile. Fin height corresponds to length L, width to thickness t, and length to depth z in Chap. 2.

□ Problem B: Temperatures in a Corner

This program illustrates application of the Gauss-Seidel iteration technique of Chap. 3. Two choices of nodal increments are available: 6 nodes, as shown in Fig. D-2, and 33 nodes, as shown in Fig. D-3. The user may select different values of thermal conductivity, inside and outside wall temperatures, and an initial guess for the nodal temperatures for starting the iteration. By running several cases the user may examine the effect of the number of nodes on the

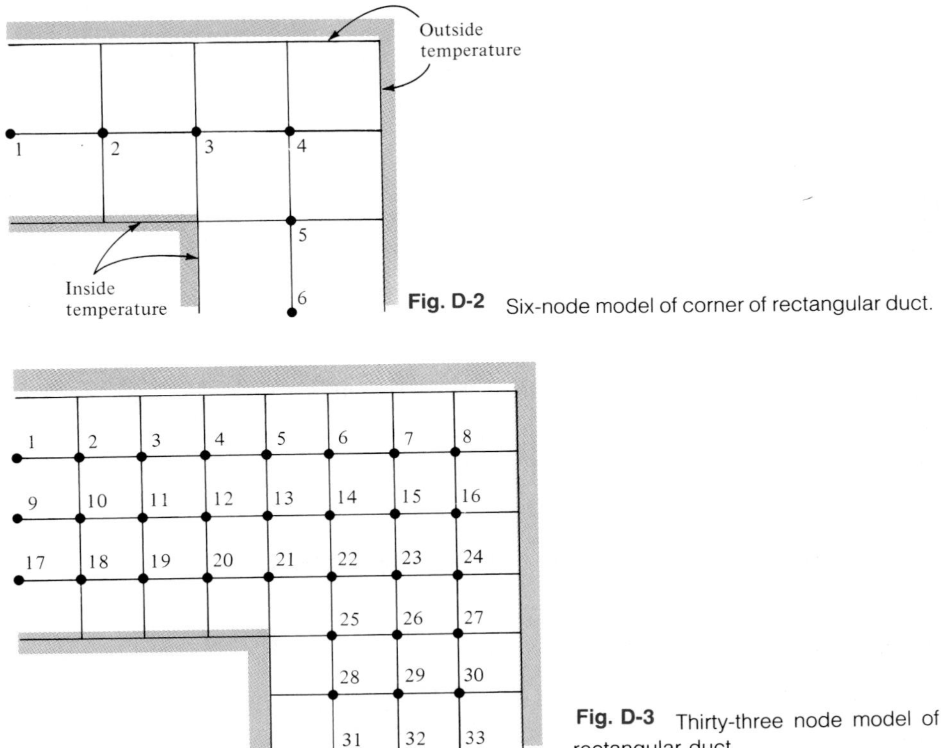

Fig. D-2 Six-node model of corner of rectangular duct.

Fig. D-3 Thirty-three node model of rectangular duct.

accuracy of the results and the influence of the initial temperature guess on the number of iterations required to reach the final solution.

☐ Problem C: Transient Temperatures in a Slab

This program illustrates how the numerical technique is applied to determine the temperature distribution in a one-dimensional slab, as shown in Fig. D-4. The program operates for the following specific set of conditions:

Total thickness = 1.2 m (2L of Fig. 4-6*a*)

Thermal conductivity = 150 W/m · °C

Density = 2500 kg/m^3

Specific heat = 1000 J/kg · °C

Overall time = 900 s

The user may also select options of a sudden exposure to a convection environment or a sudden change in surface temperature which corresponds to $h \rightarrow \infty$. To arrive at the final time of 900 s, the user should select the values of $\Delta\tau$ which are even divisions of this value: 225, 112.5, 56.25 s, etc. The program automatically rejects selections which do not meet the stability criteria of Chap. 4 as expressed in terms of acceptable values of the Fourier modulus. Finally, the program calculates an analytical value for the centerline temperature for comparison with the numerical solution.

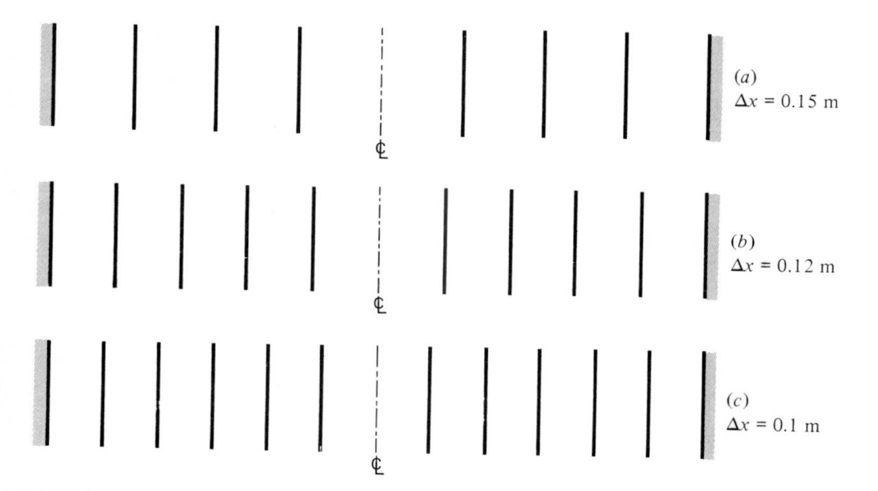

Fig. D-4 Three models for transient numerical analysis of a plane slab: (*a*) eight sublayers, (*b*) ten sublayers, (*c*) twelve sublayers. The total thickness in each case is 1.2 m.

☐ Problem D: Transient Temperatures in a Semi-infinite Solid

This problem calculates the subject temperatures for three cases:

1. Sudden change of surface temperature according to Eq. (4-8)
2. Sudden exposure to constant surface heat flux according to Eq. (4-13)
3. Sudden exposure to convection environment according to Eq. (4-15) and Fig. 4-5

The program also computes values of the error function for different arguments.

☐ Problem E: Properties of Air and Airflow over a Flat Plate

This program computes the properties of air corresponding to Table A-5 for temperatures between 250 and 500 K. It also computes the heat transfer corresponding to the equations in Table 5-2. The program menu is self-explanatory.

☐ Problem F: Forced-Convection Coefficients

Program F computes forced-convection heat transfer in accordance with the equations of Chap. 6 and Table 6-8. For air and water the program computes the fluid properties to match with Tables A-5 and A-9. The range of temperature is between 250 and 500 K for air and from 0 to 104.4°C for saturated liquid water. For other fluids the user must insert the properties (obtainable from tables in Appendix A or other sources). Otherwise the menu is self-explanatory.

As practice problems the user may wish to consider the following:

F-1 The finned annular passage in Fig. D-5 carries unused engine oil flowing at 2 m/s at a bulk temperature of 80°C. Assuming that the tube wall temperature is 60°C, calculate the coefficient of heat transfer if the length of the configuration is 2 m.

F-2 The inside of the tube in Fig. D-5 carries ethylene glycol flowing at a mean velocity of 4 m/s with a bulk temperature of 40°C. The length of the configuration is 2 m. What is the coefficient of heat transfer?

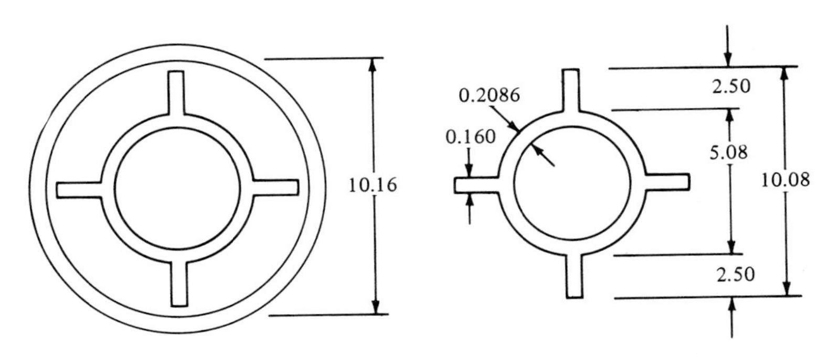

Fig. D-5 Double pipe arrangement where inner tube has four exterior longitudinal fins of rectangular profile. All dimensions are in centimeters.

☐ Problem G: Free or Natural Convection Coefficients

This problem uses the equations of Chap. 7 and Table 7-4 to calculate free-convection heat transfer with either air or water as the working fluid. The properties of air and water are built into the program and match with Tables A-5 and A-9. The temperature range allowed is 250 to 500 K for air and 0 to 104.4°C for saturated liquid water. Otherwise, the menu is self-explanatory. For practice, the user may wish to work the following design problem.

G-1 Consider the box of electronic components displayed with dimensions in Fig. D-6. Assume that all surfaces are effective surfaces for the dissipation of heat by free convection and by radiation. The effect of the mounting structure, which acts as a heat-flow path to the environment, may be neglected.

The components dissipate 325 W and, because of reliability requirements, the surface of the box may not exceed 100°C. The box is covered with a paint having an emissivity of 0.62 and must operate in the wheel well of a jet aircraft at (1) sea level (p = 1 atm) with surrounding temperature at 71°C, and (2) at an altitude where the pressure is 0.15 atm and the ambient temperature is 55°C.

Your job is to make an analysis of the heat transfer in the two still-air environments. First use the surface temperature of 100°C at the sea-level condition and observe the heat dissipation. The total obtained will be greater than, equal to, or less than 325 W.

If the temperature of 100°C does not yield a heat dissipation of 325 W, you have two alternatives: (1) reduce the environmental temperature or (2) refinish the box with a paint of higher emissivity.

First find the maximum allowable environmental temperature. Then repeat the process and find a solution (if possible) for the 71°C environmental temperature but with a higher emissivity. When this is finished, repeat the entire process for the 0.15-atm condition.

Comment on the importance of radiation in this problem.

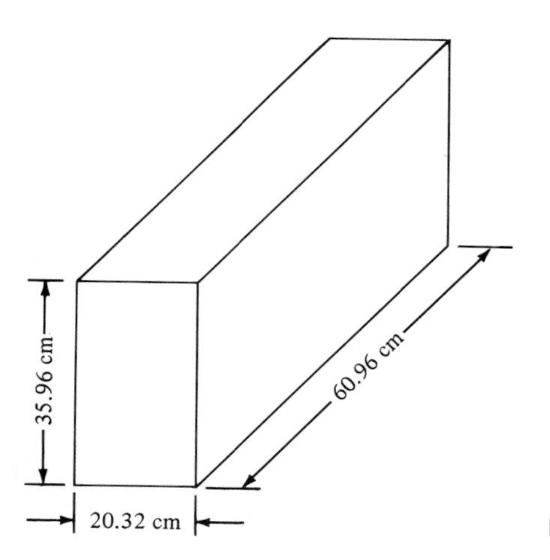

Fig. D-6 Cooling of box of electronic components.

Problem H: Radiation Functions

This program computes the radiation functions according to the Planck blackbody formula of Eq. (8-12), which are shown in Table 8-1. The menu format is self-explanatory.

Problem I: Radiation Shape-Factor Determinations

This program computes the shape factors corresponding to Figs. 8-12 to 8-16. Note in particular that

1. For perpendicular rectangles with a common edge the program computes F_{12} for the larger to smaller rectangle.
2. For short cylinders the program computes F_{21} corresponding to the nomenclature of Fig. 8-15.
3. For unequal parallel disks the program computes F_{12} corresponding to the nomenclature of Fig. 8-16.
4. For corners of parallel rectangles the program computes $F_{19'}$ according to the nomenclature of Fig. D-7. The program also tabulates the sixteen K-factors for this arrangement corresponding to Eq. (8-35).

Problem J: Radiation Network

This program computes heat-transfer rates and radiosities for a six-body problem in the shape of a room, as shown in Fig. D-8. The radiation-network method of Chap. 8 is used for the calculation in conjunction with the shape-factor

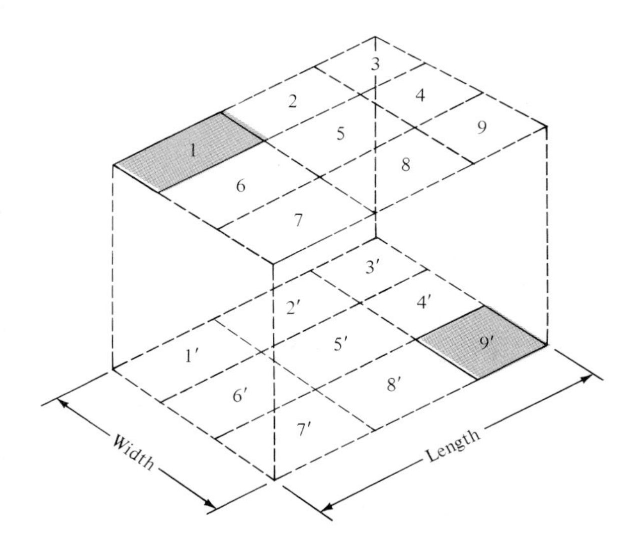

Fig. D-7 Nomenclature for shape factor for corners of parallel rectangles. Product factors correspond to Eq. (8-35).

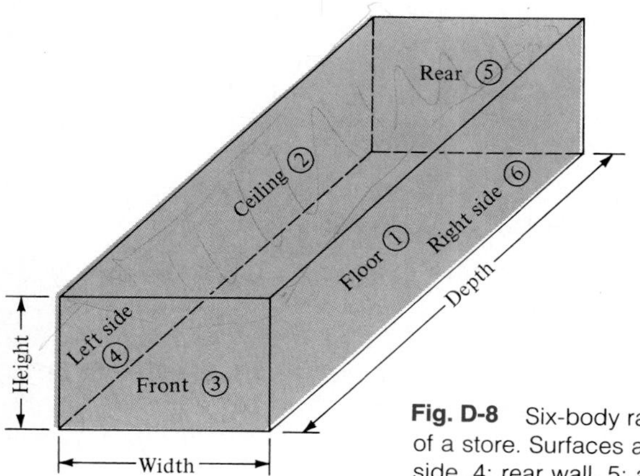

Fig. D-8 Six-body radiation network problem in the shape of a store. Surfaces are floor, 1; ceiling, 2; front wall, 3; left side, 4; rear wall, 5; and right side, 6.

determinations of Prob. I. The program can handle different kinds of problems, e.g.,

1. A six-body problem with specified temperatures and emissivities of all surfaces
2. Surrounding by a very large surface which behaves as if $\epsilon \rightarrow 1.0$ because $A \rightarrow \infty$ and the surface resistance approaches zero
3. Concave surfaces when the shape factor is converted into an equivalent flat-plate value covering the opening, with an apparent emissivity as discussed in Chap. 8 (Prob. 8-152)
4. An insulated surface which behaves as if $\rho \rightarrow 1.0$, $\epsilon \rightarrow 0$, surface resistance approaches ∞, and the calculated radiosity approaches the blackbody emissivity power of the insulated surface.

□ Problem K: Heat-Exchanger Effectiveness

This program makes use of the effectiveness-NTU method discussed in Chap. 10 and the equations of Tables 10-3 and 10-4, plotted in Figs. 10-12 to 10-17. The program menu is self-explanatory.

□ Problem L: Heat-Exchanger Flow Arrangement for Increased Heat Recovery

This design-oriented program has two objectives: (1) to show how both thermal performance and pressure-loss requirements can dictate selection of a heat exchanger, and (2) to demonstrate how heat-exchanger flow arrangement can influence the ultimate heat recovery.

Data

This program has its data built in. Basically, the hot-fluid and cold-fluid flows, specific heats, and inlet temperatures are contained in the program. In addition, the program considers an exchanger with a fixed number of tubes which, by suitable baffling on the tube side, can permit a counterflow of several 1-*n* (*n* even) arrangements.

The user may select an exchanger type and adjust the tube length between extremes of 3.5 and 7.0 m. The original tube length set by the program is 5.5 m.

All data in the program required for an intelligent solution to the design program are displayed and automatically made continuously available to the user. It clearly shows that as the number of tube passes increases, the tube-side velocity increases to yield a higher value of the overall heat-transfer coefficient and hence a lower hot-side outlet temperature. This increase in velocity also causes an increase in the tube-side pressure loss. The user will also be able to observe that as the number of tube passes increases, the turn loss becomes more and more significant.

Procedure

Follow the instructions in the menu to select the best exchanger (shortest length) yielding a maximum hot-side outlet temperature of 36.5°C and maximum tube-side pressure loss of 40 kPa. The best exchanger is the one that has the shortest length to meet the foregoing performance. When finished, summarize the results in an appropriate format.

PROBLEMS USING HEAT-TRANSFER SOFTWARE

D-1 *Fin Design Problem*

A longitudinal pin of rectangular profile is constructed as shown in Fig. D-1 with fin height = 4 cm, fin length = 100 cm, and fin width = 0.228 cm. You have a choice of two materials with the following properties:

	Al	*Mg*
k, W/m · °C	202	154
Density ρ, kg/m^3	2710	1750

The fin dimensions are to be modified so that *the minimum weight of the fin is of parament importance*. Follow the procedure below to carry out the design process.

1. Compare the heat flow, fin efficiency, and weight of the two fins in the basic configuration for h = 50 W/m^2 · °C and a base-to-ambient temperature difference of 100°C.
2. Increase the width of the magnesium fin (holding the fin height at 4 cm) until the heat flow is the same as for the aluminum fin in step 1. Note the weight and efficiency of the fin and comment.
3. Go back to the basic configuration of step 1 and increase the height of the magnesium fin (holding the width at 0.228 cm) until the flow is the same as for the aluminum fin in step 1. Again note the weight and the efficiency of the fin and comment, making sure that you include a comparison between the magnesium fins of steps 1 and 2.

4. Select step 2 or step 3 as the basis for the match of the magnesium and aluminum fins. There is an obvious choice here (remember the weight criterion). Before proceeding, comment on the fin-efficiency comparison.
5. Proceed on the basis of the choice in step 4 to check both magnesium and aluminum fins at values of the heat-transfer coefficient h = 10, 20, 30, and 40 W/m^2 · °C. Comment on which of the two fins has the superior performance.
6. The question arises as to what would happen if a deliberate attempt were made to further improve the weight picture by shaving some metal. This leads to the consideration of triangular fins. For the original value of h = 50 W/m^2 · °C and the original base width of 0.288 cm, determine the height of both the aluminum and magnesium fins needed to achieve the same heat flow as the aluminum fin in step 1. Compare the weight and fin efficiency and comment.
7. This time compare the two fins of step 6 at values of the heat-transfer coefficient of h = 10, 20, 30, and 40 W/m^2 · °C and comment on which fin has the superior performance.
8. You have accumulated a great deal of data and are now in a position to make a recommendation of which fin to use: aluminum-rectangular, aluminum-triangular, magnesium-rectangular, or magnesium-triangular. Make your recommendation.
9. Suppose that in the basic configuration of Fig. D-1 and step 1 the heat-transfer coefficient is changed to h = 350 W/m^2 · °C. Recompute the value of the fin heat flow for both the aluminum and magnesium fins for this value of the heat-transfer coefficient. Then make an attempt to get the magnesium fin to dissipate the same quantity of heat at this value of h by any method you wish, i.e., increasing the width or the height or both. When you are finished, comment on the use of a magnesium fin at the higher value of the heat-transfer coefficient.

D-2 *Temperatures and Heat Flows in a Corner*

A duct is constructed as shown in Figs. D-2 and D-3. The inside dimensions are 1.0 by 2.0 m, and the outside dimensions are 3.0 by 4.0 m. Analyze the corner section according to the following procedure:

1. After selecting the units framework, ask the computer to give you Option 1. You will be working the six-node problem.
2. Enter values of thermal conductivity, inside and outside wall temperature, and the initial starting temperature.
3. Record the value of all six temperatures and the two heat flows, i.e., the flow crossing into the wall at the inside and the flow crossing out of the wall at the outside. Comment on any difference you observe in these heat flows.
4. Return to the option section and this time choose Option 2. You will now be working the 33-node problem. Keep the same input data except for the initial starting temperature, which you should set at 0°C or 0°F.
5. Record the value of the temperatures at nodes 9, 11, 13, 15, 26, and 32. These correspond to the six nodes considered in Option 1. Also record the values of the two heat flows.
6. Repeat the process in step 5, each time changing the initial starting temperature to values of 100, 200, 300, and 400. Record the values of the initial temperatures and the number of iterations required to effect the solution.
7. Comment on the accuracy attained with the 33-node problem (Option 2) and indicate whether there is any need to refine the increment further. Comment

also on whether the selection of the starting temperature has an effect on the number of iterations.

D-3 The following procedure is to be used for analyzing the transient temperatures in the slab of Prob. C.

1. After selecting the units framework, ask the computer to give you Option 1, which considers the slab with a sudden temperature change at its faces; then ask for the eight-layer slab model.
2. Select any values you like for the environmental and initial temperatures. Enter them, but be sure to select the time increment at 225 s, in an effort to reach the total time duration of 900 s in four time increments. Observe what happens. Why was 225 s not a good choice?
3. Change the time increment successively to 112.5, 56.25, 28.125, 14.0625, 7.03125, 6, 4, 3, and 2 s. Enter these actual numbers so that the number of iterations required to reach exactly 900 s is a whole number (in this case 16, 32, 64, 128, 150, 225, 300, and 450, respectively). Record the results.
4. At the conclusion of the 2-s-increment run, respond Y or y (for "yes") to the question regarding the classical solution. Use a time of 900 s and record the result. This will be taken as the correct solution as it is determined from a solution of the governing partial differential equation.
5. Begin again with the 10-layer slab model and repeat all calculations, beginning with a time increment of 112.5 s. Take note of all time increments for which the computer gives an "excessive Fourier modulus" message. Record the results.
6. Go through an identical procedure for the 12-layer slab model, take note of all points where the time increment is excessive, and record the results.
7. Now calculate the value of the Fourier modulus at all points where the error message appeared. Comment on the interplay between the spatial and time increment.
8. Now change options by asking for Option 2, which involves a heat-transfer coefficient at the faces of the slab. Stay with the 12-layer slab model and select a high value of heat-transfer coefficient, say 3000 W/m^2 · °C or 600 Btu/ft^2 · h · °F. Begin with a time increment of 90 s (allowing 10 even increments to reach 900 s). You will receive a message saying that the calculation cannot proceed. Try successively lower time increments of, say, 56.125, 50, 45, 30, 28.125, 25, 20, 15, 12, 9, 6, and 5 s and observe when the message changes and then when the calculation proceeds to a conclusion. Take note of this time increment and comment on why the message changed.
9. Finally, for the first time increment in step 8 where the analysis went to a conclusion, change the value of the heat-transfer coefficient to successively lower values. Do this four times and record the results. Then for one of these values of the heat-transfer coefficient attempt to confirm the result by using the appropriate Heisler chart.

D-4 Soil at a certain location can be considered as a semi-infinite solid with $k = 1.0$ W/m · °C, specific heat $c = 1840$ J/kg · °C, and density $\rho = 1500$ kg/m^3. The temperature at a depth of 25 cm is 10°C at 8 P.M. on Sunday, and we assume that the temperature of the soil is uniform throughout. The temperature at the 25-cm depth is measured by a section of tubing containing a temperature-detection device. Use the computer program to investigate the following cases and record the results.

1. At 8 P.M. on Sunday the surface temperature drops to −10°C and stays there until 8 A.M. Monday. Record the initial temperature, the elapsed time, and the final temperature at 25 cm.
2. At 8 A.M. Monday, when the assumption is made that the entire soil-tubing entity is at the temperature just obtained at 25 cm, the sun begins to shine. It provides an effective constant surface heat flux of 24.85 W/m^2, and this condition continues until 4 P.M. of the same day. Again record the initial temperature, elapsed time, and final temperature.
3. The assumption that the entire soil-tubing entity is at a uniform temperature still holds when the sun dips below the horizon, and starting at 4 P.M. Monday, a cold wind blows at such a velocity to make the effective heat-transfer coefficient h = 18.3 W/m^2 · °C. The temperature of the wind is −5°C. These conditions persist until 8 A.M. Tuesday. Start with an initial temperature as calculated at the end of step 2 and determine the temperature 25 cm below the surface at 8 A.M. Tuesday. Record this value.
4. Assume that at 8 A.M. Tuesday the temperature throughout the soil-tubing entity is uniform at the value just computed at 25 cm. At this time the surface temperature drops to −12°C and stays there until 3 A.M. Friday. Will the tubing reach a temperature of 0°C? If so, approximate to the nearest hour (time and day, A.M. or P.M.) when this will happen. Record these values.
5. Before exiting this program take advantage of the fact that it will compute values of the error function and complementary error function for a specified argument. Select two adjacent arguments, values of x, corresponding to values that appear in Table A-1. Use these values as the argument and have the computer determine erf x and erfc x. Compare the computer result with the tabulated result.
6. Finally, take a value of the argument between the two values just assumed and obtain the computer result. Make a hand interpolation of the values in Table A-1 and compare with the computed value. Are you justified in using an interpolation of the tabulated values for the semi-infinite slab type of problem? Comment.

D-5 *A Radome Design Problem*

A phased-array antenna 1 m high and 2 m wide is to be installed in an Arctic region where the environmental temperature may be taken as −20°C. The electronics require that its surface, called the *radome*, be maintained at a minimum temperature of 10°C. This is to be accomplished by installing electrical heaters placed in the radome in panels 20 by 20 cm. The heaters are to be designed to operate at two power levels and must handle a heat loss to the environment in the presence of wind velocities in the 2-m direction of up to 25 m/s. Use the following procedure to examine the design possibilities.

1. An obvious dividing line is the transition from laminar to turbulent flow. Select the laminar flow and plate length for transition-to-turbulent-flow options and by trial and error assume values of the wind velocity until you observe that the transition to turbulent flow occurs at 200 cm. Because this calculation is performed for an unheated plate, you may increase this velocity slightly since the viscosity of gases increases with temperature. Record the trial-and-error value and the value you will assume.

2. You are now ready to select the heat-dissipation requirements for each 20-cm vertical strip of heaters. Each strip will require a different dissipation in the attempt to make the surface temperature constant at 10°C. Select the laminar flow and heat transfer for a constant-temperature-plate option and determine the heat-dissipation requirement for the first panel by setting the length of interest as 20 cm. Record this value, recognizing that it should be a heat flow in watts rather than a heat flux in watts per square meter.
3. Repeat the process using 40 cm as the length of interest, adjust the result to a heat dissipation, and then subtract the previous dissipation from the new value. The result will be the dissipation for the second vertical panel. Be sure to log the total dissipation for the first two panels so that when you repeat the procedure for 60 cm, you will be able to obtain the dissipation for the third panel by subtracting the total for the first two. Continue this procedure in 20-cm increments until you reach the total of 2 m (200 cm). Why do you suppose that the heat-dissipation requirement varies inversely as the downstream length?
4. The question now is: Can the dissipation requirement be eased by calling for a constant and uniform heat dissipation for all panels? Recognize that the temperature distribution on the radome from the leading edge to the trailing edge will most probably be linear under the condition of constant heat dissipation and that the computer will give you an average surface temperature and the temperature at the specified length. Thus, when you call for the laminar flow and heat transfer for a constant-heat-flux-plate option, you will obtain data for the 100- and 200-cm flow-length points. You may then assume the linear relationship and calculate the temperature at the leading edge from

$$T(x = 0) = 2T_{avg} - T(x = 200 \text{ cm})$$

Therefore, the procedure is to assume various values of heat dissipation, use a length of interest of 200 cm and a plate width of 100 cm, and by trial and error find the heat dissipation that gives a temperature $T(x = 0) = 10°C$. Record this dissipation and explain whether this value or that found in step 3 should be used.
5. It is of interest to know the boundary-layer thickness at the trailing edge ($x = 200$ cm) and where (what value of x) the boundary-layer thickness is one-half that of the trailing edge. Compute these values and record the results.
6. Consider the case of turbulent flow. In the subsequent calculations the velocity of the wind will be raised to 25 m/s. Use this velocity and repeat the computations in step 2. When you find that you get no results at the 20-cm point, comment on the reason.
7. Proceed to the 40-cm point and repeat the procedure of step 3, going successively through 60, 80, . . . , 200 cm in increments of 20 cm. In this case you will call for half of the 40-cm value of heat dissipation as the heat dissipation for the first 20-cm panel. Record these values.
8. What is the boundary-layer thickness at the 200-cm point?

D-6 A 2-m-long tube has an outside diameter of 3.175 cm and a wall thickness of 1.82 mm. Water flows through the tube at a bulk temperature of 76°C at 1.5 m/s. The wall of the tube is held at 60°C. What is the heat-transfer coefficient?

D-7 For the tube in Prob. D-6 air at 1 atm flows transverse to the tube at 8 m/s and a bulk temperature of 38.5°C. What is the heat-transfer coefficient? How good is the assumption in Prob. D-6 that the tube-wall temperature is 60°C?

D-8 In an industrial plant, carbon dioxide is cooled in a cross-flow exchanger that contains eight rows of tubes in a staggered arrangement with spacing-to-diameter ratio of 1.5 in both the transverse and longitudinal directions. Freon is condensing inside the tubes at 10°C. If the CO_2 enters the tube bank at 1 atm, 44°C, and a free-stream velocity of 12 m/s and the tube outside diameter is 2.54 cm, determine the heat-transfer coefficient.

D-9 Air at 1 atm flows at 2.5 m/s and 27°C in a rectangular 20 by 20 cm duct that is 4 m long. If its walls are maintained at a constant 77°C, find the coefficient of heat transfer.

D-10 In a medical application, distilled water (use the properties of water) flows inside a tube of 1 mm inside diameter of 0.25 m/s. The bulk temperature is 14°C, and the wall of the tube, which is 1.5 cm long, is maintained at a constant 5°C. Determine the average heat-transfer coefficient.

D-11 Wiring codes limit the current-carrying capability of wire because its I^2R dissipation must be held to a certain maximum to prevent excessive temperature and fire hazard. Consider two wires, each 1 m long, whose surface temperature must be held to 257°F (125°C). One wire has a diameter of 1 mm and a resistance of 2.2 Ω/m; the other has a diameter of 2 mm and a resistance of 0.55 Ω/m. Both wires have a surface emissivity of 0.725. Find the maximum allowable current in each wire. Take the ambient temperature as 35°C with a pressure of 1 atm.

D-12 A vertical cylinder 80 cm high and 29 cm in diameter is held throughout at 88°C. Determine the heat dissipation from the cylindrical sides if the environment is (*a*) water at 42°C and (*b*) air at 1.5 atm and 42°C.

D-13 In a jet-engine repair facility the engine to be tested is encased in a "soundproof" chamber, in the rear of which a glass window makes it possible to view the exhaust area. Assume that the exhaust gases can be considered as emitting essentially blackbody radiation at 2500 K and that the transmissivity of the glass is 0.4 between 0.4 and 0.6 μm and 0.65 between 0.6 and 1.2 μm. The window is square with sides of 0.48 m. Find the magnitude of the heat loss from the window.

1. First select Option 1, which will provide the radiation function for a particular temperature-wavelength product. Specify a temperature of 4000 K and a wavelength of 1.2 μm. Record the radiation function. Then interpolate between the upper and lower values closest to your temperature-wavelength product and note the radiation function. Comment on the accuracy you can obtain by using interpolation to the radiation functions in Table 8-1.
2. Now select Option 2 and follow the directions given in the program menu. Your aim is to obtain the radiation fraction between 0.4 and 0.6 m and between 0.6 and 1.2 μm. When you have determined them, you will be ready to complete the problem by hand.

D-14 Turn to Sec. 8-4 for the graphs of the radiation shape factors. Calculate the values of the shape factors for cases *a* through *e* below and compare with values obtained from the graphs. Then calculate the shape factor called for in part *f*.

a. Two parallel opposed 12 by 8 cm rectangles 3 cm apart.
b. Two parallel opposed 6-cm-diameter disks 3 cm apart.

c. Parallel opposed disks having diameters of 20 and 16 cm and separated by a distance of 10 cm. Be sure to specify the nomenclature for the shape factor.

d. Two concentric 20-cm-long cylinders with radii of 6 and 10 cm. Be sure to specify the nomenclature for the shape factor.

e. Two perpendicular rectangles having a common edge. The height of the smaller one is 8 cm, that of the larger one is 16 cm, and the length of the common edge is 20 cm. Be sure to specify the nomenclature for the shape factor.

f. A composite arrangement, as shown in Fig. D-7, with

Overall length of rectangles = 5.0 m

Overall width of rectangles = 4.0 m

Distance between rectangles = 2.0 m

Length of subarea 1 = 1.2 m

Width of subarea 1 = 0.8 m

Area 9′ is square, 1.0 m on side

Compute the value of $F_{1-9'}$. Also determine the size of the square 9′ to the nearest centimeter such that this shape factor has a value of 0.010.

D-15 A rectangular parallelepiped is 10, 20, and 30 cm on a side. One of the 10 by 30 sides is maintained at 800 K and has an emissivity of 0.7. The other five surfaces are at 350 K and have emissivities of 0.4. Calculate the heat lost by the hot surface (*a*) using the six-body computer program and (*b*) using the relations of Chap. 8 to work it as a two-body problem. Comment on the results.

D-16 Calculate the heat gained by each of the five cool surfaces in Prob. D-15 using both the computer program and the two-body model. Comment on the results.

D-17 The box in Prob. D-15 has one of the 10 by 20 cm ends open and exposed to a large enclosure at 250 K. Otherwise the temperatures are the same as before. Calculate the heat transfer for each of the five surfaces.

D-18 The box in Prob. D-15 has the cooler 10 by 30 cm side insulated in such a way that it behaves as if it had a very low emissivity of about 0.01. Calculate the heat transfer for all the surfaces under this condition.

D-19 The 10 by 20 and 20 by 30 cm surfaces of the box in Prob. D-15 have low emissivities of 0.01. Calculate the heat-transfer rates under these conditions.

D-20 Work Prob. D-19 as a three-body problem using the methods of Chap. 8 and compare the results with those obtained in Prob. D-19.

D-21 Two 40 by 60 cm parallel plates 20 cm apart are placed in a large enclosure at 300 K. One plate has an emissivity of 0.8 and a temperature of 900 K; the other has an emissivity of 0.4 and a temperature of 500 K. Calculate the heat transfer from each plate using the computer program.

D-22 The six-sided box in Fig. D-8 has the following properties:

Surface	*1*	*2*	*3*	*4*	*5*	*6*
T, °C	500	450	400	425	200	300
Emissivity	0.9	0.7	0.6	0.65	0.3	0.15

Calculate the heat transfer and irradiation for each surface. Take the height as 40 cm, width as 80 cm, and depth as 120 cm.

D-23 A store in the shape of Fig. D-8 is 6 m wide, 12 m deep, and 4 m high.

1. Enter the following data for the six surfaces:

Surface	*T*, °C	*Emissivity*
Front	33	0.887
Left wall	25	0.841
Right wall	26	0.792
Rear wall	26	0.825
Ceiling	31	0.649
Floor	23	0.933

If you are wondering about these data, imagine that the left and right walls face adjacent air-conditioned spaces, the floor faces the parking garage, which is very cool, the ceiling picks up the roof radiation load, and the rear wall, which faces the environment, is in the shade.

2. Carefully observe how the matrix of shape factors unfolds before you. Make one check of reciprocity. Does the shape factor between front and ceiling exhibit reciprocity with the shape factor between ceiling and front? Notice at the end of the shape-factor development that the additive property for each surface prevails.

3. The displayed interwall conductances and radiosities are of interest and should be noted. Of greater importance, however, is the radiant interchange between the six walls. Note the value of the radiation gain (in watts) for the front wall.

4. You are asked to find the value of the emissivity of the front wall that will change the radiation gain for the front wall to 500 W. Find this value to three decimal places by a trial-and-error procedure answering the questions at the end of the program.

5. The same result of 500 W can be achieved by varying the temperature of the front wall. Return the emissivity to its original value (0.887) and then use trial and error to find the front-wall temperature to the nearest tenth of a degree.

6. Let the right wall approach an insulated surface by reducing the emissivity to 0.010 and the temperature to −273°C. What will the radiosity of the wall be? What will the temperature of the insulated wall be?

D-24 *Heat-Exchanger Comparisons*

The objective of this problem is to examine the influence of different variables and heat-exchanger flow arrangements on heat-exchanger performance. Four comparison studies will be called for, with the following conditions taken as constant in each:

	Cold fluid	*Hot fluid*
Mass flow rate, kg/s	15.0	12.5
Specific heat, J/kg · °C	2450	3680

At the end of the outline for the four cases you will find a table with blanks to be filled in with the results.

1. Ask for Option 1 and then for the counterflow alternative. You will then be asked to enter the flow conditions tabulated above. The objective is to compare the effectiveness of each of the exchanger types on the basis of a particular capacity-rate ratio C and for a particular value of the NTU. After the flow data have been entered, you will be asked for values of the overall heat-transfer coefficient and the exchanger surface area. Enter these as follows:

   Overall heat-transfer coefficient U W/m^2 · °C = 200
   Heat-exchanger surface area A = 294 m^2

   You should get a value of NTU = 1.60 and you will be able to read the exchanger effectiveness. Enter this value on the result sheet. Now cycle through all the other exchanger types to find their effectiveness. When you enter the values on the result sheet, you will be able to make a comparison of effectiveness as a function of NTU for all the types.
2. Unfortunately, a comparison of exchanger types as a function of NTU for a given value of the capacity-rate ratio C may not be very meaningful because it is difficult to get all exchangers to operate at identical values of NTU. In this portion of the problem you will be asked to compare the exchanger types at the same effectiveness and find the value of NTU required to yield this prescribed effectiveness. When the value of NTU has been found, you will be able to obtain the required value of the overall heat-transfer coefficient U for the same flow conditions of step 1 and the surface area of 294 m^2. This procedure is expeditiously handled by calling for Option 2 and pretty much repeating the procedure followed in step 1. Observe that your result sheet calls for the percent increase of overall coefficient based on the counterflow case as well as its value. Note also that Option 2 calls for a trial-and-error procedure.
3. Options 1 and 2 make no mention of the terminal heat-exchanger temperatures. Suppose it is known that the hot fluid enters the exchanger at 80°C and that the cold fluid enters at 30°C. Repeat the procedure of step 2 and find the outlet temperatures for all types of exchangers for the given flow conditions and NTU value of 1.600. Enter these values in the appropriate places on the result sheet.
4. You can use Option 4 to find the required value of NTU to yield a prescribed temperature performance. Select Option 4 and then the counterflow case. With the inlet temperature of step 3 and a cold-side outlet temperature of 62.7°C, let the computer show you that the required value of NTU $\neq$ 1.600. Make the appropriate entries on the result sheet.

Table D-1 Results for Problem D-24

For all result-sheet entries:

	Cold fluid	*Hot fluid*
Flow, kg/s	15.0	12.5
Specific heat, J/kg · °C	2450	3680
C_{min}, J/°C		
Capacity ratio $C = C_{min}/C_{max}$		

Table D-1 Results for Problem D-24 (*continued*)

***Effectiveness as a function of C* and NTU**

Overall heat-transfer coefficient, $W/m^2 \cdot °C$	____
Exchanger surface area, m^2	____

Exchanger type	*Effectiveness,* ϵ
Counterflow	____
Parallel flow	____
Crossflow:	
Both fluids mixed	____
One fluid mixed, one unmixed;	
Cold fluid mixed	____
Hot fluid mixed	____
Both fluids unmixed	____
Shell and tube exchangers (1-n):	
One shell pass, two tube passes	____
One shell pass, four tube passes	____
One shell pass, six tube passes	____
One shell pass, eight tube passes	____
One shell pass, ten tube passes	____
One shell pass, twelve tube passes	____

What conclusions can you draw?

___

___

___

Overall heat-transfer coefficient to yield an effectiveness

C_{min}, J/°C is C_c. The value of C_{min} is	____
Exchanger surface, m^2	294
Required effectiveness	0.654
Counter flow NTU	1.600
Capacity ratio C	0.799

Exchanger type	*Overall coefficient U,* $W/m^2 \cdot °C$	*Required Percent Increase*
Parallel flow	____	____
Cross flow:		
Both fluids mixed	____	____
One fluid mixed, one unmixed		
Cold fluid mixed	____	____
Hot fluid mixed	____	____
Both fluids unmixed	____	____
Shell-and-tube exchangers (1-n):		
One shell pass, two tube passes	____	____
One shell pass, four tube passes	____	____

Table D-1 Results for Problem D-24 (*continued*)

Exchanger type	*Overall coefficient U, W/m² · °C*	*Required Percent Increase*
One shell pass, six tube passes	____	____
One shell pass, eight tube passes	____	____
One shell pass, ten tube passes	____	____
One shell pass, twelve tube passes	____	____

Exchanger terminal temperature	
Overall heat-transfer coefficient U, W/m² · °C	200
Exchanger surface area, m²	294
Capacity rate ratio C	0.799
Number of transfer units, NTU	1.600
Hot-side inlet temperature, °C	80
Cold-side inlet temperature, °C	30

Exchanger type	*Outlet temperatures, °C* Hot side	Cold side
Counterflow	____	____
Parallel flow	____	____
Cross flow:		
Both fluids mixed	____	____
One fluid mixed, one unmixed		
Cold fluid mixed	____	____
Hot fluid mixed	____	____
Both fluids unmixed	____	____
Shell and tube exchangers (1-n):	____	____
One shell pass, two tube passes	____	____
One shell pass, four tube passes	____	____
One shell pass, six tube passes	____	____
One shell pass, eight tube passes	____	____
One shell pass, ten tube passes	____	____
One shell pass, twelve tube passes	____	____

Value of NTU to yield a specified temperature performance	
Overall heat-transfer coefficient U, W/m² · °C	200
Exchanger surface area, m²	294
Capacity rate ratio C	0.799
Cold-side outlet temperature, °C	62.7
Hot-side inlet temperature, °C	80
Cold-side inlet temperature, °C	30
Required effectiveness	____
Hot-side outlet temperature, °C	____

Table D-1 Results for Problem D-24 (*continued*)

Exchanger Type	*Required NTU*
Counterflow	_____
Parallel flow	_____
Cross flow:	
Both fliuds mixed	_____
One fluid mixed, one unmixed:	
Cold fluid mixed	_____
Hot fluid mixed	_____
Both fluids unmixed	_____
Shell and tube exchangers (1-*n*);	
One shell pass, two tube passes	_____
One shell pass, four tube passes	_____
One shell pass, six tube passes	_____
One shell pass, eight tube passes	_____
One shell pass, ten tube passes	_____
One shell pass, twelve tube passes	_____

D-25 Entrance effects can be important in tube flow when the length is not large compared with the diameter. To study this effect, consider a tube having a length of 20 cm. For different flow rates of air, water, and engine oil (properties from Appendix A) take progressively smaller tube diameters from 2.0 cm down to 1.0 mm. Take the fluid temperatures at 20°C. Comment on the behavior of the heat-transfer coefficients under the different conditions.

D-26 A cross-flow heat exchanger with both fluids unmixed is used to heat 10 kg/s of CO_2 from 20 to 80°C using water entering at 160°C and leaving at 90°C. The overall heat-transfer coefficient is 45 W/m^2 · °C. Assuming that the CO_2 flow controls U and that $U \sim \dot{m}^{0.8}$, determine the heat transfer, exit water temperature, and exit CO_2 temperature as a function of percent CO_2 design flow rate (10 kg/s).

D-27 Suppose that the CO_2 in Prob. D-25 is heated by condensing steam at 100°C and U(design) = 180 W/m^2 · °C. Using the same assumption that $U \sim \dot{m}^{0.8}$, repeat Prob. D-25 for these new conditions.

D-28 Repeat Prob. D-24 interchanging the flow rates and specific heats of the hot and cold fluids.

D-29 Determine the heat transfers and radiosities for all the surfaces of a six-sided box like that shown in Fig. D-8 if it is 100 cm long, 60 cm wide, and 40 cm high, with the following surface conditions:

Surface	*1*	*2*	*3*	*4*	*5*	*6*
T, °C	100	200	300	400	500	600
Emissivity	0.3	0.4	0.7	0.6	0.5	0.3

D-30 A cube 10 cm on a side has five of the six surfaces maintained at 600°C with emissivities of 0.7. The sixth side is open and exposed to a large room at 25°C. Calculate the heat lost to the room (*a*) using the computer program and (*b*) working it as a two-body problem and using information from Chap. 8.

D-31 Suppose that the bottom surface (opposite the opening) in Prob. D-30 is at 600°C with $\epsilon = 0.7$ and that the other four surfaces are insulated; that is $\epsilon \rightarrow 0$. Calculate the energy lost to the room under these conditions.

D-32 A finned-tube heat exchanger is to be designed to deliver 44 kW of heat to air on the fin side. The air is heated from 20 to 30°C by hot water flowing in the tubes. The water enters at 160°C and leaves at 140°C. If the overall heat-transfer coefficient is 35 $W/m^2 \cdot °C$, calculate the area of the heat exchanger. What will the heat transfer for this area exchanger be if the water flow rate is doubled, all other factors remaining the same?

INDEX

Important physical constants

Avogadro's number	$N_0 = 6.022045 \times 10^{26}$ molecules/kg mol
Universal gas constant	$\mathcal{R} = 1545.35$ ft·lbf/lbm·mol·°R
	$= 8314.41$ J/kg mol·K
	$= 1.986$ Btu/lbm·mol·°R
	$= 1.986$ kcal/kg mol·K
Planck's constant	$h = 6.626176 \times 10^{-34}$ J-sec
Boltzmann's constant	$k = 1.380662 \times 10^{-23}$ J/molecule·K
	$= 8.6173 \times 10^{-5}$ eV/molecule·K
Speed of light in vacuum	$c = 2.997925 \times 10^{8}$ m/s
Standard gravitational acceleration	$g = 32.174$ ft/s^2
	$= 9.80665$ m/s^2
Electron mass	$m_e = 9.1095 \times 10^{-31}$ kg
Charge on the electron	$e = 1.602189 \times 10^{-19}$ C
Stefan-Boltzmann constant	$\sigma = 0.1714 \times 10^{-8}$ Btu/hr·ft^2·R^4
	$= 5.669 \times 10^{-8}$ W/m^2·K^4
1 atm	$= 14.69595$ lbf/in^2 $= 760$ mmHg at 32°F
	$= 29.92$ inHg at 32°F $= 2116.21$ lbf/ft^2
	$= 1.01325 \times 10^{5}$ N/m^2